X-Ray Spectrometry:
Recent Technological Advances

X-Ray Spectrometry: Recent Technological Advances

Edited by

Kouichi Tsuji
Department of Applied Chemistry, Osaka City University, Japan

Jasna Injuk
Micro and Trace Analysis Center, University of Antwerp, Belgium

René Van Grieken
Micro and Trace Analysis Center, University of Antwerp, Belgium

John Wiley & Sons, Ltd

Copyright © 2004   John Wiley & Sons Ltd, The Atrium, Southern Gate, Chichester,
West Sussex PO19 8SQ, England

Telephone (+44) 1243 779777

Email (for orders and customer service enquiries): cs-books@wiley.co.uk
Visit our Home Page on www.wileyeurope.com or www.wiley.com

All Rights Reserved. No part of this publication may be reproduced, stored in a retrieval system or transmitted in any form or by any means, electronic, mechanical, photocopying, recording, scanning or otherwise, except under the terms of the Copyright, Designs and Patents Act 1988 or under the terms of a licence issued by the Copyright Licensing Agency Ltd, 90 Tottenham Court Road, London W1T 4LP, UK, without the permission in writing of the Publisher. Requests to the Publisher should be addressed to the Permissions Department, John Wiley & Sons Ltd, The Atrium, Southern Gate, Chichester, West Sussex PO19 8SQ, England, or emailed to permreq@wiley.co.uk, or faxed to (+44) 1243 770620.

This publication is designed to provide accurate and authoritative information in regard to the subject matter covered. It is sold on the understanding that the Publisher is not engaged in rendering professional services. If professional advice or other expert assistance is required, the services of a competent professional should be sought.

Other Wiley Editorial Offices

John Wiley & Sons Inc., 111 River Street, Hoboken, NJ 07030, USA

Jossey-Bass, 989 Market Street, San Francisco, CA 94103-1741, USA

Wiley-VCH Verlag GmbH, Boschstr. 12, D-69469 Weinheim, Germany

John Wiley & Sons Australia Ltd, 33 Park Road, Milton, Queensland 4064, Australia

John Wiley & Sons (Asia) Pte Ltd, 2 Clementi Loop #02-01, Jin Xing Distripark, Singapore 129809

John Wiley & Sons Canada Ltd, 22 Worcester Road, Etobicoke, Ontario, Canada M9W 1L1

Wiley also publishes its books in a variety of electronic formats. Some content that appears in print may not be available in electronic books.

Library of Congress Cataloging-in-Publication Data

X-ray spectrometry : recent technological advances / edited by Kouichi Tsuji, Jasna Injuk,
　René Van Grieken.
　　　p. ; cm.
　　Includes bibliographical references and index.
　　ISBN 0-471-48640-X (Hbk. : alk. paper)
　　1. X-ray spectroscopy. I. Tsuji, Kouichi. II. Injuk, Jasna. III. Grieken, R. van (René)
　　[DNLM: 1. Chemistry, Analytical. 2. Spectrometry, X-Ray Emission – instrumentation.
3. Spectrometry, X-Ray Emission – methods. QD 96.X2 X87 2004]
QD96.X2X228 2004
543′.62 – dc22

2003057604

British Library Cataloguing in Publication Data

A catalogue record for this book is available from the British Library

ISBN 0-471-48640-X

Typeset in 10/12pt Times by Laserwords Private Limited, Chennai, India
Printed and bound in Great Britain by Antony Rowe Ltd, Chippenham, Wiltshire
This book is printed on acid-free paper responsibly manufactured from sustainable forestry
in which at least two trees are planted for each one used for paper production.

Contents

Contributors vii

Preface xi

1 Introduction 1
  1.1 Considering the Role of X-ray Spectrometry in Chemical Analysis and Outlining the Volume 1

2 X-Ray Sources 13
  2.1 Micro X-ray Sources 13
  2.2 New Synchrotron Radiation Sources ... 29
  2.3 Laser-driven X-ray Sources 49

3 X-Ray Optics 63
  3.1 Multilayers for Soft and Hard X-rays 63
  3.2 Single Capillaries X-ray Optics .. 79
  3.3 Polycapillary X-ray Optics 89
  3.4 Parabolic Compound Refractive X-ray Lenses 111

4 X-Ray Detectors 133
  4.1 Semiconductor Detectors for (Imaging) X-ray Spectroscopy... 133
  4.2 Gas Proportional Scintillation Counters for X-ray Spectrometry 195
  4.3 Superconducting Tunnel Junctions 217
  4.4 Cryogenic Microcalorimeters 229
  4.5 Position Sensitive Semiconductor Strip Detectors 247

5 Special Configurations 277
  5.1 Grazing-incidence X-ray Spectrometry 277
  5.2 Grazing-exit X-ray Spectrometry . 293
  5.3 Portable Equipment for X-ray Fluorescence Analysis 307
  5.4 Synchrotron Radiation for Microscopic X-ray Fluorescence Analysis 343
  5.5 High-energy X-ray Fluorescence.. 355
  5.6 Low-energy Electron Probe Microanalysis and Scanning Electron Microscopy 373
  5.7 Energy Dispersive X-ray Microanalysis in Scanning and Conventional Transmission Electron Microscopy 387
  5.8 X-Ray Absorption Techniques ... 405

6 New Computerisation Methods 435
  6.1 Monte Carlo Simulation for X-ray Fluorescence Spectroscopy 435
  6.2 Spectrum Evaluation 463

7 New Applications 487
  7.1 X-Ray Fluorescence Analysis in Medical Sciences 487
  7.2 Total Reflection X-ray Fluorescence for Semiconductors and Thin Films 517
  7.3 X-Ray Spectrometry in Archaeometry 533
  7.4 X-Ray Spectrometry in Forensic Research 553
  7.5 Speciation and Surface Analysis of Single Particles Using Electron-excited X-ray Emission Spectrometry 569

Index 593

Contributors

F. Adams
Department of Chemistry, University of Antwerp, Universiteitsplein 1, B-2610 Antwerp, Belgium

J. Börjesson
Department of Diagnostic Radiology, Country Hospital, SE-301 85 Halmstad, Sweden

A. Brunetti
Department of Mathematics and Physics, University of Sassari, Via Vienna 2, 1–07100 Sassari, Italy

R. Bytheway
BEDE Scientific Instruments Ltd, Belmont Business Park, Durham DH1 1TW, UK

A. Castellano
Department of Materials Science, University of Lecce, I-73100 Lecce, Italy

R. Cesareo
Department of Mathematics and Physics, University of Sassari, Via Vienna 2, I-07100 Sassari, Italy

C. A. Conde
Physics Department, University of Coimbra, P-3004-0516 Coimbra, Portugal

W. Dąbrowski
Faculty of Physics and Nuclear Techniques, AGH University of Science and Technology, Al. Mickiewicza 30, 30–059 Krakow, Poland

E. Figueroa-Feliciano
NASA/Goddard Space Flight Centre, Code 662, Greenbelt, MD 20771, USA

M. Galeazzi
University of Miami, Department of Physics, PO Box 248046, Coral Gables, FL 33124, USA

N. Gao
X-ray Optical Systems, Inc., 30 Corporate Circle, Albany, NY 12203, USA

P. Gryboś
Faculty of Physics and Nuclear Techniques, AGH University of Science and Technology, Al. Mickiewicza 30, 30-059 Krakow, Poland

P. Holl
Semiconductor Lab., MPI Halbleiterlabor, SIEMENS – Gelaende, Otto-Hahn-Ring 6, D-81739 München, Germany

J. de Hoog
Department of Chemistry, University of Antwerp, Universiteitsplein 1, B-2610 Antwerp, Belgium

Y. Hosokawa
X-ray Precision, Inc., Bld. #2, Kyoto Research Park 134, 17 Chudoji, Minami-machi, Shimogyo-ku, Kyoto 600–8813, Japan

G. Isoyama
The Institute of Scientific and Industrial Research, Osaka University, 8-1 Mihagaoka, Ibaraki, Osaka Pref. 567-0047, Japan

K. Janssens
Department of Chemistry, Universiteitsplein I, University of Antwerp, B-2610 Antwerp, Belgium

J. Kawai
Department of Materials Science and Engineering, Kyoto University, Sakyo-ku, Kyoto 606–8501, Japan

M. Kurakado
Electronics and Applied Physics, Osaka Electro-Communication University, 18-8, Hatsucho, Neyagawa, Japan

S. Kuypers
Centre for Materials Advice and Analysis, Materials Technology Group, VITO (Flemish Institute for Technological Research), B-2400 Mol, Belgium

P. Lechner
Semiconductor Lab., MPI Halbleiterlabor, SIEMENS–Gelaerde, Otto-Hahn-Ring 6, D-81739 München, Germany

P. Lemberge
Department of Chemistry, University of Antwerp, Universiteitsplein 1, B-2610 Antwerp, Belgium

B. Lengeler
RWTH, Aachen University, D-52056 Aachen, Germany

G. Lutz
Semiconductor Lab., MPI Halbleiterlabor, SIEMENS – Gelaende, Otto-Hahn-Ring 6, D-81739 München, Germany

S. Mattsson
Department of Radiation Physics, Lund University, Malmö University Hospital, SE-205 02 Malmö, Sweden

Y. Mori
Wacker-NSCE Corporation, 3434 Shimata, Hikari, Yamaguchi 743-0063, Japan

I. Nakai
Department of Applied Chemistry, Science University of Tokyo, 1-3 Kagurazaka, Shinjuku, Tokyo 162–0825, Japan

T. Ninomiya
Forensic Science Laboratory, Hyogo Prefectural Police Headquarters, 5-4-1 Shimoyamate, Chuo-Ku, Kobe 650–8510, Japan

J. Osan
KFKI Atomic Energy Research Institute, Department of Radiation and Environmental Physics, PO Box 49, H-1525 Budapest, Hungary

C. Ro
Department of Chemistry, Hallym University, Chun Cheon, Kang WonDo 200–702, Korea

M. A. Rosales Medina
University of 'Las Americas', Puebla, CP 72820, Mexico

K. Sakurai
National Institute for Materials Science, 1-2-1 Sengen, Tsukuba, Ibaraki 305-0047, Japan

C. Schroer
RWTH, Aachen University, D-52056 Aachen, Germany

A. Simionovici
ID22, ESRF, BP 220, F-38043 Grenoble, France

H. Soltau
Semiconductor Lab., MPI Halbleiterlabor, SIEMENS – Gelaende, Otto-Hahn-Ring 6, D-81739 München, Germany

C. Spielmann
Physikalisches Institut EP1, Universität Würzburg, Am Hubland, D-97074 Würzburg, Germany

L. Strueder
Semiconductor Lab., MPI Halbleiterlabor, SIEMENS – Gelaende, Otto-Hahn-Ring 6, D-81739 München, Germany

I. Szaloki
Institute of Experimental Physics, University of Debrecen, Bem tér 18/a, H-4026 Debrecen, Hungary

B. K. Tanner
BEDE Scientific Instruments Ltd, Belmont Business Park, Durham DH1 1TW, UK

M. Taylor
BEDE Scientific Instruments Ltd, Belmont Business Park, Durham DH1 1TW, UK

K. Tsuji
Osaka City University, 3-3-138 Sugimoto, Sumiyoshi-ku, Osaka 558-8585, Japan

E. Van Cappellen
FEI Company, 7451 N.W. Evergreen Parkway, Hillsboro, OR 97124-5830, USA

R. Van Grieken
Department of Chemistry, University of Antwerp, Universiteitsplein I, B-2610 Antwerp, Belgium

B. Vekemans
Department of Chemistry, University of Antwerp, Universiteitsplein 1, B-2610 Antwerp, Belgium

L. Vincze
Department of Chemistry, University of Antwerp, Universiteitsplein 1, B-2610 Antwerp, Belgium

M. Watanabe
Institute of Multidisciplinary Research for Advanced Materials, Tohoku University, 2-1-1 Katahira, Aoba-ku, Sendai 980-8577, Japan

K. Yamashita
Department of Physics, Nagoya University, Chikusa-ku, Nagoya 464-8602, Japan

M. Yanagihara
Institute of Multidisciplinary Research for Advanced Materials, Tohoku University, 2-1-1 Katahira, Aoba-ku, Sendai 980-8577, Japan

A. Zucchiatti
Instituto Nazionale di Fisica Nucleare, Sezione di Genova, Via Dodecanesco 33, I-16146 Genova, Italy

Preface

During the last decade, remarkable and often spectacular progress has been made in the methodological but even more in the instrumental aspects of X-ray spectrometry. This progress includes, for example, considerable improvements in the design and production technology of detectors and considerable advances in X-ray optics, special configurations and computing approaches. All this has resulted in improved analytical performance and new applications, but even more in the perspective of further dramatic enhancements of the potential of X-ray based analysis techniques in the very near future. Although there exist many books on X-ray spectrometry and its analytical applications, the idea emerged to produce a special book that would cover only the most advanced and high-tech aspects of the chemical analysis techniques based on X-rays that would be as up-to-date as possible. In principle, all references were supposed to be less than five years old. Due to rapid changes and immense progress in the field, the timescale for the book was set to be very short. A big effort was made to cover as many sub-areas as possible, and certainly those in which progress has been the fastest. By its nature, this book cannot cover the fundamental, well-known and more routine aspects of the technique; for this, reference is made to several existing handbooks and textbooks.

This book is a multi-authored effort. We believe that having scientists who are actively engaged in a particular technique to cover those areas for which they are particularly qualified, outweighs any advantages of uniformity and homogeneity that characterize a single-author book. In the specific case of this book, it would have been truly impossible for any single person to cover a significant fraction of all the fundamental and applied sub-fields of X-ray spectrometry in which there are so many advances nowadays. The Editors were fortunate enough to have the cooperation of truly eminent specialists in each of the sub-fields. Many chapters are written by Japanese scientists, and this is a bonus because much of their intensive and innovating research on X-ray methods is too little known outside Japan. The Editors wish to thank all the distinguished contributors for their considerable and timely efforts. It was, of course, necessary to have this book, on so many advanced and hot topics in X-ray spectrometry, produced within an unusually short time, before it would become obsolete; still the resulting heavy time-pressure put on the authors may have been unpleasant at times.

We hope that even experienced workers in the field of X-ray analysis will find this book useful and instructive, and particularly up-to-date when it appears, and will benefit from the large amount of readily accessible information available in this compact form, some of it presented for the first time. We believe there is hardly any overlap with existing published books, because of the highly advanced nature and actuality of most chapters. Being sure that the expert authors have covered their subjects with sufficient depth, we hope that we have chosen the topics of the different chapters to be wide-ranging enough to cover all the important and emerging fields sufficiently well.

We do hope this book will help analytical chemists and other users of X-ray spectrometry to fully exploit the capabilities of this set of powerful analytical tools and to further expand its applications in such fields as material and environmental sciences, medicine, toxicology, forensics, archaeometry and many others.

K. Tsuji
J. Injuk
R. Van Grieken
Osaka, Antwerp

Chapter 1

Introduction

1.1 Considering the Role of X-ray Spectrometry in Chemical Analysis and Outlining the Volume

R. VAN GRIEKEN
University of Antwerp, Antwerp, Belgium

1.1.1 RATIONALE

Basic X-ray spectrometry (XRS) is, of course, not a new technique. The milestone developments that shaped the field all took place several decades ago. Soon after the discovery of X-rays in 1895 by Wilhelm Conrad Röntgen, the possibility of wavelength-dispersive XRS (WDXRS) was demonstrated and Coolidge introduced the high-vacuum X-ray tube in 1913. There was quite a time gap then until Friedmann and Birks built the first modern commercial X-ray spectrometer in 1948. The fundamental Sherman equation, correlating the fluorescent X-ray intensity quantitatively with the chemical composition of a sample, dates back to 1953. The fundamental parameter (FP) approach, in its earliest version, was independently developed by Criss and Birks and Shiraiwa and Fujino, in the 1960s. Also various practical and popular influence coefficient algorithms, like those by Lachanche–Traill, de Jongh, Claisse–Quintin, Rasberry–Heinrich, Rousseau and Lucas–Tooth–Pine all date back to 1960–1970. The first electron microprobe analyser (EMPA) was successfully developed in 1951 by Castaing, who also outlined the fundamental aspects of qualitative electron microprobe analysis. The first semiconductor Si(Li) detectors, which heralded the birth of energy-dispersive XRS (EDXRS), were developed, mainly at the Lawrence Berkeley Lab, around 1965. Just before 1970, accelerator-based charged-particle induced XRS or proton-induced X-ray emission (PIXE) analysis was elaborated; much of the credit went to the University of Lund in Sweden. A description of the setup for total-reflection X-ray fluorescence (TXRF) was first published by Yoneda and Horiuchi and the method was further pursued by Wobrauschek and Aiginger, both in the early 1970s. The advantages of polarised X-ray beams for trace analysis were pointed out in 1963 by Champion and Whittam and Ryon put this further into practice in 1977. There have been demonstrations of the potential for micro-X-ray fluorescence (XRF) since 1928 (by Glockner and Schreiber) and Chesley began with practical applications of glass capillaries in 1947. Synchrotron-radiation (SR) XRS was introduced in the late 1970s and Sparks developed the first micro version at the Stanford Synchrotron Radiation Laboratory in 1980.

So around 1990, there was a feeling that radically new and stunning developments were

---

X-Ray Spectrometry: Recent Technological Advances. Edited by K. Tsuji, J. Injuk and R. Van Grieken
© 2004 John Wiley & Sons, Ltd   ISBN: 0-471-48640-X

lacking in XRS and scientists began to have some ambivalent opinions regarding the future role of XRS in analytical chemistry. One could wonder whether, in spite of remarkably steady progress, both instrumental and methodological, XRS had reached a state of saturation and consolidation, typical for a mature and routinely used analysis technique.

In the meantime, XRF had indeed developed into a well-established and mature multi-element technique. There are several well-known key reasons for this success: XRF is a universal technique for metal, powder and liquid samples; it is nondestructive; it is reliable; it can yield qualitative and quantitative results; it usually involves easy sample preparation and handling; it has a high dynamic range, from the ppm level to 100 % and it can, in some cases, cover most of the elements from fluorine to uranium. Accuracies of 1 % and better are possible for most atomic numbers. Excellent data treatment software is available allowing the rapid application of quantitative and semi-quantitative procedures. In the previous decades, somewhat new forms of XRS, with e.g. better sensitivity and/or spectral resolution and/or spatial resolution and/or portable character, had been developed. However, alternative and competitive more sensitive analytical techniques for trace analysis had, of course, also been improved; we have seen the rise and subsequent fall of atomic absorption spectrometry and the success of inductively coupled plasma atomic emission and mass spectrometry (ICP-AES and ICP-MS) in the last two decades.

Since 1990, however, there has been dramatic progress in several sub-fields of XRS, and in many aspects: X-ray sources, optics, detectors and configurations, and in computerisation and applications as well. The aim of the following chapters in this book is precisely to treat the latest and often spectacular developments in each of these areas. In principle, all references will pertain to the last 5–6 years. Many of the chapters will have a high relevance for the future role of XRS in analytical chemistry, but certainly also for many other fields of science where X-rays are of great importance. The following sections in this chapter will give a flavour of the trends in the position of different sub-fields of XRS based e.g. on the recent literature and will present the outline of this volume.

1.1.2 THE ROLE AND POSITION OF XRS IN ANALYTICAL CHEMISTRY

An attempt has been made to assess the recent trends in the role and position of XRS based on a literature survey (see also Injuk and Van Grieken, 2003) and partially on personal experience and views. For the literature assessment, which covered the period from January 1990 till the end of December 2000, a computer literature search on XRS was done in *Chemical Abstracts*, in order to exclude (partially) the large number of XRS publications on astronomy, etc.; still, it revealed an enormous number of publications. Figure 1.1.1 shows that the volume of the annual literature on XRS, cited in *Chemical Abstracts*, including all articles having 'X-ray spectrometry/spectroscopy' in their title, is still growing enormously and exponentially. During the last decade, the number of publications on XRS in general has nearly doubled; in 2000, some 5000 articles were published, versus 120 annually some 30 years ago. As seen in Figure 1.1.2, XRS in general seems more alive than ever nowadays.

However, the growth of the literature on specifically XRF is much less pronounced: from about 500 articles per year in 1990 to about 700 in 2000, still a growth of 40 % in the last decade. While in 1990 it looked like XRF had reached a state of saturation and consolidation, newer developments in the 1990s, e.g. the often-spectacular ones described in the other chapters of this volume, have somehow countered such fears. It is a fact that WDXRF remains the method of choice for direct accurate multi-element analysis in the worldwide mineral and metallurgy industry. For liquid samples, however, the competition of ICP-AES and ICP-MS remains formidable. It is striking that, while there are still many more WDXRF units in operation around the world than EDXRF instruments, the number of publications dealing with WDXRF is about five times lower. This clearly reflects the predominant use of the more expensive WDXRF

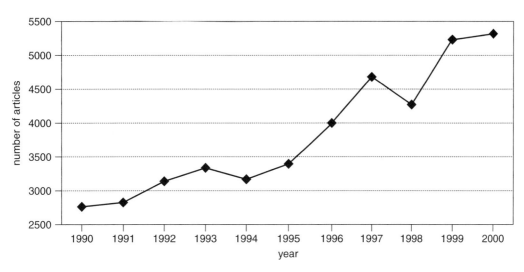

Figure 1.1.1 Total annual number of articles on X-ray emission spectrometry in the period 1990–2000 (source of data: *Chemical Abstracts*). Reproduced by permission of John Wiley & Sons, Ltd

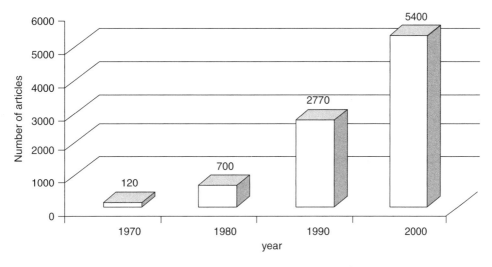

Figure 1.1.2 Number of articles on X-ray emission spectrometry since 1970 (source of data: *Chemical Abstracts*). Reproduced by permission of John Wiley & Sons, Ltd

in routine industrial analysis, where publishing is not common, while EDXRF is mostly present in academic and research institutions; there are many applications in environment-related fields where ultimate accuracies are not so mandatory.

Also the number of publications on radioisotope XRF has been increasing from 40 in 1990 to 100 in 1998, reflecting the frequent use of the technique in many field and on-line applications. In Australia alone, more than 2000 portable XRF are employed in the mining and mineral industry. It is expected that the radioisotope-based on-line installations will gradually be replaced by systems based on small X-ray tubes.

The annual number of articles dealing with various aspects of the PIXE technique (but excluding micro-PIXE) is in the range of 30 to 70 with very prominent peaks every 3 years. These are obviously related to the publication of the proceedings of the tri-annual PIXE Conferences (Figure 1.1.3), with many short articles. It is clear, and not only from the literature, that, of all X-ray emission techniques considered, PIXE is thriving the least; there is no clear growth in the literature, although PIXE might still be the method of choice for the trace analysis of large numbers of relatively small samples, like e.g. for particulate air pollution monitoring using impactor deposits of aerosols. The number of PIXE installations in the world is probably decreasing, and the future of PIXE seems to be exclusively in its micro version; some 30 institutes are active in this field at the moment. The literature on micro-PIXE is still growing; a search on Web of Science showed that the annual number of articles on micro-PIXE was around 10 at the beginning of the previous decade and around 35 in the last few years.

Since the early 1970s, SR-XRS has been experiencing remarkable growth, nowadays approaching almost 350 articles per year, with a doubling seen over the last decade. Investment in SR facilities continues to be strong and with the increasing availability of SR X-ray beam lines, new research fields and perspectives are open today. Most of the presently operational SR sources belong to the so-called second-generation facilities. A clear distinction is made from the first generation, in which the SR was produced as a parasitic phenomenon in high-energy collision experiments with elementary particles. Of special interest for the future are new third-generation storage rings, which are specifically designed to obtain unique intensity and brilliance. SR has a major impact on microprobe-type methods with a high spatial resolution, like micro-XRF, and on X-ray absorption spectrometry (XAS) as well as on TXRF. For highly specific applications, SR-XRS will continue to grow. The costs of SR-XRS are usually not calculated, since in most countries, SR facilities are free of charge for those who have passed some screening procedure. Of course, such applications cannot be considered as routine.

There are nowadays some 100 publications annually on TXRF, and this number has more or less doubled since 1990. However, it may seem that TXRF has stabilized as an analytical method for ultra-trace determination from solutions and dissolved solids due to the fierce competition from ICP-MS, in particular. There are now only a few companies offering TXRF units. But mostly for surface analysis directly on a flat solid sample, TXRF is still unique. SR-TXRF might be one of the methods of choice in future wafer surface analysis (in addition to e.g. secondary ion mass spectrometry). In addition, by scanning around the total-reflection angle, TXRF allows measurements of the density, roughness and layer thickness and depth profiling, which are, of course, of much interest in material sciences. New possibilities for improving the performance of TXRF are in using polarised primary radiation. SR has almost ideal features for employment in combination with

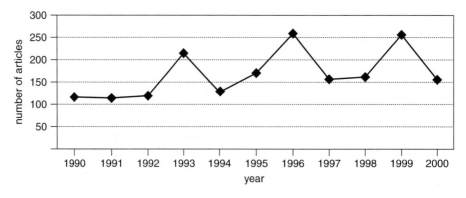

Figure 1.1.3 Annual number of articles on PIXE in the period 1990–2000 (source of data: Web of Science). Reproduced by permission of John Wiley & Sons, Ltd

TXRF. It is several orders (8–12) of magnitude higher in brightness compared to X-ray tubes, has a natural collimation in the vertical plane and is linearly polarised in the plane of the orbit of the high energy (GeV) electron or positrons. The spectral distribution is continuous, so by proper monochromatisation, the performance of selective excitation at best conditions is possible. SR offers a significant reduction in TXRF detection limits and a remarkable improvement has been achieved over the past 20 years from nanogram level in 1975 to attogram level in 1998.

Until recently, evolution of XRF into the microanalytical field was hampered because of the difficulties involved in focusing a divergent X-ray beam from an X-ray tube into a spot of small dimensions. However, the development of SR sources and the recent advances in X-ray focusing have changed the situation. Contemporary micro-XRF applications started only some 10 years ago on a significant scale, and it appears today to be one of the best microprobe methods for inorganic analysis of various materials: it operates at ambient pressure and, in contrast to PIXE and EMPA, no charging occurs. In many instances, no sample preparation is necessary. The field of micro-XRF is currently subject to a significant evolution in instrumentation: lead-glass capillaries and polycapillary X-ray lenses, air-cooled micro-focus X-ray tubes, compact ED detector systems with a good resolution even at a high-count rate and no longer requiring liquid-nitrogen cooling. Commercial laboratory instrumentation using capillary optics combined with rapid scanning and compositional mapping capability is expected to grow, and various systems are commercially available. During the 1980s, SR facilities around the world began to implement X-ray microbeam capabilities on their beam lines for localised elemental analysis. Recent trends in SR micro-XRF are towards optimisation of optics and smaller beam sizes down to the submicrometer size.

With respect to the general applications of XRS, it appears that environmental, geological, biological and archaeological applications make up a stunning 70 % contribution to the literature; undoubtedly this is far above their relative contributions in actual number of analysis, since XRF is certainly still a working horse in many types of industries, for all kinds of routine analyses, but the latter applications are published very seldom. The number of articles dealing with environmental applications of XRS, in the past decade, shows a steady growth. Interestingly, the relative contributions for the different topics covered in the environmental applications, like soils and geological material (23 %), biological materials (19 %), water (19 %), air (17 %) and waste material (8 %) have not changed considerably during the last decade.

Table 1.1.1 shows the relative share of laboratories in different countries to the literature on XRS generated in 1998 (according to *Analytical Abstracts*) and the language in which the publications were written. It appears that European countries produce almost one half of the total number of publications, while, of the non-European countries, China and Russia are leading. The low contribution of the USA is striking. There might be several reasons for this. Apparently, XRS is considered more as a routine technique by the US industry and there are almost no US academic centers working in this field. It is also true that in 1998, no volume of *Advances in X-ray Analysis* appeared and this covers the proceedings of the popular Denver X-ray

Table 1.1.1 The relative share of laboratories in different countries to the literature on XRS generated in the year 1998 (as covered by *Analytical Abstracts*) and the language in which the publications were written

Country	Relative contribution (%)	Language
China	13.4	25 % English
		75 % Chinese
Russia	10.2	70 % English
		30 % Russian
Japan	8.0	55 % English
		45 % Japanese
Other Asian	4.8	100 % English
Germany	9.3	95 % English
Italy	6.3	100 % English
UK	4.8	100 % English
Other European	26.2	100 % English
USA	5.4	100 % English
Other American	5.4	100 % English
Australia	3.6	100 % English
Africa	2.7	100 % English

Analysis Conference. Finally, the most advanced research (as described in the following chapters in this volume) may still be published in physics journals rather than in journals covered by *Analytical Abstracts*. It also appeared from our literature search that about one fourth of the XRS literature is written in less accessible languages like Russian, Chinese and Japanese.

In view of the enormous advances that are being made in XRS and that, hopefully, are covered well in the following chapters of this book, one can expect that the applications of XRS will dramatically be changed over the next few years, and that, in the literature, the distribution over fundamental aspects (probably not fully reflected yet in the literature covered by *Chemical Abstracts* and *Analytical Abstracts* discussed above) will be radically different as well.

1.1.3 VOLUME OUTLINE

All of the chapters of this volume have been written by acknowledged research and application leaders, the best that the editors could find in each of the sub-fields. A relatively large fraction of them are Japanese scientists, and this may be a bonus for readers elsewhere in the world, since only about half of the advanced XRS research in Japan is published in English and hence it is not always sufficiently widely known, e.g. in the West.

All the chapters or sets of subchapters cover topics in which remarkable progress has been made during the last decade and which offer good perspective for drastically changing the power of XRS in the near future.

Chapter 2 deals with X-ray sources, which have become more powerful and diverse in the last few years. Significant improvements have been made to the design and performances of conventional X-ray tubes, and in their miniaturisation (which is treated in a later chapter), but most impressive has been the progress in micro-X-ray sources, the development of new synchrotron sources and the first steps towards X-ray laser and laser-induced plasma X-ray sources applicable to XRS. Subchapter 2.1 (by M. Taylor, R. Bytheway and B. K. Tanner of Bede plc, Durham, UK) describes how electromagnetic rather than conventional electrostatic focusing, for shaping and steering the electron beam in the X-ray tube, allows the X-ray source dimensions to be controlled much better than in the past, to achieve a higher brilliance without target damage, to tailor the X-ray spot dimensions for optimising the input coupling with subsequent grazing-incidence X-ray optical elements, like ellipsoidal mirrors and polycapillaries (treated in a later chapter), and hence to deliver high brilliance beams of small dimension to the sample. These high-brightness micro-focus sources have been used mostly in X-ray diffraction (XRD) so far, but they are likely to have a major impact on XRS in the near future as well. Subchapter 2.2 on new synchrotron radiation sources was written by M. Watanabe (Institute of Multidisciplinary Research for Advanced Materials, Tohoku University, Japan) and G. Isoyama (Institute of Scientific and Industrial Research, Osaka University, Japan). In this subchapter, new synchrotron radiation sources are introduced and the characteristics of synchrotron radiation are summarised. New aspects and typical properties of the synchrotron radiation flux at the sample position are described for users of third-generation sources and candidates for fourth-generation sources are discussed. In Subchapter 2.3, C. Spielman (Physikalisches Institut EP1, University of Würzburg, Germany) treats a novel generation of laser-driven X-ray sources, which could produce femtosecond pulses of soft to hard X-rays, synchronisable to other events, and very high intensities, from compact laboratory X-ray sources. This section describes recent progress in the development of laser sources relevant for X-ray generation and reviews the generation of laser-produced incoherent radiation, the development of X-ray lasers and high-harmonic generation. Applications of coherent laboratory X-ray sources are still in their infancy, but these might be intriguing in the future, in XRS, X-ray microscopy, X-ray photoelectron spectroscopy and maybe X-ray interferometry, all of which have had to rely on large-scale synchrotron facilities thus far, and might open the way to attosecond science.

The third chapter is all about X-ray optics, another field that has seen an explosive growth in the last decade, in various ways, resulting in new commercial instruments and new application lines. In Subchapter 3.1, M. Yanagihara (Institute of Multidisciplinary Research for Advanced Materials, Tohoku University, Japan) and K. Yamashita (Department of Physics, Nagoya University, Japan) discuss advances in multilayer production technology, due to the progress in thin-film technology and polishing of super-smooth substrates to the sub-nanometer level, and in their performance and applications. For soft X-rays, the latter include focusing, microscopy and polarimetry; for hard X-rays, obtaining microbeams for microscopy, X-ray telescopes and multilayer-coated gratings are discussed. In Subchapter 3.2, Y. Hosokawa (X-ray Precision, Inc., Kyoto, Japan) presents the state-of-the-art for single capillaries, which make use of multiple external total reflections. He shows how a very bright and narrow X-ray microbeam can be realised using single capillaries (or X-ray guide tubes), and how this leads to a tabletop X-ray analytical microscope. Several applications are presented. In Subchapter 3.3, N. Gao (X-ray Optical Systems, Albany, NY, USA) and K. Janssens (University of Antwerp, Belgium) give a detailed treatment of the fundamentals of multi-fiber polycapillaries and the recent fused and heat-shaped monolithic versions. In the last decade, polycapillary optics have become widespread and successfully used as crucial components in commercial X-ray microanalysis and low-power compact instruments. Novel analytical applications are situated in elemental microanalysis in laboratory scale, portable and synchrotron systems, micro-X-ray absorption near-edge spectroscopy (XANES), EMPA, etc. Future developments in performance and spot size are discussed. Finally, the new compound refractive lenses, first fabricated in 1996, are presented in Subchapter 3.4 by A. Simionovici (European Synchrotron Research Facility, Grenoble, France) and C. Schroer and B. Lengeler (Aachen University of Technology, Germany). Their theory, design and properties are considered, as well as their use for imaging and microbeam production. Some focus is on parabolic refractive X-ray lenses that can be used in e.g. a new hard X-ray microscope that allows sub-micrometer resolution and for e.g. combined fluorescence spectroscopy and tomography. Applications in the realms of biology, XRF computed micro-tomography, geochemistry and environmental research are given.

The most dramatic and spectacular progress has certainly been made recently in the field of X-ray detector technology, and all this is covered in Chapter 4. In Subchapter 4.1, L. Strüder, G. Lutz, P. Lechner, H. Soltau and P. Holl (Max-Planck-Institute for Physics and Extraterrestrial Physics, pnSensor and/or the Semiconductor Lab of the Max Planck Institute, Munich, Germany) treat advances in silicon detectors. After an introduction to the basic operation principles of semiconductors and the electronics used, some important new detectors are discussed in detail and important applications in XRS and imaging are reviewed. The detectors include Silicon Drift Detectors for X-ray detection, Controlled Drift Detectors (CDD), fully depleted backside illuminated pn-CCD and Active Pixel Sensors (APS) for XRS. All these quite sophisticated detectors have left their initial fields of applications in high-energy physics, astrophysics and SR research. They are now a mature technology and open many new industrial applications. These detectors exhibit now a high quantum efficiency, excellent energy resolution, high radiation tolerance, good position resolution, high speed, homogeneous response of the full bandwidth of radiation and high background rejection efficiency. In Subchapter 4.2, C. A. N. Conde (Department of Physics, University of Coimbra, Portugal) treats the role of new gas proportional scintillation counters (GPSC) for XRS, after considering the physics of the absorption of X-rays in gases, the transport of electrons and the production of electroluminescence in gases, and the basic concepts of different types of GPSC. Their energy resolution is only 8 % for 5.9 keV X-rays, but they can be built with very large windows and be useful for very soft X-rays like the K-lines of C and O. Different types of cryogenic detectors, operating near the liquid helium temperature (implying sophisticated cooling systems) and offering unseen energy resolutions, are truly a major development of recent years. However, their commercial availability and price range is still somewhat unclear at the moment. Both superconducting tunneling junctions (STJ) and microcalorimeters are treated in detail in this volume. In Subchapter 4.3, M. Kurakado (Department of Electronics and Applied Physics, Osaka Electro-Communication University, Japan) explains the unique working principles of STJ, which usually consist of two superconductor layers and a nanometer-thick insulator layer, which is a tunnel barrier between the superconductor layers that can be passed by excited electrons or holes, i.e. quasiparticles, to give rise to a signal. Single-junction detectors and two other types of STJ detectors are discussed. Fantastic energy resolutions around 10 eV are possible. New applications are emerging, including one- and two-dimensional imaging. Other equally promising cryogenic detectors are the cryogenic microcalorimeters, treated in Subchapter 4.4 by M. Galeazzi (Department of Physics, University of Miami, Coral Gables, FL, USA) and E. Figueroa-Feliciano (NASA/Goddard Space Flight Center, Greenbelt, MD, USA). The idea of detecting the increase in temperature produced by incident photons instead of the ionisation of charged pairs, like in semiconductor detectors, was put forward almost 20 years ago, and the operating principle is rather simple, but the practical construction is quite challenging. Only in recent years has the practical construction of adequate cryogenic microcalorimeters been realised. The required characteristics, parameters and non-ideal behavior of different components and types, including large arrays, detector multiplexing and position-sensitive imaging detectors, are discussed in detail. Several expected future developments are outlined. In the last section of this chapter on detectors, W. Dabrowski and P. Gryboś (Faculty of Physics and Nuclear Techniques, University of Mining and Metallurgy, Krakow, Poland) treat position-sensitive semiconductor strip detectors, for which the manufacturing technologies and readout electronics have matured recently. Silicon strip detectors, of the same type as used for detection of relativistic charged particles, can be applied for the detection of low-energy X-rays, up to 20 keV. Regardless of some drawbacks due to limited efficiency, silicon strip detectors are most widely used for low-energy X-rays. Single-sided, double-sided and edge-on silicon strip detectors and the associated electronics are treated in great detail.

There are many special configurations and instrumental approaches in XRS, which have been around for a while or have recently been developed. Eight of these are reviewed in Chapter 5. In Subchapter 5.1, K. Sakurai (National Institute for Materials Science, Tsukuba, Japan) deals with TXRF or grazing-incidence XRF (GI-XRF). Although TXRF may have been fading away a bit recently for trace element analysis of liquid or dissolved samples, there have still been advances in combination with wavelength-dispersive spectrometers and for low atomic number element determinations. But mostly, there have recently been interesting developments in surface and interface analysis of layered materials by angular and/or energy-resolved XRF measurements, and in their combination with X-ray reflectometry. Micro-XRF imaging without scans is a recent innovation in GI-XRF as well. Future developments include e.g. combining GI-XRF with X-ray free-electron laser sources. An approach that has not been used widely so far is grazing-exit XRS (GE-XRS), related in some ways to GI-XRF. GE-XRF is the subject of Subchapter 5.2, by K. Tsuji (Osaka City University, Japan). Since the X-ray emission from the sample is measured in GE-XRS, different types of excitation probes can be used, not only X-rays but also electrons and charged particles. In addition, the probes can be used to irradiate the sample at right angles. This subchapter describes the principles, methodological characteristics, GE-XRS instrumentation, and recent applications of GE-XRF, as well as GE-EPMA and GE-PIXE. At the end of this subchapter, the future of GE-XRS is discussed, which implies the use of more suitable detectors and synchrotron radiation excitation. One interesting aspect of XRF is the enormously increased recent (commercial) interest in portable EDXRF systems. This topic is treated in the next subchapter by R. Cesareo and A. Brunetti (Department of Mathematics and

Physics, University of Sassari, Italy), A. Castellano (Department of Materials Science, University of Lecce, Italy) and M.A. Rosales Medina (University 'Las Americas', Puebla, Mexico). Only in the last few years, has technological progress produced miniature and dedicated X-ray tubes, thermoelectrically cooled X-ray detectors of small size and weight, small size multichannel analysers and dedicated software, allowing the construction of completely portable small size EDXRF systems that have similar capabilities as the more elaborate laboratory systems. Portable equipment may be necessary when objects to be analysed cannot be transported (typically works of art) or when an area should be directly analysed (soil analysis, lead inspection testing, etc.) or when the mapping of the object would require too many samples. The advantages and limitations of different set-ups, including optics, are discussed. A focused subchapter on the important new technology of microscopic XRF using SR radiation has been produced by F. Adams, L. Vincze and B. Vekemans (Department of Chemistry, University of Antwerp, Belgium). It describes the actual status with respect to lateral resolution and achievable detection limits, for high-energy, third-generation storage rings (particularly the European Synchrotron Radiation Facility, Grenoble, France), previous generation sources and other sources of recent construction. Related methods of analysis based on absorption edge phenomena such as X-ray absorption spectroscopy (XAS), XANES, X-ray micro-computed tomography (MXCT) and XRD are briefly discussed as well. Particular attention is paid to the accuracy of the XRF analyses. Subchapter 5.5 by I. Nakai (Department of Applied Chemistry, Science University of Tokyo, Japan) deals with high-energy XRF. It considers SR sources and laboratory equipment, in particular a commercial instrument for high-energy XRF that has only recently become available. The characteristics of the technique include improved detection limits, chiefly for high atomic number elements. This makes it particularly suitable for the determination of e.g. rare earths via their X-lines. Other interesting application examples pertain to environmental, archaeological, geochemical and forensic research. Low-energy EMPA and scanning electron microscopy (SEM) are the topics of S. Kuypers (Flemish Institute for Technological Research, Mol, Belgium). The fundamental and practical possibilities and limitations of using soft X-rays, as performed in the two separate instruments, are discussed. The potential of the two techniques is illustrated with recent examples related to the development of ultra-light-element based coatings for sliding wear applications, membranes for ultrafiltration and packaging materials for meat. In Subchapter 5.7, E. Van Cappellen (FEI Company, Hillsboro, OR, USA) treats ED X-ray microanalysis in transmission electron microscopy (TEM), for both the scanning and conventional mode. The section describes how EDXRS in the (S)TEM can be made quantitative, accurate and precise, and is nowadays an extremely powerful technique in materials science and has not vanished in favor of electron energy loss spectrometry (EELS) as predicted 20 years ago. Several examples are given of quantitative chemical mapping, quantitative analysis of ionic compounds and other real-world applications. Finally, J. Kawai (Department of Materials Science and Engineering, Kyoto University, Japan) discusses in detail the advances in XAS or X-ray absorption fine structure spectroscopy (XAFS), which include XANES (X-ray absorption near edge structure) and EXAFS (extended X-ray absorption fine structure). X-ray absorption techniques are now used in commercially available film thickness process monitors for plating, printed circuit and magnetic disk processes, in various kinds of industries. But they are used, both in laboratories and synchrotron facilities, for basic science as well. The X-ray absorption techniques, described extensively in this subchapter, differ in probe type (electrons and X-rays, sometimes polarised or totally reflected), detected signals (transmitted X-rays, XRF, electrons, electric currents, and many others) and application fields (high temperature, high pressure, low temperature, *in situ* chemical reaction, strong magnetic field, applying an electric potential, short measurement time, and plasma states). One shortcoming of XAS techniques, that absorption spectra of all the elements were not measurable using one beamline, has been overcome in many synchrotron facilities nowadays.

Chapter 6 reviews some advances in computerisation concerning XRS. The first subchapter, written by L. Vincze, K. Janssens, B. Vekemans and F. Adams (Department of Chemistry, University of Antwerp, Belgium) deals with modern Monte Carlo (MC) simulation as an aid for EDXRF. The use of MC simulation models is becoming more and more viable due to the rapid increase of inexpensive computing power and the availability of accurate atomic data for photon-matter interactions. An MC simulation of the complete response of an EDXRF spectrometer is interesting from various points of view. A significant advantage of the MC simulation based quantification scheme compared to other methods, such as FP algorithms, is that the simulated spectrum can be compared directly to the experimental data in its entirety, taking into account not only the fluorescence line intensities, but also the scattered background of the XRF spectra. This is linked with the fact that MC simulations are not limited to first- or second-order approximations and to ideal geometries. Moreover, by considering the three most important interaction types in the 1–100 keV energy range (photoelectric effect followed by fluorescence emission, Compton and Rayleigh scattering), such models can be used in a general fashion to predict the achievable analytical characteristics of e.g. future (SR)XRF spectrometers and to aid the optimisation/calibration of existing instruments. The code illustrated in this subchapter has experimentally been verified by comparisons of simulated and experimental spectral distributions of various samples. With respect to the simulation of heterogeneous samples, an example is given for the modeling of XRF tomography experiments. The simulation of such lengthy XRF imaging experiments is important for performing feasibility studies and optimisation before the actual measurement is performed. Subchapter 6.2 by P. Lemberghe (Department of Chemistry, University of Antwerp, Belgium) describes progress in spectrum evaluation for EDXRF, where it remains a crucial step, as important as sample preparation and quantification. Because of the increased count rate and hence better precision due to new detectors, more details became apparent in the spectra; fortunately, the availability of inexpensive and powerful PCs now enables the implementation of mature spectrum evaluation packages. In this subchapter, the discussed mathematical techniques go from simple net peak area determinations, to the more robust least-squares fitting using reference spectra and to least-squares fitting using analytical functions. The use of linear, exponential or orthogonal polynomials for the continuum fitting, and of a modified Gaussian and Voigtian for the peak fitting is discussed. Most attention is paid to partial least-squares regression, and some illustrative analytical examples are presented.

The final chapter, Chapter 7, deals with five growing application fields of XRS. J. Börjesson (Lund University, Malmö and the Department of Diagnostic Radiology, County Hospital, Halmstad, Sweden) and S. Mattsson (Department of Diagnostic Radiology, County Hospital, Halmstad Sweden) focus on applications in the medical sciences since 1995, i.e. on recent advances in *in vivo* XRF methods and their applications, and on examples of *in vitro* use of the technique. The latter deals mostly with the determination of heavy metals in tissues, in well-established ways. But there have been significant developments lately in *in vivo* analysis with respect to sources, geometry, use of polarised exciting radiation, MC simulations and calibration, and the analytical characteristics have been improved. Examples of novel *in vivo* determinations of Pb, Cd, Hg, Fe, I, Pt, Au and U are discussed. The next subchapter deals with novel applications for semiconductors, thin films and surfaces, and is authored by Y. Mori (Wacker-NSCE Corp., Hikari, Japan). Progress in the industrial application of TXRF in this field is first discussed. The use of TXRF for semiconductor analysis came into popular use in the 1990s; today, more than 300 TXRF spectrometers are installed in this industry worldwide, meaning that almost all leading-edge semiconductor factories have introduced TXRF. Since the main purpose of TXRF is trace contamination analysis, improvements in the elemental range (including light elements), detection ability (e.g. by preconcentration) and standardisation (versus other techniques) are discussed. In addition, XRF and X-ray reflectivity analysers for the characterisation of thin films made from new materials are introduced. A. Zucchiatti (Istituto Nazionale di Fisica Nucleare, Genova, Italy) wrote the next subchapter on the important application of XRS in archaeometry, covering instrumentation from portable units through PIXE and synchrotrons. Applications of the latter techniques include e.g. the study of Renaissance glazed terracotta sculptures, flint tools and Egyptian cosmetics. Also the XRF and XANES micro-mapping of corroded glasses is described. Radiation damage is a constant major concern in this field. The much larger availability of facilities and several technological advances have made archaeometry a very dynamic field for XRS today and an even greater research opportunity for tomorrow. T. Ninomiya (Forensic Science Laboratory, Hyogo Prefectural Police Headquarters, Kobe, Japan) illustrates some recent forensic applications of TXRF and SR-XRF. Trace element analysis by TXRF is used to fingerprint poisoned food, liquor at crime scenes, counterfeit materials, seal inks and drugs. Forensic applications of SR-XRF include identification of fluorescent compounds sometimes used in Japan to trace criminals, different kinds of drugs, paint chips and gunshot residues. Subchapter 7.5 deals with developments in electron-induced XRS that have mainly an impact on environmental research, namely the speciation and surface analysis of individual particles, and is written by I. Szaloki (Physics Department, University of Debrecen, Hungary), C.-U. Ro (Department of Chemistry, Hallym University, Chun Cheon, Korea), J. Osan (Atomic Energy Research Institute, Budapest, Hungary) and J. De Hoog (Department of Chemistry, University of Antwerp, Belgium). In e.g. atmospheric aerosols, it is of interest to know the major elements that occur together in one particle, i.e. to carry out chemical speciation at the single particle level. These major elements are often of low atomic number, like C, N and O. In EDXRS, ultrathin-window solid-state detectors can measure these elements but for such soft X-rays, matrix effects are enormous and quantification becomes a problem. Therefore an inverse MC method has been developed which can determine low atomic number elements with an unexpected accuracy. To reduce beam damage and volatilisation of some environmental particles, the use of liquid-nitrogen cooling of the sample stage in the electron microprobe has been studied. Finally, irradiations with different electron beam energies, i.e. with different penetration power, have been applied, in combination with the MC simulation, to study the surface and core of individual particles separately and perform some depth profiling. The given examples pertain to water-insoluble elements in/on individual so-called Asian dust aerosol particles, nitrate enrichments in/on marine aerosols and sediment particles from a contaminated river.

REFERENCE

J. Injuk and R. Van Grieken, *X-Ray Spectrometry*, **32** (2003) 35–39.

Chapter 2
X-Ray Sources

2.1 Micro X-ray Sources

M. TAYLOR, R. BYTHEWAY and B. K. TANNER
Bede plc, Durham, UK

2.1.1 INTRODUCTION

A little over a century ago, X-rays were discovered by Wilhelm Conrad Röntgen (Röntgen, 1995) as a result of the impact of a beam of electrons, accelerated through an electrostatic field, on a metallic target. Current commercial X-ray tubes work on the self-same principle, heated cathodes having replaced the original cold cathodes and water cooling enabling much higher power loads to be sustained. Electrostatic focusing of the electron beam is used in almost all tubes and of the incremental improvements in sealed tube performance over the past 50 years, the recent development of ceramic tubes (e.g. Bohler and Stehle, 1998) by Philips is the only one of note.

A discrete step in performance occurred in the late 1950s with the development (Davies and Hukins, 1984; Furnas, 1990) of the rotating anode generator. Through rapid rotation (several thousand revolutions per minute) of the target, the heat load and hence X-ray emission, could be increased as the heated region is allowed to cool in the period when away from the electron beam. Rotating anode generators are manufactured by a number of companies and provide the highest overall power output of any electron impact device.

It was recognized many years ago that for some applications, in particular those involving imaging, that a small source size was desirable. In the 1960s Hilger and Watts developed a demountable, continuously pumped X-ray tube that found use, for example, in X-ray diffraction topography (Bowen and Tanner, 1998) of single crystals. One of the electron optical configurations for this generator gave a microfocus source but with the demise of the company and the advent of synchrotron radiation, use of such very small sources was unusual.

The limitation of electron impact sources lies principally in the ability to conduct heat away from the region of electron impact, hence limiting the power density on the target. Heat flow in solids is governed by the heat diffusion equation, first derived by Fourier in 1822. This describes the temperature T at any point x, y, z in the solid and at time t. Assuming that there is a heat source described by the function $f(x, y, z, t)$ we have

$$\partial T/\partial t = \alpha^2 \nabla^2 T + f/c\rho \qquad (2.1.1)$$

where $\alpha = (K/c\rho)^{1/2}$ and K is the thermal conductivity, c is the heat capacity and ρ is the density. Thus under steady-state conditions, the key parameter in determining the temperature distribution

Table 2.1.1 Physical properties of target materials

Material	Melting temperature, T_m (°C)	Thermal conductivity, K (W cm^{-1}K^{-1})	Heat capacity, c (J g^{-1} K^{-1})	Density, ρ (g cm^{-3})	Diffusivity, α (cm s$^{-1/2}$)
Cu	1084	4.01	0.38	8.93	1.09
Al	660	2.37	0.90	2.7	0.99
Mo	2623	1.38	0.25	10.22	0.74
W	3422	1.73	0.13	19.3	0.83
Diamond (Type IIa)	3500	23.2	0.51	3.52	3.60

is the thermal conductivity. However, in transient conditions, it is the parameter α, often referred to as the diffusivity, that is the important in determining the maximum temperature at any point. Clearly, to avoid target damage, the maximum temperature T must be significantly below the melting point of the target material.

Reference to Table 2.1.1 shows that the choice of copper as the anode material is governed by more than its ease of working and relatively low cost. In rotating anode generators, even when the actual target material is tungsten or molybdenum, these materials are plated or brazed onto a copper base. Calculations performed many years ago by Müller (Müller, 1931) and Oosterkamp (Oosterkamp, 1948) showed that the maximum permissible power on a target was proportional to the diameter of the focal spot on the target. Relatively little is gained from such strategies as making turbulent the flow of coolant on the rear surface of the target. As the power of the X-ray tube is increased, there must be a corresponding increase in focal spot size and inspection of manufacturers' specifications will readily attest to this fundamental limitation. Synchrotron radiation sources, where there is no such problem of heat conduction, have proved the route past this obstacle.

2.1.2 INTER-RELATIONSHIP BETWEEN SOURCE AND OPTICS

Nevertheless, there are many applications where it is either impossible or impractical to travel to a synchrotron radiation source and thus, driven by the spectacular developments at the synchrotron radiation sources, there has been strong pressure to improve laboratory-based sources. As it is clear from the above impasse that increase in the raw power output was not the solution, attention has focused on the exploitation of X-ray optics. It was realized that there was a prodigious waste of X-ray photons associated with standard collimation techniques and that the scientific community had progressed no further than the pinhole camera. (Of the photons emitted into 2π solid angle, a collimator diameter 1 mm placed 10 cm from the source accepts only 8×10^{-5} steradians. Only 0.0013 % of the photons are used.)

The huge developments in X-ray optics over the past decade are described elsewhere in this volume. In this subchapter we confine ourselves to discussion of only two optical elements, ellipsoidal mirrors and polycapillary optics. These devices are mirrors that rely on the total external reflection of X-rays at very low incidence angles and the figuring of the optic surfaces to achieve focusing. However, as is evident from Figure 2.1.1,

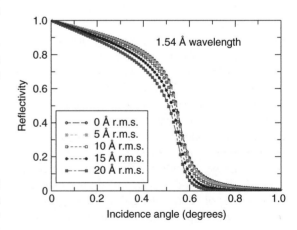

Figure 2.1.1 Reflectivity of a gold surface as a function of incidence angle and surface roughness (r.m.s. = root mean square of the amplitude of surface displacement)

because the refractive index for materials in the X-ray region of the spectrum is only smaller than unity by a few parts in 10^5, the range of total external reflection is very limited. As a consequence of this grazing incidence limitation on total external reflection, to maximize the photon collection, the optic needs to be placed very close to the X-ray source. [Although parabolic multilayer mirrors have been developed (Schuster and Göbel, 1995; Gutman and Verman, 1996; Stommer et al., 1997) that significantly enhance the flux delivered from rotating anode generators, the gains are relatively modest due to the large source size and large source to optic distance. Nevertheless, the combination of high power and insertion gain results in such devices delivering a huge intensity at the specimen.]

A small optic to source distance has an immediate consequence in that, if the beam divergence (or crossfire) at the sample is to be small, the optic to sample distance must be large. The magnification of the source is therefore high and to maintain both a small beam size at the sample and low aberrations, the source size must be very small. Thus, to achieve a high insertion gain from a grazing incidence optic, it is essential that a microfocus X-ray source be used.

2.1.3 A MICROFOCUS GENERATOR WITH MAGNETIC FOCUSING

Despite the widespread use of X-ray fluorescence analysis from extremely small electron beam spots in scanning electron microscopes, until recently all commercial X-ray generators used exclusively electrostatic focusing. This is despite the fact that electron microscope manufacturers long ago realized that magnetic focusing was superior in many ways. Electrostatically focused tubes generally exhibit side lobes to the electron beam spot and are relatively inefficient at delivering electrons to the target itself.

The first electromagnetically focused microfocus tube was described by Arndt, Long and Duncumb (Arndt et al., 1998a) in 1998. The design maximized the solid angle of collection of the emitted X-rays and thus, in association with an ellipsoidal mirror, achieved a high intensity at the sample. The observed intensity was in reasonable agreement with that calculated and compared with that achieved with non-focusing X-ray optics used with conventional X-ray tubes operated at a power more than 100 times as great.

In the patented design (US Patent No. 6282263, 2001), the electron beam, of circular cross-section, from the gun is focused by an axial magnetic lens and then drawn out by a quadrupole lens to form an elongated spot on the target. When viewed at a small take-off angle an elongated focus is seen, foreshortened to a diameter between 10 and 20 μm. Within the X-ray generator (Figure 2.1.2) the tube is sealed and interchangeable. The electron optics enable the beam to be steered and focused into either a spot or a line with a length to width ratio of 20:1. An electron mask of tungsten is included to form an internal electron aperture. The electron gun consists of a Wehnelt electrode and cathode that can be either a rhenium–tungsten hairpin filament or an indirectly heated activated dispenser cathode. The advantages of the dispenser cathode is that it is mechanically stable and, due to the lower power consumption and operating temperature, it has a greater lifetime than heated filament cathodes. It is also simpler to align in the Wehnelt electrode. The tube is run in a space-charge limited condition (as opposed to the conventional saturated, temperature limited condition), with the filament maintained at a constant temperature. As a result, the tube current is determined almost exclusively by the bias voltage between the filament and the Wehnelt electrode. The electrons are accelerated from the cathode, held at a high negative potential, towards the grounded anode. They pass through a hole in the anode before entering a long cylinder and subsequently colliding with the target. An electron cross-over is formed between the Wehnelt and anode apertures and this is imaged onto the target by the iron-cored axial solenoid. So far Cu, Mo and Rh target tubes have been run successfully. The power loading that can be achieved is such that a small amount of water-cooling proves essential.

The ability to control the electron beam spot size and shape by adjustment of the current in

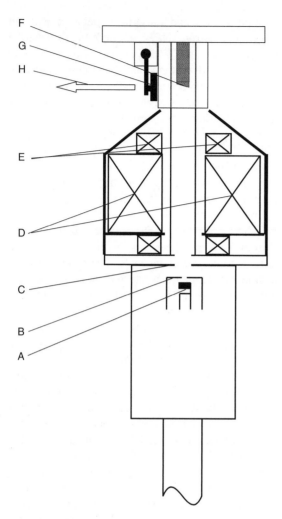

Figure 2.1.2 Schematic drawing of an electromagnetically focused microfocus X-ray tube. (A) Cathode, (B) Wehnelt grid electrode, (C) anode, (D) electromagnetic axial focusing lens, (E) electromagnetic quadrupole lens, (F) target, (G) X-ray shutter, (H) direction of X-ray beam

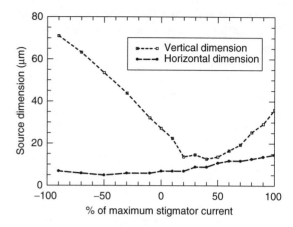

Figure 2.1.3 Source dimension as a function of the maximum current through the stigmator coils

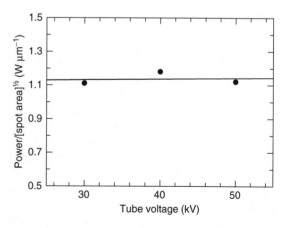

Figure 2.1.4 Power divided by the square root of the spot area under conditions for achieving maximum output intensity at three settings of tube voltage. The line is a least-squares linear fit to the data

the quadrupole stigmator coils is critical to the optimum performance of the microfocus tube. Figure 2.1.3 shows the variation of the source dimensions as a function of the current in the stigmator coils. We note that there is a significant range, close to the minimum in source area, which is almost independent of the stigmator current and where the source is approximately equiaxed.

As the tube accelerating voltage is increased, the value of the current in the focusing coils needs to be increased to achieve the minimum spot size. Increasing the tube current usually increases the spot size, due to space charge effects. It is our standard practice to set up the focusing coils with the stigmator coil current chosen to achieve an equiaxed beam. However, by adjusting the stigmator coil current to draw the source out into a line, a significant gain in intensity can be obtained without compromising on the coupling into subsequent X-ray optical elements. The output is limited by the maximum power at which target damage does not occur and Figure 2.1.4 shows that the maximum permissible power divided by the square root of the spot area remains constant whether the tube is run at 30, 40 or 50 kV. This is in agreement with the results of Müller (Müller, 1931) and Oosterkamp (Oosterkamp, 1948). The maximum loading is a fortuitously close to an easily remembered value, namely $1\,\mathrm{W}\,\mu\mathrm{m}^{-1}$.

2.1.4 SOURCE SIZE MEASUREMENT

Measurement of the spot size, so crucial for the operation of microfocus sources, is not trivial and care needs to be taken in its definition. There are two measurement methods that can be adopted. The first is to use a small pinhole (5 μm) in a heavy metal substrate (e.g. platinum) to create a 'pinhole camera' image of the source on an imaging device. The second is to use a wire or knife-edge of heavy metal (e.g. platinum or tungsten) to cast a shadow on such an imaging device.

In the former technique (Figure 2.1.5), a pinhole is used to project an image of the X-ray source onto an imaging device with source to specimen distance a and specimen to detector distance b typically, $a = 12\,\mathrm{mm}$, $b = 250\,\mathrm{mm}$.

A source size can be measured directly from the image that is collected, and scaled by a factor a/b. Resolution is affected by the size of the pinhole h, distances a and b, and the resolution of the imager. Each point of the source illuminates a spot on the imager of size $h[(b+a)/a]$. This gives a minimum distance to position the imager, where the illuminated spot is equal to the pixel size of the imager. For the Photonic Science X-Ray Eye 2i imagers (17 μm pixel size) with a 5 μm pinhole, $b_{\min}$ is 30 mm.

Each point on the image corresponds to a finite portion of the source of size $h[(a+b)/b)]$. When $b \gg a$ the size of this portion of the source is approximately equal to the pinhole size h. This limits the minimum observable source size to the size of the pinhole. In practice, the minimum size from which useful measurements can be taken is more nearly two to three times the pinhole size.

In the second method (Figure 2.1.6), a fine tungsten wire is used to cast a shadow when placed in the X-ray beam. It can be used to measure a source of size less than the wire diameter. The size of the source can be calculated from the intensity profile in two ways:

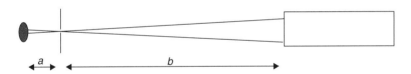

Figure 2.1.5 Schematic diagram of the pinhole geometry for determination of source size

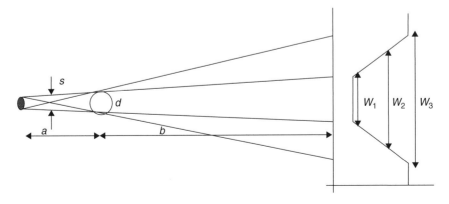

Figure 2.1.6 Shadow technique for source size measurement

(1) If the diameter of the wire d is accurately known, the source size s is given by:

$$s = \frac{W_3 - W_1}{2} \cdot \frac{d}{W_2} \qquad (2.1.2)$$

Using a digital imager and image collection software, a very smooth curve can be obtained by averaging many frames, and then averaging along the shadow.

(2) If the magnification is known, the source size s is given by the width of the sloped part of the intensity profile $(W_3 - W_1)/2$, divided by the magnification.

For both of these methods, it is necessary to identify the corners of the intensity distribution, that is where to measure W_1 and W_3. The simplest approach is to fit straight lines to the penumbra (Figure 2.1.7) and measure directly.

It is usual to quote the full width at half-maximum (FWHM) of a source but unfortunately, the majority of the microfocus source shapes are not simple statistical distributions. The standard Microsource® 80 W source is a very square shape with more intensity in the corners than the centre. The shadow measurement gives a measurement of the source dimension perpendicular to the wire axis, and integrates across the source parallel to the wire axis. Assuming a Gaussian profile we can calculate the shadow data ('Cumulative Centre' line) as shown in Figure 2.1.8.

The 'Cumulative Centre' line is a plot of the shadow that would be seen in a wire measurement and we note that the 12% (and 88%) points correspond to the FWHM dimension of the Gaussian distribution. The source dimension given by these points corresponds to the size of spot that includes 76% of the total energy. This is useful when matching source size to an optic since the aim is to capture as much of the source power with the optic as possible. By using the 12% corner points in the distribution for non-Gaussian distributions, a 'Gaussian equivalent FWHM' can be determined.

2.1.5 APPLICATIONS IN PHASE CONTRAST IMAGING

There is a general perception that a monochromatic, coherent beam is necessary for phase imaging. Such highly coherent beams are available at synchrotron radiation sources and spectacular, high-resolution phase images have been reported from third generation sources in recent years (Cloetens *et al.*, 1999, Elliot *et al.*, 2002). However, it was demonstrated by Wilkins *et al.*

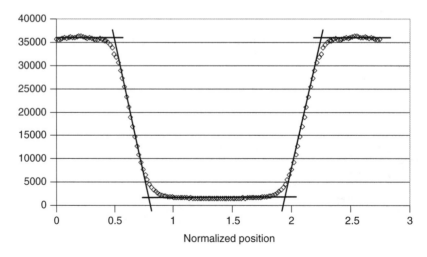

Figure 2.1.7 Measured intensity as a function of normalized position across the image of a tungsten wire with linear interpolation to determine the shadow extent

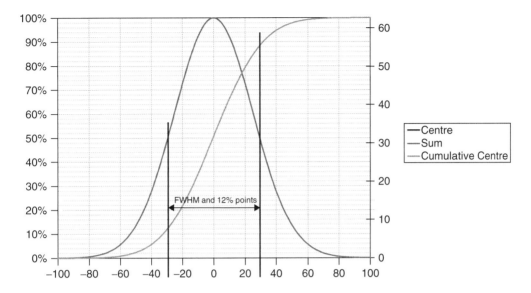

Figure 2.1.8 'Cumulative Centre' line and Gaussian profile of a model source geometry

(Wilkins *et al.*, 1996) that even for a polychromatic source, refractive index gradients distort the wavefront. The crucial insight of the Australian group was to recognize that the wavefront distortion diverges at the edge of a circular object for all X-ray wavelengths present. As a result, if the imaging medium is placed a significant distance from the specimen, then the distortion of the wavefront results in interference and hence characteristic intensity changes. Thus, this very simple phase contrast imaging method reveals regions of high electron density gradient. To avoid resolution blurring destroying the effects, a microfocus source is needed. A detailed analysis of the contrast shows that, in the differential phase contrast the wavelength appears only as a separable factor, the geometric features of the contrast being wavelength independent (Pogany *et al.*, 1997). The technique has the benefit of ultimate simplicity in that no monochromatization is thus strictly necessary. The dominance of characteristic X-ray lines gives more than adequate monochromaticity for high contrast at quite high spatial frequency. Recovery of the phase and hence the quantitative determination of the electron density is, however, much more difficult.

Figure 2.1.9(a) shows an example of phase contrast from a small fly. Very little absorption contrast is present, the image revealing the edges of the body and organs. The phase contrast imaging technique presents an interesting application in the food processing industry, where there is a need to detect and recognize low atomic number (organic) objects inside the products. Figure 2.1.9(b) shows an image of the same fly inside a tuna fish sandwich. The contrast from the sandwich is strong, but against this slow contrast variation the sharp outline of the phase contrast image is readily identifiable.

2.1.6 FOCUSING OPTICS WITH MICROFOCUS SOURCES

2.1.6.1 ELLIPSOIDAL MIRRORS

As indicated above, the key potential of microfocus sources lies in the ability to couple closely X-ray optical elements, thereby resulting in a very substantial insertion gain at the sample. The design of Arndt *et al.* (Arndt *et al.*, 1998b) is such that the distance from source to optic can be less than 10 mm. They reported a collection solid angle

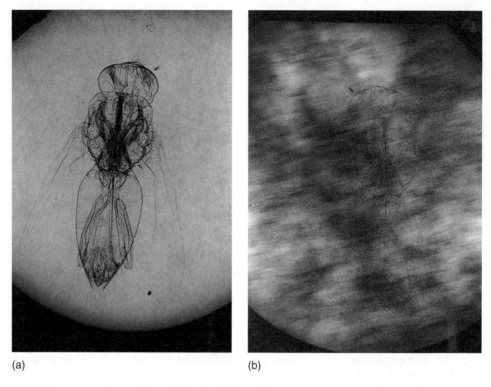

Figure 2.1.9 (a) Phase contrast image of a small fly, length 3 mm. (b) Image of the same fly inside a tuna fish sandwich

of 8×10^{-4} steradians using an ellipsoidal mirror (Figure 2.1.10) and a resulting insertion gain of over 100.

The small ellipsoidal mirrors were developed in the Czech Republic by Hudec and co-workers at the Czech Academy. They can be produced by an electroforming process (Hudec *et al.*, 1988, 1993) in which a highly polished mandrel is first made as a negative of the final required shape. Following lacquer polishing, the mandrel is coated with gold and then electroplated with nickel. The whole is then surrounded with a carbon-fibre reinforced epoxy to give structural strength. On withdrawal of the mandrel, the gold is left behind on the nickel shell (Arndt *et al.*, 1998b). The complete assembly is only a few centimetres in length and the inner diameter of the ellipsoid is approximately 1 mm (Figure 2.1.11).

Although the hole through the centre of the ellipsoid does mean that there is a halo of diverging, non-reflected X-rays emerging in addition to the focused beam (Figure 2.1.10) in practice this does not prove a problem as the focus is typically 30–60 cm from the mirror and the majority of the

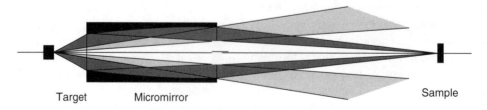

Figure 2.1.10 Ellipsoidal mirror for application in protein crystallography

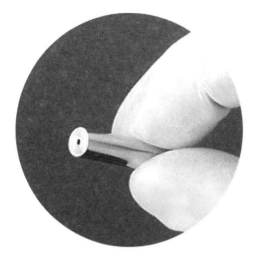

Figure 2.1.11 Photograph of a Reflex Micromirror™

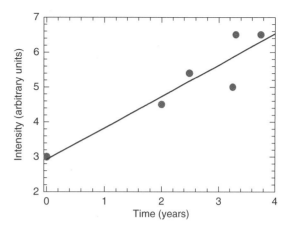

Figure 2.1.13 Improvement in Reflex s.r.o. Micromirror™ output performance over time (1997–2000). (Intensity was measured through a 500 μm pinhole at 150 mm from the mirror in all cases)

X-rays in the diverging halo can be removed by an aperture. For X-ray diffraction applications, a divergence of less than a few milliradians is necessary and to achieve this low level of crossfire at the focus, the ellipsoidal mirrors must have high magnification. As a result, the 10 μm source is magnified to typically 300 μm at the specimen. An example of the focusing is shown in Figure 2.1.12, where for a 1 mrad divergence mirror, the FWHM of the beam at the sample is 300 μm in this case.

There has been gradual improvement in the performance of the ellipsoidal mirrors over time (Figure 2.1.13). As the roughness has only a small effect on the efficiency in the total external reflection regime, the principal improvement has come from improved figuring.

In the field of protein crystallography where there is a demand for a low divergence to match crystal perfection and small beam area to match small crystal size, the results have been encouraging. Arndt and Bloomer (Arndt and Bloomer, 1999a,b) note that diffracted intensities approach that equivalent from a 5 kW rotating anode generator with X-ray mirrors, the microfocus source consuming only 80 W, a fraction of the power of

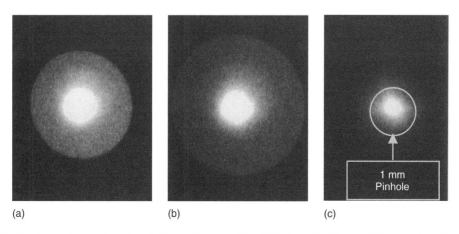

Figure 2.1.12 Focal spot size as a function of distance from an ellipsoidal mirror: (a) 15 cm, (b) 20 cm and (c) 55 cm

the rotating anode generator. Further the microfocus source is extremely small and compact compared with a rotating anode generator, which can typically weigh up to 1 tonne. Microfocus sources fit in a suitcase and weigh typically a few kilograms. In terms of structure refinement with such sources, very good agreements are found with an overall R merge from 100 Å to 1.8 Å of 3.2%. Protein data are visible out to 1.6 Å with very 'clean' diffraction patterns. Rocking curve widths are typically 0.15–0.18°.

2.1.6.2 POLYCAPILLARY OPTICS

Polycapillary optics provide an alternative method for collecting efficiently the output from a microfocus tube. They consist (Figure 2.1.14) of a bundle of hollow borosilicate glass tubes stacked together, the X-rays being totally externally reflected from the internal surfaces of the capillary tubes. By shaping the bundle appropriately, the spot size and beam divergence can be varied over a wide range.

As a capillary, like the ellipsoidal mirror, relies on total external reflection to guide the X-rays (much in the manner of an optical waveguide) the angular range of acceptance is small. Thus, as with the mirror, optimum coupling is achieved only when the polycapillary optics is placed very close to the X-ray source, again presenting a major challenge in the engineering design of the microfocus source housing. Again, as with the mirror, there is a requirement to minimize the source size and maximize the brilliance in order to maximize the intensity, as opposed to count rate, at the sample. Where the polycapillary optic gains over the ellipsoidal mirror is that by shaping the bundle so that each capillary is oriented for grazing incidence, the effective aperture angle ω can be made quite large. Although care must be taken in such comparisons as the divergence often differs, in practice there is a gain of about an order of magnitude in intensity transmitted by a polycapillary optic compared with an ellipsoidal mirror.

Over recent years the improvement in performance of polycapillary optics has been dramatic. Configured for large crossfire but small beam diameter at the sample, in combination with a microfocus X-ray tube they have major potential for X-ray fluorescence analysis.

2.1.7 APPLICATION OF MICROFOCUS TUBES AND OPTICS TO HIGH-RESOLUTION X-RAY DIFFRACTION (HRXRD)

High-resolution X-ray diffraction (Bowen and Tanner, 1998) has found a pivotal niche in the nondestructive characterization of the compound semiconductors that underpin high-speed optical

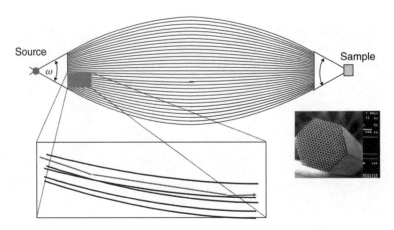

Figure 2.1.14 Schematic diagram and photograph of a polycapillary focusing optic. (Reproduced with kind permission of XOS Inc. USA)

communications systems. Since the pioneering experiments in the 1980s, the technique has moved from research laboratory to production line within the most advanced semiconductor fabrication plants in the world. The key to high-resolution X-ray diffraction measurements is the use of Bragg reflections from highly perfect crystals to control the angular divergence and monochromaticity of the beam hitting the sample. Such tailored beams can be used to measure the relative lattice parameters of substrate and heteroepitaxial layers, enabling the composition, thickness, perfection, relaxation, and uniformity to be determined without damage to the wafer.

An example of a high-resolution diffraction configuration is shown in Figure 2.1.15, where a pair of asymmetric channel-cut silicon crystals are used successively to condition the angular spread and then the wavelength spread of the X-ray beam. In this so-called DuMond arrangement, the combination of Si 022 reflections gives a beam of 11.5 arc seconds angular divergence and dispersion $\Delta\lambda/\lambda = 1.3 \times 10^{-4}$ with CuKα radiation. With the resulting beam, the Bragg peaks from substrate and layer having very closely matched lattice parameters can be separated. Further, interference phenomena, giving a precise measurement of the layer thickness can be resolved. Using current generation personal computers, the diffraction profile (rocking curve) can be simulated for a model structure, compared with the experimental data and iteratively fitted by adjustment of the layer parameters in the model. The fitting is far from trivial, due to the presence of multiple minima in the difference function between simulation and experiment, and only with the introduction of genetic algorithms by Wormington *et al.* (Wormington *et al.*, 1999) in 1999 was the problem solved for anything other than the simplest cases.

An example of a rocking curve from a psuedomorphic high electron mobility transistor (pHEMT) structure (commonly used in mobile telecommunications), consisting of three layers on a GaAs (001) oriented substrate, is given in Figure 2.1.16. The first layer is highly strained and is of $In_xGa_{1-x}As$, the second of $Al_xGa_{1-x}As$ and the third is a pure GaAs cap. Thickness and compositions deduced from fitting the X-ray data are given in Table 2.1.2.

In the application of high-resolution diffraction for in-line process control there is a constant drive towards reduction of the area of the probe beam on the sample. This must be achieved without loss of statistical accuracy or increase in measurement time. With a microfocus source and associated optic Taylor *et al.* (Taylor *et al.*, 2001) have shown that not only can the beam footprint be reduced but also the exposure time compared with larger focus sources. In the experiments reported by Taylor *et al.*, the standard air-cooled 80 W source in the current generation Bede QC diffractometer (Loxley *et al.*, 1991) was replaced by an 80 W microfocus source and an ellipsoidal mirror (Figure 2.1.17). In this setting, the monochromating is done by the aperture after the reference crystal rather than a second channel-cut crystal. Although a paraboloidal mirror would give the most efficient coupling of the beam into the beam conditioning crystal, the beam size would be large and the beam profile annular. The focusing of the ellipsoidal mirror not only reduces the spot size but also makes adjustment extremely easy, albeit at the expense of some intensity.

Figure 2.1.16 shows an example of the order of magnitude gain in count rate as well as the overall

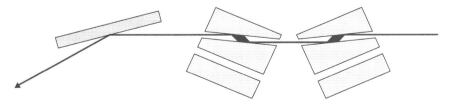

Figure 2.1.15 Schematic diagram of the high resolution double axis diffraction setting with DuMond monochromator before the specimen

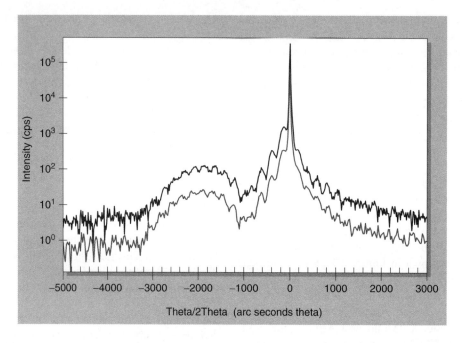

Figure 2.1.16 Double axis X-ray rocking curve from a pHEMT structure on GaAs (Taylor *et al.*, 2001). Upper curve: data collected on a Bede QC200 diffractometer with microfocus source running at 50 W, plus micro-mirror. CuKα_1 wavelength present only, beam footprint 1 mm diameter, total collection time 22 min. Lower curve: standard compact source (0.25 mm focal spot), no mirror, CuKα_1 and CuKα_2 wavelengths present. Beam footprint approximately 3 mm × 2 mm, collection time 53 min. The 10-fold increase in count rate and the beam footprint reduced by 85 % represent an increase in brightness of around 70 times. Reprinted from Materials Science and Engineering B, Vol. 80, M. Taylor, J. Wall, N. Loxley, M. Wormington and T. Lafford, 'High resolution X-ray diffraction using a high brilliance source, with rapid data analysis by auto-fitting', pages 95–98, Copyright (2001), with permission from Elsevier Science

Table 2.1.2 Thickness and composition of a pHEMT structure

Layer	Material	Thickness (nm)	Composition fraction x
3	GaAs	35.07 (+0.52, −0.3)	–
2	Al$_x$Ga$_{1-x}$As	41.57 (+0.52, −0.34)	0.286 (+0.024, −0.037)
1	In$_x$Ga$_{1-x}$As	15.75 (+0.42, −0.16)	0.117 (±0.001)
Substrate	GaAs	–	–

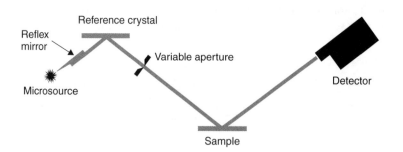

Figure 2.1.17 Schematic diagram of the high resolution configuration for high throughput and small spot size

reduction in beam area of a factor 7 between the original configuration and that adopted in the Bede QC200 system. Similar enhanced performance in the analysis of SiGe epitaxial layers has been reported by Zaumseil et al. (Zaumseil et al., 2001). Further intensity gain can be achieved by use of a polycapillary focusing optic, and it is now straightforward to locate automatically (with robot handling) a conditioned beam 300 μm diameter on a test window in a 300 mm wafer. Without the microfocus X-ray tube, the development of the fully automated X-ray tools that are now accepted for key metrological functions in the silicon industry would not have been possible.

2.1.8 APPLICATION OF MICROFOCUS TUBES AND OPTICS TO GRAZING INCIDENCE IN-PLANE X-RAY DIFFRACTION (GIIXD)

Grazing incidence in-plane X-ray diffraction is an X-ray scattering technique through which the in-plane lattice parameter and lattice orientation of very thin surface and buried semiconductor layers can be determined (Robinson and Tweet, 1992). With the incident beam close to or below the critical angle for total external reflection, a Bragg reflection is excited from planes perpendicular to the surface (Figure 2.1.18). As the depth penetration of a wave incident below the critical angle is only a few nanometres, the technique has become of major importance in studying the surface crystallography and, in particular, the construction of semiconductors. In this geometry, the simple Bragg condition for diffraction is never quite satisfied and the scattering power is therefore weak. Further, at these very low angles of incidence, much of the beam spills off the sample, not contributing to the scattering. Using normal X-ray generators, the intensity is generally so low that statistics are poor and data difficult to interpret. As a consequence, most of the developments have occurred at synchrotron radiation sources.

However, the ability to focus the beam from a microfocus source to a spot typically 300 μm is diameter with a divergence of typically 2 mrad opens up the possibility of performing these experiments in the laboratory on a realistic time-scale. Goorsky and Tanner (Goorsky and Tanner, 2002) have demonstrated extremely high intensity diffracted beams from a microfocus X-ray tube and ellipsoidal focusing optic.

In the initial experiments, the Cu target microfocus X-ray generator was run at a power of 80 W, the beam being focused onto the sample using a Reflex Micromirror^{TM} ellipsoidal mirror of focal length 300 mm and divergence at the sample of 2 mrad in both horizontal and vertical directions. The resulting spot diameter at the sample was 0.3 mm FWHM. Even so, at the extremely low angle of incidence (α_f in Figure 2.1.18) of typically 0.25°, the footprint on the sample was about 70 mm. This resulted in substantial beam spill-off for small samples. No further conditioning of the incidence beam was used. The sample was rotated

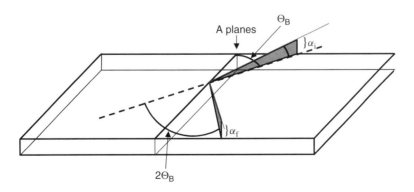

Figure 2.1.18 Geometry of GIIXD. θ_B, Bragg angle; α_f, angle of incidence of X-ray beam to surface

until it was horizontal, the vertical height being adjusted to half-cut the incident beam. A small rotation about the axis horizontal and perpendicular to the incident beam permitted the grazing angle to be tuned. The sample was rotated about the axis normal to the sample surface. A 2 mrad divergence Soller slit limited the angular acceptance of the scintillation detector in the horizontal plane in which the detector was scanned ($2\Theta_B$ axis). Over the small range of grazing incidence angles used in GIIXD, the vertical aperture of the detector was large enough to accept all beams.

The technique has proved particularly powerful for the measurement of the in-plane mosaic of the GaN-based epitaxial films used for blue light emitting devices. Measurement of the in-plane mosaic is particularly difficult and involves a model-dependent complex deconvolution of several data sets. The GIIXD technique enables the in-plane mosaic to be determined directly from a single measurement, the total experimental time (including alignment) being 5 min. Figure 2.1.19 shows an example of the measurement of the in-plane mosaic of a (001) oriented epitaxial multilayer of Fe–Au on MgO carried out at the European Synchrotron Radiation Facility and in the laboratory. The agreement is excellent. In both cases a Voigt function was necessary to fit the tails of the intensity distribution although a Gaussian was a quite reasonable approximation. The noise base of the laboratory measurements was proportionally higher than at the synchrotron radiation source as there was little discrimination against Fe fluorescence in the laboratory measurements. Use of a Si(Li) energy dispersive detector at the synchrotron source eliminated the fluorescence almost completely.

2.1.9 SUMMARY

The use of electromagnetic focusing to focus and steer the electron beam has enabled X-ray source dimensions to be controlled to a much greater tolerance than in the past. In particular the absence of side lobes on the spot and controllable astigmatism permits the achievement of a higher brilliance without target damage and the ability of tailoring the spot dimensions to optimize the input coupling to subsequent X-ray optical elements. The result is the efficient use of grazing incidence optical elements such as mirrors and polycapillaries to deliver high brilliance beams of small dimension on the sample. While the greatest impact has so far been in the field of diffraction, the availability of high brightness microfocus sources is likely to have a major impact on X-ray spectroscopy in the near future.

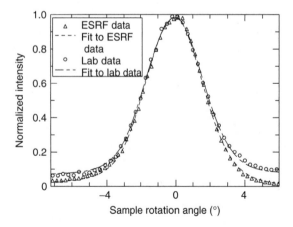

Figure 2.1.19 Comparison of GIIXD rocking curves taken at the European Synchrotron Radiation Facility and in the laboratory using a microfocus X-ray generator. Fits are to a Voigt function

REFERENCES

Arndt, U. W. and Bloomer, A. C. *Curr. Opin. Struct. Biol.*, **9**, 609 (1999a).

Arndt, U. W. and Bloomer, A. C. *Acta Cryst., Biol. Crystallogr.*, **D55**, 1672, (1999b).

Arndt, U. W., Long, J. V. P. and Duncumb, P. *J. Appl. Cryst.*, **31**, 936 (1998a).

Arndt, U. W., Duncumb, P., Long, J. V. P., Pina, L. and Inneman, A. *J. Appl. Cryst.*, **31**, 733 (1998b).

Bohler, P. and Stehle, J. L. *Phys. Status Solidi A*, **170**, 211 (1998).

Bowen, D. K. and Tanner, B. K. *High Resolution X-ray Diffractometry and Topography*, Taylor and Francis, London, 1998.

Cloetens, P., Ludwig, W., Baruchel, J., Guigay, J.-P., Pernot-Rejmánková, P., Salomé-Pateyron, M., Schlenker, M., Buffière, J.-Y., Maire, E. and Peix, G. *J. Phys. D: Appl Phys.*, **32**, A145 (1999).

References

Davies, K. E. and Hukins, D. W. L. *J. Appl. Cryst.*, **17**, 486 (1984).

Elliott, J. A., Windle, A. H., Hobdell, J. R., Eeckhaut, G., Oldman, R. J., Ludwig, W., Boller, E., Cloetens, P. and Baruchel, J. *J. Mater. Sci.*, **37**, 1547 (2002).

Furnas, T. C. *Nucl. Instrum. Methods*, , **299**, 246 (1990).

Goorsky, M. S. and Tanner, B. K. *Crystal Res. Technol.*, **37**, 647 (2002).

Gutman, G. and Verman, B. *J. Phys D: Appl. Phys.*, **29**, 1675 (1996).

Hudec, R., Valnicek, B., Ashenbach, B., Braeununger, H. and W. Burkart, *Appl. Opt.*, **27**, 1453 (1988).

Hudec, R., Arndt, U. W., Inneman, A. and Pina, L. *Inst. Phys. Conf. Ser.*, **130**, 499 (1993).

Loxley, N., Bowen, D. K. and Tanner, B. K. *Mater. Res. Soc. Symp. Proc.*, **208**, 119 (1991).

Müller, A. *Proc. R. Soc. London*, **132**, 646 (1931).

Oosterkamp, W. J. *Philips Res. Repts*, **3**, 303 (1948).

Pogany, A., Gao, D. and Wilkins, S. W. *Rev. Sci. Instrum.*, **68**, 2774 (1997).

Robinson, I.K. and Tweet, D. J. *Rept. Prog. Phys.*, **55**, **599** (1992).

Röntgen Centennial, University of Würzburg, 1995.

Schuster, M. and Göbel, H. *J. Phys. D: Appl. Phys.*, **28**, A270 (1995).

Stommer, R., Metzger, T., Schuster, M. and Göbel, H. *Nuovo Cim. D*, **19**, 465 (1997).

Taylor, M., Wall, J., Loxley, N., Wormington, M. and Lafford, T. *Mater. Sci. Eng. B*, **80**, 95 (2001).

US Patent No. 6282263 (2001).

Wilkins, S. W., Gureyev, T. E., Gao, D., Pogany, A. and Stevenson, A. W. *Nature*, **384**, 335 (1996).

Wormington, M., Panaccione, C., Matney, K. M. and Bowen, D. K. *Philos. Trans. R. Soc.*, **357**, 2827 (1999).

Zaumseil, P., Lafford, T. A. and Taylor, M. *J. Phys. D: Appl. Phys.*, **34**, A52 (2001).

2.2 New Synchrotron Radiation Sources

M. WATANABE
Institute of Multidisciplinary Research for Advanced Materials, Tohoku University, Japan

and

G. ISOYAMA
The Institute of Scientific and Industrial Research, Osaka University, Japan

2.2.1 INTRODUCTION

Synchrotron radiation is light emitted by high-energy electrons moving on a circular orbit in a uniform magnetic field. It has a continuous spectrum ranging from the infrared, through the visible, the ultraviolet, and the soft X-ray, to the hard X-ray regions.[1-4] Experiments in the soft X-ray and the hard X-ray regions began in the 1960s using light from bending magnets of synchrotrons operated for elementary particle physics. In the 1970s, synchrotron radiation users began to parasitically use storage rings constructed for colliding beam experiments, instead of synchrotrons, as a light source, because the lifetime of the electron or positron beam circulating in a storage ring is longer than a few hours and hence synchrotron radiation from a storage ring is spatially and temporally much more stable than that from a synchrotron. These synchrotrons and storage rings used parasitically are called first generation sources.

In the 1980s, synchrotron radiation users acquired their storage rings dedicated to light sources, which are called second generation sources. Figure 2.2.1 shows a schematic drawing of a storage ring as a light source. Various experiments, including spectroscopy, photoelectron spectroscopy, extended X-ray absorption fine structure (EXAFS), and X-ray diffraction, were extensively conducted with light from bending magnets of these dedicated machines. Time resolved experiments were also conducted by making use of the pulsed time structure of synchrotron radiation from these storage rings. Insertion devices such as superconducting wigglers, multipole wigglers, and undulators were developed in this decade in order to obtain higher energy light, higher flux light and brighter quasi-monochromatic light, respectively. These insertion devices consist of arrays of magnets with alternating fields and they are installed in straight sections of storage rings. Hereafter the term 'synchrotron radiation' means light not only from bending magnets but also from insertion devices.

Several third generation sources were constructed in the 1990s, and even at present some are under construction and others are in the planning stages. They are storage rings optimized for utilizing insertion devices, especially undulators. In order not to deteriorate bright light from undulators, the size and the divergence of the electron beam, the product of which is approximately equal to the emittance of the electron beam, has to be reduced considerably. Third generation sources are, therefore, low emittance rings. The vertical emittance of an electron storage ring is much smaller than the horizontal one and the diffraction

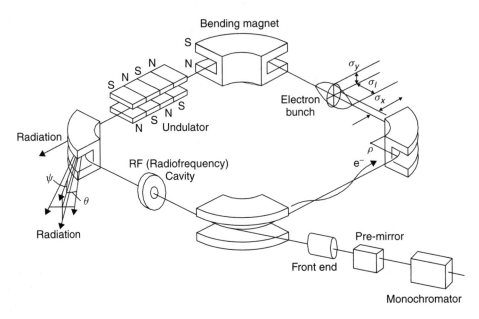

Figure 2.2.1 Schematic drawing of a storage ring as the synchrotron radiation source and a beamline. The angular coordinate is also defined for synchrotron radiation from a bending magnet

limit is achieved in the longer wavelength region of synchrotron radiation for the vertical direction.

By making use of third generation sources, experiments requiring brighter light have progressed, such as microscopy, high resolution measurement, and two-color experiments using synchrotron radiation together with laser. In addition, it is noteworthy that other synchrotron radiation sources have been newly born and some are under construction, which should be evaluated as an intermediate class between second and third generation sources. This means that the utilization of synchrotron radiation grows rapidly and spreads worldwide. Stability of the electron beam in its cross-section and position has considerably improved compared with that in the 1980s, due to cures for collective instability and development of feedback systems, so that more precise measurements can be made. A new operation mode of storage rings was recently tested, which is called top-up injection. Whenever the stored current decreases slightly, electrons are repeatedly injected to keep the current almost constant. The synchrotron radiation sources open to outside users in the world are listed in Table 2.2.1. The emittance of the storage rings indicates the horizontal emittance, as usual.

On the other hand, efforts have been made to develop free electron lasers (FEL). An FEL is comprised of an undulator and an optical resonator composed of two mirrors, and a high energy electron beam is provided from a linac or a storage ring. The first FEL was realized in 1977 in the infrared region using a low energy linac. The wavelength region is extended to the visible and the ultraviolet with FEL based on storage rings. In 2000, lasing in the vacuum ultraviolet was achieved with a special type of the linac-based FEL without an optical resonator, called self-amplified spontaneous emission (SASE). Some SASE-FELs in the soft X-ray and the hard X-ray regions are under construction or in the design stage.

Fourth generation sources are expected to appear in the 2000s.[5] One candidate for such light sources is SASE-FEL mentioned above. Others are storage rings with ultra-low emittance and energy recovery linacs recently proposed. In fourth generation sources, the emittance of the electron beam will reach the diffraction limit even in the horizontal direction and brilliance will be very

Table 2.2.1 Synchrotron radiation sources

Location	Institute	Ring	Energy (GeV)	ε_c (keV)	Emittance (π nm·rad)
Spain					
Barcelona	Catalonia SR Lab.		2.5–3		
UK					
Daresbury	SRC	SRC[2]	2.0	3.2	
Chilton	RAL	DIAMOND[3a]	3	2.5	14.3
France					
Orsay	LURE	DCI[2]	1.85	3.7	1600
	LURE	Super ACO[2]	0.8	0.67	35
		SOLEIL[3a]	2.5		3.0
Grenoble	ESRF	ESRF[3]	6	20.6	3.9
Switzerland	Paul Scherrer Inst.				
Villigen		SLS[3]	2.4	5.4	4.4
Italy					
Frascati	INFN	DAFNE[1]	0.51	0.2	1000
Trieste	Syn. Trieste	Elettra[3]	2.0	3.2	7.0
Germany					
Karlsruhe	FZK	ANKA[2]	2.5	6.2	41
Bonn	Bonn Univ.	ELSA[1]	1.6–2.7		
Dortmund	Univ. Dortmund	DELTA	1.5		11
Hamburg	HASYLAB	DORIS III[1]	4.45	16	404
	HASYLAB	PETRA[1]	12	20	25
Berlin	BESSY	BESSY II[3]	1.7 (1.9)	2.6	6.1
Denmark					
Aarhus	ISA	ASTRID[1]	0.58	0.38	160
	ISA	ASTRID II[3a]	0.6–1.4		
Sweden	Univ. of Lund	MAX I[2]	0.55	0.31	
Lund	Univ. of Lund	MAX II[3]	1.5		8.8
	Univ. of Lund	MAX III[3a]	0.7		13
Ukraine					
Kharkov	KPI	Pulse Stretcher/SR[1]	0.75–2		
Kiev	UNSC	ISI-800	0.7–1.0		
Russia					
Moscow	Kurchatov Inst.	Siberia I[2]	0.45	0.2	76
		Siberia II[2]	2.5	1.75, 7.1	880
Dubna	JINR	DELSY[3b]	1.2	1.16	10
Zelenograd	F. V. Lukin Inst.	TNK[2]	1.2–1.6		
Novosibirsk	INP	VEPP-3[1]	2.2	2.95	
	INP	VEPP-4[1]	5–7		
Jordan					
Allaan		SESAME[2b]	1.0	1.25	115 (50[c])
India					
Indore	CAT	INDUS-I[2]	0.45	0.2	73
		INDUS-II[2a]	2.5	6.3	58.1
Thailand					
Nakhon Ratchasima	NSRC	Siam Photon Source[2]	1.0	0.80	74
Singapore					
Singapore	Nat. Univ. Singapore	Helios 2[2]	0.7	1.47	1370
Taiwan					
Hsinchu	SRRC	SRRC[3]	1.3 (1.5)	1.4	19.2
Korea					
Pohang	POSTECH	PLS[3]	2 (2.5)	2.8	12.1
China					
Beijing	BSRL	BEPC[1]	(1.55) 2.2	2.28	76
Hefei	USTC	HLS[2]	0.8	0.52	53
Shanghai	SSRF	SSRF[3a]	3.5		

(continued overleaf)

Table 2.2.1 (continued)

Location	Institute	Ring	Energy (GeV)	ε_c (keV)	Emittance (π nm·rad)
Japan					
Sendai	Tohoku Univ.	TSRF[2a]	1.5	1.87	26.9
Tsukuba	AIST	TERAS[2]	0.6	0.56	
	KEK	PF[2]	2.5	4.0	36
	KEK	AR[2]	6.5	26.4	
Kashiwa	Univ. Tokyo	Super-SOR[3a]	1.8	2.42	8
Okazaki	IMS	UVSOR[2]	0.75	0.43	127
Kusatsu	Ritsumeikan Univ.	AURORA[2]	0.575	0.84	
Nishi-Harima	JASRI	SPring-8[3]	8	28.9	5.9
	Himeji Inst. Tech.	NewSUBARU[2]	1.5	2.33	67
Higashi-Hiroshima	Hiroshima Univ.	HiSOR[2]	0.7	0.88	400
Saga	Saga Pref.	[2b]	1.4	1.9	25.5
Australia					
Melbourne	Monash Univ.	Boomerang[3a]	3		12
Canada					
Saskatoon	Canadian L. S.	CLS[3a]	(2.5) 2.9	7.57	18.1
USA					
Stanford	SSRL	SPEAR 3[3b]	3	7.6	18
Berkeley	LBNL	ALS[3]	(1.5) 1.9		6.8
Stoughton	SRC	Aladdin[2]	(0.8) 1.0	(0.55) 1.1	108
Ithaca	CHESS	CESR[1]	5.5	11.0	
Argonne	ANL	APS[3]	7	19.5	8.2
Newport News	Jefferson Lab.	Helios 1[2b]	0.7		
Gaithersburg	NIST	SURF III[2]	0.4		
Upton	BNL	NSLS/VUV[2]	0.75	0.49	
	BNL	NSLS/X-Ray[2]	2.5	5	
Baton Rouge	Louisiana S. Univ.	CAMD[2]	1.5	2.56	200
Brazil					
Campinas	LNLS	LNLS-1[2]	1.37	2.08	100
	LNLS	LNLS-2[a]	2		

[1] First generation source; [2] second generation source; [3] third generation source.
[a] In planning/design stage.
[b] Under construction.
[c] With two wigglers.

high. Furthermore, the pulse duration will be shorter than that of third generation sources, so that peak brilliance will be much higher. Experiments using fourth generation sources will be directed to extensively utilize coherence, the time structure and peak brilliance.

In this subchapter, new synchrotron radiation sources are introduced briefly and characteristics of synchrotron radiation are summarized. New aspects of synchrotron radiation, most of which were topics at the 6th and the 7th International Conferences on Synchrotron Radiation Instrumentation,[6,7] are described for users exploiting third generation sources and typical properties of light at the sample position such as the number of photons are briefly mentioned.

Finally, candidates for fourth generation sources will be described.

2.2.2 THIRD GENERATION SOURCES AND OTHER NEW SOURCES

In the world, there are about 45 synchrotron radiation sources in operation, including 10 third generation sources, as given in Table 2.2.1. Third generation sources are storage rings with the horizontal emittance around or lower than 10 π nm·rad and with many long straight sections to install insertion devices, especially undulators with a large number of periods.

Figure 2.2.2 shows a schematic drawing of a planar undulator. Most of undulators are made

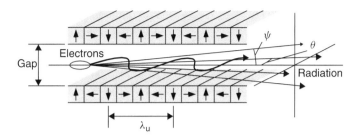

Figure 2.2.2 Electron bunch moving in a planar undulator. The angular coordinate is defined for synchrotron radiation from the undulator

with permanent magnets. The peak energy of quasi-monochromatic light from the undulator is inversely proportional to the period length, λ_u. To produce high energy photons with an undulator, λ_u has to be made short and at the same time the gap has to be reduced so as to keep the magnetic field high. The minimum gap determines the vertical physical aperture for the circulating beam in the storage ring, and the aperture has to be large enough to make the lifetime of the electron beam long. Another method is to use higher order harmonics from a planar undulator. Brightness of higher harmonics above the 5th order is, however, reduced due to errors in the magnetic field, so that light from the undulator covers only a lower portion of the spectral region obtained with that from bending magnets for the same electron energy. In order to obtain high energy photons with an undulator, the electron energy has to be made high. This is the reason why energies of third generation sources are, in general, higher than those of second generation sources.

The third generation sources are classified into two groups; those dedicated mainly to vacuum ultraviolet and soft X-ray experiments, and those to hard X-ray experiments. Typical examples of the first group are ALS, BESSY II and Elettra. Energies of these storage rings are 1.6–2 GeV, circumferences are 180–250 m, and 6–10 undulators are installed. The members of the second group are ESRF, APS and SPring-8. Their energies are 6–8 GeV, circumferences are 0.84–1.4 km, and 30–50 undulators are installed. Along with development of these third generation sources, undulator technologies have progressed significantly.[8] One of the novel undulators is an in-vacuum undulator with small λ_u, which has no vacuum chamber in the magnet gap, so that the magnet gap can be made extremely small. In addition, the storage rings can be operated in a special manner, which ensures a long lifetime even though the vertical aperture is small in straight sections for undulators. Magnetic field errors of undulators are, furthermore, reduced, so that higher harmonics up to the 11th order can be used for experiments. By making best use of these achievements, storage rings belonging to a new group have been constructed or are under construction, which have intermediate energies of 2.4–3 GeV but can provide bright light in the hard X-ray region below 20 keV as well as the vacuum ultraviolet and the soft X-ray regions. These sources in the third group of third generation sources are Swiss Light Source, DIAMOND, SOLEIL and CLS.

Some of the other new storage rings are compact synchrotron radiation sources, which use superconducting bending magnets or strong-field normal-conducting magnets to extend the photon energy region toward the hard X-ray region with relatively lower electron energies. They are AURORA, Helios 1 and 2, and HiSOR. It is remarkable that old soft X-ray rings with energies of 1 GeV belonging to second generation sources are moved to other places and modified to add the ability to generate hard X-rays with superconducting wigglers. They are SESAME transformed from BESSY I (Berlin) and the Siam Photon Source from SORTEC (Tsukuba). ANKA is an intermediate scale X-ray source with the circumference of 110 m.

2.2.2.1 CHARACTERISTICS OF SYNCHROTRON RADIATION

The motion of an electron in a storage ring is shown schematically in Figure 2.2.1, together with the definition of the angular coordinate for light from bending magnets. If electrons circulating in the storage ring have the same energy E, they oscillate incoherently around the central orbit with small amplitudes, which is called betatron oscillation. An electron with a different energy $E + \Delta E$ circulates on a different orbit slightly away from the central orbit. The energy spread of the electron beam σ_E/E is typically $5 \times 10^{-4} - 1 \times 10^{-3}$, where σ stands for the standard deviation of the Gaussian distribution. Owing to the betatron oscillation and the energy spread, the electron beam has a finite beam size σ_x and divergence σ'_x in the horizontal direction, and σ_y and σ'_y in the vertical direction. If the contribution of the energy spread to the beam size is negligible, the emittance, defined as the phase space area occupied by the electron beam, is approximately equal to $\pi \sigma_x \sigma'_x$ or $\pi \sigma_y \sigma'_y$. The shape of the cross-section is usually elliptic, and sizes are $\sigma_x = 0.05 - 0.5$ mm and $\sigma_y = 0.01 - 0.05$ mm and divergences are $\sigma'_x = 0.01 - 0.05$ mrad and $\sigma'_y = 0.005 - 0.01$ mrad for third generation sources. These values are one to two orders of magnitude larger for second generation sources. Electrons form bunches due to the radio frequency (RF, 100–500 MHz) accelerating system used to compensate for the energy loss by emission of synchrotron radiation. The bunch length σ_l is 1–30 cm, depending mainly on the RF and the momentum compaction factor $\alpha = (\Delta L/L)/(\Delta E/E)$, where L is the circumference for an electron with the energy E. The pulse duration of the electron beam and accordingly the time duration of light pulses is 30 ps–1 ns. Novel techniques to make the light pulse duration considerably shorter have been developed, which will be described later. The time interval between bunches is the inverse of the RF and it is 2–10 ns. When a storage ring is operated in the single bunch mode or in the several bunch mode with equal intervals, pulsed light with an interval of 0.1–1 μs, which is the revolution time of the storage ring or a fraction, can be used for time resolved experiments.

In many cases, recently, the intensity of synchrotron radiation is given by the number of photons N_{ph} per second for a nominal stored beam current (100–500 mA), rather than for a unit current (mA), where synchrotron radiation from a storage ring is regarded as DC light. We define three quantities, flux, angular flux and brilliance as well as their practical units most frequently used to measure the intensity of synchrotron radiation at a certain wavelength or photon energy as follows:

$$\text{flux} = dN_{ph}/[dt(d\lambda/\lambda)]$$
$$\rightarrow \text{phs/s/0.1\%bw}$$
$$\text{angular flux} = dN_{ph}/[dt\,d\theta(d\lambda/\lambda)]$$
$$\rightarrow \text{phs/s/mrad/0.1\%bw}$$
$$\text{brilliance} = dN_{ph}/[dt\,d\theta\,d\psi\,dx\,dy(d\lambda/\lambda)]$$
$$\rightarrow \text{phs/s/mrad}^2/\text{mm}^2/0.1\%\text{bw}$$

where x and y are coordinates on the cross-section of the light source, phs and 0.1%bw denote the number of photons and the fractional bandwidth of $\Delta\lambda/\lambda = \Delta\omega/\omega = 10^{-3}$, respectively. The angular flux defined above is the number of photons per horizontal angular acceptance $\Delta\theta = 1$ mrad integrated over the vertical angle ψ. This quantity is useful only for light from bending magnets. The brilliance, which is often called brightness, is the number of photons per $\Delta\theta = 1$ mrad and $\Delta\psi = 1$ mrad divided by the cross-sectional area of the electron beam. The brilliance is useful for light from undulators as well as from bending magnets. The peak brilliance is defined as the instantaneous peak value obtained at peaks of light pulses.

Light from bending magnets has a smooth and continuous spectrum characterized by the critical photon energy ε_c, which is calculated in practical units as $\varepsilon_c(\text{keV}) = 2.22 \times E^3(\text{GeV})/\rho(\text{m})$, where ρ is the orbit radius in the bending magnet. Angular fluxes for several light sources are given in Figure 2.2.3, where critical photon energies are indicated by the arrows. The angular flux spectrum has a broad maximum around the photon energy $0.4 \times \varepsilon_c$. When it is plotted in the double

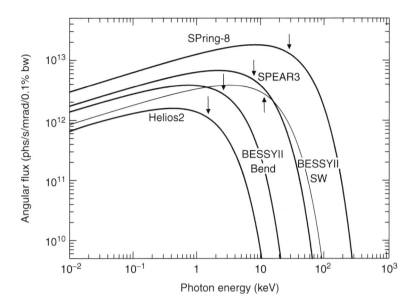

Figure 2.2.3 Angular flux spectra of synchrotron radiation from bending magnets for some storage rings. The arrows indicate critical photon energies. The angular flux spectrum is also shown for the superconducting wiggler SW of BESSY-II

logarithmic scales, the angular flux decreases gradually in the lower photon energy region, while it drops sharply in the higher photon energy region, as can be seen in Figure 2.2.3. The flux, which is the number of available photons, is given by the product of the angular flux and the acceptance angle $\Delta\theta$.

The magnetic field in bending magnets is given by $B(\text{T}) = 3.34 \times E(\text{GeV})/\rho(\text{m})$. The critical photon energy is, therefore, proportional to the magnetic field for the same electron energy. The bending magnets of BESSY-II, SPEAR 3 and SPring-8, angular flux spectra of which are shown in Figure 2.2.3, are conventional normal-conducting magnets with a magnetic field around 1 T. Since a superconducting magnet can generate a magnetic field up to 4–10 T, it is possible to produce synchrotron radiation in a photon energy region several times higher than that from conventional magnets for the same electron energy, if superconducting magnets are used as bending magnets (Helios 2) or as a superconducting wiggler (BESSY-II SW).

The angular divergence of light from bending magnets is small in the vertical direction and it is given typically by $1/\gamma$ around ε_c, where $\gamma = E/mc^2$ is the Lorentz factor with $mc^2 = 0.511$ MeV being the rest energy of an electron. The vertical angular divergence depends on the photon energy; it becomes larger as the photon energy decreases. In second generation sources, the electron beam divergence σ'_y is smaller than the angular divergence of light from bending magnets in the photon energy region below ε_c, while it is comparable or larger above ε_c. In third generation sources, however, σ'_y is smaller than the angular divergence in most of the photon energy region.

Light from bending magnets is linearly polarized with the electric vector lying on the plane of the electron motion when it is observed on the plane, while it is, in general, elliptically polarized when observed above or below the plane. When an angle of observation in the vertical direction is large, it is, not perfectly but practically, circularly polarized.

The undulator has a transverse magnetic field, which varies sinusoidally along the central axis with a period length longer than 1 cm and the number of periods ranging from a few to hundreds. The angular coordinate for light from an undulator is defined in Figure 2.2.2. Light from a planar undulator comprises a series of harmonic peaks. The wavelength λ of a peak is given by

$$\lambda = \frac{\lambda_u}{2n\gamma^2}\left(1 + \frac{K^2}{2} + \gamma^2\Theta^2\right) \quad (2.2.1)$$

where n is an integer indicating the harmonic order, $\Theta^2 = \theta^2 + \psi^2$, and K is the deflection parameter given by $K = 93.4 \times \lambda_u(m)B_0(T)$ with B_0 being the peak magnetic field. When the magnetic field in a planar undulator is not very high and the maximum deflection angle of electrons is of the order of $1/\gamma$, for which K is around one, quasi-monochromatic light with high brilliance can be obtained. The fractional bandwidth of the nth harmonic peak is given by $\Delta\lambda/\lambda \approx 1/nN$, where N is the number of periods. To calculate the wavelength for a helical undulator, the second term $K^2/2$ in the parentheses of Equation (2.2.1) should be replaced with K^2. Light from a planar undulator is linearly polarized, while that from a helical undulator is circularly polarized. If the electron energy is fixed, the deflection parameter has to be varied to change peak wavelengths. For an undulator made with permanent magnets, the peak magnetic field B_0 is varied by mechanically changing the magnet gap.

The angular divergence of light for a single electron is approximately axially symmetric even for a planar undulator, and as small as the fraction of the vertical angular divergence of light from a bending magnet. Its standard deviation is $\sigma'_r \approx 1/[\gamma(nN)^{1/2}] \approx (\lambda/L_u)^{1/2}$ for $K \approx 1$, where L_u is the total length of the undulator. The effective source size for a single electron, originating from the depth of the light source along the undulator length and the angular divergence, is calculated at the center of the undulator as $\sigma_r \approx (\lambda L_u)^{1/2}/(4\pi)$. Overall source sizes and divergences of light from an undulator are denoted by Σ_x and Σ'_x in the horizontal direction, respectively, and by Σ_y and Σ'_y in the vertical direction, where $\Sigma_x = (\sigma_r^2 + \sigma_x^2)^{1/2}$, $\Sigma_y = (\sigma_r^2 + \sigma_y^2)^{1/2}$, $\Sigma'_x = (\sigma'^2_r + \sigma'^2_x)^{1/2}$, and $\Sigma'_y = (\sigma'^2_r + \sigma'^2_y)^{1/2}$.

The intensity of light from undulators is usually expressed in terms of the brilliance. The brilliance is proportional to N^2 in an ideal case. Figure 2.2.4 shows examples of brilliance spectra of light at $\Theta = 0$ for $K = 1$ together with those from a bending magnet and a multipole wiggler. The

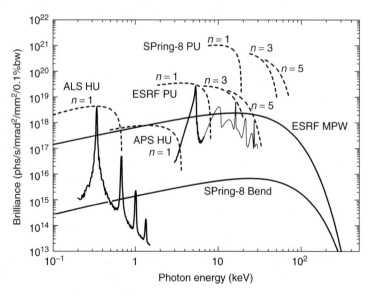

Figure 2.2.4 Brilliance spectra of synchrotron radiation from insertion devices and a bending magnet for some storage rings. PU, HU, MPW and Bend denote a planar undulator, a helical undulator, a multipole wiggler, and a bending magnet, respectively. The broken curves show envelopes of the brilliance for odd harmonics obtained by varying the undulator gap

broken curves show envelope of peak values of the brilliance for odd harmonics obtained by varying the undulator gap. As can be seen in Figure 2.2.4, the odd harmonics appear stronger than the even harmonics, and the fundamental peak with $n = 1$ is the strongest in a planar undulator, while only the fundamental peak is strong for a helical undulator. The brilliance of the fundamental peak is more than 10^3 times as high as that of light from bending magnets. As N becomes considerably large, the brilliance increases, but not as much as N^2, because the energy spread of the electron beam moderates the increase of the brilliance. The longest undulator for a storage ring is, at present, the 25 m long undulator at SPring-8. The undulator covers the photon energy range from 8 to 18 keV with the first harmonic and the brilliance is higher than 10^{20} phs/s/mrad^2/mm^2/0.1 %bw.

The flux for the undulator may be approximately calculated from the brilliance multiplied by $2\pi \Sigma_x \Sigma_y$ and $2\pi \Sigma'_x \Sigma'_y$. The flux increases linearly with the period number N, even though beam sizes and divergences as well as the energy spread of the electron beam are non-zero and finite. Figure 2.2.5 shows flux spectra for two of the undulators and the multipole wiggler appearing in Figure 2.2.4. The solid curves represent the flux spectra for the undulator with $K = 1$ and the multipole wiggler, while the broken curves show envelopes of the flux peak obtained for the undulators by changing K.

When K is much larger than unity, light from an undulator turns to a continuous spectrum and it extends to the higher photon energy region. The angular divergence becomes large and its properties become similar to those of light from bending magnets. This undulator is called a multipole wiggler. The brilliance and the flux of light from the multipole wiggler are proportional to N. As can be seen in Figure 2.2.5, the flux from the multipole wiggler exceeds the flux from the undulator at ESRF.

2.2.2.2 NEW ASPECTS OF SYNCHROTRON RADIATION

Energy Range

The third generation sources for X-rays, namely ESRF, APS and SPring-8, were designed and constructed to provide X-rays around 10 keV energy with the first harmonic light from undulators. Their electron energies are much higher than those of the second generation sources for X-rays. As a result, higher energy photons are available from bending

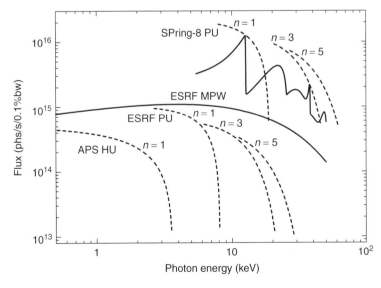

Figure 2.2.5 Flux spectra of synchrotron radiation from insertion devices. The broken curves show envelopes of the brilliance for odd harmonics obtained by varying the undulator gap

magnets of these storage rings and their critical photon energies are 20.6, 19.5 and 28.9 keV for ESRF, APS and SPring-8, respectively. The angular flux of SPring-8 is shown in Figure 2.2.3. As can be seen, experiments can be conducted using high-energy photons up to 100 keV or higher. This feature confers several advantages as follows. Materials containing heavy elements can be investigated, for instance, by means of the fluorescence trace analysis of their K absorption edges without being disturbed by other lighter elements. In diffraction experiments, it is possible to detect scattered X-rays with large wave numbers, so that the Fourier transformation can be accurately made even for amorphous materials or liquids. Samples become more transparent to the higher energy photons, so that images of thick samples can be taken easily. Signals from Compton and other scattering processes can be measured clearly, because their cross-sections become larger and comparable to the cross-section for the photoelectric process. New elements with higher nuclear levels can be excited for nuclear inelastic resonance.

Flux and Brilliance

In the early stage of undulator usage, the magnet gap of an undulator was not allowed to change during the experiment, because the beam position moved with the gap. Technologies have been developed to reduce magnetic field errors of undulators causing beam position movements and to correct beam positions precisely. As a result, the gap can be changed synchronously with scanning a monochromator during the experiment, so that the highest brilliance at the peak can be used at any photon energy. Hereafter we call it 'synchronous tuning'.

The photon number within a required spot size and a divergence for a given fractional bandwidth, which is the intensity at the sample position, should be as high as possible for the experiment. The product of the size and the divergence of the light beam at the source point is conserved at any focusing point in the optical system, which we tentatively call the optical invariant, provided the optical system is ideal. In the following discussion on the light intensity, the efficiency of optical elements is neglected. When the product of the size and the divergence of the light beam required at the sample position, which is the acceptance of light at the sample, is smaller than the optical invariant of the synchrotron radiation, the available flux is limited by the acceptance at the sample position. The flux is reduced by the ratio of the acceptance to the optical invariant. The high brilliance of the source is, therefore, essential to obtain sufficient intensity in microbeam experiments of small areas and in high energy-resolution experiments with narrow slits in the monochromators. When the acceptance at the sample position is larger than the optical invariant of the synchrotron radiation, the intensity is given by the flux and the fractional bandwidth of a monochromator. The available flux introduced to an optical system is, for example, given in Figure 2.2.5. The typical value of the flux is 10^{15} phs/s/0.1 %bw. If synchrotron radiation from a multipole wiggler is used, higher intensity light can be obtained. When a high flux is required in a somewhat larger area, light from a bending magnet is also useful. If light emitted in a horizontal angular region of 1 mrad is gathered with a mirror, the flux will be 10^{14} phs/s/0.1 %bw. A large flux is required for investigation of dilute systems, secondary processes and external field effects. When undulator radiation is used with an appropriate filter but without a monochromator, intense light may be obtained though its bandwidth is not very small. This utilization was first applied to photochemical reaction on solid surfaces and is often used for fluorescence excitation.

Coherence and Source Degeneracy

The coherence of light can be improved to a certain extent with optical devices at the expense of the flux. In practice, however, sufficient intensity is required, so that a light source should generate highly coherent light by itself. Synchrotron radiation is partially coherent. Here coherent properties of synchrotron radiation will be briefly introduced.

Light is transversely coherent when the optical invariant at a wavelength λ is nearly equal to $\lambda/4\pi$, which corresponds to the diffraction limit. If the electron beam size and the angular divergence at the source point are much smaller than those of the synchrotron radiation, σ_r and σ_r', which is realized in the vertical direction in third generation sources, synchrotron radiation from an undulator is transversely coherent. Light from bending magnets is transversely coherent in the vertical direction in the longer wavelength range even for second generation sources. The transverse coherence ensures that all the light coming from the source point can interfere at dispersive elements in a monochromator to obtain the highest spectral resolution without losing intensity but also at the sample to obtain sharp diffraction or interference patterns. Longitudinal coherence of light from an undulator is expressed in terms of the coherence length defined as $l_c = \lambda^2/\Delta\lambda$, where $\Delta\lambda$ is the spectral width of light. Since the fractional spectral width of the nth harmonic light from an undulator is approximately equal to $1/nN$, the coherence length is given by $nN\lambda$. Interference experiments such as holography can be done on a specimen, the thickness of which is of the order of the coherent length. The coherence length can be done longer using a high resolution monochromator. By coherently illuminating surfaces with microstructures, interference patterns are generated by scattered light, which are called speckle patterns, and they give information on the microstructures.

The source degeneracy of synchrotron radiation, which is an index showing the intensity of coherent light, is given as $\delta = B_p\lambda^3/4c$, where B_p is the peak brilliance expressed in mrad, and 100 %bw.[9] For ordinary undulators, δ is less than unity contrary to lasers, for which δ is enormous. Free electron lasers, which can produce high power coherent light with large δ in the shorter wavelength regions, are awaited for quantum optical experiments such as nonlinear optics.

Polarization

If light from bending magnets is confined with an appropriate horizontal slit in the beamline, linearly polarized light is obtained on the median plane of the storage ring and elliptically polarized light is obtained off the plane. Linear polarization exceeding 95 % is easily obtained, while the degree of circular polarization is around 80 %. Intense circularly polarized light recently became available from an undulator, in which the electron beam describes a helical motion. The switching frequency of the helicity is 0.1–100 Hz, depending on the type of helical undulator used. Some undulators can generate not only circularly polarized light, but also linearly polarized light with the electric vector lying even on the vertical plane as well as elliptically polarized light. The degree of polarization of undulator radiation can be kept high with synchronous tuning. Several kinds of dichroic experiments such as magnetic circular dichroism have been performed using polarized light.

Higher Order Light Rejection

Light from a bending magnet has a broad and continuous spectrum, while that from an ordinary undulator consists of the fundamental peak accompanied by its higher harmonics. Higher order light from a monochromator, due to shorter wavelength components in the incident light, is rejected in the beamline, using filters, mirrors and detuning of a monochromatizing crystal. On the other hand, there are a few trials to construct novel undulators, magnetic periods of which are quasi-periodic, so that higher harmonics do not appear at integer multiples of the fundamental photon energy. Similarly, a quasi-periodic grating has been developed.

Time Structure

In the vacuum ultraviolet region, two-color experiments such as two-photon absorption experiments and pump-probe experiments have been conducted using a synchrotron radiation pulse and a synchronized laser pulse. Some of them are experiments using a free electron laser in the ultraviolet region instead of a conventional laser. Recently, time resolved X-ray diffraction studies have been conducted, in conjunction with synchrotron radiation and a laser pulse, to investigate instantaneous heat-up phenomena in solids by laser light. Such two-color experiments will be conducted more in the hard X-ray region.

If synchrotron radiation is available with much shorter pulse duration than ordinary values of 30 ps–1 ns, many interesting fields, such as the dynamics of chemical reactions and structure analysis on the timescale of lattice vibration periods would be opened. There are two methods known to obtain shorter pulses. One is isochronous operation of the storage ring to shorten the bunch by reducing the momentum compaction factor. This operation mode has been attempted in a few rings. The storage ring called NewSUBARU has been designed and constructed to reduce the electron bunch to as short as a few picoseconds or even shorter. The other method is to slice off an extremely short electron pulse from an electron bunch. When an electron bunch comes in an undulator, a femtosecond (fs) laser pulse is synchronously injected from outside to interact with the electrons in the beam as they move side by side in the undulator, which is tuned at the laser wavelength. Electrons within the fs laser pulse are slightly accelerated or decelerated due to the interaction between them, which is the inverse process of free electron lasers to be described later. As the electron bunch moves in a downstream bending magnet, energy-modulated electrons move away from the central orbit and a fs electron pulse is produced. A fs light pulse is obtained in the beamline by screening light from the main bunch in the central orbit.

2.2.2.3 PHOTON INTENSITIES AVAILABLE AT BEAMLINE

A beamline is composed of a front end, pre-mirrors, a monochromator, refocusing mirrors, and an instrument for the measurement, which is called the end station. Soft X-ray beamlines are directly connected to storage rings without windows, while hard X-ray beamlines are equipped with Be windows separating high vacuum in storage rings and low vacuum in the beam lines. Even for hard X-rays, long beamlines have to be evacuated to suppress absorption of X-rays by the air and to prevent ionization of the air, which may react with or erode beam pipes and other devices. In beamlines for the photon energy region below 20 keV, pre-mirrors are often used to gather synchrotron radiation. These mirrors are coated with high reflectance materials as a monolayer or as multilayers and are used at grazing incidence angles. Synchrotron radiation is monochromatized with gratings below 1.5 keV, while crystals are used above 1 keV.

The available flux at the sample position is reduced by the efficiency of the optical elements and performance of a beamline. It should be noted that the polarization properties of the light source is not necessarily transferred to the end station due to the polarization characteristics of the optical elements. Figure 2.2.6 shows an example of the throughput spectrum for the undulator beamline BL39XU at SPring-8, which was obtained by synchronous tuning. The intensity around 10 keV is 4×10^{12} phs/s for a fractional bandwidth of 0.02 % at 100 mA and the beam size at a sample position is 1.5 mm wide and 0.6 mm high without using refocusing mirrors. In other energy regions, the following monochromatized photon beams are available at advanced beamlines. Around 0.4 keV, the bandwidth of light obtained is 40 meV and the intensity is 5×10^8 phs/s. Around 14 keV, the bandwidth obtained is 2.5 meV and the intensity is 1.6×10^9 phs/s. The bandwidths will be reduced to one tenth of the present values before long. A microbeam, which has a diameter of about 1 μm and a photon density of 10^{13} phs/s/mm^2, is available around 20 keV using a zone plate and a double-crystal monochromator.

2.2.3 FOURTH GENERATION SOURCES

There is no clear definition of fourth generation sources, but nevertheless it is widely accepted that next generation sources with much higher performance will show up before long. The key words may be much higher brilliance surpassing that of third generation sources and higher coherence. In the following, candidates for such light sources will be briefly introduced.

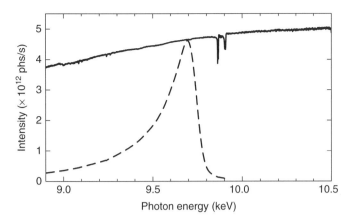

Figure 2.2.6 Throughput spectrum for the undulator beamline BL39XU at SPring-8 obtained by the synchronous tuning. The structures of the solid curve around 9.9 eV are due to monochromator glitches. The broken curve is the spectrum measured with a fixed undulator gap. The fractional bandwidth is 0.02 %. (Courtesy of M. Suzuki)

2.2.3.1 SASE-FEL

A free electron laser (FEL) is a device that produces high power coherent light by means of stimulated emission of radiation from the relativistic electron beam. This process involves many electrons interacting with coherent radiation. Let e_i be the electric field of radiation emitted by the ith electron in a bunch and E be the electric field of coherent radiation. Since the radiation power is proportional to the square of the electric field, the total power may be given by

$$P \propto \left(E + \sum_{i=1}^{n} e_i\right)^2 = E^2 + 2E \sum_{i=1}^{n} e_i + \left(\sum_{i=1}^{n} e_i\right)^2 \quad (2.2.2)$$

where n is the number of electrons in the bunch. The electric field E of radiation propagating in the z-direction with wave number k and frequency ω has such spatial and time dependences as $\exp[j(kz - \omega t)]$. The first term of the right-hand side of Equation (2.2.2) is the power of coherent radiation, the second term is the stimulated emission or absorption by an electron due to the coherent radiation, and the third term is spontaneous radiation.

When there is no coherent radiation, only spontaneous radiation given by the third term is emitted. Since wavelengths of synchrotron radiation in the X-ray region are much shorter than the bunch length in a storage ring, phases of electric fields produced by electrons will be randomly distributed and they cancel each other out on the average, so that the third term becomes

$$P \propto \left(\sum_{i=1}^{n} e_i\right)^2 = \sum_{i=1}^{n} e_i^2 + \sum_{i=1}^{n} \sum_{\substack{j=1 \\ j \neq i}}^{n} e_i e_j$$

$$\approx \sum_{i=1}^{n} e_i^2 = n e^2 \quad (2.2.3)$$

Thus an obvious result is derived that the intensity of synchrotron radiation is proportional to the number of electrons, or the beam current. When a short electron bunch from a linac passes a bending magnet and emits light in the far infrared and the sub-millimeter regions, the wavelength is comparable to or longer than the electron bunch. Electric fields emitted by individual electrons add up coherently and the radiation power is proportional to the square of the electron number, which is called coherent synchrotron radiation. It has been also observed on the storage rings MAX II and BESSY II in isochronous operation.

Next, we will look into the problem of emission of light by an electron beam in an undulator when coherent radiation exists. An electron oscillates transversely in the undulator and consequently it moves along the electric field E, so that it exchanges energy with the coherent radiation. Whether the electron emits or absorbs radiation by stimulated emission or absorption, depending on the phase of e_i relative to that of E, as can be seen in Equation (2.2.2). Since electrons are randomly distributed longitudinally in a bunch, half of them in the acceleration phase with respect to coherent radiation gain energy and the other half in the deceleration phase lose it, so that the total energy change will be zero on average, but the energy of the electron beam is modulated in the spatial period of the light wavelength. As the electron beam moves in the undulator, the higher energy electrons catch up with the lower energy ones, so that the energy modulation is converted to density modulation of the electron beam. When higher density parts of the electron beam sit on the deceleration phase, the number of decelerated electrons is larger than the number of accelerated electrons and hence a part of the kinetic energy of the electron beam is, on aggregate, transformed to coherent light power. This is the principle of FEL with an optical resonator.

The shortest wavelength realized with FEL with optical resonators is, at present, slightly shorter than 200 nm. In the shorter wavelength regions, however, reflectivity of mirrors used for optical resonator falls off rapidly and hence conventional FEL do not work in the vacuum ultraviolet and the X-ray regions. In order to realize FEL in shorter wavelength regions, a new type of FEL called SASE has been proposed,[10] where noise components in spontaneous light are amplified as seeds up to the saturation level with a high-gain amplifier FEL. A schematic drawing of SASE is shown in Figure 2.2.7. SASE is, in principle, a simple system consisting of a linac and an undulator. A high intensity, low emittance electron beam is necessary for SASE in the short wavelength region and therefore a laser photocathode RF electron gun is usually adopted. Photoelectrons emitted by irradiation of picosecond ultraviolet lasers are accelerated by a very high electric field in the RF cavity to a velocity close to that of light and thereby increase of the emittance due to space charge forces can be reduced. This electron gun produces an electron beam with a normalized emittance ε_n of a few $\pi \mu$m·rad, a few picoseconds pulse duration and a peak current up to 100 A. The normalized emittance is a constant of motion and does not change with energy. As the electron beam is accelerated in a linac, the emittance ε decreases inversely proportional to the electron energy due to adiabatic damping as $\varepsilon = \varepsilon_n/\gamma$, where γ is the Lorentz factor. In a chicane type bunch compressor shown in Figure 2.2.7, electrons in the head of a bunch are slightly decelerated compared with electrons in the central part and those in the tail are accelerated, using an upstream acceleration part. The lower energy electrons in the head make a detour, while the higher energy electrons take a short cut and catch up with the head, so that the bunch becomes shorter and the peak current is enhanced to a few kA. The high intensity and low emittance beam is then accelerated in the main linac and produces high power SASE

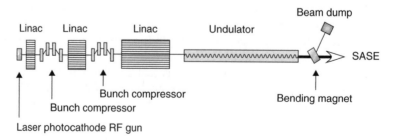

Figure 2.2.7 Schematic drawing of a typical SASE system

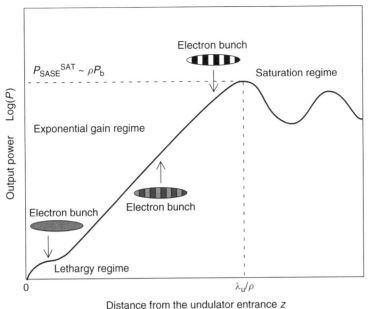

Figure 2.2.8 Evolution of the SASE power along the undulator. ρ, P_b and P_{SASE}^{SAT} are the FEL parameter, the electron beam power, and the saturation power of SASE, respectively[10]

in a long undulator. The evolution of the radiation power emitted by the electron beam passing through the undulator is shown schematically in Figure 2.2.8. At the entrance region of the undulator, where the electron distribution is uniform in the bunch, spontaneous radiation dominates. The power increases linearly with the distance from the entrance, z. Note that the vertical axis is plotted using a logarithmic scale. In this region called lethargy regime, the incipient stage of SASE is in progress though the power level is much lower. The startup of SASE is thought to develop as follows. There are many electrons randomly distributed in a bunch and these electrons produce extremely short and randomly distributed electric field spikes when they are transversely accelerated in the magnetic field of an undulator. The frequency distribution of the noise power fluctuates statistically, but if the long-term average is taken it will extend with a constant intensity to extremely high frequencies. Part of the noise within the bandwidth of the FEL gain, the center of which is located around the fundamental frequency of undulator radiation, is gradually amplified by bunching electrons at a single optical wavelength. The density modulation of the electron beam and the phase of coherent radiation affect each other to amplify the coherent radiation due to stimulated emission, given by the second term of the right-hand side of Equation (2.2.2), where the coherent radiation is provided by upstream electrons. The electron beam continues to interact with coherent light, so that the density modulation of electrons is enhanced gradually, as shown schematically in Figure 2.2.8. Meantime, the interaction comes in the exponential gain regime, where the gain per unit length is constant and the power of coherent light is exponentially amplified with z. The radiation power continues to be amplified exponentially, but eventually the power will saturate because the electron beam amplifying coherent light gradually loses kinetic energy and at the same time the energy spread increases in the amplification process. Experimental studies on SASE have been conducted progressively from the far-infrared region to the shorter wavelength regions and now SASE is successfully generated in the vacuum ultraviolet region between 80 and 180 nm at DESY.[11] A wavelength spectrum of SASE measured around 109 nm is shown in Figure 2.2.9.

A new SASE facility is in the course of construction at SPring-8, which is named SPring-8 Compact SASE Source (SCSS), to extend the SASE wavelength down to 3.6 nm in a water window.[12] The aim of this project is to blaze a trail in the short wavelength region important for biology with a relatively small-sized facility. The unique features of this facility are that C-band linacs operating at 5.7 GHz are used and an in-vacuum undulator with a period length of 15 mm is employed, in order to make the facility compact. The spectral range and the peak brilliance are shown in Figure 2.2.10.

There are two proposals, one in the United States and one in Europe for constructing a user facility for coherent X-rays based on SASE;[13] the Linac Coherent Light Source (LCLS) at the Stanford Linear Accelerator Center (SLAC) and the TeV Energy Superconducting Linear Accelerator-X-Ray Free Electron Laser (TESLA-XFEL) at DESY. The LCLS project is to construct a SASE-FEL facility in a wavelength region ranging from 0.15 to 1.5 nm using the SLAC linac. The electron beam is accelerated to 14.3 GeV with the linac and led to a 111.8 m long undulator. The magnet gap of the undulator is fixed at 6 mm, but the wavelength can be tuned by varying the electron energy. The peak power of SASE at $\lambda = 0.15$ nm is 9 GW and energy in a pulse is 2.6 mJ. The peak brilliance runs up to 1.2×10^{33} (phs/s/mrad^2/mm^2/0.1 %bw), as shown in Figure 2.2.10. Since the repetition rate of the linac is 120 Hz, the average power is 0.31 W. The average brilliance is 4.2×10^{22} (phs/s/mrad^2/mm^2/0.1 %bw). It is not very different from the brilliance of third generation light sources, but the peak brilliance is extremely high.

The TESLA-XFEL project is similar to the LCLS project, but a superconducting linac will be used. An electron beam is accelerated to 20–50 GeV with the linac and led to a 323.5 m long undulator. The period length of the undulator is 60 mm and the magnet gap is variable between 12 and 22 mm to scan the wavelength. The peak power at $\lambda = 0.1$ nm is 37 GW and energy per pulse is 3.7 mJ. The peak brilliance is comparable to that of LCLS as shown in Figure 2.2.10. The average power is 210 W and the average brilliance is 4.9×10^{25} (phs/s/mrad^2/mm^2/0.1 %bw). The average brilliance of the X-ray beam is extremely high.

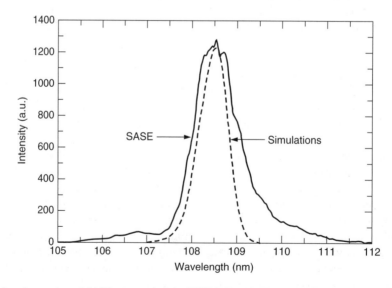

Figure 2.2.9 Wavelength spectrum of SASE measured at the TESLA Test Facility in DESY. The central wavelength is around 109 nm.[11] Reprinted from Rossbach, J. First observation of self-amplified spontaneous emission in a free-electron laser at 109 nm wavelength. *Physical Review Letters* **85** (2000) 3285. Reproduced by permission of The American Physical Society

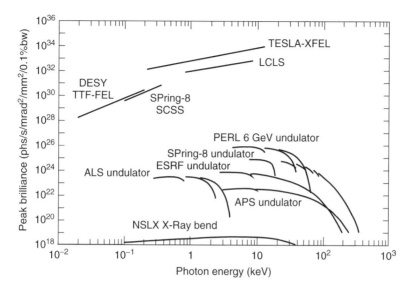

Figure 2.2.10 Peak brilliance of synchrotron radiation for various light sources. NSLX X-ray bend is for a second generation source, SPring-8 undulator, ESRF undulator, APS undulator, and ALS undulator are for third generation sources, RERL 6 GeV undulator is for an energy recovery linac, and TESLA-XFEL, LCLS, SPring-8 SCSS, and DESY TTF-FEL are for SASE

2.2.3.2 FOURTH GENERATION STORAGE RINGS AND ENERGY RECOVERY LINACS

A natural extension of the present technology for fourth generation sources would be higher performance storage rings called fourth generation storage rings.[14] The emittance of the electron beam will be reduced to a fraction of 1 πnm·rad, which is one tenth of the typical emittance of third generation light sources. Since the emittance of the electron beam in a storage ring is inversely proportional to the third power of the number of bending magnets, the circumference of the ring should be doubled. A drawback of this method to reduce the emittance is that the dynamic aperture is considerably reduced and consequently the beam lifetime becomes shorter. A countermeasure devised for the problem is the top-up injection, so that the electron beam is continuously injected to compensate for beam loss due to the short lifetime. As the electron beam size becomes smaller, spatial and temporal stabilities of the stored beam are crucial issues. These problems including short and long term drifts of the beam position and fast beam instability are, however, extensively studied in third generation sources.

Another candidate for fourth generation sources was recently proposed, which is called the energy recovery linac (ERL). Several proposals for constructing a new synchrotron light source based on ERL have been advanced throughout the world.[15] A schematic drawing of the energy recovery linac is shown in Figure 2.2.11. The electron beam is accelerated with a superconducting linac to a final energy and then passes through bending magnets and undulators to produce synchrotron radiation. After making a round, it comes back to the entrance of the linac again 180° out of phase with the accelerating RF field and decelerated in the linac, thereby delivering the kinetic energy of the electron beam to the accelerating beam through the RF field. If the electron beam is injected continuously to the linac, this process will be repeated successively, so that a continuous beam can be obtained without supplying enormous RF power to accelerate the beam.

If an electron beam with $\varepsilon_n = 1$ πµm·rad is accelerated to 5 GeV, the emittance will be reduced to 0.1 πnm·rad. The average beam current should be comparable to that of third generation sources,

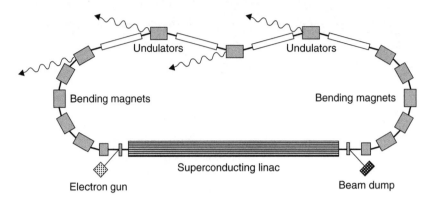

Figure 2.2.11 Schematic drawing of an energy recovery linac

so that superconducting linacs will be used to avoid RF power loss in acceleration structures. The accelerated beam produces high intensity and high brilliance synchrotron radiation in all the bending magnets and undulators installed on the return pass to the entrance of the linac for energy recovery. The outward appearance of ERL is not very different from that of a third generation source, but it is not a storage ring and consequently it is free from various technical constraints imposed on storage rings for storing an electron beam in a circular pass for a long time. The electron beam may be focused down to a few micrometers in diameter and bunches may be compressed to 100 fs. The beam lifetime is not a critical issue because electrons are used only once to produce synchrotron radiation in the return pass, so that the undulator gap may be reduced to a few millimeters and accordingly its period length may be made as short as some millimeters.

ERL has many advantages over the conventional synchrotron radiation sources based on storage rings. To turn these advantages into reality, however, there are some technical issues to be solved, including the electron source and stability of the beam, because ERL is a dynamic accelerator and there is no such built-in stabilizer as radiation damping in a storage ring. ERL can provide synchrotron radiation with an average brilliance that is two to three orders of magnitude higher than in the case of third generation sources and the estimated peak brilliance is also extremely high.

As an example, the peak brilliance of the Photo-injected Energy Recover Linacs (PERL) proposed at Brookhaven National Laboratory is shown in Figure 2.2.10. Advantages of ERL compared with SASE are such that many undulators can be used simultaneously and higher energy photons can be obtained. ERL and SASE will play complementary roles as fourth generation sources. Indeed there is a proposal to combine both at Daresbury Laboratory.[5]

REFERENCES

1. Spencer, J. E. and Winick, H. Properties of synchrotron radiation, in *Synchrotron Radiation Research* (Eds H. Winick and S. Doniach), Plenum Press, New York (1982).
2. Krinsky, S., Perlman, M. L. and Watson, R. E., Characteristics of synchrotron radiation and of its sources, in *Handbook of Synchrotron Radiation* **Ia** (Ed. E. E. Koch), North-Holland, Amsterdam (1983).
3. Watanabe, M. and Isoyama, G. Synchrotron radiation and free electron lasers, in *Dynamics during Spectroscopic Transitions* (Eds E. Lippert and J. D. Macomber), Springer, New York (1995).
4. Thompson, A. C. and Vaughan, D. (Eds) X-ray DATA booklet, Center for X-ray Optics, Lawrence Berkeley National Laboratory (2001).
5. Hasnain, S. S., Kamitsubo, H. and Mills, D. M. New synchrotron radiation sources and next generation light sources. *J. Synchrotron Rad.*, **8**, 1171 (2001).
6. Hasnain, S.S., Helliwell, J. R. and Kamitsubo, H. (Eds) Proc. of 6th Int. Conf. on Synchrotron Radiation Instrumentation. *J. Synchrotron Rad.*, **5**, Part 3 (1998).
7. Gudat, W. and Zimmermann, P. (Eds) Proc. of 7th Int. Conf. on Synchrotron Radiation Instrumentation. *Nucl. Instr. Meth. Phys. Res.*, A467–468 (2001).

8. Walker, R. P. and Diviacco, B. Insertion devices–recent developments and future trends. *Synchrotron Rad. News*, **13**(1), 33–42 (2000), and references therein.
9. Gluskin, E., McNulty, I. and Viccaro, P. J. X-ray intensity interferometer for undulator radiation, *Nucl. Instr. Meth. Phys. Res.*, **A319**, 213–218 (1992).
10. Kim, K. J. and Xie, M. Self-amplified spontaneous emission for short wavelength coherent radiation. *Nucl. Instr. Meth. Phys. Res.*, **A331**, 359–364 (1993).
11. Andruszkow, J. *et al.* First observation of self-amplified spontaneous emission in a free-electron laser at 109 nm wavelength. *Phys. Rev. Lett.*, **85**, 3825–3829 (2000).
12. Shintake, T., Matsumoto, H., Ishikawa, T. and Kitamura, H. Proc. of SPIE Conf. 4500, Optics for Fourth-Generation X-ray Sources (2001) p. 12.
13. Nuhn, H.-D. and Rossbach, J. LINAC-based short wavelength FELs: The challenges to be overcome to produce the ultimate X-ray source–the X-ray laser. *Synchrotron Rad. News*, **13**(1), 18–32 (2000).
14. Hofmann, A. and Rivkin, L. Fourth generation storage ring sources. *Synchrotron Rad. News*, **12**(2), 6–15 (1999).
15. Besn-Zvi, I. and Krinsky, S. Future light sources based upon photo-injected energy recovery linacs. *Synchrotron Rad. News*, **14**(2), 20–24 (2001), and references therein.

2.3 Laser-driven X-ray Sources

C. SPIELMANN

Physikalisches Institut EP1, Universität Würzburg, Germany

2.3.1 INTRODUCTION

Synchrotrons represent the major source of powerful X-rays and will continue to play a dominant role for X-ray science in the foreseeable future. Nevertheless, a wide range of X-ray applications in science, technology and medicine would greatly benefit from (i) X-ray pulse durations much shorter than routinely available from synchrotrons (few hundred picoseconds), (ii) synchronizability of ultrashort pulses to other events, and (iii) availability of useful fluxes from compact laboratory X-ray sources. Triggered by these demands a large number of research groups made enormous efforts to develop novel generation X-ray sources driven by high power lasers.

Advances in ultrashort-pulse high-power laser technology over the last decade (Perry and Mourou, 1994; Umstadter *et al.*, 1998) paved the way towards compact, versatile laboratory X-ray sources in a number of laboratories for spectroscopic as well as other applications. Ultrashort-pulsed X-ray radiation became available from femtosecond-laser-produced plasmas (Gibbon and Förster, 1996; Giulietti and Gizzi, 1998; and references therein).

These sources are now capable of converting up to several per cent of the driving laser pulse energy into *incoherent* X-rays emitted in a solid angle of $2\pi - 4\pi$ and delivering pulses with durations down to the subpicosecond regime. Femtosecond laser produced plasma sources matured to a point where a wide range of applications can be tackled in a wide spectral range extending from the soft to the hard X-ray regime. Already demonstrated examples include time-resolved X-ray diffraction and absorption spectroscopy (Helliwell and Rentzepis, 1997; Rousse *et al.*, 2001b) as well as medical radiology with improved contrast and resolution (Gordon III *et al.*, 1995; Grätz *et al.*, 1998).

Many laboratory X-ray applications would greatly benefit from or rely on (spatially) coherent sources with high average and/or peak power. One of the major approaches to laboratory production of coherent X-rays is the development of X-ray lasers. Whereas short-wavelength lasing has been successfully demonstrated with compact, table-top setups using several promising schemes at $\lambda > 15$ nm in the XUV range (Lemoff *et al.*, 1995; Nickles *et al.*, 1997; Korobkin *et al.*, 1998; Rocca, 1999), lasing at shorter (soft-X-ray) wavelengths could only be achieved at large-scale facilities so far (Zhang *et al.*, 1997; Klisnick *et al.*, 2002).

Another promising route to developing compact coherent X-ray sources is high-order harmonic generation (HHG) with ultrashort-pulse lasers (L'Huillier and Balcou, 1993). Extensive theoretical and experimental research provided valuable insight into the microscopic (strongly driven atomic dipole) and macroscopic (propagation effects, e.g. phase mismatch) phenomena relevant to HHG (for a recent review see Brabec and Krausz, 2000). Recent investigations revealed that ultrashort drivers with pulse durations well below 100 fs can produce HH conversion efficiencies comparable to XUV lasers in the 100–20 nm range (Rundquist et al., 1998; Sommerer et al., 1999; Constant et al., 1999).

Recently, few-cycle, sub-10 fs laser pulses produced HH radiation at 13–10 nm with somewhat lower efficiencies and with pulse durations estimated as <3 fs at a repetition rate of 1 kHz, resulting in the highest average and peak powers ever demonstrated from a coherent laboratory soft-X-ray ($E_{ph} \geq 100$ eV) source (Schnürer et al., 1999). Pulses in the 5–25 fs range have extended HHG even down to the water window, 2.3–4.4 nm (Spielmann et al., 1997; Chang et al., 1997b; Schnürer et al., 1998). Theoretical investigations suggest that few-cycle-driven harmonic emission is confined temporally to a tiny fraction of the laser period in the cut-off region of the spectrum (Christov et al., 1997; Spielmann et al., 1998). These predictions have been verified by recent experiments resulting in an isolated XUV/X-ray pulse of attosecond duration (Drescher et al., 2001; Hentschel et al., 2001).

In spite of these advances, applications of coherent laboratory X-ray sources are in their infancy. The photon fluxes available from state-of-the-art harmonic sources are still low. Compared to conventional X-ray sources the average and peak brilliance as a function of the photon energy is shown in Figure 2.3.1. Increasing the power of the few-cycle-driven harmonic source by one to two orders of magnitude holds promise for opening up a number of intriguing application fields in science and technology and pushing the frontiers of physical sciences. The former include X-ray spectroscopy, X-ray microscopy, X-ray photoelectron spectroscopy and possibly X-ray interferometry all of which have had to rely on large-scale synchrotron facilities thus far. In addition the sub-femtosecond pulse durations that have become available from harmonic sources, opened the way to attosecond science.

This subchapter is organized as follows: In the first part recent progress in the development of laser sources useful for X-ray generation will be briefly described. The subsequent sections review the generation of laser-produced incoherent radiation, the development of X-ray lasers and high harmonic generation.

2.3.2 LASERS FOR X-RAY GENERATION

Ultrafast solid-state laser technology has been subject to dramatic advances over the last 10 years. The invention of titanium-doped sapphire (Ti:S) along with the development of Kerr-lens mode locking (Brabec and Krausz, 2000) and chirped dielectric mirrors for ultrabroad-band dispersion control have resulted in a new generation of compact solid-state ultrafast oscillators, which can now routinely generate pulses in the sub-10-fs regime with peak powers exceeding 1 MW and pulse durations as short as 5 fs (for a recent overview see Steinmeyer et al., 1999). Unfortunately, the pulse energy and/or peak intensity of pulses emitted from oscillators is too low for using them for X-ray generation.

However, amplification of ultrashort pulses has also progressed rapidly. Presently the majority of compact high peak-power laser systems rely on chirped pulses amplification (CPA). In the CPA (Strickland and Mourou, 1985) technique a short seed pulse is first stretched in duration, then amplified and finally recompressed. In the ideal case the recompressed pulse duration should be close to the initial one. This technique allows extremely high peak powers at the output while avoiding high intensities in the amplifier which would cause nonlinear distortions or damage. The combination of CPA with Ti:S allowed the construction of laboratory-scale high-power laser systems with unprecedented characteristics. Pulses as short as ~20 fs have now become available with terawatt peak powers (pulse energy on the order of 100 mJ) at repetition rates of 10–50 Hz (Chambaret et al., 1996; Barty et al., 1996; Wang et al., 1999). Several systems with a peak power up to 100 TW have been set up in university laboratories all over the world (Yamakawa et al., 1998; Kalachnikov et al., 2001). Ti:S is an excellent laser material, but its limited energy storage capacity makes it hard to obtain pulses with

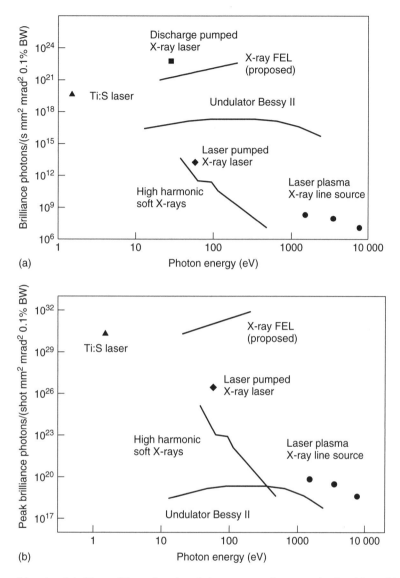

Figure 2.3.1 Average (a) and peak brilliance (b) as a function of photon energy for conventional and laser-driven X-ray sources. For the calculation of the peak brilliance the pulse duration as reported in the corresponding literature has been used

an energy of more than a few joules (Yamakawa and Barty, 2000). To further increase the pulse energy hybrid systems consisting of a Ti:S preamplifier followed by Nd:glass power amplifier chain have been developed. Pulse energies of up to 10 J for table top and 400 J have been achieved in a laboratory and a large scale facility (Danson et al., 1998; Perry et al., 1999), by drawing on this concept. Such a hybrid system offers also the possibility to generate synchronized energetic nano- and femtosecond pulses beneficial for realizing laser-driven X-ray lasers.

The other direction of current laser developments is towards higher repetition rates and shorter pulses. The development of powerful-kHz-repetition rate pump lasers and the excellent thermal properties of Ti:S opened the way to producing millijoule-energy 20-fs pulses at kilohertz frequencies (Backus et al.,

Table 2.3.1 Parameters of state-of-the-art laser sources. In the far right column potential realizations of X-ray sources are listed: high harmonic generation (HHG), X-rays from a laser produced plasma (LPPX), and X-ray laser (XRL)

	Pulse energy	Pulse duration (fs)	Repetition rate	Peak intensity (W/cm^2)	X-ray source to be realized
Oscillator	nJ	<5	100 MHz	<10^{12}	None
kHz systems	<20 mJ	>15	1–10 kHz	10^{17}	HHG, LPP
Compressed	<1 mJ	<5		10^{17}	HHG
Low rep. rate	2 J	>20	10–50 Hz	10^{20}	HHG, LPPX
Large scale facilities	<500 J	>300	Typically one shot per hour	10^{21}	LPPX, XRL

1997). The major challenge in realizing high repetition rate systems is to minimize the thermal distortions in the amplifier. Cryogenic cooling of the amplifier crystal (Durfee *et al.*, 1998) and efficient spatial filtering paved the way to sub-30 fs laser systems with a pulse energy up to 20 mJ at 1 kHz (Bagnoud and Salin, 2000) or 1 mJ at 7 kHz (Backus *et al.*, 2001). The output of these systems can now be spectrally broadened in a gas-filled capillary and subsequently compressed. Using this technique (dubbed as the hollow-fiber technique), self-phase-modulated 20-fs pulses with energies of hundreds of microjoules have been compressed in a delay line consisting of chirped mirrors to 4.5 fs (Nisoli *et al.*, 1998). They are delivered in a diffraction-limited beam and hence focusable to peak intensities in excess of 10^{17} W/cm^2. Such a source producing routinely pulses with a duration of 5–7 fs and energies of 0.5 to 0.7 mJ has been used for a wide range of applications (Sartania *et al.*, 1997).

All of these recent advances in table top X-ray sources as described in the following sections have been made possible by the above described developments of compact high power drivers. Typical values of the parameters achieved with different lasers systems are summarized in Table 2.3.1.

2.3.3 HARD X-RAYS FROM LASER PRODUCED PLASMAS

Hard X-ray diffraction methods, in which X-ray wavelength is comparable to interatomic distances of molecules, are instrumental in providing the information of the atomic arrangement in crystalline solids and liquids, and are almost exclusively responsible for our current understanding of the protein structure. Operated in an ultrafast pulsed mode, these methods may become capable of unravelling difficult dynamics in biology, chemistry, materials science, and physics. To achieve this goal of operation, however, new kinds of sources, featuring sub-picosecond pulse duration and high spectral brightness, are needed. Sub-picosecond duration is necessary because it is the shortest time interval during which chemical reactions or the motions of lattice can occur. High brightness, on the other hand, ensures an adequate signal-to-noise ratio for a reasonable acquisition time, making these sources practical. Time-resolved applications, including ultrafast X-ray diffraction (Wark, 1996; Helliwell and Rentzepis, 1997; Rousse *et al.*, 2001b) and absorption (Raksi *et al.*, 1996), X-ray photoelectron spectroscopy (Kondo *et al.*, 1996), and time-resolved X-ray microscopy will allow investigation of transient structures with unprecedented, simultaneous spatial and temporal resolution.

Presently there are several approaches for producing picosecond and even sub-picosecond hard X-ray pulses. These include Thomson scattering (Schoenlein *et al.*, 1996), laser X-ray diodes (Tomov *et al.*, 1999), optical switching of synchrotron pulses (Larsson *et al.*, 1998), laser switching of electron bunches in a synchrotron (Schoenlein *et al.*, 2000), and laser produced plasma ultrafast X-ray sources (LPPX) (Gibbon and Förster, 1996; Giulietti and Gizzi, 1998) with the latter having been made possible by the development of ultrafast tabletop terawatt lasers (Perry and Mourou, 1994). In LPPX, the intense laser pulse functions as an electron generator and accelerator. The final velocities of these energetic electrons reach tens to hundreds of keV so as to produce high yield hard X-ray pulses, and they have to be treated relativistically. In addition to the high terminal velocity, the acceleration distance is also very short, of the order of less than several tens of microns, the time spread of the electron packet is very small when it interacts with the solid target. These electrons are then stopped in the solid within a few tens of microns. The combined effect results in X-ray pulse duration on the order of a few hundred femtoseconds or less, determined mainly by how quickly these electrons are stopped and by the size of the emission volume. As a result, features of these sources include ultrashort hard X-ray pulse duration (<1 ps), high brightness and compact source size (Guo et al., 2001).

Understanding the X-ray generation mechanism is necessary for optimizing the source. Although a thorough comprehension of the process is very hard to achieve, two physical pictures responsible for the production of X-ray pulses in laser driven X-ray sources are currently being considered. In one view, an intense laser field (up to 10^{19} W/cm^2) interacts directly with the solid density plasma, which has a very short density gradient scale length, i.e. much shorter than the optical wavelength. The energy transfer from photons to electrons is achieved by inverse Bremsstrahlung absorption, resonant absorption, and vacuum heating (for an overview see e.g. Gibbon and Förster, 1996 or Giulietti and Gizzi, 1998). The electrons in the plasma are heated up within the optical pulse duration followed by a rapid cooling. These rapid temperature changes results in a very short burst of black body radiation extending into the hard X-rays region. This mechanism works only efficiently, if the laser pulse interacts with the solid, i.e. no prepulse or pedestal is allowed.

The second mechanism is proposed based on a series of experiments with dual laser pulses, a weak prepulse and an intense main pulse (Rousse et al., 1994). The prepulse preforms a plasma. It has been found that this configuration can greatly increase the efficiency of the energy transfer between the optical photons and the electrons because the preformed plasma has a density gradient scale length similar to the optical wavelength. The electrons in this plasma are further accelerated by the main pulse upon its arrival. As the main mechanism for the further heating were relativistic ($I \times B$) and resonant heating identified (Wilks and Kruer, 1997). These acceleration processes are known to accelerate these electrons in the preplasma to a few hundred keV or even a few MeV (Wharton et al., 1998). Characteristic X-rays are emitted from the solid density plasma at near room temperature via the production of inner-shell holes and subsequent inner-shell electronic transitions. Therefore, an enhanced coupling between the optical field and the electrons increases both the number of hot electrons and its temperature, and consequently more hard X-rays are produced. As a consequence of the non-thermal nature electrons are only generated at high peak intensities resulting in an X-ray source spot size comparable to that of a laser (a few micrometers).

Solid target illumination with terawatt ultrashort pulses can generate, through the above described mechanisms, ultrafast pulses of 10 keV-range X-ray line radiation and 100-keV-range continuum emission with high flux. For spectroscopic applications Kα line sources have been mainly constructed. In a vacuum chamber an intense laser pulse is tightly focused onto a moving solid target generating a hot plasma emitting X-rays. Varying the laser and target parameters (Feurer, 1999), including intensity, duration, ratio main-to-prepulse ratio, target shape etc. allowed to maximize the Kα yield from Si, Ti and Cu (Rischel et al., 1997; Blome et al., 2001). With an optimized 20 Hz laser system e.g. 10^{11} Cu Kα photons/s/4π sr (Guo et al., 2001) have been demonstrated. The demonstrated source flux is sufficient to obtain diffraction patterns with a good signal to noise ratio from perfect crystals within a reasonable accumulation time. The most promising way of further increasing the X-ray flux is the use of recently developed kHz laser systems capable of producing several tens of millijoules (Zhang et al., 1998). It is predicted that an optimized kHz Si Kα source will achieve photon fluxes ($>10^{14}$ photons/s/4π sr) substantially higher than conventional X-ray tubes. High average flux is indispensable for detecting femtogram low atomic number material impurities in silicon wafers by X-ray fluorescence analysis, currently only feasible with synchrotron radiation (Streli *et al.*, 2001). For some spectroscopic applications the problem of debris has to be addressed. Debris from the target can obscure or damage samples in the vicinity of the plasma. One solution is the use of liquid target, which minimizes the undesired effects of debris (Tompkins *et al.*, 1998). A liquid source also ensures that each laser pulse hits fresh material (Hemberg *et al.*, 2000).

Time resolved X-ray spectroscopy requires an X-ray source that is a spectrally, spatially and temporally well-characterized. Spectral and spatial measurements can be easily performed with standard methods. Accurate measurements of the duration of hard X-ray pulses represents a major challenge. The knowledge of the X-ray pulse duration is crucial because it determines the temporal resolution of the envisaged time resolved experiments. X-ray streak cameras have been built with resolutions in the range of a picosecond (Chang *et al.*, 1997a). X-ray/optical cross-correlation methods based on photoelectron spectroscopy successfully implemented for the characterization of soft X-rays cannot be extended into the hard X-ray range. Phase transitions can be extremely fast, and therefore call for an ultrashort X-ray pulse in a diffraction experiment. From these measurements the upper limit of the hard X-ray pulse duration has been estimated in the range of 100–300 fs (von der Linde *et al.*, 2001; Feurer *et al.*, 2001).

Based on these newly developed sub-picosecond X-ray sources, a series of proof-of-principle time-resolved X-ray absorption (Raksi *et al.*, 1996) and X-ray diffraction (Rischel *et al.*, 1997; Rose-Petruck *et al.*, 1999; Siders *et al.*, 1999; Cavalleri *et al.*, 2000; Rousse *et al.*, 2001a) experiments have been performed over the last few years. In all these experiments the structural dynamics have been initiated by an ultrashort laser pulse. A typical setup as used for time-resolved X-ray spectroscopy is shown in Figure 2.3.2. Raksi *et al.* investigated the photodissociation of SF6 by studying the modifications of the X-ray absorption during the transition from the molecular to the atomic state. Rischel *et al.* exposed a Langmuir–Blodgett film to a short laser pulse at an intensity well above the damage threshold, causing a disorder of the film. The transition between the ordered and disordered state was characterized by time-resolved X-ray diffraction. This transition took place on a timescale of the order of a few hundred femtoseconds. Such a fast dynamic was not accessible with previously available X-ray sources. More

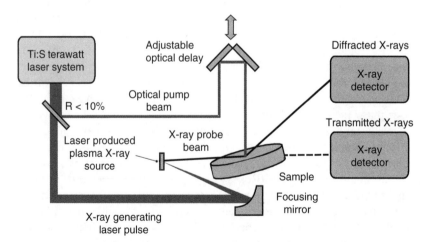

Figure 2.3.2 Typical setup for time-resolved X-ray spectroscopy. A fraction of the output of a high power laser system is split off and after an adjustable delay focused onto the sample to initiate a structural change. The major fraction of the laser energy is focused onto a solid or liquid used to generate X-rays. The pulsed X-ray radiation is used to probe the structural changes in a reflection or transmission geometry

recently Rose-Petruck *et al.* used ultrashort X-ray pulses to measure the response of a GaAs crystal to a sudden heating. The material is heated up so rapidly that no notable expansion can take place. The impulsive heat deposition launches an acoustic wave that travels into the material forming regions of compression and expansion. The lattice dynamics was characterized by observing the temporal evolution of the Bragg diffraction line. In this experiment sub-picosecond temporal and sub-milliangstrom spatial resolution has been demonstrated. After these first pioneering experiments time-resolved X-ray diffraction has been used to characterize non-thermal melting (Siders *et al.*, 1999; Rousse *et al.*, 2001a), acoustic transients in solids (Sokolowski-Tinten *et al.*, 2001), and solid–solid phase transitions (Cavalleri *et al.*, 2001). These experiments clearly demonstrate the impressive potential of table-top picosecond X-ray sources and brings us closer to the goal of watching, on femtosecond time scales, so called molecular movies.

2.3.4 X-RAY LASERS

Most lasers operate in the visible and infrared spectral regions along on an axis defined by a Fabry-Perot style of reflecting cavity. The lasing medium is pumped to produce population inversion. Amplification along this axis is provided by stimulated emission. The intensity exponentiates further with each pass in the medium. A high degree of collimation and coherence is achieved with efficient cavities. In the X-ray region the reflectivity of available cavity mirrors is relatively poor. Hence, X-ray lasers to date have been based mainly on amplification of inherent spontaneous emission (ASE) throughout the medium. This occurs in a single pass along a particular direction. This requires gain coefficients in the medium approximately 100 times larger than those for more familiar lasers operating with efficient resonators. X-Ray lasers operate on electronic transitions from outer shells to inner shells. To achieve population inversion involving these transitions is much more difficult than for visible transitions, primarily because of the very short excited lifetimes of X-ray transitions (Rocca, 1999).

Soft X-ray lasers operating at wavelengths of 4 to 40 nm and sub-nanosecond pulse duration have been demonstrated. The laser medium is a plasma of highly ionized atoms. The plasma is produced by heating solid targets with energetic visible laser pulses having an energy of several joules to kilojoules. Such high energies can only be achieved in a few large-scale laser facilities all over the world. The major objective of current X-ray laser research is to reduce the driving laser pulse energy. Using driver pulses of a few picosecond or the combination of a nanosecond and femtosecond laser pulse has reduced the required energy by nearly two orders of magnitude (Nickles *et al.*, 1997; Lin *et al.*, 1999). Further improvements in pimping efficiency resulting from optimized target configuration and travelling wave excitation can be expected to lead to soft X-ray lasers occupying a single optical table (Korobkin *et al.*, 1998). Up to this time X-ray lasers, despite their unprecedented peak brilliance (Figure 2.3.1) (Burge *et al.*, 1997), will play only a minor role in spectroscopy.

2.3.5 HIGH HARMONIC GENERATION

A bright laboratory soft X-ray source is expected to have great impacts on physics, chemistry and biochemistry likewise. Recently laboratory sources of coherent soft X-rays extending into the water window have been demonstrated (Spielmann *et al.*, 1997; Chang *et al.*, 1997b) by means of high harmonic generation in helium. The use of femtosecond near-infrared pump pulses limits the effective interaction time relevant for the generation of the highest harmonics to less than the driver pulse duration. Under specific conditions a single attosecond X-ray pulse should be emitted. These extraordinarily short pulses are likely to open the way to the atomic timescale characterization of the quantum-mechanical evolution of the electron wave function in bound atomic states as well as for electronic processes in chemical reactions. Furthermore, the concentration of the X-rays within an extremely short time interval and the high degree of spatial coherence of the femtosecond laser-driven XUV source give rise to an unprecedented peak brightness, which paves the way towards an extension of nonlinear optics in the XUV regime.

The availability of light pulses comprising just a few oscillation cycles at intensities well above 10^{14} W/cm² opens up new prospects for coherent short-wavelength generation. Matter irradiated at these intensity levels undergoes tunnel ionization (Brabec and Krausz, 2000). For pulses containing many oscillations, the number of ionized particles (e.g. atoms, molecules, or clusters) accumulates over many optical cycles, which may result in a depletion of the ground state well before the peak of the incident pulse impinges upon the target. The consequences of the dramatically shortened exposure to the laser field are numerous and far reaching. First, the peak electron emission (or ionization) rate is much higher than in the multi-cycle regime. Secondly, the freed electrons are released into significantly stronger fields as compared to the multi-cycle regime. These implications suggest that coherent XUV generation resulting from a free-bound transition (high-order harmonic or continuum generation, henceforth HHG) (L'Huillier and Balcou, 1993) will exhibit improved efficiency and generate higher XUV frequencies. After ionization, the freed electron gains energy in the laser field. This is an essential process, because emission of high-energy photons is the result of the interaction of these electrons with their parent ions.

Following Corkum (Corkum, 1993), the electron motion in the laser field can be described by classical mechanics after tunnelling. The electron reencounters its parent ion if and only if the laser field is linearly polarized and the electron is released at a suitable phase. The maximum kinetic energy is given by $W_{kin,max} \sim 3.17 U_p$ (Corkum, 1993), where U_p is the ponderomotive energy. In units of electronvolts it can be expressed as $U_p = 9.3 \times 10^{-14} I/\lambda^2$, where I is the (cycle-averaged) laser intensity and λ is the laser wavelength in units of W/cm² and µm, respectively. With some small probability the returning electron recombines in its original ground state upon emitting a photon with an energy equal to the sum of the ionization potential I_p and the electron kinetic energy W_{kin} gained in the laser field. Because the electron wavepacket evolution is driven by a spatially coherent laser field, the emission of the high-energy XUV photons from the particles in the ensemble also occurs in a spatially coherent manner, leading to a well-collimated XUV beam collinear with the pump laser beam. In the multi-cycle regime, the process is repeated quasi-periodically over many laser periods, consequently the XUV emission spectrum is made up of discrete harmonics of the laser frequency. In the photon energy range close to the highest harmonics (cutoff region) the emission period has been predicted to be confined to a small fraction of the laser oscillation cycle. This prediction implies that the temporal evolution of the XUV output in the cutoff region can be characterized as a train of bursts of subfemtosecond duration (Brabec and Krausz, 2000). It is obvious from the above considerations that the use of a quasi-single-cycle driver benefits HHG in several respects. First, higher photon energies (shorter wavelengths) can be achieved because the electrons are released into a stronger laser field. Second, the X-rays near cutoff can be emitted in a single attosecond pulse, resulting in an unprecedented concentration of electromagnetic energy. Third, efficiency is increased due to the higher ionization rate and an increased coherence length (Spielmann et al., 1998). Experimental results of these findings are summarized in Figures 2.3.3 and 2.3.4.

At first sight it looks as if soft X-rays (<1 keV) produced by HHG are only of limited use for studying transient states of matter. However, a careful inspection of the processes to be studied and the parameter of state-of-the-art soft X-ray sources reveals they are well suited for investigating the dynamics of structural changes. Soft X-ray sources based on HHG deliver pulses with unique parameters: (i) the X-ray pulses are always shorter than the driving laser pulses (Brabec and Krausz, 2000) and can be as short as a few hundred attoseconds (Drescher et al., 2001; Hentschel et al., 2001); (ii) HHG radiation spans a broad spectral range from the visible down to a few nm (up to 500 eV). Further, the centre wavelength

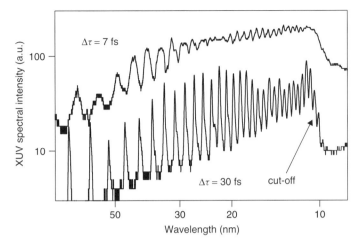

Figure 2.3.3 Typical HHG spectrum obtained with neon. The laser pulse duration was 7 and 30 fs, respectively, and the peak intensity was several times 10^{14} W/cm^2. The conversion efficiency is nearly constant over a wide range (plateau) followed by a sharp cut-off. Reproduced by permission of M. Schnürer

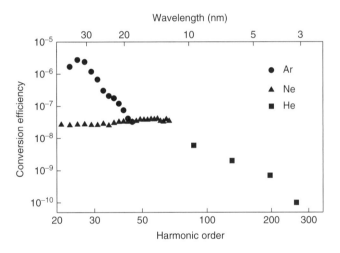

Figure 2.3.4 Energy conversion efficiencies using sub-10 fs laser pulses and an optimized target geometry. The higher the ionization potential of the noble gas the higher photon energies can be achieved via HHG. Reprinted from Schnürer, M., Cheng, Z., Hentschel, M., Krausz, F., Wilhein, T., Hambach, D., Schmahl, G., Drescher, M., Lim, Y. and Heinzmann, U. Few-cycle-driven XUV laser harmonics: generation and focusing. *Appl. Phys. B* **70**(suppl.), S227–S232, Figure 4 (2000). Reproduced by permission of Springer-Verlag GmbH & Co. KG

and width of harmonic lines is controllable with the intensity and duration of the driving laser pulses (Figure 2.3.3); (iii) the limited conversion efficiency from laser to soft X-ray radiation is compensated by an emission in a well defined beam, allowing a very effective coupling between the source and the experiment; (iv) the HHG radiation is polarized; and (v) last but not least, the necessary driving laser pulses intensity is moderate, making rather simple high repetition rate lasers usable as pump sources.

(i) Pulse duration. Using multi-TW scale lasers and solid targets, line or continuum radiation is emitted in pulses having a duration close to a picosecond. However, several processes take place on a much faster timescale; e.g. with LPPX source it was not possible to study the dynamic of the chemical shift of the silicon L-edge whilst excited with an intense femtosecond laser pulse (Nakano et al., 1998). Pulses from a HHG source are short enough to follow the dynamic. Also very fast processes take place on surfaces. To understand the basic mechanisms of adsorbate–substrate interaction studying the charge transfer between the adsorbate states and metal substrate is of fundamental importance. This process has been investigated with high spectral resolution X-ray spectroscopy (Wurth and Menzel, 2000) and revealed time constants in the order of a few femtoseconds. Soft X-ray pulses from HHG sources should allow the evolution of these reactions to be monitored in their fundamental timescale (Bauer et al., 2001; Siffalovic et al., 2001). Other experiments relying on ultrashort X-ray pulses are studies of the dynamic of photodissociation, e.g. it was found Br_2 dissociates on a timescale of 40 fs (Nugent-Glandorf et al., 2001).

(ii) Spectrum. Using few-cycle driving laser pulses the spectrum of HHG radiation has been extended down to 2 nm or 500 eV (Chang et al., 1997b; Spielmann et al., 1997). Since this wavelength range covers not only the 2p (L edges) of most 3d transition metals and 3d edges (M edges) of 4f rare earth metals it is ideally suited for the study of magnetic materials. In addition, carbon (Schnürer et al., 2000) and probably nitrogen and oxygen K absorption edge can also be covered, so that many interesting processes in biology, chemical dynamics, polymers and catalysis can be investigated. By varying the laser and target parameter the line width and position could be controlled. Even if it is a line spectrum the whole wavelength range could be covered, because the lines could be shifted by more than their initial separation by a moderate change of the laser intensity (Shin et al., 1999; Riedel et al., 2001). The width of a specific harmonic depends critically on the duration and intensity of the driving laser pulse. Very short laser pulses lead to lines broad enough to monitor chemical shifts of the absorption edge. Somewhat longer laser pulses result in narrower lines better suited for high resolution X-ray photoelectron spectroscopy (Haight, 1996; Quere et al., 2000).

(iii) Spatial properties. X-ray sources based on laser plasma interaction can be very efficient. Conversion efficiencies in the order of a percent have been demonstrated. However, the X-rays are emitted into 4π. Due to the lack of large aperture condenser optics in a typical experimental arrangement only a very small fraction of the generated photons can be used. In contrast, the coupling of HHG radiation to the experiment is nearly free from losses because it is emitted in a laser like beam (Spielmann et al., 1997). Due to the excellent spatial beam quality HHG radiation can be easily focused down to less than a micrometre spot size, paving the way to high resolution X-ray spectro-microscopy. The spatial coherence of HHG radiation makes also interferometry with soft X-ray radiation feasible (LeDeroff et al., 2000).

(iv) Polarization. HHG radiation is the only laser-based X-ray source which emits polarized radiation. The study of surfaces and magnetic materials will greatly benefit from this property (Schulze et al., 1998).

(v) Laser requirement. With a laser pulse having an energy of 1 mJ or less, enough soft X-ray photons for spectroscopy can be generated. Further improvements of the generation scheme such as quasi-phase matching (Rundquist et al., 1998) or adaptive control of the driving laser pulses (Bartels et al., 2000) will further increase the conversion efficiency. The relaxed energy requirements can be easily fulfilled by compact kilohertz laser systems which are already commercially available. The higher repetition rate results in higher average X-ray flux and allows the use of e.g. lock-in techniques enhancing the signal to noise ratio of the collected data. Also, the integration time in X-ray photoelectron spectroscopy based on time-of-flight detection will be substantially reduced by using a kilohertz system.

The significance of intense ultrashort laser pulses is likely to go well beyond the frontiers of physical science. A laboratory-scale source of coherent X-rays will undoubtedly make a major impact on time-resolved X-ray spectroscopy. Further optimization of the above described ultrashort pulse-driven harmonic generator (e.g. by the implementation of quasi-phase matching) allows the development of the first laboratory tool suitable for time-resolved spectro-microscopy. The predicted 'concentration' of the high-energy XUV photons in a single burst of attosecond duration may not only extend nonlinear optics into the X-ray regime and allow the study of the dynamics of bound electrons, but provide a tool for studying chemical reactions and wave packet dynamics on a previously inaccessible timescale. Nevertheless, at present we are just at the stage of exploring the new possibilities and phenomena, and a number of technical barriers are still left to be overcome before all of these exciting prospects can fully materialize.

REFERENCES

Backus S., Durfee III, C. G., Mourou, G., Kapteyn, H. C. and Murnane, M. M. 0.2 TW laser system at 1 kHz. *Opt. Lett.*, **22**, 1256–1258 (1997).

Backus, S., Bartels, R., Thompson, S., Dollinger, R., Kapteyn, H. C. and Murnane, M. M. High-efficiency, single-stage 7-kHz high-average-power ultrafast laser system. *Opt. Lett.*, **26**, 465–467 (2001).

Bagnoud, V. and Salin, F. Amplifying laser pulses to the terawatt level at a 1-kilohertz repetition rate. *Appl. Phys. B*, **70**, S165–170 (2000).

Bartels, R., Backus, S., Zeek, E., Misoguti, L., Vdovin, G., Christov, I. P., Murnane, M. M. and Kapteyn, H. C. Shaped-pulse optimization of coherent emission of high-harmonic soft X-rays. *Nature*, **406**, 164–166 (2000).

Barty, C. P. J., Guo, T., Le Blanc, C., Raksi, F., Rose-Petruck, C., Squier, J., Wilson, K. R., Yakovlev, V. V. and Yamakawa, K. Generation of 18-fs multiterawatt pulses by regenerative pulse shaping and chirped-pulse amplification. *Opt. Lett.*, **21**, 668–670 (1996).

Bauer, M., Lei, C., Read, K., Tobey, R., Gland, J., Murnane, M. M. and Kapteyn, H. C. Direct observation of surface chemistry using ultrafast soft-X-ray pulses. *Phys. Rev. Lett.*, **87**, 025501-1-4 (2001).

Blome, Ch., Sokolowski-Tinten, K., Dietrich, C., Tarasevitch, A. and von der Linde, D. Set-up for ultrafast time-resolved x-ray diffraction using a femtosecond laser-plasma keV x-ray-source. *J. Phys. IV*, **11**, 491 (2001).

Brabec, T. and Krausz, F. Intense few-cycle laser fields: Frontiers of nonlinear optics. *Rev. Mod. Phys.*, **72**, 545–591 (2000).

Burge R. E., Slark, G. E., Browne, M. T., Yuan, X. C., Charalambous, P., Cheng, X. H., Lewis, C. L. S., MacPhee, A. and Neely, D. Spatial coherence of X-ray laser emission from neonlike germanium after prepulse. *J. Opt. Soc. Am. B*, **14**, 2742–2751 (1997).

Cavalleri, A., Siders, C. W., Brown, F. E. H., Leitner, D. M., Toth, C., Squier, J. A., Barty, C. P. J., Wilson, K. R., Sokolowski-Tinten, K., von-Hoegen, M. H., von der Linde, D. and Kammler, M. Anharmonic lattice dynamics in germanium measured with ultrafast X-ray diffraction. *Phys. Rev. Lett.*, **85**, 586–589 (2000).

Cavalleri, A., Toth, C., Siders, C. W., Squier, J. A., Raksi, F., Forget, P. and Kieffer, J. C. Femtosecond structural dynamics in VO_2 during an ultrafast solid-solid phase transition. *Phys. Rev. Lett.*, **87**, 237401-1-4 (2001).

Chambaret, J. P., Le Blanc, C., Cheriaux, G., Curley, P., Darpentigny, G., Rousseau, P., Hamoniaux, G., Antonetti, A. and Salin, F. Generation of 25-TW, 32-fs pulses at 10 Hz. *Opt. Lett.*, **21**, 1921–1923 (1996).

Chang, Z., Rundquist, A., Wang, H., Murnane, M. M., Kapteyn, H. C., Liu, Z., Shan, B., Liu, J., Niu, L., Gong, M., Zhang, X. and Lee, R. Demonstration of a 0.54 picosecond X-ray streak camera. *Proc. SPIE*, **2869**, 971–976 (1997a).

Chang, Z., Rundquist, A., Wang, H., Murnane, M. and Kapteyn, H. C. Generation of coherent soft X-rays at 2.7 nm using high harmonics. *Phys. Rev. Lett.*, **79**, 2967–2970 (1997b).

Christov, I. P., Murnane, M. M. and Kapteyn, H. C. High-harmonic generation of attosecond pulses in the 'single-cycle' regime, *Phys. Rev. Lett.*, **78**, 1251 (1997).

Constant, E., Garzella, D., Berger, P., Mevel, E., Dorrer, C., Le Blanc, C., Salin, F. and Agostini, P. Optimizing high harmonic generation in absorbing gases: model and experiment. *Phys. Rev. Lett.*, **82**, 1668 (1999).

Corkum, P. B. Plasma perspective on strong-field multiphoton ionization. *Phys. Rev. Lett.*, **71**, 1994–1997 (1993).

Danson. C. N., Collier, J., Neely, D., Barzanti, L. J., Damerell, A., Edwards, C. B., Hutchinson, M. H. R., Key, M. H., Norreys, P. A., Pepler, D. A., Ross, I. N., Taday, P. F., Toner, W. T., Trentelman, M., Walsh, F. N., Winstone, T. B. and Wyatt, R. W. W. Well characterized 10^{19} W/cm^2 operation of VULCAN-an ultra-high power Nd:glass laser. *J. Mod. Opt.*, **45**, 1653–1669 (1998).

Drescher, M., Hentschel, M., Kienberger, R., Tempea, G., Spielmann, Ch., Reider, G. A., Corkum, P. B. and Krausz, F. X-ray pulses approaching the attosecond frontier. *Science*, **291**, 1923–1927 (2001).

Durfee III, C. G., Backus, S., Murnane, M. M. and Kapteyn, H. C. Design and implementation of a TW-class high-average power laser system. *IEEE J. Select. Topics Quantum Electron.*, **4**, 395–406 (1998).

Feurer, T. Feedback-controlled optimization of soft-X-ray radiation from femtosecond laser-produced plasmas. *Appl. Phys. B*, **68**, 55–60 (1999).

Feurer, T., Morak, A., Uschmann, I., Ziener, C., Schwoerer, H., Forster, E. and Sauerbrey, R. An incoherent sub-picosecond X-ray source for time-resolved X-ray-diffraction experiments. *Appl. Phys. B*, **72**, 15–20 (2001).

Gibbon, P. and Förster, E. Short-pulse laser-plasma interactions. *Plasma Phys. Control. Fusion*, **38**, 769–793 (1996).

Giulietti, D. and Gizzi, L. A. X-ray emission from laser-produced plasmas. *La Rivista del Nuovo Cimento*, **21**, 1–93 (1998).

Gordon III, C. L., Yin, G. Y., Lemoff, B. E., Bell, P. M. and Barty, C. P. J. Time-gated imaging with an ultrashort-pulse, laser-produced-plasma X-ray source. *Opt. Lett.*, **20**, 1056–1058 (1995).

Grätz M., Kiernan, L., Wahlström, C. G., Svaneberg, S. and Herrling, K. Time-gated x-ray tomography. *Appl. Phys. Lett.*, **73**, 2899–2901 (1998).

Guo, T., Spielmann, Ch., Walker, B. C. and Barty, C. P. J. Generation of hard X-rays by ultrafast terawatt lasers. *Rev. Sci. Instrum.*, **72**, 41–47 (2001).

Haight, R. Electron dynamics at semiconductor surfaces and interfaces. *Chem. Phys.*, **205**, 231–244 (1996).

Helliwell, J. R. and Rentzepis, P. M. (Eds) *Time Resolved Diffraction*, Clarendon Press, Oxford (1997).

Hemberg, O., Hansson, B. A. M., Berglund, M. and Hertz, H. M. Stability of droplet-target laser-plasma soft x-ray sources. *J. Appl. Phys.*, **88**, 5421–5425 (2000).

Hentschel, M., Kienberger, R., Spielmann, Ch., Reider, G. A., Milosevic, N., Brabec, T., Corkum, P. B., Heinzmann, U., Drescher, M. and Krausz, F. Attosecond metrology. *Nature*, **414**, 509 (2001).

Kalachnikov, M. P., Nickles, P. V., Schonnagel, H. and Sandner, W. On the way to 100 TW-10 Hz titanium–sapphire laser facilities. *Nucl. Instr. Meth. Phys. Res. A*, **472**, 254–259 (2001).

Klisnick, A., Kuba, J., Ros, D., Smith, R., Jamelot Chenais, G., Popovics, C., Keenan, R., Topping, S. J., Lewis, C. L. S., Strati, F., Tallents, G. J., Neely, D., Clarke, R., Collier, J., MacPhee, A. G., Bortolotto, F., Nickles, P. V. and Janulewicz, K. A. Demonstration of a 2-ps transient X-ray laser. *Phys. Rev. A*, **65**, 033810-1–4 (2002).

Kondo H., Tomie, T. and Shimizu, H. Time of flight photoelectron spectroscopy with a laser-plasma x-ray source. *Appl. Phys. Lett.*, **69**, 182–184 (1996)

Korobkin, D., Goltsov, A., Morozov, A. and Suckewer, S. Soft X-ray amplification at 26.2 nm with 1-Hz repetition rate in a table-top system. *Phys. Rev. Lett.*, **81**, 1607–1610 (1998).

Larsson, J., Heimann, P. A., Lindenberg, A. M., Schuck, P. J., Buckbaum, P. H., Lee, R. W., Padmore, H. A., Wark, J. S. and Falcone, R. W. Ultrafast structural changes measured by time-resolved X-ray diffraction. *Appl. Phys. A*, **66**, 587–591 (1998).

LeDeroff, L., Saliers, P., Carre, B., Joyeux, D. and Phalippou, D. Measurement of the degree of spatial coherence of high-order harmonic using a Fresnel-mirror interferometer. *Phys. Rev. A*, **61**, 043802-1–4 (2000).

Lemoff, B. E., Yin, G. Y., Gordon III, C. L., Barty, C. P. J. and Harris, S. E. Demonstration of a 10-Hz femtosecond-pulse-driven XUV laser at 41.8 nm in Xe IX. *Phys. Rev. Lett.*, **74**, 1574–1577 (1995).

L'Huillier, A. and Balcou, P. High-order harmonic generation in rare gases with a 1-ps 1053-nm laser. *Phys. Rev. Lett.*, **70**, 774–777 (1993).

Lin, J. Y., Tallents, G. J., MacPhee, A. G., Smith, R., Wolfrum, E., Zhang, J., Eker, G., Keenan, R., Lewis, C. L. S., Neely, D., O'Rourke, R. M. N., Pert, G. J., Pestehe, S. J. and Wark, J. S. Optimization of double pulse pumping for Ni-like Sm X-ray lasers. *J. Appl. Phys.*, **85**, 672–675 (1999).

Nakano H., Goto, Y., Lu, P., Nishikawa, T. and Usegui, N. Time-resolved soft X-ray absorption spectroscopy of silicon using femtosecond laser plasma X-rays. *Appl. Phys. Lett.*, **75**, 2350–2352 (1999).

Nickles, P. V., Shlyaptsev, V. N., Kalachnikov, M., Schnürer, M., Will, I. and Sandner, W. Short pulse X-ray laser at 32.6 nm based on transient gain in Ne-like titanium. *Phys. Rev. Lett.*, **78**, 2748–2751 (1997).

Nisoli, M., Stagiara, S., De Silvestri, S., Svelto, O., Sartania, S., Cheng, Z., Tempea, G., Spielmann, Ch. and Krausz, F. Toward a terawatt-scale sub-10-fs laser technology. *IEEE J. Select. Topics Quantum Electron.*, **4**, 414–420 (1998).

Nugent-Glandorf, L., Scheer, M., Samuels, D. A., Mulhisen, A. M., Grant, E. R., Yang, X., Bierbaum, V. M. and Leone, S. T. Ultrafast time-resolved soft x-ray photoelectron spectroscopy of dissociating Br_2. *Phys. Rev. Lett.*, **87**, 193002-1–4 (2001).

Perry, M. D. and Mourou, G. Terawatt to petawatt subpicosecond lasers. *Science*, **264**, 917–924 (1994).

Perry, M. D., Pennington, D., Stuart, B. C., Tiethohl, G., Britten, J. A., Brown, C., Herman, S., Golick, B., Kartz, M., Miller, J., Powell, H. T., Vergino, M. and Yanovsky, V. Petawatt laser pulses. *Opt. Lett.*, **24**, 160–162 (1999).

Quere F., Guizard, S., Martin, Ph, Petite, G., Merdji, H., Carre, B., Hergott, J. and De Deroff, L. Hot-electron relaxation in quartz using high-order harmonics. *Phys. Rev. B*, **61**, 9883–9886 (2000).

Raksi, F., Wilson, K. R., Jiang, Z., Ikhlef, A., Cote, C. Y. and Kieffer, J.-C. Ultrafast X-ray absorption probing of a chemical reaction. *J. Chem. Phys.*, **104**, 6066–6069 (1996).

Riedel, D., Hernandez-Pozos, J. L., Palmer, R. E., Baggott, S., Kolasinski, K. W. and Food, J. S. Tunable pulses vacuum ultraviolet light source for surface science and materials spectroscopy based on high order harmonic generation. *Rev. Sci. Instrum.*, **72**, 1977–1983 (2001).

Rischel, C., Rousse, A., Uschmann, I., Albouy, P.-A., Geindre, J.-P., Audebert, P., Gauthier, J.-C., Förster, E., Martin, J.-L. and Antonetti, A. Femtosecond time-resolved X-ray diffraction from laser-heated organic films. *Nature*, **390**, 490–492 (1997).

Rocca, J. J. Table-top soft x-ray lasers. *Rev. Sci. Instrum.*, **70**, 3799–3828 (1999).

Rose-Petruck, C., Jimenez, R., Guo, T., Cavalleri, A., Siders, C. W., Raksi, F., Squier, J. A., Walker, B. C., Wilson, K. R. and Barty, C. P. J. Picosecond-milliangstrom lattice dynamics measured by ultrafast X-ray diffraction. *Nature*, **398**, 310–312 (1999).

Rousse, A., Audebert, P., Geindre, J. P., Fallies, F., Gauthier, J. C., Mysyrowicz, A., Grillon, G. and Antonetti, A. Efficient K alpha X-ray source from femtosecond laser-produced plasmas. *Phys. Rev. E*, **50**, 2200–2207 (1994).

Rousse, A., Rischel, C., Fourmaux, S., Uschmann, I., Sebban, S., Grillon, G., Balcou, P., Forster, E., Geindre, J. P., Audebert, P., Gauthier, J. C. and Hulin, D. Non-thermal melting in semiconductors measured at femtosecond resolution. *Nature*, **410**, 65–68 (2001a).

Rousse, A., Rischel, C. and Gauthier, J. C. Colloquium: Femtosecond X-ray crystallography. *Rev. Mod. Phys.*, **73**, 17–31 (2001b).

Rundquist, A., Durfee IIII, C. G., Chang, Z., Herne, C., Backus, S., Murnane, M. M. and Kapteyn, H. C. Phase-matched generation of coherent soft X-rays. *Science*, **280**, 1412–1415 (1998).

Sartania S., Cheng, Z., Lenzner, M., Tempea, G., Spielmann, Ch., Krausz, F. and Ferencz, K. Generation of 0.1-TW 5-fs optical pulses at a 1-kHz repetition rate. *Opt. Lett.*, **22**, 1562–1564 (1997).

Schnürer, M., Spielmann, C., Wobrauschek, P., Streli, C., Burnett, N. H., Kan, C., Ferencz, K., Koppitsch, R., Cheng, Z., Brabec, T., and Krausz, F. Coherent 0.5-keV X-ray emission from helium driven by a sub-10-fs laser. *Phys. Rev. Lett.*, **80**, 3236–3239 (1998).

Schnürer, M., Cheng, Z., Hentschel, M., Tempea, G., Kálmán, P., Brabec, T. and Krausz, F. Absorption-limited generation of coherent ultrashort soft-X-ray pulses. *Phys. Rev. Lett.*, **83**, 722–725 (1999).

Schnürer, M., Streli, C., Wobrauschek, P., Hentschel, M., Kienberger, R., Spielmann, Ch. and Krausz, F. Femtosecond X-ray fluorescence. *Phys. Rev. Lett.*, **85**, 3392–3395 (2000).

Schoenlein, R. W., Leemans, W. P., Chin, A. H., Volfbeyn, P., Glover, T. E., Balling, P., Zolotorev, M., Kim, K. J., Chattopadhyay, S. and Shank, C. V. Femtosecond X-ray pulses at 0.4 Å generated by 90 degrees Thomson scattering: a tool for probing the structural dynamics of materials. *Science*, **274**, 236–238 (1996).

Schoenlein, R. W., Chattopadhyay, S., Chong, H. H. W., Glover, T. E., Heimann, P. A., Shank, C. V., Zholents, A. A. and Zolotorev, M. S. Generation of femtosecond pulses of synchrotron radiation. *Science*, **287**, 2237–2240 (2000).

Schulze D., Dörr, M., Sommerer, G., Ludwig, J., Nickles, P. V., Schlegel, T., Sandner, W., Drescher, M., Kleineberg, U. and Heinzmann, U. Polarization of the 61st harmonic from 1053-nm laser radiation in neon. *Phys. Rev. A*, **57**, 3003–3007 (1998).

Shin H. J., Lee, D. G., Cha, Y. H., Hong, K. H. and Nam, C. H. Generation of nonadiabatic blueshift of high harmonics in an intense femtosecond laser field. *Phys. Rev. Lett.*, **83**, 2544–2547 (1999).

Siders, C. W., Cavalleri, A., Sokolowski-Tinten, K., Tóth, Cs., Guo, T., Kammler, M., Horn von Hoegen, M., Wilson, K. R., von der Linde, D. and Barty, C. P. J. Detection of nonthermal melting by ultrafast X-ray diffraction. *Science*, **286**, 1340–1342 (1999).

Siffalovic, P., Drescher, M., Spieweck, M., Wiesenthal, T., Lim, Y. C., Weidner, R., Elizarov, A. and Heinzmann, U. Laser-based apparatus for extended ultraviolet femtosecond time-resolved photoemission spectroscopy. *Rev. Sci. Instrum.*, **72**, 30–35 (2001).

Sokolowski-Tinten, K., Blome, C., Dietrich, C., Tarasevitch, A., Horn von Hoegen, M., von-der-Linde, D., Cavalleri, A., Squier, J. and Kammler, M. Femtosecond X-ray measurement of ultrafast melting and large acoustic transients. *Phys. Rev. Lett.*, **87**, 225701-1–4 (2001).

Sommerer, G., Rottke, H. and Sandner, W. Enhanced efficiency in high-order harmonic generation using sub-50 fs laser pulses. *Laser Phys.*, **9**, 430–432 (1999).

Spielmann, C., Burnett, N. H., Sartania, S., Koppitsch, R., Schnürer, M., Kan, C., Lenzner, M., Wobrauschek, P. and Krausz, F. Generation of coherent X-rays in the water window using 5-femtosecond laser pulses. *Science*, **278**, 661–664 (1997).

Spielmann, C., Kan, C., Burnett, N. H., Brabec, T., Geissler, M., Scrinzi, A., Schnürer, M. and Krausz, F. Near-keV coherent X-ray generation with sub-10-fs lasers. *IEEE J. Select. Topics Quantum Electron.*, **4**, 249–265 (1998).

Steinmeyer, G., Sutter, D. H., Gallmann, L., Matuschek, N. and Keller, U. Frontiers in ultrashort pulse generation: pushing the limits in linear and nonlinear optics. *Science*, **286**, 1507–1512 (1999).

Streli, C., Wobrauschek, P., Beckhoff, B., Ulm, G., Fabry, L. and Pahlke, S. First results of TXRF measurements of low-Z elements on Si wafer surfaces at the PTB plane grating monochromator beamline for undulator radiation at BESSY II. *X-ray Spectrometry*, **30**, 24–31 (2001).

Strickland, D. and Mourou, G. Compression of amplified chirped optical pulses. *Opt. Commun.*, **56**, 219–221 (1985).

Tomov, I. V., Oulianov, D. A., Chen, P. and Rentzepis, P. M. Ultrafast time, resolved transient structures of solids and liquids studied by means of X-ray diffraction and EXAFS. *J. Phys. Chem. A*, **103**, 7081–7091 (1999).

Tompkins, R. J., Mercer, I. P., Fettweis, M., Barnett, C. J., Klug, D. R., Porter, G., Clark, I., Jackson, S., Matousek, P., Parker, A. W. and Towrie, M. 5–20 keV laser-induced x-ray generation at 1 kHz from a liquid-jet target. *Rev. Sci. Instrum.*, **69**, 3113–3117 (1998).

Umstadter, D. P., Barty, C. P. J., Perry, M. and Mourou, G. A. Tabletop, ultrahigh-intensity lasers: dawn of nonlinear relativistic optics. *Opt. Photon. News*, **9**, 41 (1998).

von der Linde, D., Sokolowski-Tinten, K., Blome, C., Dietrich, C., Zhou, P., Tarasevitch, A., Cavalleri, A., Siders, C. W., Barty, C. P. J., Squier, J., Wilson, K. R., Uschmann, I. and Forster, E. Generation and application of ultrashort X-ray pulses. *Laser Part. Beams*, **19**, 15–22 (2001).

Wang, H., Backus, S., Chang, Z., Wagner, R., Kim, K., Wang, X., Umstadter, D., Lei, T., Murnane, M. M. and Kapteyn, H. C. Generation of 10-W average-power, 40-TW peak-power, 24-fs pulses from a Ti:sapphire amplifier system. *J. Opt. Soc. Am. B*, **16**, 1790–1794. (1999).

Wark, J. S. Time-resolved X-ray diffraction. *Contemp. Phys.*, **37**, 205–218 (1996).

Wharton, K. B., Hachett, S. P., Wilks, S. C., Hey, M. H., Moody, J. D., Yanovsky, V., Offenberger, A. A., Hammel, B. A., Perry, M. D. and Joshi, C. Experimental measurements of hot electrons generated by ultraintense ($>10^{19}$ W/cm^2) laser–plasma interactions on solid-density targets. *Phys. Rev. Lett.*, **81**, 822–825 (1998).

Wilks, S. C. and Kruer, W. L. Absorption of ultrashort, ultraintense laser light by solids and overdense plasmas. *IEEE J. Quant. Electr.*, **QE33**, 1954–1968 (1997).

Wurth W. and Menzel, D. Ultrafast electron dynamics at surfaces probed by resonant Auger spectroscopy. *Chem. Phys.*, **25**, 141–149 (2000).

Yamakawa, K., Aoyama, M., Matsuoka, S., Kase, T., Akahane, Y. and Takuma, H. 100-TW sub-20-fs Ti:sapphire laser system operating at a 10-Hz repetition rate. *Opt. Lett.*, **23**, 1468–1470 (1998).

Yamakawa, K. and Barty, C. P. J. Ultrafast, ultrahigh-peak, and high-average power Ti:sapphire laser system and its applications. *IEEE J. Select. Topics Quantum Electron.*, **6**, 658–675 (2000).

Zhang, J., MacPhee, A. G., Lin, J., Wolfrum, E., Smith, R., Danson, C., Key, M. H., Lewis, C. L. S., Neely, D., Nilsen, J., Pert, G. J., Tallents, G. J. and Wark, J. S. A saturated X-ray laser beam at 7 nanometers. *Science*, **276**, 1097–1100 (1997).

Zhang, P., He, J. T., Chen, D. B., Li, Z. H., Zhang, Y., Bian, J. G., Wang, L., Li, Z. L., Feng, B. H., Zhang, X. L., Zhang, D. X., Tang, X. W. and Zhang, J. Effects of a prepulse on γ-ray radiation produced by a femtosecond laser with only 5-mJ energy. *Phys. Rev. E*, **57**, R3746–3748 (1998).

Chapter 3
X-Ray Optics

3.1 Multilayers for Soft and Hard X-rays

M. YANAGIHARA
Institute of Multidisciplinary Research for Advanced Materials, Tohoku University, Japan

and

K. YAMASHITA
Department of Physics, Nagoya University, Japan

3.1.1 INTRODUCTION

X-Rays have an advantage over visible light to achieve higher resolution in image formation. Furthermore, X-ray spectroscopy has an advantage that near the absorption edges the spectra provide element- and state-specific information. However, it has been difficult to handle X-rays using techniques similar to those applied in the longer wavelength region because grazing-incidence optics based on total reflection is essential for a reflector of X-rays. Actually the X-ray reflectivity almost vanishes at incidence angles beyond the critical angle of the total reflection. Therefore, grazing-incidence optics needs a large-sized mirror.

A multilayer coating consists of alternate layers of different materials with appropriate optical constants. Its periodic structure enhances the reflectivity due to the constructive interference of the multiple reflections from the interfaces by analogy with Bragg reflection from a crystal. Soon after the discovery of X-rays, it was proposed that synthetic layered structures might extend the spectral region to long wavelengths. As the X-ray reflection from each layer is very small, a few hundreds of layer pairs are needed to synthesize multilayer structures. Besides, for the constructive interference among the partial waves, the layer thickness should be controlled at the sub-nanometer level over large areas. Early attempts to fabricate such structures were not successful due to technological limitations. Since the 1970s, fabrication of multilayer structures of sufficient quality for X-ray optics has been widely developed due to the progress in thin-film technology and polishing of supersmooth substrates to the sub-nanometer level. In the low energy region of soft X-rays, multilayer coatings are now practically applicable at nongrazing-incidence angles. In particular, in the 13-nm wavelength region, a normal-incidence reflectivity of about 70 % is now achieved using Mo/Si multilayers. Based on these successful achievements, the multilayers have been applied as normal-incidence reflectors to microscopy, telescopes, and lasers to advance soft X-ray science. Also in the field of industry their application has been widely attempted as a key component of reductive projection lithography for next-generation large scale integrated circuit. In these applications, multilayers are coated on plane or figured substrates to form images of soft X-rays. For hard X-rays, multilayers have also been developed for grazing-incidence optical systems. The multilayer coating on a reflecting surface makes it possible to enhance the reflectivity in the proper energy band beyond the critical angle. Furthermore, depth-graded multilayers, so called 'multilayer supermirrors', extend the reflection band in the hard X-ray region of 10–100 keV. Such mirrors have been used as focusing components not only in synchrotron radiation facilities but also in usual X-ray laboratories. Using multilayers, a hard X-ray telescope has been developed for balloon-borne observations and will be put on board of future X-ray astronomy missions. Furthermore, reflection multilayers are used as coating on gratings to enhance the diffraction efficiency for soft and hard X-rays and as polarizers for soft X-rays. On the other hand, a transmission-type multilayer has been fabricated by depositing a multilayer on a very thin transparent substrate. It has been applied as a polarizer, a phase shifter, and a beam splitter in the soft X-ray region.

3.1.2 DESIGN PRINCIPLE

Optical properties of materials are closely related to the response of the electrons. By the Drude model, where a free electron oscillates with an outer electric field, the dielectric constant is given by

$$\varepsilon(\omega) = 1 - \frac{\omega_p^2}{\omega^2 + i\Gamma\omega} \quad (3.1.1)$$

where $\omega_p^2 = Ne^2/\varepsilon_0 m$ is the squared plasma frequency and Γ is a damping constant. The plasma energy $\hbar\omega_p$ of metals ranges below 20 eV. As we can neglect Γ in Equation (3.1.1) we see that ε is slightly smaller than 1 when $\omega \gg \omega_p$, that is, in the X-ray region. The index of refraction is related to the dielectric constant as

$$n = \sqrt{\varepsilon} \quad (3.1.2)$$

In the X-ray region, the index of refraction is usually denoted as

$$n \equiv 1 - \delta + i\beta \quad (3.1.3)$$

From the above discussion δ and β are very small compared to 1. β is the same quantity as the extinction coefficient and is related to the absorption coefficient μ by

$$\beta = \frac{\lambda\mu}{4\pi} \quad (3.1.4)$$

where λ is the wavelength. δ and β are given by

$$\delta \cong \frac{r_e \lambda^2}{2\pi} \sum_j N_j f_{1j} \quad \beta \cong \frac{r_e \lambda^2}{2\pi} \sum_j N_j f_{2j}$$
$$(3.1.5)$$

where the sum is taken over every atomic species of atomic density N_j and atomic scattering factor $f_j = f_{1j} - if_{2j}$. The term $r_e = e^2/4\pi\varepsilon_0 mc^2$ is the classical electron radius. The atomic scattering factor is a useful concept because the index of refraction of a material is calculated using the atomic scattering factors of its constituent atoms.[1] This is a good approximation except for the absorption edges where the near-edge structure is dominant in absorption spectra.

Now an electromagnetic wave is incident from a medium 1 with an index of refraction n_1 onto a boundary with a medium 2 with n_2 at a grazing angle of incidence θ_1. Snell's law is presented by

$$n_1 \cos\theta_1 = n_2 \cos\theta_2 \quad (3.1.6)$$

where θ_2 is the grazing angle of refraction. The Fresnel reflection and transmission coefficients, r and t, for s and p polarizations are formulated by

$$r_s = t_s - 1 = \frac{n_1 \sin\theta_1 - n_2 \sin\theta_2}{n_1 \sin\theta_1 + n_2 \sin\theta_2} \quad (3.1.7)$$

and

$$r_p = t_p - 1 = \frac{n_2 \sin\theta_1 - n_1 \sin\theta_2}{n_2 \sin\theta_1 + n_1 \sin\theta_2} \quad (3.1.8)$$

respectively. The reflectivity R is given by $R = r^*r$. Figure 3.1.1 shows the reflectivity for s and p polarizations calculated for $(\delta_2, \beta_2) = (0.2, 0.1)$ and $(0.01, 0.001)$ when $n_1 = 1$ (vacuum). The

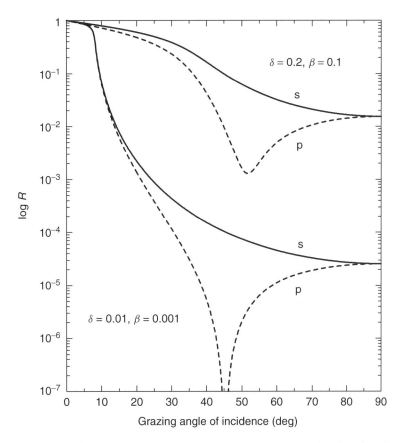

Figure 3.1.1 Calculated reflectivities for s (solid line) and p (broken line) polarization as a function of grazing angle of incidence for $\delta, \beta = 0.2, 0.1$ and $0.01, 0.001$, respectively

former is a typical example for the extreme ultraviolet region, and the latter for the X-ray region. For X-rays, the reflectivity for both s and p polarizations decreases rapidly at the critical angle of the total reflection, which is obtained from Snell's law as $\sin\theta_c = \sqrt{2\delta}$. It is practically calculated for a wavelength λ (nm) as

$$\sin\theta_c = \sqrt{2\delta} = 2.325 \times 10^{-2}\sqrt{\rho f_1/A}\lambda \quad (3.1.9)$$

where ρ (g/cm^3) is the mass density and A is the atomic weight. The reflectivity becomes very low at normal incidence. The reflectivity for p polarization shows a minimum around 45°, which is the Brewster reflection. The result is confirmed from Fresnel's and Snell's equations as n of every material is very close to 1 in the X-ray region.

The Bragg angle θ_m of the mth order reflection from a multilayer with a d-spacing is given by

$$m\lambda = 2d\sin\theta_m\sqrt{1 - \frac{2\delta - \delta^2}{\sin^2\theta_m}} \quad (3.1.10)$$

It is reduced to the usual form $m\lambda = 2d\sin\theta_m$ when θ_m is large. Figure 3.1.2 shows the boundary of the total reflection ($\theta = \theta_c$) calculated for a Pt single layer mirror using Equation (3.1.9) and the Bragg reflection in 1st and 5th order ($\theta = \theta_m; m = 1,5$) calculated for a multilayer of $d = 1.5$ nm using Equation (3.1.10) as a function of λ. The right-hand side of each curve in the $\theta - \lambda$ plane is accessible by X-ray optical systems. It is obvious that multilayers enable the applicable region to be extended to five times shorter wavelengths than

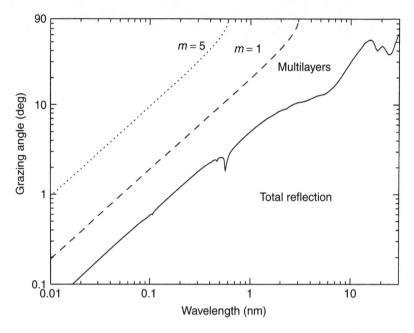

Figure 3.1.2 Applicable boundaries shown in the θ–λ plane for the total reflection of a Pt mirror (solid line) and the Bragg reflection of a multilayer with $d = 1.5$ nm in 1st (broken line) and 5th (dotted line) order. The right-hand side of each curve is accessible for X-ray optics

without them. Normal incidence optics become available in the wavelength region longer than 3 nm. The bandwidth of the Bragg peak in mth order is given as

$$\Delta E = E/mN \qquad (3.1.11)$$

where N is the number of total layer pairs. The X-ray energy (E) and the wavelength are connected by

$$E(\text{keV}) \cdot \lambda(\text{nm}) = 1.23985 \qquad (3.1.12)$$

The maximum number of layer pairs is estimated from the ratio of the d-spacing to the penetration depth of the incident X-rays.

A multilayer is designed so that all boundaries add in phase to the reflected wave. Here the 'layer-by-layer' method[2] is presented. Let the overall reflectivity and transmissivity of a multilayer be r_0 and t_0 at the moment, and then a film with a thickness l be deposited onto the surface of the multilayer. The new amplitude reflectivity and transmissivity of the multilayer structure are given by

$$r_{\text{new}} = \frac{r(1 - rr_0) + (r_0 - r)\exp(-i\alpha)}{1 - rr_0 + r(r_0 - r)\exp(-i\alpha)}$$

$$t_{\text{new}} = \frac{t_0(1 - r^2)\exp(-i\alpha/2)}{1 - rr_0 + r(r_0 - r)\exp(-i\alpha)} \qquad (3.1.13)$$

where r is the Fresnel reflection coefficient of the deposited film of an index of refraction n with respect to the vacuum and

$$\alpha = \frac{4\pi}{\lambda} l \sqrt{n^2 - \cos^2\theta} \qquad (3.1.14)$$

is the phase delay due to the propagation inside the deposited layer. As the form of Equation (3.1.13) is the same for s and p polarizations, the subscripts are omitted in it. Reflection loss due to the surface roughness is accounted for in r in the form of a Debye–Waller factor

$$\exp[-2(2\pi\sigma\sin\theta/\lambda)^2] \qquad (3.1.15)$$

where σ is the root mean square surface roughness. In order to suppress σ, the thickness must be controlled within an accuracy of the atomic scale. The reflectivity of a multilayer structure is obtained by calculating r_{new} layer by layer, starting from the Fresnel reflection coefficient of a substrate. This method is useful not only to calculate the multilayer reflectivity for a given structure, but also to optimize the thickness of each layer by monitoring the reflectivity to achieve the highest value.

Using the layer-by-layer method, the criterion for selecting the optimal pair for a high-reflectivity multilayer was established. That is, β of both materials is small and the difference in δ between the two is as large as possible. In other words, the multilayer has a periodic structure made of alternately deposited heavy and light elements. In general, materials are used in the region below their core absorption edges. Furthermore, material combinations have to be selected to form uniform layers and stable interfaces in physical and chemical aspects.

A synthetic multilayer usually has a periodic structure. Furthermore, multilayers with aperiodic structure have been developed. The bandwidth of the reflection from an X-ray multilayer is quite narrow, so that X-ray supermirrors have been widely developed to achieve a wide energy band of high reflectivity beyond the critical angle. Generally, it consists of a top layer for total reflection for long-wavelength light, a bottom multilayer with a periodic structure for short-wavelength light, and a middle multilayer with nonperiodic structure for middle-wavelength light. The number of layers has to be determined to minimize the attenuation of incident X-rays, but to maximize the reflectivity.

3.1.3 SOFT X-RAY MULTILAYERS AND THEIR APPLICATIONS

For soft X-ray multilayers around 100 eV, Mo/Si and Mo/Be are the most successful combinations. For Mo/Si multilayers a normal-incidence reflectivity of about 71 % has been achieved at 12.8 nm using magnetron sputtering.[3] The multilayer consists of 50 layer pairs and contains B_4C diffusion barriers. Its measured reflectivity is shown in Figure 3.1.3, where a peak value of 70.0 % at 13.5 nm for another multilayer is also shown. The

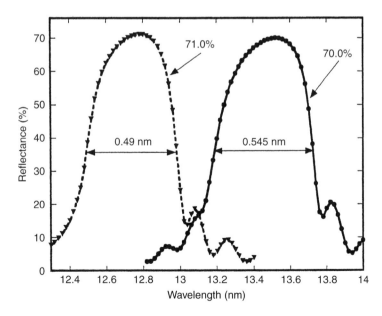

Figure 3.1.3 Normal-incidence reflectivity measured for magnetron-sputtered Mo/Si multilayers. They consist of 50 layer pairs and contain B_4C diffusion barriers.[3] Reprinted from *Opt. Eng.* **41** (2002) 1797. Reproduced by permission of SPIE

bandwidths are reasonable from Equation (3.1.11). The ideal reflectivity is about 75 %. Using a magnetron sputtered Mo/Be multilayer of 70 layer pairs a normal-incidence reflectivity of 70.2 % has been achieved at 11.34 nm.[4] The Mo-based multilayers are the basis for the development of the soft X-ray lithography.[5,6]

Up to now, it has been difficult to achieve high multilayer reflectivity below the 11-nm region due to the surface roughness in comparison with the incident wavelength. A normal-incidence reflectivity of 18.9 % has been achieved recently using a magnetron sputtered Cr/C multilayer with 150 layer pairs at a wavelength of 6.42 nm.[7] A CoCr/C multilayer of 16.1 % has been also fabricated at 6.1 nm. The motivation to develop a highly reflective multilayer in this wavelength region is X-ray photoemission spectroscopy with inner-shell excitation using a multilayer-coated Schwarzschild objective.

It is a primary subject for soft X-ray microscopy to observe living cells under natural environment in the water-window region between 4.4 nm (C K absorption edge) and 2.3 nm (O K absorption edge). In this region the absorption coefficient of C, N (K edge), and Ca (L edges) changes by a large amount within a narrow wavelength region, while that of O in the water is very low, which provides good contrast for living cells. Very recently a reflectivity of 20.5 % has been achieved at $\theta = 73°$ using a Cr/Sc multilayer of 600 layer pairs.[8] It was fabricated using an ion-assisted dual-target magnetron sputter system to avoid intermixing of the two metals.

3.1.3.1 FOCUSING AND MICROSCOPY

Figure 3.1.4 illustrates the Schwarzschild optics. It is a combination of a concave and a convex spherical mirror, where the centers of the two spheres are slightly shifted to eliminate the higher order aberration. Multilayer-coated Schwarzschild objectives have been installed at a beamline of a synchrotron radiation facility to carry out photoemission experiments with high spatial resolution.[9] The objectives provide a small radiation spot of $0.5\,\mu m$ on the sample. Three objectives are currently available at $E = 74, 95$, and $110\,eV$. The 74 and 95-eV objectives are made of Mo/Si multilayers, while the 110-eV one is made of Ru/B$_4$C. The objective reflectivity of the 95-eV objective is about 25 %, and that of the 110-eV one is 1.1 %. Figure 3.1.5 shows a cross-sectional image of a molecular-beam epitaxy-grown p-n GaAs superlattice with different periods obtained by sample scanning. As can be seen, the spectromicroscope allows layers as thin as $0.25\,\mu m$ to be imaged. The contrast of the image is based on the shift in energy between the Ga 3d level in p-type and n-type GaAs. A multilayer-coated Schwarzschild photoelectron microscope with the use of He-I and He-II resonance lines has been also developed.[10]

A transmission multilayer is available as a beam splitter in soft X-ray interferometry. A Mirau

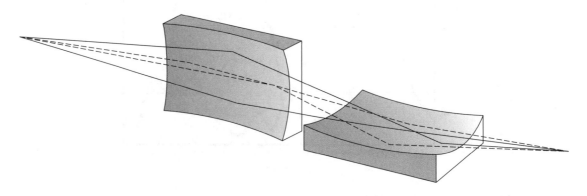

Figure 3.1.4 A schematic drawing of the Schwarzschild optics. It is a combination of a concave and a convex spherical mirror

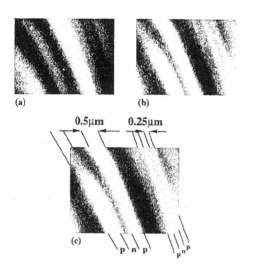

Figure 3.1.5 Cross-sectional image of an MBE-grown *p-n* GaAs superlattice with different periods obtained by sample scanning. The two top images are obtained by tuning the energy analyzer at the kinetic energy of the Ga 3d peak of (a) *n*-type GaAs and (b) *p*-type GaAs. The layers of *n*-type GaAs appear bright in (a) and dark in (b). Image (c) is the difference between the two top images.[9] Reprinted from *Rev. Sci. Instrum.* **71** (2000) 5. Reproduced by permission of American Institute of Physics

interferometric microscope has been developed by combining a multilayer beam splitter with an imaging-type soft X-ray microscope based on a multilayer-coated Schwarzschild objective.[11] When using it, some stacking defects have been observed on a multilayer reflection mask with high contrast.

The enhanced reflectivity of a multilayer is also sufficiently powerful to collect soft X-ray fluorescence from low atomic number elements such as C, N, O, and F. Photon-in/photon-out soft X-ray absorption measurements of catalytic C chemistry have been possible by the development of a high-efficiency detection system for C K fluorescence. A Cr/C multilayer was coated on the concave collector.[12] Nearly background-free C near edge X-ray absorption fine structure (NEXAFS) spectra have been obtained.

3.1.3.2 POLARIMETRY

The ratio of multilayer reflectivity for s to p polarization is usually three orders of magnitude or more at the Brewster angle ($\theta \approx 45°$ in the soft X-ray region). Thus the multilayer works as a useful polarizer for soft X-rays.[13] Figure 3.1.6 shows the polarizance calculated for a Mo/B$_4$C multilayer with $d = 5.1$ nm for photons of $E = 175$ to 190 eV as a function of θ. The polarizance is higher than 0.995 between $\theta = 44°$ and 47° independent of the photon energy. A transmission multilayer is also available as a phase shifter like a quarter-wave plate for visible light. Using reflection and transmission multilayers, ellipsometric studies have been performed in the soft X-ray region.[14–16]

When a multilayer is coated on a grating, its diffraction efficiency is enhanced.[17] Moreover, it becomes a hybrid of polarizer and grating when used around $\theta = 45°$, resulting in a polarization spectrometer in the soft X-ray region. The polarization spectrometer is applicable to investigation into polarized soft X-ray fluorescence from solids, gas phases and astrophysical plasmas. A Rowland-circle mount polarization spectrometer based on a grating coated with a Mo/B$_4$C multilayer has been constructed.[18] The polarizance of the grating was higher than 98.9 % at a wavelength of 6.7 nm at $\theta \approx 45°$. CrB$_2$ has a layer structure like MgB$_2$, which becomes a superconductor at 39 K. The B 1s emission from CrB$_2$ consists of σ and π emission due to the B 2p$_{xy}$ and 2p$_z$ orbitals, respectively. The π emission cannot be separately observed in any measurement geometry using a traditional grazing-incidence spectrometer. Figure 3.1.7 shows the B 1s σ and π emission spectra measured for a CrB$_2$ single crystal with an energy width of 0.9 eV. The most notable feature is that the π electron contributes to the upper valence band as compared with the σ electron, which is comparable with the band calculation.

At the L$_{2,3}$ edges of transition metals, large magnetic circular dichroism (MCD) has been observed. It has attracted much interest offering a means to separate electron spin and orbital moment in contributions to magnetism. MCD measurements in the soft X-ray region require incident light with high-degree circular polarization, while optical rotation measurements need only linearly polarized light. For polarization methods (Faraday, Kerr

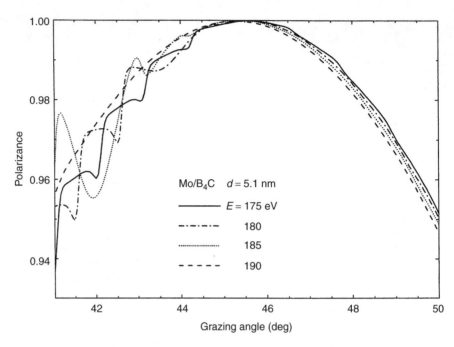

Figure 3.1.6 Calculated polarizance for a Mo/B$_4$C multilayer with $d = 5.1$ nm for photons of $E = 175$ to 190 eV as a function of θ

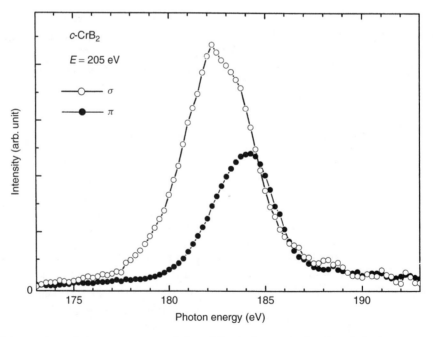

Figure 3.1.7 B 1s σ and π emission spectra measured for a CrB$_2$ single crystal using the polarization spectrometer equipped with a grating coated with a Mo/B$_4$C multilayer. The excitation energy is 205 eV. (From ref. 18)

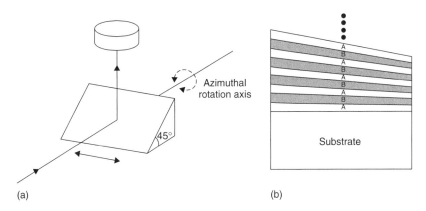

Figure 3.1.8 A schematic setup of the rotation analyzer based on a multilayer polarizer (a) and a schematic cross-section of the multilayer polarizer with a lateral gradient in period (b)

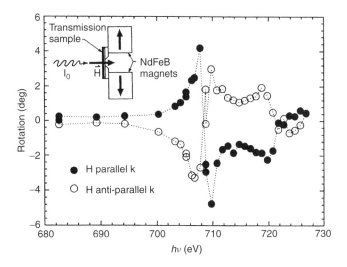

Figure 3.1.9 Faraday rotation observed for a Fe(2.0 nm)/Cr(1.9 nm) multilayer of 40 periods at the Fe $L_{2,3}$ edge region. A 3-kOe field was applied parallel or antiparallel along the beam direction and sample normal. (From ref. 19)

rotation) in the soft X-ray region, measurement of optical rotation required the development of a continuously tunable linear polarizer. The right-hand side of Figure 3.1.8 shows a schematic cross-section of the multilayer polarizer with a lateral gradient in period. $R_s = 0.01$ and $R_s/R_p \cong 3 \times 10^4$ were estimated for a W/B$_4$C multilayer around 700 eV.[19] The left-hand side of Figure 3.1.8 shows a schematic of the rotation analyzer based on the multilayer polarizer. The analyzer rotates the multilayer about the incident beam with fixed incidence angle while monitoring the reflected intensity. Continuous tunability is achieved by translating the multilayer to position the Bragg peak at the desired energy. Figure 3.1.9 shows the resonance in the Faraday rotation, observed for a transmission sample of a Fe(2.0 nm)/Cr(1.9 nm) multilayer at the Fe $L_{2,3}$ edge region. A 3-kOe field was applied parallel or antiparallel along the beam direction and sample normal as illustrated in Figure 3.1.9. The MCD absorption spectra were also measured for the same sample. Analysis of the MCD data showed good agreement with the measured Faraday rotation.

3.1.4 HARD X-RAY MULTILAYERS AND THEIR APPLICATIONS

In the 1–100 keV region, material combinations such as W/C, W/Si, W/B$_4$C, Ni/C, Cr/C, Co/C, Ru/Si, Rh/C, Mo/C, Pt/C, Pt/Si, and Pt/Ti are promising for high-reflectivity multilayers as grazing-incidence reflectors. Among them, a reflectivity of 68 % has been achieved using a Pt/C multilayer ($d = 4.3$ nm, $N = 50$) at $\lambda = 0.154$ nm and $\theta = 1.1°$.[20] The interfacial roughness is estimated to be 0.32 nm. There are many instruments using such multilayers with periodic structure. Solid Fabry–Perot etalons for X-rays have been constructed.[21] Each etalon consists of two W/C multilayers separated by a carbon spacer. The thick carbon spacer works as a resonant cavity. The structure was characterized at grazing incidence in reflection using Cu Kα radiation. The observed reflectivity was approximately 50 % of that calculated, which might result from the interfacial roughness. This interferometer is applicable to determine the absolute value of the thickness and δ of the spacer or λ and θ.

Figure 3.1.10 shows a spectral reflectivity curve calculated for a Pt/C multilayer supermirror at $\theta = 0.3°$. The cross-sectional view of the supermirror structure observed by transmission electron microscopy (TEM) is shown in Figure 3.1.11. Figure 3.1.10 shows, for comparison, reflectivity curves calculated for a Pt single layer mirror and a Pt/C multilayer mirror with $d = 4.0$ nm and $\gamma = 0.4$, where γ is the ratio of layer thickness of heavy element to d-spacing. The great advantage of the supermirror for the wide band and practical reflectivity is obvious. This supermirror has no top layer for total reflection and the d-spacing gradually decreases from the surface to the substrate as adjusted to match the energy band concerned. Figure 3.1.12 shows a measured spectral reflectivity for a W/Si multilayer supermirror with $N = 50$ at $\theta = 0.5°$,[22] for an example. It can be seen that the experimental data fit the theoretical predictions quite well. It provides a beam with uniform spectral intensity from 7 keV up to 20 keV.

3.1.4.1 MICROBEAM AND MICROSCOPY

Recently, X-ray microbeams have been greatly required for the determination of crystal structures by diffraction, trace elements analysis and imaging,

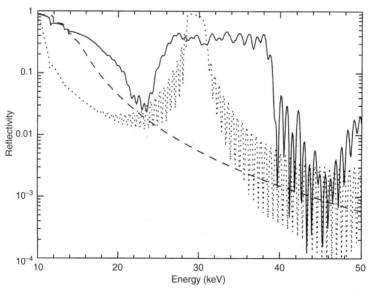

Figure 3.1.10 Spectral reflectivity curve calculated for a Pt/C multilayer supermirror at $\theta = 0.3°$. Reflectivity curves calculated for a Pt single layer mirror (broken line) and a Pt/C multilayer mirror with $d = 4.0$ nm (dotted line) are also shown for comparison

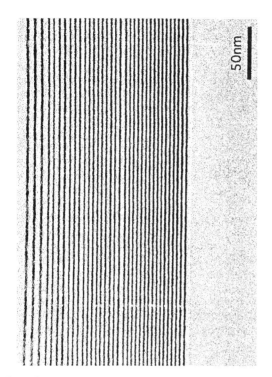

Figure 3.1.11 Cross-sectional view of a Pt/C multilayer supermirror observed by TEM. (Courtesy of N. Ohnishi, Chubu University)

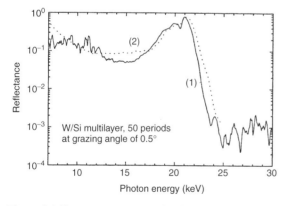

Figure 3.1.12 Measured spectral reflectivity for a W/Si multilayer supermirror with $N = 50$ at $\theta = 0.5°$.[22] Reported from *J. Synchrotron Rad.* **5** (1998) 239. Reproduced by permission of IUC

because new artificial materials just synthesized are usually small powders or composed of small grains, or the inner materials in biological cells are very small. The sizes are around $10\,\mu m$ or less. The optical elements for imaging are grazing incidence mirrors, Fresnel zone plates, and Bragg–Fresnel lens. By the use of single mirrors, such as toroidal and ellipsoidal mirrors, it is possible to obtain microbeams. However, it is desired to suppress the aberration caused by the grazing incidence optics, by the use of double reflection. Such an objective coated with a multilayer is the Kirkpatrick–Baez type, as shown in Figure 3.1.13. The Kirkpatrick–Baez objective consists of two grazing spherical or nonspherical mirrors, each of which focuses the light horizontally or vertically independently. In many cases, the surface of the grazing incidence mirrors is used at the total reflection condition of heavy metals, so that large mirrors have been required. Therefore, the multilayers and the supermirrors are used for coating materials to gather more flux.

In Figure 3.1.14 is illustrated a graded W/B$_4$C multilayer deposited on a flat substrate, which was bent to a parabola.[23] The d-spacing has to be laterally graded because the angle of incidence differs continuously from the edge of the surface. The obtained nonlinear lateral gradient differed from the theoretical calculations by less than 1%. The vertical spot size was measured by the use of 8 keV X-rays and was about $7\,\mu m$, as shown in Figure 3.1.15. Recently, a spot size of $1\,\mu m$ has been achieved. Many efforts have been made to fabricate multilayer- or supermirror-coated Kirkpatrick–Baez objectives to obtain microbeams at synchrotron radiation facilities and laboratories.[24]

3.1.4.2 X-RAY TELESCOPE

A normal-incidence telescope has been investigated in the soft X-ray region.[25] On the other hand, grazing incidence X-ray optics have been well investigated to construct X-ray telescopes. Usually the optical configuration of Wolter type I, as shown in Figure 3.1.16, is adopted for astronomical observations using the total reflection of single metal layer mirrors. The X-ray telescope on board the Einstein and ROSAT satellites revolutionized X-ray astronomy by observing a number of objects

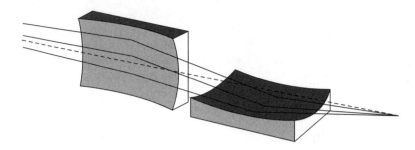

Figure 3.1.13 Schematic drawings of the Kirkpatrick–Baez optics

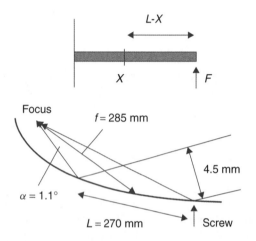

Figure 3.1.14 Schematic of a mirror bent to a parabola. A laterally graded W/B$_4$C multilayer was deposited on a flat substrate.[23] Reported from *Rev. Sci. Instrum.* **70** (1999) 3227. Reproduced by permission of American Institute of Physics

point spread function of this telescope. X-ray emissions seem to originate from Bremsstrahlung or inverse Compton of non-thermal electrons around a black hole. The telescope was of a multi-nested thin foil type assembled with four quadrant units made of 510 pieces of coaxially and confocally aligned thin foils coated with Pt/C supermirrors. The inner and outer diameter and focal length are determined to be 12 cm, 40 cm and 8 m, respectively, which correspond to a grazing angle of 0.105–0.356°. The design parameters, d and N, of supermirrors are divided into 13 groups corresponding to each grazing angle.

with good image quality, though the energy region was limited to be 0.2–4.5 keV corresponding to a grazing angle of <2°. Further development was made to cover the energy region up to 10 keV by the use of extreme grazing incidence optics at grazing angles <1°. Those telescopes were put on board X-ray astronomy satellites Chandra, XMM-Newton, and ASCA, which made it possible to observe in the energy region of 0.1–10 keV.[26–28] The next step is to develop hard X-ray telescopes sensitive in the region of >10 keV by the use of supermirrors. One of them was on board a balloon and the first hard X-ray image of the black hole candidate Cyg X-1 was observed in the 20–40 keV band as shown in Figure 3.1.17.[29] The extended feature is not a real structure, but is caused by the

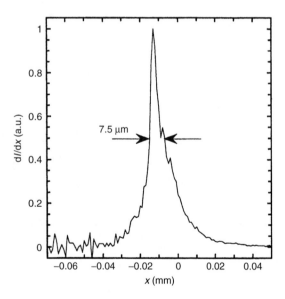

Figure 3.1.15 Focusing experiment performed for the bent mirror using X-ray of 8 keV. The vertical spot size was about 7 μm.[23] Reported from *Rev. Sci. Instrum.* **70** (1999) 3227. Reproduced by permission of American Institute of Physics

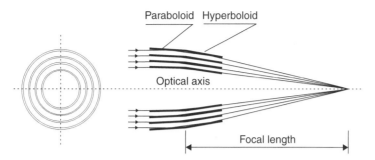

Figure 3.1.16 Schematic drawings of the Wolter type I optics

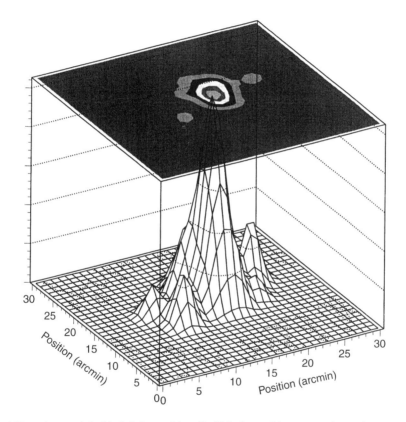

Figure 3.1.17 Hard X-ray image of the black hole candidate CygX-1 observed by a supermirror telescope on board a balloon. (From ref. 29)

3.1.4.3 MULTILAYER-COATED GRATING

Multilayer-coated gratings have been developed for hard X-rays (<10 keV) as well as for extreme ultraviolet and soft X-ray light. Usually, the conventional reflection gratings coated with a Au layer cover the energy range up to 2 keV with $\Delta E = 1$ eV. Therefore, grazing-incidence multilayer-coated gratings are developed to obtain high efficiency and resolution in the 1–10 keV region. The combination of depth structure of a multilayer and lateral structure of a grating leads to high diffraction efficiency and energy resolution.

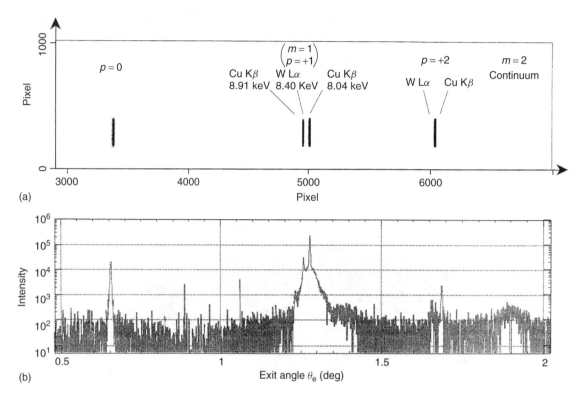

Figure 3.1.18 Raw diffraction image of a multilayer-coated grating at $\theta_i = 0.66°$ (a) and the projected profile (b). X-Rays were generated from a Cu target contaminated with W. Cu Kα, W Lα, and Cu Kβ peaks ($m = 1$, $p = +1$) are found around $\theta_e = 1.29°$. (From ref. 30)

This type of grating has to satisfy the Bragg condition and grating rule expressed by

$$2d(\sin\theta_i + \sin\theta_e) = m\lambda$$
$$2D(\cos\theta_i - \cos\theta_e) = p\lambda \quad (3.1.16)$$

where D is the period of grooves, θ_i grazing angle of incidence, θ_e exit angle, m and p the spectral order of the Bragg reflection and grating diffraction. A laminar grating with a 1200 grooves/mm and a 39-nm groove depth coated with a Pt/C ($d = 5$ nm, $N = 10$) multilayer was fabricated and evaluated with characteristic X-rays.[30] The dispersion measurement was performed with a charge-coupled device detector. The first order peak reflectivity and resolution were obtained to be 30 % and more than 100 for Cu Kα respectively, as shown in Figure 3.1.18.

3.1.4.4 DOUBLE-MULTILAYER MONOCHROMATOR

A double-multilayer monochromator was proposed from analogy with a double-crystal monochromator. This novel technique uses an identical pair of multilayers. A double-multilayer monochromator has been installed on a soft X-ray beamline.[31] Also, a hard X-ray beamline has been installed. In this case, a W/Si double-multilayer monochromator was used to suppress higher order harmonics from the synchrotron X-rays of a bending magnet.[32] It enabled suppression of all higher order harmonics to less than six orders of magnitude with a ~30 % throughput in the first Bragg peak (8 keV). The use of such a multilayer monochromator in tandem with a crystal monochromator has a great advantage over the traditional method using a large grazing-incidence mirror with respect to size and cost.

REFERENCES

1. Henke, B.L., Gullikson, E.M. and Davis, J.C. X-ray interactions: photoabsorption, scattering, transmission, and reflection at $E = 50-30,000$ eV, $Z = 1-92$, *At. Data Nucl. Data Tables*, **54**, 181–342 (1993). These data are available at http://www-cxro.lbl.gov/optical_constants/.
2. Yamamoto, M. and Namioka, T. Layer-by-layer design method for soft-x-ray multilayers. *Appl. Opt.*, **31**, 1622–1630 (1992).
3. Bajt, S., Alameda, J.B., Barbee Jr, T.W., Clift, W.M., Folta, J.A., Kaufmann, B. and Spiller, E.A. Improved reflectance and stability of Mo–Si multilayers. *Opt. Eng.*, **41**, 1797–1804 (2002).
4. Bajt, S., Behymer, R.D., Mirkarimi, P.B., Montcalm, C., Wall, M.A., Wedowski, M. and Folta, J.A. Experimental investigation of beryllium-based multilayer coatings for extreme ultraviolet lithography. *SPIE*, **3767**, 259–270 (1999).
5. Kinoshita, H. and Watanabe, T. Experimental results obtained using EUV laboratory tool at New Suraru. *Jpn. J. Appl. Phys.*, **39**, 6771–6776 (2000).
6. Sweeney, D.W., Hudyma, R.M., Chapman, H.N. and Shafer, D.R. EUV optical design for a 100-nm CD imaging system. *SPIE*, **3331**, 2–10 (1998).
7. Takenaka, H., Nagai, K., Ito, H., Ichimaru, S., Ohchi, T., Muramatsu, Y., Gullikson, E.M. and Perera, R.C.C. Highly reflective Co_xCr_{1-x}/C multilayer mirror for use in X-ray photoemission spectroscopy in the wavelength region around 6 nm. *J. Phys. IV*, **104**, 255–258 (2003).
8. Eriksson, F., Gullikson, E.M., Johansson, G.A., Hertz, H.M. and Berch, J. High near-normal reflectivity in the water window of Cr/Sc multilayer X-ray mirrors. Abstracts of the 6th International Conference on the Physics of X-ray Multilayer structures, p. 2 (2002). The current records of the multilayer reflectivity are found at http://www-cxro.lbl.gov/multilayer/survey.html.
9. Barbo, F., Bertolo, M., Bianco, A., Cautero, G., Fontana, S., Johal, T.K., La Rosa, S., Margaritondo, G. and Kaznacheyev, K. Spectromicroscopy beamline at ELETTRA: performance achieved at the end of commissioning. *Rev. Sci. Instrum.*, **71**, 5–10 (2000).
10. Kondo, Y., Ejima, T., Takatsuka, H. and Watanabe, M. Microscopic ultraviolet photoelectron spectroscopy using He-I and He-II resonance lines. *Surf. Rev. Lett.*, **9**, 521–527 (2002).
11. Haga, T., Takenaka, H. and Fukuda, M. At-wavelength extreme ultraviolet lithography mask inspection using a Mirau interferometric microscope. *J. Vac. Sci. Technol.*, **B18**, 2916–2920 (2000).
12. Fischer, D.A., Sambasivan, S., Kuperman, A., Platonov, Y. and Wood, J.L. Multilayer mirror fluorescence detection system for photon-in photon-out in-situ carbon K-edge NEXAFS. *Synchrotron Rad. News*, **15**, 16–20 (2002).
13. Dhez, P. Polarizers and polarimeters in the x-uv range. *Nucl. Instrum. Methods*, **A261**, 66–71 (1987).
14. Yamamoto, M., Mayama, K. and Yanagihara, M. Thin film ellipsometry at a photon energy of 97 eV. *J. Electron Spectrosc. Relat. Phenom.*, **80**, 465–468 (1996).
15. Schafers, F., Mertins, H.-Ch., Gaupp, A., Gudat, W., Mertin, M., Packe, I., Schmolla, F., Di Fonzo, S., Soullie, G., Jark, W., Walker, R., Le Cann, X., Nyholm, R. and Eriksson, M. Soft-X-ray polarimeter with multilayer optics: complete analysis of the polarization state of light. *Appl. Opt.*, **38**, 4074–4088 (1999).
16. Haga, T., Tinone, M.C.K., Ozawa, A., Utsumi, Y., Itabashi, S., Ohkubo, T. and Shimada, M. Fabrication of semitransparent multilayer polarizer and its application to soft x-ray ellipsometer. *SPIE* **3764**, 13–27 (1999).
17. Seely, J.F., Kowalski, M.P., Cruddace, R.G., Heidemann, K.F., Heinzmann, U., Kleineberg, U., Osterried, K., Menke, D., Rife, J.C. and Hunter, W.R. Multilayer-coated laminar grating with 16% normal-incidence efficiency in the 150-A wavelength region. *Appl. Opt.*, **36**, 8206–8213 (1997).
18. Ishikawa, S., Ichikura, S., Imazono, T., Otani, S., Oguchi, T. and Yanagihara, M. Polarization performance of a new spectrometer based on a multilayer-coated laminar grating in the 150–190-eV region. *Opt. Rev.*, **10**, 58–62 (2003).
19. Kortright, J.B., Rice, M. and Carr, R. Soft-x-ray Faraday rotation at Fe $L_{2,3}$ edges. *Phys. Rev.*, **B51**, 10240–10243 (1995).
20. Yamashita, K., Akiyama, K., Haga, K., Kunieda, H., Lodha, G.S., Nakajo, N., Nakamura, N., Okajima, T., Tamura, K. and Tawara, Y. Fabrication and characterization of multilayer supermirror for hard X-ray optics. *J. Synchrotron Rad.*, **5**, 711–713 (1998).
21. Barbee Jr, T. and Underwood, J.H. Solid Fabry–Perot etalons for X-rays. *Opt. Commun.*, **48**, 161–166 (1983).
22. Erko, A., Veldkamp, M., Gudat, W., Abrosimov, N.V., Rosslenko, S.N., Shekhtman, V., Khasanov, S., Alex, V., Groth, S., Schroder, W., Vidal, B. and Yakshin, A. Graded X-ray optics for synchrotron radiation applications. *J. Synchrotron Rad.*, **5**, 239–245 (1998).
23. Morawa, C., Pecci, P., Peffen, J.C. and Ziegler, E. Design and performance of graded multilayers as focusing elements for X-ray optics. *Rev. Sci. Instrum.*, **70**, 3227–3232 (1999).
24. Protopopov, V.V. Graded X-ray multilayer optics for laboratory-based applications. *SPIE*, **4141**, 116–127 (2000).
25. Windt, D.L., Kahn, S.M. and Sommargren, G.E. Diffraction-limited astronomical X-ray imaging and X-ray interferometry using normal-incidence multilayer optics. *SPIE*, **4851**, 441–450 (2003).
26. Weisskopf, M.C. Three years of operation of the Chandra X-ray observatory. *SPIE*, **4851**, 1–16 (2003).

27. Turner, M.J.L., Briel, U., Ferrando, P., Griffiths, R.G., Villa, G. and the EPIC Team. Science Highlights, Calibration, and Performance of EPIC on Newton-XMM: the MOS cameras. *SPIE*, **4851**, 169–180 (2003).
28. Serlemitsos, P.J., Jalota, L., Soong, Y., Kunieda, H., Tawara, Y., Tsusaka, Y., Suzuki, H., Sakima, Y., Yamazaki, T., Yoshioka, H., Furuzawa, A., Yamashita, K., Awaki, H., Itoh, M., Ogasaka, Y., Honda, H. and Uchibori, Y. The X-ray telescope on board ASCA. *Publ. Astron. Soc. Japan.* **47**, 105–114 (1995).
29. Ogasaka, Y., Tamura, K., Okajima, T., Tawara, Y., Yamashita, K., Furuzawa, A., Haga, K., Ichimaru, S., Takahashi, S., Fukuda, S., Kitou, H., Gotou, A., Kato, S., Satake, H., Nomoto, K., Hamada, N., Serlemitsos, P.J., Tueller, J., Soong, Y., Chan, K.-W., Owens, S., Brendse, F., Krimm, H., Baumgartner, W., Barthelmy, S.D., Kunieda, H., Misaki, K., Shibata, R., Mori, H., Itoh, K. and Namba, Y. Supermirror hard x-ray telescope and the results of first observation flight of InFOCuS. *SPIE*, **4851**, 619–630 (2003).
30. Tamura, K., Yamashita, K., Kunieda, H., Yoshioka, T., Watanabe, M. and Haga, K. Development of multilayer coated gratings for high-energy x-ray spectroscopy. *SPIE*, **3766**, 371–379 (1999).
31. Mekaru, H., Tsusaka, Y., Miyamae, T., Kinoshita, T., Urisu, T., Masui, S., Toyota, E. and Takenaka, H. Construction of the multilayered-mirror monochromator beamline for the study of synchrotron radiation stimulated process. *Rev. Sci. Instrum.*, **70**, 2601–2605 (1999).
32. Lingham, M., Ziegler, E., Luken, E., Loeffen, P., Mullender, S. and Goulon, J. Double multilayer monochromator for harmonic rejection in the 5–60 keV range. *SPIE*, **2805**, 158–168 (1996).

3.2 Single Capillaries X-ray Optics

Y. HOSOKAWA

X-ray Precision, Inc., Kyoto, Japan

3.2.1 INTRODUCTION

In general, it is not easy to apply X-ray spectrometry to the characterization of small areas of electronic materials, semiconductors, tissues, etc. because the interactions between X-rays and those samples are relatively low compared with those of light or electron beams. However, X-ray spectrometry, in principle, gives us a variety of information not only about the surface but also about the inside of a sample, without destruction. These advantages of the technique encourage us to expand its limits in spite of many difficulties. Especially, for samples to be measured under atmospheric conditions it is very interesting and challenging for us to perform research and development in the field of analysis techniques based on X-ray spectrometry.

X-rays commonly used for analysis are: (1) transmitted and absorbed X-rays; (2) A fluorescent X-rays; (3) scattered X-rays; (4) diffracted X-rays; (5) totally reflected X-rays; and (6) refracted X-rays. Basically, either discrete intensities of these X-rays or the entire spectrum is measured.

Since Wilhelm Röentgen discovered X-rays, research and application development based on X-rays have proceeded and advanced independently. So far, research and development have not produced any measuring instruments that integrate the use of the various X-rays mentioned above. It can be said that this is mainly due to geometrical constraints imposed by X-ray irradiation. It has been difficult to obtain X-ray microbeams because X-rays do not interact with electromagnetic fields in the way electron beams do and because no suitable and small lenses are available for X-ray optics as is the case for light due to the fact that the refractive index of X-rays is smaller than one. It can be expected that an integrated X-ray analytical instrument can only be realized if a very bright X-ray microbeam is available.

This subchapter will give a description of how a very bright and narrow X-ray microbeam can be realized using single capillaries (or X-ray Guide Tubes, XGT). In addition, a tabletop X-ray analytical microscope based on the use of an XGT and applications thereof are shown.

3.2.2 FORMING OF X-RAY MICROBEAM

There are many methods to obtain a micrometer-sized X-ray beam as shown in Table 3.2.1. In the case of imaging systems, narrow beams of about 0.1 μm diameter can be achieved in a relatively low energy area but only monochromatic X-rays are available. Non-imaging systems can transport X-rays over longer distances over a wide range of energies although relatively small in size. In laboratories, X-ray tubes are most widely used to generate X-rays to obtain an X-ray microbeam. Commercially available X-ray generators are relatively small in size and easy to use, but often their X-ray intensities are not sufficient. Therefore, in order to ensure higher X-ray intensities, it is necessary to get optimum X-ray taking solid angle and to expand the spectral range. XGTs (single

Table 3.2.1 Methods to obtain X-ray microbeams

X-ray microbeam forming systems	Methods	Designation	Features	Remarks
Imaging system	(1) Grazing incidence method	Kirkpatrick–Baez mirror	Two-dimensional total reflection mirror	Long focusing distance
		Wolter–Mirror	Rotating non-spherical total reflection mirror	
		Multichanneling plate	Multi-bundle single-total reflection capillary	
		Kumakov–Lens (Polycapillary)	Multi-bundle poly-total reflection capillary	
	(2) Diffraction method	Zone plate	Soft X-ray, $0.1\,\mu m\,\phi$	Short focusing distance
		Schwartzchild	Soft–hard X-ray	
		Johnson Crystal	Soft X-ray	
		Multilayer	Soft–hard X-ray	
		Bragg–Fresnel Zone Plate	Soft X-ray	
	(3) Refraction method	Refractive lens	Hard X-ray, $10\,\mu m\,\phi$	Long focusing distance
Non-imaging system	(1) Slit method	Slit/collimator	Continuous X-ray $50–500\,\mu m\,\phi$	Simple
	(2) Grazing incidence method	XGT (single capillary)	Wide area continuous X-ray $1–10\,\mu m\,\phi$	Fine beam
		Polycapillary	Continuous X-ray, low pass $30–100\,\mu m\,\phi$	High brightness
	(3) Diffraction method	Asymmetry reflection crystal	Single color X-ray – $100\,\mu m\,\phi$	

capillaries) and polycapillaries used in a grazing-incidence configuration in non-imaging systems are expected to meet the above requirements rather than the conventional combination of a slit and a collimator.

3.2.3 X-RAY MICROBEAMS USING A SINGLE CAPILLARY

In the case of a single capillary, X-rays are totally reflected on the inside surface if X-rays are emitted with an incident angle of less than θ_c (radian) $= 0.02 \times \sqrt{\rho}$ (g/cm^3)$/E$(keV). In this way, parallel, focused or diverging X-rays are generated depending on the geometry of the hollow tube. More elaborate discussion about the fundamental theory and applications of total reflection can be found elsewhere.[1-3]

Different morphologies of a single capillary are shown in Figure 3.2.1. Studies on the shape of these capillaries are described elsewhere.[4-8]

References 9 and 10 deal with studies and experiments on the relation between the intensity of totally reflected X-rays and the shape and smoothness of the internal surface. Reference 11 describes a study on the reduction of the X-ray background signal obtained by using capillary optics. Reference 12 deals with studies on extreme ultraviolet rays.

At present, single capillaries are produced using either the pulling method or the gas pressure method. Reference 6 describes the pulling method. In the case of the pulling method, the pulling should be controlled in such a way that the middle part of the hollow tube is expanded toward the outside.

The gas pressure method forces a gas flow through a preformed capillary which is thereby pressed against a press mold in an electric furnace. In this case, care should be taken on how to narrow both ends of the capillary. At present, efforts are being undertaken to develop better production processes.

Different X-ray output diameters are required depending on the type of measurement. A combination of various optically prealigned capillaries with different output diameters can be installed

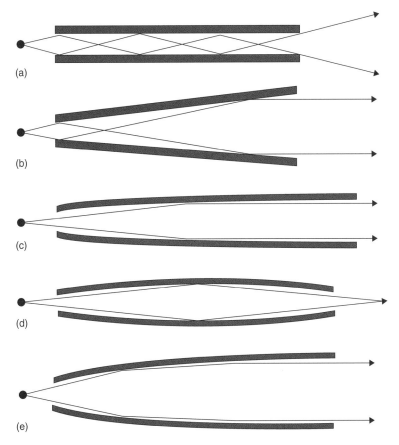

Figure 3.2.1 Single capillary shapes. (a) Cylindrical capillary: easy to make using the pulling method. Brightness and straightness of the X-ray beam are low. (b) Tapered capillary: relatively easy to make using the pulling method. Straightness of the X-ray beam is good. When used reversed, it is possible to get a fine beam but working distance is as short as several millimeters. (c) Rotated parabolic capillary: possible to obtain a long working distance beam. (d) Rotated ellipsoid capillary: if a pinpoint X-ray source is available, it is possible to get a fine and very long distance reaching beam. (e) Walter–Mirror single capillary: most efficient usage of the X-ray source is achieved with the possibility to get a fine beam. Difficult to produce using the normal pulling method

to ease the change between output diameters. Figure 3.2.2 shows two single capillaries with different output diameters (100 μm and 10 μm), which are mounted together and can be installed as such in an instrument.

3.2.3.1 AN INSTRUMENT FOR THE MEASUREMENT OF TWO-DIMENSIONAL DISTRIBUTIONS BY X-RAY SCANNING USING A SINGLE CAPILLARY

As an example of an actual application of a single capillary, a tabletop type X-ray analytical microscope that can make two-dimensional surface distribution measurements by X-ray scanning is shown in Figure 3.2.3(a).

A small-sized instrument, which can be used in a normal laboratory is realized by a combination of single capillaries and a low power X-ray generator of 50 W. An X-ray microbeam is emitted vertically through a single capillary and reaches the sample. The sample is placed on an X-Y stage and can thus be scanned in two dimensions. Both transmitted X-rays and fluorescent X-rays are detected, and a two-dimensional image is obtained by computing signals from these detectors.

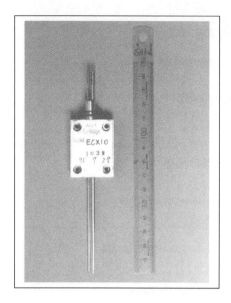

Figure 3.2.2 An example of two XGTs (single capillaries) mounted together. The XGT on the left has a nominal diameter of 100 μm and the one on the right has a nominal diameter of 10 μm. Both are optically aligned in parallel. Input diameters are 100 and 60 μm, respectively. Output diameters are about 100 and about 8 μm. Both overall lengths are 130 mm. The upper end is the X-ray input side and the lower end is the capillary output

Details of this setup are described in reference 4. Reference 5 deals with an archetype of this kind of instrument, which uses a conventional collimator.

Figure 3.2.3(b) shows an example of a soft X-ray transmitting window consisting of Mylar film with a thickness of several micrometers, which is installed so that the sensitivity for soft X-rays is improved.

3.2.4 AN EXAMPLE OF THE APPLICATION OF A TABLETOP TYPE X-RAY ANALYTICAL MICROSCOPE

Figure 3.2.4 shows the transmitted X-ray intensity measurement of the press groove of a pull-tab of a soft drink can by means of a microbeam scanned over its surface. Figure 3.2.4(a) shows a two-dimensional distribution of the intensity of transmitted X-rays. Figure 3.2.4(b) shows an intensity histogram that corresponds to (a). The horizontal axis is the intensity of transmitted X-rays and the vertical axis is the number of pixels in (a). The X-ray intensity in the gray colored area

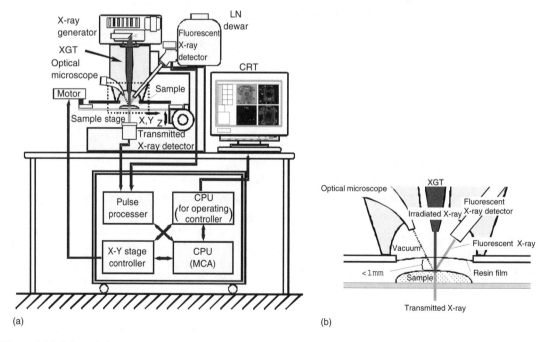

Figure 3.2.3 Schematic layout of an X-ray analytical microscope

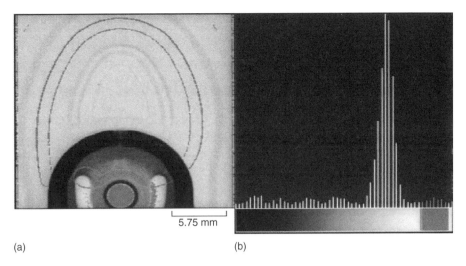

Figure 3.2.4 An example of measurement of the depth of a pull-tab of a soft drink can

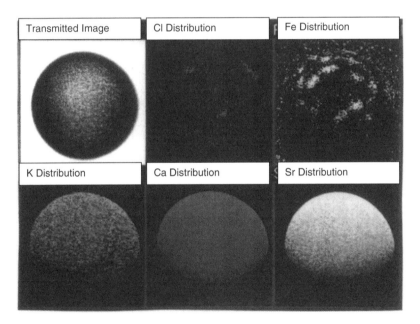

Figure 3.2.5 Mapping images of a cultured pearl obtained with an X-ray analytical microscope. The distribution of iron compounds has swirly shape. The shadow in the lower part of the mappings is due to the shadow of X-ray absorption caused by the spherical shape of the pearl

corresponds to the thickness of the pull-tab. The depth of the pull-tab groove can be measured with this technique.

Figure 3.2.5 shows the results of a nondestructive direct observation of a cultured pearl by fluorescent and transmitted X-rays. It has been said that the structure of a pearl is made up of more than 1000 layers of organic and inorganic secretions of a shellfish piled up alternately around an inorganic nucleus.

The theory that trace amounts of iron compounds are reversed in the nucleus was proven.

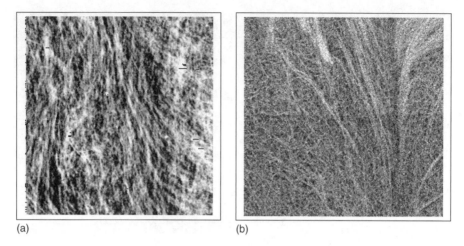

Figure 3.2.6 Scanned image of transmitted X-rays of FRP (a) and the distribution of filler on the surface obtained from the fluorescent X-rays (b)

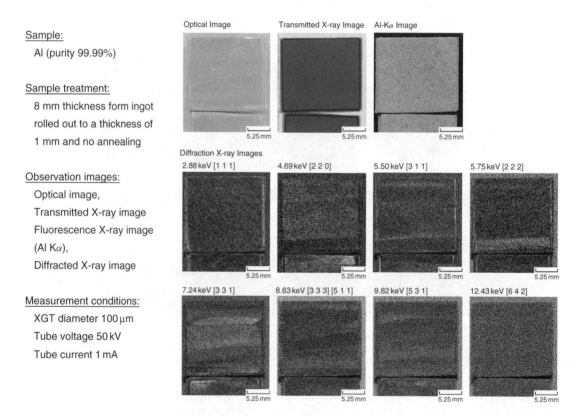

Figure 3.2.7 Distribution of aluminum crystal grains in cool rolled aluminum samples (no annealing) as observed with an X-ray analytical microscope. The horizontal bands indicate small crystals being oriented in the rolling direction

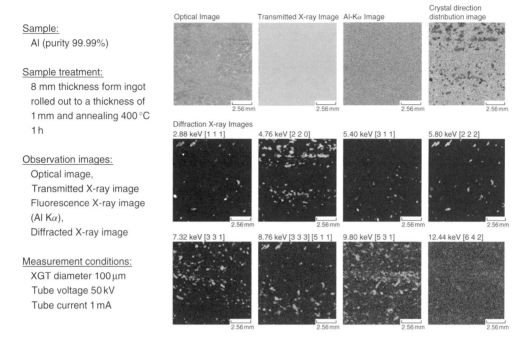

Figure 3.2.8 Distribution of aluminum crystals observed with an X-ray analytical microscope (annealing at 400 °C for 1 h)

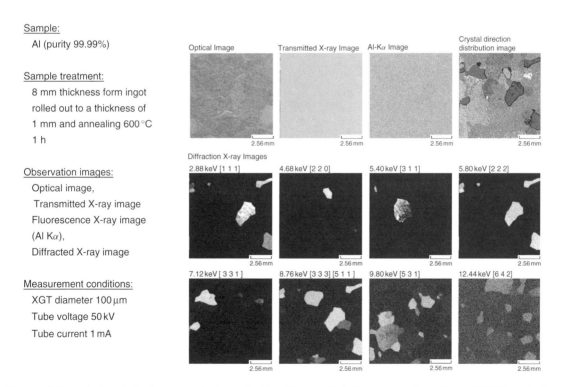

Figure 3.2.9 Distribution of aluminum crystals observed with an X-ray analytical microscope (annealing at 600 °C for 1 h). Size, orientation, and distribution of crystals can clearly be observed with X-ray analytical microscopy

Figure 3.2.6 shows an example of the application of X-ray analytical microscopy to the study of FRP (Filler Reinforced Plastics), which is a kind of the engineering plastic. A scanned image of the transmitted X-rays shows the distribution of fillers throughout the whole thickness, and a scanned fluorescent X-ray image of Ca corresponds to the two-dimensional distribution of filler in a surface layer of about 10 μm of the FRP. Fillers are commonly added in the process of injection molding to increase mechanical strength of the FRP. Unless the distribution of the fillers are uniform in the FRP, it is likely to be damaged due to stress concentration.

3.2.5 X-RAY ANALYTICAL MICROSCOPY APPLIED FOR DIRECT OBSERVATION OF ALUMINUM CRYSTAL GROWTH

Figure 3.2.7 shows images of a rolled aluminum sample obtained with X-ray analytical microscopy. It can be observed that crystal grains are small and oriented in the rolling direction.

These samples were annealed at a temperature of 400 °C and 600 °C for 1 h and observed with an X-ray analytical microscope. The results are shown in Figures 3.2.8 and 3.2.9. It can be clearly seen that the size of the crystal grains increased with increasing temperature.

Figure 3.2.10 shows the damage to an aluminum test piece submitted to repeated fatigue test cycles.

3.2.6 MULTIDIMENSIONAL ANALYSIS WITH X-RAY MICROBEAMS

By two-dimensionally scanning a sample with X-ray microbeams, various kinds of distribution analysis techniques can be proposed as shown in Table 3.2.2.

It is expected that an integrated X-ray analysis instrument can be realized that enables

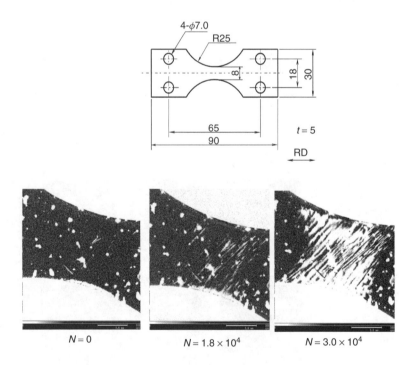

Figure 3.2.10 Images of an aluminum plate obtained with an X-ray analytical microscope. The slip bands resulting from by repeated fatigue test cycles can be seen. N corresponds to the number of cycles, pulling was done at test plane (311) or effective depth ($t_e = 11.81$ μm)

Table 3.2.2 Multidimensional analysis by means of X-ray microbeams

	Distribution analyzing probe	Abbreviation	Nature of the interactions
1	Irradiated X-ray beam induced current distribution	IXBCD	Surface electric charge
2	X-ray beam induced sample current distribution	XBICD	Inner electric charge
3	Transmitted X-ray intensities distribution	TXID	Electron density
4	Fluorescent X-ray intensities distribution	FXID	Atomic
5	Diffracted X-ray intensities distribution	DXID	Crystallization
6	Scattered X-ray intensities distribution	SXID	Electron density
7	Refracted X-ray Intensities distribution	RXID	Electron density

multidimensional X-ray analysis by visualizing the detected X-ray signals shown in Table 3.2.2.

ACKNOWLEDGEMENTS

In this subchapter, the descriptions regarding IXBCD are based on suggestions made by Dr Yoichi Ghoshi of The Japan National Institute for Environmental Studies. The topographic observations of aluminum crystals are the results of a cooperation with Professor Yoshio Miyoshi *et al.*, at the School of Engineering, Shiga Prefecture University. The author also wishes to thank the Japanese Scientific and Technological Agency and Horiba, Ltd for their support.

REFERENCES

1. A.H. Compton and S.K. Allison. *X-rays in Theory and Experiment*, Braunworth and Company, Inc., New York, 1936.
2. V.E. Cosslett and W.C. Nixon. *X-Ray Microscopy*, Cambridge University Press, London, 1960.
3. T. Namioka and K. Yamashita. *X-Ray Imaging Optics*, Baifukan, Japan 1999.
4. D.H. Bilderback. Microbeam generation with capillary optics. *Rev. Sci. Instrum.*, 2059–2063, **66**(2), 1995.
5. Y. Shuzuki and Y. Chikaura. Characterization of metalorganic chemical vapor deposition grown GaAs on Si by means of X-ray scattering radiography. *J. Appl. Phys.*, 1290, **70**, 1991.
6. Y. Hosokawa, S. Ozawa, H. Nakazawa and Y. Nakayama. An X-ray guide tube and a desktop scanning X-ray analytical microscope. *X-Ray Spectrom.*, 380–387, **26**, 1997.
7. C. Liu and J.A. Golovchenko. Surface trapped X-ray: whispering-gallery modes at $\lambda = 0.7$ Å. *Phy. Rev. Lett.*, 788–791, **79**(5), 1997.
8. Yu.I. Dudchik, *et al.* Formation of X-ray beams with the aid of a tapered micro capillary. *Tech. Phys.* 562–564, **43**(5), 1998.
9. S.V. Kuhlevsky, *et al.* Wave-optics treatment of X-rays passing through tapered capillary guides. *X-Ray Spectrom.*, 354–359, **29**, 2000.
10. B. Chen. Theoretical consideration of X-ray transmission through cylinder capillaries. *Rev. Sci. Instrum.*, 1350–1353, **72**(2), 2001.
11. P. Engstrom and C. Riekel. Background reduction in experiments with X-ray glass capillary optics. *Rev. Sci. Instrum.*, 4061–4063, **67**(12), 1996.
12. R. Bruch, H. Merabet, M. Bailey, S. Showers and D. Schneider. Development of X-ray and extreme ultraviolet optical devices for diagnostics and instrumentation for various surface applications. *Surf. Interface Anal.*, 236–246, **27**, 1999.

3.3 Polycapillary X-ray Optics

N. GAO
X-ray Optical Systems, Albany, NY, USA

and

K. JANSSENS
University of Antwerp, Antwerp Belgium

3.3.1 INTRODUCTION

A polycapillary X-ray optic consists of an array of a large number of small hollow glass tubes formed into a certain shape. The optic collects X-rays that emerge from an X-ray source within a large solid angle and redirects them, by multiple external total reflections, to form either a focused beam or a parallel beam, as illustrated in Figure 3.3.1.

The use of polycapillary optics has become widespread in various X-ray analysis applications after more than a decade of extensive research and development work led by a few groups around the world.[1-4] These optical devices have also been successfully used as crucial components in commercial X-ray analysis instruments and have dramatically enhanced the system performance. The rapid development of polycapillary optics also triggered the development of related X-ray equipment such as microfocus X-ray sources and compact X-ray spectrometers. As a result, the development of analytical X-ray instruments is being driven towards compact, low-power instruments that can be used in-line. This opens opportunities for an even larger variety of applications in different fields of science and technology, both in academic laboratories and industrial institutions.

3.3.2 FUNDAMENTALS OF POLYCAPILLARY OPTICS

The fundamental mechanism with which X-rays transmit through a curved capillary tube has been described in detail on many occasions in the literature.[5-7] As shown in Figure 3.3.2, the direction of X-ray radiation can be changed when X-rays undergo multiple total reflections within a bent capillary, as long as the incident angle at each reflection is less than a critical angle θ_c. As this critical angle is very small [of the order of a few mrad for X-rays in the 5–30 keV range: θ_c (mrad) $\sim 30/E$ (keV) for reflection on glass surfaces], the bending curvature of the capillary has to be gentle and the capillary diameter has to be small to maintain the total reflection condition. The typical radius of curvature of the individual hollow glass tubes within a polycapillary optic is about a few hundreds of millimeters and the channel diameter is anywhere from a few micrometers to a few tens of micrometers. The optimal curvature and channel size are usually determined by the particular application and the energy range the optic will be used in. The first practical use of polycapillary focusing device, reported in 1990,[5] was a multifiber polycapillary optic that consisted of an array of curved fiber bundles guided through supporting thin metal screens, successfully assembled by

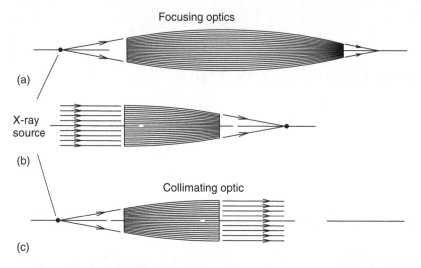

Figure 3.3.1 Polycapillary X-ray optics that produce a focused (a,b) or parallel beam (c), starting from a X-ray point source (micro-focus X-ray tube (a, c)) or a quasi-parallel X-ray source (synchrotron (b))

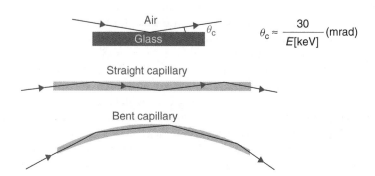

Figure 3.3.2 Schematic representation of the principles of capillary optics. θ_c is the critical angle for total reflection

Kumakhov et al.[8] A photograph of a multi-fiber optic and a cross section of a single fiber are shown in Figure 3.3.3. The total number of capillary channels can be as many as a few millions. The capture angle can be as large as 30° and the output beam size can be up to 50–100 mm.

A new generation of polycapillary devices, referred to as monolithic polycapillary optics, was announced in 1992.[8] Instead of using supporting metal screens, as employed in multi-fiber optics, the polycapillary fibers were closely packed and then fused together and formed into the desired shape through a heating process. One of the most distinguishing properties of a monolithic optic is that the cross-section of an individual channel changes along the length of the optic so that each individual channel points to the focus of the optic. As a result, the transmission efficiency of the optic is significantly increased and other beam properties, such as the focal spot size (for a focusing optic see Figure 3.3.1a and 3.3.1b) and beam divergence (for a collimating optic see Figure 3.3.1c), are improved. Photographs of monolithic optics and their cross sections are shown in Figure 3.3.4. Although a monolithic polycapillary optic is superior to a multi-fiber optic in general, a multi-fiber optic is still a better option for some applications such as large beam X-ray diffraction[9] and X-ray lithography[10] in which a high uniformity of the output beam is required.

Fundamentals of Polycapillary Optics

 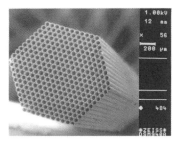

Figure 3.3.3 Photograph of multi-fiber optics and the polycapillary fiber (right)

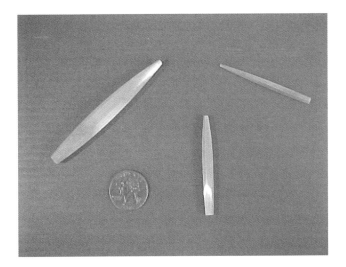

  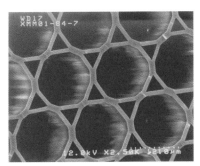

Figure 3.3.4 Photographs and SEM images showing the typical size of monolithic polycapillary optics and their cross-section

The transmission efficiency of a polycapillary optic, defined as the ratio of the number of X-rays exiting at the output end of the optic to those entering the input end of the optic, is a function of energy because of the dependency of the critical angle on the energy. Figure 3.3.5 shows simulation results for a typical focusing optic. The drop of the transmission efficiency on the high-energy end results from the decreasing critical angle θ_c and the more strict conditions for total reflection that are associated with this decrease. The particular transmission-energy curve will be determined by the particular optic design. For example, an optic designed for high-energy

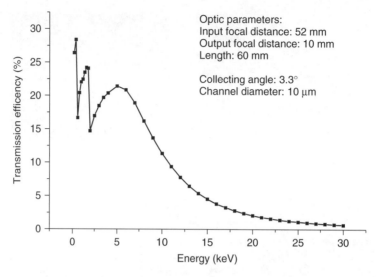

Figure 3.3.5 Transmission efficiency of a polycapillary optic as a function of the X-ray energy

X-rays can have a more favorable high-energy to low-energy transmission efficiency ratio. However, this does not necessarily mean that the absolute transmission efficiency at the high energies is higher than that at the low energies. The band-pass property shown in Figure 3.3.5 can be very useful in many applications where the high-energy radiation generates an undesired background.

3.3.3 FABRICATION AND SHAPE OPTIMIZATION

3.3.3.1 FABRICATION TECHNIQUES

Polycapillary optics are made by heating bundles of glass tubes in a furnace and pulling them to a particular shape. Today's technology allows control of the diameter vs length profile to better than 5 μm.[11] One of the recent developments is the ability to produce lenses with a non-constant curvature vs length profile; previously, the capillary tubes in focusing lenses featured a constant curvature over the entire length of the lens. In one particular example, the experimental flux density observed was 79 % higher than for a constant curvature optic at 8 keV and the focal spot size decreased from 52 to 40 μm. The development allows to produce bundles of tubes with small taper angles on the size that faces the X-ray source (Figure 3.3.1). On the focusing side (part nearest to the sample), the taper angles are larger. One of the future challenges is to make the focal spot size more constant with energy so that probe sizes do not vary by as much as a factor of two over a broad X-ray energy range.

3.3.3.2 SIMULATION TOOLS

Several Monte Carlo ray-tracing simulation programs are available that are able to calculate the trajectory of individual X-ray photons within a particular hollow glass tube that form part of a polycapillary lens.[7,12,13] Each of them has unique features that greatly help researchers to better understand the properties of the polycapillary optics and their potential application capabilities.

In one of the pioneering simulation tools developed by Xiao et al.,[7] the trajectory of an X-ray photon was described as a classically accelerated particle undergoing collisions with walls defined by the capillary. Using the small incident angle approximation, the motion of a particle was divided into the one along the capillary axis with a constant velocity and the one transverse to the axis, which has acceleration associated with the curvature of the capillary. The model simplified a complicated ray-tracing problem in three-dimensional space to a much more manageable two-dimensional problem, which significantly reduced the simulation time, and consequently made it possible to perform large numbers of simulations to optimize the optic design for a given application. The results generated from the simulation were compared to the experimental data and excellent agreement has been achieved for different types of polycapillary optics.

State-of-the-art simulation programs can handle various optic profiles, including circular, parabolic, elliptical, and polynomial shapes. The photon position and angular distributions can be obtained at any image plane, so the dimensions and divergence of the output beam can be calculated. Different source distributions can be handled, including Gaussian and line sources. The simulation has become an essential tool to provide information for the optic design, performance analysis and evaluation of potential applications.

3.3.4 FOCUSING OPTICS

A polycapillary focusing optic collects a large solid angle of X-rays from an X-ray source and focuses them to a small spot (see Figure 3.3.1a). The X-ray flux density obtained at the focus can be more than three orders of magnitude higher than that from a conventional pinhole aperture that provides the same beam size. Important performance specifications for a polycapillary focusing optic include the focal spot size, the output focal distance (working distance), and the intensity gain.

3.3.4.1 FOCAL SPOT SIZE

The best way to estimate the focal spot size is to use one of the simulation tools described earlier, because it is usually determined by a number of factors such as the focal distance, the channel diameter, the optic profile and the X-ray source. On the other hand, it is also possible to estimate the focal spot size with the following empirical relation:

$$S \approx a \cdot f \cdot \theta_{\max} + d_{\text{out}} \quad (3.3.1)$$

where S is the full width at half-maximum (FWHM) of the spot, d_{out} is the channel diameter of the capillary at the output end of the optic, and $\theta_{\max}$ is the maximum angle with which the X-rays exit from the individual capillary tubes at the given energy. In many cases, the approximation $\theta_{\max} = \theta_c$ is valid; d_{out} is usually very small compared to S and can be neglected. a is an adjustment parameter determined by the optic design and the X-ray source property, and typically ranges from 1.0 to 1.5. Since the focal spot size is proportional to the focal distance, a smaller spot size can be obtained by using lenses with shorter focal distance. The trade-off is the loss of the working space between the optic and the sample.

Equation (3.3.1) applies to both the input and the output of the optic. The input focal spot size will determine the area of the X-ray source that can be 'seen' by the optic and therefore the optimal source size for the optic. This is important in designing an optic for an existing X-ray source, or choosing the right X-ray source for an existing optic. Equation (3.3.1) indicates that the focal spot size will change with the critical angle, i.e. that it is dependent on the X-ray energy. Figure 3.3.6 shows the simulation result of the output focal spot size of a focusing optic as a function of the X-ray energy. The spot size change with energy can be a problem in some applications where highly accurate quantitative analysis is needed for a wide range of elements.

Simulation results also indicate that, to a certain extent, the output focal spot size of a polycapillary optic is a function of the source size. When the X-ray source size is much less than the input focal spot size of the optic, the output focal spot size can be much less than predicted by Equation (3.3.1). This is because, when the source size is small enough, the maximum incident angle $\theta_{\max}$, predominated by the source size, will be less than the critical angle θ_c. Consequently the X-rays exiting the optic will have a divergent angle less

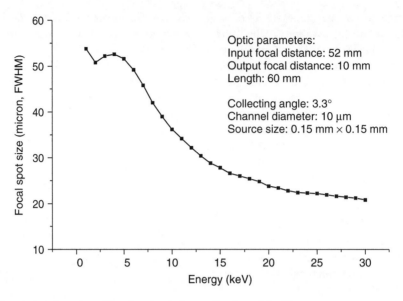

Figure 3.3.6 Focal spot size of a polycapillary focusing optic as a function of the X-ray energy

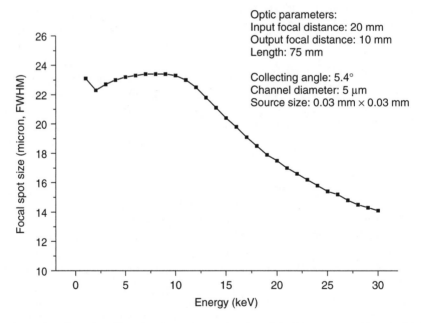

Figure 3.3.7 Focal spot size of a polycapillary focusing optic as a function of the X-ray energy with a small X-ray source

than the critical angle, and a smaller output focal spot size is obtained.

As an example, Figure 3.3.7 shows the simulation result for an X-ray source with 30 μm anode focal spot diameter. Clearly, the change of the spot size with the energy is much less than in Figure 3.3.6, where a 150 μm focal spot source was used. Other factors contributing to the result in Figure 3.3.7 include the small channel size and the optic design. It is shown that the use of very small X-ray source provides one possible solution for the problem of the spot size change with the energy that was mentioned earlier.

The most commonly used method to measure the beam size is the knife-edge scan, where a sharp knife edge is moved across the beam and the derivative of the knife-edge scan profile is used to determine the beam size. Figure 3.3.8 shows a typical knife-edge scan profile and its derivative for a polycapillary focusing optic with 5 mm output focal distance. The focal spot sizes (FWHM) of 13.8 and 23.2 μm were obtained at 17.4 and 8.0 keV, respectively.

In another often-used method, a thin wire is moved across the beam and the intensity of the fluorescent X-rays is recorded as a function of the wire position. The obtained scan curve is the convolution of the wire thickness and the beam size S_b. The latter can be estimated by means of the following formula:

$$S_b \approx \sqrt{S^2 - T_{wire}^2}$$

where S is the FWHM of the measured scan curve (see Figure 3.3.8) and T_{wire} the thickness of the wire. The result obtained is dependent upon the material the wire is made of. It must be recognized, however, that the result obtained with this method does not represent the spot size at any particular energy. Instead, the result is the beam size at an effective average energy $\langle E \rangle$ of all the X-rays that can excite the fluorescence of the element the wire is made of. Usually, $\langle E \rangle$ is close to the adsorption edge energy of the element considered. The result is also dependent upon the geometry of the X-ray tube such as take-off angle and its operating conditions.

3.3.4.2 FLUX DENSITY GAIN

The combination of the large collecting angle and small focal spot size of a focusing optic results in a high flux density, defined as the number of photons passing through a unit area per unit time. The flux density gain of an optic is defined as the ratio of the X-ray flux density obtained at the focus of the optic to that obtained with a pinhole aperture placed at a certain distance to the source. Usually the distance is the smallest one that can practically be employed for the application/X-ray source under consideration. The reason for using the term flux density rather than intensity is because the latter has been widely accepted in the X-ray analysis community to refer to the photon counting rate (detected number of photons per unit time), although this is a misuse of the term based upon the strict classical definition, as pointed out by Jenkins et al.[14] It is therefore less confusing to use the flux density gain rather than the intensity gain for a polycapillary optic. The gain clearly depends upon how far the pinhole

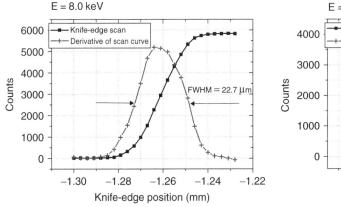
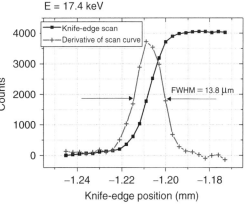

Figure 3.3.8 Focal spot size of a polycapillary focusing optic at different energies. The focal distance of the optic is 5 mm

is placed from the X-ray source. It is therefore important to specify the source-to-sample distance when reporting the intensity gain data.

The experimental result given in Figure 3.3.9 shows the typical level of the flux density gain of a polycapillary focusing optic, which is more than three orders of magnitude for W L line X-rays (8–12 keV). The polycapillary optic has an output focal distance of 10 mm and produces a spot of 40 µm (FWHM) at the W Lα energy. The data for the pinhole case were obtained using a 2 mm pinhole aperture and then scaled down to the equivalent of a pinhole of 40 µm diameter. The flux density gain is also a function of the source size because of the dependency of the optic performance on the source size (Figure 3.3.10). The rapid decrease of the intensity with the increasing source size indicates the importance of using microfocus X-ray sources in applications where polycapillary optics are employed.

Polycapillary optics can also provide 'gain' in special cases. In the above discussion for the pinhole aperture case, we assume the sample is placed right after the aperture. In practice, however, a

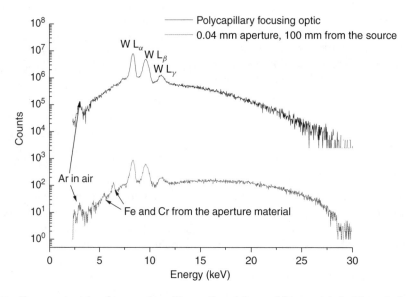

Figure 3.3.9 Scatter X-ray spectra taken from a polycapillary optic and from a 0.04 mm pinhole. The optic has an output focal distance of 10 mm. The pinhole data were acquired with a 2 mm pinhole aperture and the optic data were then rescaled. The pinhole was placed 100 mm to the X-ray source that has a tungsten target and 10 µm spot size

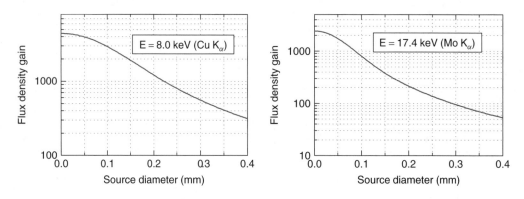

Figure 3.3.10 Flux density gain of a polycapillary focusing as a function of the X-ray source size at different energies

minimal working distance is indispensable in most applications. When a very small beam size is required and a large X-ray source is to be used, the X-ray source has to be far enough from the pinhole aperture to ensure the 'image' of the source on the sample plane matches with the desired beam size. This will dramatically reduce the collecting angle of the pinhole aperture and thereby the efficiency of the system. Using a polycapillary optic, however, can be beneficial in this case, as illustrated in Figure 3.3.8. Despite the relatively low transmission efficiency with the large X-ray source, it still provides a reasonable intensity gain. The successful application of a focusing polycapillary optic together with neutron sources[15] is a good example of this type of application.

The absolute X-ray intensity or flux density achievable with polycapillary optics strongly depends upon the characteristics of the X-ray source used. Haller *et al.*[16] reported having achieved a Cu Kα intensity of 5.4×10^9 photon/s in a spot of 30 μm (FWHM) corresponding to a flux density of 7.6×10^{12} photons/s/mm^2.

3.3.5 COLLIMATING OPTICS

A polycapillary collimating optic converts divergent X-rays from the X-ray source into a quasi-parallel beam (see Figure 3.3.1c). Although a collimating optic can be made such that all capillary channels are perfectly parallel to each other at the output end of the optic, the output X-rays have a certain divergent angle, which is determined by the critical angle and thereby the X-ray energy. The performance specifications for a collimating optic include the intensity gain, the output beam dimensions and the output beam divergent angle, which is a critical parameter for applications such as X-ray diffraction (XRD) wavelength dispersive spectrometer. The intensity gain is defined as the ratio of the output beam intensity obtained with the optic to the beam intensity obtained without the optic. Unlike in the case of focusing optics where the term *flux density* is used to describe the performance of the optic, the term *intensity*, referring to the number of photons per unit time, is used for characterizing collimating optics. Usually the total number of X-ray photons is the most important for the applications in this field. The base case (without the optic), however, must be defined carefully for collimating optics. It is a common practice to define the geometry of the base case as of a conventional aperture collimator providing a beam with the same dimension and divergent angle θ_d as delivered by the optic. The collection angle of the aperture collimator in the base case is therefore equal to the divergent angle, which is approximately twice the critical angle at a given energy. The intensity gain is then determined by the following formula:

$$\text{Gain} = \frac{\Omega_o \cdot T}{\theta_d^2} = \frac{\Omega_o \cdot T}{(2 \cdot \theta_c)^2}$$

where Ω_o and T are the collection solid angle and the transmission efficiency, respectively, of the optic. It is follows from the above expression that the larger the output beam size, the higher intensity gain will be obtained. For example, a polycapillary collimating optic with 5° collection angle, 30 % transmission efficiency and 6 mm output beam diameter will provide an intensity gain of approximately 200 at Cu Kα (8.04 keV, $\theta_c = 3.75$ mrad). If the output beam diameter is doubled, the corresponding collecting solid angle of the optic will be four times larger. Although the transmission efficiency of the optic will decrease to about 18 % due to the greater bending curvature of outer capillaries, the intensity gain of the optic will increase by a factor of 2.4 to approximately 100.

Besides the intensity gain, another critical parameter to be considered in the design of collimating optics is the divergent angle θ_d, which is usually determined by the requirement of the application. For example, when a collimating optic is used for wavelength-dispersive spectroscopy (WDS), the most efficient coupling with the diffracting crystal is obtained when the beam divergence of the optic is matched to the mosaicity of the crystal. In this case, an optic providing the maximum intensity may not be the optimal for the application. Figure 3.3.11 shows simulation results on how the intensity and divergent angle of the

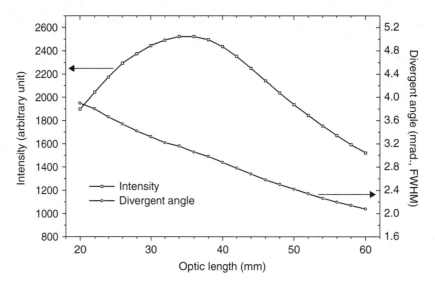

Figure 3.3.11 Simulation result of the output beam intensity and divergent angle of a collimating optic as a function of the optic length

output beam changes with the optic length at given input focal distance and energy.

The conclusion that the beam divergent angle θ_d is determined by the critical angle θ_c only holds when the maximal incident angle of the X-rays at the entrance of the optic is larger than the critical angle. This is generally true because the optic will have the maximal collecting angle under such conditions. In some cases where the maximal incident angle is less than the critical angle, however, θ_d will be smaller than θ_c because the incident angle of X-rays decreases slightly after each reflection in a collimating optic. This condition can be achieved when a very small X-ray source is used. Figure 3.3.12 show results of experimental determinations of θ_d of a collimating optic when coupled to a 20 μm X-ray source. The divergent angle was measured by acquiring the rocking curve of the beam using a highly oriented single crystal such as silicon or germanium. Figure 3.3.12(a) is the result at Cu Kα (8.04 keV), where the obtained divergent angle of 2.4 mrad (FWHM) was much less than the critical angle at Cu Kα, which is 3.8 mrad. In Figure 3.3.12(b) the result for Mo Kα (17.4 keV) are shown, where $\theta_d = 1.1$ mrad is also smaller than the critical angle at Mo Kα ($\theta_c = 1.7$ mrad). A Si (400) crystal was used in both measurements as a Bragg reflector.

3.3.6 ANALYTICAL APPLICATIONS OF POLYCAPILLARY OPTICS

3.3.6.1 ELEMENTAL ANALYSIS

Polycapillary lenses are currently being used in different forms of X-ray microanalysis: in micro-XRF instruments of various kinds, they are used to focus the primary radiation emitted by a (microfocus) X-ray tube (see Figure 3.3.1a) or by a synchrotron storage ring (see Figure 3.3.1b). Thus, only a small spot on the sample surface is irradiated and the local composition can be derived by using an energy-dispersive detector to collect the emitted fluorescent radiation. Focal spots in the range 10–20 μm diameter at an energy of 17.4 keV have recently been obtained.

In electron probe X-ray microanalysis, polycapillary optics are used to collect fluorescent X-rays from the sample and redirect them to the detector. Such applications include, but are not limited to: (1) superconducting microcalorimeter detectors[17,18] or tunnel-junction detectors,[19] in which polycapillary optics are used to increase the

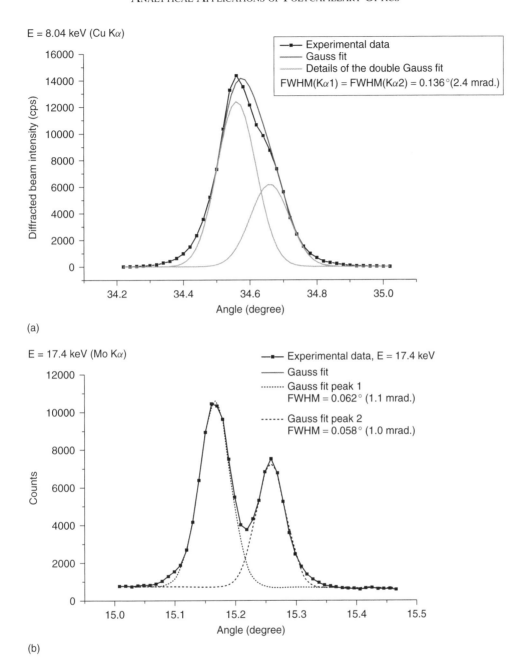

Figure 3.3.12 Measurement of the divergent angle of a collimating optic using a Si(400) crystal

effective collection solid angle; (2) environmental scanning electron microscopy (ESEM) or low-vacuum scanning electron microscopy (LV-SEM) where the use of polycapillary optics improves the contrast of X-ray imaging;[20] and (3) wavelength dispersive spectrometer, where polycapillary collimating lenses (see Figure 3.3.1c) are now being used to replace soller-slit assemblies, thus increasing the throughput of the wavelength dispersive detection system.[21]

Micro-XRF

Laboratory micro-XRF

Initially, custom-designed polycapillary lenses have been used by various groups to extend the capabilities (both in terms of absolute sensitivity and in terms of lateral resolution) of existing (i.e. commercially available) micro- and mini-XRF instruments.[22] Usually, the geometry of these systems imposes rather severe constraints on the dimensions of the polycapillary optics that may be used, so that lenses with optimal performance could not always be used effectively.

In situations were there are no or not so strict geometric constraints, Bichlmeier et al.[23] tested out several combinations of low-power micro-focus tubes, polycapillary lenses and Peltier-cooled energy-dispersive detectors and measured relative DL (detection limit) values for typical laboratory systems situated in the 20–50 ppm range for transition elements in a glass matrix. The same authors used this system for quantitative analysis of historical glass beads and gold coins and various industrial materials.[24] A detailed comparison of the performance of polycapillary lenses offered by three commercial manufacturers for use in micro-XRF spectrometers was performed by Haschke and Haller.[25] Next to an examination of the primary parameters of the lenses (spot size and gain), also secondary characteristics such as the variation of the spot size with working distance and energy, the halo-effect and the reproducibility with which lenses can be produced, were considered.

Worley et al.[26] have used a combination of two polycapillary lenses, one between X-ray tube and sample and another between sample and detector for improving the detection power of a micro-XRF instrument for analysis of radioactive samples. During XRF analysis of highly radioactive samples usually problems are encountered since the spontaneous radiation from the radioactive materials can affect the analysis in different ways. First, the energies of the radiation may overlap with that of the characteristic lines of the elements of interest. Second, the high-intensity of the radiation from the radioactive materials can increase the background and reduce the energy resolution of the detection system, reducing the detection sensitivity of the system. In some cases, the radiation can be so strong that the X-ray detector is saturated. The problem becomes more severe in the case of micro-XRF. Here the fluorescence signal is usually smaller in magnitude because it originates from a small area while the spontaneous radiation background remains unchanged.

This low peak-to-background situation can be improved in two ways: (1) a focusing X-ray optic can be used to provide a higher primary beam intensity in a small area so that the signal will be enhanced while the background remains unchanged; and (2) a spatial filter can be placed between the sample and the detector so the detector will only collect X-rays from a particular area. In this case, the net fluorescent signals will remain unchanged or become lower, but the background will be significantly reduced. Monolithic polycapillary focusing optics can be used for both purposes; Figure 3.3.13(a) shows the arrangement used in a proof-of-principle experiment involving an ^{55}Fe source in which the local Fe concentration was determined; in Figure 3.3.13(b) spectra obtained from this source with and without a polycapillary lens between the sample and detector are compared. In the case where an optic is used to preferentially guide the fluorescent X-rays to the detector, the Mn Kα signal (mainly arising from the radioactive decay of ^{55}Fe via the electron capture mechanism) is significantly reduced. Similarly, Fiorini et al.[27] have used a polycapillary conical collimator to manufacture a micro-XRF system that employs an unfocused beam and uses a conical polycapillary waveguide to restrict the area on the sample surface to a small spot from which fluorescent X-rays can reach the detector.

Portable micro-XRF

In Figure 3.3.14, a scheme of a transportable micro-XRF instrument suitable for *in situ* analysis of large paintings at a museum, etc. is shown. This figure details how polycapillary optics can be coupled to a compact X-ray tube to create a small focal spot (ca. 100 μm diameter in this case) on a large painting. This arrangement

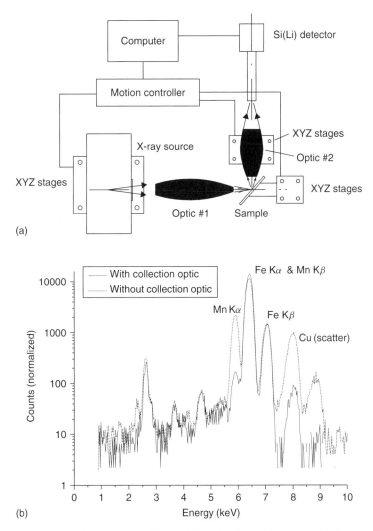

Figure 3.3.13 (a) Diagram of the experimental setup of the double-optic micro-XRF system. (b) Comparison of XRF spectra of ^{55}Fe radioisotope sample, with and without the collecting optic between the sample and the detector

is sufficiently stable to allow transport of the entire instrument between the laboratory and the museum without significant loss of lens alignment occurring. Other polycapillary-based instruments with a geometry and lay-out specifically optimized for *in situ* analyses have been described.[28] The use of the focusing optics has two advantages: on the one hand a larger number of photons from the X-ray source is captured by the optic than would be done by a pinhole of ca. 1 mm^2, thereby reducing spectrum acquisition times. On the other hand, this higher photon flux is focussed into a smaller spot, permitting the selective irradiation of small features. The use of the optic does prevent the use of tube voltage settings above 30 kV which is sometimes useful for demonstrating the presence of high atomic number elements such as Sn or Sb in pigmented materials by means of their K lines. Figure 3.3.15 illustrates how this spectrometer is useful for making the distinction between original vs restored painted areas of the same color, when the pigments originally used are significantly different in composition from those used in later periods.

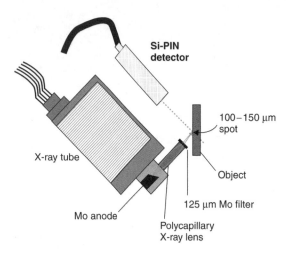

Figure 3.3.14 A portable X-ray tube with Mo target is focused to a spot size of 100–150 μm by means of a polycapillary lens. The X-ray fluorescence lines emitted by an object under study are observed with a silicon PIN diode detector

Synchrotron micro-XRF

Polycapillary optics (of the type shown in Figure 3.3.1b) are also used at bending magnet beamline L of HASYLAB (Hamburg, Germany) for focusing of monochromatic synchrotron radiation. The experimental scheme of the beamline is shown in Figure 3.3.16. Previously, at this beamline, straight or elliptical monocapillary tubes were used to collimate/concentrate the polychromatic white radiation for trace-level micro-XRF experiments.[29] Monochromatic radiation was never used since the intensity of the resulting microbeam was insufficient to allow trace-level measurements within reasonable spectrum collection times. Using lenses manufactured at Beijing Normal University (BNU, Beijing, P.R. China) and X-ray Optical Systems (XOS, Albany, NY, US), beams of, respectively, 34–53 μm and 10–40 μm were produced in the energy range 5–24 keV. The observed transmission of about 40 % at 10 keV for the BNU lens compared well with a simulation where the assumed surface roughness of the glass is 5 Å.[30] Gain factors in the range 100–400 (for the BNU lens) and 1000–2500 (for the XOS lens)[31] were obtained. Above 17 keV, however, a significant 'halo' surrounds the focal spot of focussed X-rays. The halo is due to two effects: at very high energy (>30 keV), X-ray photons can penetrate through

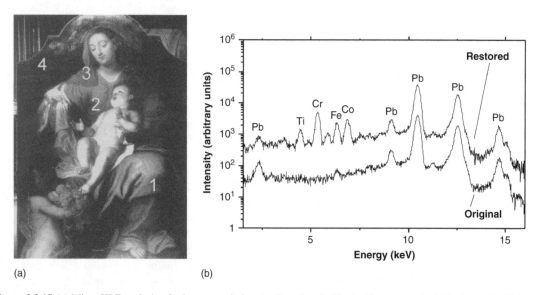

Figure 3.3.15 (a) Micro-XRF analysis of a baroque painting (attributed to P. Thys). The area marked '1' shows Fe, Pb but no Co or Cu, suggesting the use of indigo (an organic pigment), the area marked '2' contains Fe, Hg, Sr, Pb (vermillion), area '3' Ca, Mn, Zn, Pb (umber), and area '4' contains Ca, Mn, Fe, Zn, Pb (umber). (b) Two XRF spectra from adjacent spots in the area marked '1' in (a). Positions where more recent paint (containing Ti, Cr and Co) was applied during past restoration activities can clearly be distinguished from positions with original seventeenth century paint

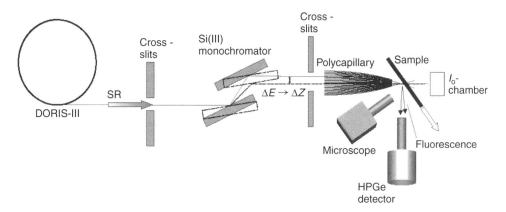

Figure 3.3.16 Polycapillary optics from Beijing Normal University installed at Hasylab beamline L for fluorescence mode X-ray near edge structure (XANES) studies. The use of polycapillary optics makes it possible to obtain a small spot with a highly stable position on the sample while the incident X-ray energy is scanned over the elemental absorption edges of interest. In this way an inexpensive Si(111) channel-cut monochromator could be used for micro-XANES instead of a costly fixed-exit monochromator

the entire body of the X-ray lens; at lower energies, they are partially transported along the capillary tubes, but near the front tip of the lens they are no longer are totally reflected and 'escape' through the remaining glass material near the lens tip. Despite these unwanted phenomena, in the focussed part of the beam, however, a significant increase of flux density is observed, as gain factors in the range 300–2500 were obtained. This increase implies that monochromatic microbeams of sufficient intensity can be produced for use in monochromatic micro-XRF and related experiments (see below). In Figure 3.3.17, the relative detection limits obtained by means of a 17.4 keV focussed X-ray microbeam from a NIST SRM1577a Bovine Liver standard sample are shown, indicating that for the transition elements, determinations down to the 10–100 ppb level are possible.

Electron Probe X-ray Microanalysis

Various types of high-resolution energy-dispersive X-ray spectrometer (EDS) are now being developed for use in X-ray microanalysis. Newbury et al.[32,33] have described the use of cryogenic microcalorimeter X-ray detectors with an energy resolution of 3 eV at 1.5 keV and count rates of up to 500 cps. Since the active area of the detector

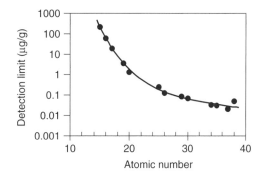

Figure 3.3.17 Relative detection limits (in μg/g) derived from a μ-XRF spectrum obtained by irradiation during 1000 s of a 1 mm thick NIST SRM 1577a bovine liver sample using a primary energy of 17.5 keV

is relatively small (0.05 – 0.2 mm²) polycapillary optics was used to increase the effective collection solid angle by a factor of 300. Thus, one of the most significant disadvantages of the small active area detection system could be overcome, leading to a microcalorimeter energy-dispersive detector that combines many of the favorable qualities of commercially available wavelength dispersive spectrometers (WDS) and semiconductor (EDS) detectors. Also a few applications are described in these papers.

A polycapillary optic can also be used as a spatial filter to eliminate background X-rays generated by the scattered electrons hitting the sample outside the area of interest. Such electron spreading is commonly seen in ESEM or LV-SEM systems where the pressure in the sample chamber needs to be raised to up to a few Torrs to accommodate non-conductive or wet samples, e.g. biological samples. Gao and Rohde[34] have demonstrated that the contrast of X-ray images was significantly improved by coupling a polycapillary focusing optic between the sample and the EDS detector in a LV-SEM.

Wavelength Dispersion of X-rays

An alternative approach for wavelength-dispersive X-ray fluorescence (WDXRF) analysis was proposed by Ebel et al. in 1983[35] where an analyzer crystal collected a divergent X-ray fluorescence beam from a small area of the sample and the lateral intensity profile of the diffracted beam was detected by means of a position-sensitive wire detector. This approach was trying to combine the high-energy resolution of the wavelength-dispersive spectrometer and the high detection efficiency of the energy-dispersive spectrometer. Its disadvantage was the low count rate, which was estimated to be an order of magnitude lower than that of an EDS. When a polycapillary lens is used to focus the primary beam into small spot, so that the fluorescent X-rays originate from a quasi-point source on the sample surface, the dispersive system will function better than when the same number of primary X-rays irradiate the sample over a large area. The principal arrangement of the experimental setup is shown in Figure 3.3.18(a). The anode focal spot of the X-ray source was placed at the input focus of the polycapillary optic. Irradiation of the sample with the focussed beam results in fluorescence X-rays that are diffracted by the crystal according to Bragg's law. The diffracted X-rays were detected by a position sensitive proportional counter (PSPC). The resulting signals were registered and amplified by the electronic system and processed by a multichannel analyzer. After proper calibration, a position on the detector corresponds to a specific energy: via Bragg's law, the angular range $\theta_1-\theta_2$ captured by the detector corresponds to an specific energy range E_1-E_2. The energy resolution of the PSXS system is determined by the X-ray energy, the spacing of the crystal used in the experiment, the spatial resolution of the PSPC, and the preamplifier noise. Effective energy resolutions from several eV (for energy below 1 keV) to several tens of eV at $\sim$10 keV were obtained, as illustrated in Figure 3.3.18(b) where a 70 mm wide LiF(200) crystal ($2d = 4.027$ Å) was used under an angle of $43°$.[24]

In electron microprobes equipped with wavelength-dispersive spectrometers, a polycapillary collimating optic can be used to collect emitted X-rays from the sample and efficiently convert them into a quasi-parallel beam that was then diffracted by a flat crystal. In comparison with the curved crystal geometry, which is commonly used in most commercial instruments, the parallel beam geometry provided by the polycapillary optic offers a much simpler scanning mechanism, more flexibility and a larger freedom of selection among analyzing crystals. Agnello et al.[21] also reported the combination of a polycapillary optic with a reflect mirror to improve the collection efficiency of a capillary-based WD spectrometer over a wide energy range. Details of this work will be described in the Detector section of this book.

3.3.6.2 STRUCTURAL ANALYSIS

In contract to X-ray fluorescence analysis where primary beams of either polychromatic and monochromatic nature can be used and where the initial divergence of these beams is of no great concern, X-ray based methods that provide structural information such as X-ray diffraction (XRD) or X-ray absorption spectroscopy (XAS) place more restrictions on the properties of the primary X-ray beams. In XRD instruments, preferably a quasi-parallel beam is used to irradiate the materials under study, in order to preserve the angular resolution in the resulting 2θ-patterns. Usually also a near-monochromatic beam is used, either by using a so-called '$K\beta$ filter' (e.g., a Zr foil in combination with a Mo anode X-ray tube) or a single-bounce graphite

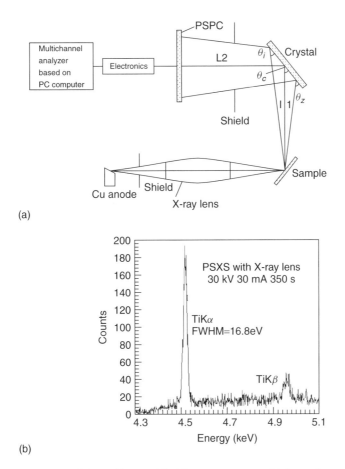

Figure 3.3.18 (a) Principal arrangement of the position-sensitive X-ray spectrometer. (b) Spectral separation in a Ti spectrum by means of the system in (a)

monochromator. In XAS, highly monochromatic synchrotron radiation ($\Delta E/E \sim 10^{-4}$) is employed that is usually obtained by employing a channel-cut or double-crystal monochromator. In both types of spectrometers, and especially in those seeking to employ X-ray beams with small cross-sections, polycapillary lenses can be employed to modify either the spatial or angular or both characteristics of the primary X-ray beams.

Micro-XANES

The experimental setup shown in Figure 3.3.16 can also be used for performing micro X-ray near edge structure (XANES) experiments. During XANES experiments, the energy of the primary, monochromatic beam is changed in small (eV) increments in a range straddling the absorption edge of an element of interest. During the energy change, the primary beam impinging on the capillary optic gradually changes height as a result of the rotation of the Si(111) reflector inside the monochromator; in view of the large collection area (of the order of 0.5–1 cm^2) of the polycapillary lens, no loss of alignment was observed during energy changes in the 5–25 keV range while also the change in position of the focused beam was kept to a minimum.

The absorption profile can be recorded either directly (by measuring the intensities that impinge on and are transmitted by the sample at each energy) or indirectly by recording the variation of the fluorescent intensity with energy, as illustrated in Figure 3.3.19. The shape and shift of the absorption profile contains information on the valence of the atomic species in question. In fluorescent mode, the valence information can be derived from trace elements down to the 50 ppm level. In Figure 3.3.19(b) an XRF spectrum of a 70 μm diameter fly-ash sample is shown containing ca. 40 ppm of Se; Figure 3.3.19(a) compares the K-edge XANES profile derived from this particle[23] to that of Sereference compounds. Most of the Se in the fly-ash particle appears to be present in the +IV (selenite) form. Other studies using this facility involve the determination of the ferric iron content in Ultra-High-Pressure eclogitic minerals,[36] the measurement of the oxidation state of U in individual particles of depleted uranium recovered from Kosovo soil[37] and the study of the Fe^{2+}/Fe^{3+} redox-equilibrium in ferro-gallic inks employed for writing documents in previous centuries.[38]

XAFS (X-ray absorption fine structure) measurements mostly are performed at synchrotron radiation facilities, in view of the fact that primary X-ray beams of a high degree of monochromaticity are required. Nevertheless, in a limited number of cases, also instruments based on laboratory sources are employed for this purpose. For in-laboratory EXAFS apparatus, usually linear spectrometers with bent crystals are used: here the monochromator crystals have to be aligned very accurately along a Rowland circle, so that the goniometer movement becomes very complicated. The mechanism of the monochromator can be very much simplified when parallel X-rays are used. Taguchi et al.[39] have used polycapillary optics to obtain a quasi-parallel X-ray beam from a conventional X-ray source and measure EXAFS spectra by a simple system utilizing an ordinary powder diffractometer.

X-ray Diffraction and Related Types of Investigations

An overview of the use of polycapillary optics in X-ray diffraction is provided by Schields et al.[40] Gubarev et al.[41] have described the design and performance of a high flux X-ray system for macromolecular crystallography that combines a microfocus X-ray generator having a 40 mm FWHM spot size at a power level of 46.5 W and a collimating polycapillary optic (see Figure 3.3.1c). The Cu Kα X-ray flux produced by this optimized

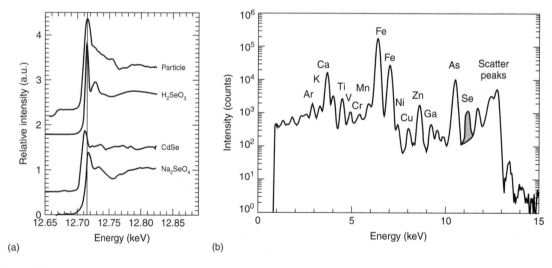

Figure 3.3.19 (a) Measured XANES spectra corresponding to various Se standard compounds and the XANES spectrum from the unknown fly ash particle of 70 μm. The concentration of Se in the particle is approximately 40 ppm. (b) The quantitative data for Se were derived from a series of XRF spectra similar to the one shown here

system through a 500 μm diameter orifice was 7.0 times greater than the X-ray flux previously reported by Gubarev.[42] The X-ray flux from the microfocus system was also 2.6 times higher than that produced by a rotating anode generator equipped with a graded multilayer monochromator and 40 % less than that produced by a rotating anode generator with the newest design of graded multilayer. Both rotating anode generators operated at a power level of 5000 W, dissipating more than 100 times the power of the microfocus X-ray system. Diffraction data collected from small test crystals are of high quality. By using the setup shown in Figure 3.3.20, the lyzozyme crystal diffraction patterns shown in Figure 3.3.21 could be obtained. 42 540 reflections collected at ambient temperature yielded a R-sym value of 5.0 % for data extending to 1.70 Å, and 4.8 % for the complete set of data to 1.85 Å. Amplitudes of the observed reflections were used to calculate difference electron density maps that revealed positions of structurally important ions and water

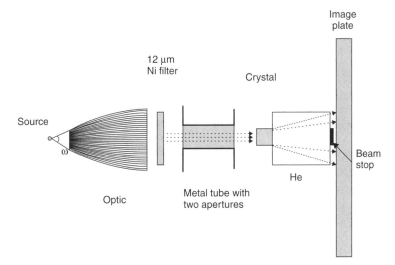

Figure 3.3.20 A microfocus source of X-rays from Oxford Instruments, Inc. is collimated by a polycapillary semilens followed by a metal tube with two apertures. A protein single crystal is then placed between the collimator and the image plate detector for oscillation photography of the diffraction pattern

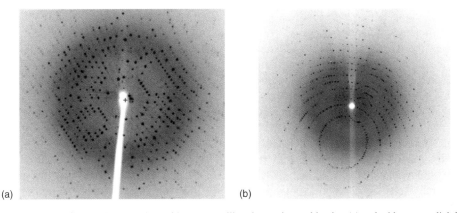

Figure 3.3.21 Lysozyme diffraction images taken with source-collimating optic combination (a) and with source-slightly focusing optic combination (b) with Cu Kα radiation. Note: the collimating lens characteristics and collimator dimensions are slightly different for these two images

molecules in the crystal of lysozyme using the phases calculated from the protein model.[43]

Other applications of collimating polycapillary optics are wafer size strain measurements [44] and X-ray lithography.[10,45] These greatly benefit from the increased intensity on the sample. Because of the higher flux, polycapillary collimating lenses allow monochromatic medical imaging to be developed for potential clinical applications.[46] Medical imaging applications will also potentially benefit from large area angular filters, which can be used to reject Compton scattered radiation originating in the patient, and thus increase the contrast of the resultant image.[47] Progress is being made in developing multi-optic large area devices. Angular filters can also be used to image diffuse sources, as high resolution gamma cameras in nuclear medicine.[48]

3.3.7 FUTURE OF THE TECHNOLOGY

It is expected that the requirements for X-ray beam with higher intensity, smaller spot, lower divergence and better uniformity will push the optic manufacturing technology towards the following directions:

- Use of polycapillary bundles with large outer diameter (10–20 mm) and small channel diameter (1–5 μm). A polycapillary optic with large outer diameter and small channel size will have a large collecting solid angle and high transmission efficiency, and thereby provide higher beam intensity. The increase of the outer diameter while keeping the channel size small implies a significant increase of the total number of channels. For example, a polycapillary bundle with 4 mm diameter and 10 μm channel diameter will have approximately 100 000 channels. This number will go up to approximately 2 600 000 for a polycapillary bundle with 10 mm diameter and 5 μm channel diameter. This is a considerable challenge for the polycapillary bundle manufacturing technology when a large open area and a smooth inner surface need to be maintained.

- Optics with varying channel size or open area. The fact that the transmission efficiency of the center channels of a polycapillary optic is higher than that of the outer channels causes the profile of the output beam to become non-uniform across the optic. This non-uniformity, which is also energy dependent, is unacceptable in X-ray lithography and some XRD applications. By using a varying channel size or open area it will become possible to adjust the output X-ray distribution For the design of such type of optics, simulation tools will again play a critical role. The approach will also allow one to choose the desired energy distribution of the output beam, which is of great importance in many applications.

- Use of improved materials and manufacturing processes to reduce the halo effect. The 'halo effect' has become a major obstacle for a polycapillary focusing optic to achieve small (i.e. <10 μm) focal spot sizes. The 'penetration halo' is usually tied with high-energy X-rays and can be prevented by operating the X-ray source at voltages not exceeding 40 kV. Thus, the 'penetration effect' may not have a significant impact on most applications. In XRF analysis, for example, unless the K lines of heavy elements are of great interest, the high-energy X-rays from the 'penetration halo' will have very low excitation efficiency and therefore will only make an insignificant contribution to the fluorescence signals. The 'escape halo' effect, on the other hand, has to be suppressed when pursuing small beam sizes. This can be achieved by using improved materials and structures in optic manufacturing. The manufacturing processes also need to be improved to better control the optic profile.

- Spot sizes smaller than 10 μm in diameter. Theoretically a focal spot of less than 10 μm can be achieved at an energy of 17.4 keV when the output focal distance is short enough, e.g. 2–3 mm. From a practical point of view, however, many other factors can limit this capability. The halo effect discussed above is one of them. The short focal distance can also be a problem for real applications, where a certain working space between the optic and the sample is needed to place other components such as a detector and a video camera. With the rapid and continuous development on the polycapillary bundle and optic manufacturing technology, we believe that achieving a 5–10 μm focal spot size is viable in the near future.

- Focusing optics producing consistent spot size at different energies. The fact that the focal spot size of the focusing optic is dependent upon the X-ray energy can cause noticeable errors in quantitative XRF analysis requiring high precision. Preliminary simulation results show that polycapillary optics having a constant spot size can be realized in the energy range 1–25 keV. The optic profile, channel size and the X-ray source size all play an important role in obtaining such property. The manufacturing requirement is also very challenging.

Polycapillary X-ray optics has been widely accepted in the X-ray analysis community and have been successfully used in commercial instruments. The development of optics and the performance enhancement will further extend the application area and the capability of the existing instruments. The large collecting solid angle and high efficiency of polycapillary optics also allows a system to achieve the same level of performance as conventional systems but with a much less powerful X-ray source. The significantly reduced power requirement makes it possible to build a compact and high-performance system that does not need intensive maintenance. Such systems have great potentials for in-line monitoring and *in situ* analysis in a wide range of areas of science and technology.

REFERENCES

1. Yan, Y. and Ding, X. *Nucl. Instrum. Meth.*, An investigation of X-ray fluorescence analysis with an X-ray focusing system. **B82**, 121–124 (1993).
2. Gao, N., Ponomarev, I., Xiao, Q. F., Gibson, W. M. and Carpenter, D. A. Monolithic polycapillary focusing optics and their applications in microbeam X-ray fluorescence. *Appl. Phys. Lett.*, **69**, 1529–1531 (1996).
3. MacDonald, C. A. Applications and measurements of polycapillary X-ray optics. *J. Sci. Technol.*, **6**, 32–47 (1996).
4. Arkadiev, V. A., Beloglazov, V. I., Bjeoumikhov, A. A., Gorny, H. E., Langhoff, N. and Wedell, R. *Poverkhnost.*, **1**, 48–54 (2000).
5. Arkadiev, V. A., Kolomitsev, A. I., Kumakhov, M. A., Ponomarev, I. Y., Khodeev, I. A., Chertov, Y. P. and Shakparonov, I. M. Wide-band X-ray optics with a large angular aperture. *Sov. Phys. Usp.*, **32**, 271 (1989).
6. Ullrich, J. B., Kovantsev, V. and MacDonald, C. A. Measurements of polycapillary X-ray optics. *J. Appl. Phys.*, **74**, 5933–5939 (1993).
7. Xiao, Q. F., Ponomarev, I., Kolomitsev, A. I. and Kimball, J. C. Numerical simulation for capillary-based X-ray optics. *SPIE Proc.*, **1736**, 227–238 (1992).
8. Kumakhov, M. A., *Nucl. Instrum. Meth.*, **B48**, 283 (1990).
9. Kardiawarman, York, B. R., Qian, C., Xiao, Q. F., Gibson, W. M. and MacDonald, C. A. X-Ray and ultraviolet sensors and applications. *SPIE Proc.*, **2519**, 197–201 (1995).
10. Klotzko, I., Xiao, Q. F., Gibson, D. M., Downing, R. G., Gibson, W. M., Karnaukhov, A. and Jezewsky, C. J. Investigation of glass polycapillary collimator for use in proximity based X-ray lithography. *SPIE Proc.*, **2523**, 175–182 (1995).
11. Gao, N. and Ponomarev, I. Y., Polycapillary X-ray optics: manufacturing status, characterization and the future of the technology. *X-ray Spectrom.*, **32**, 186–194 (2003).
12. Vincze, L., Janssens, K., Adams, F., Rindby, A. and Engström, P., *Rev. Sci. Instrum.*, **69**, 3494–3503 (1998).
13. Balaic, D. X. and Nugent, K. A., *Appl. Opt.*, **34**, 7263 (1995).
14. Jenkins, R., Gould, R. W. and Gedcke, D., *Quantitative X-ray Spectrometry*, 199–201, Marcel Dekker, New York, 1995.
15. Xiao, Q. F., Chen, H., Sharov, V. A., Mildner, D. F. R., Downing, R. G., Gao, N. and Gibson, D. M., *Rev. Sci. Instrum.*, **65**, 3399–3402 (1994).
16. Haller, M., Gao, N., Fraser, G., Loxley, N., Taylor, M. and Wall, J., Enhancement of X-ray analysis by close coupling of polycapillary optics with X-ray microsource. Poster presentation at 48th Annual Denver X-ray Conference (2000).
17. Hohne, J., Buhler, M., von Hentig, R., Hertrich, T., Hess, U., Phelan, K., Wernicke, D., Redfern, D. and Nicolosi, J. Design features of a high resolution microcalorimeter EDX system. *Mikrochim. Acta*, **138**, 259–264 (2002).
18. Newbury, D. E., Wollman, D. A., Hilton, G. C., Irwin, K. D., Bergren, N. F., Rudman, D. A. and Martinis, J. M. The approaching revolution in X-ray microanalysis: the microcalorimeter energy dispersive spectrometer. *J. Radioanal. Nucl. Chem.*, **244**, 627–635 (2000).
19. Frunzio, L., Li, L. and Prober, D. E. Detection of single X-ray photons by an annular superconducting tunnel junction. *Appl. Phys. Lett.*, **79**, 2103–2105 (2001).
20. Gao, N. and Rohde, D. Using a polycapillary optic as a spatial filter to improve micro X-ray analysis in

low-vacuum and environmental SEM systems. *Microsc. Microanal. Proc.*, **7**(Suppl. 2), 700–701 (2001).

21. Agnello, R., Howard, J. McCarthy, J. and O'Hara, D. The use of collimating X-ray optics for wavelength dispersive spectrometry. *Microsc. Microanal. Proc.*, **3**(Suppl. 2), 889–890 (1997).

22. Worley, C. G., Colletti, L. P. and Havrilla, G. J. Optimizing the focal spot size and elemental sensitivity of a monolithic polycapillary optic. Paper F-29 presented at the 1998 Denver X-ray Conference (1999).

23. Bichlmeier, S. Janssens, K., Heckel, J., Gibson, D., Hoffmann, P. and Ortner, H. M. Component selection for a compact micro-XRF spectrometer. *X-ray Spectrom.*, **30**, 8–14 (2001).

24. Bichlmeier, S., Janssens, K., Heckel, J., Hoffmann, P. and Ortner, H. M. Comparative material characterization of historical and industrial samples by using a compact micro-XRF spectrometer. *X-ray Spectrom.*, **31**, 87–91 (2002).

25. Haschke, M. and Haller, M. Examination of polycapillary lenses for their use in micro-XRF spectrometers. *X-ray Spectrom.*, **32**, 239–247 (2003).

26. Worley, C. G., Collett, L. P. and Havrilla, S. Quantitative analysis of radioactive materials by X-ray fluorescence. *Abstracts of Papers of the American Chemical Society*, **223**, 042-NUCL, Part 2, 7 April 2002.

27. Fiorini, C., Longoni, A. and Bjeoumikhov, A. A new detection system with polycapillary conic collimator for high-localized analysis of X-ray fluorescence emission. *IEEE Trans. Nucl. Sci.*, **48**, 268–271 (2001).

28. Bronk, H., Rohrs, S., Bjeoumikhov, A., Langhoff, N., Schmalz, J., Wedell, R., Gorny, H. E., Herold, A. and Waldschlager, U. ArtTAX – a new mobile spectrometer for energy-dispersive micro X-ray fluorescence spectrometry on art and archaeological objects. *Fres. J. Anal. Chem.*, **371**, 307–316 (2001).

29. Janssens, K., Vincze, L., Vekemans, B., Williams, C. T., Radtke, M., Haller, M. and Knöchel, A. The nondestructive determination of REE in fossilized bone using synchrotron radiation induced K-line X-ray microfluorescence analysis. *Fres. J. Anal. Chem.*, **363**, 413–420 (1998).

30. Vincze, L., Wei, F., Proost, K., Vekemans, B., Janssens, K., He, Y., Yan, Y. and Falkenberg, G. Suitability of polycapillary optics for focusing of monochromatic synchrotron radiation as used in trace level micro-XANES measurements. *J. Anal. At. Spectrom.*, **17**, 177–182 (2002).

31. Proost, K., Vincze, L., Janssens, K., Gao, N. and Falkenberg, G. Characterization of a polycapillary lens for use in micro-XANES experiments. *X-ray Spectrom.*, **32**, 215–222 (2003).

32. Newbury, D., Wollman, D., Nam *et al.*, Energy-dispersive X-ray spectrometry by microcalorimetry for the SEM. *Microchim. Acta*, **138**, 265–274 (2002).

33. Wollman, D. A., Irwin, K. D., Hilton, G. C., Dulcie, L. L., Newbury, D. E. and Martinis, J. M. *J. Microsc.*, **188**, 196–223 (1997).

34. Gao, N. and Rohde, D. Using a polycapillary optic as a spatial filter to improve micro X-ray analysis in low-vacuum and environmental SEM systems. *Microsc. Microanal. Proc.*, **7**(Suppl. 2), 700–701 (2001).

35. Ebel, H., Mantler, M., Gurker, N. and Wernisch, J. *X-Ray Spectrom.*, An X-ray spectrometer with a position sensitive wave detector (PSD). **12**, 47 (1983).

36. Schmid, R., Wilke, M., Ober, R., Dong, S., Janssens, K., Falkenberg, G., Franz, L. and Gaab, A. Micro-XANES determination of ferric iron and its application in thermobarometry. *Lithos*, **70**, 381–392 (2003).

37. Salbu, B., Janssens, K., Lind, O. C., Proost, K. and Danesi, P. R. Oxidation states of uranium in DU particles from Kosovo. *J. Environ. Radioact.*, **64**, 167–173 (2003).

38. Janssens, K. H., Proost, K., De Raedt, I., Bulska, E., Wagner, B. and Schreiner, M. The use of focussed X-ray beams for non-destructive characterization of historical materials. In *Proceedings of the NATO Advanced Institute on Molecular Archaeology*, NATO Science Series, Vol. 117, Kluwer, Dordrecht, 2003, pp. 193–200.

39. Taguchi, T., Xiao, Q. F. and Harada, J. A new approach for in-laboratory XAFS equipment. *J. Synchr. Rad.*, **6**, 170–171 (1999).

40. Schields, P. J., Gibson, D. M., Gibson, W. M., Gao, N., Huang, H. P. and Ponomarev, I. Y. Overview of polycapillary X-ray optics. *Powder Diffraction*, **17**, 70–80 (2002).

41. Gubarev, M., Ciszak, E., Ponomarev, I., Gibson, W. and Joy, M. A compact x-ray system for macromolecular crystallography. *Rev. Sci. Instrum.*, **71**, 3900–3904 (2000).

42. Gubarev, M. *J. Appl. Crystallogr.*, First result from a macromolecular crystallography system with a polycapillary collimating optic and a microfocus X-ray generator, **33**, 882 (2000).

43. Huang, H., Hofmann, F. A., MacDonald, C. A., Gibson, W. M., Carter, D. C., Ho, J. X., Ruble, J. R. and Ponomarev, I. *Proc. SPIE.*, **4144**, 100–109 (2000).

44. Hofmann, F. A., Gibson, W. M., Lee, S. M. and MacDonald, C. A. Polycapillary X-ray optics for thin film strain and texture analysis. In *Thin Film Stresses and Mechanical Properties VII*, R. C. Cammarata, M. A. Nastasi, E. P. Busso, W. C. Oliver, eds, Materials Research Society Proceedings, vol. **505**, pp. 3–14, 1998.

45. Turcu, I. C. E., Forber, R., Grygier, R., Rieger, H., Powers, M., Campeau, S., French, G., Foster, R., Mitchell, P., Gaeta, C., Cheng, Z., Burdett, J., Gibson, D., Lane, S., Barbee, T., Mrowka, S. and Maldonado, J. R. High power X-ray point source for next generation lithography. *SPIE Proc.*, **3767**, 21–32 (1999).

46. Sugiro, F. R. and MacDonald, C. A. Monochromatic imaging with a conventional source using polycapillary X-ray optics. *SPIE Proc.*, **4320**, 427–435 (2001).

47. Abreu, C. C. and MacDonald, C. A. Beam collimation, focusing, filtering and imaging with polycapillary X-ray and neutron optics. *Phys. Med.*, **XIII**, 79–89 (1997).

48. MacDonald, C. A., Gibson, W. M. and Peppler, W. W. X-Ray optics for better diagnostic imaging: In *Technology in Cancer Research and Treatment*, **1**, 111–123 (2002).

3.4 Parabolic Compound Refractive X-ray Lenses

A. SIMIONOVICI
ESRF, Grenoble, France

C. SCHROER, and B. LENGELER
Aachen University, Aachen, Germany

3.4.1 INTRODUCTION

In his early experiments, W. C. Röntgen could not find any noticeable refraction of X-rays by various materials nor was he able to focus X-rays with a rubber lens. He concluded that is was not possible to concentrate X-rays by refraction.[1] Today, it is well known that refraction of hard X-rays in matter – although not zero – is weaker than that of visible light by several orders of magnitude. While this is a great advantage for radiography, where an X-ray image can be interpreted as a straight projection through a non-refracting body, it led to the belief that refractive lenses for hard X-rays could not be made.[2] Therefore, a great variety of X-ray optical elements relying on other physical effects were developed, such as mirrors[3–5] and capillaries[6,7] based on external total reflection, multilayer mirrors,[8,9] Fresnel zone plates,[10] Bragg–Fresnel optics,[11,12] and bent crystal optics[13–15] based on diffraction. While all these X-ray optical elements can be used to produce a focused X-ray beam, only some of them have been shown to work as imaging devices.

While refractive optics are most common for visible light, the interplay between weak refraction and strong absorption is what makes the design of refractive lenses for hard X-rays difficult. In 1996 it was first demonstrated that refractive lenses could be fabricated despite these difficulties.[16] They were made by drilling a linear array of small holes into a lens material, such as aluminium or beryllium, focusing the X-rays in one direction. By crossing two such lenses or by crossing consecutive holes in one lens, focusing in two directions is possible.[16–18] The spherical aberration introduced by their shape prohibits their use as high quality imaging optics. Today, they are commonly used for beam conditioning at various beamlines at the European Synchrotron Radiation Facility (ESRF) in Grenoble, France.

Since then, several refractive lens designs have been developed by different groups.[16–26] In this subchapter we will focus on parabolic refractive X-ray lenses that have been designed and developed at Aachen University in collaboration with the ESRF in Grenoble.[19,20] They are high-quality imaging optics for hard X-rays, with applications in imaging and microanalysis and are routinely used at beamline ID22, ID18F and several other beamlines of the ESRF. They are used in a new hard X-ray microscope[19] that allows sub-micrometer resolution. More transparent lens materials, such as beryllium, boron, or carbon, are expected to push the resolution limit well below 100 nm.

To obtain an intense hard X-ray microprobe, a synchrotron radiation source can be imaged onto the sample by a refractive lens in the strongly demagnifying setup. This allows performing hard X-ray analytical techniques, such as fluorescence spectroscopy, diffraction, small angle scattering, and absorption spectroscopy, with lateral resolutions in the micrometer and sub-micrometer range.[27] A combination of fluorescence spectroscopy and tomography allows the determination of the spatial distribution of elements inside a sample.[28-35]

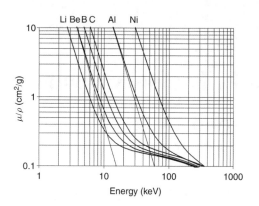

Figure 3.4.1 Mass absorption coefficient μ/ρ for Li, Be, B, C, Al, and Ni. Compton scattering dominates the attenuation below $0.2\,\text{cm}^2/\text{g}$. (Reproduced from ref. 19)

3.4.2 PHYSICS OF REFRACTIVE X-RAY LENSES

3.4.2.1 REFRACTION AND ABSORPTION IN MATTER

The refractive index for hard X-rays in matter is typically expressed in the form $n = 1 - \delta + i\beta$, where δ is the refractive index decrement that describes the refraction. For a given atomic species it is

$$\delta = \frac{N_A}{2\pi} r_0 \lambda^2 \rho \frac{Z + f'(E)}{A} \quad (3.4.1)$$

where N_A is Avogadro's constant, r_0 is the classical electron radius, λ and E the wavelength and energy of the X-rays, respectively, ρ is the mass density, $Z + f'(E)$ the real part of the atomic form factor in forward direction, and A is the atomic mass. Away from absorption edges, $f'(E)$ is small and δ is proportional to E^{-2} and ρ. Since Z/A is almost constant for all elements, δ/ρ varies very little as a function of the atomic species away from absorption edges.

Absorption inside the material is described by β that is related to the linear attenuation coefficient μ by

$$\beta = \frac{\mu\lambda}{4\pi} \quad (3.4.2)$$

β includes photoabsorption as well as the attenuation of the incident beam by inelastic (Compton) scattering. The dependence of μ/ρ as a function of the X-ray energy is shown in Figure 3.4.1. While at low energies, photoabsorption dominates μ/ρ, Compton scattering becomes the most important contribution to μ/ρ at higher energies (below $0.2\,\text{cm}^2/\text{g}$).

3.4.2.2 LENS DESIGN

Since the refractive index for any lens material is smaller than one ($n < 1$), a focusing lens needs to have a concave shape (Figure 3.4.2a). For hard X-ray energies around 20 keV, refraction is about six orders of magnitude weaker (e.g. for aluminium ($\delta = 1.35 \cdot 10^{-6}$) than that for visible light in glass ($n \approx 1.5$). To obtain a focal distance of about 1 m for visible light, a biconvex glass lens needs to have a radius of curvature R of about 1 m. To obtain the same focal distance with an aluminium lens for X-rays (20 keV), a radius of curvature of less than $R = 3\,\mu\text{m}$ would be required for a single biconcave lens. It is difficult to build a lens with such a radius of curvature, in particular with a sufficiently large aperture R_0. To circumvent this difficulty, we have chosen a radius of curvature that is large enough to be manufacturable and compensate the resulting lack of refraction of a single lens by stacking N lenses behind each other as shown in Figure 3.4.2(b). This strategy has been followed by most groups that have built refractive lenses.

In the X-ray range, there is no lens material that is as transparent as glass for visible light. Attenuation μ inside a lens is always significant,

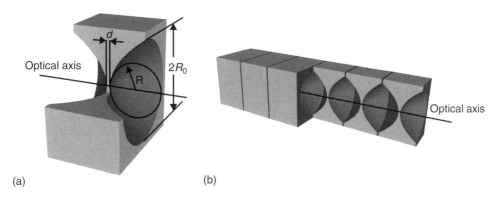

Figure 3.4.2 Design of the parabolic refractive X-ray lenses. A large number of single lenses (a) are stacked behind each other to form a refractive lens (b). (Reproduced from ref. 32)

requiring a careful choice of the lens material. As μ/ρ increases strongly with increasing atomic number Z ($\mu/\rho \propto Z^3$ for X-ray energies away from absorption edges), atomic species with low Z are good lens materials, such as Li, Be, B, C, and compounds thereof. Other important material properties are stability in the X-ray beam, low small angle scattering, and the machinability.[18,19] Although the attenuation inside aluminium is significantly higher than that for the other lens materials, it is particularly easy to machine and has therefore been very useful for the development of the lenses. As manufacturing techniques for more transparent lens materials, such as beryllium, have become available, the importance of aluminium as a lens material will decrease. However, all experimental results in this subchapter have been obtained with aluminium lenses. The energy range of operation for aluminium lenses lies between about 10 keV and 120 keV. For more transparent lens materials, such as beryllium, the energy range has been extended toward lower energies down to about 5 keV. Beryllium parabolic refractive lenses of high quality have been fabricated recently.[36]

As the radius of curvature R has to be as small as possible to limit the number of single lenses, the spherical lens approximation that is successfully applied in classical optics does not apply to most X-ray lens designs. It requires the aperture R_0 of the lens to be much smaller than the radius of curvature R. For most lens designs this would require the lens aperture to be prohibitively small. Unless the radius of curvature is made excessively large requiring a large number of single lenses, spherical aberration is very pronounced in the X-ray regime (see ref. 20 for experimental results on effects of spherical aberration). By choosing a parabolic lens shape, spherical aberration can be avoided. In the design described here, the lens surfaces are two rotational paraboloids as shown in Figure 3.4.2(a). The surface may be described by

$$y = \frac{r^2}{2R} \qquad (3.4.3)$$

where r is the radial coordinate ($r < R_0$). For parabolic lenses the aperture R_0 can be chosen independently of the radius of curvature R. For our current lens design, $R = 200\,\mu m$ and $2R_0 = 1$ mm. To minimize the attenuation inside the lens, the distance d between the apices of the parabolas should be minimized (see Figure 3.4.2(a)). For the aluminium lenses described in this chapter, $d = 5\,\mu m$.

3.4.2.3 LENS PROPERTIES

The optical properties of the parabolic refractive X-ray lenses are discussed in detail in Lengeler et al.[20] Here, we review the most important properties, such as the focal length f, the transmission T_p, and the effective aperture D_{eff}.

Using the lens maker formula, the focal distance of a single lens is $f_s = R/2\delta$. Assuming the stack of lenses to be thin compared to the focal distance (thin lens approximation), the focal length f_0 is given by

$$f_0 = \frac{R}{2N\delta} \qquad (3.4.4)$$

where N is the number of single lenses in the stack. The thin lens approximation no longer holds for lenses whose focal distance is short compared to their overall thickness. In that case, the focal distance is given by[27]

$$f = f_0 \frac{1}{1 - \frac{1}{6}\frac{l}{f_0}} \qquad (3.4.5)$$

where l is the overall length of the lens. The two principal planes of the thick lens are slightly shifted at its centre.

Since the attenuation in a refractive lens is always significant, the transmission T_p, i.e. the fraction of the photons (homogeneously) incident on the geometric aperture that is transmitted through the lens, is an important quantity, limiting the flux behind the lens. It is given by

$$T_p = \exp(-\mu N d) \frac{1}{2a_p}[1 - \exp(-2a_p)],$$

$$a_p = \frac{\mu N R_0^2}{2R} + \frac{N\delta^2 k_1^2 \sigma^2 R_0^2}{R^2} \qquad (3.4.6)$$

where k_1 is the wave number of the incident radiation, and σ the root mean square (rms) roughness of the lens surfaces.[20] The first term in a_p describes the attenuation inside the lens. The second term accounts for the surface roughness of the lenses. The aluminium lenses considered here have a roughness σ below 100 nm,[20,37] The second term in a_p is negligible compared to the first one for aluminium lenses, and we can neglect the roughness in the following.

As the attenuation increases towards the outer parts of the lens, the effective aperture D_{eff} describing the diffraction at the lens is smaller than the geometric aperture R_0. It is given by[20]

$$D_{eff} = 2R_0\sqrt{\frac{1}{a_p}[1 - \exp(-a_p)]} \qquad (3.4.7)$$

A sufficiently large geometric aperture R_0 loses its influence on D_{eff}. In that case, D_{eff} is governed only by attenuation inside the lens material:

$$D_{eff} = 2\sqrt{\frac{2R}{\mu N}} \qquad (3.4.8)$$

For a fixed focal distance f_0, the effective aperture

$$D_{eff} = 2\sqrt{4\frac{\delta}{\mu}f_0} = 2\sqrt{C\left(\frac{\mu}{\rho}\right)^{-1}f_0}$$

$$C = 4\frac{N_a}{2\pi}r_0\lambda^2\frac{Z}{A} \qquad (3.4.9)$$

depends on the lens material only through μ/ρ (C is approximately independent of the lens material). This emphasizes the fact that the major figure of merit for the lens material is μ/ρ shown in Figure 3.4.1.

3.4.2.4 IMAGING USING PARABOLIC REFRACTIVE X-RAY LENSES

A parabolic refractive X-ray lens can be used in analogy to a glass lens for visible light and has a wide range of applications. Two important applications are the microbeam production for X-ray microanalysis and the magnified imaging of a sample in a new hard X-ray microscope. For both applications, the aberration-free image transfer is crucial for the success of the method. The image quality strongly influences the microbeam characteristics, such as the lateral beam size, the depth of focus, the gain of the flux in the microbeam, and the absence of a low intensity background. For microscopy, the magnification, lateral resolution, the absence of distortion, and depth of field all depend on the quality of the optics.

As in classical optics, the parabolic refractive X-ray lenses can be used to image an object that is located a distance L_1 before the lens into an image plane at a distance $L_2 = L_1 f/(L_1 - f)$ behind the lens. The image is magnified by a factor $m = L_2/L_1 = f/(L_1 - f)$. The parabolic shape is crucial to obtaining a high quality image. Magnified imaging using refractive lenses can improve the resolution in hard X-ray microscopy to well below the sub-micrometer level. So far, using aluminium lenses, a resolution of 350 nm has been demonstrated.[19] Using more transparent lens materials, such as beryllium, the effective aperture can be significantly increased. This has increased the resolution to well below 100 nm, and a resolution of 50 nm is expected in the near future. Taking advantage of the large penetration depth of hard X-rays, magnified imaging can be combined with tomographic techniques.[38–42] This allows the non-destructive reconstruction of the three-dimensional structure of a sample with highest resolution. The possibility to demagnify an X-ray lithography mask may allow the transfer of finer lateral structures into thick resists, e.g. for MEMS (Microelectromechanical Systems) applications.[41]

3.4.2.5 MICROBEAM PRODUCTION

To obtain a small intense microbeam for microanalysis, the X-ray source is imaged onto the sample position in a strongly demagnifying geometry. This requires the distance L_1 from the source to be large compared to the focal distance f. The microbeam is then formed at a distance $L_2 = L_1 f/(L_1 - f)$ slightly larger than f behind the lens. The lateral beam size is determined by the source size, the geometric demagnification m of the source and diffraction effects at the aperture of the lens. To obtain an intense and small microbeam, the source should be as small as possible, emitting a high intensity into the solid angle that is spanned by the effective aperture of the lens at the distance L_1. This requires highly brilliant X-ray sources, such as undulator sources at third generation storage rings. For a source with a Gaussian intensity profile, such as an ESRF undulator source, the lateral beam size (full width at half-maximum (FWHM)) is[20]

$$B_{v,h} = 2\sqrt{2\ln 2} L_2 \sqrt{\frac{\sigma_{v,h}^2}{L_1^2} + \frac{a}{2k_1^2 R^2}}$$

$$a = \mu N R + 2N k_1^2 \delta^2 \sigma^2 \qquad (3.4.10)$$

where $\sigma_{v,h}$ are the rms electron beam source size in vertical and horizontal direction, respectively. Note that σ in a is the rms roughness of the lens surfaces. For an ESRF U42 high β undulator source, the source is well described by $\sigma_v = 13 \mu m$ and $\sigma_h = 300 \mu m$. The first term under the square root in Equation (3.4.10) describes the geometric demagnification of the source while the second term is the broadening of the beam due to diffraction at the lens aperture and the roughness of the lens surfaces. At distances of 40 m to 60 m from an ESRF undulator source and with a focal distance in the metre range, beam sizes of several micrometres horizontally and several hundred nanometres vertically are achieved. While diffraction at the lens aperture becomes relevant for the vertical direction, it does not significantly contribute to the relatively large horizontal beam size.

As the focal distance f is typically three to four orders of magnitude larger than the effective aperture D_{eff}, the depth of focus

$$d_{long} = 4 \frac{\min(B_v, B_h)/(2\sqrt{2\ln 2}) \cdot L_2}{D_{eff}} \qquad (3.4.11)$$

is larger than the lateral beam size by the same factor. This allows scanning samples with a thickness in the millimetre range with a beam of constant lateral extension ('pencil beam'). This is particularly important for scanning microtomography techniques such as fluorescence microtomography.[31–35] For scattering experiments it is also important, that the wave vector k of the incident radiation is sufficiently well defined, i.e. the beam divergence $\Delta k/k \approx D_{eff}/L_2$ is small enough. In most scattering experiments, a $\Delta k/k$ in the range from 10^{-3} to 10^{-4} is sufficient.

The gain in intensity in the microbeam compared to the intensity behind a pinhole of equal

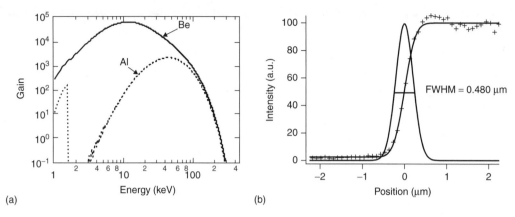

Figure 3.4.3 (a) Gain of a microbeam setup using an aluminium and beryllium lens ($f = 500$ mm, $L_1 = 60$ m, rms source size $13 \times 300 \, \mu m^2$). (b) Vertical profile through a microbeam measured by a fluorescence knife edge technique. An error function is fitted to the measured data (crosses). Its derivative gives the vertical profile of the microbeam. (Reproduced from ref. 26)

size is[20]

$$g_p = T_p \frac{4R_0^2}{B_v B_h}, \quad (3.4.12)$$

It depends on the transmission T_p and on the lateral beam size that is determined by the imaging geometry. Figure 3.4.3(a) shows the gain as a function of photon energy for aluminium and beryllium lenses in a fixed geometry.

For illustration purposes, the properties of a microbeam (energy 18.2 keV) produced by aluminium lenses ($N = 220$) at the ID22 high β undulator beamline of the ESRF are discussed. For a source to lens distance $L_1 = 41.7$ m, the minimal microbeam size was found $L_2 = 335$ mm behind the lens. From this, a focal distance of $f = 333$ m can be extracted. Using the radius of curvature $R = 209 \, \mu m \pm 5 \, \mu m$ measured by profilometry and the thin lens approximation,[4] a focal length of $f_0 = 291$ mm $\pm$ 7 mm would be expected. However, the lens with 220 individual lenses has a length of $l = 220$ mm that is comparable to f_0. The thin lens approximation does not hold in this situation and the focal distance must be calculated according to Equation 3.4.5 that yields $f = 332.6$ mm $\pm$ 7 mm in excellent agreement with the measured value. Here, the thick lens character adds about 14% to the focal distance.

The horizontal and vertical microbeam sizes were measured, using a fluorescence knife-edge technique. As knife-edge a gold strip (thickness 100 nm, width 5 mm, length several centimetres) deposited on a silicon wafer (thickness 550 μm) was used. The gold $L\alpha$ radiation ($E = 9.71$ keV) was measured by an energy dispersive detector (Si(Li)) as the gold edge was scanned through the beam. Figure 3.4.3(b) shows the scan through the microbeam in the vertical direction. A vertical beam width of 480 nm was measured. Including diffraction and roughness, a beam size of 450 nm is expected. The horizontal microbeam size is one order of magnitude larger due to the larger source size. It was measured to be 5.17 μm (FWHM) as compared to the theoretical value of 5.7 μm (FWHM). The measured transmission $T_p = 0.114\%$ yields an average lens thickness of $d = 5 \, \mu$m on the optical axis. The measured gain is 367 as compared to the theoretical value of 340 that is slightly smaller due to the discrepancy in the horizontal beam size.

The background of the microbeam contributes to the signal obtained in microprobe applications. In principle, a deconvolution with the point-spread function is necessary to remove the contribution of the background. Although the intensity of the background radiation is generally small, it may contribute significantly to the signal, since the area over which it is integrated is several orders of magnitude larger than the lateral area of the microbeam. The background is therefore an important characteristic of the microbeam.

As the dimensions of the knife-edge are large compared to the beam size, the knife-edge technique measures the integral flux over the half-plane covered by the gold knife. This allows the integral flux outside the microbeam to be measured and to compare it to that in the beam spot.[27] The knife-edge scan shown in Figure 3.4.3(b) can be evaluated in view of the background. The integral flux that falls outside a vertical interval $[-2\,\mu m, 2\,\mu m]$ around the spot is about 3.4%, the flux falling outside the interval $[-1\,\mu m, 1\,\mu m]$ is 4.6% and that falling outside $[-0.5\,\mu m, 0.5\,\mu m]$ is 10.4%. Therefore, the monochromatic microbeam produced by the refractive lens has a low background and is well suited for clean microprobe experiments. No additional pinholes are needed in front of the sample, which is an exceptional advantage over focusing devices (such as the FZP lenses) which feature a zero order transmitted beam and thus require an OSA (Order Sorting Aperture).

Using beryllium as the lens material, a gain in flux of one order of magnitude was obtained in the microbeam.[36] Recently we developed nanofocusing lenses (NFL) with focal lengths of around 10 mm at X-ray energies in the range of 10–100 keV.[43] These lenses are particularly useful for creating small microbeams at a short distance from the source, e.g. at 25 keV, a lateral beam size of 330 nm by 110 mm (FWHM) was achieved at a distance of 42 m from the source at ID22.

3.4.3 APPLICATIONS

Following the presentation of the characteristics of refractive lenses, we would like to present the results obtained at the ID22 beamline of the ESRF. ID22[41] is a beamline dedicated to microspectroscopy, micro-imaging and microdiffraction in the range 6–70 keV. Although the CRL applications span all three probes employed at ID22, we will present here only the microspectroscopy applications, obtained using X-ray fluorescence (XRF) spectroscopy. The microprobe setup of ID22 is used for performing fluorescence spectroscopy, which is the probe used in collecting data for three different techniques: direct XRF mapping, X-ray fluorescence computed tomography (XFCMT) and X-ray absorption spectroscopy (XAS).

3.4.3.1 EXPERIMENTAL SETUP

The ID22 beamline is optimized for performing micro-X-ray fluorescence (μ-XRF) and XAS with micron resolution at high energy. The basic concept in the design of this beamline was the simplicity of the optics guaranteeing the high flux/brilliance necessary for achieving a sensitive X-ray fluorescence microprobe setup, while conserving the high degree of coherence required for phase-sensitive imaging. The required optical quality of the beam was achieved by using highly polished and optically flat materials in the beam path.

The beam from the high β U42 undulator passes through polished Be and diamond windows to impinge onto a highly flat horizontally deflecting mirror with $\leq 1.5\,\mu$rad slope error and 1.5 Å rms micro-roughness, thus suppressing higher energy harmonics and lowering the transmitted heat load. A double crystal fixed-exit monochromator is used with Si crystals in either the 111 or 113 orientations. The setup includes several normalization detectors (photodiodes, ionization chambers) and a few focusing devices such as Fresnel zone plates (FZP), CRL lenses or bent mirror Kirkpatrick–Baez assemblies. Hereafter, the CRL lenses are exclusively used for focusing the beam.

The sample environment comprises a pinhole assembly serving to define the horizontal size of the beam. A high precision sample stage with seven degrees of freedom is used for positioning the sample and rotating the sample axis perpendicular to the beam. A few solid state detectors (SSD) such as Si(Li), HpGe and Si drift diodes can be included in the setup either individually horizontally/vertically or in pairs, on either side of the sample in the horizontal plane. An X-ray intensified charge-coupled device (CCD) camera ($\geq 1\,\mu$m resolution) is used in transmission behind the sample for direct imaging and/or alignment purposes. A typical setup with the drift diode positioned vertically and exclusively used for XFCMT is shown in Figure 3.4.4. As the U42 undulator is in a high ß section of the

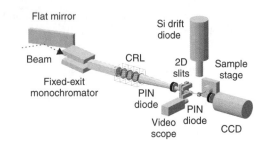

Figure 3.4.4 Experimental setup of the microprobe dedicated to fluorescence spectroscopy on beamline ID22 at the ESRF. (Reproduced from ref. 32)

storage ring lattice, the source size is $900 \times 30\,\mu m^2$ FWHM horizontal × vertical, which puts great stress on the focusing optics. At 15 keV, for a 1 m focal distance and 45 lenses, this yields beamspots of roughly $18 \times 0.75\,\mu m^2$ (H × V). However, storage ring electron beam average motions convoluted with the mechanical instabilities of the double crystal monochromator and mirror and diffraction at the lens combine into an *effective* spot size of $18 \times 1.3\,\mu m^2$ as measured by knife-edge fluorescence of a thin Au-plated Si wafer.

Depending on the energy and number of lenses, one can focus down to $0.5\,\mu m$ in vertical size and use a $10\,\mu m$ diameter pinhole to define the horizontal size. Alternatively, closing down the horizontal slits located at 13 m upstream from the lens to define a secondary source of $30\,\mu m$, one can focus down to about $2.5\,\mu m$ in horizontal size. The monochromatic flux obtained in these conditions varies from 10^9 to 10^{10} ph/s in the beamspot, depending on the energy in the range 14–25 keV.

A very interesting capability of ID22 is the PINK beam mode of operation, whereby one uses the full undulator beam, after a mirror reflection which serves as a low pass filter, and transmission through a selected attenuator foil, which serves as a high pass filter. This combination is directly used, without monochromatization, as incident beam on the CRLs, yielding a medium resolution ($\Delta E/E \approx 10^{-2}$) highly focused, high flux beam with intensities as high as 10^{12} photons/s. Recently, ID22 was extended by adding an *in vacuum* high flux, high energy U23 undulator, which covers the energy range 7–100 keV.

3.4.3.2 MICRO XRF: HIGH FLUX PINK BEAM SPECTROSCOPY OF SINGLE CELLS

In this work,[45] we aimed to achieve SXRF microanalysis of single cells by imaging the intracellular distribution of trace elements and pharmacological doses of the anticancer drug, 4′-iodo-4′-deoxydoxorubicin (IDX). Spatial distribution and concentration of trace elements in tissues are important, as they are involved in some pathological conditions and in many biological functions of living organisms like metabolism and nutrition. Llabador and Moretto[46] have reviewed the importance of microprobes in biology and emphasized their future use in cell physiology, pharmacology, and toxicology. CRLs used in a 'PINK' (polychromatic) excitation provide a fast acquisition rate and sub-ppm limits of detection. Additionally, the CRLs are currently the only focusing devices capable of sustaining the high flux, high heatload of a PINK beam without noticeable damage.

A human ovarian adenocarcinoma (IGROV1) cell line prepared using a previously published protocol[47] was used for this study. The cells, grown directly on $0.2\,\mu m$ Formvar film were incubated with complete culture medium and exponentially growing cells were exposed to $5\,\mu M$ IDX. Cell monolayers were rinsed, then cryofixed into liquid nitrogen chilled isopentane and freeze-dried at $-30°C$. Analyses were performed at 14 keV using either PINK or monochromatic excitation. PINK excitation is produced by direct, high intensity, medium bandwidth beams from the undulator impinging onto a flat mirror. The multi-strip mirror spans several full undulator harmonics and decreases the beam heatload by a horizontal deflection at a grazing angle of 2.6 mrad which gives an energy cutoff of 24 keV for the Pd strip. In order to remove the contribution of lower energy harmonics, a 2 mm Al filter was used. In the PINK beam, CRLs produce a flux of about 5×10^{10} photons/s/μm^2 and a beam spot of $10\,\mu m$ horizontally by $1\,\mu m$ vertically for 50 Al lenses, focal distance 713 mm. For the same spot size, CRLs with monochromatic excitation give a flux about

10 times less. The minimal detection limits (MDL) evaluated using a NIST standard reference material (SRM 1833) yielded about 50 ppb for elements such as Zn. Data analysis was performed using the WinAxil software[48] in order to correct for X-ray photon background and fit elemental X-ray lines detected in the sample.

In the case of a cell monolayer, the thin sample approximation can be applied. Previous Rutherford backscattering spectrometry (RBS) analyses[49] have determined a mean surface mass for freeze-dried cells around $260\,\mu\text{g/cm}^2$. With polychromatic PINK beam a close look at the undulator spectrum reveals a contribution of photons of higher energy harmonics at 16.9, 19.7, and 22.4 keV, which are greatly reduced by the lens and $10\,\mu\text{m}$ pinhole assembly installed before the sample. Taking into consideration only maxima of each contributing harmonic, the k ratio of the number of photons relative to the 14 keV harmonic is obtained from the undulator spectrum calculated using the ESRF Synchrotron Radiation Workshop (SRW) code.[50] Then, the effective ratio through the pinhole is calculated according to CRL formulas that give the intensity distribution in the plane of the sample. Only the 14 keV harmonic is effectively focused inside the $10\,\mu\text{m}$ pinhole; the other three higher energy harmonics before the mirror cutoff at 24 keV spread out over a large area and give a reduced contribution through the pinhole. The PIN diode placed before the sample and used for normalization, generates a total current I_0 for a given flux of N_0 photons/s of energy E_0 as:

$$I_0 \propto N_0(1 - e^{-\mu_0 d}) \qquad (3.4.13)$$

with μ_0 the energy deposition coefficient for Si, and d the Si PIN-diode thickness. For a given PIN diode, the current will always be a function of the incident energy of the photons through $\mu(E)$, as d is constant. So the measured current is in fact:

$$I = I_0(1 + k_1 f_1 + k_2 f_2 + k_3 f_3) \qquad (3.4.14)$$

with k ratios previously calculated and f_i obtained according to PIN-diode calculations using the energy deposition coefficient for each contributing harmonics of energy E_i. Since I measured in pA is known, the number of 14 keV photons N_0 is estimated, then those of higher and lower energy harmonics. From these values and those of fluorescence cross-sections $\sigma_F(E_i, Z)$, corrected quantitative elemental maps were generated. The following equation was used for quantization, based on the thin sample approximation:

$$N_Z = \frac{N_0 t C_Z \sigma_F \exp(-\mu_{\text{air}}(E_F)\rho_{\text{air}} d)\varepsilon_{\text{det}}\Omega}{4\pi \sin\alpha}$$
$$\times \int_0^T \exp\left[-x\rho_S\left(\frac{\mu_s(E_0)}{\sin\alpha} + \frac{\mu_s(E_F)}{\sin\beta}\right)\right]dx \qquad (3.4.15)$$

where N_Z is the number of counts in the characteristic line of element Z, N_0 is the incident photons/s, t is the integration time, C_Z is the concentration of element Z, σ_F is the fluorescence cross-section in cm^2/g, μ is the mass absorption coefficients of air or sample for the respective energies, ρ is the density (g/cm^3) of air or sample, ε_{det} the detector efficiency, Ω the detection solid angle, α and β the incident and take-off angles, x the sample depth coordinate, integrated from 0 to T (sample thickness). Spectra taken from single-cells treated with $5\,\mu\text{M}$ of IDX are shown in Figure 3.4.5. The spectrum taken using monochromatic beam with 120 s acquisition time (Figure 3.4.5a) is still of poor quality while the one from PINK-beam (Figure 3.4.5b) 100 s counting time shows well-defined X-ray peaks of intracellular elements P, S, Cl, Ca, K, Mn, Fe, Cu, Zn, and I from drug treatment.

Potassium is the major element in cells and gives elemental maps with the highest counting statistics. Compared to light microscopy cell visualization, potassium maps depict the cell boundaries roughly, particularly in the nuclear region. Iodine imaging of cells treated with $5\,\mu\text{M}$ of IDX could be performed in this study and yielded intracellular distributions of trace elements comparable to previous results obtained by micro-PIXE (proton induced X-ray emission) for higher doses of IDX of about $20\,\mu\text{M}$.[51] Particularly, the co-localization of iron and iodine within the cell nucleus is still observed. The results obtained on the iodine distribution, in comparison with potassium and iron are presented in Figure 3.4.6. This was also found

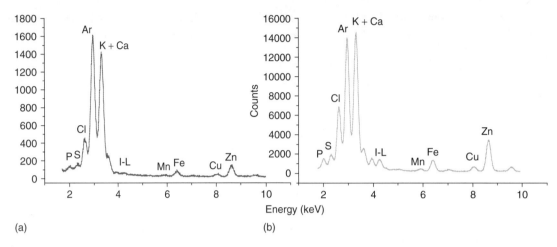

Figure 3.4.5 Spectra obtained from a IGROV1 cell treated with 5 μM of iododeoxydoxorubicin and freeze-dried. (a) Spectrum obtained using a 14 keV monochromatic focused beam, acquisition time 120 s. (b) Spectrum obtained using a focused 14 keV polychromatic PINK excitation, acquisition time 100 s. (Reproduced from ref. 45)

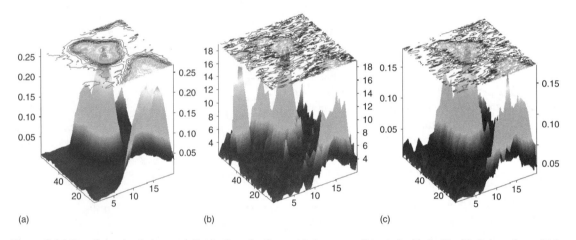

Figure 3.4.6 Two-dimensional elemental distribution of a freeze-dried cancer cell treated with 5 μM of iododeoxydoxorubicin. Cell was mapped with a 14 keV polychromatic PINK excitation, stepsize $1 \times 3\,\mu m^2$ (V × H) and 2.5 s acquisition time/step, scan size: $60 \times 60\,\mu m^2$, around 2 h total acquisition time. (a) Potassium distribution (Kα X-ray line), (b) iron (Kα X-ray line), (c) iodine (L_β X-ray line). (Reproduced from ref. 45)

using monochromatic excitation (data not shown), but more than 12 h of mapping were necessary.

From quantitative analysis, the surface concentrations displayed a maximum of $0.02\,\mu g/cm^2$ for iron and $0.15\,\mu g/cm^2$ for iodine in the nuclear region. Using a mean value of freeze-dried cells surface mass of $260\,\mu g/cm^2$ obtained by RBS, maximum concentrations for cells treated with 5 μM IDX were 10340 ppm for potassium, 274 ppm for zinc, 76 ppm for iron, and 580 ppm for iodine. These results are in agreement with previously published data on IGROV1 cells trace element content.[47,52] Values concerning detector efficiency and the K/L partial photoionization cross-sections, fluorescence yields and branching ratios are determined within 10% accuracy.[53–55] When estimating error propagation the main contributing source of error on the calculated concentration C will be given by the standard deviation on the cell thickness T. The use of appropriate certified standards simulating such thin biological matrices should reduce uncertainty on quantitative results from 30–40 % estimation to 10–15 %. The radiation dose deposited within a pixel is roughly about 10^6 G with PINK excitation and can be down to 10^4 G using monochromatic excitation which is several orders of magnitude less than scanning ion microprobe, and still acceptable for freeze-dried samples. A compromise is required between the desired resolution, mapping time, and the dose that the sample can sustain. Improvements in limits of detection can be reached working above the K edge of iodine $B_K \approx 33.2$ keV with micron-sized beams of high energy. Alternatively, using the newly commissioned setup at ID22, the samples can be raster-scanned using a piezo YZ assembly, spreading out the heatload over several successive scans. This experiment opens a new way toward possibilities of mapping intracellular distribution of drugs used at pharmacological doses. SXRF is complementary to confocal microscopy or nuclear microprobe analysis, and brings missing information not accessible by any other technique. Finally, simultaneous chemical speciation by XAS and microanalysis of living cells are exciting perspectives under investigation that will soon be reached routinely with SXRF microprobes.

3.4.3.3 MICRO XFCMT

X-Ray fluorescence computed Microtomography (XFCMT) is a nondestructive, noninvasive imaging method which was introduced over 15 years ago[56] and started to play an increasing role in microanalysis.[28–35] XFCMT is an excellent complementary technique to phase contrast imaging in that it offers the much-needed elemental sensitivity down to trace element concentrations with the same micron-sized spatial resolution. In order to retrieve the quantitative two-dimensional/three-dimensional (2D/3D) elemental information at the end of the tomographic scan, reconstruction techniques are used as opposed to the direct imaging methods associated with 2D mapping. As it requires a pencil beam as its probe, synchrotron radiation fluorescence tomography expanded and became a precise and relatively straightforward method of microanalysis only with the advent of third generation synchrotron sources such as the ESRF, APS and SPRING 8 that provide beams with high energy/high flux/high coherence characteristics.

In the following, we describe high precision experiments performed at the ID22 beamline of the ESRF on real samples featuring inhomogeneous elemental distributions, from the fields of plant physiology and astrophysics.

Depending on the desired resolution, either of the vertical or horizontal scanning geometries are used, with the associated detectors positioned vertically, respectively horizontally at 90° with respect to the beam. The horizontal focusing/scanning geometry exhibits a significant decrease in flux necessary to demagnify the rather large horizontal source size by closing down the beamline horizontal slits but has a better spectral purity, as it features Rayleigh and Compton scattering a few tens of times less than the vertical geometry, thanks to the 90° angle between incident and outgoing photons in the orbit plane and the high degree of linear polarization in the horizontal plane.

The two movements used for the tomographic scans are Z/X (vertical/horizontal, precision 0.1 µm) and R_X/R_Z (rotation around a horizontal/vertical axis, precision 0.001°). The other movements (Y, R_Y, x) are used to align the sample rotation axis in the beam. The sample is mounted on a Huber goniometer head which is pre-aligned on a visible microscope setup in order to bring its rotation axis perpendicular to the beam and to reduce precession of the tomographic rotation axis at the beam position.

Long Range Ion Transport in Higher Plants

In order to study the impact of certain genes and their mutual interaction on the phenotype of higher plants, plant physiologists investigate the long range transport of ions such as K^+, Ca^{2+} or Cl^-.[57] The long range transport and uptake of heavy metals is an important issue in environmental studies.[58] The elements under investigation are highly diffusive in the complex matrix of the plant, and the sample preparation required for standard analytical methods, such as EDX or SIMS, requires the preparation of cryosections or fracture surfaces of the plant samples. While this kind of sample preparation is difficult to achieve for certain plants, it is nearly impossible for others. In addition, there is always the danger of introducing artefacts changing the element distribution. Hard X-ray fluorescence microtomography can avoid these difficulties. It allows one to determine the element distribution on a virtual section through a bulk of an opaque sample. The region of interest does not need to be laid open. A detailed description of the experimental method can be found elsewhere.[31-33]

To illustrate the strength of this method, the distribution of potassium, iron, and zinc is determined inside a mycorrhizal root of a tomato plant grown on heavy metal polluted soil. Prior to the experiment, the sample was shock frozen and freeze dried. These two steps of sample preparation were required to avoid diffusion of the elements of interest and to reduce the attenuation of the low energy fluorescence radiation, respectively. The tomographic scan was performed using a microbeam with a flux of 4.8×10^9 photons/s at an energy of about 19.8 keV (5th harmonic of the undulator source at beamline ID22 of the ESRF tuned to 19.8 keV, filtered using a molybdenum foil with a thickness of 250 μm and a Pd coated total reflection mirror with a reflection angle of 0.15°). The lateral beam extension was 1.9 μm horizontally and 0.9 μm vertically (FWHM). The microbeam was generated by imaging the undulator onto the sample using an aluminium refractive X-ray lens with $N = 220$ single lenses ($L_1 = 41.7$ m, $f = 385$ mm, $L_2 = 388$ mm). To reduce the effective source size horizontally, a horizontal pair of slits located 13 m upstream from the lens with a gap of 0.1 mm was introduced into the optical path. In order to reduce scattering, the sample was placed in a helium-filled chamber.

For each projection, the sample was scanned in 105 translational steps of 1 μm. At each point of the scan, the fluorescence radiation and the incident and transmitted beam intensity were recorded for 2 s. 132 projections were acquired at equidistant steps over a full rotation of the sample. The data for potassium normalized to the incident flux and the attenuation (negative logarithm of the quotient of the transmitted and incident flux) are shown in Figure. 3.4.7 as a function of translational position and rotation angle in a so-called sinogram.

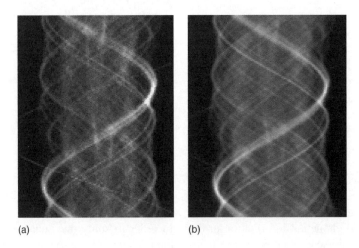

Figure 3.4.7 Tomographic data from a mycorrhizal root of the tomato plant. (a) K Kα fluorescence signal ($E = 3.31$ keV) and (b) attenuation at the energy of the incident radiation (19.8 keV). The fluorescence signal and the negative logarithm of the ratio of the transmitted and incident flux are plotted in shades of grey as a function of the translational position (horizontal axis) and rotation (vertical axis) in so-called sinograms

The attenuation signal in Figure. 3.4.7(b) can be reconstructed by standard tomographic techniques such as filtered back projection, since the underlying tomographic model is the standard radon transform.[59] The fluorescence signal (Figure 3.4.7a), however, is generated in a more complicated way. The incoming beam is attenuated along its path. At each point along the path, fluorescence is excited and radiated into the full solid angle. The part of the radiation that falls into the solid angle of the fluorescence detector is attenuated by an a priori unknown attenuation of the sample before it reaches the detector. If, as in this case, the sample has low density, secondary fluorescence and scattering effects can be neglected. The attenuation effects of the fluorescence become apparent in Figure 3.4.7(a), where the fluorescence signal is slightly weaker on the left side (far side of the detector) than on the right. This asymmetry in the sinogram can be used to estimate the attenuation inside the sample self-consistently.[34] The results of such a self-consistent reconstruction are shown in Figure 3.4.8 for potassium, iron, and zinc.

In Figure 3.4.8 the element distributions are reconstructed with subcellular resolution (<1 μm). The surface of the root is covered with soil that is rich in iron. The outermost cellular layer of the root has died off, yielding a low fluorescence signal. In the next cellular layer the mycorrhiza fungus accumulates zinc, stopping the zinc from being transported into the plant as seen by the low concentration of zinc in the central part of the root.

To determine the element distribution inside plant samples, fluorescence microtomography has great advantages over standard analytical techniques, in particular because of the minimal sample preparation required. In some cases where the samples are small and only elements with high energy fluorescence lines are of interest, the freeze drying step in the sample preparation can be omitted, and the samples can be investigated in the frozen hydrated state. This way, the only sample preparation required is shock freezing that is necessary to stop diffusion during the measurement. The attenuation inside the ice, however, is significant in particular for fluorescence energies such as those of potassium and calcium. In addition, the sample must be kept cold during the measurement.

Study of Micrometeorites in Sealed Containers

The XFCMT[60] was performed on a microfragment from the Tatahouine meteorite, which fell in 1931 and was promptly collected at the Natural History Museum in Paris. The grain analyzed had been recently retrieved from the original site and recovered in a weathered state from the first centimeters of soil, and was compared to the pristine ones retrieved in 1931 and stored at the Museum. A series of experiments in SEM and TEM were performed[61] on this meteorite, in order to study the remnants of pleomorphic bacteria present in fractures and fault lines of the meteorite.

For this measurement, a monochromatic beam of 14 keV was obtained using CRLs, associated with a 10 μm diameter pinhole to define a 2 × 10 μm² beamspot. The grain was placed in a sealed thin (≤10 μm thick) quartz capillary in

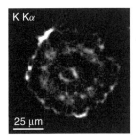
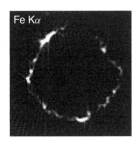
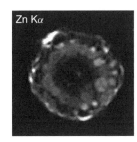
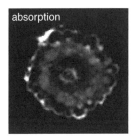

Figure 3.4.8 Virtual section across a mycorrhizal root of a tomato plant grown on heavy metal polluted soil. The distribution of potassium, iron, and zinc are shown together with the attenuation coefficient μ_0 at the energy of the incident microbeam (19.8 keV). (Reproduced from ref. 34)

order to mount it on the goniometer head. This way, the cleanliness and nondestructiveness of the measurement was guaranteed – all the more since we wanted to perform infrared (IR) spectroscopy on it afterwards. The tomography was done in the vertical scanning geometry, making use of the good spatial resolution available (2 μm). The stronger scattering which this geometry produced was evident in the energy spectra, where low-energy tails of the Rayleigh and Compton peaks at 14 and 13.6 keV, respectively produced a non-negligible background. The Axil[48] line fitting program was used to deconvolute individual elementary lines.

The biological study of this meteorite[61] was able to reveal bacteriomorphs with sizes between 0.1 and 0.6 μm, which were obtained by cultures of the soil surrounding the grains. These were observed in SEM and TEM and presumed to be real bacteria or their remnants. It was postulated that these bacteria appear at the fracture sites of the grain, following fluid circulation of carbonates from the soil due to terrestrial weathering. Therefore, our study aimed to identify and locate non-invasively carbonate phases as well as pyroxenes and chromites specific to the grain. The grain was analyzed through the quartz walls of the capillary which posed no problem for the imaging of any elements with the exception of Si which was the main constituent of the capillary walls. In Figure 3.4.9 the distribution of Fe, Cr and Ca particular to the phases is shown, with a resolution of 2 μm using the ART reconstruction algorithm.

The tomographic set of data collected complements the 3D transmission tomography, electron microscopy and IR data, allowing a rich image of the grain to be obtained and to characterize in detail its morphology and structure nondestructively, prior to the other analyses which require sample preparation and possibly alteration of the grain. This study is part of a benchmark[62] to establish the feasibility of such detailed analyses on Martian meteorites in the quarantine phase through the walls of a mini-P4 sample holder.

Reconstruction

In this work the samples were exclusively imaged using our modified version[32,63] of the algebraic reconstruction technique (ART) as described by Kak and Slaney.[59] This technique requires several simplifications and assumptions to reduce the complexity of the problem and the required calculation time. The sample is divided into a series of 'voxels' (volume pixels) of width and height equal to the beam size.

The following approximations are made throughout the calculation:

- The scattering from sample and surrounding air is neglected, as well as 'enhancement effects' due to secondary/ternary fluorescence or to Compton-excited fluorescence.
- The correction for the finite horizontal size (respectively vertical for horizontal scans) of the beam is not applied, i.e. the sample is considered

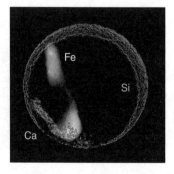 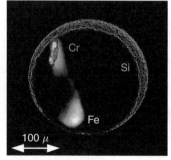

Figure 3.4.9 Fluorescence tomograms of a micrometeorite grain inside a quartz capillary. Reconstructions (using ART) of the Kα lines for Si (capillary), Fe, Cr and Ca are shown. Resolution ≈2 μm. (Reproduced from ref. 57)

to be a 2D slice of thickness equal to the width of the detector and equal to the horizontal size of the beam (respectively vertical for horizontal scans).
- The detector to sample distance is constant as well as the detector efficiency taken as 100%, regardless of the photon energy or angle of incidence.

We are continuously developing the ART method, since by its phenomenological character which closely mimics the physical interaction and by its open frame, it offers possibilities of adding the necessary procedures for direct quantitative interpretation of the reconstructed images including the full battery of appropriate corrections (self-absorption, enhancement, detector response function) seen in fluorescence spectroscopy.

Discussion

When planning an experiment, the question we have to answer is: How would the estimated resolution change for a given change in the scanning parameters, such as number of translations/rotations and counting time per step? Specifically, what is the smallest number of rotations which allow the realization of a resolution close to the scanning step. The resolution function R should be sought in the form:

$$R = R(N_t, N_\theta, \Gamma_t) \qquad (3.4.16)$$

with N_t/N_θ number of translations/rotations and Γ_t the beamsize FWHM in the direction of the translation. Rayleigh's criterion would impose a translation step of about 1.22 times the FWHM of the beam. However, non-negligible contrast is obtained for steps well below the beamsize but it is customary to use an oversampling of up to 30% of the beamsize in order to obtain the smallest achievable modulation.

In transmission tomography, based on reconstruction techniques derived from Fourier transforms, it has been proved[59] that the number of rotations N_θ necessary is such that:

$$N_\theta \cong \frac{\pi}{2} N_t \qquad (3.4.17)$$

That sampling follows the Fourier space requirements for obtaining an even resolution in the 2D (xy) space. However, when using algebraic reconstruction techniques, one deals only with the passage from the rotating referential attached to the sample (s, t) to the referential attached to the beam (xy) which is done by nearest-neighbour interpolation.

Perspectives

For the next few years, part of the activity of the ID22 group will be centered around the developments of XFCMT and its applications, both hardware and software. The hardware efforts will concentrate onto rendering the acquisition from both EDX and WDX spectrometers fast and user-friendly. We are implementing digital signal processing (DSP) systems to record the MCA spectra and lower the processing time in order to accept count rates of a few tens of kcps. Where necessary, a WDX spectrometer will be used to either record high-resolution spectra of otherwise-overlapping lines, or to eliminate saturation effects produced by the matrix or major elements. The overhead time of the acquisition system will allow counting times of 50–100 ms/pt thus opening the way to obtaining full tomographic scans in less than an hour. We are implementing a symmetric double EDX detector setup, in conjunction with a simultaneous WDX acquisition. This will effectively halve the scan duration to 180° instead of the 360° needed in regular tomographic scans and will allow tackling the 3D imaging by acquiring several adjacent slices, of thickness equal to the beamsize in the other dimension. Fully quantitative imaging requires well qualified and calibrated normalization monitors – photodiode detectors and ionization chambers in our case which we are currently implementing. We will also implement a polarization monitor, as fluorescence signals are strongly dependent on the incoming beam polarization which even at an undulator beam varies by a few percent as a function of the photon energy or the sample position with respect to the electron beam orbit.

Software-wise, we have already implemented the online spectrum fitting based on the AXIL package and we are also using Monte-Carlo simulations[64] in order to check the validity of our reconstruction and quantization.

The ART reconstruction package we use corrects for incident beam absorption and self-absorption effects from the matrix and is under development to include self-absorption effects due to the elements imaged. Besides the absorption signal which serves to generate a transmission tomogram, we are planning to exploit the Rayleigh and Compton scattering signals to obtain an 'effective' Z image useful in ascertaining the matrix. Finally, we will couple image processing capabilities such as Principal Component Analysis (PCA) and clustering in order to treat the correlation patterns in the reconstructed images.

Our aim is to deliver a fast and accurate method of 2D/3D characterization of the internal elemental structure of various samples, with quantitative capabilities, in a nondestructive, non-invasive way and without a priori knowledge of the sample constituents.

3.4.3.4 X-RAY ABSORPTION: CADMIUM SPECIATION IN MUNICIPAL SOLID WASTE FLY ASHES

This work[65] is part of a collaboration between the Chalmers University in Göteborg, Sweden and ID22, ESRF. Incineration of Municipal Solid Waste (MSW) has two main advantages: reducing the waste volume by about 90% and reducing reactivity by destruction of organic compounds. As in combustion of other fuels, the potentially toxic trace metals are concentrated in the fine ash fractions, i.e. the fly ash. The content of these metals (i.e. Pb, Ni, Cu, but mainly Cd) makes this residue ecologically harmful. An effective and safe handling of such ash requires a thorough knowledge of its chemical properties, in particular, their potential for dissolution and leaching. The dissolution and transport of metal ions from the ash matrix to soil water are key steps since dissolved ions are available for biological uptake and ground water contamination. As a consequence, knowledge of the total concentrations of heavy metals in ashes provides only limited information, as it does not show how strongly the metals are bound to ash constituent. Thus, the risk associated with the presence of heavy metals depends primarily on their speciation which is a difficult task because of their relatively high dilution and the structural and chemical complexity of the host material.

This work describes a method for the determination of Cd speciation and its possible quantitation in single MSW fly ash particles. Cd distribution within single particles was investigated by synchrotron radiation induced micro-X-ray fluorescence (μ-SRXRF) spectrometry and its speciation on single spots by synchrotron radiation induced micro-X-ray absorption spectroscopy (μ-SRXAS). Both techniques can be used in air and they are usually nondestructive, due to the lower energy deposition compared to charged particle excitation. The high brightness of the 3rd generation synchrotron radiation sources and the development of X-ray focusing optical elements make it possible to create beams of micrometer size with high intensity making μ-XRF and μ-XAS appropriate tools for the analysis of individual particles with diameters of several tens of micrometers. Elemental maps, showing the 2D projection of the 3D elemental distribution of a particle, are created by scanning the particle in a regular grid pattern by the X-ray microbeam and detecting the induced fluorescent intensities at each position (xy-μ-XRF scan). μ-XAS spectroscopy was used for the direct determination of the chemical forms of Cd in particular micro-sized areas of high Cd concentrations, by measuring the absorption of the excitation beam or the intensity of the Cd Kα X-ray line as a function of the excitation energy in the vicinity of the K absorption edge ($E_K = 26.711$ keV) of Cd ($-100 \text{ eV} < E - E_K < 400-1000 \text{ eV}$). Since the spectral scan is performed in the vicinity of an X-ray absorption edge of a chosen target element (Cd in our case), the method is therefore element-specific, which precludes interferences from compounds of other elements and means that little or no sample preparation is required. In combination with the penetrating capability of X-rays, it makes μ-XAS ideal for complex or difficult samples, like fly ashes.

The study has been performed on fly ashes from a bubbling fluidised bed (BFB) combustion unit of 2×15 MW fired with MSW. Only textile filter ashes were investigated here. Eleven single particles of different dimensions (varying from ca. 30 to 200 μm in diameter) were selected and each of them was glued on a 100 μm diameter quartz capillary before the μ-XRF analysis. μ-XAS experiments were performed only on the particles presenting high Cd concentrations (three particles) and only on those spots where Cd was predominantly accumulated (four spots totally), as shown by the XRF maps.

Both μ-SRXRF and μ-SRXAS measurements were performed in the 1st experimental hutch of the ID22 beamline (EH1) of the ESRF. For the demagnification of the synchrotron source and creating the microbeam, a compound refractive lens (CRL) consisting of 94 individual Al lenses was employed. A Au knife-edge sample was used to determine the size of the focused beam. The beam size was H × V = 10 × 8 μm^2 during fluorescence experiments and 12 × 3 μm^2 during the absorption experiments. The NIST-SRM1832 thin glass and Cu and Au thin metal foils (Goodfellow, UK) were measured in order to estimate the number of incident photons. All the μ-XRF experiments were performed by using monochromatic radiation at an excitation energy of 27 keV.

Cd K-edge μ-XAS experiments were performed on pure Cd compounds and fly ash particles as well. The reference compounds used in this study (Cd, CdCl$_2$, CdO, CdSO$_4$, CdS, CdBr$_2$) were chosen due to their probability to be found in the ash material. A small amount (0.5 mg) of each reference material was crushed and mixed with 0.3 mg of boron nitride and pressed to form pellets of 1 mm thickness. The reference XAS spectra were recorded in transmission mode. Due to the low Cd concentration and the small dimensions of the fly ash particles, their XAS spectra were recorded in fluorescence mode. Due to the relatively small thickness of the fly ash particles, no self-absorption correction was necessary for these spectra, as evidenced by the good agreement between the fluorescence and transmission XAS spectra integrated for longer. All the μ-XAS spectra were obtained by scanning the energy in the 26.65–26.95 keV range in 1 eV steps maintaining the same setup used for the microfluorescence experiments. Each energy scan was repeated four times with 2 s live time/energy point. The higher intensity Si [111] reflex was used, yielding a resolution of about 3.5 eV, to be compared with the Cd core-hole width of 7.8 eV.

The command view (CMDV) data analysis package by Ansell[66] was used to fit the data obtained on the fly ashes by linear combinations of the measured standards. Initially, factor analysis was performed in order to roughly estimate the dominant standards for further fitting the fly ashes. Then a discrete fit of the fly ashes by linear combinations of the interpolated standards was made, using the three most representative candidates evidenced by the factor analysis: CdSO$_4$, CdCl$_2$ and CdO.

Results and Discussion

The concentrations of the major and minor components in each measured voxel of the single fly ash particles were determined from scanning μ-XRF experiments as detailed in Camerani et al.[67] The large differences between the average and maximum concentrations within individual particles indicate a considerable variation in concentrations with 'hot-spots' containing about 10–100 times higher amount of a given element than the average. This finding supports the idea that the trace metals have special affinity to some ash minerals located near the particle surface. This can occur as a result of particle interactions,[68] or as a result of the variables which determine the transfer of elements to the raw gas in the furnace, such as: (i) occurrence and distribution of the elements in the input waste; (ii) physical and chemical conditions in the furnace bed, e.g. temperature, redox conditions, chlorine and oxygen content; and (iii) the kinetics in the furnace bed, e.g. residence time and mixing conditions.[69] The inhomogeneous distribution

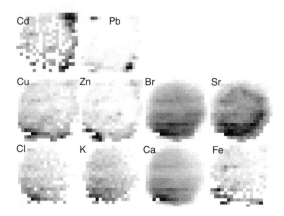

Figure 3.4.10 μ-XRF maps of various elements obtained from particle 6 (diameter ca. 160 μm), showing self-absorption of the lightest elements. Image size: 20 × 20 pixels, pixel size: 10 × 8 μm, spectrum collection time per pixel: 6 s. Darker tones indicate a higher elemental abundance; light tones indicate lower concentrations. (Reproduced from ref. 65)

of heavy metals among and within individual particles is illustrated, as an example, in Figure 3.4.10, where the spatial metal distributions of particle 6 (ca. 200 μm diameter) are shown.

Inside each of those particles, the enrichment of Cu, Zn, Pb and Cd (at the 3600, 29700, 11800 and 200 ppm level, respectively) in some well-defined hot-spots relative to the surrounding areas is clearly visible, suggesting that a part of these metals might be arranged in inorganic precipitates. Around the boundaries of those spots, zones with intermediate concentration levels are present, probably caused by diffusion of the heavy metals during the condensation process or by binding to ion exchange sites. The right-hand lower and upper quarters of particle 6 show, for example, a zone with a high abundance of Cd, possibly corresponding to the nucleation area.

Despite its low concentration, Cd could be detected by μ-SRXRF in each single particle. Due to the penetration depth of the high energetic primary and characteristic photons (few hundred micrometers range), the elemental signals originating from the whole excited depth of the particle are detected simultaneously. This makes the μ-XRF maps, shown in Figure 3.4.10, become 2D projections of the 3D distribution of these trace elements throughout the fly ash particles. So, from the μ-XRF maps alone, it is not possible to judge whether the spots of accumulation of the heavy metals are situated on the particle surface (i.e. most prone to leaching) or are present at some depth within the particle (where they might be more shielded from chemical attack by water).

To further understand the speciation of Cd in single fly ash particles, the oxidation state and the chemical surroundings of Cd ions were studied by μ-XAS in the areas of the particles showing a higher Cd concentration. μ-XAS spectra of pure Cd compounds such as Cd, CdO, $CdSO_4$, CdS, $CdCl_2$, $CdBr_2$ pellets were measured prior to the single particle analysis, normalized and reported in Figure 3.4.11(a). Comparisons of XAS spectra for fly ashes and for reference compounds show that in all the particles studied Cd is present in oxidation state +2 instead of metallic form. This can be explained by the oxidizing conditions experimented by the toxic metals after volatilization during combustion.

The μ-XAS spectra of the 'hot-spots' of the particles were also expressed mathematically as a Linear Combination (LC) of XAS fit vectors, using the measured absorption data of the Cd reference compounds. The μ-XAS spectra of one of the two 'hot-spots', within particle 6, is shown in Figure 3.4.11(b). The comparison between the linear combinations of the standard spectra and the measured XAS-spectra of the Cd hot-spots allow the concentrations to be estimated of the possible Cd compounds in those spots, e.g. in the case of the spot a in particle 6, Cd is present as an admixture of $CdSO_4$ (70%), CdO (19%) and $CdCl_2$ (11%) with over 90% confidence level in the fitting process. In all the spots analysed, Cd was always present as a combination of just $CdSO_4$, $CdCl_2$ and CdO. Thus, it can be seen that the Cd in MSW fly ashes is present as water soluble species ($CdCl_2$ and $CdSO_4$) up to 86% and 76%, respectively.

These results confirm what was found by μ-XRF mapping and agree with earlier studies.[70,71] Such metal speciation would make the management of this filter ash problematic with respect to their possible leaching. On the other hand this is not a conclusive identification because the spectra

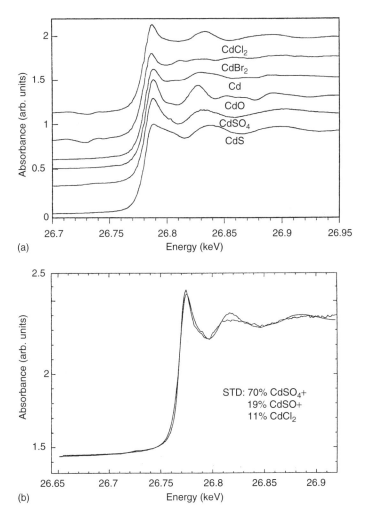

Figure 3.4.11 (a) XAS spectra of pure Cd compounds. The spectra are vertically offset to allow comparisons and were taken in fluorescence mode. (b) XAS fitted spectrum of a fly ash particle, using linear combinations of the previously measured standards. (Reproduced from ref. 65)

of other Cd sulfates and some other possibly significant compounds, like Cd silicates, were not examined due to the lack of standard materials. These investigations are planned during future work.

ACKNOWLEDGEMENTS

We would like to thank W. H. Schröder (Research Center Jülich, Germany) for the ongoing fruitful collaboration to develop X-ray fluorescence microtomography of plant samples.

The authors gratefully acknowledge support from J.M. Rigal, and A. Homs during the experiments.

The financial support of the Swedish Natural Science Research Council (NFR) for the work of M.C. Camerani is gratefully acknowledged.

REFERENCES

1. W. C. Röntgen. *Sitzungsber. physikal.-medizin. Gesellschaft*, **132**, 1895.
2. A. Michette. *Nature (London)* **353**, 510, 1991.

3. P. Kirkpatrick and A. Baez. *J. Opt. Soc. Am.* **38**, 766, 1948.
4. Y. Suzuki and F. Uchida. *Rev. Sci. Instrum.* **63**, 578, 1992.
5. O. Hignette, G. Rostaing, P. Cloetens, A. Rommeveaux, W. Ludwig and A. Freund. *Proc. SPIE* **4499**, 105, 2001.
6. D. Bilderback, S. A. Hoffman and D. Thiel. *Science* **263**, 201, 1994.
7. S. A. Hoffman, D. J. Thiel and D. H. Bilderback. *Nucl. Instrum. Methods A* **347**, 384, 1994.
8. J. Underwood, T. Barbee Jr and C. Frieber. *Appl. Opt.* **25**, 1730, 1986.
9. J. Underwood, A. Thompson, Y. Wu and R. Giauque. *Nucl. Instrum. Methods A* **266**, 296, 1988.
10. B. Lai, W. Yun, D. Legnini, Y. Xiao, J. Chrzas, P. Viccaro, V. White, S. Bajikar, D. Denton, F. Cerrina, E. Fabrizio, M. Gentili, L. Grella and M. Baciocchi. *Appl. Phys. Lett.* **61**, 1877, 1992.
11. V. V. Aristov, Y. A. Basov, G. N. Kulipanov, V. F. Pindyurin, A. A. Snigirev and A. S. Sokolov. *Nucl. Intrum. Methods A* **274**, 390, 1989.
12. P. Chevallier, P. Dhez, F. Legrand, M. Idir, G. Soullie, A. Mirone, A. Erko, A. Snigirev, I. Snigireva, A. Suvorov, A. Freund, P. Engström, J. A. Nielsen and A. Grübel. *Nucl. Instrum. Methods A* **354**, 584, 1995.
13. U. Lienert, C. Schulze, V. Honkimaki, T. Tschentscher, S. Garbe, O. Hignette, A. Horsewell, M. Lingham, H. F. Poulsen, N. B. Thomsen and E. Ziegler. *J. Synchrotron Rad.* **5**, 226, 1998.
14. G. Hölzer, O. Wehrhan and E. Förster. *Cryst. Res. Technol.* **33**, 555, 1998.
15. T. Missalla, I. Uschmann, E. Förster, G. Jenke and D. von der Linde. *Rev. Sci. Instrum.* **70**, 1288, 1999.
16. A. Snigirev, V. Kohn, I. Snigireva and B. Lengeler. *Nature (London)* **384**, 49, 1996.
17. P. Elleaume. *Nucl. Instrum. Methods A* **412**, 483, 1998.
18. B. Lengeler, J. Tümmler, A. Snigirev, I. Snigireva and C. Raven. *J. Appl. Phys.* **84**, 5855, 1998.
19. B. Lengeler, C. G. Schroer, M. Richwin, J. Tümmler, M. Drakopoulos, A. Snigirev and I. Snigireva. *Appl. Phys. Lett.* **74**, 3924, 1999.
20. B. Lengeler, C. Schroer, J. Tümmler, B. Benner, M. Richwin, A. Snigirev, I. Snigireva and M. Drakopoulos. *J. Synchrotron Rad.* **6**, 1153, 1999.
21. Y. Kohmura, M. Awaji, Y. Suzuki, T. Ishikawa, Y. I. Dudchik, N. N. Kolchevsky and F. F. Komarov. *Rev. Sci. Instrum.* **70**, 4161, 1999.
22. B. Cederström, R. N. Cahn, M. Danielsson, M. Lundqvist and D. R. Nygren. *Nature* **404**, 951, 2000.
23. B. Cederström, M. Danielsson and M. Lundqvist. *Proc. SPIE* **4145** 294, 2001.
24. Y. I. Dudchik, N. N. Kolchevsky, F. F. Komarov, Y. Kohmura, M. Awaji, Y. Suzuki and T. Ishikava. *Nucl. Instrum. Methods A* **454**, 512, 2000.
25. M. A. Piestrup, J. T. Cremer, H. R. Beguiristain, C. K. Gary and R. H. Pantell. *Rev. Sci. Instrum.* **71**, 4375, 2000.
26. V. Aristov, M. Grigoriev, S. Kuznetsov, L. Shabelnikov, V. Yunkin, T. Weitkamp, C. Rau, I. Snigireva and A. Snigirev. *Appl. Phys. Lett.* **77**, 4058, 2000.
27. C. G. Schroer, B. Lengeler, B. Benner, T. F. Günzler, M. Kuhlmann, A. S. Simionovici, S. Bohic, M. Drakopoulos, A. Snigirev, I. Snigireva and W. H. Schröder. *Proc. SPIE.* **4499**, 52, 2001.
28. J. P. Hogan, R. A. Gonsalves and A. S. Krieger. *IEEE Trans. Nucl. Sci.* **38**, 1721, 1991.
29. T. Yuasa, M. Akiba, T. Takeda, M. Kazama, A. Hoshino, Y. Watanabe, K. Hyodo, F. A. Dilmanian, T. Akatsuka and Y. Itai. *IEEE Trans. Nucl. Sci.* **44**, 54, 1997.
30. G.-F. Rust and J. Weigelt. *IEEE Trans. Nucl. Sci.* **45**, 75, 1998.
31. A. Simionovici, M. Chukalina, M. Drakopoulos, I. Snigireva, A. Snigirev, C. Schroer, B. Lengeler, K. Janssens and F. Adams. *Proc. SPIE.* **3772**, 328, 1999.
32. A. S. Simionovici, M. Chukalina, C. Schroer, M. Drakopoulos, A. Snigirev, I. Snigireva, B. Lengeler, K. Janssens and F. Adams. *IEEE Trans. Nucl. Sci.* **47**, 2736, 2000.
33. C. G. Schroer, J. Tümmler, T. F. Günzler, B. Lengeler, W. H. Schröder, A. J. Kuhn, A. S. Simionovici, A. Snigirev and I. Snigireva. *Proc. SPIE.*, **4142**, 287, 2000.
34. C. G. Schroer. *Appl. Phys. Lett.* **79**, 1912, 2001.
35. C. G. Schroer, B. Benner, T. F. Günzler, M. Kuhlmann, B. Lengeler, W. H. Schröder, A. J. Kuhn, A. S. Simionovici, A. Snigirev and I. Snigireva, *Proc. SPIE.* **4503**, 230, 2002.
36. C. G. Schroer, M. Kuhlmann, B. Lengeler, T. F. Günzler, O. Kurapova, B. Benner, C. Rau, A. S. Simionovici, A. Snigirev and I. Snigireva. *Proc. SPIE.* **4783**, 10, 2002.
37. C. G. Schroer, B. Lengeler, B. Benner, T. F. Günzler, M. Kuhlmann, J. Tümmler, C. Rau, T. Weitkamp, A. Snigirev and I. Snigireva. *Proc. SPIE.* **4145B**, 274, 2000.
38. C. G. Schroer, T. F. Günzler, B. Benner, M. Kuhlmann, J. Tümmler, B. Lengeler, C. Rau, T. Weitkamp, A. Snigirev and I. Snigireva. *Nucl. Instrum. Methods A* **467–468**, 966, 2001.
39. C. Rau, T. Weitkamp, A. Snigirev, C. G. Schroer, J. Tümmler and B. Lengeler. *Nucl. Instrum. Methods A* 467–468, **929**, 2001.
40. C. G. Schroer, B. Benner, T. F. Günzler, M. Kuhlmann, B. Lengeler, C. Rau, T. Weitkamp, A. Snigirev and I. Snigireva. *Proc. SPIE.* **4503**, 23, 2002.
41. C. G. Schroer, B. Benner, T. F. Günzler, M. Kuhlmann, C. Zimprich, B. Lengeler, C. Rau, T. Weitkamp, A. Snigirev, I. Snigireva and J. Appenzeller. *Rev. Sci. Instrum.* **73** (3), 1640, 2001.
42. C. G. Schroer, J. Meyer, M. Kuhlmann, B. Benner, T. F. Günzler, B. Lengeler, C. Rau, T. Westkamp and A. Snigirev. *Appl. Phys. Lett.* **81**, 1527, 2002.
43. C. G. Schroer, M. Kuhlmann, U. T. Hunger, T. F. Günzler, O. Kurapova, S. Feste, F. Frehse, B. Lengeler, M. Drakopoulos, A. Somogyi, A. S. Simionovici, A. Snigirev, I. Snigireva, C. Schug and W. H. Schröder, *Appl. Phys. Lett.* **82**, 1485, 2003.
44. http://www.esrf.fr/exp_facilities/ID22.

45. S. Bohic, A. Simionovici, R. Ortega and A. Snigirev. *Appl. Phys. Lett.* **78**, 3544, 2001.
46. Y. Llabador and P. Moretto. *Nuclear Microprobe in the Life Sciences*, World Scientific, Singapore, 1997.
47. R. Ortega, P. Moretto, A. Fajac, J. Benard, Y. Llabador and M. Simonoff. *Cell. Mol. Biol.* **42**, 77, 1996.
48. B. Vekemans, K. Janssens, L. Vincze, F. Adams and P. Van Espen. *X-ray Spectrom.* **23**, 278, 1994.
49. R. Ortega, PhD Thesis, Bordeaux 1 University, 1994.
50. O. Chubar and P. Elleaume. *Proc. EPAC98 Conf.* **1177**, 1998.
51. R. Ortega, G. Deves, S. Bohic, A. Simionovici, B. Ménez and M. Bonnin-Mosbah. *Nucl. Instrum Methods B* **181**, 480, 2001.
52. R. Ortega, P. Moretto, Y. Llabador and M. Simonoff. *Nucl. Instrum Methods B* **130**, 426, 1997.
53. M. O. Krause. *J. Phys. Chem. Ref. Data* **8**, 307, 1979.
54. D. T. Cromer and D. Liberman. *J. Chem. Phys.* **53**, 1891, 1970.
55. D. T. Cromer and D. Liberman. *Acta Crystallogr., Sect. A* **37**, 267, 1981.
56. P. Boisseau. PhD Dissertation, MIT, 1986.
57. A. J. Kuhn, W. H. Schröder and J. Bauch. *Planta* **210**, 488, 2000.
58. M. Kaldorf, A. J. Kuhn, W. H. Schröder, U. Hildebrandt and H. Bothe. *J. Plant Physiol.* **154**, 718, 1999.
59. C. Kak and M. Slaney. *Principles of Computerized Tomographic Imaging*, IEEE Press, New York, 1988.
60. A. Simionovici, M. Chukalina, B. Vekemans, L. Lemelle, Ph. Gillet, Ch. Schroer, B. Lengeler, W. Schröder and T. Jeffries. *Proc. SPIE* **4503**, 222, 2002.
61. Ph. Gillet, J. A. Barrat, Th. Heulin, W. Achouak, M. Lesourd, F. Guyot and K. Benzerara. *Earth Planet. Sci. Lett.* **175**, 161, 2000.
62. L. Lemelle, A. Simionovici, R. Truche, Ch. Rau, M. Chukalina and P. Gillet. *Am. Mineral. in press.*
63. M. Chukalina, A. Simionovici, A. Snigirev and T. Jeffries. *X-Ray Spectrom.* **31**, 448, 2002.
64. L. Vincze, K. Janssens, B. Vekemans and F. Adams. *Proc. SPIE.* **3772**, 328, 1999.
65. M. C. Camerani, A. Somogyi, A. S. Simionovici, S. Ansell, B.-M. Steenari and O. Lindqvist. *Environ Sci. Technol.* **36**, 3165, 2002.
66. CMDV data analysis program, © S. Ansell, 2001, http://www.esrf.fr/computing/scientific/catalog/cmdv/welcome.html.
67. M. C. Camerani, A. Somogyi, M. Drakopoulos and B. M. Steenari. *Spectrochim. Acta B*, **56**, 1355, 2001.
68. B. S. Haynes, M. Neville, J. Q. Richard and A. F. Sarofim. *Colloid Interface Sci.* **87**, 267, 1982.
69. H. Beleviand and H. Moench. *Environ. Sci. Technol.* **34**, 2501, 2000.
70. A. Golding, C. Bigelow and L. M. Veneman. *Chemosphere*, **24**, 271, 1992.
71. C. S. Kirby and J. D. Rimstidt *Environ. Sci. Technol.* **27**, 652, 1993.

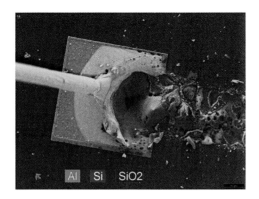

Plate 1 (Figure 4.1.28)

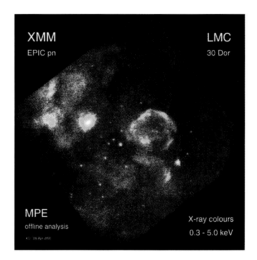

Plate 2 (Figure 4.1.50)

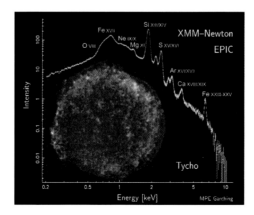

Plate 3 (Figure 4.1.51)

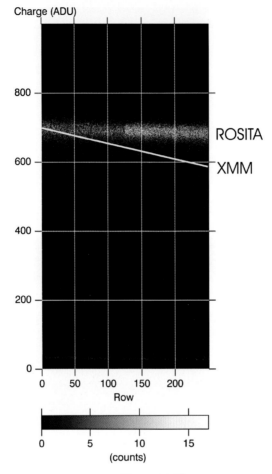

Plate 4 (Figure 4.1.59)

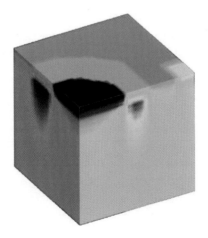

Plate 5 (Figure 4.1.61)

Chapter 4

X-Ray Detectors

4.1 Semiconductor Detectors for (Imaging) X-ray Spectroscopy

L. STRÜDER[1,3], G. LUTZ[1,3], P. LECHNER[2,3], H. SOLTAU[2,3] and P. HOLL[2,3]

[1] Max-Planck-Institute für Physik und extraterrestrische Physik, München and Garching, Germany,
[2] PNSensor, München, Germany and [3] MPI Halbleiterlabor, München, Germany

4.1.1 INTRODUCTION

Large scale use of semiconductor detectors in X-ray spectroscopy is a recent development which was prompted by their success in particle physics based on technological developments in silicon detectors and in microelectronics. Applying the planar technology first used in electronics also to detector production[1] allowed the fabrication of high granularity fast detectors. This made it possible to perform extremely precise particle position measurements at a high event rate. Simultaneously the small energy needed for creating an electron-hole pair (a minimum of 1.12 eV and an average of 3.65 eV as compared to 30 eV for ionization of gases) created the potential for precise spectroscopic measurements. This aspect came to its full use when semiconductor detectors were applied to X-rays.

This subchapter will concentrate on present use of silicon detectors which has culminated in imaging spectroscopic detectors in X-ray astronomy. Here the use of silicon charge-coupled device (CCD) detectors with their extended energy range and vastly improved spectroscopic capabilities compared to the previously used gas detectors has lead to many new important discoveries.

Before proceeding to this subject, however, an introduction to the basic operation principles of semiconductors and the electronics used for digesting the fairly small electronic signals will be given. Furthermore detectors aimed only at spectroscopy will be treated. These, as well as the focal-plane imaging X-ray detectors, rely on operating principles developed during the period of rapid developments since the first introduction of position sensitive detectors into particle physics in 1980.[2]

Now semiconductor detectors have spread to other fields of science and technology. Parts of those developments are described in this subchapter. The presentation will be done in two parts. The first part will deal with general aspects of semiconductors when used as X-ray and particle detectors and describes the functional principles of detector structures including the readout of the signals. The second part describes some important detectors in more detail and deals also with important applications in X-ray spectroscopy and imaging.

PART 1: SEMICONDUCTOR PROPERTIES AND DETECTION PRINCIPLES

4.1.2 DETECTOR RELATED PROPERTIES OF SEMICONDUCTORS

Compared with other materials, semiconductors have unique properties that make them very suitable for the detection of ionizing radiation. Furthermore, semiconductors – especially silicon – are the most widely used basic materials for electronic amplifying elements (transistors) and more recently for complete microelectronics circuits. Thus, part of the process technology that already existed in (micro) electronics could be taken or adapted for detector production.[1] Integration of detector and electronics could be envisaged. In the following, the properties of important semiconductor materials will be discussed.

The uniqueness of semiconductor material properties can best be appreciated by comparing them with the most widely used radiation detectors that are based on ionization in gas. Values for silicon will be used in this comparison.

- The small band gap (1.12 eV at room temperature) leads to a large number of charge carriers per unit energy loss of the ionizing particles to be detected. The average energy for creating an electron–hole pair (3.65 eV) is an order of magnitude smaller than the ionization energy of gases (30 eV).
- Therefore the energy of X-rays can be measured with much higher precision and down to lower energies than is possible with gas detectors.
- The high density (2.33 g/cm^3) leads to a short absorption length of low and medium energy X-rays and to a large energy loss per traversed length of the ionizing particle (3.8 MeV/cm for a minimum ionizing particle). Therefore it is possible to build thin detectors that still produce large enough signals to be measured. In addition, the very small range of delta-electrons prevents large shifts of the centre of gravity of the primary ionization from the position of the track. Thus an extremely precise position measurement (of a few micrometres) is possible.
- Despite the high material density, electrons and holes can move almost freely in the semiconductor. The mobility of electrons ($\mu_n = 1450$ cm^2/V s) and holes ($\mu_p = 450$ cm^2/V s) is at room temperature only moderately influenced by doping. Thus charge can be rapidly collected (several ns) and detectors can be used in high-rate environments.
- The excellent mechanical rigidity makes the use of foils for containment of the gas superfluous and allows the construction of self-supporting structures. Therefore very thin radiation entrance windows can be constructed and high quantum efficiency can be reached down to very low energies.
- An aspect completely absent in gas detectors is the possibility of creating fixed space charges by doping the crystals used. It is thus possible to create rather sophisticated field configurations without obstructing the movement of signal charges. This allows the creation of detector structures with new properties that have no analogy in gas detectors.
- As detectors and electronics can both be built out of silicon, their integration into a single device is possible.

The most commonly used semiconductor detector materials are germanium and silicon but also other compound materials are used, such as GaAs CdTe and CdZnT. Germanium and silicon are indirect semiconductors, their most important difference being a factor of almost 2 in the band gap and the much shorter photon absorption length for germanium due to its high Z-value. Therefore Ge is suitable for measurements of fairly high energy X-rays while direct conversion in silicon detectors with reasonable efficiency is limited to X-rays energies of a few tens of keV. Due to the small band gap and the corresponding thermal generation of electron–hole pairs germanium detectors have to be operated at cryogenic temperatures while silicon detectors can work at room temperature. High Z-semiconductors with sufficiently large band gap for room temperature operation are available as compound semiconductors. These materials are however not available with the perfection of Si and Ge, instead the large concentration of crystal defects dominates the properties of the devices.

As silicon is the most commonly used material in the electronics industry, it has one big advantage with respect to other materials, namely a highly developed fabrication technology. For X-ray detection one problem is the limit in thickness that can be depleted with the application of reasonably low voltages, as intrinsic material cannot be manufactured. This limits the maximum energy which can be detected with reasonable efficiency.

4.1.3 THE SIMPLEST SEMICONDUCTOR DETECTOR STRUCTURE: THE DIODE

The most basic semiconductor structure is a diode, a combination between n- and p-doped semiconductors. It has electrically rectifying properties and may also serve as sensor for ionizing radiation. In that case it will usually be very asymmetrically doped similar to the case shown in Figure 4.1.1 where a lowly doped n-type substrate is connected to a highly doped thin p-layer. Such a device can be operated in partial depletion (partial depletion occurs already without application of a bias) or fully depleted by applying a sufficiently high reverse bias. In the latter case a highly doped n-layer has to be provided on the surface opposite to the diode in order to prevent charge injection (holes) from the backside electrode.

Width of the depletion region, charge density, electric field and potential can be calculated straightforwardly from simple electrostatics. One result useful to remember is that the bias voltage, V_{bias}, grows linearly with the doping density of the bulk and quadratic with the depletion depth:

$$V_{bias} = \frac{qN_D}{2\varepsilon\varepsilon_0}d^2 \qquad (4.1.1)$$

where N_D is the donor concentration, d the depleted thickness, q the charge of one electron, and $\varepsilon\varepsilon_0$ the dielectric constants of silicon. Signal charge is collected not only from the space charge region where electrons and holes are separated and driven to the electrodes by the electric field, but also incompletely from the nondepleted region.

Here partial collection is due to diffusion into the space charge region.

The simple diode by itself is already a useful X-ray detector providing spectroscopic information with limited precision due to the large capacitive load it presents to the electronics measuring the signal charge. The precision of charge measurement is a central subject for all types of detectors and will be treated in the next section.

4.1.4 THE MEASUREMENT OF SIGNAL CHARGE

Signal charge is measured with a charge sensitive amplifier, an inverting amplifier with a capacitance in its feedback loop (Figure 4.1.2). The detector is presented by its capacitance C_D, the noise of the amplifier by a noise voltage U_n at its input. We will consider only the case of very high amplification. In that case the voltage at the amplifier input has to remain zero when a charge Q_s is deposited at the amplifier input. One can then read immediately off the figure that the output voltage has to be

$$V_{out} = Q_s/C_f \qquad (4.1.2)$$

All charge is transferred from the detector onto the feedback capacitance C_f. Similar consideration can be made for the effect of amplifier noise U_n. When considering the charge required at the input in order to compensate the effect of the noise voltage so as to keep the output voltage at zero. One finds as noise charge

$$Q_n = U_n(C_D + C_f + C_i) \qquad (4.1.3)$$

with C_i the amplifier input capacitance. The noise charge equals the product of serial noise voltage times the total, 'cold' capacitance.

These results hold in the limit of infinitely large and fast amplification. They are good approximations for realistic situations. The latter result demonstrates the importance of keeping the detector capacitance low in order to contain serial noise at a low level.

The signal to noise ratio can be improved by signal shaping when signal and noise have different frequency spectra. This is the case for (white) thermal noise which exhibits a flat frequency spectrum. Here the signal to noise ratio improves

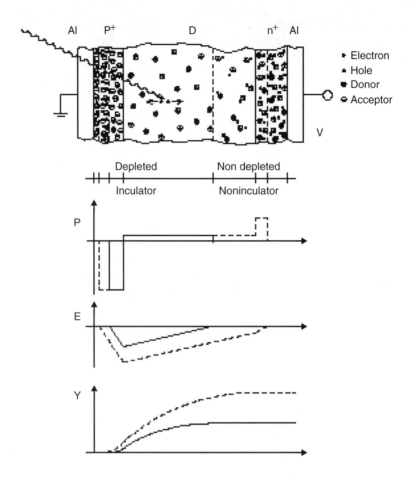

Figure 4.1.1 Schematic structure of a semiconductor diode with charge density, electric field and potential for partial (solid line) and full (dashed) depletion

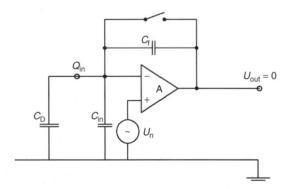

Figure 4.1.2 Schematics of a charge sensitive amplifier connected to a detector represented by its capacitance C_D. The sum of all noise sources inside the amplifier is represented by a single noise voltage source at the amplifier input

with the square root of the shaping time constant. Unfortunately all amplifying devices exhibit also low frequency ($1/f$) noise with a noise frequency spectrum corresponding to that generated by a short current pulse in the detector. Therefore the signal to noise ratio cannot be improved by shaping for this type of noise.

Besides serial noise, parallel noise has to be considered. The principal source of this noise is the detector leakage current which flows into the input of the amplifier. Integrating this current over the signal shaping time gives a 'dark charge' to be added to the signal. This added charge is subject to statistical fluctuations which in the simplest condition is described by Poisson statistics according to the number of elementary charges in this 'dark charge'. Signal shaping results in a noise rising with the square root of the shaping time.

Taking these three noise sources into account one arrives at the assumption that the dominant source of amplifier noise is due to the input transistor where

$$\text{ENC}^2 = \underbrace{\left(4kT\frac{2}{3g_m}C_{\text{tot}}^2\right)A_1\frac{1}{\tau}}_{\text{white series noise}} + \underbrace{(2\pi a_f C_{\text{tot}}^2)A_2}_{\text{low frequency noise}}$$

$$+ \underbrace{\left(qI_1 + \frac{2kT}{R_f}\right)A_3\tau}_{\text{parallel noise}} \quad (4.1.4)$$

A detailed derivation is given in Pinotti et al.[3] ENC is the equivalent noise charge, g_m the transconductance of the FET, A_1, A_2 and A_3 are constants depending on the shaper's filter function, a_f is a constant, which parameterizes the amount of low frequency noise, I_1 is the total leakage current and R_f is the equivalent resistor of the feedback.

Additional sources of noise contributing to spectroscopic performance degradation come from statistical fluctuations in the ionization (electron–hole creation) process itself (Fano noise will be described in the section on interaction of radiation with semiconductors) and in imperfections in the charge collection of the detector which can have several origins (to be described in the context of these detectors).

4.1.5 OTHER SEMICONDUCTOR DETECTOR STRUCTURES

The diode described previously is capable of measuring the energy of X-rays – although with moderate resolution and/or at low rate. Based on this structure position sensitivity can be reached by splitting the diode into small units and reading them out individually. Strip or pad-like structures are commonly used. This development was driven by particle physics where emphasis was on precision position measurement while ionization energy loss measurement was of minor interest.

Most effort went into strip detectors where a large degree of sophistication has been reached: single- and double-sided strip detectors with integrated capacitive coupled readout, integrated biasing circuits of various kinds have been developed. The detectors are able to withstand radiation fluences in the range of 10^{14} cm^{-2} hadronic particles.[4] In principle they can also be used as X-ray detectors, however, due to the large capacitance of strips (typically 1 pF/cm length) the energy resolution will be moderate. A breakthrough for X-rays has been reached with the invention of the semiconductor drift chamber by Gatti and Rehak in 1984.[5] This device, originally aimed at position measurement with the help of the drift time, has revolutionized X-ray spectroscopy; in particular, through the invention of single-sided structured devices of circular geometry[6] which in addition had the input transistor integrated in the device.[7]

The semiconductor drift chamber also initiated the development of the pn-CCD which in X-ray detection has very important advantages compared to standard CCDs which are based on MOS (metal–oxide–semiconductor) structures working in deep depletion mode. These earlier, still widely used types of CCDs are not discussed in detail here. Controlled drift detectors (CDDs) represent a combination of silicon drift detectors and pn-CCDs. They can be operated rapidly with simultaneous good position and energy measurements.

The DEPFET (depleted p-channel field effect transistor), a new device invented in 1985 by Kemmer and Lutz,[6] combines the function of detector and amplifier. It is the base structure of a new type of pixel detector under development for X-ray telescopes in astronomy.

In the following, the working principles of the devices will be described while detailed descriptions and applications are postponed to later sections.

4.1.5.1 SILICON STRIP DETECTORS

The basic principle of a strip detector is shown in Figure 4.1.3. The diode of Figure 4.1.1 is split into many strips each of which in the simplest case is

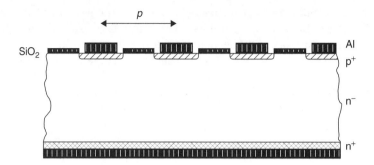

Figure 4.1.3 Cross-section of a silicon diode strip detector

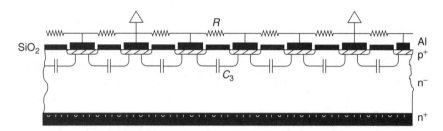

Figure 4.1.4 Strip detector with charge division readout

connected to its own amplifier. The position of the strip hit provides a coordinate, the signal height the energy of the X-rays. For a fraction of events the signal will be split between neighbouring strips due to the size of the charge cloud caused by electrostatic repulsion and diffusion on their way towards the strips. For low energy X-rays the preferred radiation entrance side is the unstructured n^+ backside. This is due to the bad charge collection properties in the gap regions between the strips where electrons experience no drift towards the backside but have to cross a potential barrier with the help of diffusion.

Typical strip pitches p are in the range of 20 to a few hundred micrometres. This may require a large number of readout channels. This number can be decreased by using capacitive charge division readout as shown in Figure 4.1.4. Here only every fourth strip is connected to an electronics channel and the charge collected at non-connected strips couples capacitively through the naturally present inter-strip capacitances to those strips connected to the electronics. The ratio of signals in neighbouring readout strips is used for position interpolation.

For proper charge collection the potential of the intermediate strips have to be held at the same potential as the readout strips. This is accomplished by high resistive connections between strips. A complication arises in the extraction of energy from the data as part of the signal collected in intermediate strips is lost to strip-backside capacitances. Corrections for this loss have to be applied.

Two-dimensional position measurement can be obtained by having (crossed) strips on both sides of the wafer and using simultaneously holes and electrons which are collected on opposite sides of the wafer. Figure 4.1.5 shows a double-sided strip detector with holes being collected on the lower side and electrons at the top side. This drawing is incomplete in the sense as it does not show the complications with n-strip isolation. The always present positive charge in the silicon oxide gives rise to an electron accumulation layer right below the oxide which shortens neighbouring n-strips. It can be interrupted with boron implantation. This implantation can either be strip like or, simpler and with better behaviour, uniformly over the whole

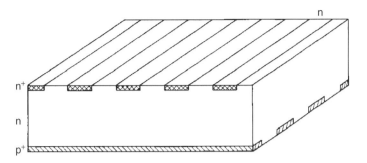

Figure 4.1.5 Schematics of double-sided strip detectors. The figure does not address the complications of electrical shortening between neighbouring n-strips due to the electron accumulation layer induced by the positive oxide charge

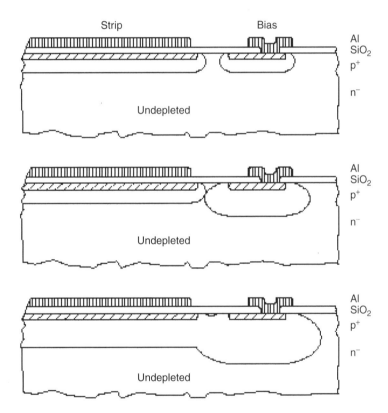

Figure 4.1.6 Strip-biasing by punch-through of capacitively coupled strip detector. When applying the bias voltage the depletion region around the bias-electrode grows and after touching that of the strip draws the strip potential along with it

surface. In the latter case the doping density has to be much smaller than that of the n-strips and roughly match that of the oxide charge so as to avoid breakdown due to large electric fields.

Capacitively coupled readout[8] shields the electronics from the dark current of the detector, not however, from the noise due to fluctuations of this current. It requires additional biasing circuitry to supply the potential to the strips. Both features can be simply integrated into the detector as shown in Figure 4.1.6. The capacitances are formed by the separation by an oxide layer between a doped strip and an aluminium strip. Biasing is performed with the help of a punch-through mechanism from one biasing strip running next to the ends of the strip electrodes. The potential of the implanted strips follows that of the bias strip within several volts. This and similar biasing methods also working for n-strips[9] are considerably simpler than biasing with polysilicon resistors, still the most widely used method.

4.1.5.2 DRIFT DETECTORS

The semiconductor drift chamber was invented by Gatti and Rehak in 1984.[5] It is based on the sideward depletion principle shown in Figure 4.1.7. The basic structure is a double-sided diode with diodes on both wafer surfaces and a small n^+ bulk contact on the side which is capable of depleting the complete wafer.

Space charge regions around the diodes are already present before applying any bias (Figure 4.1.7a). Leaving the n^+ contact at ground and applying a negative bias to the diodes makes the depletion region grow from both wafer surfaces simultaneously until they join (Figure 4.1.7b) at a quarter of the bias voltage needed for a standard diode (compare Equation.4.1.1 in Section 4.1.3). Further increase does not change the form of the potential in the diode region, only the electrons are retracted all the way towards the n^+ contact (Figure 4.1.7c).

If electron–hole pairs are created by radiation the holes will move toward one of the two p^+ contacts and the electrons towards the middle plane. Those are only very slowly removed by diffusion. Controlled drift of these electrons towards the n^+ anode can be achieved by superposition of an electric field parallel to the wafer surface. This can be done by dividing the diode into strips as shown in Figure 4.1.8 and applying potentials which rise continuously from strip to strip. Thus one arrives at the basic semiconductor drift chamber structure. The position of impinging radiation can be derived from the drift time (if the entrance time of the radiation is known), the energy from the signal height measured at the n^+ anode. The device has remarkable properties which make it very suitable

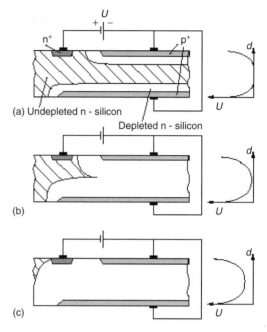

Figure 4.1.7 The principle of sideward depletion. In (a) no reverse voltage is applied, only the intrinsic depletion zones have developed. When applying a negative bias to the p^+ electrodes electrons are pushed towards the n^+ contact and the depletion voltages grow. (b) shows the full depletion, starting from the rectifying junctions on both wafer surfaces. (c) shows the configuration of 'over-depletion'

for X-ray spectroscopy. Those and many variations of the device will be described in a later section. The most important features of such devices are

- fast signal detection within a timing precision of nanoseconds
- law read node capacitance for low noise and fast signal detection.

4.1.5.3 CHARGE-COUPLED DEVICES

Standard CCDs are based on MOS structures which operate in the deep depletion mode, i.e. in thermal non-equilibrium. The principle of a three-phase MOS CCD is shown in Figure 4.1.9. Signal electrons created by ionization assemble in potential maxima at the Si–SiO$_2$ interface created by the transfer gates. Periodic variation of their potential drives the charge towards the readout anode. In the simple form shown in Figure 4.1.9

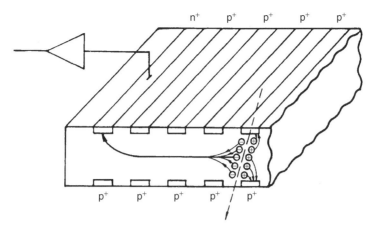

Figure 4.1.8 Semiconductor drift chamber structure derived from sideward depletion structure of Figure 4.1.7. Dividing the diode into strips and applying a continuously rising potential superimposes a horizontal field that drives the signal electron towards the n⁺ anode which is connected to the readout electronics. Upon arrival of the signal charge at the n⁺ anode the amount of charge and the arrival time can be measured

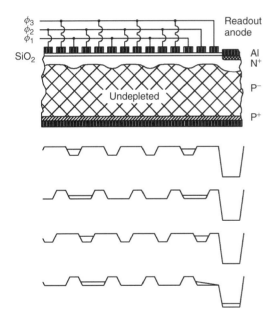

Figure 4.1.9 Principle of a three-phase MOS CCD

such a device will work badly due to the presence of potential maxima in the regions between the non-overlapping gates and because of the trapping of signal electrons in defects at the Si–SiO₂ boundary. These problems could be alleviated by designing modified structures, for example, buried channel CCDs in which the electrons are transferred in a depth of almost 1 μm. For X-rays there remains, however, at energies below 1 KeV, the disadvantage of the thick radiation entrance window and at high energy that of the fairly thin sensitive depleted region.

4.1.5.4 THE pn-CCD

A pn-CCD works on a different principle (Figure 4.1.10) from the previously described MOS-based devices (Figure 4.1.9), resulting in some rather interesting properties. While in MOS-based devices minority carriers are collected, these devices use the majority carriers, which are produced in the completely depleted and therefore also radiation-sensitive bulk material. The working principles can be explained from the silicon drift chamber. The possibility of using the drift-detector principle for the CCD has been mentioned already in the first publication by its inventors.[5]

Looking at the sideward depletion structure (Figure 4.1.7) we notice that the electron potential valley can be moved towards the top surface by biasing the lower diode negatively with respect to the top diode. Applying to the strips a periodically varying potential rather than a continuously rising one, the valley will be structured in depth. The lower diode may be formed as a single large-area

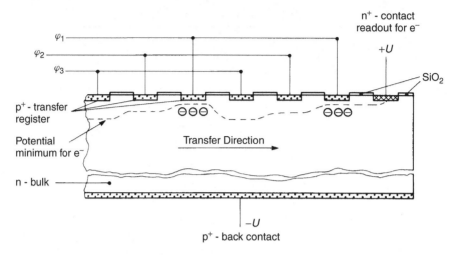

Figure 4.1.10 The fully depleted pn-CCD: cross-section along the channel

structure, and from that one arrives at the device shown diagrammatically in Figure 4.1.10. This device however still has some problems. In order to obtain sufficient modulation of the potential valley depth, the valley has to be not more than a distance of the order of the strip pitch away from the surface. This reduces (with reasonable bulk doping) the for potential barrier the holes between the top gates themselves and between the top gates and the back diode, thus enabling the thermionic injection of holes. Furthermore, spread of the signal charges along the gates is not prevented. Both problems can be solved simultaneously with an increase of doping in the surface region if this is done in a strip-like fashion, with strips perpendicular to the gate direction. These strips may be called 'channel guides', the narrow gaps between the strips have the function of channel stops.

Compared with standard CCDs, the fully depleted pn-CCDs have several important advantages when used as an X-ray detector:

- Enhanced sensitivity as the full volume is depleted. This is especially important for the measurement of X-rays at higher energies and for high energy charged particles.
- Uniform response with backside illumination as the backside consists of a unstructured large-area diode.
- High speed of operation due to charge transfer at a moderate distance from the surface.
- The possibility of building larger cell structures.
- Increased radiation hardness because radiation-sensitive MOS structures do not play an essential role in the function of the device.

4.1.5.5 THE DEPFET DETECTOR–AMPLIFIER STRUCTURE

The DEPFET structure which simultaneously possesses detector and amplification properties was proposed by Kemmer and Lutz[6] in 1987 and has subsequently been confirmed experimentally.[10] It is based on the combination of the sideward depletion method (Figure 4.1.7) – as used in a semiconductor drift chamber shown in Figure 4.1.8 – and the field effect transistor principle.

In Figure 4.1.11 a p-channel transistor is located on a fully depleted n-type bulk. As was done in the pn-CCD the potential valley has been moved close to the top side. Signal electrons generated in the fully depleted bulk assemble in a potential maximum ('internal gate') and increase the transistor channel conductivity by induction. The device can be reset by applying a large positive voltage on the clear electrode.

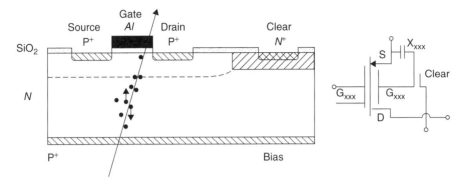

Figure 4.1.11 The DEPFET detector–amplifier principle

The DEPFET has several interesting properties:

- combined function of sensor and amplifier;
- full sensitivity over complete wafer;
- low capacitance and low noise;
- nondestructive repeated readout;
- complete clearing of signal charge: no reset noise.

These properties make it an ideal building block for an active X-ray pixel detector, or an electronic amplifying device for other silicon detectors.

4.1.6 INTERACTION OF RADIATION WITH SEMICONDUCTORS

The interaction of radiation with semiconductor materials causes the creation of electron–hole pairs that can be detected as electric signals. For charged particles, ionization may occur along the path of flight by many low-recoil collisions with the electrons. Photons have first to undergo an interaction with a target electron (photo or Compton effect) or with the semiconductor nucleus (e.g. pair conversion of photons). In any case, part of the energy absorbed in the semiconductor will be converted into ionization (the creation of electron–hole pairs), the rest into phonons (lattice vibrations), which means finally into thermal energy.

The fraction of energy converted into electron–hole pair creation is a property of the detector material. It is only weakly dependent on the type and energy of the radiation except at very low energies that are comparable with the band gap. For a given radiation energy, the signal will fluctuate around a mean value N given by

$$N = \frac{E}{w} \quad (4.1.5)$$

with E the energy absorbed in the detector and w the mean energy spent for creating an electron–hole pair (3.65 eV for silicon). The variance in the number of signal electrons (or holes) N is given by

$$\Delta N^2 = FN \quad (4.1.6)$$

with F the Fano factor ($F = 0.115$ for silicon).[11] Fano arrived at this expression by considering the probabilities of ionizing and non-ionizing collisions of charged particles in gases, making some rather arbitrary assumptions in his model. His approach has been adapted to semiconductors by Shockley.[12]

Very important aspects of the detector material in spectroscopic applications are the penetration depth of charged particles and the absorption length of photons. A very small absorption length will result in a high probability of generating the signal close to the surface, where signal charge may only be partially collected because of surface treatment (e.g. doping), coverage with insensitive material (e.g. a naturally or artificially grown insulation layer) or deterioration in the semiconductor properties, which usually appears close to the surface due to distortion of the lattice. A very large absorption length leads

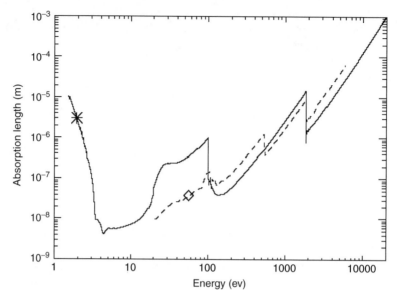

Figure 4.1.12 Energy dependence of the photon absorption length in silicon (*) and in silicon dioxide (◇)

to inefficiencies as radiation may traverse the detector without interaction. The dependence of the absorption length on photon energy for silicon is given in Figure 4.1.12.

Looking into the photon absorption process in more detail one discovers additional important aspects. Ejecting for example an electron from the K-shell will in some cases be followed by the capture of an electron from the L-shell, this process being accompanied by the emission of a photon with an energy of 1.74 keV. As seen from Figure 4.1.12, this photon has an average range of 10 μm and therefore a reasonable chance to be emitted from the detector without being detected. This 'missing energy' is responsible for the occurrence of secondary 'escape' peaks shifted downwards by this energy in the X-ray spectra.

4.1.7 THE RADIATION ENTRANCE WINDOW

The quality of the radiation entrance window is of great importance for spectroscopic detectors, and in particular for radiation with short penetration depth as for example low energy X-rays.

In the ideal case all energy should be converted into signal charge which then should reach without loss the collecting electrode. Unfortunately there are always absorbing layers present at the surface or a region in which part of the charge gets lost by recombination or other processes. All the devices to be described in the second part of this paper have a homogeneous unstructured large area entrance window which is optimized for spectroscopy. In the following the physical processes determining the spectroscopic performance are discussed.[13,14]

Figure 4.1.13 shows the situation for a p^+-n diode window which is covered with a thin layer of oxide. Photons interacting in the centre bulk (c) are completely converted and produce the proper peak in the spectrum. Photons with high energy (d) may traverse the detector without interaction. Photons interacting near the p^+ contact (b) will lose part of the signal by recombination and give rise to the shoulder on the left of the peak. Photons interacting in the oxide (a) will emit part of the charge into the silicon and create the 'flat shelf' on the left.

It remains to emphasize that both quantities, the spectral resolution as well as the background, strongly determine the usefulness of a

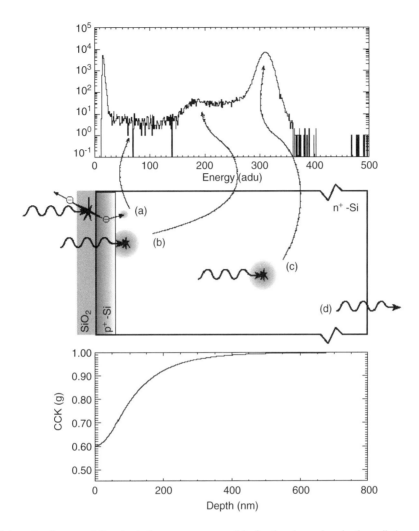

Figure 4.1.13 Schematic diagram of the physical processes responsible for the charge loss in the radiation entrance window. The spectrum is from the Cu L line (930 eV). The shape of the spectrum can be fitted with a model[13] which leads to the charge collection efficiency as a function of the depth of interaction (shown in the bottom part of the figure)

spectrometer. Beside the energy resolution the peak to background (or peak to valley, P/V) ratio is the most important performance figure since it defines the ability of the instrument to separate weak X-ray lines from the dominant lines.

The good performance of a thin optimized radiation entrance window manifests itself in a high quantum efficiency as shown in Figure 4.1.14 and in excellent spectroscopic properties (to be shown together with the respective detectors in Part 2 of this subchapter).

PART 2: SEMICONDUCTOR DETECTORS IN X-RAY SPECTROSCOPY AND IMAGING

Having given an overview of the basic detection techniques for X-rays in semiconductors we will now look at the subject from the standpoint of possible applications. X-ray astronomy has been pushing for several years the instrumentation for broadband imaging non-dispersive X-ray spectrometers: Since the launch of the European XMM-Newton

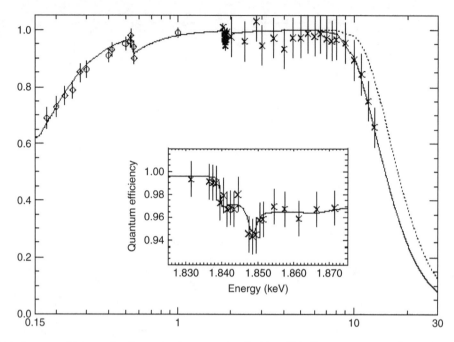

Figure 4.1.14 Quantum efficiency for X-rays in the range of 150 eV to 30 keV energy of a pn-CC of 300 μm thickness (measurements and solid line) equipped with a thin optimized radiation entrance window. Close to 100 % quantum efficiency is reached over most of the range. Remarkable is the high efficiency at low energies. The falloff at high energy is due to the limited thickness. The dotted line represents an extrapolation to 500 μm thick sensitive volume

satellite in December 1999, reliably operating X-ray CCDs have been delivering extraordinary images, recorded in a single photon counting mode, imaged through the largest X-ray telescope ever built. Related applications in other fields of basic and applied science will equally be mentioned. State of the art X-ray detectors with energy, time and/or position resolution at high quantum efficiency from the near infrared up to 30 keV are described in detail. They are all based on the concept of sideward depletion, the underlying principle of silicon drift detectors. They have been primarily developed for astrophysics experiments in space, for material analysis and for experiments at synchrotron radiation facilities. The functional principles of the silicon devices, i.e. detectors and on-chip electronics, are derived from basic solid state device physics. The spatial resolution, the spectroscopic performance of the systems, the long term stability and the limitations of the detectors are described in detail. Field applications show the unique usefulness of state of the art silicon radiation detectors.

4.1.8 THE DETECTION OF OPTICAL PHOTONS AND X-RAYS

Imaging of photons is best known in the visible domain, ranging from a wavelength of 3.500 Å up to 6.000 Å. For those applications optics and detectors are equally well developed. However, all those imaging systems do not count the incoming photons individually to measure their position, energy and arrival time.

The photon information is either integrated in the grains of a photographic film that is afterwards developed chemically or the photons are collected in individual *picture cells* (pixels) and after a given time sequentially read out. The photonic or electronic content of each grain or pixel is then 'counted' to measure the intensity of the incident photon flux. Traditionally, the energy of the photons is determined by an arrangement of various filters, transparent only for a narrow, well defined bandwidth of the incoming photons. In this sense, the image is a static, integrated reconstruction of a local photon intensity distribution.

Single 'optical' photons cannot be counted up to now in a practical manner, i.e. with reasonably large arrays. The energy of the photons is too small to detect them individually with non-cryogenic detectors: it is a fraction of 1 eV in the near infrared and up to 4 eV for the violet part of the visible spectrum.[1] In gas detectors more than 20 eV are needed for the ionization of a detector gas atom, and room temperature silicon detectors need at least 1 eV for the generation of an electron–hole pair in the optical range and 3.65 eV for ionizing particles with sufficiently large energy. For a proper electronic extraction of the very weak signal of one optical photon, read out electronics should operate below 0.1 electrons equivalent noise charge (ENC). This has not yet been reached in state of the art silicon sensor systems. The best noise values obtained so far are 0.9 e^- (rms). From approximately 11.000 Å to 3.000 Å only one electron-hole pair per photon is generated due to the ionization process and its statistics in silicon. In this sense, direct spectroscopic information in the optical range is physically not available from silicon detectors.

The X-ray imaging detector systems which are described below record simultaneously the energy, position and arrival time of each individual X-ray photon without using selective absorbers. The physical reasons for being able to make truly energy-dispersive X-ray detectors are the low electron–hole pair creation energy (average) of about 3.65 eV in silicon at room temperature and the very thin radiation entrance windows of only a few tens of partially insensitive atomic layers of silicon and native SiO_2 which can be penetrated by the (even soft) X-rays. For a good quantum efficiency at higher X-ray energies only the depleted thickness of silicon (signal interaction depth) is relevant. At 500 μm sensitive detector thickness, for example, 25 % of 25 keV X-rays are converted in electron–hole pairs and can be collected and detected (see Figure 4.1.14). For two-dimensional silicon detectors with high position and energy resolution, the fabrication by a planar process – comparable to the fabrication in state of the art microelectronics – is obligatory. Depletion thicknesses of 1000 μm are technically a limit for the detector fabrication in planar processes.

In the energy band between 0.1 keV and 30 keV state of the art imaging silicon detector systems are ideal for the direct detection with high quantum efficiency, position and energy resolution.

The astrophysical requirements have driven the developments of the high resolution X-ray detectors from 100 eV to 10 keV in the last 10 years. The X-ray Multi Mirror Mission (XMM) of the European Space Agency (ESA) was successfully launched in a highly eccentric orbit on 10 December 1999 with three large X-ray telescopes and reflecting grating spectrometers all having specially designed X-ray CCDs in their focal planes.[15,16] The wavelength dispersive gratings are read out with the more conventional back illuminated 25 μm deep depleted MOS CCDs for energies up to 4 keV.[17]

The review of silicon detectors and applications is based on basic principles and discusses physical limitations. As limiting factors the electronic noise in physical systems and the effect of the radiation entrance window will be treated. Device simulations will serve as an intuitive approach to understand the collection and motion of signal electrons in the detectors. Various types of silicon drift detectors (SDDs) will be introduced, among those circular, linear and multi-cell devices. Applications will demonstrate the broad use of SDDs. The controlled drift detector (CDD) will highlight the large variety of silicon drift devices. pn-CCDs as high resolution X-ray imagers will introduce the field of position resolved spectroscopy. The excellent properties of pn-CCDs will demonstrate the progress in the field. The conceptually most advanced device in spectroscopic X-ray imaging is represented by the backside illuminated active

---

[1] Cryogenic detectors are able to perform single photon counting in the near infrared, visible and soft X-ray domain.[20] The band gap for this kind of detectors is in the mV range as compared to 1.1 eV for Si. But they require cooling down to the order of 100 mK.

pixel sensor (APS) DEPFET detector. The concept, functional principles, measured and expected properties will be described in detail.

All experimental results shown here are from devices which have been designed, fabricated and tested at the MPI semiconductor laboratory.

4.1.8.1 X-RAY DETECTION

The absorption depth of photons in silicon oxide and silicon varies over five orders of magnitudes in the bandwidth of 1 eV to 20 keV, as can be seen in Figure 4.1.12. The average range of the photon in the silicon varies from several millimetres in the near infrared to a few tens of ångstrøm only for UV light and then increases again for higher energies to 1 mm for approximately 20 keV. The absorption is most efficient at the silicon M-, L- and K-edges at approximately 20 eV, 100 to 150 eV and 1830 eV, respectively.

The X-ray detectors should be able to absorb all incident radiation and transfer a variation of five orders of magnitude of absorption lengths into a quantum efficiency over the whole range of interest with high homogeneity and an efficiency close to 100 %. This should be independent of the photon interaction location in the detector body, where the photon to electron–hole conversion occurred.

The primary conversion process of the incident radiation into a detectable quantity can go into light, heat and electrical charges. The incident photons can be directly converted into light, e.g. in scintillators, which will then be analysed with the help of light sensitive detectors, i.e. photomultipliers or photodiodes to finally yield an electronic signal. Another technique makes use of the increase of temperature caused by the absorption of the photon energy. The temperature increase is then used to break up Cooper pairs in a superconductor or to make a current or voltage change in a microthermometer, resulting in an electronically detectable quantity. The last possibility is to convert the incident radiation directly into electrical charges. The generation of electron–hole or electron–ion pairs in semiconductors and gas counters can be directly amplified to generate an electronic pulse, proportional to the energy of the incoming photons. Up to now all three types of techniques have been used to realize two-dimensional, X-ray sensitive detector systems.[18]

Scintillators with photodiode or photomultiplier readout can go to the highest energies; cryogenic detectors as bolometers or superconducting tunnel junctions can achieve to date the best energy resolution;[19] avalanche photodiodes can achieve a time resolution for individual events of several picoseconds; with proportional gas counters sensitive areas without insensitive gaps in the order of several hundred cm^2 have been built; operation at high temperatures has been achieved with HgI detectors, but there is no detector combining all the needed properties in one single detector system with highest quality. Up to now, only state of the art X-ray CCDs and APS unify the broad band properties, with some compromises in the above list of desired physical parameters. The most advanced systems are all made on silicon as absorbing detector material and with integrated on-chip electronics.

The availability of very good starting silicon, the highly elaborated fabrication techniques and the well matched physical properties of silicon, makes silicon microsystems – detector and electronics, monolithically unified – a good candidate for satisfactory performance figures for many application scenarios.[2]

4.1.9 SILICON DRIFT DETECTORS

To obtain the lowest possible noise in radiation measurements, the total capacitance of the signal charge collecting node must be minimized. In conventional structures the sensitive area always correlates with the readout capacitance. Either the sensitive area is made very small or the sensitive thickness very large to reduce capacitance. The silicon drift-type detectors decouple collection area from the readout node size, since an electric field parallel to the wafer surface transports the signal

---

[2] Simple silicon pad and strip detectors, as well as hybrid pixel detectors (detectors and electronics on different chips, then bump bonded to form a hybrid detector) will not be treated, because their use is restricted to count X-rays. They are unable to reach Fano-limited energy resolution, mainly because of their high electronic readout noise.

charge to a small output node, whose size is independent of the sensitive area.

Other silicon drift-type detectors like pn-CCDs[15,20] and APS[10,21,22] also make use of the so-called principle of sideward depletion (Figure 4.1.7). The functional principles of SDDs and CCDs are related. The SDD could be called a phaseless pn-CCD (see Section 4.1.12), because the charge is continuously drifted out, or a pn-CCD could be called a discrete SDD, since a CCD drifts the charge packets in clocked time intervals, discretely, to the readout node.

4.1.9.1 OPERATION PRINCIPLE

The basic concept of semiconductor drift detectors has already been presented in Section 4.1.5. Here we go into more detail and consider its consequences on detector properties.

The silicon drift chamber is derived from the principle of sideward depletion (see Figure 4.1.7) by adding an electric field, which forces the electrons to the n$^+$ readout anode. This is simply achieved by implanting a parallel p$^+$ strip pattern at both sides of a semiconductor wafer (instead of the homogeneous implant on both surfaces shown in Figure 4.1.8) and superimposing a voltage gradient at both strip systems. The direction of the voltage gradient is such that the n$^+$ readout anode has the highest (positive) potential, therefore collecting all the signal electrons accumulated in the local potential minimum (bottom of the parabola in Figure 4.1.15) and drifting them to the absolute potential minimum for electrons at the n$^+$ readout node (Figure 4.1.16). The holes from the ionization process disappear directly in their local potential minimum in the p$^+$ strips. From the Poisson equation it can be easily derived that in the case of full depletion with the depletion voltage U_D and in a one-dimensional approximation across the silicon wafer (y) and with a linear superimposed drift field parallel (x) to the wafer surface, the electric potential $\Phi(x, y)$ is with d the wafer thickness and the depth:

$$\Phi(x, y) = U_D - \frac{\rho}{2\varepsilon_0 \varepsilon_{Si}}(y^2 - yd) - \frac{\Phi_{out} - \Phi_{in}}{x_{out} - x_{in}} \tag{4.1.7}$$

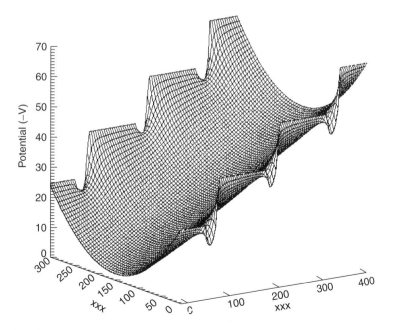

Figure 4.1.15 Simulation of the electric potential of a silicon drift detector in the bulk of the detector far from the readout node

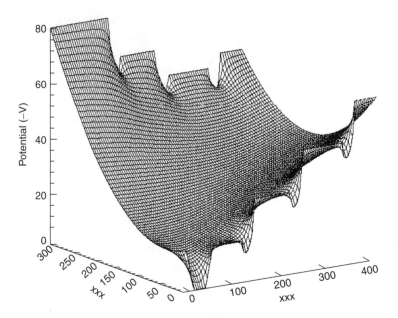

Figure 4.1.16 Simulation of the electric potential of a silicon drift detector in the vicinity of the readout node

The drift field E_d is usually applied between the outer p$^+$ drift structures (Φ_{out}) and the inner p$^+$ drift strips (field strips) or rings (Φ_{in}) in the vicinity of the readout node. Neglecting the potential perturbations close to the n$^+$ anode, the x component of the drift field can be written as

$$E_D = \frac{\Phi_{out} - \Phi_{in}}{x_{out} - x_{in}} \quad (4.1.8)$$

The drift speed for practical drift fields varies between 1 μm/ns and 10 μm/ns. The maximum drift length realized up to now is about 4 cm.[23]

4.1.9.2 ENERGY RESOLUTION

The measurement of the total energy of the incident radiation is achieved by a careful, low-noise 'counting' of all electrons arriving at the n$^+$ readout node. The number of created electron–hole pairs is proportional to the energy of the incident X-ray and that the average energy required to create one electron–hole pair is 3.65 eV at room temperature. For the following considerations, we assume to operate the first amplifying stage in an ideal source follower configuration. The detector's anode, collecting all signal charges can in principle be made very small, limited only by technological parameters. If this readout node is then directly coupled to the gate of an on-chip pre-amplifying first transistor, the total read-node capacitance can be kept as low as 100 fF (or lower), translating in a high sensitivity of the on-chip amplifier, i.e. the increase of the readout node voltage with the arrival of one electron. With

$$U_{out} = \frac{Q_{inj}}{C_{tot}} \quad (4.1.9)$$

where U_{out} is the increase of the output voltage, Q_{inj} the injected charge and C_{tot} the total readout node capacitance, an X-ray of 6 keV stimulates a voltage change of 2.6 mV. This corresponds to a sensitivity of 1.6 μV/electron. The noise of the on-chip amplifier in such configurations can be kept as low as 5 electrons rms at temperatures around −30 °C.

For optimum noise and speed performance the implementation of the first amplifying stage on the detector is essential to the use of SDDs.

The anode is connected to an amplifying junction field effect transistor (JFET) integrated directly on the detector chip (Figure 4.1.17). This way the capacitance of the detector–amplifier

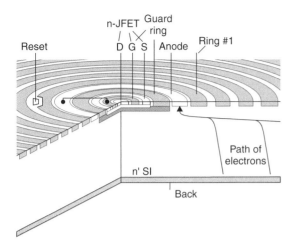

Figure 4.1.17 On-chip single-sided junction FET coupled to the readout node of a SDD

system is minimized by eliminating bond wires between detector and amplifier, thus avoiding all kinds of stray capacitances between the readout node and ground, making the system again faster and less noisy. Further advantages are evident as the effect of electrical pickup is significantly reduced and problems of microphony, i.e. noise introduced by mechanical vibration, are excluded.

With the help of Figure 4.1.17, the basics of the amplification and resetting process of the integrated JFET can be easily understood. In the centre of the schematic drawing, a single sided n-channel JFET is shown. Let us assume that electrons, generated by ionizing radiation, drift towards the readout anode. The voltage change, generated at the readout node is directly coupled to the p^+ gate of the n-channel transistor (source and drain are n^+ implants, the transistor channel is a deep n implant). The negative voltage on the p^+ gate reverse biases the junction, thus depleting into the transistor channel, resulting in a current drop through the transistor. This change of current can be precisely measured.

As it collects more and more electrons the JFET gate gets increasingly reverse biased relative to the transistor channel. At a given potential difference the gate is discharged by impact ionization in the transistor channel close to the junction of the p^+ gate and the drain at the end of the channel.[24]

During detector operation the gate adjusts its potential in a way that all signal electrons and leakage current are compensated by the breakdown mechanism. In other words, the integrated JFET resets itself automatically, there is no need for an externally clocked reset pulse, and the SDD and integrated electronics are operated with direct current voltages only.

The energy resolution is limited by the fluctuations in the generation of electron–hole pairs in the conversion process and the electronic noise generated by the input amplifier and the detector leakage current. The electronic noise has been considered in Section 4.1.4, the fluctuations in the ionization process in Section 4.1.8. The following results have been obtained for the ENC:

$$\mathrm{ENC}^2 = \underbrace{\left(4kT\frac{2}{3g_\mathrm{m}}C_\mathrm{tot}^2\right)A_1\frac{1}{\tau}}_{\text{white series noise}} + \underbrace{(2\pi a_\mathrm{f} C_\mathrm{tot}^2)A_2}_{\text{low frequency noise}}$$

$$+ \underbrace{\left(qI_\mathrm{l} + \frac{2kT}{R_\mathrm{f}}\right)A_3\tau}_{\text{parallel noise}} \qquad (4.1.4)$$

and for the Fano noise

$$\Delta N^2 = FN \qquad (4.1.6)$$

The white serial noise that is due to thermal fluctuations in the transistor channel scales with $1/\sqrt{\tau}$, the $1/f$ low frequency noise due to trapping of charge carriers in the vicinity of the channel is independent of shaping and the parallel noise due to detector and gate leakage current as well as the feedback resistor contribution grows with $\sqrt{\tau}$.

The leakage current has its physical origin in the thermal generation of electron–hole pairs in the semiconductor through energy levels in the forbidden band gap. Those levels may arise from (mainly) metal contamination in the silicon or imperfections in the silicon lattice. In the case of 'mid-band-gap' traps, the leakage current attenuates approximately a factor of two every 7 K in temperature reduction. In the case of the low frequency, or $1/f$ noise, electrically active traps capture and release the charge carriers in the transistor channel and therefore give rise to a change of the electric field in the channel, influencing the current flow. The perturbations of the electric field are described by the density of traps and their capture and emission time constants.

If the detector leakage current could be made infinitely small (e.g. by cooling the detector and front-end electronics) the time shaping constant should be made as long as possible to obtain the lowest ENC, up to the moment, when the shaping time constant independent $1/f$ noise sets a lower limit for the noise. Of course, again, this conflicts with the requirement of high count rate capabilities; long shaping times, i.e. long signal processing times lead to signal pile-up and therefore degrade the system performance. To beat pile-up, again, the only possibility is to lower the total input capacitance C_{tot} and thus lower τ to achieve the same ENC. The $1/f$ noise contribution is independent of τ, which cannot easily be overcome by operational means. This technologically intrinsic limitation of the noise level has its origins in the nonperfect crystal properties of the starting material and the fabrication process.

The achievable energy resolution ΔE_{FWHM} of a SDD can be as good as

$$\Delta E_{FWHM} = 2.355\, w \sqrt{ENC^2 + \frac{FE}{w}} \quad (4.1.10)$$

F is the Fano factor,[25,26] E the total X-ray energy, w the pair creation energy, ENC the rms fluctuation of the readout noise and 2.355 the conversion factor between the standard deviation σ (rms) of a Gaussian and the FWHM $\ln(2\sqrt{2}) = 2.355$. With $F = 0.115$, $w = 3.65$, for $E = 6\,\text{keV}$ and a readout noise of 10 electrons (e.g. close to room temperature), the best achievable energy resolution is 150 eV FWHM. Values of 140 eV to 150 eV are now routinely achieved at $-10\,°C$ with SDDs. By further reduction of the temperature, i.e. reduction of the detector leakage current, i.e. reduction of ENC from 10 to 5 electrons, the energy resolution improves to 125 eV FWHM at 6 keV. State of the art SDD systems operate very close to the above values.[7,27] For ENC = 0, the Fano limit can be derived, which is 119 eV (FWHM) for 6 keV X-rays (see also Figure 4.1.45).

4.1.9.3 POSITION RESOLUTION

In standard applications of SDDs in high energy physics experiments the position resolution of SDDs is obtained by a precise measurement of the drift time. The 'start' signal could be delivered by the bunch crossing time mark and the 'stop' time by the SDD. In our short consideration we restrict ourselves to minimum ionizing particles, traversing the SDD perpendicular to the detector's surface. According to

$$x_{drift} = \mu_n E_d t_{drift} \quad (4.1.11)$$

the position x_{drift} can be obtained easily with the electron mobility μ_n, the electrical drift field E_d and the measured drift time t_{drift}. If the readout anode is segmented in many individual nodes, the position is measured in two dimensions, with the help of the drift time (x) and the position of the anode (y) as indicated in Figure 4.1.18.

The position resolution of a SDD was derived for minimum ionizing radiation by Rehak[28] including the effects of charge spreading during the collection and drift time. For realistic assumptions in high energy physics experiments, the limit for

Figure 4.1.18 Sketch of a two-dimensional silicon drift detector. The n$^+$ readout nodes are indicated as black squares in the vertical direction on the left-hand side (y-direction). The drift of the signal charges is perpendicular to the p$^+$ field strips (x-direction)

the position measurement precision of SDDs is approximately $2\,\mu m$ rms.

With the SDD principle in mind, the designer has great flexibility in the choice of anode configurations and drift directions. For instance at the semiconductor laboratory of the Max-Planck-Institutes (MPI-HLL) large SDDs have been fabricated with linear drift geometry, i.e. parallel strips,[29] up to 4.2×3.6, $4.2\,cm^2$ and a $55\,cm^2$ cylindrical geometry on $4''$ wafers, in which electrons drift along the radial direction to one of 360 anodes placed at the wafer edge.[30] Both systems have been used as particle trackers.

For the use in imaging X-ray spectroscopy this straightforward use of SDDs is not practical. The CDD in Section 4.1.11 shows alternatives for the simultaneous measurement of position and energy with SDDs.

4.1.10 SILICON DRIFT DETECTORS FOR X-RAY DETECTION

To make the detectors suitable for spectroscopic X-ray applications, the strip system on both surfaces is replaced by a large area pn-junction on one side, which is used as a very homogeneous thin entrance window for the radiation[6,7,31] (Figure 4.1.19). A further improvement is the use of circular drift electrodes, which force the signal electrons to a very small anode in the centre of the device, from where they are transferred to the gate of an integrated JFET.

The radiation entrance window, denoted as 'back contact' in Figure 4.1.19 and 4.1.20, plays an important role in X-ray spectroscopy for the detection of light elements, i.e. for X-ray radiation between 100 eV and 1 keV and for the detection of trace elements in the presence of additional (strong) continuous and X-ray line emission:[13,14] X-ray absorbing layers on top of the radiation entrance window as e.g. SiO_2 or Si_3N_4 in SDDs would significantly lower the quantum efficiency for low energy X-rays. Depending on the photon energy, a fraction of the X-rays would be stopped in the insensitive layers.

Figure 4.1.21 shows a typical low energy (900 eV) X-ray spectrum obtained with a detector having a homogeneous entrance window. The origin of the low energy background (shoulder and flat shelf) has been discussed in Section 4.1.7. This background extending from zero to the X-ray energy is of extreme importance in spectroscopy as it will hide weak lower energy lines. It can be reduced by technological means by suppression of SiO_2–Si interface recombination. This is accomplished by soothing the interface states at the boundary by technological means, e.g. hydrogen

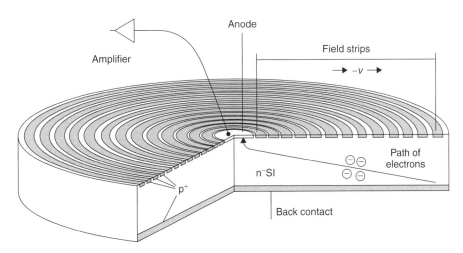

Figure 4.1.19 Cylindrical silicon drift chamber with an integrated amplifier for spectroscopic applications. The entire silicon wafer is sensitive to radiation. Electrons are guided by an electric field towards the small-sized collecting anode in the centre of the device

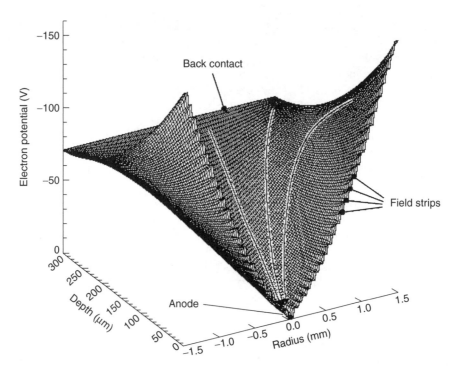

Figure 4.1.20 Simulated potential energy distribution in a circular silicon drift chamber with homogeneous radiation entrance window. The simulation includes the whole detector shown in Figure 4.1.19 including the electron collecting readout node. The arrows indicate the paths of the electrons drifting to the anode

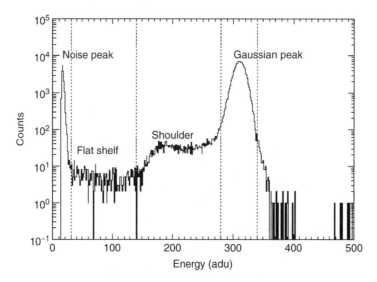

Figure 4.1.21 Typical low energy X-ray spectrum at 900 eV, showing the different components of the detector background. The 'noise peak' arises from the detector, signal processors and digitization circuits. As can be seen here, the trigger threshold is about 50 eV. The flat shelf arises from ionization in the layers above the sensitive silicon (see also Figure 4.1.13). The shoulder has its origin in interaction areas close to the p^+ back diode. The p/v ratio of this measurement is approximately 2000:1

termination of bonds. Recombination processes at the silicon surface region are controlled by appropriate doping and annealing procedures.

If the recombination is prevented or suppressed due to proper thermal treatments, the signal charges diffuse in the field free region of the p^+ implant until they eventually reach the edge of the space charge region. At that moment, the charge is swept away to the n^+ readout node.

The low energy background strongly determines the usefulness of a spectrometer. Beside the energy resolution the peak to background (or peak to valley, P/V) ratio is the most important performance figure since it defines the ability of the instrument to separate weak X-ray lines from the dominant lines.

Within the cylindrical area no charge splitting is possible,[3] resulting in a single reading of a given charge package. The low energy response down to 100 eV can be made as good as in any other semiconductor detector, by keeping all fast timing capabilities of the SDD system. In addition to the above mentioned features an electron sink electrode at the structured surface is implemented as well as an integrated voltage divider. The electron sink takes out all surface generated current components, reducing the leakage current to the pure bulk contribution of less than 1 nA per cm² for a depletion depth of 500 μm. The integrated voltage divider supplies all voltages for the drift rings. Only the innermost and outermost p^+ ring needs to be contacted. The details of these techniques are described elsewhere.[21] For a SDD of 5 mm² active area the maximum drift time from the edge of the detector is about 150 ns, while the time spread of the signal charges of one photon event is approximately 5 ns. As we have usually no event trigger signal in the field of X-ray detection, the device has no position resolution within the sensitive area of 5 mm².

The electric potential of the cylindrical silicon drift chamber is shown in Figure 4.1.20 in a two-dimensional cut perpendicular to the surface through the silicon wafer. It shows the potential energy for electrons of the SDD of Figure 4.1.19, including all field strips and the central electron collecting anode. The equipotential of the homogeneously doped radiation entrance window can be seen on the back, the field strips (rings) with their decreasing (negative) potential on the front side. There is no field free region in the device and all electrons in the sensitive area are guided within less than 150 ns towards the readout node. However the time spread of the charge cloud is only in the order of 5 ns. Overlapping charge clouds limit the single photon counting capability to about 10^6 X-rays per detector element.

The cylindrical SDD has outstanding properties: at moderate temperatures of about $-10\,°C$ (achievable by Peltier cooling), the devices have already good spectroscopic properties,[32] comparable to state of the art Si(Li) detectors, but with count rate capabilities up to 10^6 counts per second (cps) to be compared to the order of 10^4 cps with a 70 ns shaping of the classical Si(Li) detector concept, without the need of liquid nitrogen cooling (Figure 4.1.22). The energy resolution at two different shaping times and temperatures are shown in Figure 4.1.23. If the incident photons are correctly collimated within the sensitive area, the P/V ratio, i.e. the ^{55}Fe peak count rate divided by the average number of counts around 1 keV, can be as large as 7000:1. The P/V ratio determines the sensitivity limit for weak X-ray intensities, as it describes the extraction of an X-ray line from the detector background.

Very recently a 10 mm² large SDD based on the same layout was fabricated and tested. At $-10\,°C$ it has shown a similar energy resolution as the above described 5 mm² large SDD. This is an important step towards the use of SDDs in applications where larger areas must be covered.

4.1.10.1 THE SILICON DRIFT DETECTOR DROP (SD³)

The previously described 'conventional' cylindrical SDD with the on-chip amplifier located in the middle of the detector exhibits partial events, i.e. events whose signal charge is partially correctly collected and partially drained by the positive potential of the first SSJFET on the SDD. An

---

[3] In the vicinity of the on-chip FET, charge losses were observed. These are avoided by operating the FET outside the sensitive area (see Section 4.1.10.1).

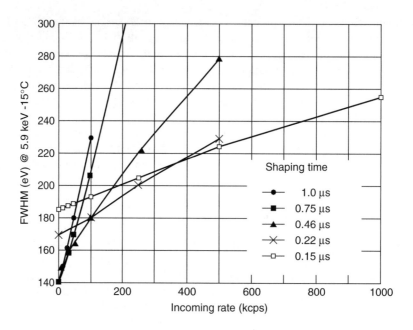

Figure 4.1.22 Silicon drift detector energy resolution as a function of the X-ray (^{55}Fe source) count rate. The measurement was done close to room temperature. The signal processing time (here peaking time) ranged from 150 ns to 1 μs

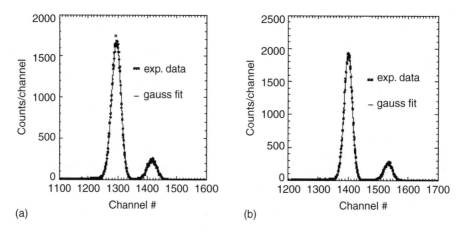

Figure 4.1.23 (a) Manganese spectrum recorded with a circular SDC at 25 °C. The shaping time was 0.25 μs, the FWHM is 178 eV. Typical values at room temperature scatter from 170 eV to 190 eV. (b) Manganese spectrum recorded with a SDC at −10 °C. The shaping time was 1 μs, the FWHM is 144 eV. The p/v ratio is as good as 7000. In this case routinely obtained energy resolutions are in between 140 eV and 150 eV (FWHM) for the circular SDDs

area with a diameter of 300 μm around the on-chip FET can be affected by this partial signal collection mechanism, leading to a decreased P/V ratio. In addition, if X-rays of higher energies are used e.g. 15 keV to 25 keV, the X-rays eventually traverse the sensitive thickness of 300 μm to 500 μm and may be converted into electron–hole pairs close to the SiO_2 – silicon interface in the FET vicinity. In case the conversion cascade of electrons and holes is close enough to the border of the on-chip FET, they may increase the number of interface traps and give rise to an increase of fixed positive oxide charges or interface states. The latter effects degrade the performance of the on-chip SSJFET.

The following three reasons, namely, partial signal collection, increased radiation hardness and reduced output node capacitance, have led us to the design of a SDD with eccentric readout FET[33] as shown in Figure 4.1.24. Energy resolutions at the MnKα line of an ^{55}Fe source were measured down to 124 eV at $-20\,^{\circ}$C with count rates around 1000 cps. Typical values scatter around 128 eV FWHM (Figure 4.1.25), being confirmed by many other users, both scientific and industrial. The SD^3 under test had a sensitive area of 5 mm^2. The electronic noise contribution at $-15\,^{\circ}$C was 5 electrons (rms).

The SD^3 detector now has the potential to directly compete with Si(Li) detectors in the detection of light elements. Cryogenics are not longer needed and microphonics is completely avoided. The SD^3 will shortly be also available in larger formats up to 30 mm^2.

4.1.10.2 WORKS OF ART INVESTIGATIONS WITH SILICON DRIFT DETECTORS

In archeometry different kinds of investigations are used for the characterization of art objects. In particular, XRF (X-ray fluorescence) spectroscopy is a nondestructive technique widely used for the identification of chemical elements in pigments, metal alloys, and other materials. The classical high resolution cryogenic detectors, like Si(Li) and HP(Ge) detectors (whose energy resolution is of the order of 130 eV FWHM at the Mn Kα line), are not completely suitable for the realization of portable instrumentation because of the need of liquid nitrogen in the cooling system.

Recently new silicon PIN diodes simply cooled by a Peltier element have been introduced. Their energy resolution (of the order of 200 eV FWHM at the Mn Kα line at $-30\,^{\circ}$C) is in some cases unsatisfactory (especially for the analysis of light chemical elements). At low energy, the main contribution to the FWHM is due to the electronic noise of the detector front-end system, which is

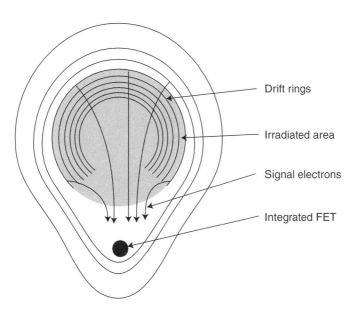

Figure 4.1.24 Silicon drift detector 'drop' with asymmetric JFET location. The drift of the electrons is shaped by adequate field structures on one of the surfaces. The read node capacitance was reduced by more than a factor of 2. In a practical application the on-chip FET will be excluded from direct X-rays by a collimator

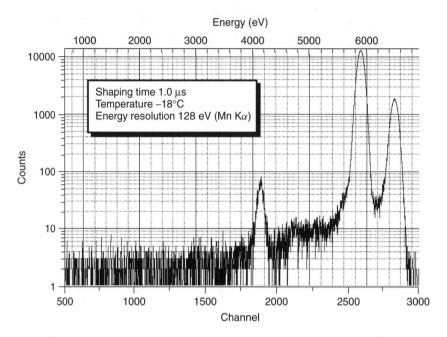

Figure 4.1.25 Measured Mn Kα and Mn Kβ spectrum with an SD^3 from an ^{55}Fe source. The peak-to-background ratio is about 9000:1, the FWHM 128 eV at a shaping time of 1 μs. The electronic noise (ENC) was 6 electrons only

associated with the detector–capacitance, directly dependent on the detection area. In addition this performance is only obtained with a PIN type detector at very low output count rates, typically below 1000 cps.

The possibility to operate the SDD at non cryogenic temperatures, i.e. at or close to room temperature and the good energy resolution (in the order of 140 eV FWHM at 6 keV) makes these detectors suitable for the realization of high resolution portable instrumentation. Recently a portable high resolution X-ray spectrometer – based on the SDD, cooled by a Peltier element – was realized at the research laboratories of Politecnico di Milano.[34] A commercial miniaturized X-ray tube was utilized as an excitation source.

The measurements on different kinds of art objects confirmed the ease of use combined with the high class performance, in particular the high energy resolution. Figure 4.1.26 shows a spectrum of an orange pigment recorded with the above described system. The almost background-free detection of the individual chemical elements helps to identify the composition of complex materials directly at the location of the work of art. Many measurements were carried out in museums and cathedrals in Italy, Germany and France.

4.1.10.3 ELEMENT IMAGING IN ELECTRON MICROSCOPES WITH SILICON DRIFT DETECTORS

Silicon drift detectors have been tailored to use them as energy dispersive spectrometers in scanning electron microscopes. The RÖNTEC-XFlash^{TM} system was developed to record at temperatures achievable with a single stage Peltier cooler, X-ray fluorescence spectra at about 10 times higher count rates than conventional energy dispersive X-ray spectrometers.[35] This results in spatially resolved element mapping with a high dynamic range, i.e. several hundred grey levels within short measurement times. Figures 4.1.27 and 4.1.28 show the analysis of a damaged pressure sensor with a spatial resolution of 1024 × 768 pixels. The measuring time was 8 min and the average X-ray count rate was 500 000 cps. The detector system used

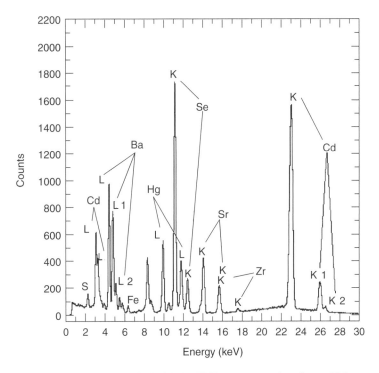

Figure 4.1.26 Spectrum of an orange pigment acquired with the XRF spectrometer based on a SDD

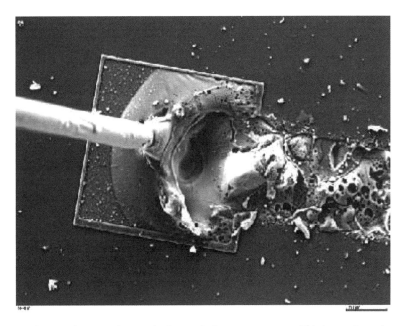

Figure 4.1.27 Scanning electron microscope image of a damaged micropressure sensor. This image shows the topological structure of the electron channel. A 20 keV acceleration voltage was used (photo: RÖNTEC)

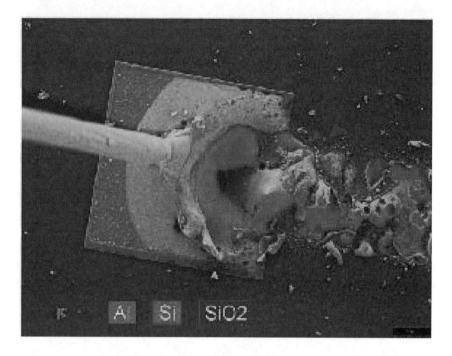

Figure 4.1.28 Element imaging with a SDD of the damage pressure sensor shown in Figure 4.1.27. This Al plated component was destroyed through electrical overload. The local power dissipation melted the Al layer such that the Si underneath appeared. Because of their nearly identical atomic mass, these elements can hardly be distinguished in the electron image (Figure 4.1.27), but the chemical components can clearly be identified in the SDD image. The spatial resolution is 1024 × 768 pixels, the measuring time was 8 min and the average count rate was 500 000 cps (photo: RÖNTEC)

is the ROENTEC XFlash™ operated at −10 °C. The RÖNTEC-ColorSEM™ system was used for the data acquisition, signal and image processing. Figure 4.1.27 shows the conventional image recorded with the electron detector. Figure 4.1.28 shows the spectroscopic SDD image with Al, Si and SiO_2 as the main chemical components of the pressure sensor. Compared to conventional energy dispersive X-ray systems, a factor of 10 is gained in the number of grey steps for comparable measurement times. In total, 90 different elements can be resolved.

4.1.10.4 SILICON DRIFT DETECTOR ARRAYS

Another concept for (coarse) position resolved X-ray spectroscopy with SDDs, for ultra high count rates is shown in Figure 4.1.29.[7] The whole sensitive surface is segmented in relatively small drift detectors, each having its own amplifying chain.[36] Every channel is connected to an individual signal processor, producing its own position resolved X-ray spectrum with count rates around 100 000 cps at a temperature of 0 °C with a resolution of better than 160 eV for the MnKα line of an ^{55}Fe source (see Figure 4.1.30). This corresponds to a count rate capability of 2×10^6 cps per cm^2, which is just right for the X-ray holography applications planned at HASYLAB (see Section 4.1.10.6).

Silicon drift detector arrays have also been used for the scintillation light readout of CsI(Tl) crystals in γ-ray cameras. The high quantum efficiency at a wavelength around 4500 Å and the low noise readout have led to an energy resolution in the detection of γ-rays better than the conventional photomultiplier readout of the scintillator[37] (see Section 4.1.10).

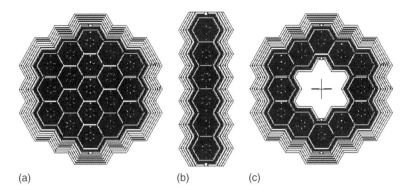

Figure 4.1.29 Examples of multicell drift detector layouts: (a) large area detector 19 cells, 95 mm^2, exists equally as a 39 cell detector unit with 195 mm^2 and as a 61 cell detector having 305 mm^2; (b) linear chain, 6 cells, 30 mm^2; and (c) closed ring with a hole in the centre, 12 cells, 60 mm^2. All plots show the layout of the detectors structured front side with the field strip system and the readout transistor in each cell's centre. The hexagonal cell shape has be chosen for an optimum ratio of area and border length

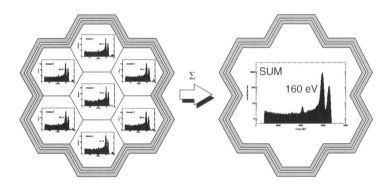

Figure 4.1.30 Spectroscopic performance of the seven-channel SDD. The sum of seven simultaneously recorded ^{55}Fe spectra yields an energy resolution of 160 eV (FWHM at 5.9 keV) at a temperature of $-20\,^\circ$C. The detector was irradiated without collimator. The low energy background is therefore dominated by split events, i.e. charge sharing between neighboring detector cells when an X-ray hits the border

4.1.10.5 COMPACT X-RAY FLUORESCENCE SPECTROMETER

The 12-channel ring detector of Figure 4.1.29 with a hole in its centre is the basic component of a compact X-ray fluorescence (XRF) spectrometer (Figure 4.1.31).

The sample is excited by X-rays passing through the central hole of the detector chip, and the SDD ring device receives the characteristic fluorescence photons emitted by the sample covering a large fraction of the solid angle around the sample. The X-rays from the source may be intensified and focused by a capillary fibre as shown in Figure 4.1.31. This compact XRF spectrometer has been implemented in a table-top system for material analysis.

4.1.10.6 X-RAY HOLOGRAPHY

X-Ray holography is a new experimental method to reveal the crystal structure of solids. The sample is irradiated by an intense beam of monochromatized synchrotron radiation X-rays under different angles. The variation of the emitted fluorescence intensity with the direction of the incident photons allows the direct three-dimensional reconstruction

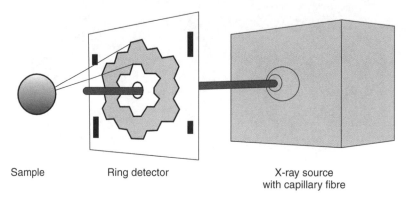

Figure 4.1.31 Principle of a compact XRF spectrometer using the SDD ring structure

of the sample's electron density distribution. This technique requires the energy-dispersive detection of an enormous number of scattered and fluorescent X-ray photons to gain sufficient statistics. Therefore it depends on a large area and fast detector system. Single cell SDDs and small multicell drift chips like the one shown in Figure 4.1.29 have already been used in X-ray holography. A new system currently in production aims towards the complete detection of all photons emitted by the sample (Figure 4.1.32). More than 1000 SDD cells grouped in 61-channel multicell drift chips and arranged in a football-like configuration cover almost the complete sphere around the sample. The pixelation of the detector allows the channels to be discarded in the direction of intense Bragg reflections that do not carry useful information.

4.1.10.7 A γ-RAY CAMERA USING MULTICHANNEL DRIFT DETECTORS

The use of SDDs for the detection and spectroscopy of photons is on the high energy end restricted to 20 to 30 keV, limited by the low atomic number of silicon and the detector thickness, which is 500 μm. For hard X-rays and γ-rays either direct converting detectors of high-Z materials like Cd(Zn)Te have to be used or indirect converting systems like scintillators coupled to photomultiplier tubes (PMTs). Recently, scintillation detectors experienced a step forward in performance by replacing the PMT by a SDD

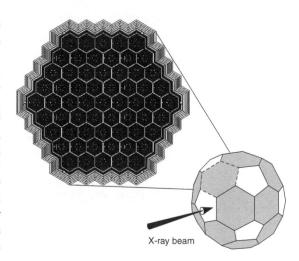

Figure 4.1.32 Experimental setup for X-ray holography. The irradiated sample is surrounded by an almost complete sphere of detectors. Each hexagon of the football-like configuration is a 61-channel SDD. The whole detector consists of more than 1000 channels

used as a low-capacitance photon detector for the scintillation light. That way not only the known practical problems of PMTs like requirement of space, incompatibility with magnetic fields, and the necessity of a high voltage are avoided, but also the quantum efficiency and energy resolution of the system are improved. The transmittance of the SDD entrance window can be tuned by deposition of anti-reflective coatings to the emitted wavelengths of a wide range of scintillators. For 122 keV X-rays a position resolution of 300 μm was measured and simultaneously an energy resolution of 15 %.[38]

Compared to direct converting pixelated Cd(Zn)Te detectors with equal position resolution the scintillator–SDD combination requires a considerably lower number of readout channels. In addition it has the advantages of comprehensive material experience, existing technologies, proved long term stability, and practically unlimited availability of high quality material.

For the readout of multichannel drift (Mc drift) chips the amplifier chip ROTOR (rotational trapezoidal readout) based on JFET-CMOS technology has been developed. ROTOR includes a preamplifier, a filter amplifier, and a peak stretcher for each detector channel, as well as one analogue multiplexer that sends the series of data of all channels to an external flash ADC. The system is able to handle the random asynchronous event occurrence of a large number of SDDs with low read noise at count rates exceeding 10^5 cps per channel. At present 165 eV (FWHM) have been achieved with an ^{55}Fe source.

4.1.11 THE CONTROLLED DRIFT DETECTOR (CDD)

For X-ray imaging purposes, a detection scheme which needs the time mark of the X-ray to be detected is disadvantageous. A new readout scheme was invented recently.[39,40] In addition to the drift field for the transport of charges parallel to the wafer surface, a potential barrier for electrons is implemented which provides a channel guide for electrons,[41] thus the lateral spread of the charges is prevented (Figure 4.1.33). This technique is very similar to the channel stop configurations for the pn-CCDs to be described in Section 4.1.12. In order to keep the generated electrons at their position for a well defined time, i.e. in the integration time for the incoming photons, an additional electron potential barrier is formed perpendicular to the channel stop implants. This additional control of the electrons in the direction of the readout node can be made by means of clocking the drift strips. At a given, externally determined moment, the potential barriers are released, defining the start signal for the drift time

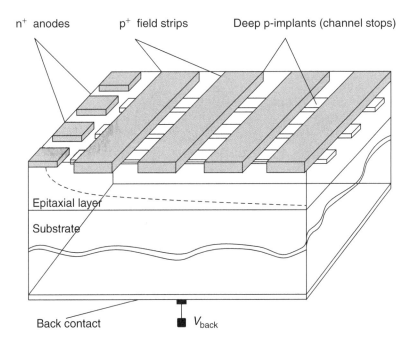

Figure 4.1.33 Cut through a CDD perpendicular to the wafer surface, and parallel to the electron drift direction

measurement. Upon arrival at the readout node, the stop mark is measured, and thus the position of the generated signal charge cloud according to Equation (4.1.11) calculated. No external trigger is needed, the trigger signal is generated by the detector system itself.

The pn-CCD as well as this concept use a homogeneous large rectifying p^+ implant for the complete depletion of the detector (Figure 4.1.33). The large p^+ (backside) contact is the radiation entrance window for the photons. As discussed already in Section 4.1.7 (Figure 4.1.13), a structured surface would lead to (a) an inhomogeneous response as a function of the incident photon energy and position and (b) incomplete charge collection of the signal charge package when produced close to the gaps of p^+ field strips.

The simulation of the CDD intuitively shows the functional principle: Figure 4.1.34 shows the electric potential during the photon integration time, which must be long compared to the signal drift and readout time. Figure 4.1.35 shows the potential distribution once the drift strips have been clocked to remove the electron potential barrier to allow the electron drift towards the read nodes.

The time measurement of the arriving electrons with a precision of 10 ns would yield a position measurement precision of 50 μm for standard drift fields, depending on the pixel size. The read out system has a 'fast channel' for the time measurement and a 'slow channel' for the charge measurement. This scheme makes the CDD extremely interesting for X-ray measurements for high photon rates and fast, low noise readout. As the SDD and the pn-CCD, the CDD has on-chip amplifiers integrated on the detector, one for each individual readout node.

Concerning the readout speed and count rate capability, the CDD is a real alternative to the CCDs (pn-CCDs). The drift towards the readout node happens with a velocity which is only controlled by the externally applied drift field. The transfer is not interrupted by a pixel wise reading of the charge content, as in the case of CCDs. For a 1 cm long drift distance a CDD typically needs 5 μs, while a pn-CCD type detector requires about 500 μs.

The development status of the CDD is progressing towards a detector system, which is showing all performance parameters as designed. To date, the CDD is existing in a prototype version.[42]

Figure 4.1.36 shows the position and energy resolved X-rays from a ^{55}Fe source recorded at 100 kHz frame rate. The same set-up was used

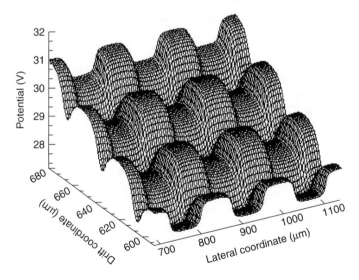

Figure 4.1.34 Simulation of the two-dimensional controlled SDD in the signal accumulation mode. The generated signal charges are confined in their local potential minima

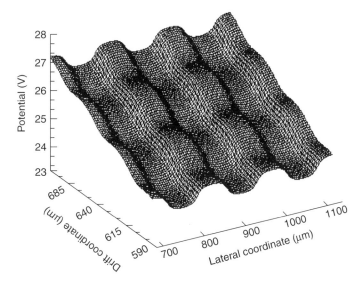

Figure 4.1.35 Potential simulation of the two-dimensional controlled SDD in the signal drift mode. The signal charges in the pixels drift towards the readout nodes

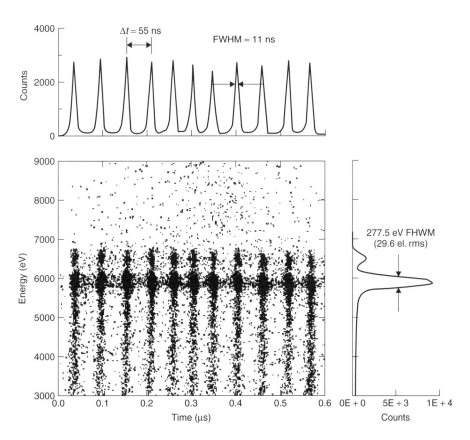

Figure 4.1.36 Position and energy resolved image of a CDD detector recorded at a frame rate of 100 kHz. At room temperature an energy resolution of 277 eV was measured

to take first images at the synchrotron in Trieste (Figure 4.1.37). In the meantime, CDDs have been produced in larger formats. They are foreseen as scattering detectors for small animal analysis in Compton camera systems.

4.1.12 FULLY DEPLETED BACKSIDE ILLUMINATED pn-CCDs

Conceptually the pn-CCD is a derivative of the SDD.[5] The development of the pn-CCDs started in 1985. In the following years the basic concept was simulated, modified and designed in detail.[43] n-channel JFET electronics was integrated in 1992,[3,44] and the first reasonably fine working devices were produced in 1993. Up to then, all presented devices were 'small' devices, i.e. 3 cm^2 in sensitive area.[15]

The flight type large area detectors for X-ray astronomy (6 × 6 cm^2) were produced from 1995 to 1997, with a sufficiently high yield to equip the X-ray satellite missions ABRIXAS and XMM[20,45,46] with defect-free focal plane pn-CCDs.

XMM was launched on 10 December 1999 from Kourou in French-Guiana. Commissioning of the scientific payload was completed in the middle of March 2000. In this overview, the basic instrument features as measured on the ground and in orbit will be shown.

4.1.12.1 THE CONCEPT OF FULLY DEPLETED, BACKSIDE ILLUMINATED, RADIATION HARD pn-CCDs

For ESA's X-ray Multi Mirror Mission (XMM), we have developed a 6 × 6 cm^2 large monolithic X-ray CCD[47] with high detection efficiency up to 15 keV, low noise level (ENC≈5e(rms) at an operating temperature of −90 °C) and an ultra-fast readout time of 4.6 ms per 3 × 1 cm^2 large subunit (Figure 4.1.38). A schematic cross-section, already showing some of the advantages of the concept is displayed in Figure 4.1.39.

The pn-CCD concept and the fabrication technology allow for an optimum adaptation of the pixel size to the X-ray optics, varying from 30 μm up to 300 μm pixel size. Up to now systems with 50 μm to 200 μm have been produced. The XMM telescope performance of 13 arcsec half energy width (HEW) translates to 470 μm position resolution in the focal plane. The FWHM of the point spread function (PSF) is about 7 arcsec. A pixel size of 150 × 150 μm^2 was chosen, with a position

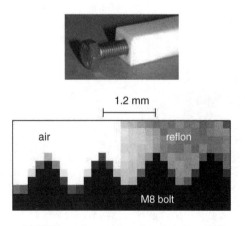

Figure 4.1.37 X-ray image of an iron bolt in a Teflon nut. The image was recorded at the synchrotron in Trieste with an energy of 15 keV. The frame rate was 100 kHz

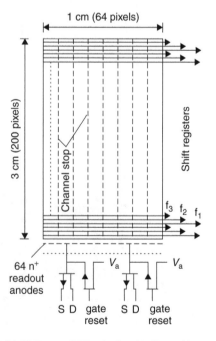

Figure 4.1.38 One pn-CCD subunit with 64 on-chip amplifiers and a size of 3 × 1 cm^2. Each column is terminated by an on-chip JFET amplifier

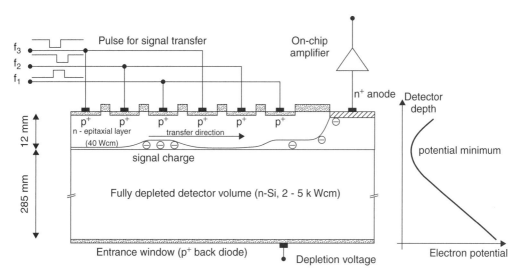

Figure 4.1.39 Schematic cross-section through the pn-CCD along a transfer channel. The device is back illuminated and fully depleted over 300 μm thickness. The electron potential perpendicular to the wafer surface is shown on the right-hand side

resolution of 120 μm, resulting in an equivalent spatial resolving capability of 3.3 arcsec. This is sufficient to fully conserve the positional information from the X-rays from the mirrors. The quantum efficiency is higher than 90 % at 10 keV because of the sensitive thickness of 300 μm.

The low energy response is given by the very shallow implant of the p^+ back contact; the effective 'dead' layer is smaller than 200 Å.[14] The good time resolution is given by the parallel readout of 64 channels per subunit, 768 channels for the entire camera. A high radiation hardness is built in by avoiding active MOS structures and by the fast transfer of the charge in a depth of more than 10 μm.

The spatially uniform detector quality over the entire field of view is realized by the monolithic fabrication of the pn-CCD on a single wafer. For reasons of redundancy 12 individually operated $3 \times 1\,\text{cm}^2$ large pn-CCDs subunits were defined. Non-homogeneities were not observed over the whole sensitive area in the energy band from 500 eV up to 8 keV, the precision of the measurements was always limited by Poisson statistics. The insensitive gap in the vertical separation of the pn-CCDs is about 40 μm, neighbouring CCDs in horizontal direction have insensitive regions of 190 μm.

The basic concept of the pn-CCD is shown in Figure 4.1.39 and is closely related to the functional principle of the SDDs. A double-sided polished high resistivity n-type silicon wafer has both surfaces covered with a rectifying p^+-boron implant. On the edge of the schematic device structure (Figure 4.1.39) a n^+-phosphorus implant (readout anode) still keeps an ohmic connection to the nondepleted bulk of the silicon. A reverse bias is now applied to both p^+ junctions, i.e. a negative voltage is applied with respect to the n^+ anode. For simplicity let us assume, that the silicon bulk is homogeneously doped with phosphorus with a concentration of 1.0×10^{12} per cm^3. The depletion in the high ohmic substrate, with a resistivity of about $4\,\text{k}\Omega\text{cm}$, develops from both surfaces, until the depletion zones touch in the middle of the wafer in the case of homogeneous doping of the wafer. The potential minimum for electrons is now located in the middle of the wafer. An additional negative voltage on the p^+ back diode shifts the potential minimum for electrons out from the centre towards the surface having the pixel structure. Typical depletion voltages on the backside are $-150\,\text{V}$. To make a CCD-type detector, the upper p^+ implant must be divided

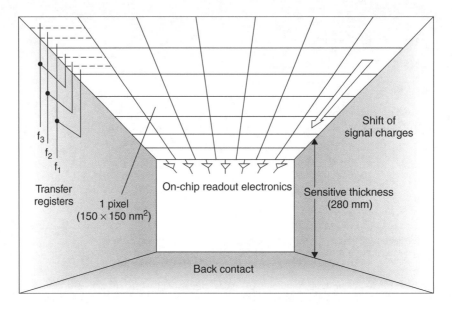

Figure 4.1.40 Inside the pn-CCD. The X-rays hit the device from the backside (bottom). The charges are collected in the pixel well close to the surface having the pixel structure. After integration, they are transferred to the on-chip amplifier

in p^+ strips as shown in Figure 4.1.39 and 4.1.40. Adequate voltages should now be applied to the three shift registers, such that they form local electron potential minima at a distance of approximately 10 μm from the surface. Three p^+ strips (shift registers) with the potentials (Φ_1, Φ_2 and Φ_3) comprise one pixel. Charges are collected under Φ_3, the potential minimum for electrons. A reasonable change with time of the applied voltages transfers the charges in the local electron potential minimum in a discrete way towards the n^+ readout node. In reality the side having the p^+ shift registers has an additional phosphorus doped epitaxial layer, 12 μm thick, with a concentration of approximately 10^{14} donors per cm³. The interface of the epi-layer and the high resistivity bulk silicon fixes the electron potential minimum to a distance of about 10 μm below the surface.

As can be seen in Figure 4.1.38, one pn-CCD subunit consists of 64 individual transfer channels each terminated by an on-chip JFET amplifier. Figures 4.1.41, 4.1.42 and 1.4.43 show the charge transfer mechanism in a depth of approximately 10 μm below the shift registers. The p^+ backside contact is not shown: it expands quite uniformly an additional 260 μm towards a negative potential of −150 V. The sequence of changing potentials shows nicely the controlled transfer from register Φ_3 to register Φ_2, one-third of a pixel.

This concept is seen from a different point of view in Figure 4.1.40, seen from the inside of a pn-CCD: X-rays hit the detector from the rear side (back contact). The positively charged holes move to the negatively biased back side, electrons to their local potential minimum in the transfer channel, located about 10 μm below the surface having the pixel structure. The electrons are fully collected in the pixels after 5 ns at most, the collection of holes is completed in no more than 15 ns because of their reduced mobility. As can be seen in Figure 4.1.40, each CCD line is terminated by a readout amplifier. The on-chip single sided JFET has already been described in Section 4.1.4 as the first amplifying element in the SDD.

The focal plane layout of XMM is depicted in Figure 4.1.44. Four individual quadrants each having three pn-CCD subunits are operated in parallel. The camera housing and its mechanical, thermal and electrical properties are described elsewhere.[48]

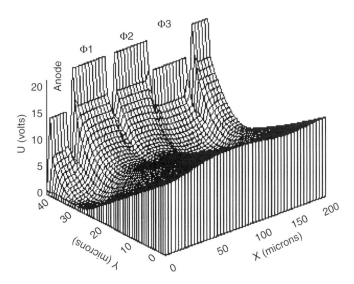

Figure 4.1.41 Negative potential of a pn CCD shift register. In this operating condition the signal charges are stored under the register Φ_3 only. The p^+ backside potential is only shown up to the depth of $40\mu m$ for clarity. It expands to $-150\,V$ at $300\mu m$ distance from the pixel surface

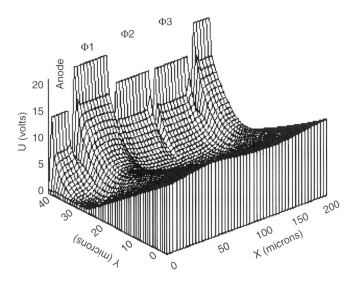

Figure 4.1.42 Negative potential of a pn-CCD shift register. In this operating condition the signal charges are stored under the registers Φ_2 and Φ_3. The electrons now share a larger volume for a short time. Note that the electrons are still nicely confined in the potential well

4.1.12.2 LIMITATIONS OF THE CCD PERFORMANCE

The performance of the CCDs is subject to several limitations: physical and technical. Some of these limitations, which are subject to most detectors described in this subchapter, have already been discussed in Sections 4.1.6 and 4.1.7. As a result, a distortion of the spectrum to lower energies has been found (Figure 4.1.13 and 4.1.21;

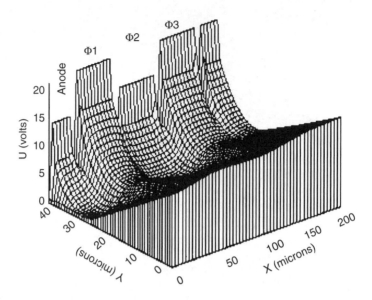

Figure 4.1.43 Negative potential of a pn-CCD shift register. In this operating condition the signal charges are stored under the register Φ_2 only. The charge was transferred by one-third of the pixel length in approximately 150 ns

the quantum efficiency as function of energy is presented in Figure 4.1.14). The low energy tail in the spectrum (p/V ratio) and the drop in quantum efficiency at low energy is due to partial charge collection of photons converting in or near the entrance window. This drop is still much lower than in other (MOS) CCDs. The loss at high energy is due to the limited thickness of the silicon wafer. Figure 4.1.14 also indicates the improvement achievable by increasing the sensitive thickness to 500 μm.

As the bias voltage required for depletion grows with the square of the thickness but only linearly with doping density (Equation 4.1.1) the maximum depletable thickness achievable thickness at reasonable bias is limited. As an example, 4.5 kΩcm n-type silicon ($N_D = 10^{12}$ donors per cm^3) and a bias voltage of 500 V result in a depletion depth of 800 μm.

Important aspects on the subject of energy resolution were also treated in Sections 4.1.4 and 4.1.6 and for drift detectors in Section 4.1.9. These aspects concern electronic noise (Section 4.1.4), ionization statistics (Section 4.1.6) and their combined effect (Section 4.1.9). Those results apply also to CCDs. According to Figure 4.1.45, the

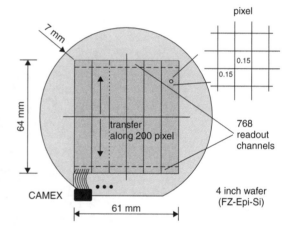

Figure 4.1.44 The focal plane of the pn-CCD camera on XMM and ABRIXAS consist of 12 independent, monolithically integrated pn-CCDs with a total area of $6 \times 6 \text{cm}^2$. In total 768 on-chip amplifiers process the signals and transfer them to a VLSI JFET-CMOS amplifier array. 12 output nodes of the CAMEX arrays are fed into 4 ADCs, i.e. one ADC per quadrant

Fano noise is dominant for energies above 1 keV for an electronic ENC of 5 electrons. For 1 electron noise (ENC) this threshold is lowered to 50 eV only.

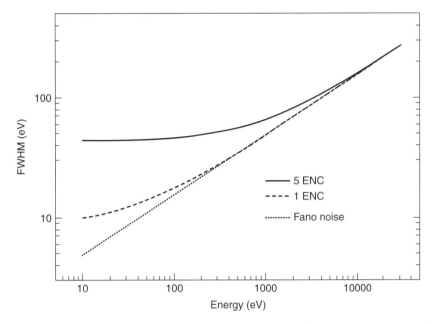

Figure 4.1.45 Energy resolution as a function of the photon energy. The Fano noise is taken into account as well as a 5 electron and 1 electron electronics ENC, respectively

Added to these contributions is an essential property of CCDs, the loss of charge during transfer from the place of origin towards the readout node.

Charge Transfer Noise

In CCDs, where the signal electrons are transferred over many pixels, the charge shifting mechanism from pixel to pixel must be excellent. Leaving charges behind during the transfer means reducing signal amplitude. This loss can be corrected, but adds noise to the signal amplitude measurement.

$$\text{ENC}^2_{\text{trans}} = n_{\text{lost}} = q\frac{E}{w}(1 - \text{CTE})N_{\text{trans}} \quad (4.1.12)$$

In a simple model the lost charges n_{lost} can be parameterized according to Equation (4.1.12), where the left behinds can be considered as a backward flow loss of electrons. As the loss process is of statistical nature, it is treated similar to a signal leakage current. N_{trans} denotes the number of transfers in the CCD and CTE is the charge transfer efficiency, a number close to one, of the order 10^{-5}, depending on CCD type, radiation damage, temperature, etc.

For the pn-CCD (as for any other SDD-type system) the electronic noise can be reduced by reducing the read node capacitance, by lowering the leakage current and the $1/f$ noise constants and by optimizing the shaping time constant τ. The total read noise, if not correlated, can be added quadratically

$$\text{ENC}^2_{\text{tot}} = \text{ENC}^2_{\text{el}} + \text{ENC}^2_{\text{fano}} + \text{ENC}^2_{\text{trans}} + \ldots \quad (4.1.13)$$

and delivers the total ENC. State of the art systems of today exhibit electronic noise figures around 3 electrons and a readout speed per pixel below 1 μs and operating temperatures higher than $-90\,°\text{C}$.

In pixelated detectors not all the signal charges of one single X-ray photon will be collected in a single pixel, so the electronic content of several pixels must be added. This increases, for the so-called split events, the noise floor by $\sqrt{N}$, with N as the number of pixels affected.

4.1.12.3 DETECTOR PERFORMANCE (ON GROUND AND IN ORBIT)

The best values for the readout noise of the on-chip electronics is 3 electrons rms at 180 K for the most recent devices; typical values scatter around 5 electrons rms for the XMM system. This includes all noise contributions described in Equation (4.1.4). The charge transfer properties of the pn-CCDs on XMM are reasonably good, in the order of a several percent signal loss from the last to the first pixel over a distance of 3 cm charge transfer. As the charge transfer losses describe the position dependent energy resolution, it is one of the key parameters for the spectroscopic performance, especially after radiation damage may have occurred.

Figure 4.1.46 shows a ^{55}Fe spectrum of a pn-CCD in a flat field measurement resulting in a typical energy resolution of 130 eV at an operating temperature of $-120\,°C$.[15] The XMM flight camera was operated at $-90\,°C$ during calibration on the ground with a resolution of about 145 eV (FWHM) over the entire area of 36 cm^2.

The main effect on the degradation of energy resolution was the reduction of the charge transfer efficiency (CTE) at warmer temperatures. Leakage currents and on-chip JFET properties only played a minor role. The impact of the material properties of silicon and related impurities and their consequences for the operation of scientific grade X-ray pn-CCDs including the effects of radiation damage, is treated in detail in the literature.[49,50]

The equivalent dose of 10 MeV protons over the expected life time of XMM is 4×10^8 p/cm^2. Figures 4.1.47 and 4.1.48 show the results of the irradiation tests with 10 MeV protons: the expected decrease of energy resolution over the 10 year dose is from 145 eV to 158 eV at an operating temperature of $-100\,°C$. At the actual operating temperature of $-90\,°C$ the expected effect of trapping and detrapping at A-centres, generated by the radiation, is even more reduced.

In a single photon counting mode the quantum efficiency was measured with respect to a calibrated solid state detector. Figure 4.1.14 shows measurements from the synchrotron radiation facilities in Berlin and Orsay. At 525 eV a 5 % dip can be seen from the absorption at the oxygen edge in

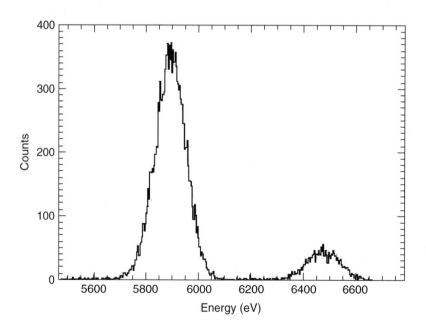

Figure 4.1.46 Mn Kα spectrum of an ^{55}Fe source. The measured FWHM is 130 eV at $-120\,°C$

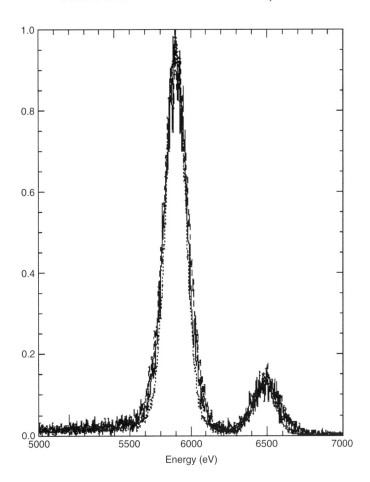

Figure 4.1.47 ^{55}Fe energy spectrum after different 10 MeV proton fluencies of 0 p/cm^2 (dotted line), 4.1×10^8 p/cm^2 (solid line), 6.1×10^8 p/cm^2 (dashed line), measured at the low (and after irradiation unfavourable) temperature of 142 K. The expected dose over a life time of 10 years is 4.0×10^8 MeV p/cm^2

the SiO$_2$ layers. The same happens at the SiK-edge at 1840 eV showing the fine structure of a typical XAFS spectrum (see inset of Figure 4.1.14). For all energies the quantum efficiency is nicely represented by a model using the photoabsorption coefficients from the atomic data tables. The quantum efficiency on the low energy side can be further improved with respect to the measurements shown in Figure 4.1.14 by increasing the drift field at the p$^+$ junction entrance window[14] and by using <100> silicon instead of <111> silicon. The useful dynamic range of the pn-CCD camera on XMM was adjusted from 100 eV to 15 keV (Figure 4.1.49).

Split events, i.e. events with electrons in more than one pixel, originating from one single photon, were reconstructed and summed to one photon event. In total, about 70% of all events are single pixel events, 28% are two pixel events and 2% are events with three and four pixels involved. In the case of the XMM pn-CCDs one single X-ray photon never spreads the generated signal charge over more than four pixels.

The readout electronics of the pn-CCD system is described in the literature.[47,51] A charge sensing amplifier followed by a multicorrelated sampling stage, multiplexer and output amplifier (CAMEX64B JFET/CMOS chip) guide the

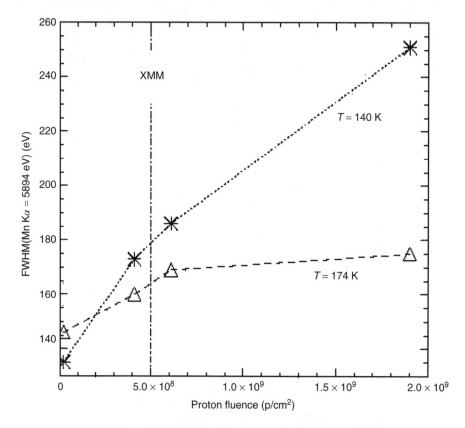

Figure 4.1.48 FWHM of the Mn Kα spectrum (5894 eV) with dependence on proton fluence and temperature. Before proton exposure the lower operating temperature of 140 K gains better results. After a 10-MeV proton fluence of more than 2.0×10^8 cm^{-2} the higher temperature of 174 K results in a better energy resolution. The FWHM is degraded from 135 eV (140 K) to 160 eV and 175 eV (174 K) after 4.1×10^8 p/cm^2 and 1.9×10^9 p/cm^2, respectively. A FWHM of 160 eV is expected after the 10 year XMM mission

pn-CCD pixel content as a voltage signal to a 10 MHz 12-bit flash ADC system. The whole system, i.e. CCD and CAMEX64B amplifier array dissipate a power of 0.7 W for the entire camera (768 readout channels), a value which is acceptable in terms of thermal budget on XMM realized through passive cooling. A further increase of the readout speed can be made only at the expense of further increase of power, or a degradation of the noise performance.

The charge handling capacity of the individual pixels was tested with the 5.5 MeV α particles from a radioactive ^{241}Am source. Around 10^6 electrons can be properly transferred in every pixel. The spatial resolution of the camera system was intensively tested in the PANTER facility with the flight mirror module in front of the focal plane. The first light image of the Large Magellanic Cloud in (Figure 4.1.50), as well as the quantitative analysis of the point spread function have shown a perfect alignment of the telescope system on the ground: the spatial resolution of the entire telescope system measured on the ground corresponds exactly to the performance in orbit.

The operating temperature of XMM in orbit is $-90\,^\circ$C. This temperature optimizes on one side the requirement of 'warm' operating conditions to avoid contamination and to release stress to the mechanical structures. On the other side, it matches the need of 'cold' temperatures because of leakage current reduction and efficient charge transfer.

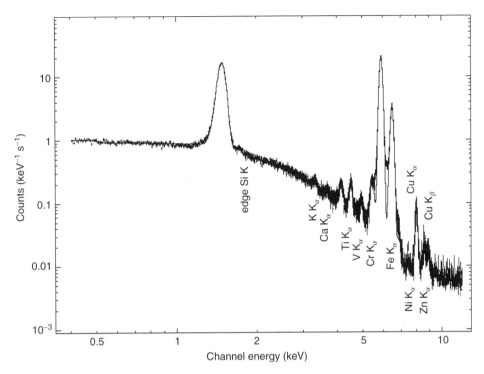

Figure 4.1.49 Calibration spectrum with the internal radioactive source including the background with the filter wheel in closed position. The continuous background below the Mn lines arises mainly from photoelectrons stimulated from the ^{55}Fe source in the Al target. The iron Kα line between Mn Kα and Mn Kβ is not resolved. The additional lines are due to X-ray fluorescent background from the camera structure

Up to now, almost 4 years after launch, no instrumental surprise has occurred: the energy resolution is equal to the ground measurements, as is the case for the charge transfer efficiency. To date, the electrical stability of the instrument is perfect. The first light image in (Figure 4.1.50) and the observation of Tycho Brahes supernova remnant, which includes chemical analysis from the X-ray spectra (Figure 4.1.51), qualitatively summarize the above statements.

4.1.12.4 FRAME STORE pn-CCDS FOR ROSITA AND XEUS

Future missions and other applications require pn-CCDs with smaller pixels and even faster readout. Two potential applications are the German/European ROSITA mission to be installed on the International Space Station (ISS) in 2007 and ESA's XEUS mission to be launched around 2015.

As in conventional CCDs, pn-CCDs can equally be designed in a frame store format. This optimizes the photon integration to charge signal transfer time, but requires more space on a chip because the store area does not serve as an active area but as an analogue storage region (Figures 4.1.52 and 4.1.53). A cross-section through image and storage area is shown in Figure 4.1.54. In Table 4.1.1 the expected characteristics of the devices for ROSITA and XEUS are compared to those achieved in XMM.

The area to be processed in a quasi defect free manner increases by the size of the store area. A $7.5 \times 7.5\,cm^2$ large image area can be realized monolithically on a 6″ wafer (Figure 4.1.52). If pn-CCDs should be also used for the $14 \times 14\,cm^2$ focal plane, there may be a possible extension of the focal plane camera on XEUS (Figure 4.1.55). By that technique the whole field of view could be covered with a minimum of insensitive gaps in

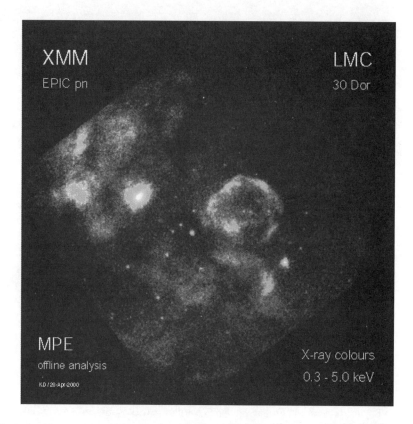

Figure 4.1.50 The Large Magellanic Cloud in X-ray colors. First light image of the pn-CCD camera. The field of view of 30 arcmin corresponds approximately to our perception of the size of the moon. The image shows the area of 30 Doradus a supernova remnant as an extended source of X-rays. The 'north-east' of 30 Dor shows an emission of X-rays up to 5 keV (blue), while the 'south-west' rim appears much softer in X-rays (yellow and red). The supernova 1987A is the bright source 'south west' of 30 Doradus. About 40 new X-ray objects have been found in this exposure. The exposure time was about 10 h

between the buttoned devices. The central part, the inner diameter of 7 cm, would be homogeneously sensitive.

The major change in concept, besides the smaller pixel size, is the dramatic increase in frame rate because of the modified readout philosophy: by doubling the processed area and dividing it in an image and store section, we will get towards the required readout speed for the large collecting area of the XEUS mirrors. We expect to get a frame rate of the whole camera of 200 per second.

As the pixel size shrinks, the number of read nodes and transfers increases. At the same time, the system will be requiring more readout time and being more sensitive to radiation damage due to the higher number of transfers. If spreading of signal charges over more than one pixel is needed for the improvement of position resolution (Figures 4.1.56 and 4.1.57), the effective read noise per event will be higher by a factor $\sqrt{n}$ (n is the number of pixels involved). The readout noise of every pixel involved must be quadratically added to get the total noise for one photon event.

To date, the signals of one row (64 pixels) are processed in parallel in 23 μs. The extension to 128 channels on the CAMEX amplifiers, to match the new pixel pitch, was realized in a new fabrication of a CAMEX128 for the low noise operation. In addition, the signal processing time must be shortened by a factor of two to obtain the same readout time per row. The increased readout speed will certainly have an impact on the power

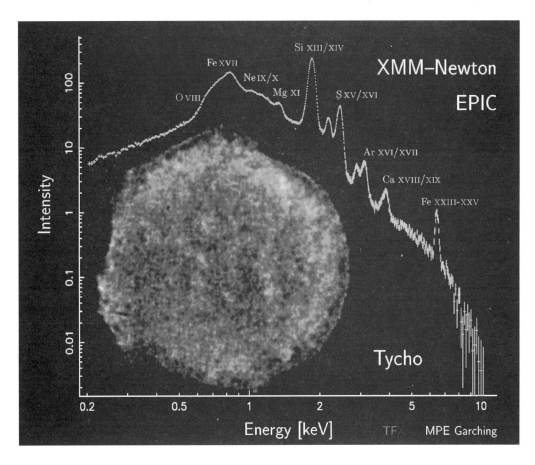

Figure 4.1.51 Supernova remnant Tycho in the Cassopeia region, discovered in 1572 by Tycho Brahe. The radius of Tycho is 12 light-years. The temperature of several million degrees gives rise to X-rays from 0.2 keV to 10 keV. The X-ray spectrum shows an abundance of many elements among those O, Mg, Si, S, Ar, Ca, Fe and Ni. The distribution of the elements is not homogeneous; the explosion dynamics are currently under study

consumption which is actually below 1 W for the 36 cm^2 array.

If 128 channels are read out with 12.8 MHz, 10 µs would be required for the parallel readout of one pixel line. For the parallel transfer from the image to the storage area 100 ns are needed for one transfer. A device of 1000 × 1000 pixels would be divided (as in the XMM-EPIC case) in two identical halves of the image area, i.e. 500 × 1000 pixels each (Figure 4.1.52). For the parallel 500 shifts, 50 µs would be needed for the transfer from the image to the shielded storage area. The readout time for the storage area while integrating X-rays in the image part, would then be 500 × 10 µs = 5 ms. That means, within 5 ms the whole focal plane would be read out. The out-of-time probability for the X-ray events will then be 1:100. In this operation mode, 200 image frames can be taken in 1 s with a full frame time resolution of 5 ms.

The above discussion shows the present technological limits for applications in space. Of course, for XRF applications smaller (and therefore faster) devices can be made.

According to the progress of the development for both detector systems – pn-CCDs and APS – a decision about the final choice for the XEUS wide field imager has to be taken at a later stage.

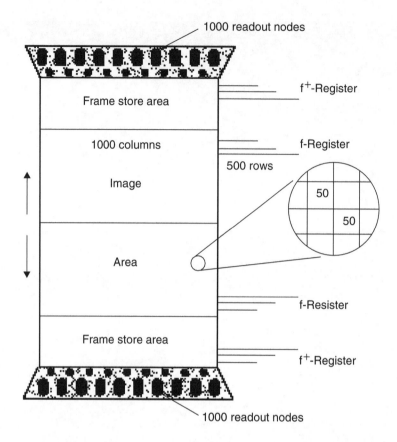

Figure 4.1.52 Example for a pn-CCD operated in a frame store mode. The imaging area may have a pixel size of $50 \times 50\,\mu m^2$ or $75 \times 75\,\mu m^2$ and a store area of $75 \times 50\,\mu m^2$

4.1.12.5 NEW DEVICES

Recently, the first prototypes of the pn-CCDs for the ROSITA mission have been tested with excellent results. The low energy response was significantly improved (Figure 4.1.58). The trigger threshold was as low as 50 eV, the peak-to-background ratio is 50:1 and the FWHM for C_K X-rays (277 eV) is below 80 eV. This width is still large compared with the theoretical limit (around 40 eV) and reveals some additional improvements to be done in the near future. At AlKα (1498 eV) the FWHM is also 80 eV.

The charge transfer efficiency was improved at least by a factor of 10 at the critical lower energies (Figure 4.1.59). That enables us to get closer to the limits given by silicon as a detector material.

4.1.12.6 OTHER APPLICATIONS

In recent years many other applications with pn-CCDs have been realized. The fields of application are very different as well as the appreciated advantages of the pn-CCD. Some examples are:

(1) *X-ray microscopy*. At BESSY II a new generation of X-ray microscope has been installed, equipped with a pn-CCD system. The high efficiency for X-rays below 1 keV and the high radiation hardness were the key properties to switch from MOS-type CCDs to fully depleted back illuminated CCDs.[52]
(2) *Plasma diagnostics*. X-Ray spectroscopy is frequently used for plasma diagnostics in fusion reactors. The temperature can be

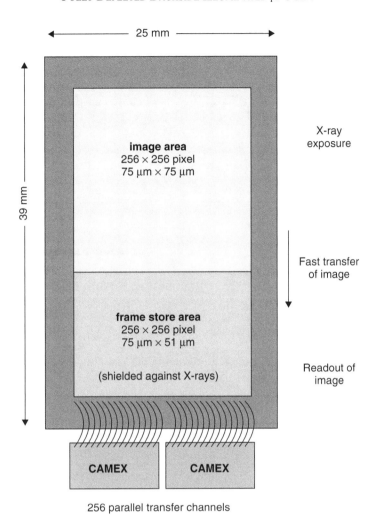

Figure 4.1.53 Smaller (prototype) version of the ROSITA pn-CCD, having a pixel size of $75 \times 75\,\mu m^2$ and a store area of $75 \times 51\,\mu m^2$ and a format of 256×256 pixels in the imaging area

determined quite precisely (the black body radiation of the plasma exhibits X-ray energies up to 10 keV with high flux) and the contamination of the plasma can be analysed.[53] It is foreseen to install a pn-CCD system at the new fusion reactor in Greifswald, Germany. The preferred properties are the high quantum efficiency at the high energies and the fast readout.

(3) *Quantum optics*. Multi-ionization processes have been studied with a pn-CCD system at the MPI for quantum optics. They have improved sensitivity by the high efficiency in the VUV range at energies around 25 eV. The radiation, stimulated with a femtosecond laser was recorded in an integration mode, i.e. not in a single photon counting mode.[54]

(4) *Electron emission channelling spectroscopy*. The emission channelling spectroscopy technique allows the direct determination of the lattice sites of radioactive impurity atoms that are incorporated into single crystalline solids.[55] Electrons from a few keV to several hundreds of keV were recorded with high precision in energy and position.

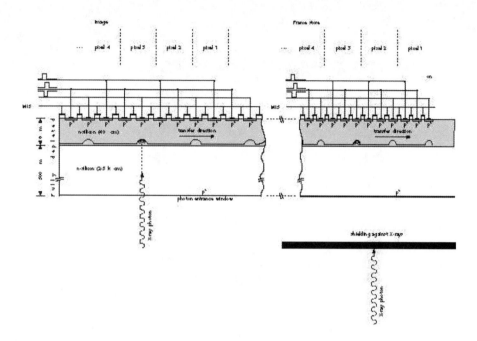

Figure 4.1.54 The frame store pn-CCD as designed for the German ROSITA mission will have a format of 256 × 256 pixels with a size of 75 × 75 μm² in the image area and 256 × 256 pixels with a size of 75 × 51 μm²

Table 4.1.1 Comparison of expected properties of pn-CCDs in XEUS and ROSITA with those reached at XMM

Property	XMM	ROSITA	XEUS
Status	Operating	Prototyping	Research
Type	Full frame	Frame store	Frame store or APS
Format	400 × 384	256 × 256	1024 × 1024
Pixel size (μm²)	150 × 150	75 × 75	50 × 50 or 75 × 75
Readout noise	5 electrons	3 electrons	1 electron
Sensitive thickness (μm)	295	450	450
Frame rate (frames/s)	50	20	200 (1000)
Readout speed (ns/pix)	350	100	50
Energy resolution at Mn Kα (5.9 keV) (eV)[a]	140	130	125
Energy resolution at C Kα (eV)	130	80	45
Energy range (keV)	0.15–15	0.1–20	0.05–20

[a]The energy resolution (FWHM) refers to incident X-rays of the MnKα line at 5.9 keV and CKα measured at temperatures around −100 °C.

(5) *CAST*. The CAST experiment is dedicated to the search of solar axions. Its main components are a large superconducting magnet of 10 T, an X-ray telescope mounted behind the magnet and a pn-CCD camera system to detect the X-ray radiation between 1 keV and 8 keV behind the telescope caused by the conversion of axions into X-rays through the Primakov effect.

(6) *Transition radiation*. A novel K-edge imaging method has been developed aiming at the efficient use of transition radiation generated by a high energy electron beam for applications in material science, biology and medicine.[55]

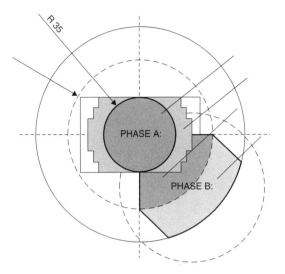

Figure 4.1.55 Possible focal plane layout with a diameter of 15 cm, composed by a central detector and four surrounding detectors with circular shaped pixels. This example shows the use of pn-CCDs, but can equally be applied to APS detectors

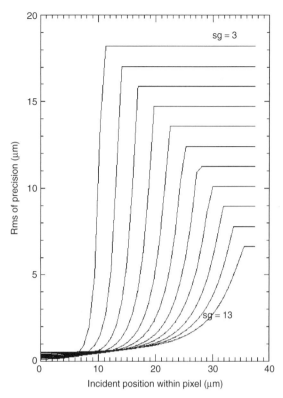

Figure 4.1.57 The same calculation is made as for Figure 4.1.56 for a pixel size of 75 μm. 1000 electrons have been generated and processed with a noise level of 5 electrons (rms). The pixel edge is located at the x-coordinate 0, while the centre of the pixel is located at 25 μm (see Figure 4.1.55) and 37.5 μm (this figure)

4.1.13 ACTIVE PIXEL SENSORS FOR X-RAY SPECTROSCOPY

Large format arrays covering a wide energy bandwidth from 1 eV to 25 keV will be used in the focal plane of X-ray telescopes of the next generation.[56] As the readout speed requirements increase drastically with the collecting area, but noise figures have to be on the lowest possible level, CCD-type detectors do not seem to be able to fulfil all the experiment needs. Active pixel sensors have the capability to arbitrarily select areas of interest and to operate at readout noise levels below 1 electron (rms).

One prominent candidate for the use of an APS is XEUS (X-ray Evolving Universe Spectroscopy

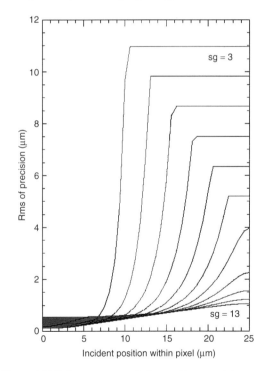

Figure 4.1.56 Improvement of the position precision as a function of the Gaussian spreading of the electron charge cloud. 1000 electrons have been generated and processed with a noise level of 5 electrons (rms). The typical sigma of the Gaussian ('sg') electron distribution is 7 μm. The assumed pixel size is 50 μm

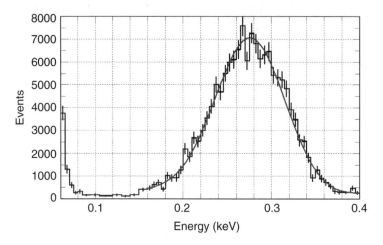

Figure 4.1.58 Carbon spectrum recorded with a frame store pn-CCD. The peak energy is at 277 eV corresponding to 73 electrons generated by the incoming low energy X-ray. The measured FWHM is around 80 eV. Due to partial absorption of signal carriers, the measured peak is shifted by 30 eV towards lower energies, if compared to the peak position of the Mn Kα line at 5.9 keV. A peak shift correction was applied

mission).[57] The launch is supposed to be around 2015. It represents a potential follow-on mission to the ESA cornerstone XMM currently in orbit. The XEUS mission is considered as part of ESA's Horizon$^+$ 2000$^+$ program within the context of the International Space Station (ISS).

4.1.13.1 THE WIDE FIELD IMAGER FOR XEUS – AN INTRODUCTION

The wide field imager (WFI) on XEUS is one out of three scientific instruments in the focal plane of the X-ray telescope with a field of view of about 5 arcmin, corresponding to 73 mm on the focal plane detector. The large collecting area (up to 30 m^2) and high angular resolution (2–5 arcsec) of the X-ray optics requires new detector technologies.[22] The physical quantities of interest are imaging (position resolution), and spectroscopy (energy resolution) with a high detection probability (quantum efficiency) in a single photon counting mode at a high photon rate (time resolution without pile-up). The first choice for the WFI is mainly driven by its count rate capabilities and the flexibility of operation. As the collecting area of XEUS in phase A is already a factor of 20 larger than XMM and a factor of 100 in phase B, it becomes clear, that a new device concept is needed rather than improvements of existing schemes. Active pixel sensors will be in the focus of our considerations, being able to match the relevant physical parameters of the WFI. The concept of the p-channel Depleted Field Effect Transistor (DEPFET) allows to measure position, arrival time and energy with a sufficiently high detection efficiency in the range from 0.1 to 30 keV. As a fallback solution, fully depleted backside illuminated frame store pn-CCDs are considered (see previous section).

4.1.13.2 THE DEVICE AND SYSTEM CONCEPT OF THE DEPFET

In all CCD-type concepts, charges are transferred slowly over large distances, they are intrinsically sensitive to radiation damage or to metallic contamination of the base material, because of the presence of traps in the bulk silicon. In addition, because of the relatively slow charge transfer, X-rays may hit the CCD during the readout time. This gives rise to events whose position is erroneously assigned in transfer direction – the so-called *out-of-time* events.

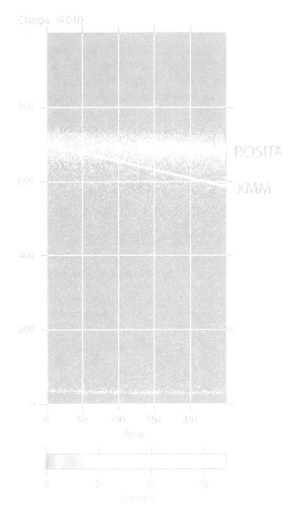

Figure 4.1.59 Improvement of charge transfer efficiency from the XMM devices to the ROSITA devices. At Al Kα (1.49 keV) the XMM charge losses over 3 cm of transfer were about 15 %, while in the new devices the loss is about 1.5 %

The XMM-EPIC pn-CCD system is limited with pile-up at count rates in the order of 20 counts per HEW and second. But with the anticipated collecting area up to 30 m² several hundreds of counts per HEW and second are expected for comparable observations. That means that a factor of 20 or more in the XEUS phase A and a factor of about 100 in phase B in frame speed is needed as compared to the pn-CCD camera on EPIC-XMM, to exploit the capabilities of the XEUS mirror system and therefore its astrophysical significance.

Perspectives of the DEPFET System

The DEPFET detector system belongs to the family of APS. That means, that every pixel has its own amplifier and can be addressed individually by external means. This results in a high degree of operational freedom and performance advantages.

The major advantages of DEPFET type devices are:

(1) Operation with high spectroscopic resolution at temperatures as high as −50 °C, keeping the total readout noise well below 5 electrons (rms) for a single reading of the signal charges.
(2) The charge does not need to be transferred parallel to the wafer surface over long distances. That makes the devices very radiation hard, because trapping (radiation induced defects), the major reason for degrading the charge transfer efficiency, is avoided.
(3) The ratio between photon integration time and read out time can be made as large as 1.000:1 for a full frame mode, that means that the so called out-of-time events are suppressed to a very large extent.
(4) As the integration time per event will be in the order of 1 ms and the read out time per line about 1 µs, more than 1000 cps per HEW (2 arcsec, i.e. 7 × 7 pixel) can be detected with a pile-up below 6 %.
(5) No additional frame store area is needed; the device is as large as the processed area.
(6) Any kind of windowing and sparse readout can be applied easily, different operation modes can be realized simultaneously.
(7) The DEPFET transistor amplifier structure offers the possibility for a repetitive nondestructive readout (RNDR). Under those conditions the readout noise can be reduced to below 1 electron (rms) by a repetitive reading of the physically same signal charge. This readout mode can be applied in selected areas, while the rest of the device is operated in the standard readout mode.

The standard DEPFET and DEPMOS devices are p-channel devices on n-type material. The use of p-type base material is very interesting for the DEPFET devices. The reasons for that is, that the use of n-channel JFETs and MOSFETs becomes possible by using holes as the signal charges. This offers an increased transconductance g_m of the transistors by a factor of three, improving the ENC at least by a factor of 1.5.

Device Concept and Functional Principle

Our DEPFET concepts are based on a detector–amplifier structure, which consists of a FET working on a depleted high resistivity substrate. The cross-section of such a device is shown in Figure 4.1.60.

The device, which was proposed by Kemmer and Lutz in 1986,[6] makes use of the sideward depletion principle.[5] Assuming that n-type semiconductor material is used, one can deplete a detector chip in such a way, that there remains a potential minimum for electrons under the channel of a FET[10] being capable of storing the signal charges for a long time – if needed, up to several seconds according to the operating temperature. It is straightforward to use such a device as detector, where signal charges (electrons) are collected in the potential minimum, from where they can steer the transistor current, acting as a so-called 'internal gate'. The signal charges change the transistor current by inducing charges inside the p-type channel of the DEPFET. The result is a simultaneous integration of the first amplifier stage

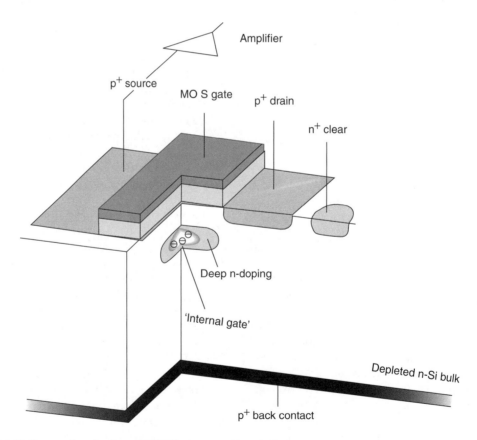

Figure 4.1.60 Cross-section of a linear DEPFET structure with periodic reset mechanism, based on the MOS version of the detector–amplifier structure. Electrons stored underneath the gate induce holes in the transistor channel, giving rise to an increased current. Upon request, the charges can be cleared through the n⁺-clear contact

on the detector chip with a detection fill factor of 1.

The potential distribution in the device, calculated by the 3D POSEIDON[60] code, is shown in Figure 4.1.61. In this figure, the backside, where the charge hits the detector is located on the down-side.

The potential maximum of the internal gate (minimum for electrons) is clearly visible and is separated from the external gate by the p-channel. The potential difference in the pixel area to its direct surroundings is about 1 V, sufficient to collect more than 100 000 electrons in one pixel. The amount of storable electrons can be tailored according to the needs of the experiments (Figure 4.1.60).

Since the electrons are collected in a potential maximum (signal charges as well as leakage current) the device has to be reset from time to time by emptying the corresponding internal gate. One straightforward way of doing it, is applying a positive voltage to an adjacent n^+ contact, which acts as a drain for electrons.

In a first approach devices were built, where periodically (hundreds of μs) all charges are removed from the potential minimum beneath the transistor. This is done by applying for a short time (hundreds of ns) a positive voltage at the substrate contact. The result of a two-dimensional simulation shows the continuous rise of the bulk potential between the region under the transistor and the substrate contact for this particular case (Figure 4.1.62). After the clear procedure, signal electrons can be collected and stored again in the electron potential minimum under the transistor channel. As the signal charges have to be removed explicitly and as the internal gate is continuously filled up with thermally generated electrons, the clear procedure can be applied upon request or in a repetitive manner. The clear mechanism acts locally where the clear pulses have been applied. The time required for a complete clear of the internal gate is estimated to be below 100 ns.

The information about the amount of signal charges stored can be recorded by measuring the rise of the transistor current. This measurement does not disturb the stored charges, therefore the readout process can be repeated several times and opens the option of a multiple nondestructive readout.

Hence, if a row of DEPFETs is activated (Figure 4.1.63) by the selective application of the external gate voltages, the charge content can be measured, a clear pulse could be applied and the charge measurement repeated without having signal electrons in the potential minimum. That means, clearing is done after having read the signal charge. The difference between both measurements is the net signal of electrons in the internal gate.

System Performance

The key parameters of the DEPFET system are listed in Table 4.1.2. Their values have been derived from prototype measurements or, if

Figure 4.1.61 Three-dimensional potential distribution of a circular DEPFET structure. The dark area is the most negative potential (drain), the red area the most positive potential, where the electrons are stored. The clear is the small yellow contact on the right-hand side

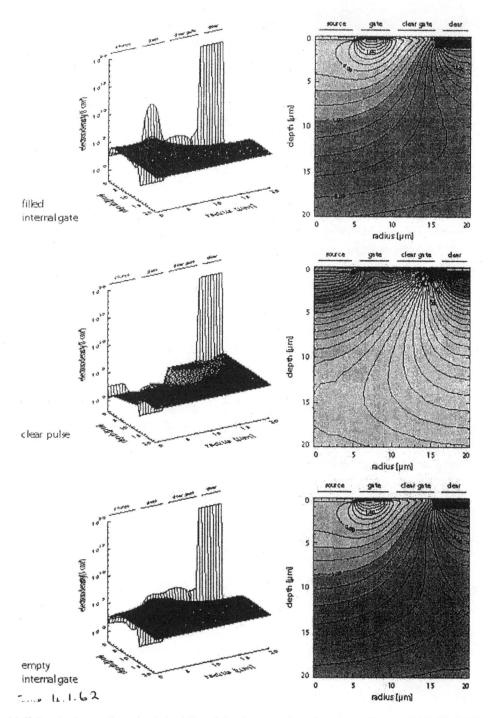

Figure 4.1.62 Result of a two-dimensional simulation of the clear procedure. One can see the potential inside the detector chip while there is a positive voltage pulse (+15 V) applied to the substrate contact; the simulation was done with the program TOSCA for a DEPFET with cylindrical symmetry where the source is in the centre of the structure. The reader is looking from the top of the device into the bulk

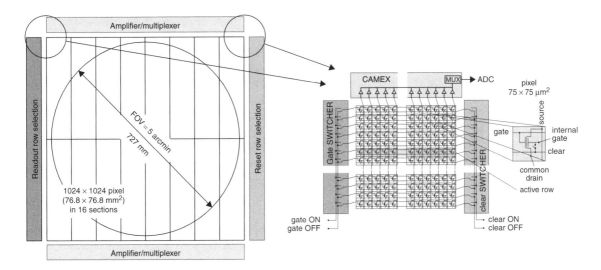

Figure 4.1.63 Layout of the focal plane pixel matrix system, consisting of the detector chip and surrounding read out and control electronics. The figure shows the sensitive area and its logical division. The extended chips limits (for electrical and thermal coupling) are not visible

transferable, from measurements with the XMM pn-CCDs. The main properties are summarized in the following sections.

Energy Resolution and Noise

Beside the statistical fluctuations of the ionization process (Fano fluctuations) the electronic noise is the dominant limitation of the energy resolution. Therefore, in order to understand the basic noise sources, the physical models of the devices are of great importance.

Considering the noise behaviour of the DEPFET, the so called 'total detector capacitance' present in conventional detector–amplifier combinations can be neglected. Only the capacitance of the internal gate is relevant. This leads to very low ENC figures for the series noise contribution. The parallel noise of the structure has its origin in the volume generation of charges inside the fully depleted substrate and surface generated currents. As there is a low resistance between source and gate, the gate leakage current normally can be neglected.

To examine the noise characteristics, measurements of the energy resolution were done with the help of an ^{55}Fe source. Figure 4.1.64 shows a spectrum with the Mn Kα and the Mn Kβ line at 5898 eV and 6498 eV.

The obtainable energy resolution with a DEPFET detector is shown in Figure 4.1.45. In the standard full frame mode with pixel read times of about 1 μs the FWHM, including readout noise and Fano fluctuations, is shown.

Position Resolution

Due to the diffusion of the signal charges during their drift from the conversion point inside the silicon into the potential minimum of the pixel, the spatial measurement precision can be improved substantially, for relatively large pixel sizes. The improvement is significant, if the signal charge cloud diameter is in the order of the pixel size. Taking into account the thickness of the silicon wafer and the according charge collection times, i.e. collection times for the generated electrons, the charge cloud, containing 96% (4 σ) of all signal charges will have a diameter of about 30 μm. This would improve the spatial resolution with a pixel size of 50 μm to less than 10 μm for the events which are all contained in one single pixel and to a

Table 4.1.2 Expected performance figures of the DEPFET focal plane detector

Integration and readout	
Readout time per row (128 channels)	2.5 μs
Total readout time	1.25 ms
Integration: readout time	800:1
Window mode	150 μs for e.g. 64 × 64 pixels
Response to radiation	
QE @ 50 eV	70 %
QE @ 100 eV	85 %
QE @ 272 eV (C Kα)	90 %
QE @ 1.740 eV (Si Kα)	100 %
QE @ 8050 eV (Cu Kα)	100 %
QE @ 10 000 eV	96 %
QE @ 20 000 eV	45 %
Depletion depth	500 μm
Rejection efficiency of MIPs	100 %
Response to radiation	
QE @ 50 eV	70 %
QE @ 100 eV	85 %
QE @ 272 eV (C Kα)	90 %
QE @ 1.740 eV (Si Kα)	100 %
QE @ 8050 eV (Cu Kα)	100 %
QE @ 10 000 eV	96 %
QE @ 20 000 eV	45 %
Depletion depth	500 μm
Rejection efficiency of MIPs	100 %
Spectroscopy	
Fano noise at 5.9 keV	118 eV FWHM
System noise	3–5 electrons (rms)
System noise with RNDR	≈1 electrons (rms) for $n = 16$
^{55}Fe resolution	125 eV
C Kα resolution	50 eV
P/V ratio at C Kα	100:1
Radiation hardness	
No change up to (@220K)	1×10^{10} p with 10 MeV per cm^2
Focal plane geometries	
Device size	7.5 × 7.5 cm^2
Device format	1000 × 1000
Pixel size	75 × 75 μm^2
Position resolution	better than 30 μm
Fill factor of focal plane	1
Operating temperature	200–240K

QE, quantum efficiency.

spatial resolution substantially below that (≤5 μm) for all other events (80 %) (Figure 4.1.56).

A theoretical and experimental study on the position resolution using the charge spreading technique and their impact on energy resolution must be considered. However, it seems reasonable that a pixel size of 75 μm to 100 μm is adequate for the anticipated angular resolution and focal length of XEUS. Tests with the beam trajectory monitor for the TTF-FEL at DESY confirmed the feasibility of sub-micron position accuracy by centroiding the charge cloud of the incident photons[61] at the border of two adjacent pixels.

Figures 4.1.56 and 4.1.57 demonstrate the effect of charge spreading and position reconstruction of the incident photon. The x-axis indicates the position of the photon hit: at $x = 0$, the photon hits the pixel exactly at the boundary to the neighbouring pixel. Here the position resolution is at its optimum. As the physical situation is symmetrical with respect to the centre of the pixel, the x-axis ends at half the pixel size. On the ordinate we plotted the position resolution (rms). The parameter 'sg' (sigma of the Gaussian) scales the lateral signal spread before arriving in the pixel well. The upper curve indicates a sg = 3 μm and increases to sg = 13 μm at the bottom. For a 500 μm thick detector the typical 'sg' is between 7 μm and 9 μm.

Quantum Efficiency and Radiation Background

As the XEUS mission intends to achieve high sensitivity from the very low energies (around 50 eV) up to 30 keV the detector entrance window as well as the sensitive thickness must be optimized. The practical thickness of such a detector is limited to 500 μm because the Compton background of the spacecraft increases with detector thickness. On the low energy side the studies on <100> oriented silicon will continue, in order to improve the spectroscopic response down to 50 eV. The limiting quantity for the low energy response is clearly the optical blocking filters. As a baseline we propose a 500 Å thick monolithically integrated Al filter on the radiation entrance side.

For X-rays in the range of 0.1 keV up to 30 keV the response is shown in Figure 4.1.14. As the silicon has the same thickness and a similar radiation entrance window, the quantum efficiency should not differ.

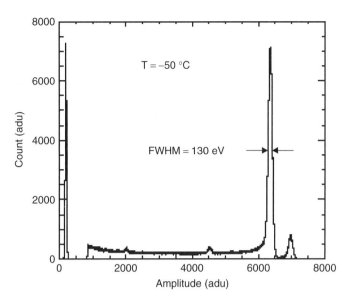

Figure 4.1.64 Manganese spectrum measured with a DEPFET structure at −25 °C. The electronic noise contribution is only 3 electrons (rms). It is obtained with a time shaping constant of 6 μs. The device was illuminated from the front side. The energy resolution of the Mn Kα line is ∼130 eV

4.1.13.3 THE REPETITIVE NONDESTRUCTIVE READOUT (RNDR)

In cases where the count rates do not exceed the pile-up limit and/or the area of interest is restricted to a smaller window, e.g. $2 \times 2 \, \text{cm}^2$ the same signal charge can be read out several times. The field of interest for RNDR in the focal plane can be chosen relatively free, leaving the rest of the detector in its conventional readout mode.

Because the electrons are confined in the electric field below the sensing gate of the DEPFET amplifier (floating gate amplifier) and are not mixed with other charges, the measurement of the amount of signal charges can be repeated as often as required. The noise, as shown in Equation (4.1.4), can be reduced by

$$\text{ENC}(n) = \text{ENC}_0/\sqrt{n} \quad (4.1.14)$$

where n is the number of readings of the signal charges and ENC_0 the noise of a single reading.

We expect a single read noise of the DEPFET structure of 4 electrons at −50 °C with a shaping time of 1 μs. After the single reading, the signal charge is transferred to the neighbouring DEPMOS or DEPFET cell. The charge is read out again and compared to the previous reading. Repeating that procedure 16 times, spending 16 μs for the reading of two pixels (Figure 4.1.65), we could achieve a single electron noise floor, corresponding to an energy resolution of less than 10 eV (FWHM). This would allow the usable X-ray bandwidth to be expanded down to 50 eV. Simulations and a design for a DEPMOS nondestructive readout device was recently proposed[62] and is being fabricated.

4.1.14 CONCLUSION

Since the invention of the SDD a large variety of new detector structures based on the principle of sideward depletion have been developed. Those detectors have left their initial fields of applications in high energy physics, astrophysics and synchrotron radiation research. They are now a mature technology and open many new industrial applications. Experiments in basic research have driven the performance parameters towards the optimum for the specific applications: high quantum efficiency, excellent energy resolution, high radiation

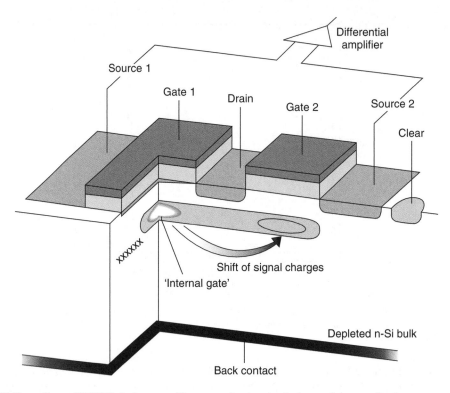

Figure 4.1.65 Two adjacent DEPFET devices are able to transfer the signal charges from one floating gate amplifier to the neighbouring one, reading the same signal charges several times. The read noise is reduced by the square root of n, where n is the number of readings

tolerance, good position resolution, high speed, large and almost defect-free devices, homogeneous response of the full bandwidth of radiation and high background rejection efficiency. It will be the aim of future developments to approach the physical limits in radiation detection and to add in additional intelligence into the local detector systems to face the steadily increasing amount of data and power dissipation.

ACKNOWLEDGEMENTS

We have profited from many discussions with scientists of the Max-Planck-Institute für Physik and extraterrestrische Physik and of the MPI-Halbleiterlabor and the company pnSensor. In particular, we are indebted to R. Richter, N. Meidinger, R. Hartmann and H. Bräuninger. We are grateful to J. Trümper for his constant support and numerous discussions over the last 15 years. We appreciate the involvement of the Universität Bonn (N. Wermes, P. Fischer) in the DEPFET work and the Institut of Astronomie und Astrophysik Tübingen (IAAT) in the pn-CCD work. The frequent discussions with our friends and colleagues from the Politecnico di Milano and BNL, New York, were always stimulating. Special thanks to Antonio Longoni, Carlo Fiorini, Marco Sampietro, Emilio Gatti, Andrea Castoldi, Chiara Guazzoni and Pavel Rehak.

REFERENCES

1. Kemmer, J. Fabrication of low noise silicon radiation detectors by the planar process. *Nucl. Instrum. Methods A* **169**, 499–502 (1980).
2. Belau, E. *et al*. Silicon detectors with 5 µm spatial resolution for high energy particles. *Nucl. Instrum Methods* **217**, 224–228 (1983).
3. Pinotti, E., Bräuninger, H., Findeis, N., Gorke, H., Hauff, D., Holl, P., Kemmer, J., Lechner, P., Lutz, G., Kink, W.,

Meidinger, N., Metzner, G., Predehl, P., Reppin, C., Strüder, L., Trümper, J., Zanthier, C.v., Kendziorra, E., Staubert, R., Radeka, V., Rehak, P., Bertuccio, G., Gatti, E., Longoni, A., Pullia, A. and Sampietro, M. The pn-CCD on-chip electronics. *Nucl. Instrum. Methods A* **326**, 85–92 (1993).

4. Andricek, L., Hauff, D., Kemmer, J., Koffeman, E., Lukewille, P., Lutz, G., Moser, H.G., Richter, R.H., Rohe, T., Soltau, H., and Viehl, A., Design and test of radiation hard p + n silicon strip detectors for the ATLAS SCT. *Nucl. Instrum. Methods A* **439**, 427–441 (2000).

5. Gatti, E. and Rehak, P. Semiconductor drift chamber – an application of a novel charge transport scheme. *Nucl. Instrum. Methods* **225**, 608–614 (1984).

6. Kemmer, J. and Lutz, G. New semiconductor detector concepts *Nucl. Instrum. Methods A* **253**, 365–377 (1987).

7. Lechner, P., Eckbauer, S., Hartmann, S., Krisch, S., Hauff, D., Richter, R., Soltau, H., Strüder, L., Fiorini, C., Gatti, E., Longoni, A. and Sampietro, M. Silicon drift detectors for high resolution room temperature X-ray spectroscopy. *Nucl. Instrum. Methods A* **377**, 346–351 (1996).

8. Caccia, M. *et al.* A Si strip detector with integrated coupling capacitors. *Nucl. Instrum. Methods A* **260**, 124–131 (1987).

9. Kemmer, J. and Lutz, G. New structures for position sensitive semiconductor detectors. *Nucl. Instrum. Methods A* **273**, 588–598 (1988).

10. Kemmer, J. *et al.* Experimental confirmation of a new semiconductor detector principle. *Nucl. Instrum. Methods A* **288** 92–98 (1990).

11. Fano, U. Ionization yield of radiations II: The fluctuations of the number of ions. *Phys. Rev.* **72**, 26–29 (1947).

12. Shockley, W. and Read, W.T. Statistics of the recombination of holes and electrons. *Phys. Rev.* **87**, 835–842 (1952).

13. Hartmann, R., Hauff, D., Lechner, P., Richter, R., Strüder, L., Kemmer, J., Krisch, S., Scholze, F. and Ulm, G. Low energy response of silicon pn-junction detectors. *Nucl. Instrum. Methods A* **377**, 191–196 (1996).

14. Hartmann, R., Strüder, L., Kemmer, J., Lechner, P., Fries, O., Lorenz, E. and Mirzoyan, R. Ultrathin entrance windows for silicon drift detectors. *Nucl. Instrum Methods A* **387**, 250–254 (1997).

15. Soltau, H., Holl, P., Krisch, S., Zanthier, C.v., Hauff, D., Richter, R., Bräuninger, H., Hartmann, R., Hartner, G., Krause, N., Meidinger, N., Pfeffermann, E., Reppin, C., Schwaab, G., Strüder, L., Trümper, J., Kendziorra, E. and Krämer, J. Performance of the pn-CCD X-ray detector system designed for the XMM satellite mission. *Nucl. Instrum. Methods A* **377**, 340–345 (1996).

16. Holland, A. *et al.* X-ray spectroscopy using MOS CCDs. *Nucl. Instrum. Methods A* **377**, 334–340 (1996).

17. van den Berg, M.L. *et al.* Back illuminated CCDs made by gas immersion laser doping. *Nucl. Instrum. Methods A* **377**, 312–320 (1996).

18. Knoll, G.F. *Radiation Detection and Measurement*, 3rd Edn, John Wiley & Sons, Inc., New York, 2000, pp. 1–80.

19. Rando, N., Peacock, A., Favata, F. and Perryman, M. S-Cam: an imaging spectrophotometer based on superconducting tunnel junctions. *Exp. Astron.* **10**, 499–517, (2000).

20. Meidinger, N., Bräuninger, H., Briel, U., Hartmann, R., Hartner, G., Holl, P., Kemmer, J., Kendziorra, E., Krause, N., Lutz, G., Pfeffermann, E., Popp, M., Reppin, C., Richter, R., Soltau, H., Stötter, D., Strüder, L., Trümper, J. and von Zanthier, C. The PN-CCD detector for XMM and ABRIXAS. *Proc. SPIE* **3765**, 192–203 (1999).

21. Lechner, P., Andricek, L., Findeis, N., Klein, P., Kemmer, J., Meidinger, N., Lutz, G., Schuster, K., Sterzik, M., Strüder, L., von Zanthier, C. and R. Richter New DEPMOS applications. *Nucl. Instrum. Methods A* **326**, 284–289 (1993).

22. Strüder, L. Wide field imaging spectrometer for ESA's future X-ray mission: XEUS. *Nucl. Instrum. Methods A* **436**, 53–67 (1999).

23. Takahashi, J. *et al.* Silicon drift detectors for the STAR/SVT experiment at RICH. *Nucl. Instrum. Methods A* **439**, 497–506 (2000).

24. Fiorini, C. and Lechner, P. Continuous charge restoration in semiconductor detectors by means of the gate-to-drain current of the integrated front-end JFET. *IEEE Trans. Nucl. Sci.* **46**, 761–764 (1999).

25. Lechner, P. and Strüder, L. Ionization statistics in silicon X-ray detectors – new experimental results. *Nucl. Instrum Methods A* **354**, 464–474 (1995).

26. Lechner, P., Hartmann, R., Soltau, H. and Strüder, L. Pair creation energy and Fano factor of silicon in the energy range of soft X-rays. *Nucl. Instrum. Methods A* **377**, 206–209 (1996).

27. Strüder, L., Lechner, P. and Leutenegger, P. Silicon drift detector – the key to new experiments. *Naturwissenschaften* **85**, 539–543 (1998).

28. Rehak, P. *et al.* Progress in semiconductor drift detectors. *Nucl. Instrum. Methods A* **248**, 367–378 (1986).

29. Gatti, E., Longoni, A., Sampietro, M., Giacomelli, P., Vacchi, A., Rehak, P., Kemmer, J., Holl, P., Strüder, L., and Kubischta, W. Silicon drift chamber prototype for the upgrade of the UA ~ 6 experiment at the CERN pp collider. *Nucl. Instrum. Methods A* **273**, 865–868 (1988).

30. Holl, P., Rehak, P., Ceretto, F., Faschingbauer, U., Wurm, J.P., Castoldi, A. and Gatti, E. A 55 cm^2 cylindrical silicon drift detector. *Nucl. Instrum. Methods A* **377**, 367–374 (1996).

31. Kemmer, J., Lutz, G., Belau, E., Prechtel, U. and Welser, W. Low capacity drift diode. *Nucl. Instrum. Methods A* **253**, 378–381 (1987).

32. Hartmann, R., Hauff, D., Krisch, S., Lechner, P., Lutz, G., Richter, R.H., Seitz, H., Strüder, L., Bertuccio, G., Fasoli, L., Fiorini, C., Gatti, E., Longoni, A., Pinotti, E. and Sampietro, M. Design and test at room temperature of the first silicon drift detector with on-chip electronics. *IEDM Tech. Digest* 535–539 (1994).

33. Lechner, P., Pahlke, A. and Soltau, H. Novel high resolution silicon drift detector. *X-ray spectrom.* in press.

34. Longoni, A., Fiorini, C., Leutenegger, P., Sciuti, S., Fronterotta, G., Strüder, L. and Lechner, P. A portable XRF spectrometer for non-destructive analysis in archeometry. *Nucl. Instrum. Methods A* **409**, 395–400 (1998).
35. RÖNTEC GmbH. X-flash detector. Product Information, vol. 98/99, 1999.
36. Gauthier, Ch., Goulon, J., Moguiline, E., Rogalev, A., Lechner, P., Strüder, L., Fiorini, C., Longoni, A., Sampietro, M., Walenta, A., Besch, H., Schenk, H., Pfitzner, R., Tafelmeier, U., Misiakos, K., Kavadias, S. and Loukas, D. A high resolution, 6-channel – silicon drift detector array with integrated JFET's designed for EXAFS: first X-ray fluorescence excitation spectra recorded at the ESRF. *Nucl. Instrum. Methods A* **382**, 524–532 (1996).
37. Fiorini, C., Longoni, A., Boschini, L., Perotti, F., Labanti, C., Rossi, E., Lechner, P. and Strüder, L. Position and energy resolution of a new gamma-ray detector based on a single CsI(Tl) scintillator coupled to a silicon drift detector array. *IEEE Trans. Nucl. Sci.* **46**, 858–864 (1999).
38. Fiorini, C., Longoni, A., Labanti, C., Rossi, E., Lechner, P., Soltau, H. and Strüder, L. A monolithic array of silicon drift detectors for high resolution gamma-ray imaging. *IEEE Trans. Nucl. Sci.* **49**, 995–1000 (2000).
39. Castoldi, A., Guazzoni, C., Longoni, A., Gatti, E., Rehak, P. and Strüder, L. Controlled drift detector: a novel silicon device for the measurement of position and energy of X-rays. *IEEE Trans. Nucl. Sci.* **44**, 1724–1732 (1997).
40. Castoldi, A., Guazzoni, C., Longoni, A., Gatti, E., Rehak, P. and Strüder, L. The controlled drift detector. *Nucl. Instrum. Methods A* **439**, 519–528 (2000).
41. Castoldi, A., Guazzoni, C., Longoni, A., Gatti, E., Rehak, P. and Strüder, L. Analysis and characterization of the confining mechanism of the controlled drift detector. *IEEE Trans. Nucl. Sci.* **46**, 1943–1947 (1999).
42. Castoldi, A., Guazzoni, C., Longoni, A., Gatti, E., Rehak, P. and Strüder, L. Room temperature 2-D X-ray imaging with the controlled drift detector. *IEEE Trans. Nucl. Sci.*, **49**, 989–994 (2002).
43. Strüder, L. and Holl, P., Lutz, G. and Kemmer, J. Development of fully depletable CCDs for high energy physics applications. *Nucl. Instrum. Methods A* **257**, 594–602 (1987).
44. Radeka, V., Rehak, P., Rescia, S., Gatti, E., Longoni, A., Sampietro, M., Bertuccio, P., Holl, P., Strüder, L. and Kemmer, J. Implanted silicon JFET on completely depleted high resistivity devices. *IEEE Electron Device Lett.* **10**, 91–95 (1989).
45. Soltau, H., Kemmer, J., Meidinger, N., Stötter, D., Strüder, L., Trümper, J., Zanthier, C.v., Bräuninger, H., Briel, U., Carathanassis, D., Dennerl, K., Haberl, F., Hartmann, R., Hartner, G., Hauff, D., Hippmann, H., Holl, P., Kendziorra, E., Krause, N., Lechner, P., Pfeffermann, E., Popp, M., Reppin, C., Seitz, H., Solc, P., Stadlbauer, T., Weber, U. and Weichert, U. Fabrication, test and performance of very large X-ray CCDs designed for astrophysical applications. *Nucl. Instrum. Methods A* **439**, 547–559 (2000).
46. Strüder, L., Bräuninger, H., Briel, U., Hartmann, R., Hartner, G., Hauff, D., Krause, N., Maier, B., Meidinger, N., Pfeffermann, E., Popp, M., Reppin, C., Richter, R., Stötter, D., Trümper, J., Weber, U., Holl, P., Kemmer, J., Soltau, H., Viehl, A. and Zanthier, C.v. A 36 cm^2 large monolithic pn-CCD X-ray detector for the European XMM Satellite Mission. *Rev. Sci. Instrum.* **68**, 4271–4274 (1997).
47. Strüder, L., Bräuninger, H., Meier, M., Predehl, P., Reppin, C., Sterzik, M., Trümper, J., Cattaneo, P., Hauff, D., Lutz, G., Schuster, K.F., Schwarz, A., Kendziorra, E., Staubert, A., Gatti, E., Longoni, A., Sampietro, M., Radeka, V., Rehak, P., Rescia, S., Manfredi, P.F., Buttler, W., Holl, P., Kemmer, J., Prechtel, U. and Ziemann, T. The MPI/AIT X-ray imager – high speed pn-CCDs for X-ray detection. *Nucl. Instrum. Methods A* **288**, 227–235 (1990).
48. Kendziorra, E., Bihler, E., Kretschmar, B., Kuster, M., Pflüger, B., Staubert, R., Bräuninger, H., Briel, U., Pfeffermann, E. and Strüder, L. PN-CCD camera for XMM: performance of high time resolution – bright source modes. *Proc. SPIE* **3114**, 155–165 (1997).
49. Krause, N., Soltau, H., Hauff, D., Kemmer, J., Stötter, D., Strüder, L. and Weber, J. Metal contamination analysis of the epitaxial starting material for scientific CCDs. *Nucl. Instrum. Methods A* **439**, 228–238 (2000).
50. Meidinger, N., Schmalhofer, B. and Strüder, L. Particle and X-ray damage in pn-CCDs. *Nucl. Instrum. Methods A* **439**, 319–337 (2000).
51. Buttler, W. *et al.* Short channel, CMOS compatible JFET in low noise applications. *Nucl. Instrum. Methods A* **326**, 63–70 (1993).
52. Wiesemann, U., Thieme, J., Früke, R., Guttmann, P., Niemann, B., Rudolph, D. and Schmahl, G. Construction of a scanning transmission X-ray microscope at the undulator U41 and BESSY II, *Nucl. Instrum. Methods A* **467–468**, 861–863 (2001).
53. Bertschinger, G., Biel, W., Herzog, O., Weinhammer, J., Kunze, H.-J., and Bitter, M. X-ray spectroscopy at the TEXTOR-94 Tokamak. *Physica Scripta* **T83**, 132–141 (1999).
54. Lindner, F., Stremme, W., Schätzel, M.G., Grasbon, F., Paulus, G.G., Walther, H., Hartmann, R. and Strüder, L. High-order harmonic generation at a repetition rate of 100 kHz. *Phys. Rev. A* **68**, 013814 (2003).
55. Ronning, C., Vetter, U., Uhrmacher, M., Hofsäss, H., Bharut-Ram, K., Hartmann, R. and Strüder, L. Electron emission channelling spectroscopy using X-ray CCD detectors. *Nucl. Instrum. Methods A* **512**, 378–385 (2003).
56. Strüder, L. and Trümper, J. Silicon pixel detectors for future X-ray missions. Leicester University, 1997, pp. 89–97.
57. XEUS Astrophysics working group. X-ray Evolving – Universe Spectroscopy – The XEUS scientific case, ESA SP-1238, 1999.
58. Cesura, G., Findeis, N., Hauff, D., Hörnel, N., Kemmer, J., Klein, P., Lechner, P., Lutz, G., Richter, R.H. and Seitz, H. New pixel detector concepts based on junction

field effect transistors on high resistivity silicon. *Nucl. Instrum. Methods A* **377**, 521–528 (1996).

59. Klein, P., Cesura, G., Fischer, P., Lutz, G., Neser, W. and Richter, R.H. and Wermes, N. Study of a DEPJFET pixel matrix with continuous clear mechanism *Nucl. Instrum. Method A* **392**, 254–259 (1997).

60. Gajewski, H. *et al. TOSCA Handbook*, Weierstrass Institute for Applied Analysis and Stochastics, Berlin, 1997.

61. Hillert, S., Schmüser, P., Ischebeck, R., Miller, U., Roth, S., Hansen, K., Karstensen, S., Leenen, M., Ng, J., Holl, P., Kemmer, J., Lechner, P., and Strüder, L. Test results on the silicon pixel detector for the TTF–FEL beam trajectory monitor. *Nucl. Instrum. Methods A* **458**, 710–719 (2001).

62. Lutz, G., Richter, R.H. and Strüder, L. Novel pixel detectors for X-ray astronomy and other applications. *Nucl. Instrum. Methods A* **461**, 393–404 (2001).

63. Gajewski, H., Kaiser, H.-Chr., Langmach, H., Nürnberg, R. and Richter, R.H. *Mathematical Modeling and Numerical Simulation of Semiconductor Detectors*, Springer-Verlag, Berlin, 2001.

4.2 Gas Proportional Scintillation Counters for X-ray Spectrometry

C. A. N. CONDE
Universidade de Coimbra, Coimbra, Portugal

4.2.1 INTRODUCTION

X-Ray spectrometry is usually performed with one of two techniques: energy dispersive and wavelength dispersive. Energy dispersive techniques use radiation detectors that give a pulse proportional to the energy dissipated in the detector medium. Most detectors used in practical spectrometry applications fit into two groups: cooled semiconductor detectors and room temperature gas detectors. While semiconductor detectors generally provide superior energy resolution, gas detectors present a better performance for applications requiring room temperature operation and/or large areas, or for X-ray spectrometry below ~2 keV.[1]

Since gas detectors are usually less bulky and cheaper than cooled semiconductor detectors, they are common in portable X-ray fluorescence analysis systems. Two types of gas detectors can be used: the standard gas proportional (ionization) counter (PC) (Figure 4.2.1) and the gas proportional scintillation counter (GPSC) (Figure 4.2.2) which was developed later.

In both cases, the detector gives a pulse with an amplitude proportional to the number n of primary electrons produced by an X-ray photon in the gas, which is itself approximately proportional to the energy E of the photon. In addition to the ionization processes in the gas that lead to the production of primary electrons, there are also excitation processes that lead to the production of light, the so-called primary scintillation. However, as the number of primary electrons is quite small (the average value is $\bar{n} \approx 273$ electrons for a 5.9 keV photon in Xe) it cannot be properly distinguished from the noise of the electronic devices used in the early amplification stages.

Thus a sort of amplification is required before the electronics can process the detector signals. In PCs, the primary electrons are made to drift towards a strong electric field region, usually in the vicinity of a small diameter (typically 25 μm) anode wire (Figure 4.2.1). In this region electrons engage in ionizing collisions that lead to an avalanche with an average multiplication gain M of the order of 10^3 to 10^4. If M is not too large, space charge effects can be neglected, and the average number of electrons in the end of the avalanche, $N_a = M\bar{n}$, is then nearly proportional to the energy E of the absorbed X-ray photon, hence the name proportional (ionization) counter given to this device.

An alternative solution (Figure 4.2.2), when the filling gas is a noble gas like Xe, is to drift the primary electrons from the absorption or drift region, under a weak electric field (<1 V cm^{-1} Torr^{-1}), towards a not too strong electric field region (in the 1–6 V cm^{-1} Torr^{-1} range) – the so-called scintillation region – so that electrons can excite, but not ionize, the gas atoms/molecules.

The excited atoms decay, in a more or less complex process, emitting light (the so-called secondary scintillation), which is detected by a photosensor, usually a photomultiplier tube.

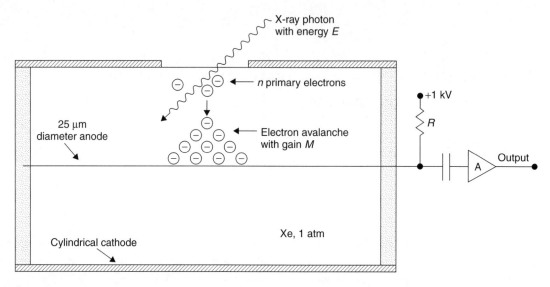

Figure 4.2.1 Scheme of a gas proportional (ionization) counter (PC)

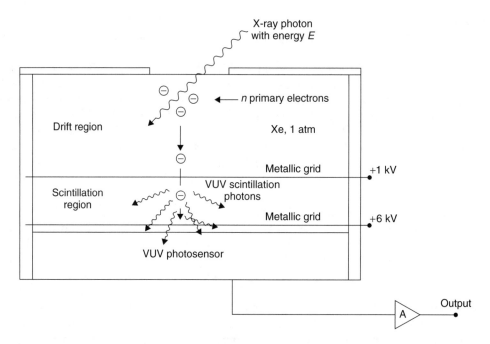

Figure 4.2.2 Scheme of a gas proportional scintillation counter (GPSC)

The intensity of the secondary scintillation is two or three orders of magnitude stronger than that of the primary scintillation. However, since the secondary scintillation is produced while the electrons drift, its rise time is much slower (a few μs) than that for the primary scintillation (a few ns). For a fixed electric field, the number N_{ph} of secondary scintillation photons produced by a single primary electron is nearly constant and can reach values as large as 500 photons per electron.

The average total number, N_t, of secondary scintillation photons produced by an X-ray photon is then $N_t = N_{ph}\bar{n}$, so the photosensor signal amplitude is nearly proportional to E, hence the name of gas proportional scintillation counter[1] (GPSC) for this device.

The energy resolution of gas counters depends mostly on the fluctuations of the physical processes involved in the detection. For PCs there are fluctuations both in n and M; for GPSCs, since the gain is achieved through a scintillation process with almost no fluctuations, only fluctuations in n need to be considered. Thus a better energy resolution can be achieved for a GPSC than for a PC; typical values for 5.9 keV X-rays are 8 % and 14 %, respectively.

The purpose of this subchapter is to delve into the physics and applications of GPSCs. We shall first consider the physics of the absorption of X-rays in gases, then the transport of electrons and the production of electroluminescence in GPSC gases, the basic concepts in GPSCs, different types of GPSCs and the applications of these devices to X-ray spectrometry.

4.2.2 THE PHYSICS OF THE ABSORPTION OF X-RAYS IN GASES

4.2.2.1 ABSORPTION OF X-RAYS IN Xe

The probability of absorption of an X-ray photon in a gas depends on the absorption cross-section, which for the heavier gases is nearly equal to the photoelectric effect cross-section, σ_p. This is approximately proportional to Z^m/E^3, where Z is the gas atomic number, m a number between 4 and 5 and E the X-ray energy. Therefore, the heavier the noble gas the stronger is the absorption, so pure Xe ($Z = 54$) or Xe-based mixtures are often used in gas-filled X-ray detectors. For Xe, the photoelectric cross-section (Figure 4.2.3) is rather large and varies from about 5×10^{-17} cm^2/atom at the ionization threshold (12.1 eV) to about 3.0×10^{-20} cm^2/atom at 10 keV, to which correspond, at room temperature and 1 atm (20 °C and number density, $N = 2.420 \times 10^{19}$ atoms/cm^3) absorption lengths, $L_p = 1/\sigma_p N$, ranging from 8.3 μm to ~1.38 cm (right axis). Partial, shell or subshell, cross-sections are plotted as dotted lines. For some subshells the partial cross-sections are grouped: M_2 and M_3, M_4 and M_5, N_2 and N_3, N_4 and N_5.

Also plotted are the coherent (Rayleigh) σ_R and incoherent (Compton) σ_C X-ray scattering cross-sections, which are much lower than the photoelectric ones. The corresponding absorption lengths, L_R and L_C, refer to the right-hand side axis.

In the following we will refer almost exclusively to pure Xe, the most common filling gas for GPSCs, though in a few cases we will consider Xe based mixtures. The binding energies for the Xe subshells are represented in Figure 4.2.4.

Once an X-ray photon is absorbed and a photoelectron ejected a multitude of processes take place, depending on the atomic subshell involved, which lead finally to the production of n primary electrons. The residual single charged ion, with a vacancy in the photoionized subshell, decays to lower energy states through the emission of X-rays (fluorescence) or electrons (Auger/Coster–Kronig and shake off), increasing then its charge state. Figure 4.2.4 shows the first stages of a typical cascade: following the photoionization of the K shell, a K_α fluorescence X-ray is emitted. The vacancy in the L_2 subshell is filled with an electron from M_1 and another electron is ejected from M_5 (Auger effect). Then the M_1 vacancy is filled with an electron from M_5 and another electron is ejected from N_5 (Coster–Kronig effect). The M_5 hole is then filled with an electron from N_3 and an electron is ejected from N_5 (Auger effect), etc. In the present example the decaying processes stop when the Xe^{9+} ion in its ground state is formed.

When an energetic electron is ejected it might also eject another electron from an outer-shell in a so-called 'shake-off' process, which was not considered in the previous example.

---
[1] This name was first suggested by J. B. Birks from Manchester University (UK), where the initial GPSC work took place, to the authors of the first article.[2] Since there is no charge multiplication in a GPSC, it cannot be considered as a PC that scintillates, so the name 'gas scintillation proportional counter', that is sometimes found in the scientific literature, is not appropriate.

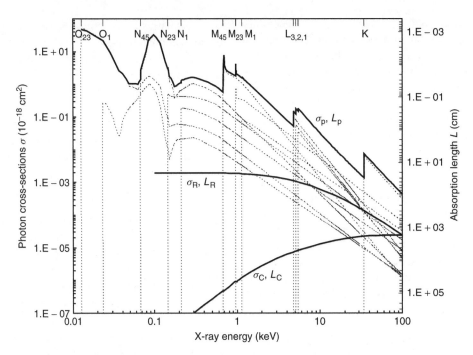

Figure 4.2.3 Photoelectric σ_p, Rayleigh σ_R and Compton σ_C cross-sections for X-rays in Xe. Corresponding absorption lengths L_p, L_R and L_C for Xe at $20\,°C$ and 760 Torr refer to the right-hand side axis. Partial shell and subshell cross-sections are plotted as dotted lines. (Based on Figure 1 of Dias et al.[3] and references therein and in http://www.photcoef.com/212154.html and http://physics.nist.gov/cgi-bin/Xcom/xcom3_1)

Highly charged ions like Xe^{9+} in the previous example, or higher charge, can arise when the decaying processes are completed. Meanwhile the ejected electrons and fluorescence X-rays proceed ionizing further Xe neutral atoms. The final stage is reached when there are no more electrons or photons with sufficient energy (larger than 12.1 eV) to ionize neutral atoms; as a result n primary electrons are produced.

All these processes can be simulated in full detail with Monte Carlo techniques provided all relevant integral and differential cross-section data and transition rates are available.[3] This way n can be calculated for the simulated X-ray absorption event. Repeating the calculations a large number of times it is possible to calculate with good accuracy its average value, $\bar{n}$, the so-called W value ($W = E/\bar{n}$), the variance σ_n of n, the so called Fano factor, $F = \sigma_n/\bar{n}$, i.e. the relative variance of n and the size and spatial distribution of the primary electron cloud.[3–5]

4.2.2.2 NONLINEARITY EFFECTS

The study of the variation of $\bar{n}$, W and F with the X-ray energy E is of great importance, since it affects the energy calibration and energy resolution of any gaseous X-ray detector, either of the PC or GPSC type. It was shown[5] that $\bar{n}$ varies almost linearly with E except near the L and K photoionization thresholds, where $\bar{n}$ exhibits discontinuities approaching 1 % (Figures 4.2.5 and 4.2.6). Agreement between Monte Carlo calculations and experimental measurements[5,6] is clearly seen.

These effects result from the fact that when photoionization of a new shell becomes energetically possible, this shell is more likely to be photoionized than the other (outer) shells and so a new set of decaying channels becomes possible with new fluorescence X-rays and Auger/Coster–Kronig electron spectra.[5]

Note that, besides discontinuities, these effects can originate false peaks in an otherwise flat X-ray

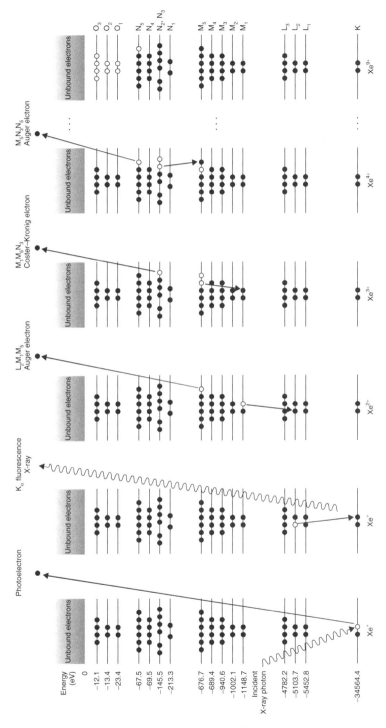

Figure 4.2.4 Typical decay cascade of a Xe$^+$ ion following the photoionization of a K shell in Xe. (Binding energies taken from http://ie.lbl.gov/atom.htm)

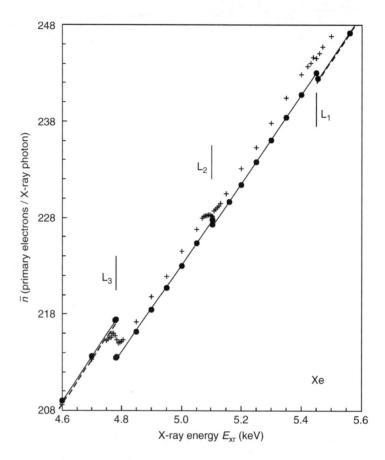

Figure 4.2.5 Mean number $\bar{n}$ of primary electrons produced by an X-ray photon near the L subshell binding energies, showing nonlinearity effects (Figure 2 of Dias et al.[5]). Solid circles are Monte Carlo values; other symbols are derived from experimental results. For details see Dias et al.[5]

energy distribution since X-rays with different energies can produce the same $\bar{n}$.

The calculated W-values for Xe under a variety of conditions are plotted in Figure 4.2.7 as a function of X-ray energy. The two lower points at 4.7 keV and 5.9 keV are absolute experimental values.[7] Recently[8] the absolute W-value for Xe at 825 Torr and 5.9 keV X-rays was measured with improved accuracy ($21.61^{+0.14}_{-0.10}$ eV) showing a small disagreement with Monte Carlo values (Figure 4.2.7). Since Xe–Ne mixtures are expected to be important for soft X-ray detection, their W-values have been measured experimentally[9] and calculated by Monte Carlo techniques;[10] evidence was found for Penning effects.

4.2.2.3 STATISTICAL FLUCTUATIONS

For a given X-ray energy E, the distribution function of n, $f(n)$, can be calculated by Monte Carlo simulation as shown in Figure 4.2.8 for 6 keV photons. The peak position of its components for the L, M, N and O shells is shifted to the right as the shell order increases. Thus, its true shape is not a pure gaussian, as is usually assumed, and the peak position m, i.e. the most probable number of primary electrons is not coincident with the average number, $\bar{n}$. These effects are more pronounced close to a shell, if the fluorescence X-ray escapes (as shown in Figure 4.2.9 of Dias et al.[5]). However, if as usually we approximate $f(n)$ to a gaussian, we can define its variance $\sigma_n = F\,\bar{n}$.

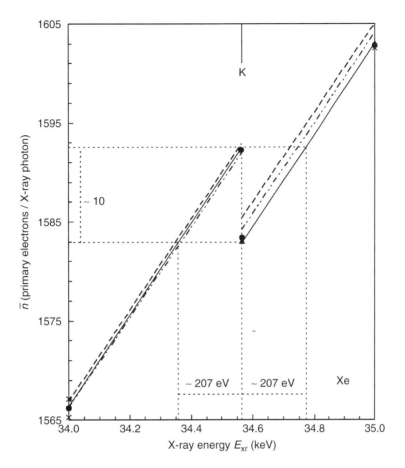

Figure 4.2.6 Mean number $\bar{n}$ of primary electrons produced by an X-ray photon near the K shell binding energy showing nonlinearity effects (Figure 3 of Dias et al.[5]). Solid symbols are Monte Carlo values; other symbols are derived from experimental results. For details see Dias et al.[5]

The full width at half-maximum, $\Delta n = 2.355\sqrt{\sigma_n}$ can then be calculated, as well as the Fano factor, F. The experimental energy resolution, R, has then a lower limit given by its intrinsic value, R_{int}:

$$R_{int} = 2.355\sqrt{\frac{FW}{E}}$$

This expression allows, in principle, the measurement of F from measurements of R, provided W is known. However, since the experimental energy resolution is for a GPSC (Figure 4.2.2) also dependent on a variety of other factors (noise, non-parallelism of the grids, photosensor imperfections, fluctuations in the number of scintillation photons, etc.) experience can only give an upper limit for F. For X-rays with energies close to the binding energies of the shells, the same effects that produce discontinuities in $\bar{n}$ and variations in W (Figure 4.2.7), produce variations in F.[5]

4.2.2.4 TAILING EFFECTS FOR LOW ENERGY X-RAYS

Very low energy X-rays (below about 1 keV or 2 keV) are absorbed very near the detector window (Figures 4.2.1 and 4.2.2) since their absorption lengths, L (Figure 4.2.3) are very small (about 20 μm for 100 eV X-rays and 200 μm for 1 keV

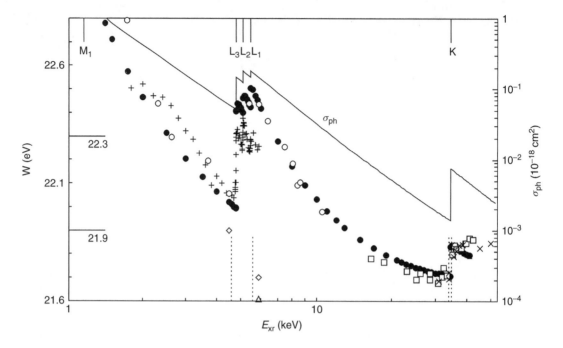

Figure 4.2.7 W value for X-rays in gaseous Xe as a function of energy (Figure 4 of Dias et al.[5]). Solid symbols are Monte Carlo values; other symbols are derived from experimental results or are absolute measurements ($\diamond, \triangle$). Continuous line represents the Xe photoelectric cross-section. For details see Dias et al.[5]

X-rays). Thus, some of the electrons in the primary electron cloud (its size may be of the order of tens of μm) can be scattered back to the detector window, and so be lost. Then, the number of primary electrons that can contribute to the X-ray detector pulse is smaller than those initially produced, and a tail arises in the function, $f(n)$.[12] This tail can be reduced by increasing the intensity of the electric field near the window decreasing thus backscattering, or else by using mixtures like Xe–Ne with longer absorption lengths.[13] An alternative is to use a driftless detector,[14] which has the inconvenience of requiring drift time compensation.

Typical calculated and experimental results for 277 eV X-rays in pure Xe are presented in Figure 4.2.9. Full circles correspond to Monte Carlo calculated events that do not have primary electrons lost to the window (A_0); open circles correspond to all events (A_1). The gray continuous line with no symbols corresponds to an experimental spectrum obtained with a GPSC.

4.2.3 TRANSPORT OF ELECTRONS IN Xe

4.2.3.1 DRIFT OF ELECTRONS IN Xe

The n primary electrons, once produced at below the 12.1 eV ionization threshold, may acquire further energy from the electric field in the place they are located (usually the absorption region in Figure 4.2.2) and at the same time lose energy through elastic an inelastic collisions. Noble gases being monoatomic have neither rotational nor vibrational states, but only electronic states. The first four states of Xe lie at 8.32, 8.44, 9.45 and 9.57 eV above the ground state. Therefore for electrons with energies below 8.32 eV no inelastic (excitation) collisions can take place and so the only way electrons can lose energy is through recoil of a Xe atom in elastic collisions. Classical mechanics teaches us that if an electron, with energy E and mass $m = 5.4858 \times 10^{-4}$ u, collides with a Xe atom (average mass $M = 131.293$ u) at

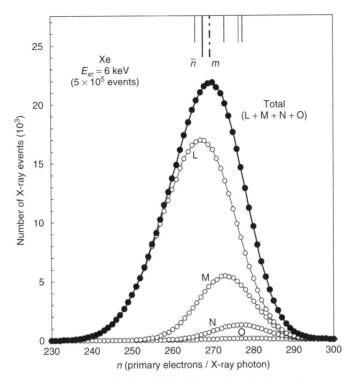

Figure 4.2.8 Monte Carlo calculated distribution function, $f(n)$, of the number of primary electrons, n, produced by 6 keV X-ray photons in pure Xe, showing the contributions from the L, M, N, O shells. The shape of the total distribution is not a pure gaussian and the position of the peak, m, is not coincident with $\bar{n}$. (Based on Dias et al.[5] and Dias[11])

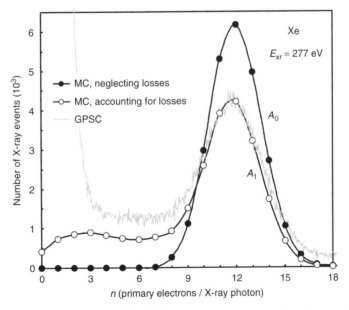

Figure 4.2.9 Tailing effects in a very low energy (277 eV) X-ray spectrum in gaseous Xe. The calculated distributions (solid and open circles) and the experimental spectrum (gray continuous line) obtained with a GPSC are represented (Borges et al. (unpublished); based on Borges et al.[15])

rest, it loses in the recoil, at most, an energy ΔE given by:

$$\Delta E = \frac{4mM}{(m+M)^2} E = 1.6713 \times 10^{-5} E$$

which is a very small fraction of the incident energy. If there is no electric field, an electron with an initial energy of 8.32 eV, losing on the average half of the above ΔE value, requires about 7×10^5 elastic collisions before it can reach thermal energies. However, if the electron is subject to an electric field it may acquire between collisions sufficient energy to compensate for the losses during elastic collisions, and so the electrons will drift along the field lines. For sufficiently strong electric fields the electron might occasionally reach energies above the 8.32 eV threshold for excitation an so Xe* species are produced. For even stronger fields ionization can take place.

All these processes can be simulated in detail by Monte Carlo techniques[3,4,16] if accurate values of the appropriate cross-sections (integral and differential) are known. In Figure 4.2.10 we plot the integral elastic, excitation and ionization cross-sections, together with the total electron–Xe collision cross-section for energies ranging from 10^{-2} to 10^4 eV.

Figure 4.2.11 depicts the calculated radial (R) versus axial (Z) position of a single electron starting at the origin with zero energy in Xe at 760 Torr subject to a 2.5 V cm^{-1} Torr^{-1} reduced electric field along the OZ axis. Each dot corresponds to the electron position after every twentieth collision. As shown, the radial diffusion is already 0.3 mm after a drift of 1.0 mm. In Figure 4.2.12 we plot the calculated energy of a single electron versus the axial distance, under the previous conditions. Since the energy lost in recoil is very small, the energy of the electron is almost proportional to Z. However, once the electron reaches an energy above the 8.32 eV threshold it will have a good chance of exciting a Xe atom, and then its energy will drop down to a very low value; from this low value energy will increase again almost linearly with the distance until a new excitation takes place; these processes repeat themselves until the electron reaches the end of the region. Thus,

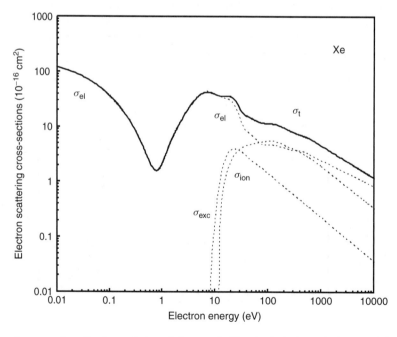

Figure 4.2.10 Integral cross-sections for the scattering of electrons in Xe: elastic (σ_{el}), excitation (σ_{exc}), ionization (σ_{ion}) and total (σ_t) (based on Figure 2 of Dias et al.[3])

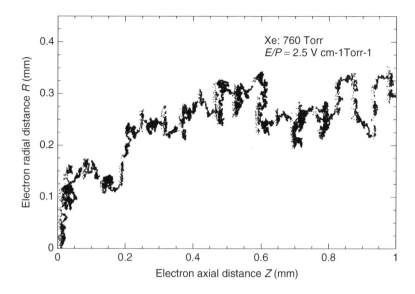

Figure 4.2.11 Calculated axial (Z) and radial (R) position of a typical single electron (every twentieth collision) drifting in gaseous Xe under a reduced electric field intensity of $2.5\,\text{V}\,\text{cm}^{-1}\,\text{Torr}^{-1}$. (Calculations by Barata[17] based on Dias et al.[3,18], Dias[4,19] and Santos[16])

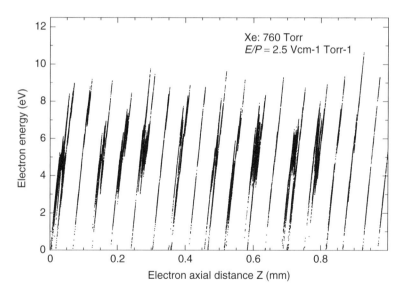

Figure 4.2.12 Calculated energy versus axial distance (every twentieth collision) for a typical single electron drifting in Xe under a reduced electric field intensity of $2.5\,\text{V}\,\text{cm}^{-1}\,\text{Torr}^{-1}$. (Calculations by Barata[17] based on Dias et al.,[3,18] Dias[4,19] and Santos[16])

the saw-tooth shape of the plot. In the example depicted in this figure, the single electron drifting across the 1.0 mm distance produced 17 excitations of Xe atoms.

On the other side, if the electric field intensity is very weak, as in Figure 4.2.13 ($0.5\,\text{V}\,\text{cm}^{-1}\,\text{Torr}^{-1}$), the electron will never reach the excitation threshold.

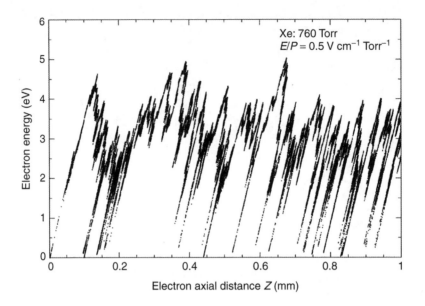

Figure 4.2.13 Calculated energy versus axial distance (every twentieth collision) for a typical single electron drifting in Xe under a reduced electric field intensity of 0.5 V cm^{-1} Torr^{-1}. (Calculations by Barata[17] based on Dias et al.,[3,18] Dias[4,19] and Santos[16])

4.2.3.2 THE SECONDARY SCINTILLATION PROCESSES IN Xe

Once an excited Xe* species is formed, if the pressure is high enough (above about 50 Torr), there is a good chance[16,20,21] that the following three body process takes place:

$$Xe^* + 2\,Xe \rightarrow Xe_2^{**} + Xe$$

which leads to the formation of the excimer Xe_2^{**}. If the pressure is not too high the excimer decays to the repulsive molecular ground state:

$$Xe_2^{**} \rightarrow 2\,Xe + h\nu_0$$

with emission of a photon $h\nu_0$ in the first VUV radiation continuum.

However, if pressure is above a few hundred Torr vibrational relaxation is favoured:

$$Xe_2^{**} + Xe \rightarrow Xe_2^* + 2\,Xe$$

followed by the decay

$$Xe_2^* \rightarrow Xe + Xe + h\nu$$

For Xe these secondary scintillation photons $h\nu$ are peaked at the VUV wavelength of 170 nm, i.e. within the transmission range of high purity quartz. However, if other rare gases were used the wavelength would be shorter (150 nm for Kr and 127 nm for Ar).[21]

We can assume that for each Xe* species formed by the drifting electrons under the influence of an electric field, a secondary scintillation photon is emitted. The same Monte Carlo techniques that were used to calculate Figures 4.2.11–4.2.13 can be used to calculate the number of secondary scintillation photons produced in the scintillation region. Usually what is calculated or measured[22] is the so-called reduced secondary scintillation intensity, Y/p, i.e. the number of secondary scintillation photons produced per unit of path length by a single electron Y, divided by the pressure p. In Figure 4.2.14 we plot calculated[23] results (full circles) together with (normalized) experimental values (open symbols) in agreement with each other. As shown the plot starts at the ~1 V cm^{-1} Torr^{-1} threshold and it is a straight line up to 5 V cm^{-1} Torr^{-1}. A numerical expression for

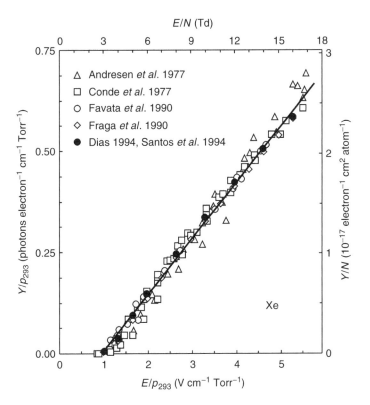

Figure 4.2.14 Reduced secondary scintillation intensity (Y/p_{293}) in Xe at 20 °C, as a function of the reduced electric field intensity. (Based on Figure 18 of Dias[4] and Figure 4 of Santos et al.[23].) Solid circles are Monte Carlo results, open symbols are experimental results (see references in Dias[4] and Santos et al.[23])

Y/p is:[23]

Y/p_{293} (photons electron^{-1} cm^{-1} Torr^{-1})

$= 0.1389 E/p_{293} - 0.1325$

Above about 5 V cm^{-1} Torr^{-1} ionization starts and so, the growth is exponential. We must point out that a single primary electron drifting across a scintillation region with 5000 V applied to it, can produce more than 500 photons.

Since the decay time of the secondary scintillation processes is very fast (tens of nanoseconds or shorter) the risetime of the secondary scintillation pulse depends almost exclusively on the drift time of the primary electrons cloud across the scintillation region. The drift velocity of electrons in Xe is plotted in Figure 4.2.15. So, we can conclude that the secondary scintillation rise time is of the order of a few microseconds.

4.2.4 THE GAS PROPORTIONAL SCINTILLATION COUNTER

4.2.4.1 BASIC CONCEPTS

We are now in a position to discuss the use of the secondary scintillation to measure the number of primary electrons and so to measure the energy of X-ray photons or other ionizing radiation. To do so we need to fully absorb the incident radiation in a region of the Xe gas where the reduced electric field intensity is below the 1 V cm^{-1} Torr^{-1} threshold for scintillation (Figure 4.2.14) and then drift all the n primary electrons to another region (the scintillation region) where they produce large amounts of secondary scintillation, but no (or little) ionization. There, the field intensity should be close to or below, the 5 or 6 V cm^{-1} Torr^{-1} threshold for ionization. The secondary scintillation will

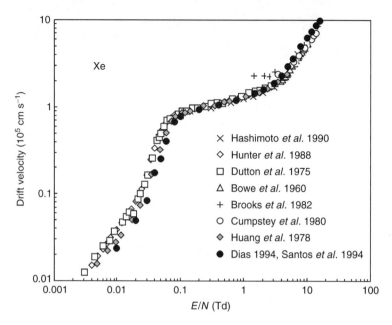

Figure 4.2.15 Drift velocities of electrons in Xe. (Based on Figure 17 of Dias et al.[3] and Figure 10 of Santos et al.[23]). Solid circles are Monte Carlo results, open symbols are experimental results (see references in Dias et al.[3])

be detected by a VUV photosensor (usually a photomultiplier tube), which produces a pulse lasting as long as there are electrons drifting across the scintillation region (a few microseconds). This pulse has an amplitude independent of the absorption position, i.e. it is proportional to n and so approximately proportional to E. Such a detector, as stated in the Introduction, is the gas proportional scintillation counter.

An earlier review article for these detectors was written by Policarpo[24] in 1977; more recent reviews have been produced by Akimov[25] and dos Santos et al.[26]

Up to the present, GPSC have been implemented mainly in three different geometries:

(i) cylindrical geometry counter;
(ii) spherical anode geometry counter;
(iii) uniform electric field geometry counter.

The cylindrical geometry GPSC[2,27] is like a standard proportional (ionization) counter (Figure 4.2.1) filled with a noble gas (Xe) and coupled in the end to a photomultiplier, to sense the secondary scintillation produced. However, since the applied voltage is below the one for starting avalanche processes it is not a proportional counter. The scintillation region is a small volume of the gas around the anode where the reduced electric field intensity lies in the 1 to 5 V cm^{-1} Torr^{-1} range; the rest is absorption region, so almost all the ionizing radiation is absorbed in this region.

In the spherical anode GPSC[28] the voltage applied to the anode is also close to the threshold for starting avalanches, which is well above the scintillation threshold. The scintillation region is also limited to a small volume close to the anode; the rest, which is the largest part of the detector volume, is absorption region.

In the uniform electric field GPSC,[29,30] already described summarily in the Introduction, there is a high transmission metallic grid delimiting the absorption region (Figure 4.2.2). This region is usually a few cm thick so that most radiations are absorbed there. The n primary electrons there produced are transferred into the scintillation region by an electric field, producing then the secondary scintillation pulse, detected by the photosensor. Since in this geometry the absorption and scintillation regions are well separated it is possible to optimize the electric field in each one and large amounts of secondary scintillation be produced. This geometry is the most used one.

4.2.4.2 THE ENERGY RESOLUTION OF A GAS PROPORTIONAL SCINTILLATION COUNTER

The energy resolution of a GPSC depends mainly on the fluctuations of n (with variance σ_n), as well as on the fluctuations in the number of secondary scintillation photons reaching the photosensor and in the gain A of the photosensor. However, since the number of secondary scintillation photons produced by a single primary electron is large and with small fluctuations, the experimental energy resolution for a narrow beam of X-rays[14,31] can be simplified and written as:

$$R = 2.355\sqrt{\frac{FW}{E} + \frac{k}{A}}$$

where k is a constant and the term k/A represents the fluctuations associated with the gain A, which is generally much smaller than FW/E.

However, if the beam enters the detector through the full window diameter, there are fluctuations in the number of scintillation photons reaching the photosensor due to solid angle effects. Indeed for X-rays entering the detector window near its border, a large number of scintillation photons will miss the photosensor due to its finite size. Thus, these events produce pulses with amplitudes smaller than central events. For a full opening window, these solid angle effects produce pulses with variable amplitudes which deteriorate the energy resolution, limiting the performance of large area GPSC.

An obvious solution is to use very large area photosensors, which has the inconvenience of a high cost. Another solution is to concentrate the primary electrons produced in the drift region into the scintillation region,[32] a technique which has its own limitations.

4.2.4.3 SOLID ANGLE COMPENSATION TECHNIQUES

Two other techniques have been recently developed that allow the construction of GPSCs with windows approaching in size the diameter of the photosensor, with little deterioration of the energy resolution.[33–36] Such techniques make use of solid angle compensation.

The first one, the curved grid technique,[33,35–37] uses a curved grid G_1 followed by a flat grid (G_2) (Figure 4.2.16a) delimiting the scintillation region. Therefore, the electric field there is no longer uniform: in the centre it is weaker than in the border of the curvature, which means that the intensity of the secondary scintillation increases with the radial distance to the axis. However, the solid angle through which the photosensor sees the secondary scintillation photons decreases with the radial distance.

With a properly calculated grid curvature[33,35,36] it is possible to achieve a radial increasing of the scintillation yield, that compensates the radial decreasing of the solid angle, leading thus to a constant pulse amplitude whatever the entrance window position of an X-ray photon.

Both spherical[33,37] and ellipsoidal[38] grids have been used. Energy resolutions of 8% were obtained for 5.9 keV X-rays and windows 25 mm in diameter, with Xe filled GPSC using standard 50 mm diameter photomultiplier tubes.[37]

The second technique (Figure 4.2.16b), the masked photosensor technique,[34–36] uses a uniform electric field scintillation region, produced by the parallel grids G_1 and G_2, and a graded mask covering the photosensor. This mask absorbs more secondary scintillation in the centre than in the border. With proper grading it is possible to compensate the radial decreasing of solid angle by a radial increasing of the mask transmission. Again, this leads to constant pulse amplitude whatever the entrance window position of the X-ray photon. With this technique energy resolutions of 10% were obtained for 5.9 keV X-rays entering the full 40 mm wide window of a GPSC which uses a 50 mm diameter photomultiplier.[34]

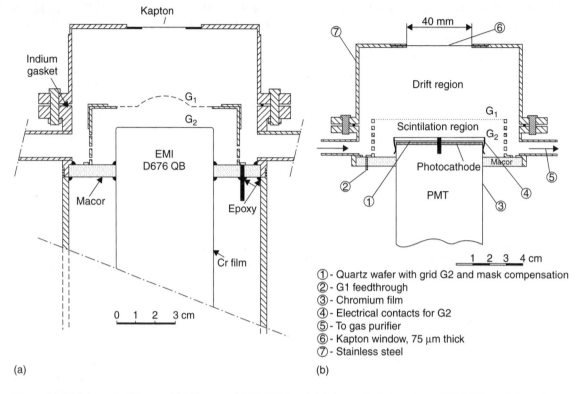

(a) (b)

Figure 4.2.16 Schematic diagram of (a) the curved grid GPSC and (b) the masked-photosensor GPSC (Figure 2 from Conde et al.[33] and Figure 4 from Veloso et al.,[34] respectively). Reproduced by permission of The Institute of Electrical and Electronics Engineers, Inc.

4.2.4.4 PHOTOSENSORS FOR GAS PROPORTIONAL SCINTILLATION COUNTERS

Traditionally, photomultipliers have been the preferred photosensors for GPSCs. As the secondary scintillation for Xe lies in the VUV (170 nm) and is rather intense (about 10^5 VUV photons per 5.9 keV X-ray) the photomultiplier needs to have a high purity quartz window (Spectrosil B) and a small number of dynodes[8] like the EMI D676 QB. However, since photomultipliers are expensive, bulky and fragile, efforts have been made to find other alternatives. Following the early use of photoionization chambers with TMAE[39] recent work has put emphasis on the use of microstrip plate detectors with CsI and photodiodes.

In a compact implementation[40] depicted in Figure 4.2.17, a standard microstrip plate (MSP)

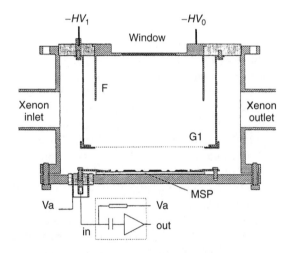

Figure 4.2.17 A compact GPSC using a CsI covered microstrip plate (MSP) as photosensor (Figure 4.2.2 of Veloso et al.[40]). Reproduced by permission of The Institute of Electrical and Electronics Engineers, Inc.

takes the role of the second grid of a GPSC, with the MSP anodes biased positively at a few hundred volts. With G_1 at a negative $-HV_1$ voltage of a few kV, a scintillation region is produced between G_1 and the MSP. An appropriate voltage applied to the window $(-HV_0)$ defines an absorption/drift region between G_1 and the window. If the MSP is covered with a CsI film (Figure 4.2.18) photoelectrons can be released from the anodes by the VUV scintillation and then be charge multiplied in the MSP anodes, producing a large amplitude pulse. However, this implementation has a drawback: since charge multiplication takes place in the Xe environment, it is accompanied by light emission that releases further electrons from the CsI covered cathodes, leading to a positive feedback process which limits the maximum allowable charge gain. To avoid this drawback the MSP can be separated from the Xe GPSC scintillation region with a thin quartz window and the MSGC filled with a non-scintillating gas like P-10.[31] The achieved energy resolutions for 5.9 keV X-rays are 12 %[31] and 11.4 %[41] for the first implementation (MSP within the Xe environment), and 10.5 % for the second implementation (MSP in a separate P-10 gas environment).

Photodiodes, whether of the vacuum type[42] or solid state (Si) type,[43,44] have also been considered as alternative photosensors for GPSCs. However, for X-ray detection the Si photodiode noise is too high and the performance is rather worse than that for a standard photomultiplier based GPSC. The recent development of VUV sensitive large area avalanche photodiodes (LAAPD), which are photodiodes with an intrinsic gain resulting from an avalanche process, allowed the development of high performance GPSCs.

When the LAAPD is placed in the Xe environment, just in front of the scintillation region,[45] an energy resolution of 7.9 % for 5.9 keV X-rays was obtained. However, LAAPDs have the inconvenience of having a gain rather sensitive to temperature fluctuations.

4.2.5 APPLICATIONS OF GAS PROPORTIONAL SCINTILLATION COUNTERS TO MATERIAL ANALYSIS

The superior performance of GPSCs over standard PCs makes them suitable for material analysis by X-ray fluorescence techniques, wherever PCs have

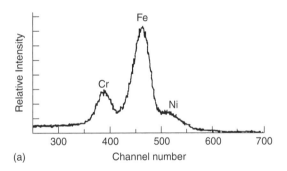

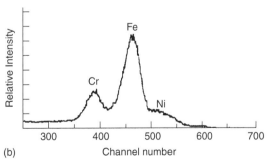

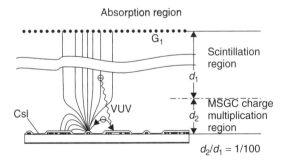

Figure 4.2.18 Detail of a CsI covered MSP as the photosensor in a GPSC (Figure 2 of Veloso et al.[40]). Reproduced by permission of The Institute of Electrical and Electronics Engineers, Inc.

Figure 4.2.19 X-ray fluorescence spectra for a stainless steel sample excited with a ^{244}Cm X-ray source, using a photomultiplier based GPSC and (a) 11 mm and (b) 25 mm diameter window collimations (Figure 8 of dos Santos et al.[46]). Reproduced by permission of The Institute of Electrical and Electronics Engineers, Inc.

been used: low cost room temperature portable instruments, large area detectors and very soft X-ray detection. Figures 4.2.19 and 4.2.20 depict X-ray fluorescence spectra for stainless steel and calcopyrite samples excited with a ^{244}Cm X-ray source[46] using a photomultiplier based GPSC with a curved grid. As shown, the performance does not deteriorate when the window collimation increases from 11 to 25 mm diameter. The detector can separate clearly the Cr and the Fe peaks.

Figure 4.2.21 shows X-ray fluorescence spectra obtained with a LAAPD based GPSC[47] of non-homogeneous geological samples containing Si (Figure 4.2.21a) and anthracite (Figure 4.2.21b) excited with a ^{55}Fe X-ray source and a calcopyrite sample (Figure 4.2.21c) excited with a ^{109}Cd source. The good separation power of the GPSC for low atomic number elements like Si, S, Ca and Ti is evident.

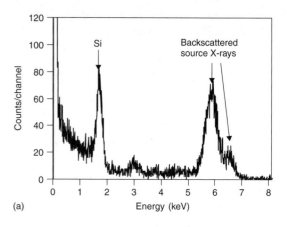

(a)

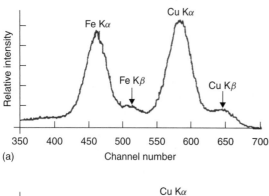

(a)

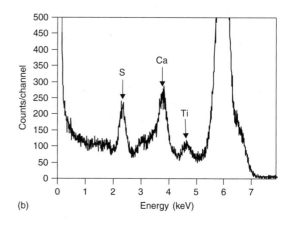

(b)

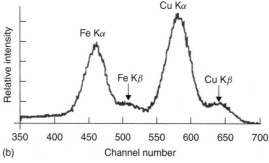

(b)

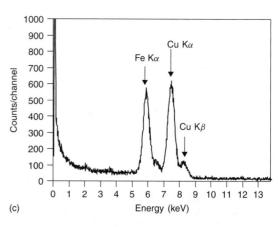

(c)

Figure 4.2.20 X-ray fluorescence spectra for a calcopyrite sample excited with a ^{244}Cm X-ray source, using a photomultiplier based GPSC and (a) 11 mm and (b) 25 mm diameter window collimations (Figure 9 of dos Santos et al.[46]). Reproduced by permission of The Institute of Electrical and Electronics Engineers, Inc.

Figure 4.2.21 X-ray fluorescence spectra obtained with a LAAPD based GPSC for Si (a) and anthracite (b) samples excited with a ^{55}Fe X-ray source, and for a calcopyrite sample (c) excited with a ^{109}Cd source (Figure 9 of Lopes et al.[47]). Reproduced by permission of The Institute of Electrical and Electronics Engineers, Inc.

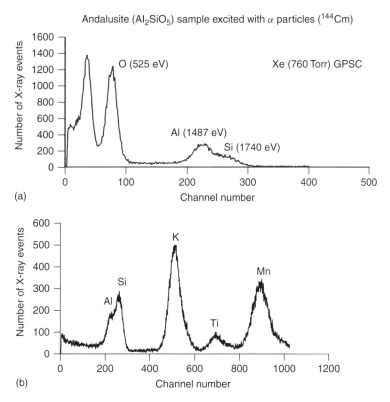

Figure 4.2.22 X-ray fluorescence spectra of an andalusite sample, obtained with a Xe filled GPSC with a very thin polyimide window: (a) excitation with a ^{244}Cm alpha particle source; (b) excitation with 5.9 keV X-rays from a ^{55}Fe source (Borges (unpublished) and Borges et al.[48])

For very soft X-ray detection and large areas, there are no other detectors (even cooled semiconductor detectors) that can match the performance of GPSC. In Figure 4.2.22 we present the spectra of an andalusite sample obtained with a GPSC filled with pure Xe and having a polyimide window (PG-W from Metorex). In Figure 4.2.22(a) the sample was excited with alpha particles from a ^{244}Cm source; in Figure 4.2.22(b) excitation was produced by 5.9 keV X-rays from a ^{55}Fe source. As shown, the oxygen K_α peak is clearly separate from the carbon peak (this peak arises from impurities and the carbon in the polyimide window). The Mn line results from coherent backscattering of the 5.9 keV X-rays.

Detecting low concentration elements is, sometimes, not easy since the corresponding peaks might be difficult to distinguish from background. However, digital risetime discrimination techniques applied to GPSC spectra might improve the peak-to-background ratio, as already demonstrated.[49]

4.2.6 CONCLUSIONS

It was shown that Xe-filled gas detectors, like standard proportional counters and gas proportional scintillation counters, suffer from nonlinearity effects, approaching 1%, and that the peak shape is not a pure gaussian. These effects must be taken into account when doing spectra fitting. GPSCs offer an energy resolution much better than that of PCs, i.e. 8% for 5.9 keV X-rays and can be built with very large windows (2.5 cm diameter or larger) and are very useful for very soft X-rays like the K lines of C and O. New photosensors have already been developed to replace the photomultiplier allowing more compact GPSCs.

ACKNOWLEDGEMENTS

This work was supported by Project CTAE/1920, Fundação para a Ciência e a Tecnologia (FCT), Lisbon, and was done in Grupo de Instrumentação Atómica e Nuclear (GIAN), Centro de Instrumentação (Unidade 217/94), Departamento de Física of the University of Coimbra.

I thank Teresa H.V.T. Dias for very interesting comments and for preparing Figures 4.2.3, 4.2.5–4.2.10, 4.2.14 and 4.2.15. Thanks are due to Filipa I.G.M. Borges for preparing Figure 4.2.22. I thank Diogo S.A.P. Freitas for preparing Figures 4.2.1 and 4.2.2, and for doing the word processing work. Thanks are also due to João A.S. Barata who did the calculations and prepared Figures 4.2.11–4.2.13 and to Hugo N. da Luz for help in preparing Figure 4.2.4.

REFERENCES

1. Knoll, G. F. *Radiation Detection and Measurement*, 3rd Edition, John Wiley & Sons, Inc., New York, 2000.
2. Conde, C. A. N. and Policarpo, A. J. P. L. A gas proportional scintillation counter. *Nucl. Instrum. Methods*, **53**, 7–12 (1967).
3. Dias, T. H. V. T., Santos, F. P., Stauffer, A. D. and Conde, C. A. N. Monte Carlo simulation of X-ray absorption and electron drifting in gaseous xenon. *Phys. Rev.*, **A48**(4), 2887–2902 (1993).
4. Dias, T. H. V. T. Physics of noble gas X-ray detectors: a Monte Carlo study, in *Linking the Gaseous and Condensed Physics of Matter: the Behavior of Slow Electrons* (Eds L. G. Christophorou, E. Illenberger and W. F. Schmidt), NATO ASI Series B: Physics **326**, 553–559, Plenum Press, New York, 1994.
5. Dias, T. H. V. T., dos Santos, J. M. F., Rachinhas, J. P. B. M., Santos, F. P., Conde, C. A. N. and Stauffer, A. D. Full-energy absorption of X-ray energies near the Xe L- and K-photoionization threshold in xenon gas electrons: simulation and experimental results. *J. Appl. Phys.*, **82**(6), 2742–2753 (1997).
6. dos Santos, J. M. F., Morgado, R. E., Távora, L. M. N. and Conde, C. A. N. The energy non-linearity of a xenon gas proportional scintillation counter at the absorption edge in xenon. *Nucl. Instrum. Methods Phys. Res.*, **A350**, 216–220 (1994).
7. Borges, F. I. G. M. and Conde, C. A. N. Experimental W-values in gaseous Xe, Kr and Ar for low energy X-rays. *Nucl. Instrum. Methods Phys. Res.*, **A381**, 91–96 (1996).
8. Vinagre, F. L. R. and Conde, C. A. N. A technique for the absolute measurement of the W-value for X-rays in counting gases. *Nucl. Instrum. Methods Phys. Res.*, **A450**, 365–372 (2000).
9. Vinagre, F. L. R. and Conde, C. A. N. Absolute W-value measurements for 5.9 keV X-rays in Xe–Ne mixtures at atmospheric pressures. *J. Appl. Phys.*, **88**(9), 5426–5432 (2000).
10. Santos, F. P., Dias, T. H. V. T., Rachinhas, J. P. B. M., Conde, C. A. N. and Stauffer, A. D. Monte Carlo simulation study of the Fano factor, W value, and energy resolution for the absorption of soft X rays in xenon–neon gas mixtures. *J. Appl. Phys.*, **89**(12), 8202–8213 (2001).
11. Dias, T. H. V. T. Radiation detection with noble gases: modeling and measurements, in *Proceedings of CAARI '98 – 15th International Conference on the Applications of Accelerators in Research and Industry* (Eds J. L. . Duggan and I. L. . Morgan), *Am. Inst. Phys. CP Ser.* **475**, 854–857 (1999).
12. Santos, F. P., dos Santos, J. M. F., Dias, T. H. V. T. and Conde, C. A. N. Pulse-height-spectrum distortion in xenon gaseous detectors for soft X-rays: experimental results. *IEEE Trans. Nucl. Sci.*, **42**(4), 611–614 (1995).
13. Borges, F. I. G. M., Santos, F. P., Dias, T. H. V. T., Rachinhas, J. P. B. M., Conde, C. A. N. and Stauffer, A. D. Xenon–neon gas proportional scintillation counter for X-rays below 2 keV: a Monte Carlo simulation study. *IEEE Trans. Nucl. Sci.* **49**, 917–922 (2002).
14. Simons, D. G. and De Korte, P. A. J. Soft X-ray energy resolution and background rejection for a driftless GPSC. *Nucl. Instrum. Methods Phys. Res.*, **A277**, 642–656 (1989).
15. Borges, F. I. G. M., dos Santos, J. M. F., Santos, F. P., Dias, T. H. V. T., Rachinhas, P. J. B. M. and Conde, C. A. N. Xenon–neon gas proportional scintillation counters for X-rays below 2 keV: experimental results. *IEEE Nucl. Sci. Symp.*, **50**, 842–846 (2003).
16. Santos, F. P. *Detectores Gasosos para Raios X: Simulação e Estudo Experimental*. PhD Thesis, Universidade de Coimbra, Portugal, 1994.
17. J. A. S. Barata, *Simulação por Técnicas de Monte Caro da Deriva de Electrões em Xénon*. Report, Universidade da Beira Interior, Covilhã, Portugal, 1998.
18. Dias, T. H. V. T. A unidimensional Monte Carlo simulation of electron drift velocities and electroluminescence in argon, krypton and xenon. *J. Phys. D: Appl. Phys.*, **19**, 527–545 (1986).
19. Dias, T. H. V. T. *Simulação do Transporte de Electrões em Gases Raros*. PhD Thesis, Universidade de Coimbra, Portugal, 1986.
20. Leite, M. and Salete S. C. P. Radioluminescence of rare gases. *Port. Phys.*, **11**, 53–100 (1980).
21. Suzuki, M. and Kubota, S. Mechanism of proportional scintillation in argon, krypton an xenon. *Nucl. Instrum. Methods*, **164**, 197–199 (1979).
22. Conde, C. A. N., Ferreira, L. R. and Ferreira, M. F. A. Secondary scintillation output of xenon in a uniform field gas proportional scintillation counter. *IEEE Trans. Nucl. Sci.*, **24**(1), 221–224 (1977).

23. Santos, F. P., Dias, T. H. V. T., Stauffer, A. D. and Conde, C. A. N. Three-dimensional Monte Carlo calculation of the VUV electroluminescence and other electron transport parameters in xenon. *J. Phys. D: Appl. Phys.*, **27**, 42–48 (1994).
24. Policarpo, A. J. P. L. The gas proportional scintillation counter. *Space Sci. Instrum.*, **3**, 77–107 (1977).
25. Akimov, Yu. K. Scintillation in noble gases and their application (Review). *Instrum. Experim. Tech.*, **41**(1), 1–32 (1998).
26. dos Santos, J. M. F., Lopes, J. A. M., Veloso, J. F. C. A., Simões, P. C. P. S., Dias, T. H. V. T., Santos, F. P., Rachinhas, J. P. B. M., Requicha Ferreira, L. F. and Conde, C. A. N. Development of portable gas proportional scintillation counter for X-ray spectrometry. *X-Ray Spectrom.*, **30**, 373–381 (2001).
27. Garg, S. P., Murthy, K. B. S. and Sharma, R. C. Development of a semi-empirical expression for light gain factors in gas scintillation proportional counters. *Nucl. Instrum. Methods Phys. Res.*, **A357**, 406–417 (1995).
28. Policarpo, A. J. P. L., Alves, M. A. F., dos Santos, M. C. M. and Carvalho, M. J. T. Improved resolution for low energies with gas proportional scintillation counters. *Nucl. Instrum. Methods*, **102**, 337–348 (1972).
29. Conde, C. A. N., Santos, M. C. M., Fátima M., Ferreira, A. and Sousa, C. A. Argon scintillation counter with uniform electric field. *IEEE Trans. Nucl. Sci.*, **22**(1), 104–108 (1975).
30. Palmer, H. E. and Braby, L. A. Parallel plate gas scintillation proportional counter for improved resolution of low-energy photons. *Nucl. Instrum. Methods*, **116**, 587–589 (1974).
31. Veloso, J. F. C. A. and dos Santos, J. M. F., Conde, C. A. N. Gas proportional scintillation counters with a CsI-covered microstrip plate UV photosensor for high-resolution X-ray spectrometry. *Nucl. Instrum. Methods Phys. Res.*, **A457**, 253–261 (2001).
32. Manzo, G., Peacock, A., Andersen, R. D. and Taylor, B. G. High pressure gas scintillation spectrometry for X-ray astronomy. *Nucl. Instrum. Methods*, **174**, 301–315 (1980).
33. Conde, C. A. N., dos Santos, J. M. F. and Bento, A. C. S. S. M. New concepts for the design of large area gas proportional scintillation counters. *IEEE Trans. Nucl. Sci.*, **40**(4), 452–454 (1993).
34. Veloso, J. F. C. A., dos Santos, J. M. F. and Conde, C. A. N. Large-window gas proportional scintillation counter with photosensor compensation. *IEEE Trans. Nucl. Sci.*, **42**(4), 369–373 (1995).
35. Conde, C. A. N., dos Santos, J. M. F. and Bento, A. C. S. S. M. Gas proportional scintillation counter for ionizing radiation with medium and large size radiation windows and/or detection volumes. US Patent 5,517,030 (14 May, 1996) (available in *http://www.uspto.gov/patft/index.html*).
36. Conde, C. A. N., dos Santos, J. M. F. and Bento, A. C. S. S. M. Gas proportional scintillation counter for ionizing radiation with medium and large size radiation windows and/or detection volumes. European Patent EP 0616722 BI (30 December, 1998) (available in *http://ep.espacenet.com*).
37. dos Santos, J. M. F., Bento, A. C. S. S. M. and Conde, C. A. N. The performance of the curved grid gas proportional scintillation counter in X-ray spectrometry. *Nucl. Instrum. Methods Phys. Res.*, **A337**, 427–430 (1994).
38. Silva, R. M. C., dos Santos, J. M. F. and Conde, C. A. N. An ellipsoidal grid gas proportional scintillation counter. *Nucl. Instrum. Methods Phys. Res.*, **A422**, 305–308 (1999).
39. Anderson, D. F. A xenon gas scintillation proportional counter coupled to a photoionization detector. *Nucl. Instrum. Methods*, **178**, 125–130 (1980).
40. Veloso, J. F. C. A., Lopes, J. A. M., dos Santos, J. M. F. and Conde, C. A. N. A microstrip gas chamber as a VUV photosensor for a xenon gas proportional scintillation counter. *IEEE Trans. Nucl. Sci.*, **43**(3), 1232–1236 (1996).
41. Freitas D. S. A. P., Veloso, J. F. C. A., dos Santos, J. M. F. and Conde, C. A. N. Dependence of the performance of CsI-covered microstrip plate VUV photosensors on geometry: experimental results. *IEEE Trans. Nucl. Sci.* **49**, 1629–1633 (2002).
42. Van Standen, J. C., Mutterer, M., Pannicke, J., Schelhaas, K. P., Foh, J. and Theobald, J. P. Vacuum photodiode as light sensing element for gas scintillation counters. *Nucl. Instrum. Methods*, **157**, 301–304 (1978).
43. Campos, A. J. A silicon photodiode based gas proportional scintillation counter. *IEEE Trans. Nucl. Sci.*, **31**(1), 133–135 (1984).
44. Lopes, J. A. M., dos Santos, J. M. F., Morgado, R. E. and Conde, C. A. N. Silicon photodiodes as the VUV photosensor in gas proportional scintillation counters. *IEEE Trans. Nucl. Sci.*, **47**(3), 928–932 (2000).
45. Lopes, J. A. M., dos Santos, J. M. F. and Conde, C. A. N. A large area avalanche photodiode as the VUV photosensor in gas proportional scintillation counters, *Nucl. Instrum. Methods Phys. Res.*, **A454**, 421–425 (2000).
46. dos Santos, J. M. F., Soares, A. J. V. D., Monteiro, C. M. B., Morgado, R. E. and Conde, C. A. N. The application of the curved-grid technique to a gas proportional scintillation counter with a small-diameter photomultiplier tube. *IEEE Trans. Nucl. Sci.*, **45**(3), 229–233 (1998).
47. Lopes, J. A. M., dos Santos, J. M. F., Morgado, R. E. and Conde, C. A. N. Silicon a xenon gas proportional scintillation counter with a UV-sensitive, large-area avalanche photodiode. *IEEE Trans. Nucl. Sci.*, **48**(3), 312–319 (2001).
48. Borges, F. I. G. M., Santos, F. P., dos Santos, J. M. F., Dias, T. H. V. T., Rachinhas, P. J. B. M. and Conde, C. A. N. The performance of a gas proportional scintillation counter for X-ray spectrometry in the 0.1–3 keV range. *IRRMA-V – 5th International Topical Meeting on Industrial Radiation and Radioisotope Measurement Applications* submitted.
49. Simões, P. C. P. S., dos Santos, J. M. F., and Conde, C. A. N. Digital risetime discrimination for peak enhancement analysis. *X-Ray Spectrom.*, **26**, 182–188 (1997).

4.3 Superconducting Tunnel Junctions

M. KURAKADO

RIKEN, Saitama, Japan

4.3.1 INTRODUCTION

In this subchapter, recent references regarding superconducting tunnel junction (STJ) detectors are limited to those related to spectrometric applications. References [1–4] are useful concerning studies of STJ detectors themselves. Historical and introductive descriptions also can be found in References 5 and 6.

The limit of energy resolution due to the statistical fluctuation ΔN of the number N of initial signal charges, $(\Delta N/N)E$, is given by $2.355(E\varepsilon F)^{1/2}$ (FWHM), where E is the radiation energy, ε is the mean energy required to excite one signal charge and F is the Fano factor of the detector. Since the energy gaps ($E_g = 2\Delta$) of metal superconductors are of the order of 1 meV, detectors made of metal superconductors potentially provide high resolution. Popular superconductors for STJ detectors are Nb ($T_c \approx 9$ K), Ta ($T_c \approx 4.5$ K), V ($T_c \approx 5.4$ K) and Al ($T_c \approx 1.2$ K), where T_c is the superconducting transition temperature ($2\Delta_{T=0} \approx 3.5 k_B T_c$).

An STJ usually consists of two superconductor layers and a 1–2 nm thick insulator layer, which is a tunnel barrier between the superconductor layers (Figure 4.3.1). Excited electrons or holes, i.e. quasiparticles, can pass through the tunnel barrier by means of quantum mechanical tunneling (Figure 4.3.2).

The dc Josephson current that flows at the bias voltage $V_B = 0$ is suppressed by applying a magnetic field parallel to the junction plane to stabilize the bias when an STJ is put to use as a detector. A quasiparticle produces signal current and signal charge when it passes through the tunnel barrier before recombination with other quasiparticles in the same electrode or escapes from the electrode to the electric lead. Quasiparticles excited by radiation must diffuse in the electrode and collide with the tunnel barrier for many times to tunnel the barrier, leading to the dependency of the tunneling probability on characteristics of the electrode such as the thickness, the mean free path in the electrode and the width of the lead. Consequently, two electrodes produce different magnitudes of signals that cause double peaks in the case of X-ray detection when the X-rays can penetrate the upper electrode. The rise times of signals, i.e. charge

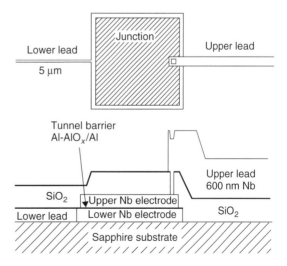

Figure 4.3.1 Example of the structure of an STJ detector

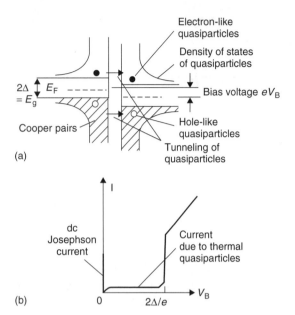

Figure 4.3.2 Energy structure, tunnel effect and current–voltage characteristics of an STJ. For low energy X-rays, the insulator layer covering the STJ is usually removed

signals or current signals, provide good methods to distinguish the signals produced in an electrode from the others.

In addition to the single-junction detectors (Figure 4.3.1), there are two other type of STJ detectors as shown in Figure 4.3.3. Type (a) consists of an absorber superconductor and two STJs that are composed of a smaller energy gap superconductor and collect quasiparticles excited in the absorber by means of the quasiparticle trapping effect. In this subchapter, this type of detector is referred to as a one-dimensional-imaging STJ detector. In the case of type (b), i.e. series-junction detectors, radiation is absorbed in a single-crystal substrate and the resulting non-thermal phonons are detected by many STJs connected in series on the surface of the substrate.

4.3.2 STATISTICAL LIMIT OF ENERGY RESOLUTION

Figure 4.3.4 shows the main relaxation processes of the energy deposited from radiation to superconducting Sn.[7,8] Radiation excites electrons which

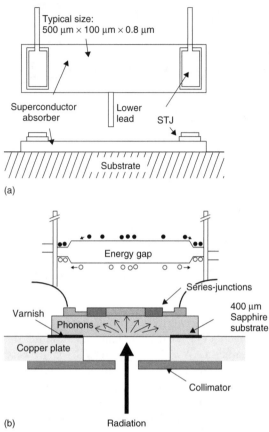

Figure 4.3.3 Two kinds of STJ detectors other than single-junction detectors: (a) one-dimensional-imaging STJ detector; and (b) series-junction detector

initiate the cascade of quasiparticle excitation in the superconductor. The dominant interaction of quasiparticles with energy $E > \sim(E_D E_F)^{1/2}$ is the electron–electron interaction, where E_D is the Debye energy, i.e. maximum energy of phonons, and E_F is the Fermi energy. Such a quasiparticle breaks Cooper pairs and thus excites other quasiparticles. The dominant interaction of quasiparticles with energy $E < \sim(E_D E_F)^{1/2}$ is the electron–phonon interaction. Such a quasiparticle emits a phonon of energy Ω and decreases its own energy by Ω. The distribution of phonons emitted by quasiparticles with energies $E \gg E_D$ is given by $\alpha^2(\Omega)F(\Omega)$, where $\alpha^2(\Omega)$ is the effective electron–phonon coupling function and $F(\Omega)$ is the density of states of phonons. The inset in

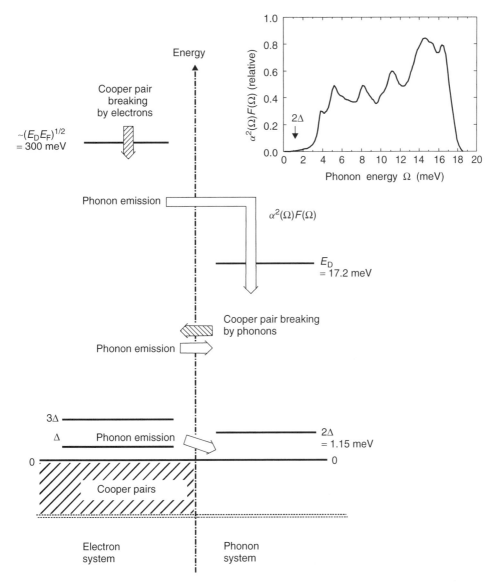

Figure 4.3.4 Cascade excitation processes of quasiparticles and phonons in a superconducting Sn at 0 K. The inset is the $\alpha^2(\Omega)F(\Omega)$, where $\alpha^2(\Omega)$ is the effective electron–phonon coupling function and $F(\Omega)$ is the density of states of phonons

Figure 4.3.4 shows $\alpha^2(\Omega)F(\Omega)$ of Sn. A phonon with energy $\Omega \geq 2\Delta$ breaks a Cooper pair and thus excites two quasiparticles. A quasiparticle with energy $E \geq 3\Delta$ can emit a phonon with energy $\Omega \geq 2\Delta$. Repetition of the phonon emission by quasiparticles and excitation of quasiparticles by phonons increases the number of quasiparticles and phonons, leading to a small ε value.

Figure 4.3.5 shows results of numerical simulations of the cascade excitation processes of quasiparticles and phonons in a bulk superconducting Sn at 0 K.[8] As can be seen in Figure 4.3.5(a), the mean energy ε required to excite one quasiparticle beyond the energy gap is $\approx 1.7\Delta (\approx 1 \text{ meV})$ and the statistical fluctuation of the number N of excited quasiparticles defined by $F \equiv \langle (N - \langle N \rangle)^2 \rangle / \langle N \rangle$

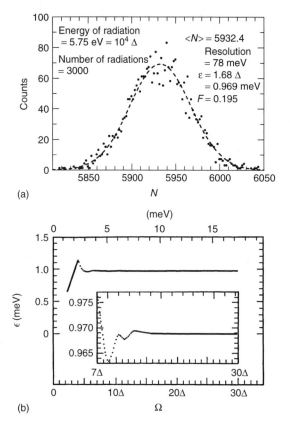

Figure 4.3.5 Results of computer simulations on the quasi-particle excitation in a bulk superconducting Sn at 0 K: (a) a spectrum of the number of excited quasiparticles for 3000 radiations with energy 5.75 eV; and (b) ε value obtained as a function of the energy Ω of an incident phonon. Spectrum (a) was obtained by Monte Carlo simulation

Figure 4.3.5(b) shows the ε value calculated as a function of the energy Ω of a phonon.[8] Non-thermal phonons with energy $\Omega \geq 2\Delta$ can excite quasiparticles as efficiently as radiations in the superconducting Sn. ε increases linearly, i.e. $\varepsilon = \Omega/2$, for $2\Delta < \Omega < 4\Delta$ because the phonon excites only two quasiparticles, and decreases for $\Omega > 4\Delta$ because the phonon can break more than 2 Cooper pairs and excite more than 4 quasiparticles. The ε for $\Omega > 6\Delta$ is almost constant and equal to that obtained for radiation, i.e. 1.7Δ. It should be noted that thermal phonons with energy $\Omega < 2\Delta$ do not contribute to signals and that STJ detectors are insensitive to those phonons, in contrast to bolometers.

The ultra-low noise performances of most of the recent STJ detectors can be attributed to the amplification of signal charge Q in the STJs themselves, i.e. multi-tunneling of quasiparticles (Figure 4.3.6).[9] The STJs having a trapping structure (Figure 4.3.6(a)) showed prominent amplification of signal charge. With each process 1 or

i.e. Fano factor, is about 0.2, where $\langle N \rangle$ stands for the mean value of N. The ε and F mean that the statistical limit of energy resolution is about 2.5 eV at 5.9 keV for Sn junctions ($E_g = 2\Delta \approx 1.15$ meV). The data of $\alpha^2(\Omega)F(\Omega)$ in Figure 4.3.4 is characteristic of Sn. Generally $\alpha^2(\Omega)F(\Omega)$ is proportional to Ω^2 for $\Omega \ll E_D$ in most superconductors. A simulation approximating $\alpha^2(\Omega)F(\Omega)$ by Ω^2 also gave $\varepsilon \approx 1.7\Delta$, suggesting that the relation holds for many metal superconductors. Furthermore, the ε value means that about 60 % of the energy lost by radiation in the superconductor contributes to signal charge, whereas only about 30 % contributes in the cases of semiconductor detectors.

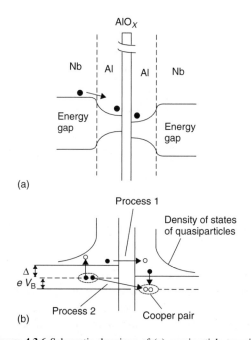

Figure 4.3.6 Schematic drawings of (a) quasiparticle trapping and (b) multi-tunneling of quasiparticles. (a) Energy gap versus distance for an STJ with quasiparticle traps. (b) Quasiparticle excitation energy versus density of states and the main quasiparticle processes in multiple tunneling

2 shown in Figure 4.3.6(b), one electron is transferred from the left electrode to the one on the right. The iteration of the processes 1 and 2 does not increase the number of excited quasiparticles but increases the signal charge. The amplification of signal charge terminates when the quasiparticle escapes from the junction or recombines with one of the other quasiparticles in the same electrode.

The statistical fluctuation of signal charge associated with the tunneling processes deteriorates the energy resolution R in addition to the initial fluctuation of N, noise and the position dependency of the signal height. The amplification factor $\langle n \rangle \equiv Q/(E/1.7\Delta)e$ and the fluctuation R_t associated with the multi-tunneling process are given by:

$$\langle n \rangle = P_1(1+P_2)/(1-P_1P_2)$$
$$R_t = 2.355(E\varepsilon G)^{1/2}$$

where $G = (1 - P_1 + 3P_2 + P_1P_2)/P_1(1+P_2)^2$, and P_1 and P_2 are probabilities of processes 1 and 2 of Figure 4.3.6(b), respectively.[10] Therefore, $R = 2.355[E\varepsilon(F+G)]^{1/2}$. When $\langle n \rangle \gg 1$, $G \approx 1$ and thus $R \approx 2.45 \times 2.355(E\varepsilon F)^{1/2}$, where $F = 0.2$, corresponding to $R \approx 10\,\text{eV}$ (7 eV) at 5.9 keV for Nb-based (Ta-based) STJs. Therefore, multi-tunneling is not preferable to attain ultimate energy resolutions but useful to attain ultra-low noise performances.

4.3.3 SINGLE-JUNCTION DETECTORS

Recently, STJ detectors realized energy resolutions about one order higher compared with semiconductor detectors.

An STJ, consisting of (from bottom to top) Nb (240 nm)/Al (200 nm)/AlO$_x$/Al (200 nm)/Nb (150 nm) and having an area of $100 \times 100\,\mu\text{m}^2$, showed an energy resolution of 29 eV and noise of 10.5 eV for 5.9 keV X-rays.[11] The STJ was cooled to 200 mK by an adiabatic demagnetization refrigerator (ADR) because 2Δ of the thick Al layers is considerably smaller than that of Nb. The Al layers acted as traps for quasiparticles. The quasiparticles generated by an X-ray absorbed in a Nb layer are quickly collected to the adjacent Al layer and trapped there, leading to efficient tunneling, amplified signals, reduced recombination in the Nb layer and thus decreased position dependency of signal heights.

As can be seen from Figure 4.3.7, showing calculated Gaussian spectra corresponding to (a) conventional semiconductor detectors and (b) an STJ detector with a resolution of 29 eV and noise of 10 eV at 5.9 keV, the improvement in resolving power for characteristic X-rays is remarkable.

A $100 \times 100\,\mu\text{m}^2$ Al/AlO$_x$/Al STJ, with an absorption efficiency of about 1% for 6 keV X-rays, attained the best energy resolution of 12 eV for 5.9 keV X-rays.[12,13] The STJ detector was cooled to 70 mK and the noise was 7 eV. The Al STJ was fabricated on a Si substrate coated with a metallic layer, which acted as a buffer layer to reduce the influence of X-ray absorption in the substrate on the STJ response. An insulator layer (polyimide) is deposited onto the buffer layer. The high energy resolution is attributed to the buffer layer underneath the STJ. The rise time of the outputs of the charge-sensitive preamplifier was about 10 μs. The charge signal of the 5.9 keV X-rays corresponded to 61% of the initially created quasiparticles ($\langle n \rangle = 0.61$), suggesting a weak amplification of signal charges owing to the multi-tunneling of quasiparticles.

A $141 \times 141\,\mu\text{m}^2$ Nb (265 nm)/Al (50 nm)/AlO$_x$/Al (50 nm)/Nb (165 nm) STJ revealed a high stability of the signal heights against temperature variation from a temperature (>50 mK) to 500 mK because of the enhanced energy gap of the thin Al layers arising from their proximity to the thicker Nb layers.[14] The pulse decay time of current signals was 4.5 μs. The amplification factor $\langle n \rangle$ was about 20. The STJ was evaluated by using low energy X-rays (70–700 eV) from synchrotron radiation. During the measurements the STJ was cooled by an ADR, and the temperature was not regulated and allowed to drift up to about 500 mK. Most of the low energy X-rays were absorbed by the upper electrode. The STJ detector showed, for example, an energy resolution of 5.9 eV and a noise of 4.5 eV for 277 eV X-rays that correspond to

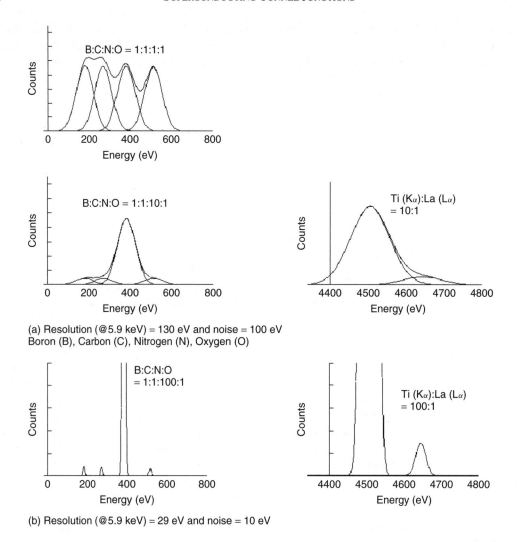

Figure 4.3.7 Sample spectra of characteristic X-rays obtainable with detectors with different resolutions: corresponding to (a) a semiconductor, and (b) a Nb/Al/AlO$_x$/Al/Nb STJ detector. The spectra were calculated assuming Gaussian distributions for the peaks

the K X-rays of carbon at a count rate of several hundred counts per second (cps) and a resolution of 13 eV, and a noise of 11.9 eV at 23 kcps. The energies of the synchrotron X-rays corresponded to the K X-rays of light elements, i.e. Be, B, C, N and O, and they produced sharp peaks in the pulse height spectrum.

Katagiri *et al.* developed a high count rate X-ray detector system with a fast current readout.[15] The STJ is a $200 \times 200\,\mu\text{m}^2$ Nb/Al/AlO$_x$/Al/Nb junction, which has a current rise time of 100 ns, a current decay time of 160 ns and resistance of 20 Ω at 0.4 K. The fast current readout system is equipped with a superconducting coil to achieve a stable biasing of the STJ. The coil works as an inductive load to signal pulses although it does not affect dc biasing by a shunt resistance. The energy resolution of bottom layer signals was about 230 eV up to 100 kcps, $\sim$265 eV at 250 kcps, and $\sim$400 eV at 500 kcps for 4 keV X-rays.

Verhoeve *et al.* studied STJs as infrared to ultraviolet photon detectors.[16] Optical photons with energy of 0.62–6.2 eV were measured by Ta (100 nm)/Al (30 nm)/AlO$_x$/Al (30 nm)/Ta (100 nm) STJs with areas of 10×10, 20×20, 30×30, 50×50 and $100 \times 100\,\mu m^2$.[16] The base electrode Ta is epitaxial. The STJs were cooled in a ^{3}He cryostat with a base temperature of 0.3 K. The rise time of the charge sensitive preamplifier was longer (2.8–70 μs) and the amplification factor $\langle n \rangle$ due to the multi-tunneling of quasiparticles was larger (6–190) for larger STJs. The $20 \times 20\,\mu m^2$ STJ (rise time = 12.5 μs and $\langle n \rangle = 30$) showed the highest energy resolution of 0.19 eV and a noise of 0.14 eV for 2.5-eV photons.

Angloher et al. obtained an energy resolution of 12 eV and a noise of 4 eV for 5.9 keV Mn K$_{\alpha 1}$ X-rays.[17,18] On a $100 \times 100\,\mu m^2$ Al/AlO$_x$/Al junction covered by a thin natural Al oxide, a superconducting Pb absorber ($2\Delta_{Pb} > 2\Delta_{Al}$) was formed (90 μm × 90 μm × 1.3 μm). The junction was prepared on a Si$_3$N$_4$ membrane with a thickness of 0.3 μm (Figure 4.3.8). The detector was operated at about 70 mK in a dilution refrigerator. The Pb absorber is coupled to the STJ via phonons. X-rays absorbed by the absorber break Cooper pairs, and the resulting quasiparticles emit phonons in relaxation and recombination processes. The phonons with energy larger than $2\Delta_{Al}$ can efficiently excite quasiparticles in the Al STJ. Phonons leaving the STJ towards the substrate can be reflected at the backside of the membrane and re-enter the STJ, increasing the signal charge. The rise time of the charge pulses was about 80 μs. Due to the high absorption efficiency of lead (51.8 %) and weak absorption efficiencies of Al (0.7 %) and Si$_3$N$_4$ (0.5 %) for 5.9 keV X-rays, multiple peaks are strongly suppressed and practically only single peaks appeared in the pulse height spectrum. The noise measured with pulser signals was 4 eV. The signal charge for 5.9 keV X-rays was 1.82 pC, i.e. $\langle n \rangle = 0.58$. With a similar detector, fluorescence lines from Si (K$_\alpha$ 1.740 keV) and W (M$_\alpha$ 1.776 keV) were clearly separated with an energy resolution of 9.7 eV.

4.3.4 ONE-DIMENSIONAL-IMAGING STJ DETECTORS

The thickness of the absorbers of one-dimensional-imaging STJ detectors (Figure 4.3.3(a)) is typically 500–1000 nm, which is thicker than the typical thickness of a superconductor electrode of an STJ (200 nm), leading to higher absorption efficiencies than single-junction detectors. The signal charge collected by each STJ depends on the incident position of radiation in the absorber. This type of STJ detector, therefore, can have high energy and lateral resolutions, e.g. 60 eV and about 5 μm, respectively, for 5.9 keV X-rays.[19]

Li et al. obtained an energy resolution of 13 eV for 5.9 keV X-rays in an area of 20 μm × 100 μm.[20] The size of the Ta absorber is 200 μm × 100 μm × 0.57 μm. The calculated absorption efficiency of the absorber is 28 % for 6 keV X-rays. X-rays from a ^{55}Fe source (Mn K$_\alpha$ (5.9 keV) and Mn K$_\beta$ (6.5 keV)) were measured at 210 mK.

Wilson et al. detected optical and ultraviolet photons with a detector that consists of a Ta absorber of 100 μm × 10 μm × 0.6 μm and two $100\,\mu m^2$ Al STJs.[21] The detector was cooled to 220 mK in a two stage ^{3}He cryostat. An amplification factor $\langle n \rangle$ of 23 and an energy resolution of about 1 eV were obtained.

Verhoeve detected 4.1-eV photons with a detector that consists of a Ta absorber of 400 μm × 50 μm × 0.1 μm and two $50 \times 50\,\mu m^2$ Ta STJs with 60 nm thick Al trapping layers. An energy resolution of 0.4 eV was obtained.[22]

4.3.5 SERIES-JUNCTION DETECTORS

Series-junction detectors (Figure 4.3.3(b)), were proposed to increase the effective areas of STJ

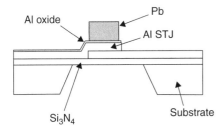

Figure 4.3.8 Schematic drawing of the single-junction detector of references 17 and 18

detectors.[5,6] The effective capacitance $C_{eff}(\equiv Q/V_S)$ can be suppressed by the series-connection of STJs, where Q is the total signal charge and V_S is the resulting signal voltage. $C_{eff} = C + nC'$, where n is the number of junctions in series, C is the electric capacitance of the junction, and C' is the input capacitance of the preamplifier. For a given total junction area S, C_{eff} takes its minimum value $2(Sc_0C')^{1/2}$ when $n = (Sc_0/C')^{1/2}$, which is not proportional to S but proportional to $S^{1/2}$, allowing larger S. The radiation energies are converted to phonons in a single-crystal substrate, and the non-thermal high energy phonons are absorbed by the series-junctions on the substrate. Therefore, the thickness of a series-junction detector is given by the thickness of the substrate. The thickness of a substrate is usually several hundred μm. Substrates containing heavy elements, e.g. Ge substrates, are preferable for detection of high energy X-rays. Calculated absorption efficiencies for some materials are shown in Figure 4.3.9.

One of demerits of series-junction detectors is the position dependence of signal heights. Kamihirata et al. measured incident positions of α particles (Figure 4.3.10). The thickness of the sapphire (Al_2O_3) substrate was 400 μm. Two-dimensional position detection was performed with a detector that is constructed by 4 series-junctions.[23] Each series-junction consists of 160 Nb STJs in series. The diameter of each STJ is 110 μm. The layer structure of the STJs is Nb (200 nm)/Al (70 nm)/AlO_x/Al (70 nm)/Nb (150 nm). The detector was cooled to about 0.35 K and was irradiated with α particles (5.486 MeV: 85 %; 5.443 MeV: 13 %) from ^{241}Am through 5 holes in a collimator as can be seen in Figure 4.3.10(a). The intensity of the available ^{241}Am source was only about 1000 Bq, and thus the source was closely attached to the collimator, resulting in a loose collimation of α particles.

By making use of the position information, the position dependence of the pulse heights was corrected, i.e. the initial energy resolution of about 10 % was improved to 0.79 %, which corresponds to 47 eV for 6 keV X-rays.[24]

4.3.6 APPLICATIONS

Niedermayr et al. studied the interaction of Ar^{9+}, O^{7+}, N^{6+}, and C^{5+} with SiH, Au, and SiO_2 targets using a 141 μm × 141 μm Nb (265 nm)/Al (50 nm)/AlO_x/Al (50 nm)/Nb (165 nm) STJ.[25] X-ray spectra resulting from the interaction of O^{7+} with a SiH surface at 10 keV/q were reported. The STJ was cooled by a two-stage ADR, which has a base temperature of 60 mK with a hold time above 20 h (<0.4 K). The STJ required no temperature regulation as long as it was kept below about 400 mK. Six satellite lines of oxygen (6 electrons in the L shell for the neutral configuration) are separated by 10 eV and range from 524.9 eV to 573.9 eV. They correspond to the 2p–1s transition of $O^{6+}, O^{5+}, \ldots, O^{1+}$. The resolution of the detector was not good enough to completely separate these lines, but it was possible to fit the data to six Gaussian lines.

Friedrich et al. reported on synchrotron-based X-ray fluorescence and absorption spectroscopes with a 3 × 3 array of 200 μm × 200 μm STJs, of which the layer structure is the same as that of reference 25.[26] The energy resolution of the STJs is about 15 eV below 1 keV and the total count rate capability of the array is about 100 kcps. As a model compound for Mn-containing proteins, they used an MgO crystal doped with 840 ppm of Mn. The spectral characteristics of the 9 STJs were quite uniform, although each pixel required a

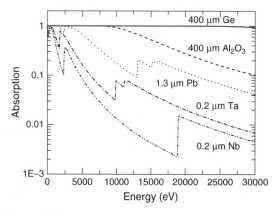

Figure 4.3.9 Calculated absorption efficiencies for some materials

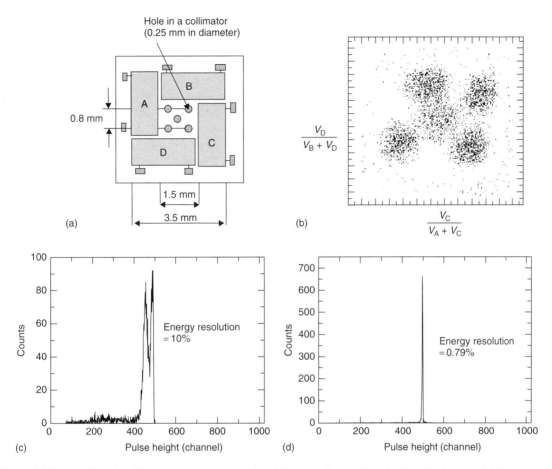

Figure 4.3.10 Two-dimensional imaging and correction of position dependency of a series-junction detector. (a) Structure and setup; (b) two-dimensional image; (c) original spectrum; and (d) spectrum after correction

slightly (±5%) different calibration to align the peaks. The weak Mn L fluorescence (~640 eV) was clearly measured even for an acquisition time of 10 s. The Mn L fluorescence was well separated from the strong O K fluorescence at 525 eV.

Frank reviewed time-of-flight mass spectrometry (TOF-MS) with low-temperature detectors.[27] Conventional mass spectrometers for biomolecules use microchannel plates (MCPs) to measure the arrival times of molecular ions. The sensitivities of MCPs decrease for large ion masses above a few tens of kDa because MCPs rely on secondary electron emission. On the contrary, low temperature detectors can measure low-energy solid-state excitation, such as phonons, and are more sensitive to weakly ionizing, slow-moving particles.

In addition to the high sensitivity for large ion masses, the high energy resolution of low temperature detectors is useful for charge discrimination and for studies of ion fragmentation and internal energies.

Sato et al. studied an STJ detector for fast timing measurements.[28] The detector was a 20 μm × 20 μm Nb (200 nm)/Al (10 nm)/AlO$_x$/Nb (150 nm) STJ in liquid helium. Instantaneous switching to the voltage state of an STJ following a decrease in the superconducting critical current (dc Josephson current) I_c induced by a heavy ion beam was observed. The output voltage was reset by sweeping the bias current $I_b (0.95 I_c < I_b < I_c)$ from the fixed value I_b to 0 V and again to I_b. The kinetic energy of the ^{40}Ar beam at the entry to the

STJ detector was about 44 MeV/nucleon. The time width of the obtained time spectrum was about 1.7 ns, which merely corresponded to the time resolution of the data acquisition system.

4.3.7 COOLING SYSTEMS

Usually, three kinds of cooling systems are adopted to cool low temperature detectors, i.e. Helium 3 (^{3}He) cryostats, ^{3}He-^{4}He dilution refrigerators, and ADRs. Helium 3 cryostats are relatively small and low cost, for example a cryostat with a diameter of 25 cm and height of 53 cm can maintain about 0.35 K for 92 h by one shot cooling.[29] On the other hand, the attainable temperature is usually higher than 0.3 K. Temperatures lower than 0.3 K usually require ^{3}He-^{4}He dilution refrigerators or adiabatic demagnetization refrigerators. The attainable temperatures of the refrigerators are usually lower than 0.1 K. Dilution refrigerators are suited for long periods of continuous cooling, for example longer than a month is possible. Less mechanical operations are required for ADRs. Recently, the development of pulse tube cryocoolers that are of low vibration and can attain about 4 K has been making it possible to operate the low temperature detectors without a supply of liquid helium and liquid nitrogen.

Shirron et al. are developing a three-stage continuous cooling ADR. The temperature stability is 8 µK rms or better over an entire cycle, and the cooling power is 2.5 µW at 60 mK using a superfluid helium bath (1.2 K) as the heat sink.[30]

Höhne et al. developed a compact ADR using a pulse tube cooler for scanning electron microscopes.[31]

Luukanen et al. developed superconductor–insulator–normal metal–insulator–superconductor (SINIS) tunnel junction refrigerators for low temperature detectors. They are Peltier type on-chip refrigerators.[32,33] Electronic cooling from 260 mK to 80 mK with a cooling power of 20 pW at 80 mK was demonstrated.

4.3.8 CONCLUSION

Superconducting tunnel junctions are showing their high possibilities as detectors for X-ray spectrometry: high energy resolution even for optical photons, high count rate capability, and one- and two-dimensional imaging. They will open new fields of application for X-ray spectrometry.

REFERENCES

1. H. Ott and A. Zehnder (Eds), *Nucl. Instrum. Methods Phys. Res. A* **370**, 1 (1996).
2. S. Cooper (Ed.), *Proceedings of the Seventh International Workshop on Low Temperature Detectors*, Max-Planck-Institute of Physics (1997); urg@mppmu.mpg.de
3. P. de Korte (Ed.), *Nucl. Instrum. Methods Phys. Res. A* **444**, 1(2000).
4. F. S. Porter et al. (Eds), *AIP Conf. Proc.* **605**, 1 (2002).
5. M. Kurakado, *X-ray Spectrom.* **28**, 388 (1999).
6. M. Kurakado, *X-ray Spectrom.* **29**, 137 (2000).
7. M. Kurakado and H. Mazaki, *Nucl. Instrum. Methods* **185**, 141 (1981).
8. M. Kurakado, *Nucl. Instrum. Methods* **196**, 275 (1982).
9. K. E. Gray, *Appl. Phys. Lett.* **32**, 392 (1978).
10. D. J. Goldie, P. L. Brink, C. Patel, N. E. Booth and G. L. Salmon, *Appl. Phys. Lett.* **64**, 3169 (1994).
11. C. A. Mears, S. E. Labov, M. Frank, M. A. Lindeman, L. J. Hiller, H. Netel and A. T. Barfknecht, *Nucl. Instrum. Methods A* **370**, 53 (1996).
12. G. Angloher, B. Beckhoff, M. Bühler, F. v. Feilitzsch, T. Hertrich, P. Hettl, J. Höhne, M. Huber, J. Jochum, R. L. Mößbauer, J. Schnagl, F. Scholze and G. Ulm, *Nucl. Instrum. Methods A* **444**, 214 (2000).
13. G. Angloher, M. Huber, J. Jochum., F. von Feilitzsch, R. L. Mößbauer, and G. Sáfrán, *J. Low Temp. Phys.* **123**, 165 (2001).
14. M. Frank, L. J. Hiller, J. B. le Grand, C. A. Mears, S. E. Labov, M. A. Lindeman, H. Netel, D. Chow and A. T. Barfknecht, *Rev. Sci. Instrum.* **69**, 25 (1998).
15. M. Katagiri, T. Nakamura, M. Ohkubo, H. Pressler, H. Takahashi and M. Nakazawa, *AIP Conf. Proc.* **605**, 177 (2002).
16. P. Verhoeve, N. Rando, A. Peacock, A. Dordrecht, A. Poelart and D. J. Goldie, *IEEE Trans. Appl. Supercond.* **7**, 3359 (1997).
17. G. Angloher, M. Huber, J. Jochum, A. Rüdig, F. v. Feilitzsch and R. L. Mößbauer, *AIP Conf. Proc.* **605**, 23 (2002).
18. M. Huber, G. Angloher, F. v. Feilitzsch, T. Jagemann, J. Jochum, T. Lachenmaier, J.-C. Lanfranchi, W. Potzel, A. Rüdig, J. Schnagl, M. Stark and H. Wulandari, *AIP Conf. Proc.* **605**, 63 (2002).
19. H. Kraus, F. v. Feilitzsch, J. Jochum, R. L. Mössbauer, T. Peterreins and F. Pröbst, *Phys. Lett. B* **231**, 195 (1989).
20. L. Li, L. Frunzio, C. M. Wilson, K. Segall, D. E. Prober, A. E. Szymkowiak and H. Moseley, *AIP Conf. Proc.* **605**, 145 (2002).

21. C. M. Wilson, K. Segall, L. Frunzio, L. Li, D. E. Prober, D. Schiminovich, B. Mazin, C. Martin, and R. Vasquez, *Nucl. Instrum. Methods A* **444**, 449 (2000); *IEEE Trans. Appl. Supercond.* **11**, 645 (2001).
22. P. Verhoeve, *Nucl. Instrum. Methods A* **444**, 435 (2000).
23. S. Kamihirata, M. Kurakado, A. Kagamihata, K. Hirota, H. Hashimoto, R Katano, K. Taniguchi, H. Sato, Y. Takizawa, C. Otani, and H. M. Shimizu, *AIP Conf. Proc.* **605**, 149 (2002).
24. M. Kurakado, S. Kamihirata, A. Kagamihata, K. Hirota, H. Hashimoto, H. Sato, H. Hotchi, H. M. Shimizu and K. Taniguchi, *Nucl. Instrum. Methods A* **506**, 134 (2003).
25. T. Niedermayr, S. Friedrich, M. F. Cunningham, M. Frank, J. P. Briand and S. E. Labov, *AIP Conf. Proc.* **605**, 363 (2002).
26. S. Friedrich, T. Niedermayr, T. Funk, O. Drury, M. L. van den Berg, M. F. Cunningham, J. N. Ullom, A. Loshak, S. P. Cramer, M. Frank and S. E. Labov, *AIP Conf. Proc.* **605**, 359 (2002).
27. M. Frank, *Nucl. Instrum. Methods A* **444**, 375 (2000).
28. H. Sato, T. Ikeda, K. Kawai, H. Miyasaka, T. Oku, W. Ootani, C. Otani, H. M. Shimizu, Y. Takizawa, H. Watanabe, K. Morimoto and F. Tokanai, *Nucl. Instrum. Methods A* **459**, 206 (2001).
29. M. Kurakado and Y. Ikematsu, *Cryogenics* **37**, 331 (1997).
30. P. J. Shirron, E. R. Canavan, M. J. DiPirro, M. Jackson, J. Panek and J. G. Tuttle, *AIP Conf. Proc.* **605**, 379 (2002).
31. J. Höhne, U. Hess, M. Bühler, F. v. Feilitzsch, J. Jochum, Rv. Hentig, T. Hertrich, C. Hollerith, M. Huber, J. Nicolosi, K. Phelan, D. Redfern, B. Simmnacher, R. Weiland and D. Wernicke, *AIP Conf. Proc.* **605**, 353 (2002).
32. A. Luukanen, A. M. Savin, T. I. Suppula, J. P. Pekola, M. Prunnila and J. Ahopelto, *AIP Conf. Proc.* **605**, 375 (2002).
33. A. Luukanen, M. M. Leivo, J. K. Suoknuuti, A. J. Manninen and J. P. Pekola, *J. Low Temp. Phys.* **120**, 281 (2000).

4.4 Cryogenic Microcalorimeters

M. GALEAZZI
University of Miami, Coral Gables, FL, USA

and

E. FIGUEROA-FELICIANO
NASA/Goddard Space Flight Center, Greenbelt, MD, USA

4.4.1 INTRODUCTION

Detectors traditionally used in X-ray spectroscopy can be divided into four groups: solid-state detectors, proportional counters and position-sensitive detectors coupled to either Bragg crystals or grazing-incidence diffraction gratings. Solid-state detectors are the most common because they are easy to use and inexpensive to operate. However their energy resolution is limited by the statistical fluctuations in the number of electron–hole (e-h) pairs that are created. The energy resolution of such a detector is given by $\Delta E_{rms} = \sqrt{FEw}$ where w is the energy necessary to create one e-h pair, F is the Fano factor and E is the energy. In practice the energy resolution of a solid-state detector is not better than 100–120 eV FWHM at 6 keV. In a proportional counter the energy necessary to create an electron–ion pair is of the order of 30 eV, and the energy resolution is even worse.

Grazing incident gratings can provide very high resolution for soft X-rays, on the order of 1 eV FWHM, but their low dispersion means throughputs are very small at high resolution. The grating efficiency also falls off rapidly at higher energies. The case of Bragg crystals is different. The energy resolution is again very good and the dispersion and efficiency are higher. However, only one resolution element is reflected, so each energy band must be measured independently, one after the other, and the net throughput for measuring a broad band spectrum is very low. Moreover the energy interval covered by a single crystal is limited, in practice, to a factor of two or so.

Almost 20 years ago the idea of detecting the increase in temperature produced by incident photons instead of the ionization of charged pairs was proposed (Fiorini and Niinikoski, 1984; Moseley *et al.*, 1984). If all the energy that is deposited is thermalized (converted into thermal phonons), there is no branching of the incident energy and no inherent statistical limitation to the resolution. Moreover the detection of phonons means that the choice is no longer restricted to materials with good electron transport properties, such as germanium or silicon, but different materials, more desirable for the particular experiment can be used. Most of the developments on cryogenic microcalorimeters have been discussed at the biannual International Workshops on Low Temperature Detectors. We refer the reader to the workshop proceedings as a useful source of information to complement this subchapter (De Korte and Peacock, 2000; Porter *et al.*, 2002).

4.4.2 MICROCALORIMETERS

A schematic view of a cryogenic microcalorimeter is shown in Figure 4.4.1. It is composed of three parts, an absorber that converts the energy of the incident X-rays into heat, a sensor that detects the temperature variations of the absorber and a weak thermal link between the detector and a heat sink. The operating principle is simple. When an X-ray hits the absorber its energy is thermalized, that is, is distributed among thermal phonons and electrons, and the temperature of the detector first rises and then returns to its original value due to the weak thermal link to the heat sink. The temperature change is proportional to the energy of the incident X-ray and is detected by the sensor. The sensor, or thermometer, is typically a resistor whose resistance has a strong dependence on the temperature at the working point, but other, non-resistive, thermometers have also been investigated.

Although the operating principle is simple, the construction of a cryogenic microcalorimeter may prove challenging. The temperature rise of the detector is inversely proportional to its heat capacity, which must therefore be as small as possible in order to have a detectable temperature variation.

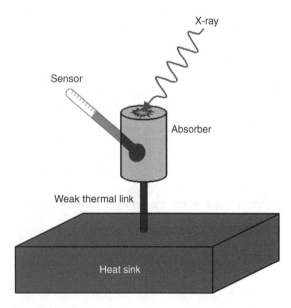

Figure 4.4.1 Schematic view of a cryogenic microcalorimeter

This is achieved in two different ways, reducing the size of the detector and/or reducing its working temperature. Typical working temperatures in the range of tens of mK and volumes of the order of 10^{-3} mm^3 are used for best energy resolution. The temperature variations to be detected may be of the order of μK, so good sensors are also necessary to detect them. Although these are stringent requirements, several groups working in the field have obtained impressive results.

Requirements for a good microcalorimeter are an absorber with small heat capacity able to convert the energy of the incident radiation into phonons and/or electrons in thermal equilibrium quickly and with high efficiency, and a sensor with low heat capacity and high sensitivity to temperature variations. We discuss these characteristics, together with the performance of cryogenic microcalorimeters in the following.

4.4.2.1 THE ABSORBER

The choice of the right absorber is one of the more critical parts of the design of a microcalorimeter. The main characteristics that must be taken into account are the quantum efficiency in the energy range of interest, the collection area of the detector, and the efficiency and speed of the absorber in thermalizing the incident energy. This must be combined with a sufficiently small heat capacity for good energy resolution (as we will see in Section 4.4.3.1, the energy resolution is proportional to the square root of the heat capacity). Different materials have been tested and used with relatively good results. These include metals, semiconductors, superconductors and insulators.

Metals usually have very good thermalization properties: the conversion of the incident energy into thermal phonons and electrons is fast (a few μs or less) and very efficient. The stopping power is also generally very good, if a high Z material is chosen. Despite these advantages they are often undesirable as absorbers because of a large heat capacity at low temperatures due to the electronic component. Nevertheless, some metals with a small electronic density of states and consequently small electronic heat capacity can be used with very good results. Among these bismuth is certainly the one that is receiving the highest attention and is giving the best results (Lindeman et al., 2002a; Wollman et al., 2000). For small absorbers, traditional metals like copper have also been used with very good results (Bergmann Tiest et al., 2002a).

Semiconductors and insulators have no electronic contribution to the heat capacity, which can then be very small. However, part of the incident energy is converted first into e-h pairs that can be very slow to recombine. It is therefore difficult to have good energy resolution since the thermalization efficiency is limited and is affected by the statistical and positional variation in the e-h pair creation and trapping. Some small or zero gap semiconductors such as HgTe can have very good performance due to the negligible energy tied up in charge carrier production (McCammon et al., 2002a; Stahle et al., 2002a).

A promising alternative to metals and semiconductors are superconductors. Their heat capacity can be small due to the absence of an electronic contribution at temperatures below ~0.1 T_C, while the stopping power is very good for high Z materials like lead or rhenium. In superconductors the incident energy is first converted into quasiparticles that then recombine releasing phonons. The details of this process are complicated and depend strongly on the characteristics of the material used (Cosulich et al., 1993). Superconductors as absorbers may sometimes be characterized by incomplete thermalization and very long tails in the pulses that affect the detector speed and energy resolution. The best results at low energy (below 10 keV) have been obtained using tin (Alessandrello et al., 1999; Silver et al., 2002), but promising results have also been obtained using rhenium (Galeazzi, 1998a). Recently, good high-energy (~50 keV) results have also been obtained using lead (Bleile et al., 2002).

In conclusion cryogenic microcalorimeters can obtain good results using a large variety of materials. The choice of the absorber can therefore be optimized depending on the requirements of the experiment.

4.4.2.2 THE SENSOR

The main characteristic of a sensor is its response to temperature variations. This characteristic is described by the sensitivity α, defined as:

$$\alpha = \frac{d \log R}{d \log T} = \frac{T}{R}\frac{dR}{dT} \qquad (4.4.1)$$

where T and R are the temperature and resistance of the sensor, respectively. The sensitivity α is dimensionless and describes the fractional resistance variation versus the temperature variation. Higher α means higher sensitivity to temperature variations, corresponding to larger output signals.

The most commonly used sensors are resistors in which the resistance has a strong temperature dependence at the working point. Figure 4.4.2 shows the schematic of a typical circuit used to bias a detector. The detector can be current or voltage biased, depending on whether the load resistor R_L is bigger or smaller than the sensor resistance R, and the temperature variation is thus read out as a voltage or current variation. We can distinguish two main categories of such sensors in wide use: semiconductor thermistors and transition edge sensors (TES), the characteristics of which are reported in Table 4.4.1.

Two different kinds of semiconductor thermistors are used: neutron transmutation doped (NTD) germanium thermistors and ion implanted silicon thermistors. Their resistance versus temperature behavior is approximately $R = R_0 \exp(\sqrt{T_0/T})$,

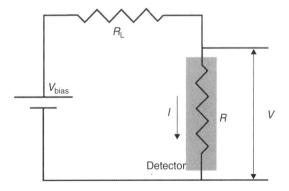

Figure 4.4.2 Schematic of the bias circuit for a microcalorimeter. If $R \ll R_L$, the detector is current biased, if $R \gg R_L$ the detector is voltage biased

Table 4.4.1 Principal characteristics of commonly used resistive sensors

Sensor	α	Temperature range	Resistance	Usual bias	Read-out electronics
Thermistors	Negative/small	Wide	Large	Constant current	Cold FET
TES	Positive/ large	Narrow	Small	Constant voltage	SQUID

where R_0 and T_0 are constant parameters characteristic of the sensor. Their sensitivity is negative and relatively small, usually around 5–6. They are often biased with an almost constant current and characterized by a very high resistance (tens of MΩ at the working point) where they are well matched to junction field effect transistors (JFET) operated near 100 K (Gatti and Parodi, 2000; McCammon et al., 2002a). To implement the readout electronics in the fabrication process and to build a large number of readout channels in a compact geometry, the use of single electron transistors (SET) for the readout of semiconductor thermistors (Schoelkopf et al., 1998) and superconducting devices with transistor-like properties for the readout of both thermistors and TES (Pepe et al., 2000) are also being investigated.

Transition edge sensors are superconducting thin films that are biased so that the temperature is actually in the phase transition between the superconducting and normal state, where α can be very high. The transition is generally very narrow, so the working temperature for a given TES is fixed. Depending on the application, it is possible to tune the transition temperature of a TES during its fabrication using the proximity effect in superconductor–normal metal bilayers (Martinis et al., 2000). For small reductions in the transition temperature, implantation of a thin superconducting film with ferromagnetic materials has also been successfully used (Young et al., 1999). The most commonly used TES are Al-Ag, Mo-Au, Mo-Cu, Ti/Au, and Ir-Au bilayers or W thin films (Hoehne et al., 1996; Cabrera et al., 2000; Wollman et al., 2000; Hilton et al., 2001; Bergmann Tiest et al., 2002a; Lindeman et al., 2002a; Tan et al., 2002). Their α is generally very high (a factor of 10 to 100 higher than in thermistors) and positive, while their resistance is very low. They are generally biased at constant voltage and their current is read out using superconducting quantum interference devices (SQUID) as current transducers (Gallop, 1991).

There are also other thermometer types that can be classified as nominally non-dissipative or non-resistive. These include devices whose capacitance, inductance, or magnetization change with temperature. They can be analyzed the same way as resistive thermometers by assuming an effective sensitivity $\alpha = (\text{d} \log X/\text{d} \log T)/Q$, where X is the thermometric parameter and Q^{-1} is the fractional dissipation. These thermometers tend to be intrinsically insensitive, but Q^{-1} can be 10^5 or more, making them potentially advantageous. They can also have a significant amplifier noise contribution, which markedly changes their optimization. Recently particularly interesting results have been obtained with magnetic thermometers, where the magnetic properties of the sensor are temperature dependent and can be read out with a dc SQUID used as magnetic flux transducer (Enss, 2002). We will discuss these detectors in more detail in Section 4.4.4.5.

4.4.3 PERFORMANCE

4.4.3.1 ENERGY RESOLUTION

In the standard theory of microcalorimeters (Mather, 1982; Mather, 1984; Moseley et al., 1984), four different noise contributions can affect the energy resolution: the Johnson noise of the sensor and any excess noise it has, the phonon shot noise produced by the random flow of energy carriers through the weak thermal link, and the electrical noise of the read out electronics (Mather, 1982). Current technology allows the construction of read out electronics whose noise is in general negligible with respect to the other contributions. As we already pointed out, for semiconductor thermistors, this is done using low temperature (around 100 K)

JFET for the first stage of amplification plus room temperature electronics for the second stage. For TES, low temperature dc SQUID are used. In the case of JFET, the total voltage noise is typically a few $nV/\sqrt{Hz}$ (Gatti and Parodi, 2000; McCammon et al., 2002a). With SQUID electronics it is possible to obtain current noise levels of a few $pA/\sqrt{Hz}$ using commercially available dc SQUID (Gallop, 1991).

The Johnson noise is a voltage noise across the detector with voltage spectral density $\sqrt{4Rk_BT}$ where k_B is the Boltzmann constant. The phonon noise essentially consists of statistical fluctuations in the temperature of the detector due to the link between the detector and the heat sink. These temperature fluctuations are then converted into a voltage or current noise by the sensor. Skipping here the mathematical passages that can be found in Mather (Mather, 1982), the intrinsic energy resolution of an ideal cryogenic microcalorimeter operating in the linear, small signal regime, can be written as:

$$\Delta E_{rms} = \xi \sqrt{\frac{k_B T^2 C}{\alpha}} \quad (4.4.2)$$

where T and C are, respectively, the temperature and the heat capacity of the detector and ξ is dimensionless and depends on the detector characteristics and bias power and has an optimized value of 1–2.

We would like to discuss some observations on the result in Equation (4.4.2). First, as intuitively pointed out before, the energy resolution depends on the heat capacity of the detector. The energy resolution also depends directly on the temperature of the detector, making the requirement of working at very low temperature even more important. Moreover, the energy resolution depends on the detector sensitivity α, which stresses the importance of working with good sensors. We also want to point out the fact that the energy resolution does not depend on the energy of the incident radiation. This is because the energy resolution is noise-limited and not statistics-limited, as is the case for most radiation detectors. We also note that the resolution does not depend on the thermal conductivity G between the detector and the heat sink. This affects the detector speed, as discussed in the next section.

The competition for the best energy resolution has been very strong in recent years. Interestingly enough, different techniques are obtaining very similar results. At 6 keV, that is often considered the reference energy, recent as-yet unpublished results show an energy resolution of 4.3 eV FWHM using silicon implanted thermistors and HgTe absorbers (Stahle, 2002b); 4.7 eV FWHM has also been obtained with a larger detector with similar characteristics (Stahle et al., 2002a). With a Ti/Au TES and Cu absorber 3.9 eV FWHM for a 5 min run and 4.5 eV FWHM for longer runs has been obtained (Bergmann Tiest et al., 2002a); 4.5 eV FWHM has also been obtained with a Mo/Cu TES that also acts as the absorber (Irwin et al., 2000), and a resolution below 5 eV FWHM has been obtained with NTD germanium thermistors and tin absorbers (Alessandrello et al., 1999; Silver et al., 2002). In Figure 4.4.3 we show the spectrum of the Mn Kα line from Bergmann Tiest et al. (2002a). At lower energies, 2 eV FWHM at 1.5 keV has been obtained with an Al/Ag TES (Wollman et al., 2000) and 2.4 eV FWHM at 1.5 keV and 3.7 eV FWHM at 3.3 keV has been obtained with a Mo/Au TES (Lindeman et al., 2002a).

4.4.3.2 COUNT RATE

To discuss the count rate capabilities of a microcalorimeter we must first introduce the effect of the detector nonlinearity. So far we implicitly assumed that whenever an incoming particle warms the detector the temperature change is so small that the characteristics of the detector do not change. That is not true in the case of relatively large temperature changes, since the thermometer sensitivity, the detector heat capacity, and the thermal conductivity to the heat sink may be temperature dependent, generating detector nonlinearity. This nonlinearity is usually small and can be easily taken into account in the analysis of the data, but it plays an important role in the detector count rate. If the count rate is high there will often be pile-up of pulses on the tails of other pulses. While this pile-up could be taken care of in a linear detector, in a microcalorimeter the starting temperature of the second pulse is higher than

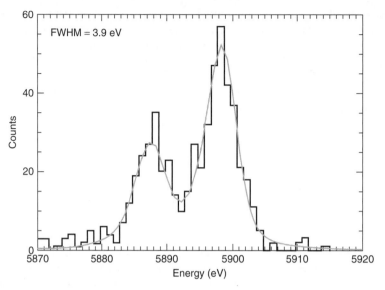

Figure 4.4.3 Spectrum of Mn Kα lines obtained with a Ti/Au TES and Cu absorber. The black line represents the experimental data, the gray line the best-fit to the data. (Courtesy of W. Bergmann Tiest, SRON National Institute for Space Research, The Netherlands; Bergmann Tiest *et al.*, 2002a). Reproduced by permission of the American Institute of Physics

the equilibrium temperature of the detector. Due to the different starting temperature, the pulse amplitude could reasonably be different (usually smaller) than the amplitude of a normal pulse generated by the same amount of energy, worsening the energy resolution of the detector. To get the best energy resolution, the count rate is therefore limited by the time necessary for the detector to return to the original temperature. The time it takes the temperature to return to $1/e$ of the maximum temperature variation during the pulse is called the time constant of the detector.

A microcalorimeter with heat capacity C and thermal conductivity to the heat sink G is characterized by an intrinsic time constant $\tau = C/G$. This is the characteristic time necessary for the microcalorimeter to return to the working temperature after an X-ray warms it. The temperature rise, determined by the time necessary for the incident energy to be converted into the thermal signal, is in general much faster and is a characteristic of the absorber. Since, as we have seen, the energy resolution of a microcalorimeter does not depend on the thermal conductivity G, it is in principle possible to build a detector with very high G and consequently very small time constant. In practice, because of technical limitations, and because the classical detector model that we have used so far does not completely describe a microcalorimeter (see Section 4.4.3.5) typical time constants of cryogenic microcalorimeters are in the range 0.1–10 ms or even longer in the case of very large detectors.

Instead of increasing the value of G, much faster detectors can be obtained using the effect known as electrothermal feedback. For resistive sensors, when there are no events in the detector, the temperature of the microcalorimeter is higher than the temperature of the heat sink due to the power dissipated by the bias signal. This power is simply equal to $I^2 R$ in the case of constant current bias and to V^2/R in the case of constant voltage bias. When an X-ray is absorbed into the detector the resistance of the sensor changes and therefore also the bias power changes. This effect can be quantified by introducing a parameter with the dimension of a thermal conductivity

$$G_{\text{ETF}} = \frac{P\alpha}{T} \frac{R - R_{\text{L}}}{R + R_{\text{L}}} \quad (4.4.3)$$

that depends on the power P dissipated into the detector at equilibrium, the detector sensitivity α, the temperature T, and on characteristics of the bias circuit (R and R_L). A detector is then characterized by an effective time constant $\tau_{eff} = C/G_{eff}$, where $G_{eff} = G + G_{ETF}$. The value of G_{ETF} can be either positive or negative, depending on the sign of the parameter α and on the ratio between R and R_L, and the effective time constant of the detector can be either longer or shorter than the intrinsic time constant τ. From Equation (4.4.3) we see that if the detector is current biased ($R_L > R$) G_{ETF} is positive when α is negative and negative when α is positive. The opposite is true if the microcalorimeter is voltage biased. In the case of voltage biased TES, with very large α, the parameter G_{ETF} is not only positive, thus reducing the time constant, but, choosing the appropriate working point, it can be much larger than G (Irwin, 1995). In this way it is possible to obtain detectors with time constant much faster than C/G. Using this strong electrothermal feedback effect cryogenic microcalorimeters with an effective time constant of 100 μs and count rates of about 500 Hz have already been built (Wollman et al., 2000).

For detectors with lower sensitivity, as semiconductor thermistors, the possibility of using an external electrical feedback system that actively reduces the bias power during an X-ray event has been proposed and tested (Galeazzi, 1998b; Meier et al., 2000). In practical terms this has the same effect as using electrothermal feedback. Reduction of more than a factor of 30 in the detector time constant (Galeazzi, 1998b) and time constants of the order of 100 μs (Silver et al., 2002) have been successfully obtained using this technique.

Microcalorimeter arrays are also being developed (Kelley et al., 1999; McCammon et al., 2002a). In particular, arrays of 1000 elements are under construction (Stahle et al., 2002c), that would allow the measurement of net count rates of the final detector of about 500 kHz.

4.4.3.3 QUANTUM EFFICIENCY

The detection efficiency of microcalorimeters is generally very good. The quantum efficiency can be more than 99 % below 10 keV when using high Z materials. This is due to the fact that the thickness of the absorber is not an intrinsic limit to the detector performance. Nevertheless, even if it is possible to make the detector thick enough to have high stopping power, to keep the heat capacity small, the area of the detector may have to be reduced, affecting the collecting area. Another important factor that bears on the effective quantum efficiency of cryogenic microcalorimeters is that they work at very low temperature. If the X-ray source is at a temperature that is higher than that of the detector, the detector must in general be shielded to reduce the shot noise due to the impinging infrared photons and the heat load due to the black body emission of the source. In this case, infrared filters are installed between the detector and the source to reflect the infrared radiation that would warm up the detector and the heat sink. These filters are generally thin organic films covered by aluminum. Well-designed filter systems have transmission efficiency above 95 % for energies above 1 keV, while the efficiency tends to drop below 200 eV (McCammon et al., 2002a).

4.4.3.4 COLLECTING AREA

To keep the heat capacity small, the detector collecting area is complementary to quantum efficiency. The collecting area of an experiment is therefore mainly affected by the requirements on energy resolution and quantum efficiency and is usually limited to a fraction of a square millimeter per detector. The fabrication of arrays is necessary to substantially increase the collection area while maintaining the best possible energy resolution. Arrays have the other advantage of being position sensitive. The main drawback in using arrays is the need for a readout channel for each pixel in the array. Currently arrays with up to 36 pixels are being used. The best results are obtained by the NASA/Goddard Space Flight Center – University of Wisconsin collaboration with silicon implanted thermistors (Kelley et al., 1999; McCammon et al., 2002a). The collaboration has built arrays in 6 × 6 and 2 × 18 geometries for the

X-ray quantum calorimeter (XQC) and X-ray spectrometer (XRS) experiments. The XQC arrays are optimized to work below 2 keV, while the XRS arrays are designed to have high quantum efficiency up to 10 keV and a slightly worse energy resolution. The XQC arrays have a collecting area of $0.36\,\text{cm}^2$, while the XRS arrays have a collecting area of $0.12\,\text{cm}^2$.

One-thousand-element arrays are currently under development and are expected to be realized in the next few years (Stahle et al., 2002c). These arrays are being built mainly to increase the imaging capabilities of the experiments, but also to increase the collecting area. Both TES and thermistors are being investigated for these larger arrays.

4.4.3.5 NON-IDEAL EFFECTS

After 15 years of experiments, microcalorimeters have reached limits close to those predicted by the ideal model (Mather, 1982; Moseley et al., 1984). However, the standard non-equilibrium theory of microcalorimeters fails to completely predict the performance of real devices due to additional non-ideal properties that play an important role at low temperatures. The resistance of the thermometer becomes dependent on readout power as well as temperature, and equilibration times between different parts of the detector can be significant. Thermodynamic fluctuations between internal parts are then an additional noise source. Excess noise of unknown origin is also limiting the energy resolution performance in some cases.

Absorber and Hot-electron Decoupling

So far we have described a microcalorimeter as a monolithic device. With current devices this is not always a sufficient approximation. The absorber is often glued to the sensor, or it is connected to it through a finite thermal conductivity. This decoupling has two main consequences, the first is to change the response of the thermometer to X-rays. The second is to introduce thermal fluctuations between the absorber and thermometer that lead to an additional noise source. Both effects tend to worsen the detector performance (Galeazzi and McCammon, 2002).

The hot electron effect is very similar. This effect is well known in metals at low temperatures and has recently been studied in semiconductors in the variable range-hopping regime (Liu et al., 2002). It is due to the interaction between electrons in the thermometer which is much stronger than the interaction between electrons and phonons. Consequently the thermometer can be described as two systems, electrons and phonons, thermally connected by a finite thermal conductivity. The resistance of the thermometer depends directly on the temperature of the electrons, and the bias power dissipated into the electron system flows to the phonon system and then to the heat sink. This increases the temperature of the electrons above the temperature of the phonons, reducing the thermometer sensitivity to X-rays and worsening the detector performance.

The resulting detector is a very complicated system. For example, the thermal conductivity between electrons and phonons depends on the volume of the thermometer. Bigger thermometers have a lower hot-electron effect, but this increases the heat capacity of the detector. Complex models have been developed to describe the performance of a realistic microcalorimeter (Figueroa-Feliciano, 2001; Galeazzi and McCammon, 2002). These models have also been used recently to optimize the design of new detectors and will be described in more detail in Section 4.4.4.4.

Thermometer Non-ohmic Behavior

In an ideal resistive thermometer, the resistance only depends on the thermometer temperature. We have already seen that this is not completely true due to the hot-electron effect. The resistance depends on the electron temperature, that can be different from the phonon temperature, which because of this decoupling can lead to a reduction in the detector performance. In addition there are other physical effects that introduce a dependence of the thermometer resistance on the bias current or voltage. This is particularly true, for example, in TES (Tan et al., 2002). It is a known property of TES that their transition temperature can be changed by applying an external magnetic field. Similarly, when a bias current is passed through the TES to readout the resistance, it generates a magnetic field around the TES. This field also changes the transition temperature of the TES. When an X-ray hits a microcalorimeter the change of sensor resistance changes the current through the sensors which, in turn, changes the transition temperature and thus the resistance value. This non-ohmic behavior of the sensor plays against the temperature dependence of the resistance, reducing the thermometer sensitivity. A similar effect, but on a much smaller scale has also been seen in semiconductor thermistors due to field effects in the doped region (Zhang et al., 1998). The theoretical models recently developed to predict the performance of microcalorimeters include the non-ohmic behavior of the sensor.

Excess Noise

We already pointed out that the energy resolution of microcalorimeters does not depend on any statistical effect on the number of pairs or particles generated by X-ray absorption, but it is limited by noise in the detector.

A major problem that historically affected silicon semiconductor thermistors is $1/f$ noise (McCammon et al., 2002b). An unbiased thermistor typically exhibits the expected noise, but increasing the bias current often results in an increase in the amount of noise. This excess noise can be modeled as $1/f$ fluctuations in the resistance, depend only on the doping density and the resistivity (or equivalently the electron temperature). Spectra taken with low base temperature and high bias show the same noise as spectra at higher base temperatures but lower biases, so that the electron temperature was the same. The presence of $1/f$ noise can introduce degradation of up to 50 % in the energy resolution of the microcalorimeter. Recently the NASA/Goddard Space Flight Center, in collaboration with the University of Wisconsin has developed a deep implant and diffusion technique for silicon (Galeazzi et al., 2002a) with which they can implant very uniform devices more than five times thicker than before. The devices produced with this technique do not show any sign of $1/f$ noise and have demonstrated improved and impressive energy resolution performance (Stahle et al., 2002a).

Excess noise in TES is also the subject of a massive investigation. Single pixel microcalorimeters with TES have achieved energy resolution of 4.5 eV at 6 keV (Irwin et al., 2000; Bergmann Tiest et al., 2002a). This result is impressive, but is more than a factor of two worse than the predicted performance. The reason for this is the presence of an extra noise source of unknown origin in the detectors. Very little is known about this extra noise and currently many different laboratories around the world are trying to eliminate or at least understand it.

The current picture is very confusing. When the TES is in the superconducting or normal state, the noise levels agree with those expected. When the TES is in the transition the measured noise is appreciably higher than expected. Measurements in a number of different laboratories all show the presence of excess noise, but the characteristics of the noise appear different (Bergmann Tiest et al., 2002a; Lindeman et al. 2002a; Tan et al., 2002). So far it has been difficult to build a common, consistent picture between the results from different groups.

4.4.4 CURRENT DEVELOPMENTS

4.4.4.1 DETECTOR MICROFABRICATION

The first X-ray microcalorimeters were handmade devices where an absorber was epoxied to a thermometer and the device was suspended by the readout leads. Apart from being time consuming, this approach also has reproducibility problems: it is hard to cut absorbers to exactly the same size each time, use the same amount of epoxy to attach the absorber, have exactly the same suspended lead length, etc. Reproducibility is very important when making microcalorimeter arrays. The more uniform the energy resolution, time constant, and quantum efficiency are, the easier is the calibration and data analysis. Ideally one would like to have an array of identical detectors.

Lithography provides a way to get very uniform devices across an entire array. The large amount of knowledge and instruments developed for the computer industry can be leveraged for these very specialized detectors. Film depositions using standard solid-state processing have thickness variations across the wafer of less than 5%, and feature size variations across a wafer of less than a micron. For microcalorimeter arrays, this technology produces very uniform arrays, and most research groups are using lithography in some or all of their process. The University of Wisconsin – NASA/Goddard Space Flight Center collaboration has made several ∼30 pixel arrays using photolithography for the thermometer and weak link fabrication, using wafer dicing saws for cutting the absorbers and manually attaching them to the thermometer with epoxy (Kelley et al., 1999; McCammon et al., 2002a). Impressive uniformity was achieved, with 5% spread in quantum efficiency and time constants, and 10% in energy resolution. In Figure 4.4.4 a photograph of a microcalorimeter that is part of the XRS detector built using photolithographic technique is shown. Full use of lithography promises even higher uniformity in detectors currently under development (Hilton et al., 2001; Beeman et al., 2002; Finkbeiner et al., 2002; Kudo et al., 2002; Stahle et al., 2002c; Ukibe et al., 2002).

4.4.4.2 LARGE ARRAYS AND DETECTOR MULTIPLEXING

For an imaging instrument, the size of the pixels used in the instrument's detector is optimized for the focal length and the point-spread function (PSF) of the optics. Given a specific pixel size, the number of pixels in the detector determines its field of view. For most applications, one desires the largest field of view possible.

Even if the application does not use imaging, a larger area is usually desirable to increase the number of detectable X-rays. As discussed in Section 4.4.3.1, the size of individual pixels is restricted by the heat capacity of the pixel and its effect on energy resolution.

In both cases, one needs to increase the number of pixels to increase the detector area. Since most microcalorimeters are now fabricated using solid-state lithography techniques, fabricating large arrays of microcalorimeters is in principle not much more difficult than fabricating a small array. However, reading out these large arrays is a problem. To reduce noise pickup, the readout devices (SQUID, JFET, or SET) need to be in close proximity to the detector. The wiring and layout becomes challenging when designing a system with hundreds of channels. These devices also dissipate power and the wiring adds to the thermal load on the cold stage of the refrigerator, impacting the performance of the refrigeration system.

To work around these issues, several groups are actively designing multiplexing schemes to read out several (∼10–30) microcalorimeters per electronic channel. For low-impedance devices, SQUID multiplexers are being developed, for high-impedance devices, both SET and JFET multiplexers are under study. Both time-division multiplexing and frequency-division multiplexing can be used, and practical implementation

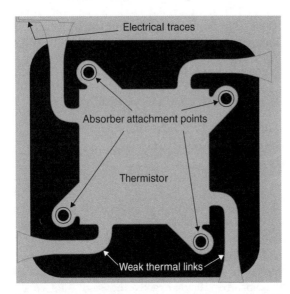

Figure 4.4.4 Photograph of a microcalorimeter built with photolithography technique, before absorber attachment. The thermistor is $300\,\mu m \times 300\,\mu m$ is size and $1.5\,\mu m$ thick and it is suspended using a Deep Reactive Ion Etching (RIE) technique. The absorber attachments are SU8 photoresist tubes (US Patent No. 4882245, 1989; Su8) also fabricated using a photolithography technique. (Courtesy of Caroline K. Stahle, NASA/Goddard Space Flight Center)

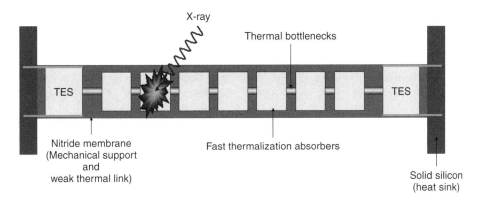

Figure 4.4.5 Schematic view of a position sensitive microcalorimeter. The heat released by the X-ray is split between the two TES. The difference in temperature change and time necessary to warm up between the two TES is used to determine the position and the energy of the incident X-ray

issues will likely decide which approach is best. Most groups working toward large (~1000 pixel) arrays plan to use some form of multiplexing for their readout.

Benford et al. (2002) have successfully fabricated and run a time-multiplexed SQUID system to read out an array of 16 TES bolometers for submillimeter observations. Irwin et al. (2002) are working on the second generation of this technology that will be compatible with X-ray TES arrays. Kiviranta et al. (2002) and Cunningham et al. (2002) are developing frequency-division SQUID multiplexers for the XEUS and Constellation-X X-ray astronomy missions, respectively. A different approach based on reading out a resistance bridge (with a TES as one of the four resistances) with a frequency-division multiplexer is under investigation by Miyazaki et al. (2002). For a more traditional approach, Kraus et al. (2002) have investigated the issues encountered in wiring 66 SQUID channels in a low-temperature system for the CRESST II dark matter search.

4.4.4.3 POSITION SENSITIVE IMAGING DETECTORS

Since the number of readout channels is the limiting factor for going to large numbers of pixels, position sensitive detectors – where one thermometer reads out several pixels or effective pixels – becomes attractive.

The operating principle of position sensitive microcalorimeters is similar to that of position sensitive proportional counters and semiconductor devices. An imaging calorimeter uses one or more thermometers to analyze the signal produced by a photon absorption event in an absorber. For the same energy photons, the signal received by the thermometers varies in some detectable way depending on the position in the absorber where the event occurred. In other words, the absorber exhibits position dependence. If one can use the information in the signal shape to determine the energy and location of photon absorption one has an imaging calorimeter. The imaging calorimeter can be a one-dimensional 'strip' absorber with one or more thermometers, or a two-dimensional 'plane' absorber with two or more thermometers. A schematic view of a one-dimensional position sensitive microcalorimeter with two sensors is shown in Figure 4.4.5.

This technique can be used in conjunction with readout multiplexers, to allow current designs that will have 1000 multiplexed readout channels to go to ~10 000 pixel arrays. Figueroa-Feliciano et al. (2002) are developing an imaging detector that has seven 0.250 mm pixels flanked by two transition-edge sensors in a one-dimensional column called a position-sensitive TES (PoST). They have obtained a preliminary result of 32 eV FWHM at 1.5 keV. Trowell et al. (2002) are working on a similar detector that uses a single long absorber, and position determination is carried out from an analysis of the pulse shape. Another approach by Ohno *et al.* (2002) uses a single SQUID channel with a segmented TES that has different time constants for absorption in the different TES segments.

4.4.4.4 MODELING OF NON-IDEAL BEHAVIORS AND DETECTOR OPTIMIZATION

We already discussed how microcalorimeter performance is limited by non-ideal effects that were not included in the standard theory of bolometers and microcalorimeters developed 20 years ago by Mather (Mather, 1982). These effects have been known for a few years now, but only very recently a complete theoretical description of microcalorimeter performance has been derived. The thermometer non-ohmic behavior has been added to the standard model to describe the performance of TES detectors by Lindeman (Lindeman, 2000). Two independent models that include all non-ideal effects of know origin have also been developed. These include the hot-electron effect, the absorber decoupling, the thermometer non-ohmic behavior, and all correlated extra noise sources. One of the two models uses block diagram algebra to solve the detector equations, obtaining an analytical solution (Galeazzi and McCammon, 2002), the other uses matrix notation to numerically solve the linearized differential equations of the microcalorimeter (Figueroa-Feliciano, 2001). The two models are in good agreement with each other, and predict the behavior of real silicon thermistor devices to high accuracy (Galeazzi *et al.*, 2002b).

Of particular interest from the experimental point of view is the fact that these models can be used to optimize the design on microcalorimeters for best performance. Up to now the detector design was carried out in a semi-empirical way, based on the standard theory and on experimental tests. The array for the XRS detector on the Astro-E2 satellite is the first detector whose design was completely optimized based on the required performance and on the characteristics of the material used. The characteristic heat capacity and thermal conductivity of all the detector components have been measured and the values have been used as input to the models to design the detector geometry for best performance. Since the detector was designed to be used in space, mechanical modeling has also been carried out in parallel to ensure the mechanical integrity of the microcalorimeter (Stahle *et al.*, 2002a). As we have shown before, the results are impressive: the energy resolution for a large detector is 4.7 eV FWHM and it is reproducible over large arrays. This is almost a factor of two better than a previous generation of the same detectors optimized in the semi-empirical way (Stahle, 2002b). The spectrum of the Mn Kα line acquired with such detectors is reported in Figure 4.4.6.

4.4.4.5 NON-RESISTIVE THERMOMETERS

As discussed in Section 4.4.2.2, the thermometer used in a microcalorimeter is not limited to a variable resistor. Any quantity that can be measured and that varies strongly with temperature can be used for thermometry. Several devices show great promise and some have achieved energy resolutions comparable with current resistive microcalorimeters.

A magnetic microcalorimeter (Enss, 2002) uses a paramagnetic material with an applied magnetic field as its thermometer (Figure 4.4.7). The magnetization of this material is a strong function of temperature, and can be measured with high resolution using a SQUID magnetometer. Due to the need for fast thermalization in the thermometer, the materials that show most promise are metals doped with rare earth ions to provide the paramagnetic moments. The paramagnetic material is thermally connected to an X-ray absorber, and weakly connected to a heat reservoir as in all other microcalorimeters. The device is usually placed directly on a SQUID to provide the best coupling between the SQUID and the paramagnetic material. The best energy resolution of 6 eV at 6 keV has been obtained by Fleischmann *et al.* (2003).

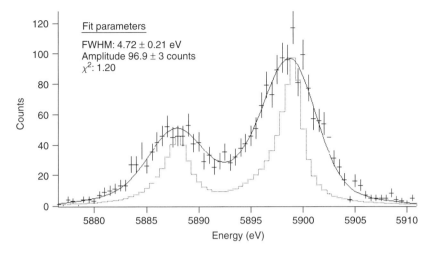

Figure 4.4.6 Spectrum of Mn Kα lines obtained with a Si thermistor and HgTe absorber optimized using the microcalorimeter non-ideal model. The datapoints represent the measured spectrum, the continuous line the best-fit to the data, and the dotted line represents the zero-resolution spectrum normalized in amplitude to the best-fit data (Hölzer et al., 1997)

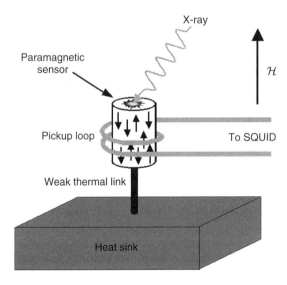

Figure 4.4.7 Schematic view of a magnetic microcalorimeter. The X-ray heats the detector, changing the magnetic properties of the paramagnetic sensor. The change is read out through the pickup loop

Other measurement techniques have also been proposed. Kinetic inductance and thermoelectric detectors are further examples of non-resistive thermometers that are under development and will be discussed in the next section.

4.4.4.6 OTHER IDEAS

Many other developments in microcalorimetry are underway. Here we outline some of the work that shows promise for the future.

As discussed in Section 4.4.3.5, current TES calorimeters show excess noise of unknown origin. To understand and/or remove this excess noise, different detector geometries are under study. One idea is that the boundary of the superconducting film, which due to lithographic processing has some degree of irregularity, could be the cause of the excess noise. In a circular or so called 'Corbino' geometry, one bias electrode is at the center of a circular film, and current flows radially outward to a ground electrode encompassing the circumference of the film. This geometry has no boundary except the electrodes, which are superconducting and are not part of the transition-edge sensor itself (Luukanen, 2002). Another idea is that when the TES is in its transition, it is actually divided into superconducting and normal state zones. In a square TES with constant current density along the direction of current flow, there is no preferred place for these zones to develop, and they may nucleate in a rather random fashion, possibly shifting in position on the

TES film, creating noise. A non-square geometry (a trapezoid, for example) would have a current density gradient, and the normal zone would tend to form in the highest current density region first and then propagate monotonically toward the low density region. Trapezoidal TES detectors with different aspect ratios (Lindeman et al., 2002b) and detectors with a 'zebra' pattern of normal and superconducting strips (Bergmann Tiest et al., 2002b) are being fabricated and tested.

In a resistive calorimeter that has weak thermal coupling between the electron and the phonon systems, a significant temperature differential can be established between the electron and phonon systems when the detector is biased. This is the so called 'hot-electron effect' that was discussed in Section 4.4.3.5. If the phonon system is thermally anchored to the heat sink (no weak link between the thermometer and the refrigerator), the electron–phonon decoupling acts as the weak link of the calorimeter and the electron temperature becomes the thermometer. This mode of operation has been successfully used for some time for dark matter and optical photon detectors using tungsten TES (Cabrera et al., 2000; Cabrera et al., 2002). Recently, Moseley and McCammon (2002) proposed similar operation of silicon thermistors for submillimeter devices. It may be possible to fabricate viable X-ray devices that operate in this hot-electron regime. The benefits of this technique are the much simpler fabrication (without need for the weak thermal link) and the robustness that comes with non-suspended devices.

To obtain faster devices and better energy resolution, constant temperature microcalorimeters are also being investigated (Moeckel et al., 2002). The idea is an extension of the external electronic feedback system that was described in Section 4.4.3.2. If the external feedback is sufficiently fast and with enough gain, it can compensate the power dissipated in the thermometer by X-rays keeping the thermometer temperature constant. In addition to being linear, this system has the potential of being faster than conventional detectors. This is especially true in devices that are limited by absorber decoupling. If properly optimized, they can also achieve better energy resolution.

The kinetic surface inductance of a superconductor is a function of the number of quasiparticles present in the superconductor. Since the number of quasiparticles depends on the temperature of the superconductor, one can use kinetic inductance as a thermometric parameter. These kinetic inductance detectors (KID) can be operated near or well below the superconducting transition temperature T_C, and can be read out by either a SQUID (Sergeev et al., 2002) or by measuring the resonant frequency of a circuit of which the KID is an inductive component. This last method may provide a simple way to multiplex these detectors (Mazin et al., 2002).

Thermometers using the thermoelectric effect are also being investigated (Gulian et al., 2002): a rapid change in temperature in a material causes a transient voltage across it that can be measured by a SQUID amplifier. They have proven the concept with gold/iron devices and are now investigating lower temperature systems such as lanthanum–cerium hexaborides as possible detector materials that could achieve resolutions comparable with thermistor devices.

REFERENCES

Alessandrello, A., Beeman, J. W., Brofferio, C., Cremonesi, O., Fiorini, E., Giuliani, A., Haller, E. E., Monfardini, A., Nucciotti, A., Pavan, M., Pessina, G., Previtali, E. and Zanotti, L., High energy resolution bolometers for nuclear physics and X-ray spectroscopy. Phys. Rev. Lett. **82**, 513 (1999).

Beeman, J., Silver, E., Bandler, S., Schnopper, H., Murray, S., Madden, N., Landis, D., Haller, E. E. and Barbera, M., The constellation-x focal plane microcalorimeter array: An ntd-germanium solution. AIP Conf. Proc. **605**, 211 (2002).

Benford, D. J., Ames, T. A., Chervenak, J. A., Grossman, E. N., Irwin, K. D., Khan, S. A., Maffei, B., Moseley, S. H., Pajot, F., Phillips, T. G., Renault, J.-C., Reintsema, C. D., Rioux, C., Shafer, R. A., Staguhn, J. G., Vastel, C. and Voellmer, G. M., First astronomical use of multiplexed transition edge bolometers. AIP Conf. Proc. **605**, 589 (2002).

Bergmann Tiest, W., Hoevers, H. F. C., Mels, W. A., Ridder, M. L., Bruijn, M. P., de Korte, P. A. J. and Huber, M. E., Performance of X-ray microcalorimeters with an energy resolution below 4.5 eV and 100 μs response time. AIP Conf. Proc. **605**, 199 (2002a).

Bergmann Tiest, W., Bruijn, M., Krouwer, E., Mels, W., Ridder, M. and Hoevers, H., Microcalorimeter performance:

absorber geometries, responsivity, noise and energy resolution. *First International Workshop on TES*, Boulder, Colorado, April 2002b. Available online at http://origins.colorado.edu/~deiker/tes/index.cgi.

Bleile, A., Egelhof, P., Kraft, S., McCammon, D., Meier, H. J., Shrivastava, A., Stahle, C. K. and Weber, M., Calorimetric low-temperature detectors for high resolution X-ray spectroscopy on stored highly stripped heavy ions. *AIP Conf. Proc.* **605**, 409 (2002).

Cabrera, B., Clarke, R., Miller, A., Nam, S. W., Romani, R., Saab, T. and Young, B., Cryogenic detectors based on superconducting transition-edge sensors for time-energy-resolved single-photon counters and for dark matter searches. *Physica B* **280**, 509 (2000).

Cabrera, B., Martinis, J. M., Miller, A. J., Nam, S. W. and Romani, R., TES spectrophotometers for near ir/optical/uv. *AIP Conf. Proc.* **605**, 565 (2002).

Cosulich, E., Gatti, F. and Vitale, S., Further results on mu-calorimeters with superconducting absorber. *J. Low Temp. Phys.* **93**, 263 (1993).

Cunningham, M. F., Ullom, J. N., Miyazaki, T., Drury, O., Loshak, A., van den Berg, M. L. and Labov, S. E., DC and AC biasing of a transition edge sensor microcalorimeter. *AIP Conf. Proc.* **605**, 317 (2002).

De Korte, P. and Peacock, T. (Eds), LTD8, *Nucl. Instrum. Methods Phys. Res. A* **444** (2000).

Enss, C., Magnetic sensors for X-ray and gamma-ray detectors. *AIP Conf. Proc.* **605**, 5 (2002).

Figueroa-Feliciano, E., Theory and Development of Position-Sensitive Quantum Calorimeters. *PhD thesis*, Stanford University, 2001.

Figueroa-Feliciano, E., Chervenak, J., Finkbeiner, F. M., Li, M., Lindeman, M. A., Stahle, C. K. and Stahle, C. M. First results from position-sensitive quantum calorimeters using mo/au transition-edge sensors. *AIP Conf. Proc.* **605**, 239 (2002).

Finkbeiner, F. M., Brekosky, R. P., Chervenak, J. A., Figueroa-Feliciano, E., Li, M. J., Lindeman, M. A., Stahle, C. K., Stahle, C. M. and Tralshawala, N., Fabrication of close-packed TES microcalorimeter arrays using superconducting molybdenum/gold transition-edge sensors. *AIP Conf. Proc.* **605**, 215 (2002).

Fiorini, E. and Niinikoski, T. O., Low-temperature calorimetry for rare decays. *Nucl. Instrum. Methods Phys. Res.* **224**, 83 (1984).

Fleischmann, A., Daniyarov, T., Rotzinger, H., Enss, C. and Seidel, G., Magnetic calorimeters for high resolution X-ray spectroscopy. *Physica B* **329**, 1594 (2003).

Galeazzi, M., Status of the rhenium beta experiment for neutrino mass study. *Nucl. Phys. B (proc. suppl.)* **66**, 203 (1998a).

Galeazzi, M., An external electronic feedback system applied to a cryogenic micro-calorimeter. *Rev. Sci. Instrum.* **69**, 2017 (1998b).

Galeazzi, M., Boyce, K. R., Brekosky, R., Gygax, J. D., Kelley, R. L., Liu, D., McCammon, D., Mott, D. B., Porter, F. S., Sanders, W. T., Stahle, C. K., Stahle, C. M., Szymkowiak, A. E. and Tan, P., Non-ideal effects in doped semiconductor thermistors. *AIP Conf. Proc.* **605**, 83 (2002a).

Galeazzi, M., Figueroa-Feliciano, E., Liu, D., McCammon, D., Sanders, W. T., Stahle, C. K. and Tan, P., Performance modeling of microcalorimeter detectors. *AIP Conf. Proc.* **605**, 95 (2002b).

Galeazzi, M. and McCammon, D., A microcalorimeter and bolometer model. *J. Appl. Phys.* **93**, 4856 (2003).

Gallop, J. C., *SQUIDS, the Josephson Effects and Superconducting Electronics*. Adam Hilger, Philadelphia, 1991.

Gatti, F. and Parodi, L., The low-noise read-out electronics of the 187Re experiment. *Nucl. Instrum. Methods Phys. Res. A* **444**, 129 (2000).

Gulian, A. M., Wood, K. S., Fritz, G. G., Van Vechten, D., Wu, H.-D., Horwitz, J. S., Badalyantz, G. R., Harutyunyan, S. R., Vartanyan, V. H., Petrosyan, S. A. and Kuzanyan, A. S., Sensor development for single photon thermoelectric detectors. *AIP Conf. Proc.* **605**, 31 (2002).

Hilton, G. C., Martinis, J. M., Irwin K. D., Bergren N. F., Wollman D. A., Huber M. E., Deiker S. and Nam S. W., Microfabricated transition-edge X-ray detectors. *IEEE Trans. Appl. Superconduct.* **11**, 739 (2001).

Hoehne, J., Forster, G., Absmaier, C., Colling, P., Cooper, S., Feilitzsch, F. V., Ferger, P., Igalson, J., Kellner, E., Koch, M., Loidl, M., Nagel, U., Pröbst, F., Rulofs, A. and Seidel, W., Progress on fabrication of iridium–gold proximity-effect thermometers. *Nucl. Instrum. Methods Phys. Res. A* **370**, 160 (1996).

Hölzer, G., Fritsch, M., Deutsch, M., Härtwig, J. and Förster, E., $K\alpha_{1,2}$ and $K\beta_{1,3}$ X-ray emission lines of the 3d transition metals. *Phys. Rev. A* **56**, 4554 (1997).

Irwin, K. D., An application of electrothermal feedback for high resolution cryogenic particle detection. *Appl. Phys. Lett.* **66**, 1998 (1995).

Irwin, K. D., Hilton, G. C., M. Martinis, J., Deiker, S., Bergren, N., Nam, S. W., Rudman, D. A. and Wollman, D. A., A Mo-Cu superconducting transition-edge microcalorimeter with 4.5 eV energy resolution at 6 keV. *Nucl. Instrum. Methods Phys. Res. A* **444**, 184 (2000).

Irwin, K. D., Vale, L. R., Bergren, N. E., Deiker, S., Grossman, E. N., Hilton, G. C., Nam, S. W., Reintsema, C. D., Rudman, D. A. and Huber, M. E., Time-division squid multiplexers. *AIP Conf. Proc.* **605**, 301 (2002).

Kelley, R. L., Audley, M. D., Boyce, K. R., Breon, S. R., Fujimoto, R., Gendreau, K. C., Holt, S. S., Ishisaki, Y., McCammon, D., Mihara, T., Mitsuda, K., Moseley, S. H., Mott, D. B., Porter, F. S., Stahle, C. K. and Szymkowiak, A. E., Astro-E high resolution X-ray spectrometer. *Proc. SPIE* **3765**, 114 (1999).

Kiviranta, M., Seppä, H., van der Kuur, J. and de Korte, P., Squid-based readout schemes for microcalorimeter arrays. *AIP Conf. Proc.* **605**, 295 (2002).

Kraus, H., Bazin, N., Cooper, S. and Henry, S., The 66-channel SQUID readout system for CRESST II. *AIP Conf. Proc.* **605**, 333 (2002).

Kudo, H., Sato, H., Nakamura, T., Arakawa, T., Goto, E., Shoji, S., Homma, T., Osaka, T., Mitsuda, K., Fujimoto, R., Iyomoto, N., Audley, M. D., Miyazaki, T., Oshima, T., Yamazaki, M., Kushino, A., Kuroda, Y., Onishi, M. and Goto, M., Fabrication of an X-ray microcalorimeter with an electrodeposited X-ray microabsorber. *AIP Conf. Proc.* **605**, 235 (2002).

Lindeman, M. A., Microcalorimetry and the Transition Edge Sensor. *PhD thesis*, University of California at Davis, 2000.

Lindeman, M. A., Brekosky, R. P., Figueroa-Feliciano, E., Finkbeiner, F. M., Li, M., Stahle, C. K., Stahle, C. M. and Tralshawala, N., Performance of Mo/Au TES microcalorimeters. *AIP Conf. Proc.* **605**, 203 (2002a).

Lindeman, M. A., Chervenak, J. A., Figueroa-Feliciano, E., Finkbeiner, F. M., Galeazzi, M., Li, M. J. and Stahle, C. K., TES physics: probing the phase transition. *First International Workshop on TES*, Boulder, Colorado, April 2002b. Available online at http://origins.colorado.edu/~deiker/tes/index.cgi.

Liu, D., Galeazzi, M., McCammon, D., Sanders, W. T., Smith, B., Tan, P., Boyce, K. R., Brekosky, R., Gygax, J. D., Kelley, R., Mott, D. B., Porter, F. S., Stahle, C. K., Stahle, C. M. and Szymkowiak, A. E., Hot-electron model in doped silicon thermistors. *AIP Conf. Proc.* **605**, 87 (2002).

Luukanen, A., Kinnunen, K. M., Nuottajärvi, A. K., Pekola, J. P., Bergmann Tiest, W. M. and Hoevers, H. F. C., First results from the CORTES – a TES utilizing a Corbin disk geometry. *First International Workshop on TES*, Boulder, Colorado, April 2002. Available online at http://origins.colorado.edu/~deiker/tes/index.cgi.

Martinis, J. M., Hilton, G. C., Irwin, K. D. and Wollman, D. A., Calculation of T_C in a normal-superconductor bilayer using the microscopic-based Usadel theory. *Nucl. Instrum. Methods Phys. Res. A* **444**, 23 (2000).

Mather, J. C., Bolometer noise: nonequilibrium theory. *Appl. Opti.* **21**, 1125 (1982).

Mather, J. C., Electrical self-calibration of nonideal bolometers. *Appl. Opt.* **23**, 3181 (1984).

Mazin, B. A., Day, P. K., Zmuidzinas, J. and Leduc, H. G., Multiplexable kinetic inductance detectors. *AIP Conf. Proc.* **605**, 309 (2002).

McCammon, D., Almy, R., Apodaca, E., Bergmann Tiest, W. M., Cui, W., Deiker, S., Galeazzi, M., Juda, M., Lesser, A., Sanders, W. T., Zhang, J., Figueroa-Feliciano, E., Kelley, R. L., Moseley, S. H., Mushotzky, R. F., Porter, F. S., Stahle, C. K. and Szymkowiak, A. E., A high spectral resolution observation of the soft X-ray diffuse background with thermal detectors. *Astrophys. J.* **576**, 188 (2002a).

McCammon, D., Galeazzi, M., Liu, D., Sanders, W. T., Smith, B., Tan, P., Boyce, K. R., Brekosky, R., Gygax, J. D., Kelley, R., Mott, D. B., Porter, F. S., Stahle, C. K., Stahle, C. M. and Szymkowiak, A. E., $1/f$ noise and hot electron effects in variable range hopping conduction. *Phys. Status Solidi B* **230**, 197 (2002b).

Meier, O., Bravin, M., Bruckmayer, M., Stefano, P. D., Frank, T., Loidl, M., Meunier, P., Pröbst, F., Safran, G., Seidel, W., Sergeyev, I., Sisti, M., Stodolsky, L., Uchaikin, S. and Zerle, L., Active thermal feedback for massive cryogenic detectors. *Nucl. Instrum. Methods Phys. Res. A* **444**, 350 (2000).

Miyazaki, T., Yamazaki, M., Futamoto, K., Mitsuda, K., Fujimoto, R., Iyomoto, N., Oshima, T., Audley, D., Ishisaki, Y., Kagei, T., Ohashi, T., Yamasaki, N., Shoji, S., Kudo, H. and Yokoyama, Y., Ac calorimeter bridge; a new multi-pixel readout method for TES calorimeter arrays. *AIP Conf. Proc.* **605**, 313 (2002).

Moeckel, N., Galeazzi, M., Lindeman, M. A. and Stahle, C. K., A constant temperature TES microcalorimeter with an external electronic feedback system. *AIP Conf. Proc.* **605**, 111 (2002).

Moseley, S. H., Mather, J. C. and McCammon, D., Thermal detectors as X-ray spectrometers. *J. Appl. Phys.* **56**, 1257 (1984).

Moseley, S. H. and McCammon, D., High performance silicon hot electron bolometers. *AIP Conf. Proc.* **605**, 103 (2002).

Ohno, M., Noguchi, Y., Fukuda, D., Takahashi, H., Nakazawa, M., Ataka, M., Ukibe, M., Hirayama, F., Ohkubo, M. and Shimizu, H. M., Development of the X-ray microcalorimeter with a superconductive iridium layer. *AIP Conf. Proc.* **605**, 259 (2002).

Pepe, G. P., Ammendola, G., Peluso, G., Barone, A., Parlato, L., Esposito, E., Monaco, R. and Booth, N. E., Superconducting device with transistor-like properties including large current amplification. *Appl. Phys. Lett.* **77**, 447 (2000).

Porter, F. S., McCammon, D., Galeazzi, M. and Stahle C. K. (Eds), Low temperature detectors. *AIP Conf. Proc.* **605** (2002).

Schoelkopf, R. J., Wahlgren P., Kozhevnikov A. A., Delsing P. and Prober D. E., The radio-frequency single-electron transistor (RF-SET): a fast and ultrasensitive electrometer. *Science* **280**, 1238 (1998).

Sergeev, A., Karasik, B., Gogidze, I. and Mitin, V., Ultrasensitive hot-electron kinetic-inductance detectors. *AIP Conf. Proc.* **605**, 27 (2002).

Silver, E., Bandler, S., Schnopper, H., Murray, S., Madden, N., Landis, D., Goulding, F., Beeman, J., Haller, E. E. and Barbera, M., X-ray and gamma-ray astronomy with NTD germanium-based microcalorimeters. *AIP Conf. Proc.* **605**, 555 (2002).

Stahle, C. K., Allen, C. A., Boyce, K. R., Brekosky, R. P., Brown, G. V., Cottam, J., Figueroa-Feliciano, E., Galeazzi, M., Gygax, J. D., Jacobson, M. B., Kelley, R. L., Liu, D., McCammon, D., Moseley, S. H., Porter, F. S., Rocks, L. E., Sanders, W. T., Stahle, C. M., Szymkowiak, A. E. and Vaillancourt, J. E., The next generation of silicon-based X-ray microcalorimeters. *Proc. SPIE* **4851** (2002a).

Stahle, C. K., *Personal communication*, 2002b.

Stahle, C. K., Lindeman M. A., Figueroa-Feliciano, E., Li, M. J., Tralshawala, N., Finkbeiner, F. M., Brekosky, R. P. and Chervenak, J. A., Arraying compact pixels of transition-edge microcalorimeters for imaging X-ray spectroscopy. *AIP Conf. Proc.* **605**, 223 (2002c).

Su8, http://aveclafaux.freeservers.com/SU-8.html.

Tan, P., Cooley, L. D., Galeazzi, M., Liu, D., McCammon, D., Nelms, K. L. and Sanders, W. T., Mo–Cu bilayers as transition edge sensors for X-ray astrophysics. *AIP Conf. Proc.* **605**, 255 (2002).

Trowell, S., Holland, A. D., Fraser, G. W., Goldie, D. and Gu, E., Development of a distributed read-out imaging TES X-ray microcalorimeter. *AIP Conf. Proc.* **605**, 267 (2002).

Ukibe, M., Kimura, T., Nagaoka, T., Pressler, H. and Ohkubo, M., Fabrication of bridge-type microcalorimeter arrays with Ti–Au transition-edge-sensors. *AIP Conf. Proc.* **605**, 207 (2002).

Wollman, D. A., Nam, S. W., Newbury, D. E., Hilton, G. C., Irwin, K. D., Bergren, N. F., Deiker, S., Rudman, D. A. and Martinis, J. M., Superconducting transition-edge-microcalorimeter X-ray spectrometer with 2 eV energy resolution at 1.5 keV. *Nucl. Instrum. Methods. Phys. Res. A* **444**, 145 (2000).

Young, B. A., Saab, T., Cabrera, B., Cross, J. J., Clarke, R. M. and Abusaidi, R. A., Measurement of T_C suppression in tungsten using magnetic impurities. *J. Appl. Phys.* **86**, 6975 (1999).

Zhang, J., Cui, W., Juda, M., McCammon, D., Kelley, R. L., Moseley, S. H., Stahle, C. K. and Szymkowiak, A. E., Non-ohmic effects in hopping conduction in doped silicon and germanium between 0.05 and 1 K. *Phys. Rev. B* **57**, 4472 (1998).

4.5 Position Sensitive Semiconductor Strip Detectors

W. DĄBROWSKI and P. GRYBOŚ
AGH University of Science and Technology, Krakow, Poland

4.5.1 INTRODUCTION

Progress in technology of semiconductor devices opens new possibilities for manufacturing semiconductor detectors. From the technological point of view a semiconductor X-ray detector, being in principle a diode, is a relatively simple structure. A very straightforward way to build a position sensitive X-ray detector is to manufacture an array of independent semiconductor diodes on a common substrate. Such arrays can be made on silicon with precision much better compared to what is needed for X-ray detectors. The difficulties are associated with requirements concerning properties of semiconductor materials, which are quite different, compared to typical materials used in the electronics industry. The two basic requirements with respect to semiconductor materials used for X-ray detectors are: (a) high density that guarantees short absorption lengths; and (b) high resistivity that allows obtaining sufficiently thick depletion layers of reverse biased junctions. Although silicon is a relatively poor material for X-ray detection for its low density, semiconductor position sensitive detectors are based in the majority on silicon.

As mentioned, silicon based technologies are suitable for manufacturing detector arrays with individual elements of almost any shape, however, each element of an array needs to be read out by an individual electronic channel. This requirement imposes some constraints on the possible arrangement of such detectors. The most commonly known and used types of position sensitive detectors are:

- strip detectors;
- pad detectors;
- pixel detectors.

Within each group one can find large variety of geometrical configurations. The boundaries between the three groups are not sharp and distinctions between the three types are made according to the schemes of the readout systems. The strip detectors are basically one-dimensional (1-D) position sensitive devices. The readout electronics is connected at the ends of strips and it is located outside the sensitive area of the detector. The pad detectors and pixel detectors are both two-dimensional (2-D) arrays. The pad detectors are often considered as 2-D arrays of relatively large elements, with sizes of the order of millimetres, while the pixel detectors are considered as 2-D arrays of small elements, with sizes of the order of a few hundreds of microns. A more fundamental distinction between these two types of detectors is established by the scheme of the readout electronics. In a pad detector each element is read out by an individual electronic channel. The readout electronics is located outside the sensitive area and the connections between the pads and the readout electronics are made with metal traces (Weilhammer *et al.*, 1996; Lin *et al.*, 1997). Thus, the electronics for readout of pad detectors has a similar architecture as the electronics for readout of strips. In the pixel detectors the readout electronics circuit for each sensitive element is located in the direct vicinity of that element. Depending on how this is realized we distinguish monolithic pixel detectors and hybrid pixel detectors. In monolithic pixel detectors the readout electronics is integrated on the same substrate as the sensor array (Turchetta et al., 2001). Hybrid pixel detectors are built of two separate devices; the sensor array is made as one device while the readout integrated circuit is made as another one and the two devices are connected together using a technique of flip chip bonding (Breibach et al., 2001; Cihangir and Kwan, 2001; Lozano et al., 2001). In either case the readout electronics overlaps with the sensitive area of the detector.

The technologies used for manufacturing silicon strip detectors and readout electronics can be considered these days as very mature ones. Silicon strip detectors have been widely used in high energy particle physics experiments for almost 20 years and much development has been done in this area. Special techniques suitable for implementation of standard monolithic planar processes on high resistivity silicon have been developed and are available as industrial standards. It is important to note that techniques of designing and manufacturing silicon strip detectors allow building full custom devices specific and optimized for given applications. Silicon strip detectors of the same type as used for detection of relativistic charged particles can be used for detection of low energy X-rays, up to 20 keV. A great advantage of silicon is its very mature, essentially industrial, technology and regardless of some drawbacks due to limited efficiency, silicon strip detectors are most widely used for low-energy X-rays.

Detector structures with position sensitive strips have been realized successfully in various laboratories using other semiconductor materials: high purity germanium (HPGe) (Rossi et al., 1997; Rossi et al., 1999; Amman and Lucke, 2000; Vetter et al., 2000) and compound semiconductors, like gallium arsenide (GaAs) (Chen et al., 1996; Smith, 1996), cadmium telluride (CdTe) and cadmium zinc telluride (CdZnTe) (Eisen et al., 1999; Gostilo et al., 2001; Kalemci and Matteson, 2002), and mercuric iodide (HgI_2) (Schieber et al., 1998). These materials are more suitable for detection of harder X-rays, above 20 keV, because of higher atomic numbers. High purity germanium, having good spectroscopic properties, could be in principle a perfect material for strip detectors for higher X-ray energies. The complications associated with cooling such detectors, however, means that there is not much development in that direction. On the other hand, significant progress has been made recently in the area of so-called room temperature semiconductor detectors, which are based mostly on semi-insulating compounds with high atomic numbers and high energy band gaps. Poor charge transport properties of those materials and difficulties in producing large crystals are still limiting factors and much effort is focused on technological improvements.

The basic concept of a strip detector is very simple and is based on splitting a continuous electrode for individual strips, assuming that signals from each strip will be read out by an individual electronic circuit. Starting from a flat semiconductor detector one can either make one electrode as a multiple strip pattern and obtain a single-sided strip detector or divide both electrodes into strips and obtain a double-sided strip detector. In order to make such a detector working as a position sensitive device it is required that the strips are well separated electrically so that the current signal induced in a given strip does not flow to the neighbouring strips. The methods of strip separation depend on the detector structures and will be discussed in detail later.

In order to use position sensitivity of strip detectors it is required that the signal from each strip is recorded and processed independently by an individual electronic circuit, like for a single detector. Thus, strip detectors require multi-channel front-end electronics and the only practical way is to make such electronics as multi-channel integrated circuits. Such solutions have become possible due to advances in Very Large Scale Integration (VLSI) electronics, and in particular, due to advances in techniques of designing, prototyping and manufacturing of Application Specific Integrated Circuits (ASICs).

Generic structures of a single-sided and a double-sided semiconductor strip detector are shown in Figure 4.5.1. One can read out every strip, as shown schematically in Figure 4.5.1, or one can

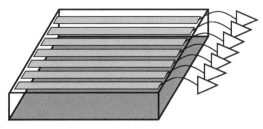

Single-sided strip detector

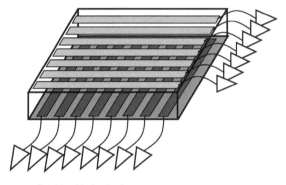

Double-sided strip detector

Figure 4.5.1 Schematic structure of a single-sided and a double-sided semiconductor strip detector

read out only every second or every third strip. In such a configuration the signals collected by the intermediate are coupled to the readout strips through the interstrip capacitance.

4.5.2 SPATIAL RESOLUTION OF SEMICONDUCTOR STRIP DETECTORS

The spatial resolution of a semiconductor strip detector depends on the intrinsic spatial resolution of the semiconductor material, strip pitch of the detector and signal-to-noise ratio of the readout electronics. For an ideal position sensitive detector of X-rays, the measured position is expected to correspond to the position at the detector surface at which the photon entered the detector. There are two effects, which impose some physical limitations on the spatial resolution of a semiconductor detector, namely, scattering of photons in the semiconductor material and diffusion of generated charge carriers during their transport in the sensitive volume of the detector. These effects are independent of the segmentation of the detector electrodes but the segmentation, i.e. strip pitch in a strip detector, should be designed taking into account these effects. Even if there were no technological limitations on precision of electrode patterns one could not obtain spatial resolution better than determined by the intrinsic effects of scattering and diffusion. From a practical point of view it is very important that one understands the intrinsic resolution of the semiconductor material as this allows one to optimize the design of the detector accordingly, taking into account other aspects like, for example, very important limitations due to readout electronics.

For most of the later considerations it is assumed that the photons enter the detector either from the top or from the bottom side, perpendicularly to the surface, as shown schematically in Figure 4.5.2. There are some particular applications, in which the detector is illuminated from the edge side, along the strip, and they will be discussed separately. The measured position of a photon entering the detector at point x_{entry} is given by the position x_{meas} reconstructed upon the electrical signals recorded at the readout strips. Depending on the readout method used the reconstructed positions will be a set of discrete numbers corresponding to strip positions or some numbers elaborated in a more sophisticated way using the signal amplitudes measured on the strips clustered around the position of photon entry. Thus, assuming that the detector surface is illuminated uniformly with photons, as shown in Figure 4.5.2, one can build up a distribution of residuals $(x_{\text{meas}} - x_{\text{entry}})$ for a given strip and define the spatial resolution as:

$$\sigma_x = \sqrt{(x_{\text{meas}} - x_{\text{entry}})^2} \qquad (4.5.1)$$

There are two issues, which should be addressed with respect to such a measure of spatial resolution, namely:

- the resolution cannot be measured directly according to Equation (4.5.1) since one cannot measure or predefine the real entry position x_{entry} of a photon:

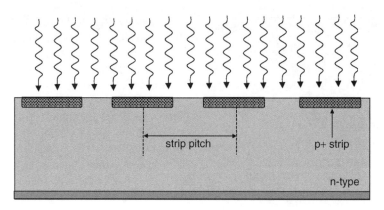

Figure 4.5.2 Schematic cross-section of a strip detector and definition of the strip pitch

- the distribution of residuals ($x_{\text{meas}} - x_{\text{entry}}$) can have a highly non-Gaussian shape and in such cases the true rms (root mean square) value is only partially meaningful for accuracy of position measurements.

An advantage of such a definition is that it is very generic and it allows one to take into account all the effects which affect the spatial resolution, i.e. scattering of photons in the semiconductor material, diffusion of charge carriers during the collection process, architecture and noise performance of the readout electronics. In applications of semiconductor strip detectors for imaging measurements the quality of a system with respect to its spatial resolution is usually described by the Modulation Transfer Function (MTF) (Besch, 1998).

4.5.2.1 INTRINSIC SPATIAL RESOLUTION OF SEMICONDUCTOR DETECTORS

Let us first briefly review the basic properties of semiconductor materials, which are potential candidates to be used for position sensitive detectors, namely, Si, Ge, GaAs, CdTe and CdZnTe. The fundamental parameter, which defines the properties of a given semiconductor material as a sensor material, is the absorption coefficient as a function of X-ray energy. The parameter, which describes directly attenuation of a photon beam in a semiconductor material, is the absorption length. Total attenuation of an X-ray beam is determined by photoabsorption as well as by incoherent and coherent scattering. The photons scattered either in incoherent or in coherent processes are eventually absorbed in the detector within distances which are usually small compared to the dimensions of the detector. For a non-segmented detector the attenuation length defines the efficiency of the detector of given thickness. In segmented detectors, like strip detectors, one has to consider cases when a scattered photon is absorbed in a different segment than the one into which the primary photon entered.

The absorption length for X-rays of given energy in a given semiconductor material defines not only the detector efficiency but also the distance over which the generated charge carriers have to be transported to the collecting electrodes. For the considered energy range of X-rays the secondary electrons are stopped within submicron distances so that one can assume that the charge carriers are generated in a single point. During transport to the collecting electrodes the carriers diffuse in all directions. In a non-segmented detector the effect of diffusion in the plane parallel to the collecting electrodes has no influence on the current signal induced in the electrodes. In a position sensitive detector, like a strip detector, the diffusing cloud of charge carriers can be divided between neighbouring strips. The number of strips into which the charge is divided depends on the range of diffusion as well as on the strip pitch. This effect has to be taken into account when designing a strip detector for a given application.

The absorption length for the basic materials (Si, GaAs and CdTe) as a function of X-ray energy from 1 keV to 100 keV, is shown in Figure 4.5.3. The plots have been generated using the database XCOM (Berger *et al.*, 1999). The plots for Ge and CdZnTe are not shown as they mostly overlap with GaAs and CdTe, respectively, except the regions around the absorption edges. The plots shown in Figure 4.5.3 are fundamental for usual considerations of detector efficiency vs detector thickness and detector material. In this subchapter we discuss these plots with respect to the issues specific for position sensitive detectors.

Diffusion

For each particular detector and particular spectrum of X-rays one can estimate intrinsic spatial resolution due to diffusion by taking into account the following aspects:

- attenuation profile of the X-ray beam in the sensitive volume of the detector;
- distribution of the electric field in the sensitive volume of the detector;
- transport of charge carriers to the collecting electrodes.

Since photons are absorbed at various depths distributed according to the absorption law and have to travel over various distances one has to average the effects over some number of photons interacting with the detector material. This can be done either analytically in simple cases or by Monte Carlo (MC) simulation. Several examples of MC simulation employed for evaluation of the performance of position sensitive semiconductor detectors, strips or pixels, can be found in recent publications (Dąbrowski *et al.*, 2000; del Risco Norrlid *et al.*, 2001; Fowler *et al.*, 2002; Mathieson *et al.*, 2002). In MC simulation one can combine the effect of diffusion with the transport of charge carriers, taking into account accurate distribution of the electric field in particular structures of strip or pixel detectors.

For a rough evaluation of the diffusion effect in various semiconductor materials we have used the following simplified analytical approach. Let us assume that the detector is well overdepleted so that the electric field is almost constant over the full thickness of the detector. Examples of silicon strip detectors show that using high resistivity material, with resistivity of the order of 10 kΩcm,

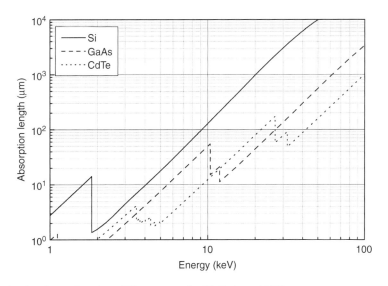

Figure 4.5.3 Absorption length as a function of X-ray energy for Si, GaAs and CdTe

one can bias the detectors with a voltage by a factor of 3–5 higher than the full depletion voltage (Andricek et al., 2000). The GaAs and CdTe detectors for X-rays are usually made using semi-insulating materials with resistivity of the order of 1×10^5 to 1×10^6 Ωcm. Such detectors behave more like solid-state ionization chambers and one can assume that the electric field is constant in the total sensitive volume. In the strip detector structures the electric field distribution around the strips may be highly non-uniform. This non-uniformity affects the trajectories and velocities of charge carriers in the regions around the readout strips. However, these regions are only a small fraction of the total sensitive volume of the detector. These details are important for construction of the strip detectors, however, they do not affect significantly the range of diffusion spread.

In addition, we assume that the electric field is lower than the value at which the drift velocity saturates. For most detectors it is possible to fulfil both conditions, i.e. to bias the detector with voltage significantly higher than the full depletion voltage and to keep the electric field below the velocity saturation point. In numerical calculation both effects can be easily taken into account but for analytical evaluation the above approximations simplify the analysis.

Let us assume that photons enter the detector perpendicularly to its surface from the side of the collecting electrode. The spread of charge carriers due to diffusion taking place during transport of the carriers to the readout electrode is characterized here by the full width at half maximum (FWHM) of the charge carrier distribution arriving at the collecting electrode when started from a point at the depth corresponding to the absorption length for X-rays of given energy. The FWHM of the charge carrier distribution in the plane parallel to the electrode is then given as:

$$\text{FWMH} = 4\sqrt{\ln 2 \, t_{n,p} D_{n,p}} \quad (4.5.2)$$

where $D_{n,p}$ is the diffusion constant for electrons or holes and $t_{n,p}$ is the drift time of charge carriers over the distance equal to the absorption length.

Assuming that the electric field E is constant in the depletion region one can express the drift time by a simple formula

$$t_{n,p} = \frac{\mu_{n,p} E}{\lambda} \quad (4.5.3)$$

where λ is the absorption length.

Of course, the charge generated by photons absorbed at smaller depths diffuse less but a significant fraction, about 37 % of incident photons which are absorbed at the depths larger then the absorption length, result in wider distributions of the collected charge.

Taking the absorption length as a function of energy, as shown in Figure 4.5.3, one can obtain the corresponding values of the FWHM of charge distribution due to diffusion. The values of the assumed electric field and the parameters of semiconductor materials are summarized in Table 4.5.1.

For silicon it has been assumed that the carriers collected by the readout electrode are holes as the most commonly used silicon strip detectors are built as p$^+$ strips in n-type bulk material. For GaAs and CdTe it has been assumed that the collected carriers are electrons. It is known that in these materials hole trapping is a very basic problem, which affects spectroscopic performance of such detectors and special efforts have to be undertaken to make those detectors as devices sensing the electron signal only (He, 2001).

The FWHM of distribution of charge collected on the readout electrode as a function of incident X-ray energy for the four considered semiconductor materials is plotted in Figure 4.5.4. One can notice that the FWHM as a function of X-ray energy follows roughly the absorption length as

Table 4.5.1 Parameters used for calculation of the FWHM of charge spread due to diffusion

	Si	GaAs	CdTe	Ge@77K
Collected charge carriers	holes	electrons	electrons	holes
Electric field (V/cm)	4×10^3	2×10^3	1×10^4	1×10^3
Mobility of collected carriers (cm^{2}V^{-1}s^{-1})	450	8500	1150	42 000
Diffusion coefficient of collected carriers (cm^{2}s^{-1})	11.6	201	29.8	280

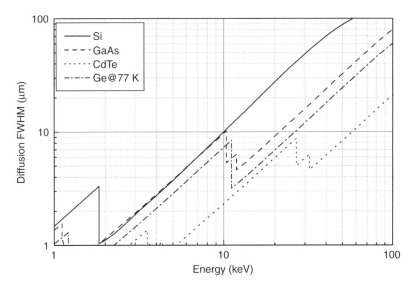

Figure 4.5.4 FWHM of charge distribution collected at the readout electrodes for a point charge generated at the depth equal to the absorption length

a function of energy. This is not surprising if one remembers that the drift time is proportional to the absorption length and the ratio of the diffusion coefficient and the mobility for a given semiconductor material is constant and is given by the Einstein relation as:

$$\frac{D_{n,p}}{\mu_{n,p}} = \frac{kT}{q} \qquad (4.5.4)$$

where k is the Boltzmann constant, T is the absolute temperature and q is the electronic charge.

At given temperature a high mobility, like in the case of GaAs, results in a short charge collection time but proportionally higher diffusion coefficient results in large diffusion spread in a short time. One can expect to gain in terms of reducing the relative effect of diffusion with respect to drift by lowering the temperature. This is shown in Figure 4.5.4 for germanium at a temperature of 77 K. The absorption length for Ge is not shown in Figure 4.5.3 since it overlaps almost completely with the plot for GaAs. In Figure 4.5.4 the diffusion FWHM for Ge at 77 K is systematically lower than for GaAs, though the reduction is not large because the diffusion spread is proportional only to the square root of the drift time.

The FWHM of charge spread due to diffusion, as shown in Figure 4.5.4, gives a first indication about the lower limit of the strip pitch in a strip detector. Using a strip pitch much smaller than the diffusion FWHM might be in principle an advantage for the spatial resolution as one could measure the signals collected on several clustered strips and then evaluate the centre of gravity of the charge distribution. That way one could measure the position with a precision much better than the strip pitch. Such methods are commonly used in particle physics applications where position measurement accuracy better than 2 μm rms has been achieved using silicon strip detectors with 25 μm strip pitch and 50 μm readout pitch (Colledani et al., 1996). However, one has to keep in mind that a small strip pitch results in dividing the total charge generated by a photon between too many strips and the signal measured at each individual strip becomes low. This is a serious limitation for measurements of low energy X-rays as the total signals generated in the semiconductor detectors are small anyway and obtaining a satisfactory signal-to-noise ratio is a non-trivial problem, even before division of charge between several strips. For semiconductor materials other than silicon the processes used for manufacturing strip detectors does not allow reducing easily the strip pitch below 100 μm and in most cases taking diffusion into account does not affect the detector design. On the other hand, for silicon there is practically no limitation on precision of the strip layout and one should take into account the results shown in Figure 4.5.4 in order to make a reasonable layout of strips given the energy of X-rays to be measured and performance of the readout electronics.

Scattering

In many cases photon scattering is more important than diffusion. The scattered photons are mostly absorbed in the detector volume but in positions displaced from their initial trajectories so that the charge can be collected in strips not necessarily nearest to the point at the surface where the photon has entered the detector. A complete evaluation of the scattering effects for a given detector geometry can be done only by full MC simulation that takes into account angular distributions of scattered photons. The two basic scattering mechanisms differ with respect to angular distribution of scattered photons. In the case of incoherent (Compton) scattering most of the photons are scattered by large angles from the trajectories of primary photons and such scattered photons can produce false position measurements. The angular distribution of photons scattered in the coherent (Rayleigh) processes has a very dominant peak in the forward direction so that most of the scattered photons diverge by very small angles from the trajectories of primary photons and they do not affect the position measurements.

The contribution of scattering effects to the spatial resolution of a semiconductor detector depends on the ratio of the cross-sections for incoherent and coherent scattering to the cross-section for the photoabsorption. The cross-section for the photoabsorption decreases with increasing energy E of X-rays as $1/E^3$, while the cross-section for Compton scattering shows relatively flat dependence on the photon energy. Thus, the relative contribution of the incoherent scattering increases strongly with increasing energy of X-rays. The cross-section for the coherent scattering depends on the photon energy in a similar way as for the photoabsorption so that the relative contributions of these two effects remains at the same level over a wide range of X-ray energy.

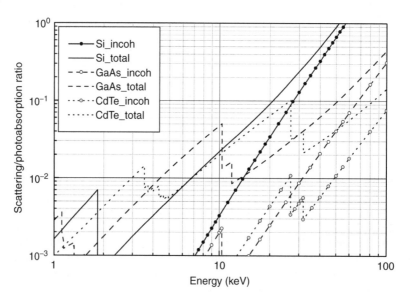

Figure 4.5.5 Ratio of the incoherent scattering coefficient to the photoabsorption coefficient and of the total scattering coefficient to the photoabsorption coefficient

The ratio of the incoherent scattering coefficient to the photoabsorption coefficient and the ratio of the total scattering coefficient, including the incoherent and the coherent scattering, to the photoabsorption coefficient as a function of X-ray energy for the three considered semiconductor materials (Si, GaAs and CdTe) are shown in Figure 4.5.5. The plots have been generated using the database XCOM (Berger et al., 1999). In a first approximation one can consider only the incoherent scattering for the coherent scattering due to its angular distribution predominating in the forward direction contributes only in a limited way to diverging photons from their primary trajectories. As discussed above, the relative cross-section of the Compton scattering increases strongly with increasing energy of X-rays. The lighter the semiconductor material the bigger the relative increase of the Compton scattering compared to the photoabsorption. For example, in silicon the ratio of the probability of incoherent scattering to the probability of photoabsorption is 0.1 for X-ray energy of about 25 keV and reaches 1 for X-ray energy of about 55 keV.

From the plots shown in Figure 4.5.5 one can obtain a rough estimate of how much the scattering may degrade the accuracy of position measurements. One should note that the scattered photons are eventually converted into photoelectrons within distances from the primary photon trajectories that are of the order of the corresponding absorption lengths. In order to evaluate how the scattering affects the spatial resolution one needs to take into account the angular distribution of scattered photons. This can be done employing full MC simulation. Such simulations are helpful at the stage of designing a detection system so that one can adjust properly the strip geometry and parameters of the readout electronics according to the intrinsic spatial resolution of the semiconductor material used. Several such examples concerning designs of particular detection systems can be found in the literature (Dąbrowski et al., 2000; del Risco Norrlid et al., 2001).

Most of the work in this area has been done for silicon as one can easily manufacture silicon position sensitive detectors with small readout electrodes (strips), comparable with the range of scattered photons. Results of such simulations are presented elsewhere (Dąbrowski et al., 2000). The simulation was performed for a 300 μm thick silicon detector with a surface dead layer of 2 μm for three values of X-ray energy (8 keV, 17 keV and 22 keV). For each photon entering the detector material perpendicularly to the surface the spatial distribution of the energy deposited in silicon was calculated using the EGS4 package (Bielajew et al., 1994). Then this distribution was projected on a line perpendicular to the strips. In such a way we could estimate the effect of scattering on the spatial resolution of a 1-D strip detector. Neglecting for the moment other effects associated with charge transport, diffusion and detector segmentation, the position at which the energy is deposited corresponds to the position reconstructed. The distributions of energy deposition for photons of 8 keV, 17 keV and 22 keV, entering a silicon detector perpendicularly to its surface, are shown in Figure 4.5.6. The non-Gaussian tails due to Compton scattering extend approximately up to 150 μm, 800 μm and 1000 μm for 8 keV, 17 keV and 22 keV, respectively. Thus, although the fractions of incoherently scattered photons are small they contribute significantly to the total rms value of residuals.

Assuming that one can measure precisely the position of the deposited charge $x_{\text{deposition}}$, i.e. neglecting a finite strip pitch, diffusion and noise of the readout electronics, the intrinsic spatial resolution can be defined according to Equation (4.5.1) as:

$$\sigma_x = \sqrt{(x_{\text{deposition}} - x_{\text{entry}})^2} \qquad (4.5.5)$$

From the results of MC simulation shown in Figure 4.5.6 one obtains the intrinsic spatial resolution as 4.8 μm, 35.9 μm and 71.1 μm for 8 keV, 17 keV and 22 keV, respectively. These values can be now compared with the spreads due to diffusion. From the plots shown in Figure 4.5.4 one can find the FWHM values of charge spread due to diffusion as 7.3 μm, 21 μm and 32 μm for X-ray energy of 8 keV, 17 keV and 22 keV, respectively. The corresponding standard deviation values are

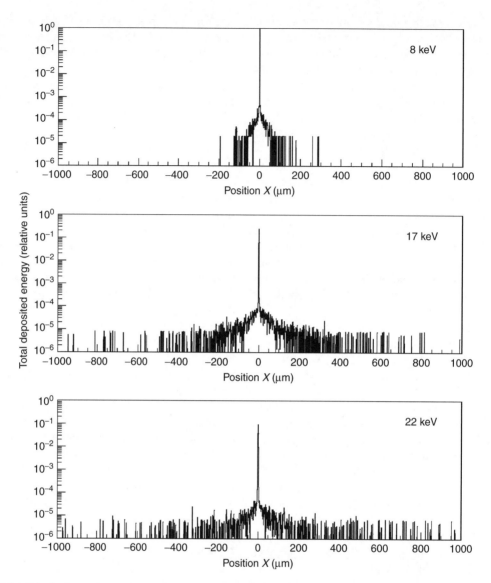

Figure 4.5.6 Distributions of energy deposition for photons of 8 keV, 17 keV and 22 keV, entering a silicon detector perpendicularly to its surface. Reported from *Nucl. Instrum. Methods Phys. Res., Sect. A* **442** (2000) 348, with permission from Elsevier Science

by a factor of 2.355 lower and one can notice that only for the lowest considered energy of 8 keV is the effect of scattering comparable with the effect of diffusion. For higher X-ray energies the effect of scattering predominates.

Up to now we have been assuming that the incident photons enter the sensitive detector volume perpendicularly to its surface. This is a good approximation for many applications since position sensitive measurements usually require well collimated X-ray beams. However, the collimation is never perfect and there is a trade-off between the aperture of a collimator and losses of X-ray intensity. Therefore, considering the intrinsic spatial resolution of a semiconductor detector, in addition to the scattering effects, one has to pay attention

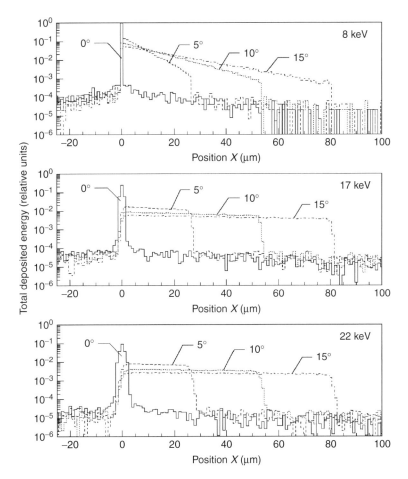

Figure 4.5.7 Distributions of energy deposition for photons of 8 keV, 17 keV and 22 keV, entering a silicon detector at the angle of 0, 5, 10 and 15°. Reported from *Nucl. Instrum. Methods Phys. Res., Sect. A* **442** (2000) 348, with permission from Elsevier Science

to the effect of parallax. In a rough approximation, for given incident angle this effect is proportional to the absorption length as the charge generated along the trajectories of primary photons is then projected on the detector surface with the read-out electrodes. In order to illustrate this effect we have performed MC simulations for the same photon energy values as considered above and for various incident angles (Dąbrowski *et al.*, 2000). Figure 4.5.7 shows the distributions of deposited energy for the incident angles of 0, 5, 10 and 15°. One can notice that the effect of parallax is far more significant compared to Compton scattering. In the case of this particular detector the distributions are limited by the detector thickness, which is only 300 μm. For the photon energy of 22 keV the distribution becomes almost flat since the detector thickness of 300 μm corresponds to the initial part of the absorption curve, which is approximately linear.

Following the same procedure as in the case of 0° one can estimate the intrinsic spatial resolution for the distributions shown in Figure 4.5.7. The results are summarized in Table 4.5.2. It is interesting to note that for X-ray energy of 22 keV the spatial resolution does not change very much with the incident angle and becomes even better for the incident angles different from 0°, although

Table 4.5.2 Intrinsic spatial resolution of 300 μm thick silicon detector vs X-ray energy and incident angle

	8 keV	17 keV	22 keV
0°	4.8 μm	35.9 μm	77.1 μm
5°	6.9 μm	35.1 μm	64.6 μm
10°	11.0 μm	45.1 μm	70.4 μm
15°	15.6 μm	49.9 μm	67.6 μm

the shapes of charge distributions are very different depending on the angle. This effect is due to limited detector thickness being, in that particular case, much smaller compared to the absorption length of 22 keV photons in silicon. Thus, some fraction of incoherently scattered photons escape the detector volume before being absorbed and this fraction depends on the incident angle of the primary photons and angular distribution of the scattered photons. Particular combinations of those two angular distributions result in variation of the intrinsic spatial resolution. For poorly collimated X-ray beams a thin detector helps to reduce the affect of parallax on the spatial resolution, of course, in expense of losing detection efficiency.

4.5.3 STRUCTURES OF STRIP DETECTORS

Concerning the physical and topological structures there are significant differences between silicon strip detectors and strip detectors made of other semiconductor materials. The use of advanced silicon technologies allows one to design and manufacture quite advanced and sophisticated structures of silicon strip detectors. On the other hand, strip detectors made of other semiconductor materials, like Ge, GaAs, CdTe, CdZnTe, are relatively simple structures. The same applies to strip detectors built on lithium-drifted silicon Si(Li). Lithium compensation technique is used for manufacturing thick silicon detectors, however, the technological step of lithium drift is not compatible with other technological steps employed in manufacturing advanced structures of strip detectors.

4.5.3.1 SILICON STRIP DETECTORS

Most commonly used silicon strip detectors are based on reverse biased strongly asymmetric junctions built on low doped silicon of resistivity between 1 kΩcm and 20 kΩcm. In majority these are p^+-n junctions built on n-type high resistivity bulk material. Development of high precision silicon microstrip detectors of this type has been driven by applications in high energy particle physics for measurements of particle tracks. Those applications require thin detectors and thickness of 300 μm has become almost a standard in this area. Silicon strip detectors can be made on thicker wafers using exactly the same technological steps as for thin detectors and detectors of thickness up to 2 mm have been manufactured successfully (Ota, 1999; Phlips et al., 2001). In order to make use of the full thickness of a detector based on the p^+-n junction one needs to bias the structure with a sufficiently high voltage, the so-called full depletion voltage, to induce the depletion layer over the full physical thickness of silicon, up to the n^+ ohmic contact on the backside. The full depletion voltage increases as the square of detector thickness and is given as:

$$V_{depl} = \frac{qN_d d^2}{2\varepsilon_{Si}} \quad (4.5.6)$$

where q is the electronic charge, d is the detector thickness, N_d is the donor concentration in n-type bulk and ε_{Si} is the permittivity of silicon.

For example, for the above mentioned 2 mm thick detector built on silicon of 20 kΩcm resistivity the full depletion voltage is 650 V. The required high bias voltage for thick detectors becomes at a certain point a limitation since it exceeds the breakdown voltage of the structure. It is worth noting that the breakdown voltage of a silicon strip structure is much lower, by an order of magnitude, compared to the breakdown voltage of homogenous silicon bulk because of local high electric field around the edges of p^+ strips. The breakdown voltage usually decreases with narrowing the strip width. Significant progress has been made recently with respect to increasing breakdown voltage in silicon strip detectors and structures with strip pitch of the order of 100 μm with breakdown voltage up to 500 V are produced routinely (Andricek *et al.*, 2000). This allows not only to build thicker detectors but also to use thinner detectors with bias voltages much higher than the full depletion voltage, which helps to reduce the charge collection time and the charge spread due to diffusion.

Given the fact that there is practically no technological limitation on the precision of strip layout of silicon strip detectors one can easily match the strip pitch to the intrinsic spatial resolution so that the resulting spatial resolution is minimally affected by the detector layout. Due to low absorption coefficient, silicon devices are suitable for high precision position sensitive measurements of low energy X-rays, up to approximately 20 keV. For this low energy range the noise of readout electronics and the achievable signal-to-noise ratio are the limiting factors to be taken into account. The readout electronics will be discussed later, however, already at this point it should be mentioned that the detector geometry, in particular the strip pitch that determines the interstrip capacitance, affects the signal-to-noise ratio. Thus, in order to optimize the layout of a silicon strip detector for a given experiment one has to take into account all the discussed issues.

It is worth noting that silicon strip detectors and readout electronics are usually designed as full custom devices. Such an approach is affordable mainly because silicon strip detectors and readout ASICs are manufactured using mature industrial processes. A designer does not need to interfere with details of technology and, in fact, in most cases is not allowed to modify the technology. Of course, there are some drawbacks of such an approach but great advantages are relatively low cost and fast turn around time for prototyping.

Single-sided Silicon Strip Detectors

Starting from a p^+-n silicon detector structure one can make a single-sided strip detector by splitting off one of the electrodes, either on the junction side or on the ohmic side, into individual strips. Since the mobility of electrons and holes in silicon differ only by a factor of 3 the collection times for carriers of each type are of the same order of magnitude. For each type of charge carriers the lifetime is much longer than the collection time so that each electrode senses the motion of electrons and holes. The detailed shapes of the induced current pulse depend on the distribution of the electric field in the depletion region but in the end the total charge induced in a given electrode is equal to the charge of carriers collected by this electrode. Thus, assuming that the integration time in the front-end electronics is longer than the charge collection time, the total charges of current signals induced in the p-side (junction side) and the n-side (ohmic side) are equal.

Regarding a silicon strip detector there is a significant difference in the strip structure on the junction side and on the ohmic side. If the strips are made on the junction side separation of strips is obtained naturally, without any additional structure, as shown in Figure 4.5.8(a). The neighbouring strips are separated by two reverse biased p^+-n junctions connected back-to-back. One should note that the surface of silicon between the strips is covered by silicon dioxide in order to protect the surface against influence of humidity and impurities adsorbed at the surface. The oxide layer grown by oxidation of the silicon crystal is always charged positively with holes trapped at the Si–SiO$_2$ interface and holes trapped in the bulk of the oxide. The positive charge in the oxide causes accumulation of electrons at the silicon surface so that the regions between the strips become n^+-type. This is one of the reasons, for which strip detectors are made on n-type silicon. For an opposite configuration, i.e. n^+ strips in p-type low doped silicon, the positive charge in the surface oxide induces an inversion n^+ layer in p-type silicon underneath the oxide that short-circuit the n^+ strips.

The structure shown in Figure 4.5.8(a) illustrates schematically a DC-coupled detector. The p^+ strips are covered with metal strips for two reasons: to reduce the strip resistance and to provide contacts for connection of the readout electronics, which usually is done by wire bonding. In this configuration the potential of strips is defined by the potential of the inputs of the readout electronics, usually close to the ground of the system. Positive

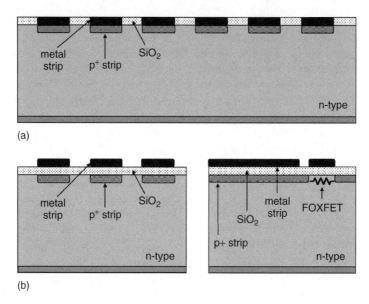

Figure 4.5.8 Techniques of signal coupling in strip detectors: (a) DC-coupled single-sided strips; (b) AC-coupled single-sided strips, cross-sections perpendicular to the strips and along the strip are shown. The FOXFET bias structure is shown

high voltage bias is then applied to the backside contact of the detector. In this configuration the DC leakage current from each strip has to flow to the input of the preamplifier connected to that strip. Such a configuration is possible, however, one has to ensure that the front-end electronics is designed in such a way that it can sink that leakage current.

In order to cut off the DC leakage current one can use capacitive coupling between the strips and the front-end electronics. In silicon detectors the coupling capacitors can be integrated in the detector structures resulting in AC-coupled strip detectors, as shown schematically in Figure 4.5.8(b). The p^+ strips and the metal strips are separated by a silicon dioxide layer and these structures form capacitors extended all along the strips. In this configuration each p^+ strip has to be biased separately through an individual resistor. The resistors are required to be of high enough value so that they do not contribute to the noise of the front-end system and they should be integrated in a small area at the ends of the strips. There are several technologies for manufacturing such resistors. The resistor can be made of low doped polysilicon layer, a diffusion layer or as channel of a MOSFET (Metal Oxide Semiconductor Field Effect Transistor) structure built between the strip and the bias line. For our purpose it is important to note that resistors of values up to a few MΩ can be realized using polysilicon and resistors of higher values can be realized using the FOXFET technique. In many applications to X-ray measurements the signal-to-noise ratio is very critical and it is required to reduce all noise sources, including the noise due to bias resistors. For such applications detectors with FOXFET bias structure, as shown in Figure 4.5.8(b), are preferable.

The readout strips can be made on the ohmic side, but in such a case one needs to implement additional structure in order to separate n^+ strips, which otherwise would be short-circuited via n-type bulk material. The interstrip resistance would be additionally lowered by the accumulation of n^+ layers formed in the interstrip regions covered by silicon dioxide.

There are two techniques used to separate the readout strip on the ohmic side, so-called 'p-stops' and 'field plates'. The p-stop structure, shown schematically in Figure 4.5.9(a), employs additional p-type strips laid out between the n^+ readout strips. This structure can be implemented either in DC-coupled or AC-coupled detectors.

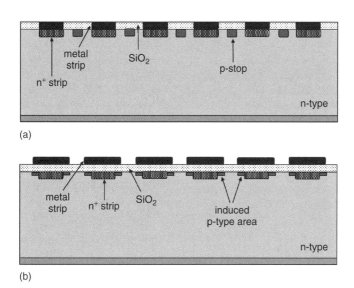

Figure 4.5.9 Strip isolation techniques on the ohmic side of a strip detector: (a) p-stops; (b) field plates

The field plate structure, shown schematically in Figure 4.5.9(b), can be implemented only in AC-coupled detectors. In this structure the metal strips are wider than the n^+ strips in silicon so that the edges of the metal strips are extended over the n-type bulk material. If the metal strips are biased negatively with respect to n^+ strips the electric field induces p-type layers at the surface of the n-type bulk, underneath the extended wings of the metal strips. These induced p-type surface layers provide separation of neighbouring readout strip. The potential of the metal strips has to be adjusted in such a way that inversion layers are induced in the low doped interstrip regions but not in the high doped n^+ strips regions.

From this short overview it is clear that the strip structure on the ohmic side is more complicated than the structure on the junction side and as far as single-sided detectors are concerned there is no particular reason to make the strip structure on the ohmic side. A small advantage of n-type readout can be due to faster signals on the n-side as these signals are induced mostly by electrons being collected faster. However, collection of holes in silicon is fast enough so that in most cases there is practically no difference between the signals collected on the n-side and the p-side, except in cases when very fast front-end electronics is used.

A strip detector can also be made starting from the Si(Li) detector structure. The strip structure on the p^+ contact side can be realized by pattering metal strips and plasma etching groves in the area between the strips in order to separate them electrically. Given that the thickness of the p^+ contact is small, about $1\,\mu m$, one can obtain a relatively fine pitch, down to $100\,\mu m$. The lithium-diffused contact, on the side opposite to the p^+ contact, is rather thick, typically between $100\,\mu m$ and $500\,\mu m$. For such a thick lithium contact layer the strip structure can be obtained by sawing deep groves in the contact but the pitch of such a structure can be only in the range of millimetres. Recently a technique of making thin lithium contacts, about $30\,\mu m$, in Si(Li) detectors has been reported (Protic *et al.*, 2001a,b). Such a contact can be then divided into strips by plasma etched grooves. The possibility of producing a strip structure with a pitch of $500\,\mu m$ has been demonstrated using this technique.

Double-sided Silicon Strip Detectors

Once the technique of separating the readout strips on the ohmic side is elaborated, the basic concept of a double-sided strip detector is relatively simple.

It is a structure with the strips on each side of the detector. The relative tilt of the strip on the n-side and on the p-side depends on the requirements concerning the spatial resolution for each coordinate and possible constraint due the overall assembly of the detectors. Double-sided detectors are 2-D devices, which are very attractive for many applications. One has to note, however, a very severe limitation of double-sided strip detectors with respect to the intensity of radiation that can be measured with these devices. The problem is illustrated schematically in Figure 4.5.10.

Let us assume that two photons, marked as black stars in Figure 4.5.10, hit the detector at the same time. Simultaneity of two events is determined by the double pulse time resolution of the readout system. The two photons will produce signals in two x-strips and two y-strips. Reconstructing the hit positions by the signals recorded at the strips one arrives at four possible combinations, of which two correspond to real hits and the other two are fake hits, sometimes called 'ghosts'. Such ambiguity introduces a very serious limitation on radiation intensity, which can be measured by means of double-sided detectors. Employing a fine strip pitch one can obtain a very precise 2-D device for measurements of X-rays, however, this structure only suits applications with rather low radiation intensity.

Edge-on Silicon Strip Detectors

As discussed earlier, there are two basic limitations associated with silicon strip detectors, namely low absorption for higher X-ray energies

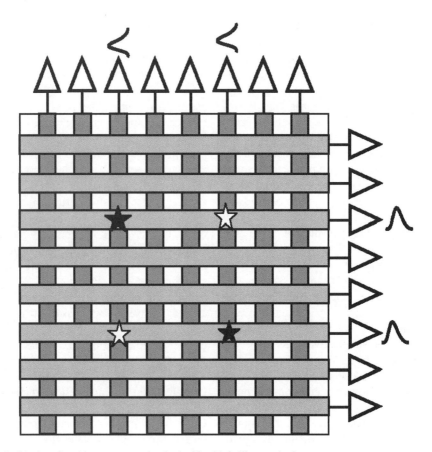

Figure 4.5.10 Ambiguity of position reconstruction in double-sided silicon strip detectors

and large cross-section for scattering increasing with increasing energy of X-rays. Both of these drawbacks can be overcome to a large extent by employing single-sided silicon strip detectors in the so-called edge-on configuration. The idea is to illuminate the detector from the edge so that photons enter the detector along the strips as shown schematically in Figure 4.5.11. The generated charge carriers are collected by the strips and the backplane in the plane perpendicular to the direction of the incident photon. The active area of such a detector is small but when combined with a mechanical scanning system it can be used as a large area high count rate 2-D device.

In the edge-on configuration the effective thickness of the detector is defined by the strip length and the strips can be made sufficiently long to cover 3 to 5 absorption lengths of X-rays of given energy. One cannot, however, achieve 100 % efficiency in this configuration because of two effects: absorption in the dead layer; and scattering of X-rays.

In the edge-on configuration the dead layer appears to be one of the serious problems associated with detector construction. In the conventional configuration, with photons entering the detector from the strip side or the backplane side, the thickness of the dead layer is of the order 1–2 μm and is practically negligible for X-ray energies above 4 keV. In the edge-on configuration the strips cannot be extended to the edge of the detector and the dead layer is much thicker. After cutting a silicon wafer the edge of the detector is mechanically damaged within a distance of a few tens of micrometres, depending on the cutting technique. Due to mechanical damage the edges of the detector exhibit low resistivity and the high voltage applied to the backplane appears on the strip side of the detector. In order to avoid the surface leakage current flowing to the strips one needs to surround the strips with a guard ring structure. For a higher bias voltage one needs a higher distance between the strip ends and the edge of the detector. Usually a multiple guard ring structure is employed in order to make the potential drop between the strips and the edge of the detector as smooth as possible. For typical silicon strip detectors of 300 μm thickness the guard ring structure extends over a distance of about 500 μm from the detector edge.

Another limitation of efficiency in edge-on detectors is due to scattering effects. The cross section of such a detector perpendicular to the photon paths is very small as it is determined by the thickness of silicon. Thus, the majority of scattered photons, except those scattered in the very forward direction, escape the sensitive volume of the detector. In a first approximation one can estimate the efficiency from the ratio of the total scattering coefficient to the photoabsorption coefficient for given X-ray energy, shown in Figure 4.5.4. The ratio reaches 1 for X-ray energy of 50 keV, which means that maximum detection efficiency can be about 50 % for that energy.

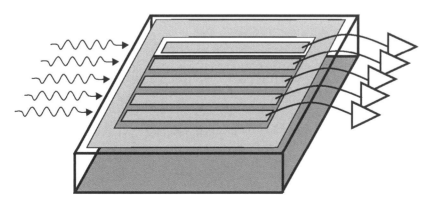

Figure 4.5.11 Schematic structure of the edge-on silicon strip detector

The edge-on configuration helps, however, to reduce the effect of scattering on the spatial resolution. The scattered photons mostly escape the detector so that they do not contribute to the spatial resolution. A very small fraction of photons are scattered into the sensitive volumes associated with the neighbouring strips but in a first approximation this effect can be neglected. Accurate estimate of effect of scattering on the spatial resolution can be done by full MC simulation for given X-ray energy and given detector geometry, as discussed before. One should note that in the edge-on configuration the effect of diffusion on the spatial resolution is decoupled from the absorption length since the absorption length can be in the range of millimetres and the charge carriers diffuse only while travelling across the distance equal to the detector thickness.

From the illumination side an edge-on detector looks like a linear array of pixels, of which dimensions are defined by the strip pitch in one direction and by the thickness of silicon in the other direction. One can build a large area 2-D detection system by combining such a detector with a mechanical scanning slit. In scanning direction one can improve the spatial resolution by using a slit which is narrower that the detector thickness. For example one can use a 100 μm wide scanning slit while the detector is typically 300 to 500 μm thick. Using a narrow slit helps to improve the spatial resolution, however, it requires more steps in the scanning and increases measurement time. An interesting idea has been proposed recently to improve the spatial resolution of silicon edge-on detectors by employing double-sided strips and using a timing structure of signals read out on the p-side and the n-side strips (Cederström *et al.*, 1999). The strips on both sides of the detector are proposed to be laid out in parallel. The shapes of the signals induced in the n-side and the p-side strips depend on the distance from the photon absorption point to the n-side and the p-side strips, respectively, due to the different mobility of electrons and holes. It has been shown by simulation that the significance of the timing information is sufficient to extract additional information about position, however, implementation of this idea requires quite advanced and sophisticated readout electronics.

It is worth noting that edge-on silicon strip detectors have been used successfully for imaging systems using X-rays of 60 keV. A complete clinical apparatus for digital radiology using a 6 cm long strip in edge-on configuration and a scanning slit system has been built successfully (Hilt *et al.*, 2000). Edge-on detectors combined with mechanical scanning systems are also proposed to be used as position sensitive detectors for digital mammography using X-rays in the range of 20 keV (Arfelli, 2000; Mali *et al.*, 2001).

4.5.3.2 *OTHER STRIP DETECTORS*

Strip detectors can be, in principle, made on other semiconductor materials like Ge, GaAs, CdTe and CdZnTe. A significant difference between silicon and other semiconductor materials is due to different technological processes used for manufacturing the detectors. The position sensitive structures on Ge, GaAs, CdTe and CdZnTe are developed at various research institutes but none of these developments has reached an industrial standard yet. A consequence of such a situation is that minimal strip pitch achievable in strip detectors using those materials is rather large compared to what can be obtained in silicon technology. The limitation on the minimal strip pitch is not only due to precision of the strip patterns but also due to electrical separation of the strips. The electrical character of electrode contacts in those detectors built on semi-insulating materials is not always well defined. They can be either ohmic or Schottky contacts and electrical separation of strips relies on very high intrinsic resistivity of the sensor material.

High purity Ge is a good sensor material with respect to many parameters and detection properties of this material are well understood. The single-sided and double-sided Ge strip detectors have been built, however, the smallest strip pitch reported is 2 mm (Protic *et al.*, 2001). This has been achieved by implementing amorphous semiconductor technology for making the strip contacts. Such contacts exhibit blocking characteristics under either polarity so that they can be used on each side of a double-sided strip detector. A problem, which appears in such Ge strip detectors, is incomplete charge collection from the regions between the strips. A common technique used to ameliorate this effect is to add additional strips between the charge sensing strips. The intermediate strips are biased at some potential with respect to the charge collecting strips so that they shape the electric field at the surface in such a way that the charge carriers are forced to move to the collecting strips.

The idea to use a timing structure of signals induced at the two sides of a double-sided detector, as mentioned earlier for edge-on silicon strip detectors, has been implemented practically in Ge double-sided detectors to make 3-D devices. Given that Ge detectors can be rather thick devices, of the order of cm, the collection times of holes and electrons are long and it is relatively easy to measure the difference in the collection time of electrons and holes, depending on the depth of photon interaction in the detector. The spatial resolution of about 0.5 mm in depth has been reported when using this technique (Wulf et al., 2001).

While the high purity Ge detectors require cooling in order to reduce thermally generated leakage current, materials like CdTe, CdZnTe and GaAs offer the possibility of building detectors working at room temperature since the leakage current is suppressed due to a relatively high energy band gap (1.6 eV for CdTe and 1.4 eV for GaAs). A basic problem of those detectors, from the spectroscopy point of view, is poor efficiency of holes collection so that the amount of the collected charge depends on the depth at which a photon interacts with the sensor material. Therefore, at the present stage of technology one can consider only single-sided strip detectors built of these materials. The transport properties of these materials are being continuously improved. Especially significant progress has been made with respect to CdTe and CdZnTe resulting in room temperature detectors with spectroscopic performance comparable with this of cooled high purity Ge detectors. The resistivity of these semi-insulating materials is sufficiently high so that sufficiently high inter-strip resistance is obtained without implementing additional separation structures. Employing CdTe and CdZnTe of very high resistivity the interstrip resistance in the range of a few GΩ has been obtained for detectors with strip pitch as low as 125 μm (Gostilo et al., 2001).

One should mention at this point the coplanar strip structure used in CdTe detectors, not necessary for position measurements. This is one of the techniques employed to eliminate signals induced by motion of holes in order to improve the spectroscopic response of CdTe detectors. The electrode is divided into two groups of interleaved strips, each group biased at different potential. The signals from the two strip structures are then subtracted and that way the resulting signal corresponds to the signal induced only by electron motion in the region near the readout electrodes. This structure has been shown to work effectively for improving energy resolution of CdTe detectors (He, 2001; Luke et al., 2001). It has been also employed for constructing a 3-D position sensitive device based on a single polarity signal (Macri et al., 2001). Because of technological limitations such structures are usually realized with rather large pitch of about 1 mm.

4.5.4 READOUT ELECTRONICS

Progress in the development of position sensitive detectors based on single-sided or double-sided strip structures is linked closely with progress in the development of the readout electronics. Advanced VLSI electronics and techniques of designing and prototyping ASICs allow the integration of a large number of channels in a single chip. As discussed before, the technology of silicon strip detectors offers the possibility of designing application specific detectors with respect to size and strip geometry. Such an approach opens new possibilities for optimizing experimental techniques. However, one has to keep in mind that custom-designed strip detectors require custom-designed readout electronics, which in fact has become a standard in the areas employing silicon strip detectors. As the progress on other strip detectors built of Ge and CdTe results in smaller pitch, more advanced applications of those detectors require custom-designed integrated readout electronics as well.

There are two aspects of readout electronics, which should be addressed individually in each particular project, namely optimization of the front-end electronics and the overall readout architecture. Depending on application one may require good spatial resolution as well as good energy resolution. For low-energy X-rays the charges generated in semiconductor detectors are small and the noise of the readout electronics can be a limiting factor for energy measurements as well as for position measurements. Concerning the readout architecture one has to take into account several aspects, like whether energy measurements are required simultaneously with position measurements, whether 1-D, 2-D or 3-D position measurements are required and the intensity of radiation. The readout electronics realized as an ASIC gives opportunity to optimize the front-end for each particular type of the detector and for each particular application.

4.5.4.1 FRONT-END ELECTRONICS

Conventionally, front-end electronics is understood as a preamplifier only. Employing ASIC technique one can integrate more functions in a readout integrated circuit (IC) connected directly to a strip detector. Usually a multi-channel front-end IC, in addition to preamplifiers, comprises also shaper circuits. In most advanced solutions front-end ICs comprise also circuitry responsible for data conversion and data storage. Regardless of overall complexity of an IC the basic problems associated with optimization of the front-end circuit are similar to those in conventional circuits.

The noise performance of a front-end system is described by the Equivalent Noise Charge (ENC) defined as the charge applied to the input in the form of a short δ-like current pulse, which gives at the output the signal amplitude equal to the rms value of noise. For a typical configuration of the front-end electronics, comprising an integrator followed by a band-pass filter, the ENC is given as

$$\mathrm{ENC} = \sqrt{\frac{F_v v_n^2 C_t^2}{\tau_p} + F_i i_n^2 \tau_p + F_{vf} v_{nf}^2 C_t^2} \quad (4.5.7)$$

where C_t is the total input capacitance including the capacitance of the input transistor, detector capacitance, feedback capacitance and any stray capacitance of the connection between the detector strip and the input of the preamplifier, v_n^2 is the spectral density of the equivalent input voltage white noise, dominated by thermal noise of the channel of the input transistor, i_n^2 is the spectral density of the equivalent input current noise, dominated by the thermal noise of the feedback resistor in the preamplifier, thermal noise of the bias resistor in the detector and the shot noise of the detector leakage current, v_{nf}^2 is the spectral density of the equivalent input voltage flicker noise, dominated by the flicker noise of the input transistor, τ_p is the peaking time, i.e. the time at which the signal at the filter output reaches the maximum, and F_v, F_i and F_{vf} are factors dependent on the filter type. There are three immediate observations resulting from Equation (4.5.7), which are important for optimization of the front-end circuit, namely:

- contribution of the voltage white noise to the ENC is proportional to the total input capacitance and inversely proportional to the square root of the peaking time;
- contribution of the current noise is independent of the input capacitance and it is proportional to the square root of the peaking time;
- contribution of the voltage flicker noise to the ENC is proportional to the total input capacitance and independent of the peaking time.

One can optimize the front-end system taking into account various requirements and constraints of a given application. There are two parameters, the detector capacitance and the detector leakage current, which are determined by the detector and they can vary over wide ranges. The total strip capacitance seen by the input of the preamplifier is the sum of the capacitance of the strip to the backplane and the capacitance of a given strip to the neighbouring strips. Both components are proportional to the strip length. For a small strip pitch the interstrip capacitance dominates. Given the large variety of possible geometrical arrangements of strip detectors one can have the detector capacitance in the range from a fraction of pF, for strip lengths of a few millimetres, up to tens of pF for strip lengths of a few centimetres. In the first approximation the leakage current per strip is proportional to the depletion volume corresponding to a given strip and so it is proportional to the strip area. The leakage current is, in addition, a strong function of temperature and lowering temperature is a way to reduce the leakage current and the shot noise associated with it.

Other parameters in Equation (4.5.7) are determined by the front-end electronics. The first electronics parameter to be considered is the peaking time. One can distinguish two categories of applications with respect to the peaking time, low count rate and high count rate experiments. In the first category we have experiments with such low rate of X-rays that there is practically zero probability of having a pile-up of consecutive pulses regardless of the duration time of pulses in the shaper circuit. Another class of applications is constituted by experiments in which the maximum peaking time is limited by the intensity of X-rays.

For low count rate experiments and for given parameters of the detector and noise parameters of the front-end electronics one can find an optimum value of the peaking, which yields a minimum value of the ENC. In practice such an approach leads to reducing all parallel noise sources, including the shot noise of the detector leakage current, as much as possible and shifting the peaking time towards longer values in order to reduce the contributions from the voltage noise sources. For a system with current noise reduced to zero, Equation (4.5.7) gives infinite peaking time for minimizing the ENC. In practice there are always some current noise sources in the detector and in the front-end electronics but for extreme cases one can use peaking time in the range of tens of μs. In many applications, however, peaking time in the range of a few μs is already considered as a long one.

In high count rate experiments the maximum value of the peaking time is a starting point for optimization of the front-end system. In such applications usually one has to make a compromise between the count rate performance of the system and the signal-to-noise ratio. One can take advantage of designing a fully custom detector system and consider the detector segmentation at the same time as the concept and parameters of the readout electronics. For the required total active area of a detector one can divide it into smaller or larger elements, e.g. shorter or longer strips in the case of a strip detector. A smaller strip area helps for the count rate problems in a twofold way: for a given intensity of X-rays the count rate per strip is smaller, and a smaller strip capacitance helps to reduce the contribution of the voltage noise sources to the ENC.

Another aspect associated with the count rate performance concerns construction of the charge-sensitive preamplifier, in particular the scheme of discharging the feedback capacitor. A simplified block diagram of a single front-end channel is shown schematically in Figure 4.5.12. The input stage is a charge-sensitive amplifier, which converts the current signals induced in the strip into voltage steps. The feedback capacitance is charged up by these signals and, for DC-coupled strips, also by the detector leakage current. In order to avoid the DC level at the preamplifier output to shift off the dynamic range one needs to discharge the feedback capacitance, either after every pulse or in some solution after a number of pulses.

There are two basic techniques used for discharging the feedback capacitance, continuous discharge and switching reset, shown schematically in Figure 4.5.12. Continuous discharging can be realized either by a resistor in parallel to the capacitor or by a controlled current source. In either case the discharging component contributes to the parallel noise at the preamplifier input. In order to limit this noise source one should use a large value resistor or a low discharging current but then the decay time constant of the preamplifier output signal is long and one still faces limitations on the pulse rate due to pile-up. Thus, this option can be used either in low count rate systems or in systems where short

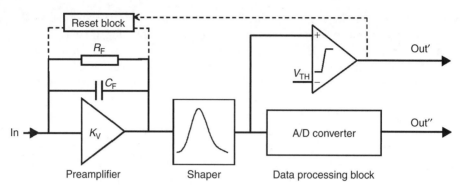

Figure 4.5.12 Block diagram of single readout channel

shaping is employed in the shaper circuit following the preamplifier so that the contribution of the parallel noise to the ENC is limited and one can use lower resistors or higher currents for discharging the feedback capacitors.

Another way of discharging the feedback capacitor is to employ a switching circuit that periodically resets the capacitor. Such a solution is commonly used in ASICs for readout of silicon strip detectors in the collider type particle physics experiments, in which, if the signals appear, they appear synchronously in all the channels. The trigger signal for discharging the capacitors is provided by the central clock of the experiment. In X-ray measurements the signals appear randomly in time and independently in each channel. After receiving a signal from the strip, the circuit has to generate the trigger signal for discharging the capacitor. In order to generate such a trigger signal one needs to implement a threshold discriminator in every channel. Various schemes used for discharging the feedback capacitor are discussed in detail elsewhere (De Geronimo et al., 2001).

The lowest level of noise optimization concerns the details of the preamplifier design. The readout electronics realized in the form of an ASIC offers the possibility to optimize the preamplifier circuit for each particular type of detector and for each particular application. A first step in optimization of the input transistor is to evaluate the level of non-reducible current noise, which comes from the detector and possibly from the feedback resistor. Given the level of the current noise and the required peaking time one can minimize the voltage noise by proper choice of dimensions of the input transistor and proper bias current. This analysis has to be done separately for preamplifiers with bipolar junction transistors (BJTs) used in the input stage and for preamplifiers employing field effect transistors (MOSFETs) as input devices. In principle one should consider also the junction field effect transistor (JFETs), however, VLSI technologies including JFET structures are very special and not easily available. On the other hand, the main drawback of MOSFETs due to high $1/f$ noise becomes less important as the quality of CMOS processes improves.

There are several basic facts concerning noise in these devices which should be kept in mind, namely:

- the voltage noise of a BJT is independent of the transistor area and is inversely proportional to the collector bias current;
- the current noise in a BJT is generated by the base current and for given current gain factor β it increases proportionally to the collector bias current;
- the voltage noise of a MOSFET depends on the drain bias current I_d and the dependence changes from $1/\sqrt{I_d}$ for transistors working in strong inversion to $1/I_d$ for transistors working in weak inversion;
- the voltage noise of a MOSFET depends on transistor channel dimensions as $\sqrt{(L/W)}$ where L and W are the channel length and the channel width, respectively;

- the input capacitance of a MOSFET is proportional to the transistor gate area $W \times L$.

Due to the presence of parallel noise in the BJTs they offer advantages compared to MOSFETs only in applications where short peaking times are required. In Bertuccio *et al*. (1997), one can find a detailed discussion of various design issues for front-end circuits based on BJTs. In short, for given detector capacitance and given peaking time there is an optimum value of the collector bias current in the input BJT that gives minimum of noise.

In MOSFET based front-end circuits the noise optimization requires proper sizing of the input device according to the detector capacitance. As mentioned above the voltage noise can be reduced by increasing the W/L ratio, however, for a given minimum value of the channel length L allowed by the technology this requires increasing the channel width W and so the input capacitance of the transistor. Thus, there is an optimum W/L ratio for which one obtains minimum of noise.

The range of peaking times, in which BJTs offer better noise performance compared to MOSFETs, depends on the detector capacitance. For a typical detector capacitance in the range 1 pF to 10 pF the breakpoint is around 100 ns, i.e. for peaking times below that value BJTs offer better noise performance and for peaking times above that value MOSFET are preferable. Another aspect, which may point to a BJT or a MOSFET, is the power consumption. Generally, BJTs offer lower voltage noise compared to MOSFETs at the same bias current and additional advantage in applications in which the total power dissipation is a limitation. It is worth noting that with decreasing feature size of modern CMOS processes the range of peaking times, in which MOSFETs offer superior performance over BJTs, extends towards shorter values of the peaking time.

4.5.4.2 *READOUT ARCHITECTURES*

Semiconductor strip detectors require that each strip is equipped with an individual electronics channel. The ASIC technique allows one to build multi-channel front-end systems, as discussed in the previous sections. However, keeping in mind that experimental set-ups may comprise hundreds or thousands of strips, one cannot imagine to build systems with as many complete spectroscopic channels working independently. For practical and cost reasons the number of output channels has to be reduced at a certain stage. The multiplexing can be done either in the front-end ASIC or in an external circuitry. Possible solutions depend on the intensity of X-rays and on the requirements concerning the energy resolution. In some solutions one can also implement buffering of data and zero suppression in the front-end ASICs.

From the point of view of system architecture used to read out silicon strip detectors one can distinguish two basic classes of systems: systems for position measurements only and systems for simultaneous position and energy measurements. In various imaging techniques employing monoenergetic X-rays it is sufficient to measure spatial distributions of X-rays of energies above a given threshold, or within a given energy window. For such applications one can use the binary readout architecture. In this scheme each front-end channel is equipped with a threshold discriminator or a window discriminator and delivers only binary (yes/no) signals. If simultaneous measurements of energy and position are required one has to preserve information on pulse amplitudes. As today, it is not feasible to integrate an individual high resolution analogue-to-digital converter (ADC) in each channel of the front-end ASIC. The constraints are associated with the total area of an IC and with the power consumption. However, if the requirements concerning resolution of the ADC are not very demanding one can implement a simple low resolution ADC in each channel. One of the proposed solutions is based on the so-called time-over-threshold method. Otherwise one has to multiplex some number of channels into a single ADC which can be either integrated in the front-end ASIC or can be an external device. The three possible system architectures are shown in Figure 4.5.13.

In the binary architecture the information delivered by a strip detector is suppressed to a

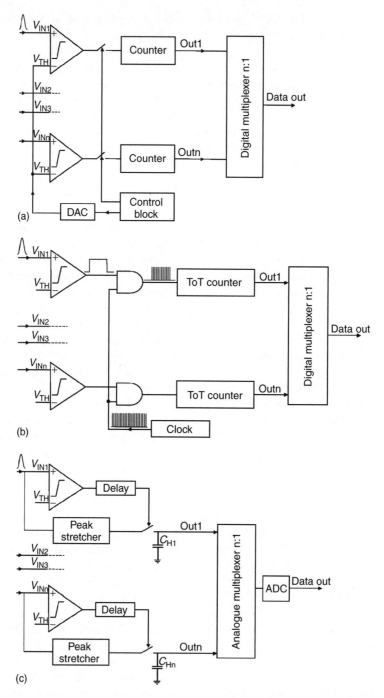

Figure 4.5.13 Three possible schemes of readout of strip detectors: (a) binary architecture; (b) analogue architecture employing time-over-threshold method; (c) analogue architecture employing multiplexing of analogue signals

minimum already in the front-end circuit. This is a significant advantage in systems comprising hundreds or thousands of channels as one can store the data in the front-end ASIC for a period of time required for performing the measurement. ASICs with binary readout architecture have been developed and used successfully to read out silicon strip detectors in diffractometry measurements (Comes et al., 1996; Dąbrowski and Dąbrowski, 2001; Dąbrowski et al., 2002). The RX64 IC (Dąbrowski et al., 2002) is a complete binary readout ASIC, which comprises 64 functionally independent channels, each built of the front-end circuit, threshold discriminator and 20-bit counter. The capacity of counters is sufficiently large so that one can store all the data from one measurement session and then read out the data into an external data acquisition system. In this scheme the intensity of X-rays is limited only by possible pile-ups in each individual front-end channel. Using this IC for readout a strip detector with 100 μm strip pitch, a count rate of 100 000 random pulses per second and per strip has been achieved (Dąbrowski and Dąbrowski, 2001).

Although the binary system does not provide direct information on the energy of X-rays the noise performance and energy resolution are equally important as in the spectrometric systems based on analogue readout schemes. One needs a good signal-to-noise ratio in the front-end circuit to be able to set the discrimination threshold at a level which is sufficiently high to suppress the rate of noise counts to a negligible level, and at the same time, sufficiently low to provide full efficiency for X-rays of given energy. In more complex measurements one may have X-rays of several energies, or like in experiments employing X-ray tubes, a continuous spectrum in addition to a distinct energy line. In such applications a window discriminator may be required instead of a single threshold discriminator.

An issue, which is very specific for binary readout architectures, and is as important as noise performance, is the matching of analogue parameters, like gain, noise and discriminator offset, for all the channels in a multi-channel ASIC. This is due to the fact that the only practical way to control the discriminator threshold in such an IC is to apply a common threshold to all channels. Thus, variations of gain and/or discriminator offset with respect to the nominal values affect the noise counts and efficiency in a similar way as the noise of a given channel.

Even with a simple circuit with a single threshold discriminator one can extract spectroscopic information by scanning the threshold and measuring that way the integral distribution of pulse amplitudes. An example of such measurements performed by means of the ASIC described in Dąbrowski et al. (2002) is shown in Figure 4.5.14. The plot shows complex X-ray spectra derived from integral ones measured simultaneously in 64 channels of the readout ASIC. One can notice that the spread between channels is really smaller than the noise of each particular channel. The differences of intensity in different channels are due to a particular distribution of X-ray intensity across the strips of the detector. Such measurements are essential for diagnostics of the system and for optimizing threshold setting for position measurements.

In some applications of strip detectors for position sensitive measurements the requirements concerning energy resolution are not very demanding. In fact, for detection systems working at room temperature usually one cannot achieve the energy resolution like in dedicated high resolution X-ray spectrometers. The requirements for the resolution of an ADC used for measurements of signal amplitudes are moderate and 6 to 8 bits can be sufficient. A simple scheme to extract information on signal amplitudes is based on the time-over-threshold principle (Becker et al., 1996; Manfredi et al., 2000). The idea is illustrated schematically in Figure 4.5.13(b). The analogue signal from the front-end circuit is applied to a simple threshold discriminator like in the binary scheme. The duration time of the discriminator response is measured in a simple way by counting pulses from a clock generator over the period equal to the duration of the discriminator response. The width of the discriminator response depends on the relative amplitude with respect to the threshold and so contains some information about the signal amplitude. A response function of such a system is nonlinear,

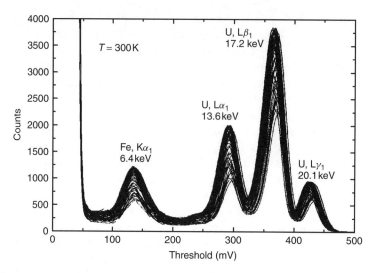

Figure 4.5.14 Spectra of Pu-238 radioactive source and Fe Kα line measured simultaneously in 64 strips of a silicon strip detector read out by a 64-channel ASIC

however, for a given pulse shape from the shaper circuit, it is well defined.

A main advantage of such a scheme is its simplicity and low power consumption that allows one to implement it in every channel. One can, however, notice easily basic limitations of this scheme, like the measurement range being limited by the discrimination level, low accuracy for small signals just above threshold, sensitivity to time jitter of the discriminator, especially for low amplitudes. The scheme has been implemented in an ASIC used for readout of silicon strip detectors in a particle physics experiment. Let us note that in the scheme shown in Figure 4.5.13(b) there is no capability to store data for more than one event in the front-end ASIC. Thus, one needs either to multiplex all the channels into one serial output, or elaborate a scheme of sparse readout (Feuerstack-Raible, 2000).

A fully analogue readout scheme, employing a true ADC, is shown schematically in Figure 4.5.13(c). In this scheme each channel is equipped with a peak detector and a sample and hold (S&H) circuit. Such schemes are commonly used in synchronous experiments where an external trigger signal, common for all the channels, is available. In applications to X-ray measurements one needs to implement a threshold discriminator in each channel to generate a trigger signal for the S&H circuit in each channel independently. The analogue signals from some number of channels are then multiplexed into one ADC, which in most cases is an external device, although one can consider to integrate it in the front-end ASIC. In such a scheme the intensity of X-rays is limited by the multiplexing rate and the speed of the ADC and not so much by the shaping in the front-end circuit. There is an obvious trade-off between the intensity of X-rays and the number of channels multiplexed into one ADC. In experiments with high X-ray intensity one can reduce the number of channels per ADC and increase the number of ADCs in the system.

Another aspect specific for such an architecture concerns control of the multiplexer and the ADC operation. One can either run the multiplexer and the ADC continuously (Fiorini *et al.*, 2001), allowing for some probability of pile-ups in the S&H circuits, or trigger the multiplexer and the ADC upon a signal occurring in the detector (Overdick *et al.*, 1997). The most commonly used scheme is based on an OR (logical sum) gate taking inputs from all the channels. Then each signal occurring in any of the channels triggers the readout sequence.

One should keep in mind the fact that readout ASICs with architectures as shown in Figure 4.5.13

are custom designed and for a given number of channels in one IC and given readout scheme there is no possibility to reconfigure the IC topology. The progress in the VLSI technology opens new possibilities for integrating more functions in single ASICs, however, they become even more specific for given applications. Using submicron CMOS processes one can easily imagine to integrate an ADC in the front-end ASIC or even several ADCs in one ASIC to increase the throughput of data. Another direction of development, particularly important for analogue readout architectures, is to integrate in the front-end ASICs analogue memory buffers that can serve as derandomizers. Amplitudes of signals occurring randomly in time are stored in such a buffer and read out with a constant rate adjusted to the speed of the multiplexer and the ADC. Providing the derandomizing buffer is sufficiently deep to remove completely statistical fluctuations from trains of incoming pulses one can increase the intensity of X-rays by an order of magnitude for the same multiplexing and data conversion rate. Several derandomizing schemes for analogue memories have been developed for ASICs used in the particle physics experiments (Anghinolfi et al., 1997) and first implementation to X-ray measurements has been reported (De Geronimo et al., 2001).

REFERENCES

Amman, M. and Luke, P.N., Three-dimensional position sensing and field shaping in orthogonal-strip germanium gamma-ray detectors. *Nucl. Instrum. Methods A* **452**, 155–166 (2000).

Andricek, L., Hauf, D., Kemmer, J., Koffeman, E., Lukewille, P., Lutz, G., Moser, H.G., Richter, R.H., Rohe, T., Soltau, H. and Viehl, A., Design and test of radiation hard p + n silicon strip detectors for the ATLAS SCT. *Nucl. Instrum. Methods A* **439**, 427–441 (2000).

Anghinolfi, F., Dąbrowski, W., Delagnes, E., Kaplon, J., Koetz, U., Jarron, P., Lugiez, F., Posch, C., Roe, S. and Weilhammer, P., SCTA – A rad-hard BiCMOS analogue readout ASIC for the ATLAS Semiconductor Tracker. *IEEE Trans. Nucl. Sci.* **44**, 298–302 (1997).

Arfelli, F., Synchrotron light and imaging systems for medical radiology, *Nucl. Instrum. Methods A* **454**, 11–25 (2000).

Becker, R., Grillo, A., Jacobsen, R., Johnson, R., Kipnis, I., Levi, M., Luo, L., Manfredi, P.F., Nyman, M., Re, V., Roe, N. and Shapiro, S., Signal processing in the front-end electronics of BaBar vertex detector. *Nucl. Instrum. Methods A* **377**, 459–464 (1996).

Berger, M.J., Hubbell, J.H., Seltzer, S.M., Coursey, J.S. and Zucker, D.S., XCOM: Photon Cross Section Database (version 1.2), Available: http://physics.nist.gov/xcom. National Institute of Standards and Technology, Gaithersburg, MD (1999). Originally published as Berger, M.J. and Hubbell, J.H. Photon Cross Sections on a Personal Computer, NBSIR 87-3597; and as Berger, M.J. and Hubbell, J.H., NIST X-ray and Gamma-ray Attenuation Coefficients and Cross Sections Database, NIST Standard Reference Database 8. National Institute of Standards and Technology, Gaithersburg, MD (1987).

Bertuccio, G., Fasoli, L. and Sampietro, M., Design criteria of low-power low-noise charge amplifiers in VLSI bipolar technology. *IEEE Trans. Nucl. Sci.* **44**, 1708–1718 (1997).

Besch, H.J., Radiation detectors in medical and biological applications. *Nucl. Instrum. Methods A* **419**, 202–216 (1998).

Bielajew, A.F., Hirayama, H., Nelson, W.R. and Rogers, D.W.O., History overview and improvements of EGS4, SLAC Report, SLAC-PUB-6499 (1994).

Breibach, J., Lubelsmeyer, K., Masing, Th. and Rente, C., Investigation of a bump bonding interconnect technology for GaAs pixel detectors. *Nucl. Instrum. Methods A* **470**, 571–575 (2001).

Cederström, B., Danielsson, M., Lundqvist, M. and Nygren, D., High-resolution X-ray imaging using the signal time dependence on a double-sided silicon detector. *Nucl. Instrum. Methods A* **423**, 135–145 (1999).

Chen, J., Geppert, R., Irsigler, R., Ludwig, J., Pfister, J., Plotze, T., Rogalla, M., Runge, K., Schafer, F., Schmid, Th., Soldner-Rembold, S. and Webel, M., Beam test of GaAs strip detectors. *Nucl. Instrum. Methods A* **369**, 62–62 (1996).

Cihangir, S. and Kwan, S., Characterization of indium and solder bump bonding for pixel detectors. *Nucl. Instrum. Methods A* **476**, 670–675 (2001).

Colledani, C., Dulinski, W., Turchetta, R., Djama, F., Rudge, A. and Weilhammer, P., A submicron precision silicon telescope for beam test purposes. *Nucl. Instrum. Methods A* **372**, 379–384 (1996).

Comes, G., Loddo, F., Hu, Y., Kaplon, J., Ly, F., Turchetta, R., Bonvicini, V. and Vacchi, A., CASTOR: a VLSI CMOS mixed analog-digital circuit for low noise multichannel counting applications. *Nucl. Instrum. Methods A* **377**, 440–445 (1996).

Dąbrowski, W., Białas, W., Gryboś, P., Idzik, M. and Kudłaty, J., A readout system for position sensitive measurements of X-ray using silicon strip detectors. *Nucl. Instrum. Methods A* **442**, 346–354 (2000).

De Geronimo, G., O'Connor, P., Radeka, V. and Yu, B., Front-end electronics for imaging detectors. *Nucl. Instrum. Methods A* **471**, 192–199 (2001).

De Geronimo, G., Kandasamy, A. and O'Connor, P., Analog peak detector and derandomizer for high rate spectroscopy. Presented at the IEEE Nuclear Science Symposium, San Diego, paper N10-4 (2001).

del Risco Norrlid, L., Ronnqvist, C., Fransson, K., Brenner, R., Gustafsson, L., Edling, F. and Kullander, S., Calculation of the modulation transfer function for the X-ray imaging detector DIXI using Monte Carlo simulation data. *Nucl. Instrum. Methods A* **466**, 209–217 (2001).

Eisen, Y., Shor, A. and Mardor, I., CdTe and CdZnTe gamma ray detectors for medical and industrial imaging systems. *Nucl. Instrum. Methods A* **428**, 158–170 (1999).

Feuerstack-Raible, M., Overview of microstrip readout chips. *Nucl. Instrum. Methods A* **447**, 35–43 (2000).

Fiorini, C., Longoni, A. and Butler, W., Multichannel implementation of ROTOR amplifier for the readout of silicon drift detectors arrays. Presented at the IEEE Nuclear Science Symposium, San Diego, paper N10-3 (2001).

Fowler, R.F., Ashby, J.V. and Greenough, C., Computational modelling of semiconducting X-ray detectors. *Nucl. Instrum. Methods A* **477**, 226–231 (2002).

Gostilo, V., Ivanov, V., Kostenko, S., Lisjutin, I., Loupilov, A., Nenonen, S., Sipila, H. and Valpas, K., Technological aspects of development of pixel and strip detectors based on CdTe and CdZnTe. *Nucl. Instrum. Methods A* **460**, 27–34 (2001).

Gryboo, P. and Dąbrowski, W., Development of fully integrated readout system for high count rate position-sensitive measurements of X-rays using silicon strip detectors. *IEEE Trans. Nucl. Sci.* **48**, 466–472 (2001).

Gryboś, P., Dąbrowski, W., Hottowy, P., Szczygieł, R., Świętek, K. and Wiącek, P., Multichannel mixed-mode IC for digital readout of silicon strip detectors. *Microelectron. Reliabil.* **42**, 427–436 (2002).

He, Z., Review of the Shockley–Ramo theorem and its application in semiconductor gamma-ray detectors. *Nucl. Instrum. Methods A* **463**, 250–267 (2001).

Hilt, B., Fessler, P. and Prévot, G The quantum X-ray radiology apparatus. *Nucl. Instrum. Methods A* **442**, 355–359 (2000).

Kalemci, E. and Matteson, J.L., Investigation of charge sharing among electrode strips for a CdZnTe detector. *Nucl. Instrum. Methods A* **460**, 527–537 (2002).

Lin, W.T., Chang, Y.-H., Chen, A.E., Hou, S.R., Lin, C.-H., Kulinich, P., Ryan, J., Steinberg, P., Wadsworth B. and Wyslouch, B., Development of a double metal AC-coupled silicon pad detector. *Nucl. Instrum. Methods A* **389**, 415–420 (1997).

Lozano, M.E., Cabruja, A., Collado, J. Santander and Ullan, M., Bump bonding of pixel systems. *Nucl. Instrum. Methods A* **473**, 95–101 (2001).

Luke, P.N., Amman, M., Lee, J.S., Ludewight, B.A. and Yaver, H., A CdZnTe coplanar-grid detector array for environmental remediation. *Nucl. Instrum. Methods A* **458**, 319–324 (2001).

Macri, J.R., Dufour, P., Hamel, L.A., Julien, M., McConnell, M.L., McClish, M., Ryan, J.M. and Widhom, M., Study of 5 and 10 mm thick CZT strip detectors. Presented at the 12th International Workshop on Room-Temperature Semiconductor X- and Gamma-Ray Detectors, San Diego, paper R8-8, (2001).

Mali, T., Cindro, V. and Mikuz, M., Silicon microstrip detectors for digital mammography–evaluation and spatial resolution study. *Nucl. Instrum. Methods A* **460**, 76–80 (2001).

Manfredi, P.F., Leona, A., Mandelli, E., Perazzo, A. and Re, V., Noise limits in a front-end system based on time-over-threshold signal processing. *Nucl. Instrum. Methods A* **439**, 361–367 (2000).

Mathieson, K., Bates, R., O'Shea, V., Passmore, M.S., Rahman, M., Smith, K.M., Watt, J. and Whitehill, C., The simulation of charge sharing in semiconductor X-ray pixel detectors. *Nucl. Instrum. Methods A* **477**, 192–197 (2002).

Ota, N., Thick and large area PIN diodes for hard X-ray astronomy. *Nucl. Instrum. Methods A* **436**, 291–296 (1999).

Overdick, M., Czermak, A., Fisher, P., Herzog, V., Kjensmo, A., Kugelmeier, T., Ljunggren, K., Nygard, E., Pietrzik, C., Schwan, T., Strand, S.-E., Straver, J., Weilhammer, P., Wermes, N. and Yoshioka, K., A 'Bioscope' system using double-sided silicon strip detectors and self-triggering readout chips. *Nucl. Instrum. Methods A* **392**, 173–177 (1997).

Phlips, B.F., Johnson, W.N., Kroeger, R.A. and Kurfess, J.D., Development of thick intrinsic silicon detectors for hard X-ray and Gamma ray detection. Presented at the IEEE Nuclear Science Symposium, San Diego, paper N12-12 (2001).

Protic, D., Krings, T. and Schleichert, R., Development of double-sided microstructured Si(Li)-detectors, Presented at the 12th International Workshop on Room-Temperature Semiconductor X- and Gamma-Ray Detectors, San Diego, paper R16-2 (2001a).

Protic, D., Th. Stohlker, H.F., Beyer, J., Bojowald, G., Borchert, A., Gumberidze, A., Hamacher, C., Kozhuharov, X., Ma and Mohos, I., A microstrip germanium detector for position-sensitive X-ray spectroscopy. *IEEE Trans. Nucl. Sci.* **48**, 1048–1052 (2001b).

Rossi, G., Morse, J., Rabiche, J.-C., Protic, D. and Owens, A.R., X-ray response of germanium microstrip detectors with energy and position resolution, *Nucl. Instrum. Methods A* **392**, 264–268 (1997).

Rossi, G., Morse, J. and Protic, D., Energy and position resolution of germanium microstrip detectors at x-ray energies from 15 to 100 keV. *IEEE Trans. Nucl. Sci.* **46**, 765–773 (1999).

Schieber, M., Zuck, A., Braiman, M., Nissenbaum, J., Turchetta, R., Dulinski, W., Husson, D. and Riester, J.L., Evaluation of mercuric iodide ceramic semiconductor detectors. *Nucl. Phys. B* **61B**, 321–329 (1998).

Smith, K.M., GaAs detector status, *Nucl. Instrum. Methods A* **383**, 75–80 (1996).

Turchetta, R., Berst, J.D., Casadei, B., Claus, G., Colledani, C., Dulinski, W., Hu, Y., Husson, D., Le Normand, J.P., Riester, J.L., Deptuch, G., Goerlach, U., Higueret, S. and Winter, M., A monolithic active pixel sensor for charged particle tracking and imaging using standard VLSI CMOS technology. *Nucl. Instrum. Methods A* **458**, 677–689 (2001).

Vetter, K., Kuhn, A., Deleplanque, M.A., Lee, I.Y., Stephens, F.S., Schmid, G.J., Beckedahl, D., Blair, J.J., Clark, R.M., Cromaz, M., Diamond, R.M., Fallon, P., Lane, G.J.,

Kammeraad, J.E., Macchiavelli, A.O. and Svensson, C.E., Three-dimensional position sensitivity in two-dimensionally segmented HP-Ge detectors. *Nucl. Instrum. Methods A* **452**, 223–238 (2000).

Weilhammer, P., Nygard, E., Dulinski, W., Czermak, A., Djama, F., Gadomski, S., Roe, S., Rudge, A., Schopper, F. and Strobel, J., Si pad detectors. *Nucl. Instrum. Methods A* **383**, 89–97 (1996).

Wulf, E.A., Ampe, J., Johnson, W.N., Kroeger, R.A., Kurfess, J.D. and Phlips, B.F., Depth measurements in a strip detector. Presented at the IEEE Nuclear Science Symposium, San Diego, paper N4–2 (2001).

Chapter 5
Special Configurations

5.1 Grazing-incidence X-ray Spectrometry

K. SAKURAI

National Institute for Materials Science, Ibaraki, Japan

Recent advances in grazing-incidence X-ray spectrometry (XRS) cover several important analytical directions, such as ultra trace element analysis using total reflection X-ray fluorescence (TXRF) and surface and interface analysis of layered materials by angular and/or energy resolved X-ray fluorescence measurements, as well as their combination with X-ray reflectometry. Another significant recent innovation in grazing-incidence XRS is micro X-ray fluorescence imaging without scans.

5.1.1 WHY GRAZING-INCIDENCE X-RAY SPECTROMETRY?

For many years, since the discovery by Roentgen in 1895, X-rays have been extensively used as a tool for the nondestructive investigation of fairly thick materials. This is because of the rather high transmission power of X-rays, which is usually expressed by penetration depth, i.e. the depth that the incident X-rays attenuate as $1/e$ of the initial intensity. The penetration depth depends on the kind of materials as well as X-ray energy; for most materials, it is in the order of μm–cm for 5–50 keV X-rays. If such transmission power can be controlled freely, it should be possible to create a number of novel opportunities for materials analysis using XRS. The use of external total reflection for a flat and smooth surface, which was first discovered by Compton in 1923,[1] is one of the most promising ways. The critical angle of the total reflection is usually very small, as listed in Table 5.1.1. The penetration depth becomes extremely shallow, typically 1–100 nm near the critical angle, leading to the technique even becoming surface sensitive. Since the X-rays impinge almost parallel to the surface, the technique is generally called grazing-incidence XRS.

Table 5.1.1 Critical angles of various materials for 0.155 nm (8.0 keV) X-rays (mrad)

Silicon	3.92
Glass	3.80
Aluminium	4.13
Titanium	5.23
Chromium	6.54
Iron	6.72
Nickel	7.01
Copper	7.02
Silver	7.70
Tungsten	9.61
Gold	9.70
Platinum	10.5

X-Ray Spectrometry: Recent Technological Advances. Edited by K. Tsuji, J. Injuk and R. Van Grieken
© 2004 John Wiley & Sons, Ltd ISBN: 0-471-48640-X

5.1.2 MODERN TOTAL REFLECTION X-RAY FLUORESCENCE

The most popular grazing-incidence XRS has undoubtedly been TXRF spectrometry, ever since the first experiment reported by Yoneda and Horiuchi in 1971.[2] In short, such experiments can be considered as X-ray fluorescence analysis of particulate samples, deposited on a flat and smooth substrate, with the excitation radiation reaching the substrate at an angle below the critical angle of total reflection. The advantage of using a total reflection mirror as a sample support is a remarkable upgrading of the detection power for trace elements due to the significant reduction of scattering background from the substrate. The technique has been continuously developed,[3,4] and historical progresses have been compiled in several textbooks,[5,6] and also in the series of proceedings of international conferences,[7–13] which have been successively held every 2 years since 1986. Recently, an interesting account reviewing the TXRF activities of a leading Austrian research group over the past 30 years has been published.[14]

In the early days of TXRF, white X-rays from a tube were usually used for excitation, as in the case of ordinary X-ray fluorescence. Since the critical angle is a function of the X-ray energy, higher-energy X-rays are most likely to penetrate the substrate to generate background. In 1984, Iida and Gohshi demonstrated that monochromatizing excitation radiation was effective in reducing unnecessary X-ray photons efficiently, resulting in an upgrading of the detection power for TXRF.[4] In their experiment, a Si(111) monochromator was used for selecting Cu $K\alpha_1$. TXRF experiments essentially require a very narrow collimated beam even when white X-rays are used, because of shallow angle irradiation. Their idea was that introducing a crystal monochromator, which limits angular divergence of the beam, does not cause a significant intensity loss for the required energy X-rays (Cu $K\alpha_1$). Subsequently, the use of a monochromator for primary photons became the most common, and many kinds of improved optics have been developed. The recent trend is the use of multilayers[15] and/or capillary optics.[16,17]

Another possible method for reducing the background and improving the detection limit is to use polarized radiation in order to decrease the scattered X-rays. When the primary beam has a linear polarization and the detector is placed in the plane of the polarization vector, the scattering intensity is proportional to $2\cos^2\phi(r/R)^2 + \sin^2\phi(r/R)^4$, while fluorescent X-rays are simply linear to $(r/R)^2$, where r and R are the detector radius and the distance from the beam to the center of the detector, respectively, and ϕ is the scattering angle (the angle between the primary and scattering beams).[18] This means when ϕ approaches 90°, by reducing the solid angle of the detector, the ratio of fluorescent and scattering intensity can be improved.

The advent of the synchrotron radiation (SR) source has had an extremely significant impact on the development of the TXRF technique. This X-ray source has unique properties, such as a high intensity, a very low angular divergence, energy tunability, and a high degree of linear polarization. Following the first experiment performed in Japan,[19] a number of TXRF studies have been done using SR at almost all facilities worldwide. One of the most important applications is the ultra trace determination of the surface contamination of semiconductor wafers. In the late 1990s, a detection limit in the order of 10^8 atoms/cm^2 was achieved for transition metals at Hamburg (HASYLAB/DESY), Stanford (SSRL) and Tsukuba (Photon Factory). At that time, such a trace level was almost the limit for other promising forms of chemical analysis, such as inductively coupled plasma mass spectrometry (ICP-MS) and atomic absorption spectroscopy (AAS), and therefore the detection power of SR-TXRF was a breakthrough. The successful analysis of ultra trace Ni on a Si wafer, for which the detection limit was 13 fg, was published in 1997 by an Austrian group.[20]

So far, synchrotrons have been used mainly by universities, and usually most beamlines are shared for different experiments. However, for specific industrial applications, such as wafer analysis, it is extremely important to construct a dedicated beamline and experimental station. In Stanford, semiconductor companies in Silicon Valley had a successful pioneering project.[21] They developed a dedicated instrument with an ultra clean environment, and recently reported very competitive results.[22] The 3rd-generation SR source[23] is obviously attractive for such applications. In ESRF, Grenoble, a new sophisticated beamline and spectrometer (Figure 5.1.1) for wafer analysis by TXRF has come onstream,[24,25] making it the future world center for this kind of activity. Another direction for industry is to use a compact synchrotron.[26] The advantage would be flexibility in the design of analytical instruments and the rather short distance from source to sample, leading to high efficiency.

When the sensitivity is such that ultra trace elements can be detected, parasitic X-rays can easily come into the spectrum as a result of contamination. Therefore, the use of a clean room is highly desirable for both sample preparation and measurement. As is often the case with multipurpose beamlines, it might be difficult to have a clean hutch. Figure 5.1.2 shows one example of a clean TXRF spectrometer used at BL39XU, SPring-8.[27,28] A compact clean booth with an air-filter unit is fitted to the spectrometer. The vacuum chamber is made of resin, and no metallic parts are used around the sample. If such clean equipment is not used, the sample surface is easily contaminated by air-particulates from the environment at the experimental hutch, as shown in Figure 5.1.3. In some cases, even one particle attached onto the area being analysed can distort the experiments.

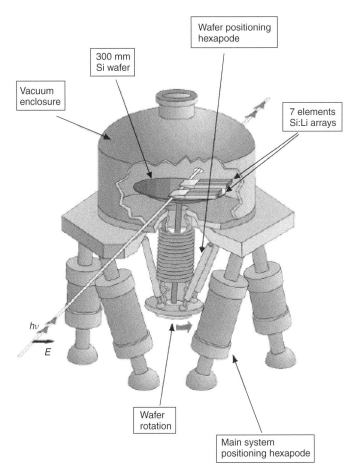

Figure 5.1.1 TXRF measuring chamber at ID27, ESRF (Grenoble, France). (Reprinted from Comin *et al.*,[24] Figure 2, with permission from Elsevier Science)

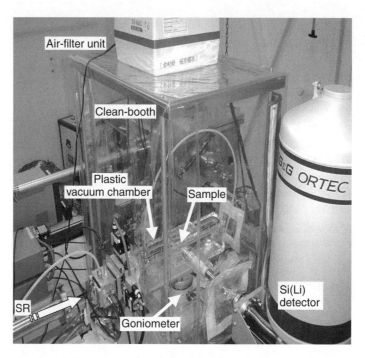

Figure 5.1.2 Clean environment for TXRF experiments at BL39XU, SPring-8 (Harima, Japan)

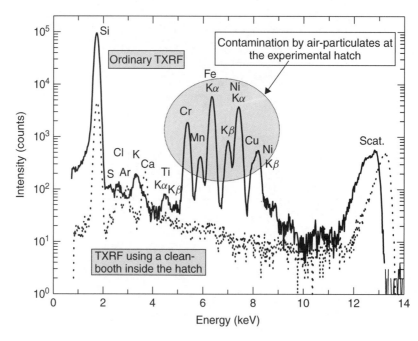

Figure 5.1.3 Influence of parasitic X-rays in TXRF experiments. Spectra of a blank Si wafer measured at BL39XU, SPring-8 (Harima, Japan). Clean instruments are crucial to prevent contamination of the sample caused by air-particulates in the environment at the beamline

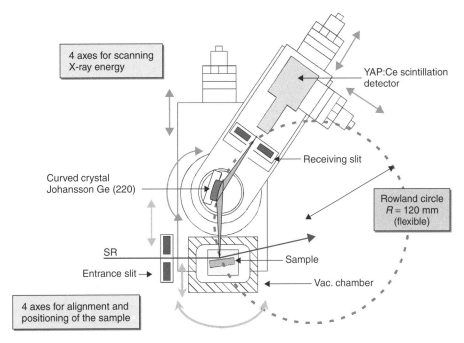

Figure 5.1.4 Schematic view of the Johansson TXRF spectrometer. (Reprinted with permission from Sakurai et al.,[29] Figure 1. Copyright (2002) American Chemical Society)

One of the biggest challenges overcome very recently in terms of instruments is the development of an efficient wavelength-dispersive spectrometer for TXRF applications.[29] The employment of a compact Johansson-type spectrometer (Figure 5.1.4) rather than a conventional Si(Li) detector, as well as the use of a quasi-monochromatic undulator X-ray source, can completely change the quality of X-ray fluorescence spectra. Typical spectra are shown in Figure 5.1.5. The energy resolution becomes 20 times better, which effectively contributes to reducing the low-energy tail of the scattering background and to separating neighboring X-ray fluorescence peaks. One can note that even some chemical effects have become visible in Kβ spectra. Another advantage of this wavelength-dispersive system is its capability with respect to high-counting-rate measurements, which makes possible the detection of weak signals from trace materials. The absolute and relative detection limit for nickel are 0.31 fg and 3.1 ppt for a 0.1 μl droplet of pure water, respectively, which is nearly 50 times better than the current best data achieved by conventional energy-dispersive TXRF using a Si(Li) detector system.

Another significant direction of modern TXRF techniques is light element analysis. The use of SR is again significant for this application. Figure 5.1.6 shows typical TXRF spectra for Na, Mg and Al on a Si wafer, obtained at BL III-4 as well as at BLIII-3, SSRL, Stanford.[30,31] The detectors used are HPGe and Si(Li) with an ultra thin (~200 nm) polymer window, which provides very low attenuation for the entering X-rays. However, Raman scattering causes background enhancement as indicated in the figure. The detection limit was around 0.1 pg, even for Na and Mg as listed in Table 5.1.2. Even B was successfully detected by using the top layer of carbon of the multilayer as a reflector. The activities are now moving to the undulator beamline at BESSY2, in Berlin, leading to further advanced results; detection limits for C and N were 0.5 and 0.8 pg, respectively.[32] A further challenge comes with the introduction of high-resolution detectors, such as STJ and other cryogenic detectors. Recently, Beckhoff has investigated the technical details for obtaining excellent X-ray spectra in the low energy region.[33]

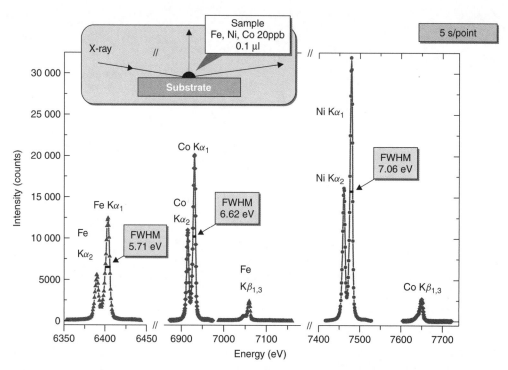

Figure 5.1.5 Wavelength-dispersive TXRF spectra for trace elements (Ni, Co, and Fe, 20 ppb each) in a 0.1 µl drop. (Reprinted with permission from Sakurai et al.,[29] Figure 2. Copyright (2002) American Chemical Society)

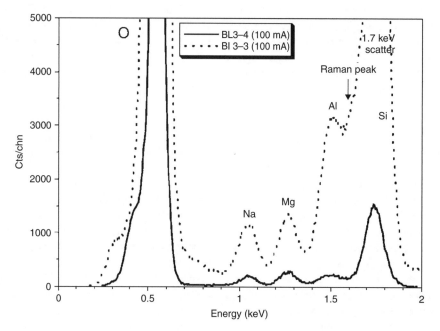

Figure 5.1.6 TXRF spectra of 50 pg Na, Mg and Al as a droplet on a Si wafer. Solid and dashed lines were obtained at BLIII-4 using filtered white radiation by a Si filter and at BLIII-3 with a multilayer monochromator, respectively. The spectra were normalized to 100 mA and 100 s counting time. (Reprinted from Streli et al.,[31] Figure 8, with permission from Elsevier Science)

Table 5.1.2 Sensitivities (S) and extrapolated detection limits (LLD) for low-Z elements performed at beamlines III-4 and III-3, SSRL (Stanford, USA)[a]

	BL III-4		BL III-3	
	S (cps/ng)	LLD (pg)	S (cps/ng)	LLD (pg)
Na	591	0.541	4769	0.127
Mg	1045	0.277	4740	0.189
Al	614	0.815	11989	0.081

[a]Beamline III-4, filtered white radiation by a Si filter; beamline III-3, equipped with a multilayer monochromator. The data are normalized to 100 mA beam current.
Reproduced with the permission of Streli et al.[31], Table 2.

5.1.3 SURFACE AND INTERFACE ANALYSIS OF LAYERED MATERIALS

Modern grazing-incidence XRS often uses the angular dependence of X-ray fluorescence intensity to explore the depth/height distribution of the elements, besides determining the average concentration. The angular dependence of X-ray fluorescence intensity from the element, $I_f(\theta)$ is given using the depth profile of the element $C(z)$ and the X-ray intensity distribution in the sample $I(\theta, z)$ as follows:

$$I_f(\theta) \propto \int I(\theta, z) \times C(z) \mathrm{d}z \quad (5.1.1)$$

This indicates that the unknown $C(z)$ can be basically calculated back from Equation (5.1.1) using both the experimentally observed $I_f(\theta)$ and the theoretically given $I(\theta, z)$. After some pioneering depth profiling work in the 1980s,[34,35] a lot of research has addressed this problem.

If the sample is just a uniform substrate, the situation is rather simple. When primary X-rays impinge on the surface at grazing incidence, reflection and refraction take place simultaneously. Refracted X-rays propagate as an evanescent wave,[36] and the intensity at depth z ($z > 0$) can be expressed as follows using a glancing angle θ, X-ray wavelength λ, and real and imaginary parts of the refractive index δ and β:

$$I(\theta, z) = S(\theta) \times \exp\left(-\frac{z}{D(\theta)}\right) \quad (5.1.2)$$

where $S(\theta) = \dfrac{4\theta^2}{(\theta + A)^2 + B^2}$

$$D(\theta) = \frac{\lambda}{4\pi B^2}$$

$$A = \sqrt{\frac{\sqrt{(\theta^2 - 2\delta)^2 + 4\beta^2} + (\theta^2 - 2\delta)}{2}}$$

$$B = \frac{\beta}{A}$$

Furthermore, during the total reflection, the primary and reflected X-rays are interfered with above the surface, and therefore a modulation of the X-ray intensity is caused at the same time. The X-ray intensity at distance z ($z < 0$) from the surface is given as follows:

$$\begin{aligned}
I(\theta, z) &= \frac{S(\theta)}{2\theta^2}\{\theta^2 + A^2 + B^2 + (\theta^2 - A^2 - B^2) \\
&\quad \times \cos[\tau(\theta)] + B\theta \sin[\tau(\theta)]\} \\
&= \frac{S(\theta)}{2\theta^2}\{\theta^2 + \sqrt{(\theta^2 - 2\delta)^2 + 4\beta^2} \\
&\quad + [\theta^2 - \sqrt{(\theta^2 - 2\delta)^2 + 4\beta^2}]\cos[\tau(\theta)] \\
&\quad + B\theta \sin[\tau(\theta)]\}
\end{aligned} \quad (5.1.3)$$

where

$$\tau(\theta) = \frac{4\pi\theta|z|}{\lambda}$$

That is, a standing wave is formed above the surface, and the period is $2\theta/\lambda$. The first surface study using the standing wave phenomena during the total reflection was reported by Bedzyk et al.[37,38] This is no doubt an extremely significant amount of research on surface physics, but one should note that the effects are often observed even in ordinary TXRF measurements.[39–42]

On the other hand, in the case of thin films, Equation (5.1.2) is not valid any more, due to the strong modulation of the X-ray electric field in a film caused by multiple reflection of incident X-rays at each interface. Assume that depth z is in the nth layer from the top of the surface. Following Parratt's very famous paper,[43] it is possible to calculate the reflection coefficient $R_{n,n+1}$ at the interface between the nth and $(n+1)$th layer and the electric field E_n at the center of the nth layer of which the thickness is d_n. Then:

$$I(\theta, z) = |a_{n-1} E_n (b_n + b_{n-1} R_{n,n+1})|^2 \quad (5.1.4)$$

where

$$a_n = \exp\left(-i\frac{kf_n d_n}{2}\right)$$

$$b_n = \exp[-ikf_n(z - \Sigma d_j)]$$

$$f_n = \sqrt{(\theta^2 - 2\delta_n) - i(2\beta_n)}$$

$$k = \frac{2\pi}{\lambda}$$

Figure 5.1.7 shows an example of the calculated internal X-ray electric field expressed as a function of both depth and glancing angle.[44] The sample assumed is a Cu[100 Å]/Ag[230 Å]/Au[500 Å]/Si thin film, and iron[3 Å], chromium[6 Å] and titanium[18 Å] are put at the surface and each interface, respectively. Interference oscillation is understood visually, and one can see that the intensity at 7 mrad is highest at the surface and exponentially decreases. This angle is therefore still in the evanescent wave region because of the shallow penetration. However, at 8 mrad, the distribution begins to change because of the interference effect. The oscillation becomes clear at 9 mrad and further changes at 10 mrad. In these ways, the interference effect is expected to cause oscillation of intensity at the surface and interfaces as the

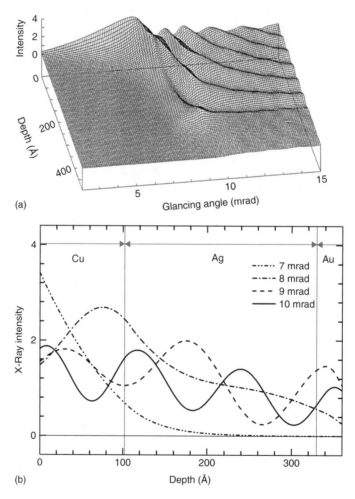

Figure 5.1.7 Calculated internal X-ray intensity distribution in Cu/Ag/Au thin film. (a) Three-dimensional representation. (b) Depth profile for 7, 8, 9 and 10 mrad incidence. (Reprinted from Sakurai and Iida,[44] Figures 5 and 6, with permission from Kluwer Academic/Plenum publisher)

angle increases. The corresponding experimental data are shown in Figure 5.1.8. One can see that the integrated fluorescent signal intensity peaks at a different angle with oscillation. Furthermore, when a certain interface is enhanced, the X-ray intensity at the neighboring interface is decreased. When the chromium signal reaches maximum, iron becomes weak, and as the titanium peaks, chromium becomes weak. Grazing-incidence XRS is obviously surface-sensitive at a low angle, but the technique can be extended further to explore interfaces by carefully tuning the glancing angle so as to obtain maximum enhancement.

When the thin film is a periodic multilayer, modern grazing-incidence XRS can analyze the depth position of trace impurities by using a standing wave generated around the Bragg diffraction condition. The technique has been often used for the determination of the atomic position of impurities in the crystal as well as of adsorbed molecules on the surface.[45,46] This method is based on dynamical diffraction theory, and therefore the main applications are performed for perfect crystals, but experiments are also possible for multilayers.[47]

Figure 5.1.9 shows an angular profile of the specular reflection, as well as X-ray fluorescence from the trace metal (Fe) for a Ni/C multilayer ($2d = 9.76$ nm), which is a promising optical device for soft and hard X-rays. A discontinuous change of Fe Kα intensity is apparent at around the Bragg peaks (16.5, 31.9 mrad). As shown in Figure 5.1.10, a comparison with the calculation based on the dynamical diffraction clarifies that the Fe impurity is almost uniformly distributed in the Ni layers, and is not present in the C layers. The result is quite interesting when considering the origin of the trace impurities in the fabrication process. In this way, the X-ray standing wave technique is used for analyzing the depth position of trace metals within 1 periodic unit of the multilayers.

Another important trend in modern grazing-incidence XRS is combined analysis of fluorescent X-rays and X-ray reflections.[48] In short, X-ray reflectometry is a $\theta/2\theta$ scan in a very low angle region, and not such a difficult experiment, but it can provide plenty of information on the surface and layered structures, i.e., layer thickness, surface and interface roughness, density of the near-surface

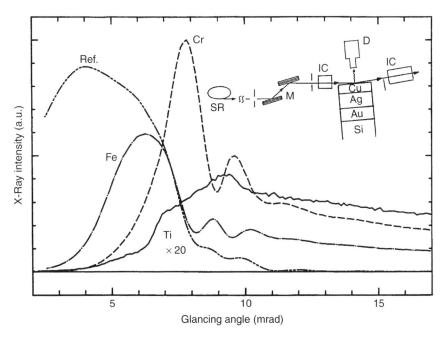

Figure 5.1.8 Experimental angular plot of reflectivity and integrated X-ray fluorescence intensities of iron, chromium and titanium from the surface and interfaces of a Cu/Ag/Au thin film. (Reproduced with permission of Plenum Press,[44] Figure 4)

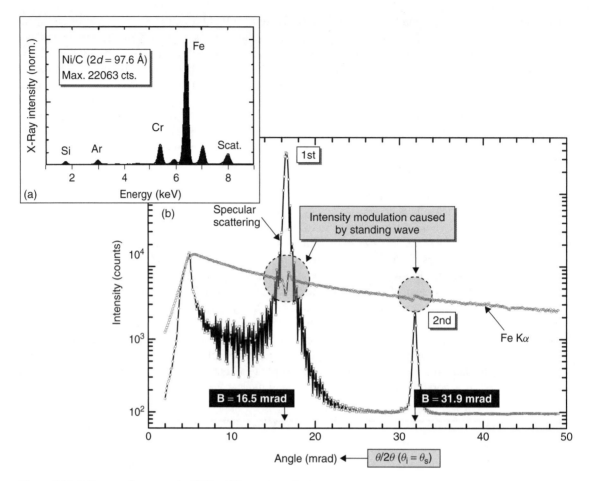

Figure 5.1.9 X-Ray standing wave for Ni/C multilayer ($2d = 9.76$ nm). (a) X-Ray fluorescence spectra. (b) Angular profile of X-ray reflectivity and iron Kα fluorescence

region, density profiles along the depth, and details of the periodic and non-periodic multilayer structures. A very important feature of the X-ray reflectivity technique is that it is not very sensitive to crystal structure, dislocations and defects, so it can be used for probing single crystals, polycrystalline samples, amorphous materials, and even liquid samples. This is because the signal depends only on the electron density of the studied material. However, this also has a negative aspect, namely the technique cannot distinguish between interface roughness and interface grading (i.e. diffusion, implantation, etc.), which both give almost the same influence on the X-ray reflectivity curve. Combining the technique with grazing-incidence XRS could help in the understanding of such problems. Further detailed analysis comes with the extension to the combination with non-specular reflections (diffuse scattering),[49,50] which are weak scattering observed around the strong specular reflection spot, and also grazing-incidence small angle X-ray scattering (GISAXS).[51,52] In the early 1990s, de Boer and his co-workers contributed successive systematic publications on both the detailed theoretical formulation and experimental applications of the angular dependence of X-ray fluorescence as well as specular and non-specular reflections.[53–56] One of their most significant works is the study on the influence of the interface roughness on X-ray fluorescence intensity from

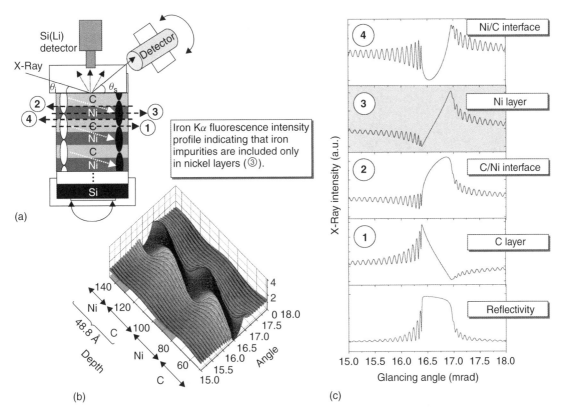

Figure 5.1.10 Calculation of angular dependence of X-ray fluorescence. (a) Basic idea of the X-ray standing wave technique. (b) Three-dimensional map for X-ray electric field intensity in the multilayer. (c) Angular profile for X-ray fluorescence for impurity iron segregated at: (1) center of the C layer; (2) C/Ni interface; (3) center of the Ni layer; (4) Ni/C interface

layered materials.[57] The theoretical work employs distorted-wave Born approximation (DWBA) up to the second-order, and X-ray fluorescence intensity has been expressed using morphological parameters such as interface rms roughness, lateral correlation length, jaggedness parameter, and the perpendicular correlation length. For a Gaussian distribution of interface heights, the refractive-index profile is an error function, with the result that the Helmholtz equation cannot be exactly solved. The breakthrough has come with finding a suitable approximation (see Figure 5.1.11) instead of trying to introduce a profile that can be solved exactly (like a tangent hyperbolicus). The proper selection of the starting point is significant here, and it was found that the use of graded interfaces (i.e. the roughness is modelled with a pack of smooth slices) allows a correct modelling for calculation of the X-ray fluorescence. The influence of the second order of DWBA is also an important problem for cases where the lateral correlation length is fairly large.

5.1.4 MICRO X-RAY FLUORESCENCE IMAGING

Recently, microscopic imaging has been performed by grazing-incidence XRS, which usually measures spatially average information. So far, imaging of X-ray fluorescence has been based on step-scans with a collimated beam (∼μm, or smaller if 3rd generation SR is available). However, it requires long measuring time, especially when pixel numbers increase to enhance the quality of the image. A novel approach toward

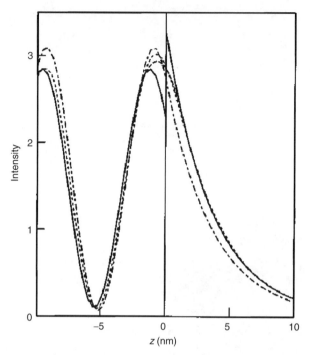

Figure 5.1.11 Possible errors in the calculation of X-ray electric field intensity distribution along the depth. The difference is caused when considering the influence of roughness. X-Ray intensity vs depth for Cu Kα radiation with a perpendicular wave vector $k = 0.375 \, \text{nm}^{-1}$ on a gold sample with a rms roughness of 1.5 nm. Dash-dotted line, no roughness; solid line, calculated using conventional Nevot–Croce factors; long-dashed line, the present approximation; short-dashed line, calculated for error–function profile using the slice method. (Reproduced with permission of de Boer,[57] Figure 2)

much more rapid X-ray fluorescence imaging comes with a combination of grazing-incidence geometry (~2°) using a rather wide beam (~cm) and parallel optics for detecting X-rays by a two-dimensional detector.[58,59] A combination of a charge-coupled device (CCD) camera and collimators can be used for micro X-ray fluorescence imaging of μm scale resolution. Figure 5.1.12 shows typical examples of X-ray fluorescence imaging obtained with a normal bending magnet synchrotron source (BL-4A, Photon Factory, Tsukuba, Japan). Although the spatial resolution is only around 20 μm, it should be noted that imaging with approximately 1M-pixels can be performed in only 1–2 min, or even less. One can see detailed patterns of the precipitation of metallic crystals, aggregation to the specific part in the tissue, and segregation at the mineral interfaces.[60] Although it is possible to perform the experiments with a laboratory X-ray source, the use of tunable monochromatic or quasi-monochromatic synchrotron X-rays is promising with respect to the selective excitation of the elements contained in the specimen. A further advantage would be the availability of more specific imaging, in addition to information on normal chemical composition, like chemical states and local structure, by making use of the X-ray absorption fine structure.[61]

5.1.5 FUTURE OUTLOOK

Recent trends in grazing-incidence XRS have been overviewed. In spite of the century-long history of X-rays, this field seems to be still growing rapidly. New X-ray synchrotron sources, such as X-ray free electron laser (XFEL) based on SASE (Self Amplified Spontaneous Emission),[62] and ERL (Energy Recovery Linac),[63] are now in the design stage, and will commence initial operation no later than 2010. Those sources are like

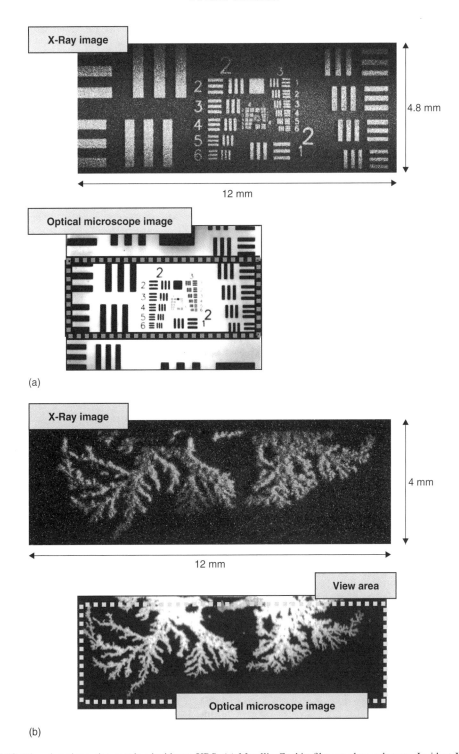

Figure 5.1.12 Micro imaging using grazing incidence XRS. (a) Metallic Cr thin film on glass substrate. Incident X-ray energy 7.2 keV. Exposure time 50 s. (b) Metallic Ag crystals precipitated from salt solution. Incident X-ray energy 7.3 keV. Exposure time 10 min

a laser in the visible light, and have a further high peak as well as average brilliance with an extremely short pulse structure. Future grazing-incidence XRS will use the coherence and the pulse structure of the source. The technique will not remain as a one or two-dimensional probe, at least three-, or maybe four- (including time-axis) dimensional experiments will become performed. The TXRF technique will be widely used for many industrial and environmental applications because of its extremely high detection power. New chemistry in ppt or in ppq range will be opened up. It will be crucial to perform the experiments with ultra clean instruments in a dust-free environment. Another important direction of future grazing-incidence XRS is more advanced surface analysis, by combining several grazing-incidence X-ray techniques and other microscopic probes for surface phenomena. Its capability with respect to exploring buried nanometer-scale structures will no doubt assume a very important role in the future development of nanotechnologies.

REFERENCES

1. A. H. Compton, *Philos.Mag.* **45** (1923) 1121.
2. Y. Yoneda and T. Horiuchi, *Rev. Sci. Instrum.* **42** (1971) 1069.
3. P. Wobrauschek and H. Aiginger, *Anal. Chem.* **47** (1975) 852.
4. A. Iida and Y. Gohshi, *Jpn. J. Appl. Phys.* **23** (1984) 1543.
5. H. Schwenke and J. Knoth, Total reflection XRF, in *Handbook of X-Ray Spectrometry*, edited by R. E. van Grieken and A. A. Markowicz, Marcel Dekker, New York, 1993, Chapter 9, p. 464.
6. R. Klockenkämper, *Total-Reflection X-Ray Fluorescence Analysis*, John Wiley & Sons, New York, 1997.
7. *Spectrochim. Acta* **B44** (1989).
8. *Spectrochim. Acta* **B46** (1991).
9. *Spectrochim. Acta* **B48** (1993).
10. *Anal. Sci. (Japan)* **11** (1995); *Adv. X-ray Chem. Anal. (Japan)*, **26s** (1995).
11. *Spectrochim. Acta* **B52** (1997).
12. *Spectrochim. Acta* **B54** (1999).
13. *Spectrochim. Acta* **B56** (2001).
14. C. Streli, *X-Ray Spectrom.* **29** (2000) 203.
15. J. Knoth, H. Schneider and H. Schwenke, *X-Ray Spectrom.* **23** (1994) 261.
16. M. A. Kumakhov, *X-Ray Spectrom.* **29** (2000) 343.
17. J. X. Ho, E. H. Snell, C. R. Sisk, J. R. Ruble, D. C. Carter, S. M. Owens and W. M. Gibson, *Acta Cryst.* **D54** (1998) 200.
18. K. Sakurai, A. Iida and Y. Gohshi, *Anal. Sci.* **4** (1988) 3.
19. A. Iida, A. Yoshinaga, K. Sakurai and Y. Gohshi, *Anal. Chem.* **58** (1986) 394.
20. P. Wobrauschek, R. Görgl, P. Kregsamer, C. Streli, S. Pahlke, L. Fabry, M. Haller, A. Knöchel and M. Radtke, *Spectrochim. Acta* **B52** (1997) 901.
21. S. Brennan, W. Tompkins, N. Takaura, P. Pianetta, S. S. Laderman, A. Fischercolbrie, J. B. Kortright, M. C. Madden and D. C. Wherry, *Nucl. Instrum. Methods A* **347** (1994) 417.
22. P. Pianetta, K. Baur, A. Singh, S. Brennan, J. Kerner, D. Werho and J. Wang, *Thin Solid Films* **373** (2000) 222.
23. D. Mills, *3rd Generation Hard X-Ray Synchrotron Radiation Sources: Source Properties, Optics, and Experimental Techniques*, John Wiley & Sons, Ltd, London 2002.
24. F. Comin, M. Navizet, P. Mangiagalli and G. Apostolo, *Nucl. Instrum. Methods B* **150** (1999) 538.
25. G. Apostolo, R. Barrett, M. Robichon, M. Navizet and F. Comin, *ESRF Highlights* (2000) 92.
26. http://www.ritsumei.ac.jp/se/d11/index-e.html
27. K. Sakurai, S. Uehara and S. Goto, *J. Synchrotron Rad.* **5** (1998) 554.
28. K. Sakurai, H. Eba and S. Goto, *Jpn. J. Appl. Phys.* **Suppl 38-1** (1999) 332.
29. K. Sakurai, H. Eba, K. Inoue and N. Yagi, *Anal. Chem.* **74** (2002) 4532.
30. C. Streli, *J. Trace Microprobe Tech.* **13** (1995) 109.
31. C. Streli, P. Wobrauschek, P. Kregsamer, G. Pepponi, P. Pianetta, S. Pahlke and L. Fabry, *Spectrochim. Acta B* **56** (2001) 2085.
32. C. Streli, P. Wobrauschek, B. Beckhoff, G. Ulm, L. Fabry and S. Pahlke, *X-Ray Spectrom.* **30** (2001) 24.
33. B. Beckhoff, R. Fliegauf and G. Ulm, *Spectrochim. Acta* **B58** (2003) 615.
34. J. M. Bloch, M. Sansone, F. Rondelez, D. G. Peiffer, P. Pincus, M. W. Kim and P. M. Eisenberger, *Phys. Rev. Lett.* **54** (1985) 1039.
35. A. Iida, K. Sakurai, A. Yoshinaga and Y. Gohshi, *Nucl. Instrum. Methods A* **246** (1986) 736.
36. R. S. Becker, J. A. Golovchenko and J. R. Patel, *Phys. Rev. Lett.* **50** (1983) 153.
37. M. J. Bedzyk, D. H. Bilderback, G. M. Bonmmarino, M. Caffrey and J. S. Schidkraut, *Science* **241** (1988) 1788.
38. M. J. Bedzyk, G. M. Bonmmarino and J. S. Schidkraut, *Phys. Rev. Lett.* **62** (1989) 1376.
39. A. Krol, C. J. Sher and Y. H. Kao, *Phys. Rev. B* **38** (1988) 8579
40. W. B. Yun and J. M. Bloch, *J. Appl. Phys.* **68** (1990) 1421.
41. A. Iida, *Adv. X-Ray Anal.* **35** (1992) 795.
42. R. Klockenkämper, J. Knoth, A. Prange and H. Schwenke, *Anal. Chem.* **64** (1992) 1115A.
43. L. G. Parratt, *Phys. Rev.* **95** (1954) 359.
44. K. Sakurai and A. Iida, *Adv. X-Ray Anal.* **39** (1997) 695.

45. B. W. Batterman, *Phys. Rev. Lett.* **22** (1969) 703.
46. P. L. Cowan, J. A. Golovchenko and M. F. Robbins, *Phys. Rev. Lett.* **44** (1980) 1680.
47. T. W. Barbee Jr and W. K. Warburton, *Mater. Lett.* **3** (1984) 17.
48. K. N. Stoev and K. Sakurai, *Spectrochim. Acta* **B54** (1999) 41.
49. Y. Yoneda, *Phys. Rev.* **113** (1963) 2010.
50. S. K. Sinha, E. B. Sirota, S. Garoff and H. B. Stanley, *Phys. Rev. B* **38** (1988) 2297.
51. J. R. Levine, J. B. Cohen, Y. W. Chung and P. Georgopoulos, *J. Appl. Cryst.* **22** (1989) 528.
52. A. Naudon and D. Thiaudiere, *Surf. Coat. Technol.* **79** (1996) 103.
53. D. K. G. de Boer, *Phys. Rev. B* **44** (1991) 498.
54. D. K. G. de Boer, *Phys. Rev. B* **49** (1994) 5817.
55. D. K. G. de Boer, *Phys. Rev. B* **51** (1995) 5297.
56. D. K. G. de Boer, A. Leenaers and W. van den Hoogenhof, *X-Ray Spectrom.* **24** (1995) 91.
57. D. K. G. de Boer, *Phys. Rev. B* **53** (1996) 6048.
58. K. Sakurai, *Spectrochim. Acta B* **54** (1999) 1497.
59. K. Sakurai and H. Eba, Japanese Patent No. 3049313 (2000).
60. K. Sakurai and H. Eba, *Anal. Chem.* **75** (2003) 355.
61. M. Mizusawa and K. Sakurai, *J. Synchrotron Rad.* (submitted).
62. B. Sonntag, *Nucl. Instrum. Methods A* **467–468** (2001) 8.
63. G. N. Kulipanov, A. N. Skrinsky and N. A. Vinokurov, *Nucl. Instrum. Methods A* **467–468** (2001) 16.

5.2 Grazing-exit X-ray Spectrometry

K. TSUJI
Osaka City University, Osaka, Japan

5.2.1 INTRODUCTION

Grazing-exit X-ray spectrometry (GE-XRS) is a method related to total reflection X-ray fluorescence (TXRF) (Figure 5.2.1a). In TXRF, the primary X-rays irradiate the sample surface at grazing angles of incidence. In contrast, in GE-XRS (Figure 5.2.1(b)), characteristic X-rays are measured at grazing-exit angles, usually less than 1°. Becker et al.[1] demonstrated the equivalence of grazing incidence and grazing exit X-ray measurements according to microscopic reversibility and reciprocity, indicating that GE-XRS can be applied to surface and thin-film analyses with low background intensity, in a manner similar to grazing-incidence X-ray spectrometry (GI-XRS) and TXRF.

Compared to GI-XRS, GE-XRS has unique advantages. In GE-XRS, different types of excitation probes can be used, not only X-rays but also electrons and charged particles. In addition, the probes can be used to irradiate the sample at right angles. This experimental geometry enables a localized analysis depending on the diameter of the probe. This subchapter describes the principles, methodological characteristics, GE-XRS instrumentation, and recent applications to X-ray fluorescence (XRF), electron probe microanalysis (EPMA), and particle induced X-ray emission (PIXE). At the end of this subchapter, the characteristics of GE-XRS and its future are discussed.

5.2.2 PRINCIPLES OF GRAZING-EXIT X-RAY SPECTROMETRY

5.2.2.1 REFRACTION OF X-RAYS

According to the Fresnel relations, the X-rays emitted from inside atoms are refracted on the surface, as shown in Figure 5.2.2. Since the refractive index of X-rays is slightly less than 1, the X-rays are refracted to the larger refraction angle (θ_{exit}) than the incident angle (θ_{in}). The refracted X-rays are obstructed by the edge of the slit, which is placed between the sample and the X-ray detector. Consequently, only the X-rays emitted from the surface region are detected through the slit. This refraction effect is important in improving surface sensitivity in GE-XRS geometry.

Principally, the sample for GE-XRS must be flat in a manner similar to that in TXRF. In TXRF, the entire sample surface must be flat; in GE-XRS, however, the requirement for a flat surface is not as severe. GE-XRS can be applied if the sample has a flat region of 1 mm or less, although this requirement depends, of course, on the diameter of the excitation probe. Further influence of surface flatness has been reported elsewhere.[2]

5.2.2.2 X-RAY EMISSION INTENSITIES UNDER GRAZING-EXIT CONDITIONS

Urbach and de Bokx have proposed a formalism for the GE-XRS calculations by applying asymptotics to plane-wave expressions.[3] Here, another

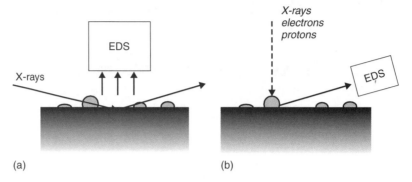

Figure 5.2.1 Experimental arrangements of GI-XRS (a) and GE-XRS (b)

approach using a reciprocity theorem is introduced. The reciprocity theorem in optics is described as '*a point source at P_0 will produce at P the same effect as that of a point source of equal intensity placed at P will produce at P_0*'.[4] It has been confirmed that this reciprocity theorem is also applicable in the X-ray region.[1] Thus, the intensity of X-ray fluorescence under grazing-exit conditions can be calculated in the same way as the calculations of the GI-XRS intensities, indicating that we can utilize the calculation formula for TXRF intensity after minor modification. Using a multi-layered model, GE-XRS intensities ($I_{exit}(\theta_{exit})$) are expressed as a function of exit angle (θ_{exit}), as follows:

$$I_{exit}(\theta_{exit}) \propto \int_0^{d_j} I_p(z) |E_j(\theta_{exit}, z)|^2 \, dz \quad (5.2.1)$$

where z is the depth in the jth layer, d_j is the thickness of the jth layer, and $E_j(\theta, z)$ is the electric field that is produced at depth z by assuming that the X-rays, which have the same energy as the detecting X-ray fluorescence, irradiated the sample surface at the grazing angle, which corresponds to the detecting exit-angle. The calculation procedure of $E_j(\theta, z)$ is described elsewhere.[5] $I_p(z)$ is the X-ray intensity emitted from depth z, and depends on the type of excitation probes (X-rays,[6] electrons[7] and charged particles).

It is possible to evaluate the information depth using Equation (5.2.1). Here, the information depth is defined as the depth where the intensity is reduced to be $1/e$ of the X-ray (fluorescence) intensity in the GI-XRS geometry when applying the reciprocity theorem. Figure 5.2.3 shows the information depth of Si Kα for the Si wafer as a function of the exit angle. It is clearly shown that the information depth is only a few nm below the critical angle, which is given approximately by the following equation:

$$\theta_c(\text{deg}) \approx \sqrt{2\delta} \approx 1.33\lambda(\text{nm})\sqrt{\frac{Z\rho}{A}} \quad (5.2.2)$$

Here, λ is wavelength of X-rays, Z is the atomic number, A is the atomic weight, and ρ is the density of the sample (g/cm^3). This equation is originally given for the critical angle for total reflection. However, the critical angle for GE-XRS can also be evaluated from this equation by applying the wavelength (λ) of the detecting X-ray fluorescence. Figure 5.2.3 suggests that surface analysis is possible at grazing-exit angles less than this critical angle. However, we must note that the critical angle in GE-XRS depends on the wavelength of the fluorescence of the observed X-ray.

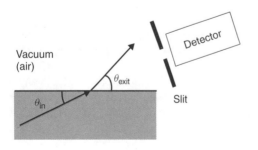

Figure 5.2.2 Refraction of X-rays on the surface

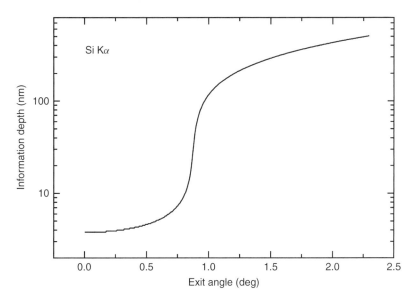

Figure 5.2.3 Information depth of Si Kα emitted from a Si wafer as a function of exit angle

5.2.2.3 GRAZING-EXIT X-RAY SPECTROMETRY INSTRUMENTATION

Commercially available GE-XRS apparatus is not common except for the Laboratory Grazing-Emission X-ray Fluorescence Spectrometer.[8] However, it is not difficult to construct GE-XRS apparatus, which consists of an excitation source, sample holder, slit, and X-ray detector. We can utilize a variety of excitation probes, such as X-rays, electrons, and charged particles. Thus, GE-XRS has been applied to XRF, EPMA (scanning electron microscopy–energy-dispersive X-ray SEM-EDX), and PIXE. In any case, an exit slit system is necessary to restrict X-ray emissions to a specific exit angle, because the X-ray intensity strongly depends on the exit angle. Since the detected X-ray intensity is weakened due to use of the slit, a high-power excitation source is desirable. It is possible to use both the EDX detector and the WDX (wavelength-dispersive X-ray). In many cases, the EDX is used because of its advantages, such as compact size and simultaneous detection of many elements. However, the combination of GE-XRS and the WDX detector is methodologically quite reasonable,[8] and is especially useful for the analysis of low Z elements.

In GE-XRS, the control of exit angle is very important, requiring an accuracy of approximately 0.01°. This angle control can easily be performed by using a stepping motor driven stage. Two types of experimental arrangement are available to change the exit angle. In the first setup, the sample stage is tilted to change the exit angle.[9] The characteristic X-rays are measured by a fixed X-ray detector. This method is easily applied to commercially available apparatus, such as EPMA or SEM-EDX. However, if the analyzing position is not set exactly on the center of rotation (tilt), it moves as the sample is tilted. In the case of localized analysis (particle analysis), this is a severe problem.[9] In the second setup, the sample is fixed and the X-ray detector is moved to change the exit angle. In this case, the analyzing position and excitation conditions (incident angles of excitation probe) are stable even if the exit angle is changed. Therefore, the latter experimental arrangement is suitable for localized analysis.

5.2.3 GRAZING-EXIT X-RAY FLUORESCENCE (GE-XRF)

Grazing-exit X-ray fluorescence was studied under the grazing incidence of primary X-rays.[10] It is shown that the method of grazing-incidence and grazing-exit XRF analysis is quite surface-sensitive.[11] In the GE-XRF performed by Noma et al., synchrotron radiation irradiated thin-film (Cr/Au/Cr/Si wafer) samples at a right angle, and the X-ray fluorescence was measured at grazing-exit angles.[12] As shown in Figure 5.2.4, an oscillation structure, caused by the interference of emitted X-rays, was observed in the exit-angle dependent curve of the XRF intensity. By fitting the experimental curve with the theoretical curve, the thickness and roughness at the interfaces were evaluated. They also measured X-ray diffraction under grazing-exit conditions.[13] Sasaki et al. applied a GE-XRF method to the structure analysis of proteins[14] and organic films[15] using synchrotron radiation.

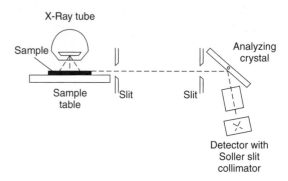

Figure 5.2.5 Schematic diagram of a GE-XRF instrument using a WDX detector.[8] Reproduced by permission of American Institute of Physics

X-Ray fluorescence at grazing-exit angles using a WDX detector is called 'Grazing Emission XEF (GE-XRF)',[8,16,17] Figure 5.2.5 is a schematic view of the GE-XRF setup. X-Rays from an X-ray tube directly irradiated the sample in a large area. X-Ray emissions from the sample were collimated by a double slit system and detected by the WDX detector. This method is especially useful for the trace analysis of low-Z elements.[18–21] The detection limits of Si in several types of organic matrices (water, beer, urine, etc.) were at the ppb level.[22] The detection limits of trace metals (Na–Sr) in mineral water were determined in an intercomparison survey by several analytical methods.[23] As shown in Table 5.2.1, GE-XRF with the WDX detector covered the weaknesses of TXRF, that is, the determination of low-Z elements.

Pérez and Sánchez also developed a GE-XRF setup using the EDX detector.[24] They carefully collimated incident and exit X-rays in the same area. In their setup, a minimum control of an exit angle of 0.03 mrad was achieved.

Micro-X-ray analysis is now one of the trends in X-ray analysis. Grazing-exit X-ray spectrometry was applied to micro-XRF using a synchrotron X-ray microbeam.[25] Micro-X-ray analysis can be performed in laboratories by using X-ray capillary optics. Two-dimensional scanning of the sample is used to obtain X-ray elemental mapping. Additionally, depth information is obtained by applying GE-XRS. Finally, three-dimensional X-ray analysis is performed.[26]

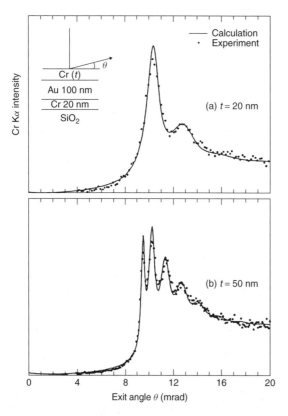

Figure 5.2.4 Angular dependence of Cr Kα fluorescence intensity emitted from Cr (20, 50 nm)/Au (100 nm)/Cr (20 nm) layered structure.[12] Reprinted from Noma, T., Iida, A. and Sakurai, K. *Phys. Rev. B* **48** (1993) 17524. Copyright (1993) by the American Physical Society

Table 5.2.1 Detection limits for the elements determined in the intercomparison survey by given techniques[23]

Technique	Detection limits (μg/l)							
	Na	Mg	K	Ca	Ni	Cu	Zn	Sr
Titration	–	–	–	700	–	–	–	–
Flame photometry	–	–	500	–	–	–	–	–
FAAS	3	2–3	3–12	3–22	4–12	1–4	1–10	17
GFAAS	–	–	–	–	0.1	0.1	–	–
ICP-AES	0.1	0.3–50	20–500	0.3–30	25	5	5	1
ICP-MS	–	7	–	20	0.1–3	0.01–3	0.08–1	0.04
TXRF	–	–	–	–	2	2	2	–
TXRF	–	–	100	50	4	3.5	3.5	4.5
GEXRF	40	3	5	20	–	–	–	–

FAAS, flame atomic absorption spectrometry; GFAAS, graphite furnace atomic absorption spectrometry; ICP-AES, inductively coupled plasma-atomic emission spectrometry; TXRF, total reflection X-ray fluorescence; GEXRF, grazing emission X-ray fluorescence.

5.2.4 GRAZING-EXIT ELECTRON PROBE MICROANALYSIS (GE-EPMA)

Hasegawa et al. measured characteristic X-rays at small take-off angles during RHEED (reflection high energy electron diffraction) experiments[27] and demonstrated that surface sensitivity was enhanced. Usui et al. applied the same technique to the analysis of super-conductive films of YBCO using SEM.[28] In both cases, the electron beam was irradiated at glancing incident angles; therefore, localized analysis was difficult and not proposed.

5.2.4.1 GRAZING-EXIT ELECTRON PROBE MICROANALYSIS SETUP

In general, GE-EPMA can be performed using commercially available EPMA (or SEM-EDX) apparatus. However, since the exit angle has to be precisely controlled, some improvements of the equipment are necessary. As described previously, there are two methods to change the exit angle. An EPMA apparatus has a function of sample inclination. Thus, by using this function, GE-EPMA measurement can be performed in the first setup just after a simple slit is placed between the sample and the X-ray detector[29] Figure 5.2.6 shows an example of the second type of experimental setup.[30] The SEM and the EDX detector were combined with a stainless steel flexible tube. The EDX detector was placed on a stepping motor driven stage, controlled by a computer. The minimum

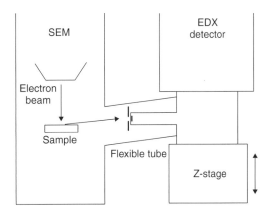

Figure 5.2.6 Schematic diagram of a GE-EPMA instrument.[30] Reproduced by permission of American Institute of Physics

step of this stage was 0.5 μm. A Ta slit (0.2 mm in the width) was attached on the top of the EDX detector at a distance of about 100 mm from the sample. The exit angle for the characteristic X-rays was changed by moving the position of the EDX detector.

Awane et al. changed the exit angle by moving the sample stage up and down, as shown in Figure 5.2.7.[31] Since the EDX detector is usually fixed on the vacuum chamber of the SEM (or EPMA), their method would be easy to apply. Although there is still a problem in that the sample position moves when the exit angle is changed, they have reported interesting results, as described below.

The calibration of the exit angle is necessary before the GE-EPMA measurements are made. For

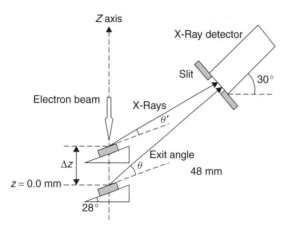

Figure 5.2.7 Experimental arrangement of GE-EPMA.[31] Reprinted from Aware, T. et al. J. Surf. Anal. **9** (2002) 171. Reproduced by permission of The Surface Analysis Society of Japan

this purpose, it is useful to measure the angle dependence of Si $K\alpha$ intensity for a Si wafer, which has a flat surface and a well-known density. This angle-dependent curve can be compared with the theoretical curve calculated by Equation 5.2.1. Finally, we can experimentally determine the exit angle.

5.2.4.2 SURFACE ANALYSIS BY GE-EPMA

Surface analysis can be performed by conventional EPMA using low-energy electrons. The lower the electron energy, the smaller the depth of penetration of the electrons into the sample. However, the selection of the analytical lines of the characteristic X-rays is restricted due to low-energy electron excitation. In some cases, this is a serious problem, especially when an EDX detector having a poor energy resolution is used. The use of GE-EPMA enables surface-sensitive analysis without reducing the electron beam energy.

It is well known that the surface of Fe–Cr metals is covered with a chemically stable oxide layer, which is primarily composed of Cr oxide. Therefore, the chemical composition of the surface layer should be different from the bulk composition. The exit angle dependences of Fe $K\alpha$ and Cr $K\alpha$ were measured and found to be almost constant at exit angles above 10°. In the exit-angle range from 0 to 1.5°, all characteristic X-ray intensities decreased as the exit angles were reduced; however, the Cr $K\alpha$ intensity decreased more slowly than with Fe $K\alpha$. The ratio of Cr $K\alpha$ intensity to that of Fe $K\alpha$ increased significantly at the grazing angle.[9] This indicates that Cr is enriched near the surface. This result agreed well with those obtained by other methods of surface analysis.

Takahashi, in JEOL, observed the surface of a contaminated semiconductor device, which had been touched with a finger, under grazing-exit conditions.[32] At a conventional exit angle of 40°, it was difficult to obtain the X-ray mapping of the contaminating elements (Ca, K, and Cl) due to poor surface sensitivity. However, under grazing-exit conditions, X-ray mapping was clearly obtained. The mapping results for K under conventional and grazing conditions are shown in Figure 5.2.8 with SEM images.

Yamanaka et al. studied the growth process of Ga on the Si surface by high-voltage SEM-EDX.[33] They reported a spatial resolution of 10–20 nm under grazing-exit conditions. Since the X-rays from the substrate are not detected at grazing angles, it is possible to detect a single particle as small as 10–20 nm. Characteristic X-rays are produced within a significantly larger interaction volume than the impinging beam, as shown in Figure 5.2.9.[34] At conventional detection angles (typically $\sim$ 40° for EPMA), the lateral resolution is determined by the dimensions of the interaction volume. In the case of GE-EPMA, however, only the X-rays that are emitted near (few nm) the surface are detected. Therefore, the lateral resolution of GE-EPMA is limited by the cross-section of the most superficial layer of the interaction volume. The lateral interaction volume ($\leftrightarrow$ in Figure 5.2.9) was evaluated for several metals at different electron accelerating voltages by Monte Carlo simulation.[7] The results are shown in Table 5.2.2, indicating that the lateral resolution would be improved considerably under grazing-exit conditions, especially for low-Z elements.

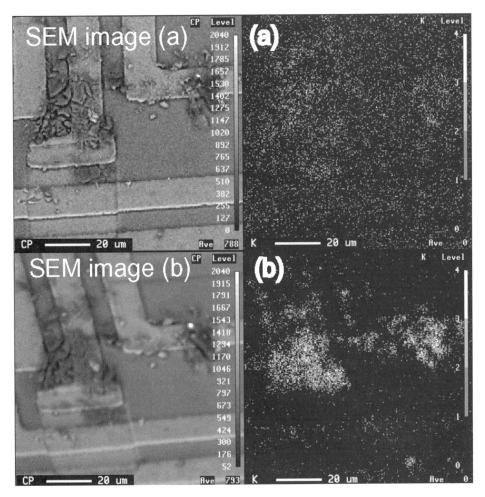

Figure 5.2.8 Comparison of X-ray mappings of K Kα taken by conventional EPMA (a, electron beam current: 2 nA) and GE-EPMA (b, 180 nA).[32] SEM images obtained by both methods are also shown. Reproduced by permission of IOP Publishing Limited

5.2.4.3 PARTICLE ANALYSIS BY GE-EPMA

In the semiconductor device manufacturing process, the detection of contaminants (small particles) on Si wafers is very important, because such particles affect the physical properties and functions of the semiconductor devices. It is not easy to measure the elemental composition of a very small particle by conventional EPMA, because the electron beam easily passes through the particle and produces strong X-rays from the Si substrate, as shown in Figure 5.2.9 Therefore, X-rays from both the particle and the substrate are detected simultaneously. In many cases, it is difficult to distinguish the X-rays emitted from the particle from those emitted from the substrate.

An artificial particle (Fe_2O_3, 1 μm in diameter) was deposited on a Au layer deposited on the Si flat substrate. The electron beam irradiated a single Fe_2O_3 particle. At a large exit-angle of 40°, Au Mα and Si Kα were observed in addition to Fe Kα, as shown in Figure 5.2.10(a).[35] This is because the electron beam penetrated into the Au–Si substrate. The same particle was measured at a grazing-exit angle of about 0°. As shown in

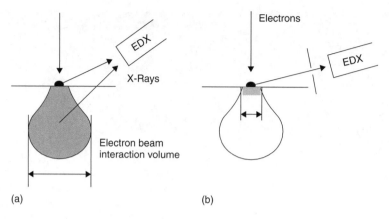

Figure 5.2.9 Interaction volume and analyzing region under conventional conditions (a) and grazing-exit conditions (b)

Table 5.2.2 Comparison of theoretically estimated interaction volumes (nm) obtained with conventional vs GE-EPMA (beam diameter = 50 nm)

		Electron accelerating voltage (kV)		
		5	10	20
Si	Conv.	311	899	1415
	GE	68	122	171
Cu	Conv.	124	294	828
	GE	84	121	148
Au	Conv.	85	184	421
	GE	62	112	190

Figure 5.2.10(b), the characteristic X-rays emitted from the substrate completely disappeared, making single-particle analysis possible with a low background intensity. Elemental analysis for atmospheric aerosols deposited on a flat substrate, taken under conventional and grazing-exit conditions, were also reported.[34]

In addition, interference between the direct X-ray beam emitted from the particle and the reflected X-ray beam on the flat substrate can be observed at grazing-exit angles.[36] This interference is useful for the enhancement of the X-ray intensities from the particle. Bekshaev and Van Grieken have theoretically studied the interference pattern to obtain the information of particle structure and composition.[37]

Awane et al. measured an inclusion (0.3 μm in diameter), which appeared on the surface of etched stainless steel samples[31] They adjusted the exit angle by moving the sample position, as shown in Figure 5.2.7. At the conventional exit angle of 30°, strong X-rays of Fe Kα, Cr Kα and Ni Kα emitted from the matrix (stainless-steel) were detected (Figure 5.2.11(a) and 5.2.11(b)), making it difficult to analyze the inclusion. When the

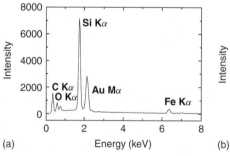
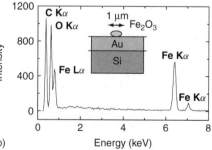

Figure 5.2.10 X-Ray spectra taken for a single particle of Fe_2O_3 deposited on a Au (100 nm) layer on a Si substrate at exit angles of 40° (a) and approximately 0° (b).[35] Reproduced by permission of Springer-Verlag KG

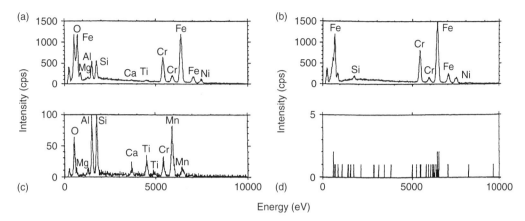

Figure 5.2.11 X-Ray spectra taken under conventional conditions (exit angle of 30°) (a, b) and grazing exit conditions (c, d) for a typical single inclusion (a, c) and the matrix (b, d).[31] Reprinted from Awane, T. *et al. J. Surf. Anal.* **9**, (2002) 171. Reproduced by permission of The Surface Analysis Society of Japan

electron beam was irradiated on the matrix at a grazing-exit angle, no X-rays were observed from the matrix, as shown in Figure 5.2.11(d), indicating that the X-ray spectrum shown in Figure 5.2.11(c) was obtained for the inclusion at the same grazing angle, although the surface was not perfectly flat. In conclusion, GE-EPMA showed that Fe and Ni were not in the inclusion.

5.2.4.4 THIN-FILM ANALYSIS BY GE-EPMA

A Cr (50%) – Ti (50%) ultra-thin film (10 nm) deposited on a Si substrate was measured.[38,39]

Figure 5.2.12(a) shows the X-ray spectrum taken at an exit angle of 45°. Due to large continuous X-ray background intensity, it was difficult to recognize the characteristic peaks of Ti Kα and Cr Kα. However, these continuous X-rays were significantly reduced at a grazing-exit angle of 0.75°, with the result that Ti Kα and Cr Kα were clearly detected with low background intensity, as shown in Figure 5.2.12(b). Reduction of continuous background intensity is important for improving detection limits.[40] By applying GE-EPMA, the detection limits of Ti and Cr were improved by factors of 4 to 10.[38]

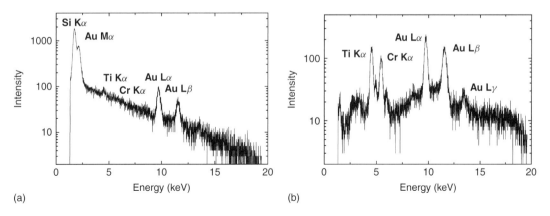

Figure 5.2.12 X-Ray spectra taken for a Cr (50%) – Ti (50%) ultra-thin film (about 10 nm) deposited on a Si substrate at an accelerating voltage of 20 kV at exit angles of 40° (a) and approximately 0.75° (b).[38]

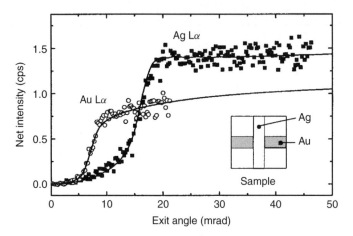

Figure 5.2.13 X-Ray intensities of Ag Lα (■) and Au Lα (○) as a function of the exit angle. A schematic diagram of the sample is shown in the inset. The curves calculated for Ag (40 nm, 8.0 g/cm^3) and Au (80 nm, 15.0 g/cm^3) are indicated by solid lines.[40] Reproduced by permission of Elsevier Science

The determination of the thickness and density of thin films is also important for their characterization. It is possible to evaluate thickness and density of thin films by other analytical X-ray techniques, such as X-ray reflectivity measurement and grazing-incidence XRF (angle-dependent TXRF). In these methods, the entire surface of the sample is irradiated by primary X-rays; and the average thickness and density for the entire surface are obtained. The advantage of GE-EPMA is in localized thin-film analysis. Figure 5.2.13 shows the angle dependence of Au Lα and Ag Lα for the Au–Ag thin films shown in the inset.[41] The electron beam was fixed on each film (Au or Ag), and then dependence on the exit angle was measured. From the curve fitting method, the thicknesses of the Au and Ag films were determined to be 80 and 40 nm, respectively.

5.2.5 GRAZING-EXIT PARTICLE INDUCED X-RAY EMISSION (GE-PIXE)

Compared to EPMA, PIXE is considered to be suitable for trace analysis because the continuous background intensity is originally low. However, some continuous background intensity is still observed, especially in the low-energy region. By applying grazing-exit measurement, it is possible to reduce this continuous background intensity and enhance the surface sensitivity.

5.2.5.1 GRAZING-EXIT PARTICLE INDUCED X-RAY EMISSION SETUP

Figure 5.2.14 shows an experimental setup of GE-PIXE, which was developed originally for total reflection PIXE experiments[42] at Amsterdam Free University in The Netherlands. The proton beam, which was produced in H$_2$ plasma and accelerated to an energy level of 2.5 MeV, irradiated the surface of the sample. The characteristic X-rays

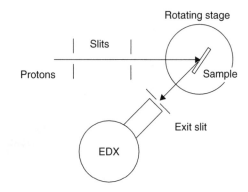

Figure 5.2.14 Schematic diagram of the top view of the GE-PIXE experimental setup.[42] Reproduced by permission of Elsevier Science

were measured through a slit by an EDX detector at a detector angle of 45° with respect to the proton beam to reduce Bremsstrahlung X-rays. To change the exit angle, the sample stage was rotated using a stepping motor with a minimum step of 0.057° in a vacuum chamber.

5.2.5.2 SURFACE AND THIN-FILM ANALYSIS BY GE-PIXE

Au (50 nm) – Cu (500 nm) double layers on glass were measured by the GE-PIXE setup shown in Figure 5.2.14.[43] At a large exit angle of 45°, the Cu Kα was the dominant X-ray peak due to the relatively large thickness of the Cu layer. The same sample was measured at a grazing-exit angle of about 0.5°. The background intensity in the low energy region was considerably reduced. In addition, the X-rays emitted from the thin Au layer were dominantly observed. This indicates that surface-sensitive PIXE analysis is possible under grazing-exit conditions.

Rodríguez-Fernández et al. proposed SPIX (surface sensitive particle-induced X-ray analysis),[44–46] which is essentially the same as GE-PIXE. They measured Cl implanted into a Si wafer by SPIX.[47] Cl Kα intensities were plotted as a function of tilt angle, as shown in Figure 5.2.15. By fitting with theoretical curves, the depth of the implanted Cl layer was determined to be 110 nm with a depth resolution of 8 nm, indicating the potential of SPIX performance. In this measurement, the sample tilt angle was determined with a precision of ±0.1° without an exit slit. Therefore, they have calculated the angle dependent curves after folding in the extended detector geometry.[48]

5.2.5.3 PARTICLE ANALYSIS BY GE-PIXE

The analysis of atmospheric aerosols is one of the most important applications of PIXE. Low-Z elements, such as Ca, K, Na, etc., are significant elements in environmental analysis. However, the continuous X-ray background intensity in low-energy regions is still too high in the conventional PIXE spectrum, making light element analysis difficult in some cases. The PIXE spectra, shown in Figure 5.2.16(a), were taken for aerosols collected on a Si wafer at an exit angle of 4.4°.[43] Due to large continuous X-ray background intensity, it was difficult to identify small amounts of elements in aerosols. The same sample was measured at a grazing-exit angle of 0.4°. As shown in Figure 5.2.16(b), the continuous X-ray background intensity was reduced significantly, and Ca, Ti, and Zn could be detected with the low background intensity.

5.2.6 CHARACTERISTICS OF GE-XRS AND ITS FUTURE

Grazing-exit X-ray spectrometry has been applied to conventional X-ray analytical methods, such as XRF, EPMA and PIXE. In the measurement of X-rays at grazing-exit angles, the detection of X-rays emitted from deep inside the sample is reduced, leading to the improvement of the degree of surface sensitivity and the detection limits. Therefore, GE-XRS would provide unique opportunities for surface analysis, thin-film analysis, and particle analysis, using conventional analytical X-ray methods. Compared with GI-XRS, the most important advantage of GE-XRS is 'localized surface analysis'. In GE-XRS, the excitation beam can

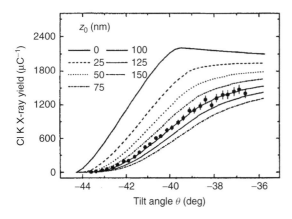

Figure 5.2.15 Calculation of Cl Kα X-ray yield produced by 1.2 MeV ^{1}H$^+$ ion bombardment as a function of the tilt angle for Cl layers located at different depths z_0 (nm) in Si.[47] Reproduced by permission of Elsevier Science

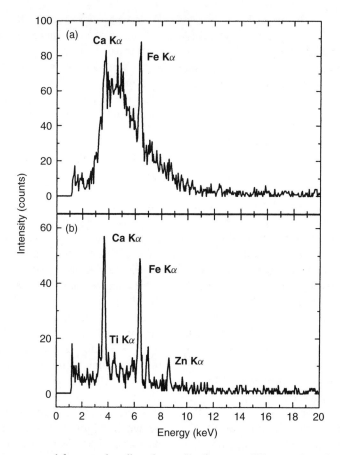

Figure 5.2.16 PIXE spectra measured for aerosols collected on a Si substrate at different exit angles of 4.4° (a) and 0.4° (b). The lifetime for EDX measurements was 180 s.[43] Reprinted with permission from Tsuji, K., Spolnik, Z., Wagatsuma, K., Van Grieken, R. and Vis, R.D. *Anal. Chem.* **71**, 5033 (1999). Copyright (1999) American Chemical Society

irradiate the sample at a large angle of 90°. Therefore, localized analysis is possible, depending on the diameter of the excitation beam.

The X-ray intensity detected by GE-XRS is weak because the X-rays are detected at a small exit solid angle through the slit. This is one of the drawbacks of GE-XRS. To overcome this drawback, a multi-X-ray detector system would be useful. In the present experimental configuration, X-rays are measured only from a small azimuth angle while characteristic X-rays are emitted in all directions. Thus, the use of a ring-type silicon drift X-ray detector (SDD),[49] shown in Figure 5.2.17, will enable the detection of characteristic X-rays with high efficiency even at grazing-exit angles, which can be changed by moving either the

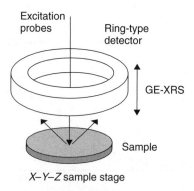

Figure 5.2.17 Ring-type SDD for high efficiency GE-XRS

detector or the sample. In the case of GE-XRF, synchrotron radiation is an ideal X-ray source for the micro X-ray beam. In laboratory GE-XRF, new developments of X-ray optics for micro X-ray beams would be a key technology. In both cases, not only elemental mapping, but also three-dimensional XRF near the surface region, by controlling the sample stage and the exit angle, will be attractive subjects for further research.

REFERENCES

1. Becker, R.S., Golovchenko, J.A. and Patel, J.R., *Phys. Rev. Lett.* **50**, 153 (1983).
2. Tsuji, K., Sasaki, A. and Hirokawa, K., *Jpn. J. Appl. Phys.* **33**, 6316 (1994).
3. Urbach, H.P. and de Bokx, P.K., *Phys. Rev. B* **53**, 3752 (1996).
4. Born, M. and Wolf, E., *Principles of Optics*, Pergamon, Oxford (1991).
5. de Boer, D.K.G., *Phys. Rev. B* **44**, 498 (1991).
6. Tsuji, K. and Hirokawa, K., *J. Appl. Phys.* **75**, 7189 (1994).
7. Tsuji, K., Delalieux, F., Wagatsuma, K. and Sato, S., *X-Ray Spectrom.* submitted.
8. de Bokx, P.K. and Urbach, H.P., *Rev. Sci. Instrum.* **66**, 15 (1995).
9. Tsuji, K., Murakami, Y., Wagatsuma, K. and Love, G., *X-Ray Spectrom.* **30**, 123 (2001).
10. Sasaki, Y. and Hirokawa, K., *Appl. Phys. A* **50**, 397 (1990).
11. Tsuji, K., Sato, S. and Hirokawa, K., *J. Appl. Phys.* **76**, 7860 (1994).
12. Noma, T., Iida, A. and Sakurai, K., *Phys. Rev. B* **48**, 17524 (1993).
13. Noma, T. and Iida, A., *J. Synchrotron Rad.* **5**, 902 (1998).
14. Sasaki, Y.C., Sezuki, Y. and Ishibashi, T., *Science* **263**, 62 (1994).
15. Sasaki, Y.C., Sezuki, Y., Tomioka, Y., Ishibashi, T., Satoh, I. and Hirokawa, K., *Phys. Rev. B* **50**, 15516 (1994).
16. de Bokx, P.K., Kok, Chr., Bailleul, A., Wiener, G. and Urbach, H.P., *Spectrochim. Acta B* **52**, 829 (1997).
17. Wiener, G., Kidd, S.J., Mutsaers, C.A.H., Wolters, R.A.M. and de Bokx, P.K., *Appl. Surf. Sci.* **125**, 129 (1998).
18. Claes, M., de Bokx, P., Willard, N., Veny, P. and Van Grieken, R., *Spectrochim. Acta B* **52**, 1063 (1997).
19. Spolnik, Z.M., Claes, M., Van Grieken, R., de Bokx, P.K. and Urbach, H.P., *Spectrochim. Acta B* **54**, 1525 (1999).
20. Spolnik, Z.M., Claes, M. and Van Grieken, R., *Anal. Chim. Acta* **401**, 293 (1999).
21. Kuczumow, A., Claes, M., Schmeling, M., Van Grieken, R. and de Gendt, S., *J. Anal. At. Spectrom.* **15**, 415 (2000).
22. Claes, M., Van Dyck, K., Deelstra, H. and Van Grieken, R., *Spectrochim. Acta B* **54**, 1517 (1999).
23. Hołyńska, B., Olko, M., Ostachowicz, B., Ostachowicz, J., Węgrzynek, D., Claes, M., Van Grieken, R., de Bokx, P., Kump, P. and Necemer, M., *Fres. J. Anal. Chem.* **362**, 294 (1998).
24. Pérez, R. and Sánchez, H.J., *Rev. Sci. Instrum.* **68**, 2681 (1997).
25. Noma, T. and Iida, A., *Rev. Sci. Instrum.* **65**, 837 (1994).
26. Tsuji, K. and Delalieux, F., *J. Anal. At. Spectrom.* **17**, 1405 (2002).
27. Hasegawa, S., Ino, S., Yamamoto, Y. and Daimon, H., *Jpn. J. Appl. Phys.* **24**, L387 (1985).
28. Usui, T., Kamei, M., Aoki, Y., Morishita, T. and Tanaka, S., *Physica C* **191**, 321 (1992).
29. Tsuji, K., Wagatsuma, K., Nullens, R. and Van Grieken, R., *Anal. Chem.*, **71**, 2497 (1999).
30. Tsuji, K., Spolnik, Z. and Ashino, T., *Rev. Sci. Instrum.*, **72**, 3933 (2001).
31. Awane, T., Kimura, T., Suzuki, J., Nishida, K., Ishikawa, N. and Tanuma, S., *J. Surf. Anal.* **9**, 171 (2002).
32. Takahashi, H., *Inst. Phys. Conf. Ser.* **165**, 435 (2000).
33. Yamanaka, T., Shimomura, N. and Ino, S., *Appl. Phys. Lett.* **77**, 3983 (2000).
34. Tsuji, K., Wagatsuma, K., Nullens, R. and Van Grieken, R., *J. Anal. At. Spectrom.*, **14**, 1711 (1999).
35. Tsuji, K., Spolnik, Z., Wagatsuma, K., Nullens, R. and Van Grieken, R., *Mikrochim. Acta*, **132**, 357 (2000).
36. Tsuji, K., Spolnik, Z., Wagatsuma, K., Zhang, L. and Van Grieken, R., *Spectrochim. Acta B* **54**, 1243 (1999).
37. Bekshaev, A. and Van Grieken, R., *Spectrochim. Acta B* **56**, 503 (2001).
38. Spolnik, Z., Tsuji, K., Saito, K., Asami, K. and Wagatsuma, K., *X-Ray Spectrom.* **31**, 178 (2002)
39. Spolnik, Z., Zhang, J., Wagatsuma, K. and Tsuji, K., *Anal. Chim. Acta* **455**, 245 (2002).
40. Tsuji, K., Spolnik, Z. and Wagatsuma, K., *Spectrochim. Acta B* **56**, 2497 (2001).
41. Tsuji, K., Saito, K., Asami, K., Wagatsuma, K., Delalieux, F. and Spolnik, Z., *Spectrochim. Acta B* **57**, 897 (2002).
42. Tsuji, K., Huisman, M., Spolnik, Z., Wagatsuma, K., Mori, Y., Van Grieken, R. and Vis, R.D., *Spectrochim. Acta B* **55**, 1009 (2000).
43. Tsuji, K., Spolnik, Z., Wagatsuma, K., Van Grieken, R. and Vis, R.D., *Anal. Chem.* **71**, 5033 (1999).
44. Rodríguez-Fernández, L., Lennard, W.N., Xia, H. and Massoumi, G.R., *Appl. Surf. Sci.* **103**, 289 (1996).
45. Rodríguez-Fernández, L., Lennard, W.N., Massoumi, G.R., Xia, H., Huang, L.J., Hung, Y. and Chang, S., *Nucl. Instrum. Methods Phys. Res. B* **136–138**, **1191** (1998).
46. Xia, H., Rodríguez-Fernández, L., Lennard, W.N., Massoumi, G.R., Dimov, S. and Chryssoulis, S., *Nucl. Methods Phys. Res. B* **149**, 1 (1999).
47. Xia, H., Lennard, W.N., Rodríguez-Fernández, L. and Massoumi, G.R., *Surf. Sci.* **384**, 291 (1997).
48. Lennard, W.N., Kim, J.K. and Rodríguez-Fernández, L., *Nucl. Instrum. Methods Phys. Res. B* **189**, 49 (2002).
49. Lechner, P., Fiorini, C., Hartmann, R., Kemmer, J., Krause, N., Leutenegger, P., Longoni, A., Soltau, H., Stotter, F., Stotter, R., Struder, L. and Weber, U., *Nucl. Instrum. Methods Phys. Res. A* **458**, 281 (2001).

5.3 Portable Equipment for X-ray Fluorescence Analysis

R. CESAREO[1], A. BRUNETTI[1], A. CASTELLANO[2] and M. A. ROSALES MEDINA[3]

[1] Department of Mathematics and Physics, University of Sassari, Sassari, Italy, [2] Department of Materials Science, University of Lecce, Lecce, Italy, and [3] University of 'las Americas', Puebla, Mexico

5.3.1 INTRODUCTION

Energy-dispersive X-ray fluorescence (EDXRF) analysis is a nondestructive, multi-elemental and simple technique, which is based on the irradiation of a sample by a low intensity X-ray beam, and by the detection of secondary X-rays emitted by the sample.[1-3]

The energy of these secondary X-rays characterises the elements present in the sample where the intensity is proportional to their concentration. The 'complex' of secondary X-rays and primary scattered radiation is called the 'X-ray spectrum'.

The sample can be in any state (solid, liquid, gaseous or of various size and nature), and it will absolutely not be altered by the analysis, for this reason it can be analysed many times. These features make EDXRF especially suitable for *in situ* and on-line analysis.

In the past, portable EDXRF equipment was composed of radioactive sources and proportional gas counters.[4] But the high energy resolution of these detectors limited the range of possible applications. Then, in the 1980s, nitrogen-cooled detectors substituted proportional gas counters. However, the need for liquid N_2 and the intrinsic delicacy of these detectors limited the use of such equipment.

Only in the last few years, has technological progress produced miniature and dedicated X-ray tubes,[5-9] thermoelectrically cooled X-ray detectors of small size and weight,[6,8-14] small size multi-channel analysers[8,15] and dedicated software.[16-19] This progress allowed the construction of completely portable small-sized EDXRF systems that have similar capabilities as the more elaborate laboratory systems, and which, by definition, can be used anywhere by one person for *in situ* analysis, but without the problems connected with nitrogen cooling, big size X-ray tubes and high costs.

Portable EDXRF equipment is absolutely necessary in many cases, when objects to be analysed cannot be transported (typically works of art) or when an area should be directly analysed (soil analysis, lead inspection testing, etc.) or when the mapping of the object would require too many samples.

Typical examples where portable EDXRF systems are needed are the following:

- archaeometry (measurements of frescoes, paintings, alloys in churches, museums, in open air, etc.;
- analysis of lead-based paints;
- analysis of soil contaminants;
- geological surveys;
- *in vivo* analysis of toxic elements (lead, cadmium, platinum, etc.);
- quality control in industry (for ex. alloys analysis);

- quality control of X-ray tubes (maximum energy and spectrum).

5.3.2 INSTRUMENTATION

A typical EDXRF system (Figure 5.3.1) is composed of:[1-3]

- an excitation source (a radioactive source or a X-ray tube);
- an X-ray detector with electronics;
- a single- or a multichannel analyser;
- software for elements identification and quantitative analysis.

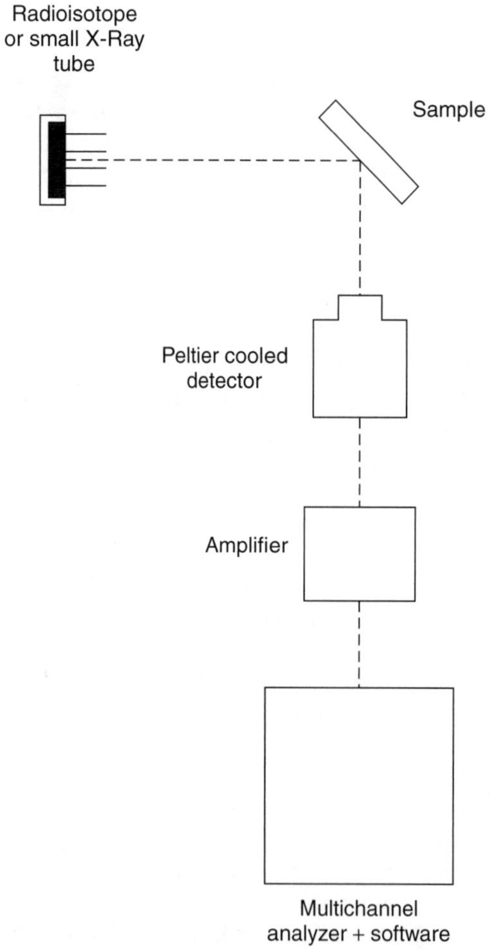

Laboratory EDXRF systems are generally equipped with high power X-ray tubes, which can be collimated and used with secondary targets, nitrogen-cooled high resolution Si (Li) or Ge detectors[15,20-22] and multichannel analysers with sophisticated software for quantitative automatic evaluation.

Portable EDXRF systems have, of course, different requirements, such as: low weight, small size, low counting time and the possibility to be equipped with batteries. They generally have higher minimum detection limits and are less flexible than laboratory systems.

5.3.2.1 X-RAY SOURCES

For portable EDXRF equipment both radioactive sources (emitting α, β, γ, X or bremsstrahlung radiation) or X-ray tubes can be employed. Among the commercial PEDXRF equipment, about one half use radioisotopes and one half X-ray tubes. Radioisotopes are generally used emitting X-rays (or γ-rays in the X-ray energy region). However, in special cases also α sources are employed for analysis of low Z elements.

X-Ray tubes used for portable EDXRF systems have a Be window, small size and low power.

Figure 5.3.1 Scheme of portable EDXRF analysis equipment composed of a radioactive source (or small size low-power X-ray tube), a Peltier- cooled semiconductor detector and a MCA with dedicated software

Radioactive Sources

General requirements for radioactive sources to be employed in portable EDXRF equipment are:

- sufficient long half-time ($> \approx 1$ year);
- emission of only one or two lines of proper energy and sufficient intensity.

Radioactive sources are especially suitable for portable X-ray equipment and are still employed, in spite of the current availability of a variety of small size X-ray tubes. They have the advantage of having very small sizes (Figure 5.3.2) and of being intrinsically stable but the disadvantage of emitting radiation of fixed energy and low intensity (orders of magnitude lower than for X-ray tubes). Radioactive sources are, therefore, not flexible versus energy and not suitable for analysis of low amounts of chemical elements.

X and γ radioactive sources generally employed for EDXRF analysis are shown in Table 5.3.1.[23–24] As observed above, in special cases α radioactive sources are employed or mixed α-X radioactive sources, where the analysis of very low Z elements is required.

X-Ray Tubes

The general requirements for a X-ray tube suitable for portable EDXRF equipment are the following:

- low power of a few watts (HV: variable from 5 kV to a maximum of 40 kV; current variable from 10 μA to a maximum of about 0.5 mA);
- air or internal cooling;
- small size to allow portability;
- anode suitable to analytical problems;
- thin Be window;
- good shielding and collimation, to guarantee only forward irradiation and reduce radiation dose.

A great variety of X-ray tubes of various types (maximum voltage, current, anode), size and cost is currently available for portable EDXRF analysis, depending on the problem, and more specifically on the element or elements to be analysed. X-Ray tubes for portable EDXRF equipment are constructed by many companies[5–9] and some of them are shown in Figure 5.3.3.

In 2001 a very compact, battery-powered X-ray tube, incorporating in the same case a radiation

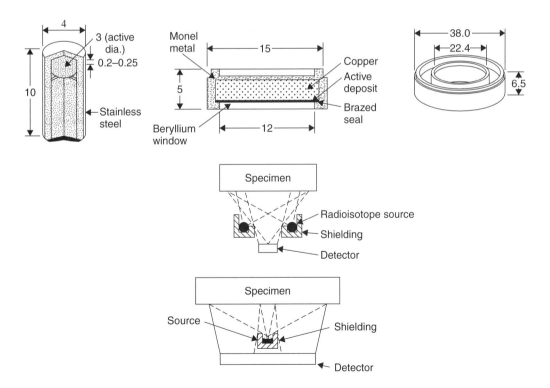

Figure 5.3.2 Radioactive sources for EDXRF analysis (from top left, point disc and annular source) and typical geometries with annular (centre) and central (bottom) source

Table 5.3.1 Typical radioisotopes for EDXRF analysis

Isotope	$t_{1/2}$ (years)	Energy of emitted radiation (keV)	Typical number of emitted photons/s sterad	Elements which can be analysed
^{55}Fe	2.7	5.9 and 6.5 (Mn K lines)	10^6–10^7	from Si to Ti
^{238}Pu	86.4	13.5 and 16.8 (U L lines)	5×10^6	from Ca to As (K)
^{109}Cd	453 days	22.1 and 25.0 (Ag K lines)	2×10^6	from Ca to Mo (K lines) and heavy elements (L lines)
^{241}Am	433	59.5	10^7	from Fe to Gd (K lines) and heavy elements (L lines)
^{57}Co	270 days	122 and 136	2×10^6	Heavy elements (K lines)
^{153}Gd	241 days	41.3 and 47.3 (Eu K lines), 97.4 and 103	5×10^6	Heavy elements (K lines)

source with HV power and control electronics, was designed and constructed for AMPTEK by Photoelectron Co.[24] (Figure 5.3.3). It has a bulk silver anode, HV adjustable from 10 to 35 kV and beam current from 0 to 100 µA. The dimensions of the external case are $19 \times 7 \times 3.3 \, \text{cm}^3$ and it has an overall weight of about 450 g. In 2001, Moxtek[9] developed a very compact, battery-operated transmission-anode X-ray tube with the following characteristics: Pd or Ag anode, HV from 10 to 30 kV, current from 0 to 0.1 mA, dimensions of $18 \times 7 \times 3 \, \text{cm}^3$ and a weight of about 450 g. More recently a 40 kV, 0.1 mA, battery-operated X-ray tube was produced by Moxtek ('Bullet™'). This tube operates at a power of 4 W with a maximum input of 12 W, and is available in a side-window configuration (W anode), or with a 2 µm Ag or Pd transmission target. The transmission target X-ray tube has a Be window of 0.25 mm thickness and a volume of $30 \, \text{cm}^3$ approximately (Figure 5.3.4).

Leaving out analysis of trace elements (with concentration lower than 1 ppm), which requires high-current tubes with proper secondary targets, low-power X-ray tubes are generally adequate. They may be selected primarily as a function of the atomic number of the element or elements to be analysed. Table 5.3.2 gives characteristics of X-ray tubes useful for EDXRF analysis. It is important to observe that small portable X-ray tubes are not available for excitation of K-lines of heavy elements. In this case, the only option currently given, is the use of radioactive sources.

A quite different type of X-ray generator, based on the properties of pyroelectric crystals was recently proposed by Brownridge and Roboy[26] and developed by Amptek.[6]

Also in 2003, carbon nanotube based field emission X-ray tubes were developed by Applied Nanotechnologies.[27] The conventional X-ray tube is based on a metal filament that emits electrons which are accelerated to bombard the metal target. ANI X-ray tubes utilise carbon nanotube field emitters as the electron source.

X-Ray Optics

In some applications, very small areas must be irradiated for EDXRF analysis. This happens when very small samples are analysed, such as grains or microfragments, or when analysis with a very reduced space resolution is needed.

The simplest way to produce an X-ray microbeam is to use a pinhole collimator between the X-ray source and the sample. But, unfortunately, only a small fraction of the original photon flux can pass through the pinhole. This results in low count rates, which limits the sensitivity of the technique.

Higher flux densities can be achieved by employing polycapillary lenses, consisting of several hundred thousand glass fibres. The channels are all directed towards one focus point, requiring straight fibres in the centre of the lens and strongly bent fibres near the surface. The strongly bent fibres are less efficient in transporting X-ray photons than the straight ones. This difference in efficiency increases with the increase in photon energy due to the energy dependence of the critical angle for total reflection. Polycapillary optics cause a

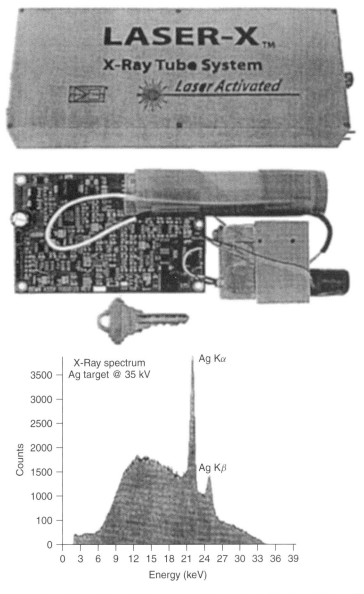

Figure 5.3.3 Battery-operated 'Laser' X-ray tube with Ag anode, which works at 30 kV and 0.1 mA. The case containing the tube and the high voltage supply is also shown (centre), and a typical X-ray spectrum emitted by the tube is illustrated at the bottom of the figure (with permission of Amptek, Inc.)

significant change in the shape of the excitation spectrum: high energy photons are filtered out and the excitation spectrum is distorted towards low energies. This effect leads to smaller focal spot sizes in the high energy region.

When applied in combination with air-cooled, low-power X-ray tubes, capillary collimators and in particular polycapillary lenses have proven to be very suited for focusing the X-ray beam in small areas of 10–100 μm diameter.[28]

A polycapillary optical element for EDXRF analysis has a typical length of 40–70 mm, like a cylinder thicker in the middle, with a diameter of 5–10 mm. It is able to focus an X-ray beam

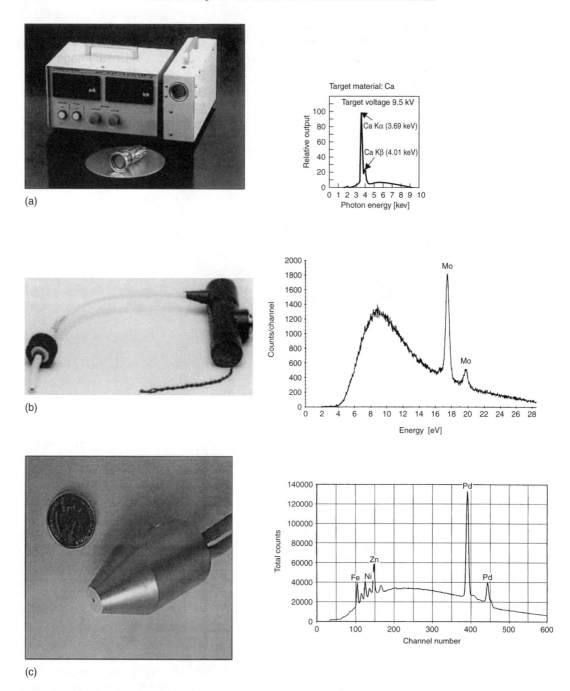

Figure 5.3.4 Typical X-ray tubes for portable EDXRF equipment. (a) A Ca anode, 8 kV and 0.1 mA from Hamamatsu (8.8 cm length × 3.7 cm diameter and 35 g weight), and typical X-ray spectrum; (b) Mo anode, 30 kV and 0.2 mA from Oxford (10 cm length × 3.3 cm diameter and 200 g weight), and typical X-ray spectrum; (c) W anode, battery-operated X-ray tube from Moxtek, 40 kV and 0.1 mA (30 cm^3 volume), and typical X-ray spectrum (with permission of Hamamatsu, Oxford and Moxtek)

Table 5.3.2 Characteristics of small-size, portable X-ray tubes and elements which can be analysed

Element or elements to be analysed	Anode material	Kilovoltage (kV)	X-ray emission
Al, Si, P, S, Cl	Calcium	5–8	3.7 keV (Ca K lines) plus Bremsstrahlung
Al, Si, P, S, Cl	Silver (L lines) or Pd	5–10	3 keV (Ag L lines) plus Bremsstrahlung
Cl, Ar, K, Ca	Titanium	10	4.5 keV (Ti K lines) plus Bremsstrahlung
From Ca to Y (K lines) and from W to U (L lines)	Molybdenum	30	17.5 keV (Mo K lines) plus Bremsstrahlung
From Ca to Mo (K lines) and from W to U (L lines)	Silver	30	22 keV (Ag K lines) plus Bremsstrahlung
From Ca to Sn (K lines) and from W to U (L lines)	Tungsten	35	Bremsstrahlung plus 8.3 and 9.8 keV (W L lines)
From Fe to Ba (K lines) and from W to U (L lines)	Tungsten	50	Bremsstrahlung plus 8.3 and 9.8 keV (W L lines)
Rare earths, from lanthanum to hafnium (L lines)	Molybdenum	20	17.5 keV plus Bremsstrahlung

up to energies of about 20 keV. For a portable EDXRF apparatus the polycapillary optical system should be rigidly connected to the X-ray tube (Figure 5.3.5).

5.3.2.2 X-RAY DETECTORS

The general features for a detector suitable for high quality portable EDXRF equipment should be the following:

- sufficient good energy resolution (better than 200–250 eV at 5.9 keV);
- sufficient good efficiency in the energy interval to be analysed;
- thin Be window;
- Peltier cooling, for reducing the size.

The typical, high resolution X-ray detector has been for a long time the nitrogen-cooled Si(Li) or HpGe detector, with an energy resolution of about 120–150 eV at 5.9 keV[20–22] (Figure 5.3.6). However, the nitrogen cooling makes these detectors difficult to be used for portable instruments.

Proportional gas counters have typically been used as substitutes for nitrogen-cooled detectors in portable EDXRF equipment until about 1994–1995,[4,29,30] when thermoelectrically cooled Si-PIN and CdZnTe started to be produced by AMPTEK.[8]

Currently, equipment based on radioactive sources and proportional gas counters is no longer in use.

In the last few years, small size thermoelectrically cooled semiconductor detectors have become available, such as HgI_2,[12] Si-PIN,[8] Si drift,[6,10,11] CdTe and CdZnTe (CZT).[8,13,14] These detectors are cooled to about $-30\,^\circ$C by means of a Peltier circuit, and are contained in small boxes including a high quality preamplifier and the Peltier circuit (Figure 5.3.7).

The HgI_2 detector was the first to be constructed; it has an energy resolution of about 180–200 eV at 5.9 keV (Figure 5.3.6), and an efficiency of about 100 % in the whole range of X-rays (Figure 5.3.8). It has the disadvantage of producing many disturbing 'escape peaks' when irradiated.

The Si-PIN detector is currently the most employed in portable EDXRF equipment. A typical model has a Si thickness of 300 μm, exhibits an energy resolution of 160–200 eV (Figure 5.3.6) and is useful up to about 25–30 keV (Figure 5.3.8), because the efficiency is decreasing at an energy larger than 15 keV, due to the limited thickness. Another model has a thickness of 500 μm, an energy resolution of about 200–220 eV and can be used up to 50 keV approximately. High energy resolutions are obtained using high shaping time, in

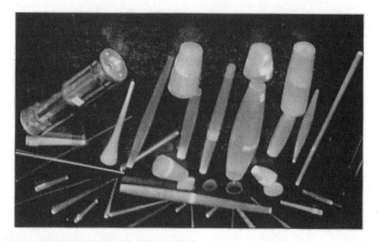

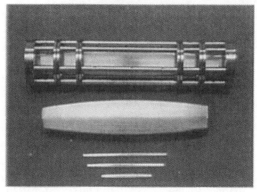
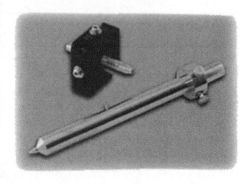

Figure 5.3.5 Typical capillary collimators from Roentgen Optics (Moscow), IfG (Berlin) and X-ray Optical Systems (Albany, NY, USA). (Published with permission)

the order of 15–20 μs. Therefore, these detectors are subject to an energy resolution degradation at a count rate higher than a few thousands of photons/s.

The Si drift detector (SDD) was developed by Gatti et al.[10] The bulk detector is produced by KeteK,[31] and the complete detector with electronics by EIS[6] and by Roentec.[11] It typically has a Si thickness of 300 μm an active area of 5–10 mm², and an energy resolution of approximately 130–150 eV at 5.9 keV (Figure 5.3.6). The integration of the first stage of the front-end electronics enables the operation of the SDD at extremely short shaping times even at low temperatures, making the SDD the optimal solution for high count rate applications. Shaping times of 0.25 to 2 μs can be used.

CdTe and CZT detectors have a typical thickness of 2 mm, and thus an efficiency of about 100 % in the whole X-ray energy range (Figure 5.3.8).

The energy resolution of the best CZT detector is about 250, 500 and 750 eV at 5.9, 59.6 and 122 keV, respectively (Figure 5.3.6). A comparison between various thermoelectrically cooled detectors can be found elsewhere.[32]

A summary of the useful Peltier-cooled compared to the nitrogen-cooled detectors is given in Table 5.3.3. It should be observed that the efficiency in the low-energy region is dependent on the thickness of the Be window (Figure 5.3.8). The minimum standard window compatible with the portability and hardiness of the equipment is 8 μm Be, having a transmission of about 80 % at 2 keV.

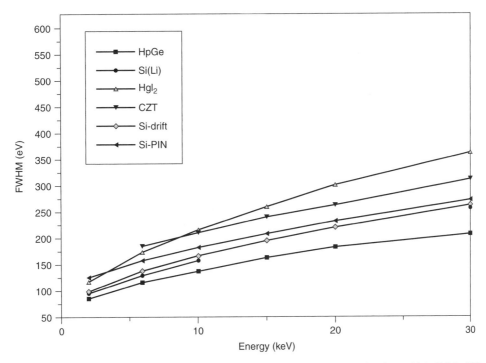

Figure 5.3.6 Energy resolution (full width at half maximum, FWHM) of 3.5 mm thick HpGe, 3 mm thick Si(Li), 300 μm thick Si-drift, 300 μm thick Si-PIN, 1 mm thick HgI$_2$, and 2 mm thick CdZnTe

Typical X-ray spectra of a Si-PIN, Si drift, and CdZnTe detector are shown in Figure 5.3.9. The following considerations can be made, concerning Peltier-cooled semiconductor detectors for portable EDXRF equipment:

- the SDD presents the best energy resolution among the thermoelectrically cooled detectors and works also at high count rates, due to the low shaping time; it is useful up to about 25–30 keV;
- Si-PIN detectors are the most employed in the energy interval 1–25 keV, in spite of the fact that it works at low count rates;
- HgI$_2$, CdTe CZT are the best thermoelectrically cooled detectors in the X-ray region of 30–120 keV.

5.3.2.3 SINGLE- OR MULTICHANNEL ANALYSERS

When one or a few elements must be analysed, a single-channel analyser may be employed, with a timer-scaler. The advantage of the small size of this solution, which was relevant in the past, is currently replaced by the multichannel analyser (MCA) board coupled to a PC,[6,15,16] or by a pocket MCA coupled to a laptop computer (Figure 5.3.10).[8] All the modern MCA models are not only able to register and store the X-ray spectra, but are also coupled to dedicated software to compute chemical elements and relative concentration.

5.3.2.4 SOFTWARE FOR THE AUTOMATIC IDENTIFICATION AND QUANTITATIVE EVALUATION OF ELEMENTS

The X-ray spectra recorded by the EDXRF instrumentation must be analysed in order to extract the desired information, i.e. the chemical element contents.

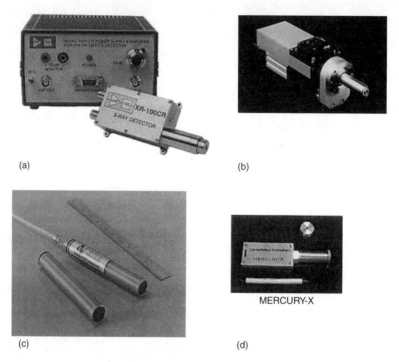

Figure 5.3.7 Typical thermoelectrically cooled, small size semiconductor detectors for portable EDXRF analysis. (a) A Si-PIN detector by Amptek, (b) a Si-drift from Roentec (X- Flash 1000), (c) a CZT from eV and (d) a HgI$_2$ from Constellation Tech. (Published with permission)

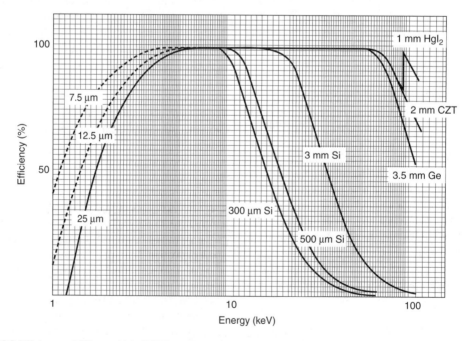

Figure 5.3.8 Efficiency of 300 μm thick Si-PIN or SDD, 500 μm Si-PIN, 3 mm thick Si(Li), 3.5 mm thick HpGe, 2 mm thick CZT and 1 mm thick HgI$_2$. At low energies the efficiency is determined by the thickness of the Be window (7.5, 12.5 and 25 μm thickness, respectively)

Table 5.3.3 Comparison between the best performances of different commercial X-ray detectors

Detector and performances	Si(Li)[a]	HpGe[b]	Si-PIN[c] 300 μm	Si-PIN[d] 500 μm	Si-drift[e]	HgI$_2$[f]	CZT[g]
Energy resolution FWHM at 5.9 keV	129	115	158	250	129	180	190
Energy resolution at 59.6 keV	360	300	–	–	–	480	500
Useful energy range (keV)[h]	1–60	1–120	1–25	1–35	1–25	2–120	2–120
Efficiency (%) at 2 cm from the source	0.002	0.004	0.0014	0.006	0.0006	0.001	0.0018
Shaping time (μs)	6–12	6	20	20	5	12	3
Cooling system	Liquid N$_2$	Liquid N$_2$	Peltier	Peltier	Peltier	Peltier	Peltier

[a] Si(Li) with an area of 10 mm^2 and a thickness of 3 mm by VacuTec[21] or ThermoNoran.[22]
[b] HpGe with an area of 20 mm^2 and a thickness of 3.5 mm by PGT Sahara.[15]
[c] Si-PIN with an area of 7 mm^2 and a thickness of 300 μm by AMPTEK.[8]
[d] Si-PIN with an area of 25 mm^2 and a thickness of 500 μm by AMPTEK[8] or Moxtek.[9]
[e] Si-drift with an area of 3 mm^2 and a thickness of 300 μm by Roentec Inc.[11] or EIS.[6]
[f] HgI$_2$ with an area of 5 mm^2 and a thickness of 1 mm by Constellation Techn.[12]
[g] CZT with an area of 9 mm^2 and a thickness of 2 mm by AMPTEK[8].
[h] The minimum and maximum detectable energy depends on the thickness of the window (Figure 5.3.8) or of the detector, respectively.

Generally an X-ray spectrum is composed of a background, due essentially to multiple scattering and noise, and a set of peaks (both fluorescence and scattered) superimposed to the background. The quantity of interest is the net area under each photoelectric peak. To determine it, two possible approaches can be used:

- a model of the spectrum including the background contents as well as the peaks part;
- extraction of the background on the basis of a model or a priori hypothesis.

The first approach requires a complete knowledge of the sample, that is generally not available. Thus the second approach is generally used.[16–18,33] After extraction and peaks identification, a qualitative description of the chemical elements contents can be given, due to some kind of proportionality of the peak area to the element content.

However, if a quantitative determination of the concentrations is required, then peak area must be further analysed. In fact, the area of a X-ray fluorescence peak (number of recorded photons during the analysis time) can be altered by enhancements and self-absorption phenomena from the matrix of the sample or from the environment (collimators, detector, air, etc.).

The concentration of a given element can be obtained in several ways, and different techniques have been developed. The most used in EDXRF is based on the so-called 'fundamental parameter determination' (FP). It is a set of mathematical equations of fluorescence emission based on fundamental physical parameters and on the instrumental parameters. The first practical algorithm was proposed by Criss and Birks[34] as a modification of the equations derived by Sherman[35] and Shiraiwa and Fujino.[36] A good review article of this and other methods can be found elsewhere.[18] However, due to the particular nature of the portable X-ray fluorescence systems, the FP method cannot be fully applied, but requires special procedures for taking into account the nature of the sample to be analysed or additional information from Compton and Rayleigh scattered peaks.

A different approach was proposed by Piorek et al.,[19] who made use of an identification technique based on a χ^2 test of the experimental data versus reference data. This technique was applied to commercial alloys and the results are satisfactory.

Future improvement of software for EDXRF analysis with portable equipment will mainly depend on increasing future computational power. Software which is currently only in top level desktop computers and uses sophisticated Monte Carlo codes will be possibly used in normal laptops and give automatic analysis in a few seconds.

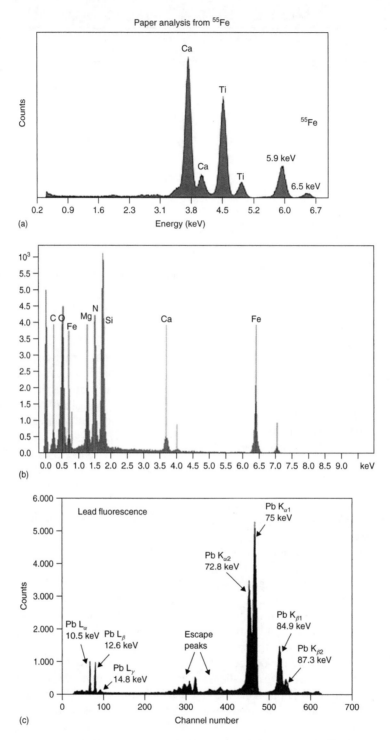

Figure 5.3.9 X-ray spectra from thermoelectrically cooled detectors. (a) Paper analysis with a ^{55}Fe source and Si-PIN detector;[8] (b) low-energy X-ray spectrum collected with a SDD;[11] (c) lead fluorescence with a CdZnTe[8] (with permission of Amptek and Roentec)

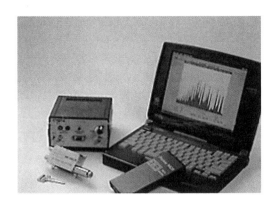

Figure 5.3.10 Amptek portable EDXRF analysis equipment with the pocket MCA. (With permission of Amptek Inc.)

Employing codes will also reduce the interaction with the user to a minimum.

In Section 5.3.2.5 some consideration is given to the software used by different manufacturers of portable EDXRF equipment.

5.3.2.5 PORTABLE EDXRF EQUIPMENT

In the last few years there has been an increasing number of commercial portable EDXRF equipment available. Some use radioactive sources while others use miniature X-ray tubes. The majority employ Si-PIN detectors, while several employ Si drift and HgI_2. Very few use other detectors, such as CdTe or CdZnTe (for higher-energy X-rays).

Much of this commercial equipment uses multichannel analysers with dedicated software for the automatic processing of the data.

Table 5.3.4, lists the manufacturers and the characteristics of their portable EDXRF equipment.

In addition to commercial EDXRF equipment, several research groups have designed and constructed their own portable equipment, by assembling in various combinations X-ray sources, detectors and MCAs.

The following is an exhaustive list of commercial EDXRF portable equipment, listed in an arbitrary order:

EIS

Portable EDXRF analysers were designed and constructed by EIS,[6] for applications in various fields (archaeometry, chemistry, industrial). They are composed of gas cooled 40–50 kV, 1 mA X-ray tubes with W anode and Be window, of a SDD with about 130–140 eV energy resolution at 5.9 keV, and of a MCA (Figure 5.3.11). The equipment is mounted on a tripod. The EIS equipment is characterised by the very good energy-resolution and reduced thickness of the Be window (Figure 5.3.12).

No dedicated software is available yet.

NITON

NITON Co. (Bedford, MA, USA)[37] has developed several models of portable EDXRF instruments, such as XL-300, XL-500, XL-700 and XL-800, according to specific applications. The model XL-300 is specifically intended for lead analysis; the model XL-500 (Prospector) is intended for mining applications; XL-700 is specifically dedicated to thin film and filters analysis; and the model XL-800 for alloy analysis. All models are composed of a radioactive source and a thermoelectrically cooled detector. The employed radioactive sources are: ^{109}Cd (10 mCi) as standard source, while ^{241}Am, ^{55}Fe is available upon request.

Detectors such as thermoelectrically cooled Si-PIN or CdZnTe are employed.

All components are included in a container (size 206 × 75 × 45 mm and weight about 1 kg) (Figure 5.3.13).

As the instrument performs multiple measurements of a sample, the liquid crystal display provides the following information: the reading number; the serial number of the measurement; the elements that have been detected; and the spectrum. A summary screen is displayed after the instrument completes a protocol. Measurements for each detected element are automatically displayed in μg.

The mode of operation, 'paint', 'thin' or 'bulk' is selected from the main menu displayed on the LCD screen before a test is performed. Measurements for each detected element

Table 5.3.4 Characteristics of portable EDXRF equipment and manufacturers

Manufacturer	Radio active source	X-ray tube anode $kV_{max} I_{max}$	Detector and energy resolution (eV)	MCA and software	Typical Application	Weight of the measuring head	Cost (Euros)
EIS	–	W 35 kV 1 mA	Si-drift (140 eV)	MCA	Various	≈5	35 000
NITON[a] XL-800	^{109}Cd ^{241}Am ^{55}Fe	–	Si-PIN (180 eV) CZT	MCA + dedicated software	Alloys	1	Not available
NITON[a] XL-300				MCA + dedicated software	Lead in paint	1	Not available
AMPTEK	–	Mo 35 kV 0.1 mA	Si-PIN (160 eV)	MCA	Various	3	13 500[b]
Thermo MT 9000 XRF field Analyser	^{55}Fe ^{109}Cd ^{241}Am	–	HgI$_2$	MCA + dedicated software	Various, soils and minerals	1.9	Not available
ThermoN Metallurg. Pro	^{55}Fe ^{109}Cd ^{241}Am	–	HgI$_2$	MCA + dedicated softwares	Alloys	1.9	Not available
EDAX[a] Map-4[c]	^{57}Co	–	CZT	MCA + dedicated software	Lead in paint	2.5	Not available
EDAX[a] CT3000[c]	^{109}Cd ^{241}Am ^{57}Co	–	Si-PIN	MCA + dedicated softwares	Alloys	2.5	30 000–50 000
Roentec ARTAX	–	W, Mo 50 kV 1 mA	Si-drift X-flash	MCA	Works of art	High	78 000
XRF Co.	^{153}Gd	–	CZT	MCA + dedicated software	Lead in paint	Not available	≈10 000
Oxford Instrum. Horizon 600[a]	–	W, Ag, Mo, …30 kV	Peltier-cooled detectors	MCA + dedicated software	Alloys	2	30 000
Warington Inc. LeadStar	^{57}Co	–	CdTe	MCA + dedicated software	Lead in paint	<1 kg	16 000

[a] Battery-operated.
[b] Obtained by adding the price of the single components.
[c] EDAX no longer markets and sells radio isotope based products. EDAX now supplies the 'Alloy Checker', which is an X-ray tube based system weighing approximately 2.2 kg. It replaces the CT3000 system and the parameters are almost identical.

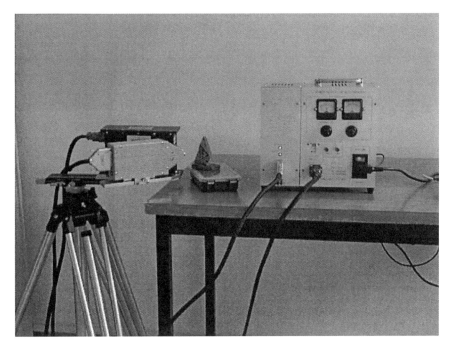

Figure 5.3.11 EIS portable EDXRF equipment, composed of a small size W anode X-ray tube (40–50 kV, 1 mA) and a high resolution SDD. (With permission of EIS)

are automatically displayed and this information is recorded for later transfer to a PC. Detection limits at 99.7 % confidence for a 1 min measurement for several sample types are reported in Table 5.3.5.

The following applications were specifically studied by NITON:

- lead in paint detection;
- soil contamination;
- site profiling;
- confirming containment programmes;
- on-site analysis of dust wipe samples;
- screening of sludges and liquids;
- working exposure monitoring;
- air monitoring;
- coating and filter analysis.

At the end of 2002, Niton Instruments released a new generation of portable EDXRF systems: (a) the pistol shaped XL_t, an X-ray tube based analyser; and (b) the banana shaped XL_i, an isotope-based analyser using film[241,37].

AMPTEK

AMPTEK Inc.,[8] developed over the last few years the Si-PIN detector, which completely changed the approach of portable EDXRF equipment. Over the years detectors with increasingly better performances were constructed and currently Si-PIN detectors with 158 eV at 5.9 keV are available. Also, high resolution CdZnTe detectors were recently constructed by AMPTEK.

AMPTEK researchers applied these detectors to portable EDXRF systems, using various radioactive sources, X-ray tubes and the Pocket AMPTEK multichannel analyser (Figure 5.3.10).

In 2001, a LASER X-ray tube was designed and constructed by AMPTEK, completely battery equipped. This X-ray tube is contained in a single compact enclosure, which includes the X-ray tube, the power supply and the control electronics. The LASER X box has a weight of 450 g and overall dimensions of $185 \times 71 \times 33$ mm.

LASER X has been designed to simplify the EDXRF process by providing a grounded anode, variable current and voltage with operation ease. It

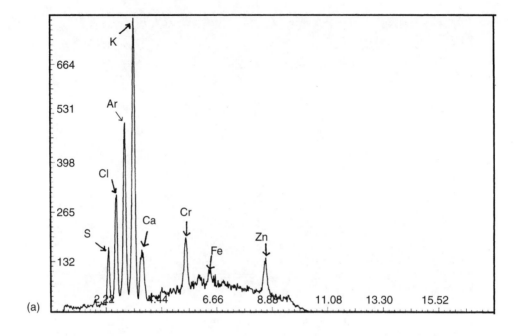
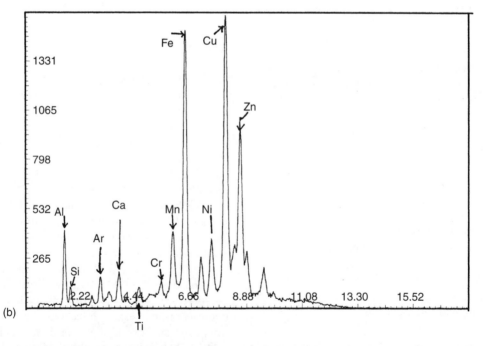

Figure 5.3.12 Typical X-ray spectra obtained with the equipment in Figure 5.3.11. (a) X-ray fluorescence spectrum of the head of a match, containing P, S, Cl, Ar (from the air), K, Ca, Mn, Fe, Zn. (b) X-ray fluorescence spectrum of an Al-alloy, containing Al, Cl, Ar (from the air), K, Ca, Ti, Cr, Mn, Fe, Ni, Cu, Zn

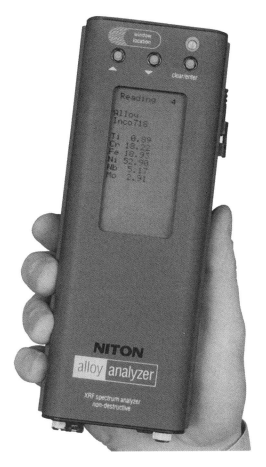

Figure 5.3.13 Niton portable equipment for EDXRF analysis. (With permission of Niton Europe GmbH)

Table 5.3.5 Detection limits (DLs) for the Niton 701 at 99.7 % confidence level for a 1 min measurement

Element	Thin samples (μg/cm^2)	DL (μg) for 25 mm diameter filter
L X-rays		
Lead	0.5	4
Uranium	0.6	<3
Thorium	0.6	<3
Mercury	0.7	4
K X-rays		
Arsenic	1	8
Chromium	12	20
Cobalt	<5	<10
Copper	1	4
Iron	10	15
Manganese	<12	<25
Molybdenum	<1	<3
Nickel	<3	7
Rubidium	<1.5	<3
Strontium	<1.5	<3
Zirconium	<0.8	<3
Zinc	0.8	3

features a 35 kV, 100 μA power supply with a Ag target and a Be end window of 250 μm.

Many possible applications for portable EDXRF equipment are indicated by AMPTEK, including the study of works of art, alloys analysis and analysis of lead.

Roentec ArtAX

The ArtAX μ-XRF portable system[11] is mainly composed of the following parts: (a) a X-ray fine focus tube, air cooled, with a W or Mo anode (other materials are available on request) and Be window of 0.2 mm thickness, which works at 50 kV, 1 mA and has a focal spot size of 1×1 mm; (b) a Si-drift X-Flash X-ray detector of 5 mm^2, and typical energy resolution of 140–170 eV at 5.9 keV. The detector head carries a color charge-coupled device camera and a sample illumination; (c) a XYZ stage with stepper motors for regulating the sample to source detector distance; (d) XRF software for spectra accumulation and storing, element identification, and peak and background net area calculation.

Monocapillary and polycapillary lenses for focusing the X-ray beam until 75 μm space resolution are available on request.

The ArtAX XRF system (Figure 5.3.14) is specifically designed for qualitative and quantitative analysis of works of art.

CT3000 by EDAX

The EDAX model CT3000[38] is specifically dedicated to alloy analysis. It is composed of a standard X-ray source of 15 mCi of ^{109}Cd (on request also 100 mCi ^{241}Am or 12 mCi ^{57}Co or 20 mCi ^{55}Fe), and of a Si-PIN detector. Qualitative analysis can be completed in seconds, while quantitative analysis is available in minutes, with a detection limit of the order of 5 ppm with accuracy ± 10 %. This

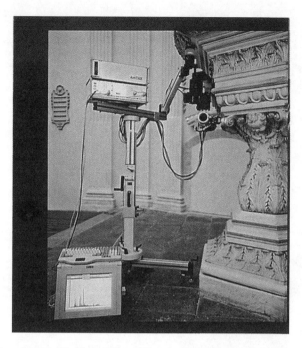

Figure 5.3.14 Roentec ArtAX equipment. The X-ray tube and detector, and the *XYZ* stage with stepper motors, is shown on the left and the control unit on the right. (With permission of Roentec GmbH)

instrument is contained in a case weighing approximately 2.5 kg (including the battery) and has a size of 25.4 × 20.3 × 10.2 cm (Figure 5.3.15). Dedicated software is included, based on a specialised FP method, that carries out automatic calibration, background subtraction, and quantitative evaluation based on fundamental parameters. Using the 'chemistry' mode of the 'Positive Metals Identification' (PMI) software, a quick and accurate result is obtained. The results of a measurement, the spectrum and the intensity data for each essay could be easily downloaded into a computer using RS232C serial cable. With two rechargeable NiCd batteries, the analyser could be used for 8 h of continuous operation.

Typical alloy analysis includes stainless steel, cobalt alloys, nickel alloys, copper–nickel alloys, bronzes, brasses, aluminium alloys, titanium alloys, molybdenum alloys and lead alloys.

As of 2003, EDAX no longer markets and sells radio isotope products. EDAX now constructs and supplies the 'Alloy Checker', an X-ray tube based system weighing approximately 2.2 kg (Figure 5.3.15). It replaces the CT3000 system and the specifications are almost identical.

9000 XRF Field Analyser and Metallurgist Pro by Thermo Measure Tech and ThermoNoran

Thermo Measure Tech[39] constructs a portable EDXRF instrument (9000 XRF Field Analyser, Figure 5.3.16) mainly composed of: (a) radioactive excitation sources such as ^{55}Fe for analysis of S to Cr, ^{109}Cd, for analysis of Ca to Rh (K X-rays) and Ba to U (L lines), and ^{241}Am for analysis of Cu to Tm (K X-rays) and W to U (L lines); (b) thermoelectrically cooled HgI_2 detector; (c) 2000 channels multichannel analyser with spectrum smoothing, energy calibration and automatic peak identification; (d) dedicated software packages. The 9000 XRF unit includes a probe, with X-ray source and detector, and the MCA. The probe weighs 1.9 kg and has dimensions of 12.7 × 7.6 × 21.6 cm,

(a)
- "Unlimited" mode assays take as little as 1 to produce a definitive lead classification with 95% confidence and no substrate correction.
- A full spectrum analyzer which provides both K and L shell results simultaneously and can store over 6000 assays, including full energy spectra.

(b)

Detection Limits (mg/kg)

Antimony	14
Arsenic	5
Barium	6
Cadmium	10
Chromium	95
Mercury	10
Lead	7
Nickel	15
Silver	15
Selenium	4
Thallium	11

(c)

Typical Alloy Analysis

Stainless Steel
Hast Alloys
Tool Steels
Cobalt Alloys
Nickel Alloys
Copper–Nickel alloys
Brasses
Bronzes
Aluminum alloys
Zirconium alloys
Titanium alloys
Molybdenum alloys
Lead alloys

Figure 5.3.15 EDAX portable EDXRF models. (a) MAP-4 for lead-based paint analysis; (b) CT2000 for metal analysis; (c) CT3000 for alloy analysis. (With permission of EDAX)

and is suitable for either *in situ*, field or laboratory measurements. The 9000 XRF Analyser can be operated from a battery/AC line (Figure 5.3.16). Source selection and acquisition are automated through stored, user-defined analysis procedures.

Each data acquisition procedure includes selection of source, acquisition time, computation method and display of analysis results. Components of the microprocessor based analyser include a user interface keypad, LCD display, 2000 MCA, data storage memory and RS-232 communication port. The system may be powered by either an internal battery pack or external AC power with results up to 300 analyses (each with up to 25 elements). The Application Generator is a PC software package. Its menu-driven operator interface provides easy access to a complete range of software tools for developing procedures.

The 9000 XRF Field Analyser uses the FP method. Three operations are sequentially performed (Figure 5.3.16): (a) measurement of the X-ray lines intensity from a pure element sample; (b) determination of the fluorescent yield of the analyte (from tabulated data); (c) determination of the matrix terms that can alter the original intensity emitted (enhancement and self-absorption phenomena).

Several calibration models are also available. Analysis results from each model are compared with results from the others to determine the

Figure 5.3.16 9000 Field Analyser for portable EDXRF analysis and working operations. (With permission of Thermo MeasureTech)

optimum analysis conditions for the application. Typical detection limits for soil analysis with various radioactive sources are shown in Figure 5.3.17.

The Metallurgist Pro equipment uses ^{55}Fe, ^{109}Cd or ^{241}Am sources with a HgI$_2$ thermoelectrically cooled detector (Figure 5.3.18). The equipment is specifically dedicated to alloy analysis, and is able to analyse up to 21 elements at the same time. It is preprogrammed with a comprehensive library of over 225 alloys, with the capability for the user to input 25 custom alloys.

Table 5.3.6 shows the minimum detection limits of the equipment for various elements in alloys.

Horizon 600 Alloy Sorter from Oxford

Portable EDXRF equipment was recently constructed by Oxford,[5] called the Horizon 600 Alloy Sorter. This equipment, shown in Figure 5.3.19, is specifically dedicated to analysis of alloys. It is able to analyse elements from calcium ($Z = 20$) to uranium ($Z = 92$), and up to 20 elements simultaneously. It is mainly composed of a digital 'cold cathode' X-ray tube working up to 30 kV, and equipped with a programmable primary beam filter. The detector is a high resolution, thermoelectrically cooled solid state semiconductor detector with the pulses from the pulse amplifier processed in a 1024 MCA. Dedicated software automatically gives the elements of the analysed alloy with their corresponding concentration. The equipment is fully battery operated and has a weight of approximately 2 kg and a size of $24 \times 20 \times 12\,\text{cm}^3$.

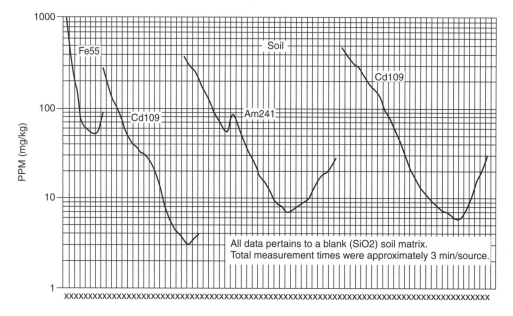

Figure 5.3.17 Typical minimum detection limits for soil analysis with the 9000 Field Analyser. (With permission of Thermo MeasureTech)

Table 5.3.6 Minimum detection limits (in %) of the Metallurgist Pro for various elements in different alloys, in a measuring time of 19 s

Element	Mild steel	Stainless	Inconel	Cu–Zn	Aluminium
Ti	0.015	0.02	0.05	0.03	0.015
V	0.008	0.02	0.02	0.01	0.01
Cr	0.06	0.3	0.4	0.06	0.07
Mn	0.15	0.25	0.2	0.04	0.06
Fe	0.5	0.5	0.25	0.07	0.03
Co	0.2	0.3	0.2	0.05	0.02
Ni	0.15	0.4	0.5	0.08	0.015
Cu	0.08	0.2	0.25	0.35	0.015
Zn	0.05	0.1	0.25	0.3	0.012
Se	0.02	0.02	0.02	0.05	0.003
Zr	0.03	0.04	0.04	0.04	0.003
Nb	0.005	0.008	0.06	0.008	0.003
Mo	0.005	0.015	0.1	0.008	0.003
Ag	0.6	0.6	0.9	0.8	0.8
Sn	0.15	0.15	0.2	0.2	0.2
Ta	0.1	0.15	0.2	0.9	0.025
W	0.15	0.2	0.2	0.9	0.025
Au	0.1	0.15	0.15	0.4	0.02
Pb	0.035	0.04	0.04	0.04	0.005
Bi	0.035	0.04	0.04	0.05	0.005
As	0.03	0.04	0.04	0.07	0.006
Re	0.08	0.1	0.1	0.35	0.015
Y	0.01	0.01	0.01	0.01	0.002

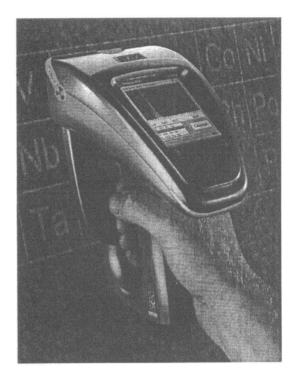

Figure 5.3.19 The Horizon 600 Alloy Sorter. (Provided courtesy of Oxford Instruments)

Figure 5.3.18 The Metallurgist Pro model. (With permission of Thermo Noran)

ICS-4000 by XRF Corporation

XRF Corporation[40] manufactures analytical instrumentation to detect γ- and X-radiation in the field. The ICS-4000 spectrometer is specially designed for fast, accurate field analysis in a variety of health physics, nuclear medicine and environmental monitoring applications. This device incorporates a sensitive CdZnTe detector with a powerful MCA.

The XRF/Pb model is specifically devoted to EDXRF analysis of lead in painted surfaces, dust and soil. This hand-held equipment is basically composed of a ^{153}Gd radioactive source, a CdZnTe detector and a MCA (Figure 5.3.20). An annual source replacement is recommended.

X-MET 880, 970 and 2000 from Metorex

The X-MET equipment from Metorex[41] is specifically designed and constructed for alloy analysis and control of coating processes. The equipment is mainly composed of a radioactive source (^{55}Fe, ^{109}Cd or ^{241}Am) of a high-resolution Si-PIN detector and a 2048 channel MCA for data acquisition. Dedicated software can utilise either fundamental parameters or empirical methods. Data collection, analysis and management is completely automated. The software can store thousands of analytical results.

The X-MET equipment have a weight of 5.8 kg and dimensions of $36 \times 29 \times 10$ cm^3 (Figure 5.3.21). The measuring head has a weight of about 2 kg.

Figure 5.3.20 The ICS-4000 portable EDXRF analyser for lead analysis. (With permission of XRF Co.)

Figure 5.3.21 The X-MET 880 Alloy Analyser. (With permission of Metorex International Oy)

The X-MET 880 model is a portable tool for analysis in the field. It is capable of measuring 80 elements in the periodic table from atomic number 13 (aluminium) through to atomic number 92 (uranium). Analyses includes stainless steels, chrome molybdenum steels, tool steels, alloy steels, nickel, cobalt, copper, titanium, aluminium, magnesium, zinc and lead based alloys.

The X-MET 970 model is specifically suited for real-time analysis of thin and ultra-thin coatings.

Typical applications include: phosphate on zinc coated steel, where P is analysed in a range of 0.5–2 g/m^2, in a measuring time of 5 s; chromate on zinc coated steel, where Cr is measured in a range of 0–70 mg/m^2, in a measuring time of 120 s; chromium on aluminium, where Cr is analysed in a concentration range of 0–70 mg/m^2 in a measuring time of 120 s; and titanium on aluminium, where Ti is analysed in a range of 0–40 mg/m^2, in a measuring time of 120 s.

The X-MET 2000 is specifically dedicated to alloys analysis. The following alloys can be typically analysed: nickel alloys, containing Ti, Cr, Mn, Fe, Co, Ni, Cu, Nb, Mo, W; stainless steels, containing Ti, V, Cr, Mn, Fe, Co, Ni, Cu, Nb, Mo, W; cobalt based alloys, containing Cr, Mn, Fe, Co, Ni, Mo, W; low alloy steels, containing Cr, Mn, Fe, Ni, Cu, Mo; tool steels, with V, Cr, Mn, Fe, Co, Ni, Mo, W; aluminium alloys, with Mn, Fe, Ni, Cu, Zn; copper alloys, with Mn, Fe, Ni, Cu, Zn, Sn, Pb; titanium alloys, containing Ti, V, Fe, Cu, Zr, Nb, Mo, Sn to name a few, there are additional alloys as well.

Lead Star and μ-Lead by Warrington

Two models of portable EDXRF instruments were developed by Warrington Inc. for lead analysis of paint.[42] The first one, μ-LEAD, is based on a 10 mCi ^{57}Co source, and a scintillation detector with filtering techniques. The second one, the LeadStar, uses a CdTe detector. Both employ MCAs and dedicated software. Figure 5.3.22 shows the LeadStar instrument.

Besides commercial portable EDXRF equipment, various additional equipment has been designed, constructed and developed for scientific purposes by various research groups around the world. They will be discussed in the next section.

5.3.3 APPLICATIONS

As shown above, portable EDXRF equipment can be employed in many fields and for many applications. In the following, typical applications are shown both of commercial equipment and of

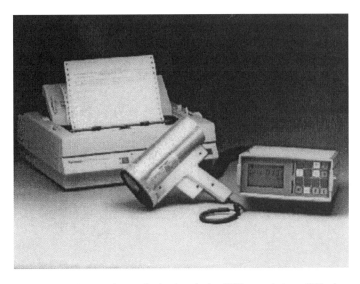

Figure 5.3.22 The LeadStar instrument by Warrington for lead analysis. (With permission of Warrington Inc.)

equipment assembled by various research groups in the following fields:

- archaeometry;
- environmental analysis;
- analysis of lead in paint;
- analysis of industrial alloys;
- soil analysis;
- mineral analysis;
- *in vivo* and human analysis of trace elements;
- analysis of soil on Mars (Pathfinder mission).

Some selected applications in some of these fields are discussed in the following.

5.3.3.1 ARCHAEOMETRY

Portable EDXRF equipment is, of course, especially suited for analysis of works of art, that require nondestructive and on site analysis.

In the last 5 years Cesareo and co-workers[43–49] have designed and constructed various portable EDXRF equipment for applications in the field of archaeometry:

- For analysis of S and Cl in frescoes, they developed equipment composed of a Ca anode X-ray tube, working at 5 kV, 0.1 mA (which is not able to excite Ca), a thin-window Si-PIN detector and an AMPTEK pocket MCA. This equipment can also excite lower Z elements, such as Al, Si and P[43–45] (Figure 5.3.23).
- For analysis of ancient gold alloys, composed of gold, silver and copper, and bronzes, generally composed of copper, tin and lead, they employed a W anode 30–35 kV, 0.2 mA X-ray tube, a Si-PIN detector and an AMPTEK pocket MCA[46–48] (Figure 5.3.24).
- For analysis of pigments in paintings and frescoes. The same system can be employed, used for analysis of alloys (Figure 5.3.24).

The most recent and interesting application of portable equipment used in the study of works of art was the analysis of the famous frescoes in the Chapel of the Scrovegni painted by Giotto in Padua during the period 1303–1305.[49]

Begun in 1303 and consecrated on 25 March 1305, the chapel, dedicated to Our Lady of the Annunciation, was commissioned by Enrico Scrovegni in suffrage for the soul of his father, Reginaldo, accused of usury. It was E. Scrovegni who commissioned Giotto to execute the frescoes in the interior of the chapel; this cycle of paintings

Figure 5.3.23 Portable equipment for the analysis of sulfur in frescoes and monuments. A Ca anode X-ray tube is employed, working at 6 kV and 0.1 mA, a thin window Si-PIN detector, and a SILENA MCA

signals 'a point of no return in the entire history of western painting'.

In the period of EDXRF analysis, i.e. July 2001–January 2002, the frescoes were under restoration, and the following problems were examined:

- detection of the presence of S or Cl on the surface of the frescoes, which is a signal of pollution, and its removal;
- detection of the presence of elements showing previous restoration areas (signaled, for example, by the presence of titanium or zinc);
- determination of the pigments used by Giotto, with special notice to gold haloes.

The variety of X-ray spectra of pigments is shown in Figure 5.3.25.

The possible presence of sulfur and chlorine was determined with two different types of equipment: one using the Ca anode X-ray tube, described previously, and the second one using a Pd anode X-ray tube working at 6–8 kV, to selectively excite Pd L lines, which, having an energy of about 2.8 keV, are suited to the excitation of sulfur and chlorine. The use of the Ca anode X-ray tube gives rise to a 'cleaner' spectrum with respect to the Pd L X-ray tube, but the counting rates are much lower, due to the large output window of the first tube (X-ray tubes output irradiate an area of about 1 cm^2).

The fresco pigments were also analysed with the same Pd X-ray tube working at 10 kV, and with a W X-ray tube working at 30–35 keV. The following results were obtained:

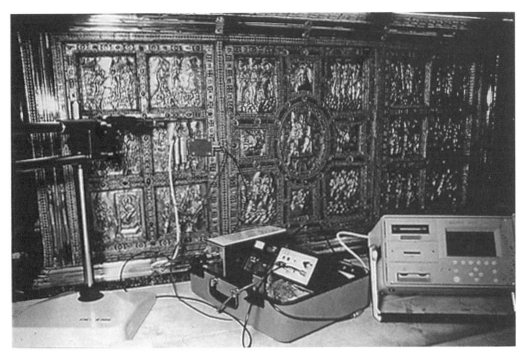

Figure 5.3.24 Portable equipment used for the analysis of alloys (bronzes, brasses, gold) and pigments. A W anode X-ray tube is employed, working at 35 kV, 0.3 mA and a Si-PIN or Si-drift detector. The equipment is shown during measurements on gold composition of the altar of S. Ambrogio, in Milan

- Sulfur was detected everywhere, at a concentration level from about 1% to about 10%, depending on the exposition and on the undergoing pigment; sulfur content was for example lower in the case of azurite pigments, higher in the white and green pigments.
- The S content strongly decreased after using a cleaning process based on ion-exchange resins (Figure 5.3.26)
- Chlorine was detected only once, in an area that was possibly recently cleaned.
- Titanium was detected in many areas, indicating a recent restoration; additionally this element was then 'cleaned'.
- Gold of the haloes shows a complicated X-ray spectrum, corresponding to a number of layers (Figure 5.3.27).

About 40 haloes were analysed, many of them in good condition (gold haloes), others damaged, and others completely black.

X-ray spectra of a good condition gold halo compared with a black halo are shown in Figure 5.3.27. From left to right peaks are visible due to the following elements: sulfur K lines at 2.3 keV, due to pollution effects; argon K lines, at 2.95 keV, due to the air; tin Lα lines, at 3.45 keV; calcium K lines, at 3.7 keV, mainly due to the plaster; iron Kα and Kβ lines, at 6.4 and 7.06 keV; nickel Kα and Kβ lines, at 7.5 and 8.3 keV, due to background effects; copper Kα and Kβ lines, at 8.04 and 8.94 keV; tungsten L lines, at 8.35, 9.8 and 11.3 keV, respectively, due to the X-ray tube anode; gold L lines, at 9.67, 11.5 and 13.4 keV; silver K lines, at 22.1 and 25.2 keV, due to fluorescence effects in the detector; lead L lines, at 10.5, 12.6 and 14.8 keV; strontium Kα lines, at 14.15 keV, due to the plaster; tin Kα and Kβ lines, at 25.2 and 28.7 keV, respectively.

There are several cases of partial or total peaks overlap: sulfur K with lead M, tin L with calcium K, gold Lα with tungsten Lβ.

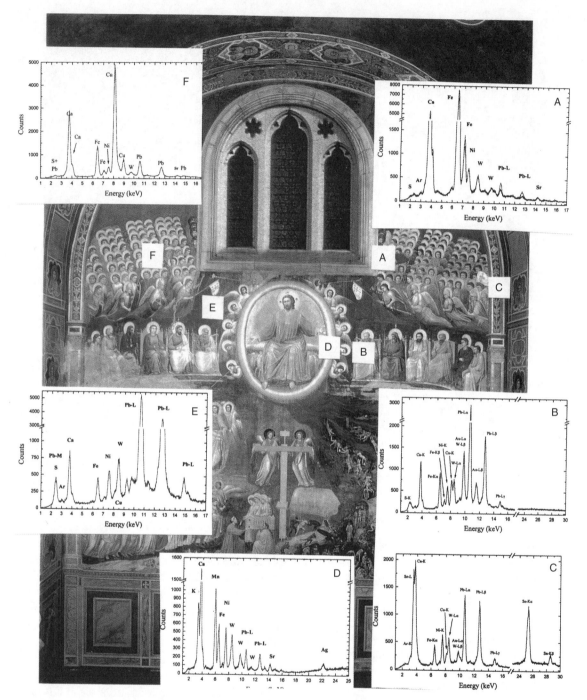

Figure 5.3.25 A general view of *The Last Judgement* by Giotto in the Chapel of the Scrovegni, Padua, Italy. There is a variety of pigments producing a variety of X-ray spectra: A–a red flag; B–a gold halo in good condition; C–a black halo; D–medallion to the left of God; E–white flag; F–green flag

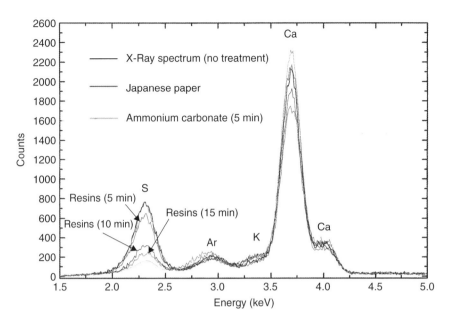

Figure 5.3.26 X-ray spectra of sulfur in frescoes, obtained with the Hamamatsu Ca anode X-ray tube, and with the Si-PIN detector

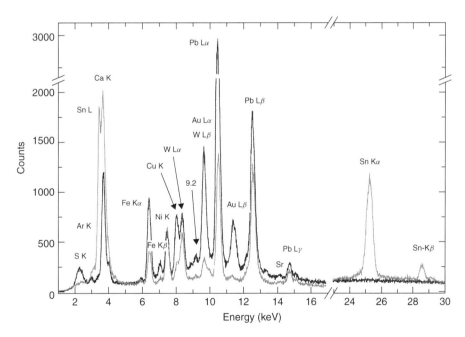

Figure 5.3.27 EDXRF spectra of a golden halo (darker line) and a black halo. The spectra are relatively complicated including: S K lines; Ar K lines; Sn Lα line; Ca K lines; Mn Kα line (in traces); Fe Kα and Kβ lines; Ni Kα line; Cu Kα line; W Lα and Lβ lines; Pb Lα, Lβ and Lγ lines; Sr Kα line and Sn Kα and Kβ lines. The differential attenuation of Pb L lines is clearly visible in both X-ray spectra: in the golden halo the Pb Lβ line is more attenuated by the Au leaf than the Lα line; in the black halo the Pb Lβ line is less attenuated by the effect of the Sn sheet. The difference in the copper content between the black and golden halo is also remarkable

Not considering X-rays of the elements argon, nickel, tungsten and silver, which are not due to the interaction of the X-ray beam with the fresco (argon is dependent on the air; nickel on the X-ray tube window, tungsten on the tube anode; and silver, at least partially, on the Si-PIN detector), the other X lines are related to the fresco pigments. However, they must be assigned to the proper fresco layer.

In the following, only haloes in good condition will be considered. First of all, the ratio of the X-rays of all elements with respect to gold L X-rays was calculated, and the Pb ($L\alpha/L\beta$) ratio (Table 5.3.7). If an element belongs to the gold alloy, then its ratio with respect to gold should remain approximately constant.

From these ratios and from the mean values it may be deduced that not one of the elements belongs to the gold alloy. Therefore, the gold should be of high purity.

Further, it may be deduced that lead/gold is not varying too much, and, therefore, lead can be assigned to the 'second level'. In this hypothesis, the Pb L lines should be attenuated in a different manner by the gold leaf. This effect is, in fact, clearly visible in Figure 5.3.27, where the differential attenuation of Pb $L\alpha$ and Pb $L\beta$ lines by gold and in the absence of gold is clearly visible.

Calculating this effect for all gold haloes in good conditions, the mean thickness of the gold layer may be calculated, which turns out to be 1.6 ± 0.5 μm. From this result it may be concluded that the gold leaf is extremely thin (minimum value 1.0, maximum value 2.6 μm) and of relatively constant thickness.

The attenuation of lead X lines (lead is present as white lead, i.e. basic carbonate of lead) by this gold leaf is about a factor of 2. The thickness of this layer of white lead can be calculated from the Pb/Au counts ratio and from the Au thickness as about 5 μm of Pb equivalent thickness, corresponding to a much larger thickness of the pigment.

The situation concerning the attribution of copper to the correct layer is complicated. Looking at the X-ray spectra, it turns out that X-rays of Cu are clearly more intense when the halo is superimposed to an azurite background, which is at a deeper layer than lead. Excluding these cases, the ratio Cu/Au will be lower and also more constant $\approx 0.6 \pm 0.25$. It is therefore reasonable that Cu X-rays come both from the azurite and from the glue between the lead carbonate and gold. This hypothesis seems to be confirmed by Figure 5.3.27. From the ratio Cu/Au ≈ 0.6 it turns out that the copper equivalent thickness between lead and gold is about 1 μm. Larger, by a factor of 2-3, should be the Cu equivalent thickness corresponding to azurite.

Finally calcium, iron and strontium should come, at least partially, from the plaster. However, in this hypothesis Ca K lines are attenuated by lead carbonate, copper and gold, giving rise to an attenuation factor of about 5×10^5, which is too high to give reasonable Ca counts in the X-ray spectra. These should also come from calcium carbonate deposit on the surface of the fresco, due to pollution.

In conclusion, the golden haloes in Giotto's frescoes are composed of the following layers:

- a superficial layer of sulfur, in the form of calcium sulfate due to pollution effects over the centuries;
- a layer of pure gold, with a mean thickness of 1.6 μm;
- a layer of glue containing copper, with an equivalent Cu thickness of about 1 μm;
- a layer of white lead, with an equivalent Pb thickness of about 5 μm;
- a possible layer of azurite, containing copper;
- the plaster containing iron and strontium.

5.3.3.2 ENVIRONMENTAL ANALYSIS: ANALYSIS OF LEAD IN PAINTS

One of the most serious public health hazards, which particularly affects children, is related to

Table 5.3.7 Pb($L\alpha/L\beta$) ratio and ratio of the intensity of elemental X-rays with respect to gold intensity

Pb($L\alpha/L\beta$)	Pb $L\alpha$/Au $L\beta$	Fe/Au	Cu/Au
1.74 ± 0.09	5.6 ± 2.2	4.1 ± 3.0	1.3 ± 1.4

lead-based poisoning from the paint found in many old houses. For example, the estimated number of apartments in New York City alone that may be affected by this hazard is greater than 300 000 and they are mainly occupied by low-income families. Recent legislation in the US provides specific requirements for new inspection procedures in federally funded housing programmes as well as for the disclosure of information and inspections during the sale and transfer of all private residential houses constructed prior to 1978. Inspection companies and state and federal agencies, which are charged with the responsibility of supervising compliance with this legislation, employ EDXRF hand-held instruments for on-site detection of lead concentration in paint. An abatement plan to eliminate paint poisoning hazards can be required for concentrations of lead in paint equal to or exceeding 1 mg/cm^2. This application also demonstrates the necessity of using K X-rays in the determination of Pb concentration. The Pb L X-rays can only serve as supplementary information, due to their strong attenuation by paint overlays. The high energy of the Pb K X-rays (75 and 85 keV) imposes also additional restrictions on the choice of the excitation source and detector.

Excitation radioactive sources can be used, such as ^{57}Co and ^{109}Cd; in this case they are more practical than X-ray tubes for portable instruments. Further, Peltier-cooled CdZnTe or HgI$_2$ near room temperature are favoured in this application.

5.3.3.3 ANALYSIS OF INDUSTRIAL ALLOYS

The last 20 years are marked by the continuously increasing use of EDXRF analysers for alloy assaying and identification, specifically in this field. This trend has been initiated by the advent of inexpensive memory and microprocessor chips. Another factor contributing to the success of EDXRF in alloy sorting is its ability to make the decision about alloy grade automatically.

Important properties of alloys are the direct consequence of their chemical composition. Up to about 50 chemical elements are involved to make thousands of alloys known to be in use. However,

Table 5.3.8 EDXRF performance in identification of commercial alloys

Alloy	Identification results (%)
Ni–Cu	100
Cu	90–100
Cr-Mo steels	95–100
Ti alloys	95–100
Al alloys	90–100

only about 10 to 20 elements can be found in any single alloy and only about 10 elements need to be monitored in a sample of an alloy in order to positively identify it.

The most obvious approach to alloy identification would be to measure the chemical composition of an alloy, and then compare its composition with tabulated data of elemental concentration ranges for each alloy. This method of identification is not very efficient because of the relatively long time required. Faster and more effective algorithms have been developed, using a mathematical formalism of the χ^2 distribution, based on comparison of the net X-ray intensities of selected elements in the unknown sample, with those of the known reference samples.[19] Typical identification results and performances are shown in Tables 5.3.8 and 5.3.9, where Metorex X-MET 2000 equipment was employed, which uses a ^{109}Cd radioactive source.

5.3.3.4 ANALYSIS OF SOILS AND ROCKS

Soil analysis is a typical subject for commercial EDXRF equipment (Table 5.3.4), but many researchers are working in this field also with self-made equipment.

Field analysis of soils and rocks by portable EDXRF analysis was systematically carried out by Potts *et al.*[50–54] They employed commercial equipment with ^{55}Fe, ^{109}Cd and ^{241}Am sources and a HgI$_2$ detector for analysing soil and rocks *in situ*. The authors first developed a correction procedure for surface irregularity effects. In fact, very few samples on which *in situ* measurements are made are perfectly flat as calibration samples. Discrepancies in measurements will then occur. For real samples, the most common situation is that there is an air gap between the sample surface and the analyser head. A simple correction procedure was investigated, based on the intensity of the scatter peaks observed when the sample is excited with (monoenergetic) radioactive sources. The principle of the procedure is that if the intensity of this scatter peak is affected by surface irregularity effects to the same extent as fluorescence intensity, and if the scatter peak intensity from a perfectly flat surface is known, then it is possible to calculate a corrected fluorescence intensity representative of that that would be observed for a flat surface in contact with the portable EDXRF analyser. The measurements showed that the correction is effective for air gaps of up to 3 to 4 mm.

Potts et al.[52] further studied the effects of granulometry. Accepting that almost all silicate rock samples are crystalline, it is clear that for medium and coarse grained rocks, the volume of sample from which the X-ray fluorescence signal originates may not have a mineral assemblage that fully represents the average composition of the sample. This effect will cause additional uncertainty in the measurement due to the finite number of minerals of each type within the excited volume. A series of portable EDXRF measurements was made, therefore, of a range of flat polished blocks of rock, having grain sizes that varied from fine to medium to coarse. Results indicated that a 10 % sampling precision can generally be achieved for a single determination on fine to medium grained rocks. When a coarse grained Shap granite was investigated, the corresponding figure for the number of separate determinations that must be averaged to achieve 10 % sampling precision was 21. The conclusion was that grain size may have a considerable effect on the precision of analytical results.

Argyraki et al. used the same equipment for the determination of lead in contaminated soil,[53] i.e. a medieval lead smelter site at Bowle Hill, Wirksworth, Derbyshire, active over 500 years ago.

Portable EDXRF analysis is particularly suited for the analysis of lead, because the detection limits of portable equipment (typically about 50 ppm) are a factor of 10 times lower than the trigger level for this element (500 ppm for domestic parks and allotments) so that sensitive measurements can be made down to concentrations at which a risk assessment would be necessary to assess the impact of the lead contamination.

The site at Bowle Hill is currently a grass field covering the upper part and top of a scarp slope which overlooks a valley. A detailed investigation by Imperial College using Auger sampling and Laboratory ICP-AES was made for comparison. The following observations resulted:

- PXRF data gave systematically low results because of the presence of moisture.
- Further discrepancies in the PXRF results are due to the fact that the analysed sample surface is not flat.
- Care is required in comparing laboratory measurements on samples removed from the site with *in situ* measurements because of differences in the nature of the samples, even if determinations are made at the same location.
- It is essential to collect a certain proportion of samples in duplicate.

Table 5.3.9 Typical performance of the EDXRF analyser in alloy analysis[41]

Alloy	Ti	Cr	Mn	Fe	Co	Ni	Cu	Zn	NbMo	Sn	Pb
Low alloy steel	0.01	0.02–0.04	0.1	0.2–0.25	0.25–0.50	0.1	0.05–0.25	0.10–0.15	0.006	0.15	0.15
Stainless	0.02	0.2	0.1	0.2	0.1	0.2	0.06	0.2	0.01	0.3	0.05
Steels	0.03	0.3	0.2	0.3		0.5	0.1		0.03		0.3
Ni/Co Alloys	0.1	0.2–0.5	0.1–0.3	0.12–0.5	0.1–0.5	0.2–0.5	0.05–0.3	0.3	0.04–0.08	0.3	0.15
Cu-alloys	0.02	0.1	0.02–0.06	0.02–0.06	0.05	0.05–0.08	0.15–0.4	0.03–0.07	0.01	0.008–0.2	0.2–0.3
Al-alloys	0.02	0.05–0.2	0.1	0.04–0.1	0.05	0.04	0.05	0.06	0.003–0.005	0.005–0.2	0.01–0.02
Ti-alloys	0.2–0.6	0.1	0.1	0.06	0.05	0.05	0.02	0.02	0.01	0.005	0.01

- Provided *in situ* results are corrected for moisture content and for surface irregularity effects, there is no significant bias between *in situ* portable EDXRF analysis and laboratory measurements for lead. However, greater variation was observed in the portable EDXRF results, because of the much smaller mass of soil from which the analytical signal was derived.

Similar measurements were carried out by Potts *et al.*[54] to evaluate sources of arsenic contamination at a heritage industrial site.

Kump *et al.*[55] developed a quantification procedure for *in situ* EDXRF analysis of soil, based on the spectrum analysis file from AXIL.[56] Then a Montana soil SRM 2710 was analysed in the laboratory both with a Si(Li) and a Si-PIN detector. Typical results are shown in Table 5.3.10.

5.3.3.5 ANALYSIS OF TRACE ELEMENTS *in vivo* AND IN HUMANS

The use of X-ray fluorescence for the determination of heavy metals such as lead, cadmium, platinum, mercury and gold *in vivo* is now a fairly widespread technique.[1,56–59] *In vivo* X-ray fluorescence was first used to measure both the concentration and distribution of iodine in the thyroid. This particular application had the advantage from the analytical point of view that the iodine concentration in the thyroid is comparatively high (≈ 400 ppm) and the thyroid is a relatively superficial organ. An increasing range of applications followed, the first of which was lead in fingerbone and teeth. Bone lead measurements have subsequently been extended to include the tibia and calcaneous, while the other elements studied include mercury and strontium in bone, and cadmium, platinum and lead in the kidney.

The use of external radioisotope sources has been very successful for the determination of metals in organs with little overlying tissue, such as lead in the tibia. However, for organs at larger depths, much higher skin doses and longer irradiation times are required.

Concerning lead analysis, the tibia is typically tested because of the favourable and superficial position of this bone. Lead concentration in the tibia ranges between a few ppm to about 100 ppm.

Two X-ray fluorescence techniques are traditionally employed for the *in vivo* measurement of lead in bone: K X-ray fluorescence (KXRF) and L X-ray fluorescence (LXRF). The KXRF technique has been more widely used and validated. It measures lead approximately 37 mm into the bone and therefore, it provides data on the total amount of lead throughout the bone. The LXRF technique, which has been used mainly in paediatric studies, measures lead only 2 to 3 mm into the bone.

Typical, modern and portable equipment for KXRF analysis uses a radioactive source for K X-ray excitation (^{57}Co, ^{152}Gd or ^{109}Cd, using the 88 keV emission, but also ^{99m}Tc and ^{133}Xe were previously employed) and a semiconductor detector that can be a nitrogen-cooled HpGe, or Peltier cooled, high resolution and large area CdZnTe or HgI$_2$.

For LXRF analysis portable equipment can be employed, which uses a low-power Mo anode (or other material) X-ray tube and a large area Si-PIN detector.

Table 5.3.10 Analysis results and reference data for SRM 2710 Montana soil (g/g)

Element	Reference data	Measured by Si(Li)	Measured by Si-PIN
Al	0.0644	0.121	0.095
Si	0.289	0.31	0.31
S	0.0024	0.0049	0.0036
K	0.021	0.020	0.020
Ca	0.0125	0.0115	0.0121
Ti	0.00283	0.0023	0.0023
Mn	0.01	0.0105	0.01
Fe	0.0338	0.0371	0.0345
Ni	0.000014	–	0.00028
Cu	0.000295	0.0032	0.0031
Zn	0.00695	0.0073	0.0066
As	0.00063	0.00095	0.00093
Pb	0.0055	0.0045	0.0053
Rb	0.00012	0.000114	0.000077
Sr	0.00033	0.00030	0.00026
Y	0.000023	0.000037	0.000017
Zr	–	0.00012	0.00009
Nb	–	0.0000072	–

5.3.3.6 ANALYSIS OF SOIL ON MARS; THE α-PROTON X-RAY SPECTROMETER (MARS PATHFINDER APXS)

A very interesting application of portable equipment for X-ray analysis is the portable apparatus transported on the Martian surface to determine the chemical composition of Martian soil and rocks[60–62] (Figure 5.3.28). On 4 July, 1997, the portable EDXRF apparatus constructed for analysing rocks landed on Mars.

The principle of the APXS technique is based on three interactions of α particles from a radioisotope source with matter: (a) simple Rutherford backscattering; (b) production of protons from (α,p) reactions on light elements; and (c) generation of characteristic X-rays upon recombination of atomic shell vacancies created by α bombardment. Measurement of the intensities and energy distributions of these three components yields information on the elemental chemical composition of the sample. In terms of sensitivity and selectivity, data are partly redundant and partly complementary: α backscattering is superior for light elements (C,O), while proton emission is mainly sensitive to Na, Mg, Al, Si and S, and X-ray emission is more sensitive to heavier elements (from Na to Fe and beyond). Actually, charged particle excitation is preferred to any other kind of excitation since it produces the best signal to noise ratio due to the absence of any Compton scattering.

The APXS equipment consists of two parts: the measuring head and the electronic box. The measuring head contains nine ^{244}Cm sources in

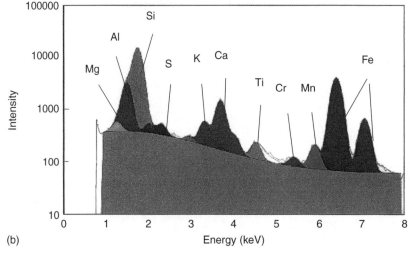

Figure 5.3.28 (a) APXS instrument for Martian soil analysis, composed of a ^{244}Cm α source and a Si-PIN detector; (b) a related X-ray spectrum. (With permission of John Wiley & Sons, Ltd.)

a ring-type geometry and three detectors for the measurement of the three components: a telescope of two Si detectors for the measurement of α-particles and protons and a Si-PIN X-ray detector with its preamplifier.

One of the most exciting aspects of the Mars Pathfinder APXS experiment is the way it will be deployed to analyse Martian surface soil and rock samples. While usually the APXS instrument is deployed after the landing, and therefore it will analyse whatever single sample happens to be under the instrument, the APXS on the Mars Pathfinder is mounted on one end of a rover that will give it unlimited mobility around the landing site. The deployment mechanism will place the APXS vertically for soil analysis and horizontally against a rock.

A laboratory unit identical to the flight unit was used in the laboratory to derive the elemental library and to establish the accuracy and the detection limits of the Mars Pathfinder APXS instrument (Figure 5.3.28).

Figure 5.3.28 also shows X-ray spectra of the Allende meteorite obtained in the laboratory instrument using ^{244}Cm α excitation source. The resolution of the Si-PIN detector is even better than the previous version based on HgI_2 detectors: it is good enough to separate the K lines from almost all elements and for the heavier elements even the K lines. There were other advantages: Si, besides being a much easier material to handle and procure for space applications produces significantly better signal to noise ratio, especially in the 1–10 keV X-ray region.

The X-ray spectra provide information on elements heavier than Na, but matrix effects play an important role. The approach taken is an interactive one: in a first step, data from the α and proton spectra are combined and the complex sample spectrum is decomposed into its individual components, using a least square fitting procedure with a library of standard spectra, and applying appropriate corrections for matrix effects. In a second step the X-ray spectra are analysed, using a library of standard spectra and the results from the first step for matrix corrections. This step yields improved data for the ratios of the elements, which are used in a second least squares fit. There are several commercial programs available for qualitative and semiquantitative analysis. After the X-ray peaks and their intensities have been identified and determined, interelement effects are corrected with an empirical correction procedure and/or model calculations based on fundamental parameters.

ACKNOWLEDGEMENTS

This work was partially supported by the National Research Council (CNR) Special Project for the Conservation of Cultural Heritage.

REFERENCES

1. Cesareo, R. Photon induced X-ray fluorescence in medicine, in *Nuclear Analytical Techniques in Medicine* (Ed. R. Cesareo), Elsevier, Amsterdam, 1988, pp. 1–121.
2. Van Grieken, R. and Markowicz, A. (Eds) *Handbook on X-ray Spectrometry: Methods and Techniques* Marcel Dekker Inc., New York, 1992.
3. Rhodes, J.R. Design and application of X-ray emission analyzers using radioisotope X-ray and gamma ray sources. *Am. Soc. Testing Mater.* **485**, 243–285 (1971).
4. Watt, J.S. Radioisotope X-ray analysis, in *Handbook on X-ray Spectrometry: Methods and Techniques* (Eds. R. Van Grieken and A. Markowicz), Marcel Dekker Inc., New York, 1992.
5. *Oxford Analytical Systems Division*, Scotts Valley, CA, USA.
6. EIS-XRS, Rome, Italy; eissrlrm@tin.it
7. *Hamamatsu Photonics*, Hamamatsu City, Japan; www.hamamatsu.com
8. AMPTEK Inc., Bedford, MA, USA; www.amptek.com
9. Moxtek Inc., Oren, UT, USA; www.moxtek.com
10. Fiorini, C. and Longoni, A. Application of a new non cryogenic X-ray detector in portable instruments for archaeometric analysis. *Rev. Sci. Instrum.* **69**, 1523–1527 (1998).
11. Roentec GmbH, Berlin, Germany; www.roentec.de
12. Constallation Techn., Largo, FL, USA; www.contech.com
13. eV (a Division of II-VI Inc.), Saxonburg, PA, USA; www.evproducts.com
14. Arlt, R., Ivanov, V. and Khusainov, A., Advances in high-resolution CdTe and CdZnTe detectors, in *Hard X-ray and Gamma-ray Detector Physics, Optics and Applications*, SPIE.

15. PGT Princeton Gamma Tech; www.pgt.com/Nuclear/Xray_detectors.html
16. Vekemans, B. et al. Comparison of several background compensation methods useful for evaluation of energy-dispersive X-ray fluorescence spectra. *Spectrochim. Acta* **50B**, 149–169 (1995).
17. Brunetti, A. and Steger T. X-Ray spectra background fitting by projection onto convex sets, *Nuclear Instrum. Methods*, **A441**, 504–509 (2000).
18. Lachance, G.R. and Claisse, F. *Quantitative X-Ray Fluorescence Analysis*, J. Wiley & Sons, Ltd., New York, 1995.
19. Piorek, S., Puusaari, V., Piorek, E. and McCann, B. Identification and quantitative analysis of alloys using X-ray fluorescence analyzer with a Si-PIN detector, in *Proceedings of the European Conference of Energy Dispersive X-Ray Spectrometry*, (Eds. J.E. Fernandez and A. Tartari), Editrice Compositori, Bologna, 1999, p. 145.
20. Perkin Elmer Instrum., Ortec Products; www.ortec-online.com
21. VacuTec Messtechnik GmbH, Dresden, Germany.
22. ThermoNORAN X-ray detectors, Middleton, WI, USA; www.thermo.com
23. *Radiation Sources*, The Radiochemical Centre, Amersham.
24. Isotope Products Lab., Valencia, CA, USA; www.isotope-products.com
25. Photoelectron Corporation; www.photoelectron.com
26. Brownbridge, J.D. and Roboy, S. Investigations of pyroelectric generation of X-rays. *J. Appl. Phys.* **86**, 640–647 (1999).
27. Yue, G.Z. et al. Generation of continuous and pulsed diagnostic imaging X-ray radiation using a carbon-nanotube-based field-emission cathode. *Appl. Phys. Lett.* **81**, 355–357 (2002).
28. Institute for Roentgen Optics, Moscow, Russia; Institut für Gerätebau GmbH, Berlin, Germany; X-ray Optical Systems, Inc., Albany, NY, USA.
29. LND Inc., Oceanside, NY, USA; www.lndinc.com
30. Centronic Ltd., Croydon.
31. Ketek GmbH, München, Germany; www.ketek-gmbh.de
32. Iwanczyk, J.S., Patt, B.E., Wang, Y.J. and Khusainov, A. Comparison of HgI_2, CdTe and Si-PIN X-ray detectors. *Nucl. Instrum. Methods* (1996).
33. Brunetti, A. Removal of the continuum of X-ray spectra using morphological operators, *IEEE Trans. Nucl. Sci.*, **45**, 2281–2287 (1998).
34. Criss, J.W. and Birks, L.S., Calculation methods for fluorescence X-ray spectrometry. *Anal. Chem.* **40**, 1080 (1968).
35. Sherman, J. A theoretical derivation of the composition of mixable specimen from fluorescent X-ray intensities. *Adv. X-Ray Anal.* **1**, 231–251 (1958).
36. Shiraiwa, T. and Fujino, N. Theoretical calculation of fluorescent X-ray intensities in fluorescent X-ray spectrochemical analysis. *Jpn. J. Appl. Phys.* **5**, 886–899 (1966).
37. Niton Co., Bedford, MA, USA; www.niton.com
38. EDAX Inc., Mahwah, NJ, USA; www.edax.com
39. Thermo MeasureTech; www.thermomt.com
40. XRF Co.; www.xrfcorp.com
41. Metorex Int. Oy, Espoo, Finland; www.metorex.com
42. Warrington Products; www.warringtonusa.com.
43. Castellano, A. and Cesareo, R. A portable instrument for energy-dispersive X-ray fluorescence analysis of sulfur. *Nucl. Instrum. Methods B* **129**, 281–284 (1997).
44. Cesareo, R., Castellano, A., Boccolieri, G. and Marabelli, M. A portable apparatus for energy-dispersive X-ray fluorescence analysis of sulfur and chlorine in frescoes and stone monuments. *Nucl. Instrum. Methods Phys. Res. B* **155**, 326–330 (1999).
45. Cesareo, R., Cappio Borlino, C., Stara, G., Brunetti, A., Castellano, A., Buccolieri, G., Marabelli, M., Giovagnoli, A.M., Gorghinian, A. and Gigante, G.E. A portable EDXRF apparatus for the analysis of sulphur and chlorine on frescoes and stony monuments. *J. Trace Microprobe Technol.*, **18**, 23–33 (2000).
46. Cesareo, R., Castellano, A., Marabelli, M., Bandera, S., Fiorini, C., Longoni, A. and Gigante, G.E. The golden altar of S. Ambrogio in Milan: non destructive XRF analysis with a portable apparatus, in Proc. Int. Conf. Sci. Technol. Safeguard Cult. Herit. Med. Basin, Elsevier, Amsterdam, 1999, pp. 541–545.
47. Gigante, G.E. and Cesareo, R. Non-destructive analysis of ancient metal alloys by *in situ* EDXRF transportable equipment. *Radiat. Phys. Chem.* **51**, 689–700 (1998).
48. Cesareo, R., Castellano, A., Gigante, G.E., Marabelli, M., Rosales, M.A., Santopadre, P. and Silva da Costa, M. Portable instruments for EDXRF-analysis in archaeometry, in Proc. 5th Int. Conf. Non Destruct. Test. Study Conserv. Works Art, Hungarian Chemical Society Budapest, 1996, pp. 183–192.
49. Cesareo, R., Castellano, A., Buccolieri, G., Quarta, S., Marabelli, M., Santopadre, P., Ieole, M. and Brunetti, A. Portable equipment for energy-dispersive X-ray fluorescence analysis of Giotto's frescoes in the Chapel of the Scrovegni. *Nucl. Instrum. Methods in Phys. Res. B.* (2003).
50. Potts, P.J., Webb, P.C., Williams-Thorpe, O. and Kilworth, R. Analysis of silicate rocks using field-portable X-ray fluorescence instrumentation incorporating a mercury iodide detector; a preliminary assessment of analytical performance. *Analyst* **120**, 1273–1278 (1995).
51. Potts, P.J., Webb, P.C. and William-Thorpe, O. Investigation of a correction procedure for surface irregularity effects based on scatter peak intensities in the field analysis of geological and archaeological rock samples by portable XRF spectrometry. *J. Anal. Atomic Spectrom.* **12**, 769–776 (1997).
52. Potts, P.J., William-Thorpe, O. and Webb, P.C. The bulk analysis of silicate rocks by portable X-ray fluorescence: the effects of sample mineralogy in relation to the size of the excited volume. *Geostand. Newsl.* **21**, 29–41 (1997).
53. Argyraki, A., Ramsey, M.H. and Potts, P.J. Evaluation of portable XRF analysis for the *in situ* determination of lead in contaminated land. *Analyst* **122**, 743–749 (1997).
54. Potts, P.J., Ramsey, M.H. and Carlisle, J. Use of portable X-ray fluorescence in the characterization of arsenic

contamination associated with industrial buildings at a heritage arsenic works site near Redruth, Cornwall, in *J. Environ. Monit.* **4**, 1017–1024 (2002).
55. Kump, P., Necemer, M. and Rupnik, A. Development of the quantification procedures for in-situ XRF analysis, in *IAEA Meeting on Portable EDXRF Equipment and Applications*, in press.
56. Vekemans, B., Janssens, K., Vincze, L. Adams, L. and Van Espen, P. Analysis of X-ray spectra by iterative least squares (AXIL). New developments. *X-Ray Spectrom.* **23**, 278–285 (1994).
57. Sommervaille, L.J., Chettle, D.R., Scott, M.C., Aufdenrheide, A.C., Wallgren, J.E., Wittmers Jr, L.E. and Rapp Jr, G.P. Comparison of two in vitro methods of bone lead analysis and the implications for *in vivo* measurements. *Phys. Med. Biol.* **31**, 1267 (1986).
58. Green, S., Bradley, D.A., Palethorper, J.E., Mearman, D., Chettle, D.R., Lewis, A.D., Mountford, P.J. and Morgan, W.D. An enhanced sensitivity K-shell X-ray fluorescence technique for tibial lead determination. *Phys. Med. Biol.* **38**, 389–396 (1993).
59. Ahlgren, L., Liden, K., Mattsson, S. and Tejning, S., X-ray fluorescence analysis of lead in human skeleton *in vivo*. *Scand. J. Work Environ. Health* **2**, 82–86 (1976).
60. Rieder, R., Wänke, H., Economou, T. and Turkevich, A., Determination of the chemical composition of martian soil and rocks: the alpha-proton X-ray spectrometer; http://astro.uchicago.edu/papers/economou/pathfinder/apxs.html.
61. Economou, T.E., Iwanczyk, J.S. and Rieder, A., A HgI_2 X-ray instrument for the Soviet Mars 94 Mission. *Nucl. Instrum. Methods A* **322**, 633–638 (1992).
62. Economou, T., Turkevich, A., Rieder, R. and Wänke, H. Chemical composition of martian surface and rocks on Pathfinder mission. *Lunar and Planet. Sci. Conf.* **XXVII**, 1111–1112 (1996).

5.4 Synchrotron Radiation for Microscopic X-ray Fluorescence Analysis

F. ADAMS, L. VINCZE and B. VEKEMANS
University of Antwerp, Antwerp, Belgium

5.4.1 INTRODUCTION

X-Ray fluorescence (XRF) spectrometry had an important impact as one of the first commercially available instrumental techniques for elemental analysis (Van Grieken and Markowicz, 2002). During most of this time it suffered from an important constraint: the apparent impossibility to limit the beam size for its use as a microbeam method for the analysis of small and/or heterogeneous samples. Early attempts to use the method for spot analysis or profiling heterogeneous samples were based on the use of pinholes used in conjunction with X-ray tubes but the resulting flux throughput was too low for most practical applications. Direct focusing of X-rays was made quasi impossible by the opposition between a high absorption coefficient and a near unity (and negative) index of refraction. It was only over the last 20 years that methods of practical use for beam confinement gradually appeared (see Subchapter 2.1) on the basis of diffraction, refraction or reflection.

This evolution led to two new methods in XRF that are based on the confinement of the interaction volume of the primary X-ray beam with the material being analysed. In total reflection XRF (TXRF), by irradiating a flat sample with a parallel X-ray beam below the angle of total reflection, the in-depth penetration of the primary X-rays is confined to a few tenths of a nanometer below the surface allowing very sensitive surface analysis. Alternatively, the method can be exploited for bulk trace analysis of liquids, e.g. aqueous solutions brought on a clean inert surface. This methodology is discussed in detail in Subchapter 5.1 of this book. A second confined impinging beam technique is micro-XRF (μ-XRF) analysis which is based on the localised excitation and analysis of a microscopically small area on the surface of a larger sample. It provides information on the distribution of major, minor and trace elements in heterogeneous materials or can be used for the analysis of objects of reduced dimensions. μ-XRF is currently exploited with laboratory X-ray sources (see Chapter 2) but is considerably more powerful when applied with X-ray emitted from a synchrotron radiation (SR) source (Janssens et al., 2000). It is the application of this latter method with SR that is the topic of this subchapter.

Due to their high intensity and directionality SR sources are ideal for the generation of microscopically confined X-ray beams. The high directionality of the radiation allows the straightforward realisation of monochromatic micron size X-ray beams from the emitted radiation of the storage ring. In addition, the polarisation can be used to reduce the relative contribution of scattered radiation reaching the detector and thus to enhance considerably the signal-to-background ratio of fluorescence spectra, decreasing significantly detection limits. X-Ray generated from SR sources are coherent sources (Margaritondo et al., 1998; Lengeler, 2001) but the exploitation of this characteristic is hampered by two contrasted length scales: the microscopic

X-ray wavelength of the order of nm and the much larger cross-section of the beam (spot size) (Pfeiffer et al., 2002).

SR-based μ-XRF offers a number of advantages compared to other microprobe techniques: it combines high spatial resolution with high sensitivity, can be used in atmospheric conditions and is relatively insensitive to beam damage to the sample. As we will show Later in this subchapter, the simplicity of the method and the quite good understanding of the physics of the processes involved makes it more adaptable for quantitative analysis than a number of other beam methods of analysis.

This subchapter will describe the actual status of μ-XRF with SR sources with respect to lateral resolution, achievable detection limits and sensitivity for high energy third generation storage rings, particularly the European Synchrotron Radiation Facility (ESRF, Grenoble), previous generation sources and other sources of recent construction. Storage ring sources and their characteristics are fully described in Subchapter 2.2. Related methods of analysis based on absorption edge phenomena such as X-ray absorption spectroscopy (XAS) and X-ray absorption near-edge spectroscopy (XANES), X-ray microcomputed tomography (MXCT) and microscopic X-ray diffraction (XRD) will also be briefly discussed. We refer to Subchapter 5.3 for details on XAS and its applications.

5.4.2 SYNCHROTRON MICROSCOPIC X-RAY FLUORESCENCE ANALYSIS

Conceptually, the instrumentation for X-ray microanalysis is extremely simple. It consists of a mechanical sample stage with precision computer controlled microstepping motors for X, Y, Z and (optionally) rotational movement of the sample in the beam path, a semiconductor type detector for the measurement of the generated fluorescence radiation, different visualisation tools for observation and positioning of the sample and, finally, a range of diagnostic and control tools.

A particular advantage of SR μ-XRF in comparison with the conventional X-tube sources is the extremely high brilliance that is obtained. In addition, in the plane of the storage ring the radiation is linearly polarised with the E-vector parallel and the B-vector normal to the ring plane. The radiation is highly collimated along a direction tangential to the movement of the electrons in the ring thus facilitating the delivery of the radiation to a predefined sample area.

The high intensity and directionality implies that SR is ideally suited for the generation of X-ray microbeams with very high intensity, exceeding now considerably 10^{10} photons/s/μm^2. The polarisation of the incident radiation can be used to reduce the relative contribution of scattered radiation reaching the detector, as scattering cross-sections are dependent on the polarisation whereas the photoabsorption cross-sections are not. When performing measurements in the plane of the SR source this increases the signal-to-background ratios by more than two orders of magnitude, depending on the source characteristics and the particularities of the XRF set-up (degree of polarisation). Thanks to the high directionality of the beam, quasi-monochromatic X-ray microbeams can be generated from the white SR spectrum through the use of X-ray monochromators. By tuning the energy with a monochromator over a given energy range, the strong energy dependence of the inner shell photoelectric cross-sections can be exploited to either increase specificity of measurements or, else, to obtain speciation information in the XAS application mode (see later).

A direct exploitation mode with broad band polychromatic excitation is also possible. It has the advantage that (nearly) all elements in the sample are excited with quite comparable efficiency providing a more uniform spectrometer response over the range of elements of interest. Since losses in flux due to the monochromatisation process do not occur, the elemental efficiency of polychromatic set-ups is also higher than when monochromatic excitation is used, making it more appropriate for general-purpose materials characterisation. In such circumstances, however, quantification of the detected X-ray intensities is more complicated. Also, the signal-to-noise ratios, and hence, detection limits do not match those obtained with monochromatic excitation as is illustrated in

Subchapter 6.1 (Figure 6.1.12) for the measurement of rare earth elements. For quantitative analysis the use of monochromatic radiation is, hence, in general, a preferable approach.

Currently, a large number of existing storage rings are employed in μ-XRF experiments. They combine the advantages of XRF as an elemental analytical tool with the unique possibilities of SR. Of special significance are the new third generation storage rings that are specifically designed to obtain unprecedented intensity, high radiation energy and brilliance. A number of these are now routinely operational, the ESRF, the Advanced Photon Source (APS, Argonne, USA) and SPring-8 (Harima, Japan) are the most important examples. Compared to earlier second generation rings these SR facilities are characterised by their high energy of 6 to 8 GeV.

Significant in these devices also is the systematic use of insertion devices that are placed in the straight sections of the storage ring (wigglers and undulators). Wigglers are magnetic structures that create multiple oscillations around the beam path and hence increase both the energy and the intensity of the radiation. Undulators are designed to create smaller and more frequent deflections, giving rise to interference effects and in such conditions the coherent radiation is concentrated around specific energies. Other newly built SR sources follow the design characteristics of these sources and now commonly include the use of insertion devices (Subchapter 2.2).

A double crystal monochromator comprising a pair of crystals is a standard item in most monochromatic μ-XRF set-ups because the exit direction is kept constant during energy scanning. The energy resolution of the double crystal monochromator is of the order of $\Delta E/E = 10^{-4}$ or less, and this is sufficient for absorption edge applications.

In its primary utilisation mode as a tool for elemental analysis by XRF analysis a high photon flux rather than such a high-energy resolution is required and an energy resolution of the order of $\Delta E/E = 10^{-2}$ is sufficient for the purpose. Synthetic multilayers, made by vacuum deposition of alternate thin layers of two materials with a different electron density provide this ('pink beam') resolution while, through a wide energy band-pass, yielding a photon flux one to two orders of magnitude higher than available with a high resolution double crystal monochromator.

Flux throughput through pinholes is insufficient for most practical analytical purposes and techniques for generating intense X-ray microbeams are, hence, mostly based on the use of various types of X-ray optics (see Chapter 3). Grazing incidence bent mirrors in various configurations and geometries using crystals and multilayers, several types of glass mono-capillaries, complex polycapillary lens systems (Subchapters 3.3 and 3.4), diffractive lenses (Fresnel zone plates), Bragg–Fresnel lenses, one- or two-dimensional waveguides and refractive lenses (Subchapter 3.4) have been developed and tested for use in micron size to sub-micron focusing at several synchrotron beam lines. At present, it is possible to obtain a sufficient beam intensity on microscale samples to allow reliable sub ppm level determinations of a large number of elements. In particular circumstances spot sizes of less than 100 nm can be achieved (Bilderback and Hoffman, 1994).

In a number of μ-XRF installations capillary optics are used as focusing devices (Vincze et al., 2002a,b) because of their inherent constructional simplicity. These pseudo-focusing devices provide a good lateral resolution and can be used for both polychromatic and monochromatic X-ray, but suffer from the short working distance (typically <100 μm) between capillary tip and the sample. Another popular design is based on Kirkpatrick–Baez focusing mirrors. Such systems are achromatic (the spot position is independent of the incident energy) and provide a long working distance. They are quite popular for XAS applications, as the beam position remains fixed while scanning over a specific edge.

The measurement part of the μ-XRF set-up is mostly a semiconductor detector, either a conventional Si(Li) or an intrinsic Ge detector or other types of solid state detectors such as HgI_2, GaAs, silicon drift, etc. (see Chapter 4). Wavelength dispersive spectrometers are seldom used because of the slow measurement process and the resulting loss in sensitivity. The limited energy resolution of the

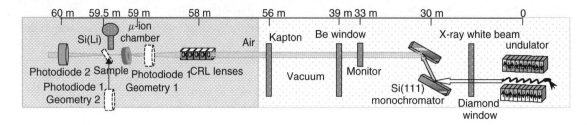

Figure 5.4.1 Schematic layout of the ID 18F experimental station. Photodiode 1 is either placed into the sample position if the sample is measured (geometry 1) or in front of the mini-ionisation chamber (geometry 2). Adapted from Somogyi et al. (2003)

energy dispersive detection gives rise to complicated spectra with multiple spectra interferences. Also, the high count rates must be adequately taken into account, e.g. by using digital pulse processing. To take full profit of the polarisation and, hence, to increase signal-to-noise ratios the X-ray detectors are positioned in the plane of the storage ring at typically an angle of 90° to the incident beam direction. Techniques were developed for fast and reliable nonlinear least squares deconvolution of X-ray spectra that circumvent spectral interferences (Janssens et al., 1996). Multivariate statistical techniques for data reduction of image scans are available (Janssens et al., 1996; Janssens et al., 2000). The topic is discussed in detail in Subchapter 6.2.

For the measurement of elemental distribution maps, spectra are taken as the sample is moved over the beam path. Contrarily to the vacuum requirements of most other microprobe methods, samples are normally observed in air, allowing the measurement of samples in their natural (e.g. wet state) conditions. Two-dimensional mapping of the repartition of elements in larger objects than the X-ray beam is possible with relative detection limits in the ppb region and absolute detection limits for many elements well below the femtogram.

All major SR sources are involved with this type of application with at least one instrumental set-up. Major SR sources have usually several possibilities for μ-XRF and related methods, e.g. in the ESRF there are the following beamlines that are at least partly devoted to microanalysis at a number of insertion devices (ID) or bending magnets (BM): ID 13 (microfocus beamline for microdiffraction, microdiffuse scattering including micro-small angle scattering); ID 18F (X-ray fluorescence); ID 21 (X-ray microscopy); ID 22 (microfluorescence, imaging and diffraction); ID 27 (TXRF facility for semiconductor wafer contamination analysis); and BM 29 (X-ray absorption fine structure for EXAFS and XANES).

The ID 18F beamline at ESRF is a typical set-up for XRF and is illustrated in Figure 5.4.1 (Somogyi et al., 2001a,b; Somogyi et al., 2003). For another representation of the same instrument see Figure 6.1.6 in Subchapter 6.1. The design goal of this instrument was to make available a dedicated instrument for μ-XRF and to improve procedures of μ-XRF in order to reach 5 % average accuracy of quantification down to sub-ppm concentration levels for elements of $Z > 13$. To do this it is necessary to insure high reproducibility of the measurement geometry and instrumental parameters of the set-up, a very good short and long-term stability and precise monitoring (<1 %) of the intensity of the incoming beam (Somogyi et al., 2003)

The energy of the excitation radiation is tuned between 6 and 28 keV and monochromatic radiation of $\Delta E/E = 10^{-4}$ is created by a fixed exit double crystal Si (111) monochromator. Compound refractive lenses are used for focusing to a routinely achievable spot size of 1–2 μm vertically and 12–15 μm horizontally. Pinhole collimation reduces this beam size further when required. Incoming, focused and transmitted beams are monitored with ionisation chambers and photodiodes. A miniature ionisation chamber with an aperture of 50 μm diameter as an entrance window was developed at the ESRF for measuring the intensity of the focused beam close to the sample (Somogyi et al., 2003). The characteristic X-ray line intensities are detected with a Si(Li) detector of 30 mm² active area, 3.5 mm active thickness placed in 90° geometry to the incoming linearly polarised X-ray beam. Fast scanning XRF measurements (>0.1 s live time/spectrum) are possible.

The degree of polarisation is estimated at $>99.7\%$ as could be deduced from the ratio of fluorescence lines with primary and multiple Compton intensity (see Figure 6.1.11 in Subchapter 6.1). The available relative detection limits (DLs) are <0.1 ppm for elements of $Z > 25$. DLs down to a few ppb are possible for a number of elements on the basis of 1000 s live time measurements and ppm DLs can be reached for measurements of a few seconds (Figure 5.4.2). The absolute DLs are <1 fg for elements of $Z > 25$ (Figure 5.4.3). The flux in the focused beam is 10^9–10^{10} photons/s depending on the energy of the incoming beam (Vekemans et al., 2003). Figures 5.4.2 and 5.4.3 show DLs obtained at ESRF with the instrument shown in Figure 5.4.1 for reference samples on the basis of the use of a set of compound refractive lenses (see Subchapter 3.4).

The design characteristics of other SR installations can be found in the literature, e.g. the X-26A beamline of the National Synchrotron Light Source in Brookhaven National Laboratory (Smith, 1995), beamline L at HASYLAB (Falkenberg et al., 2001) and the 131D beamline 'GSECARS' at the APS, Argonne (Newville et al., 1999).

5.4.3 MICRO X-RAY FLUORESCENCE ANALYSIS AS AN ACCURATE METHOD OF MICROANALYSIS

Most beam methods for microanalysis (e.g. electron probe microanalysis, micro-Auger spectroscopy, secondary ion mass spectrometry) cannot be considered as accurate analytical methods except when applied to quite simple samples mainly because of matrix effects. The application of beam methods for quantitative analysis should, hence, need to rely on the use of reference materials for calibration. Very few are currently available (Adams, 2000).

As is explained in Subchapter 6.1, the physical basis of the of the X-ray matter interaction is quite well understood and the physical constants governing the interaction and the extent of radiation absorption can be derived accurately from measurements obtained in well-defined conditions. It is thus, in principle, possible to correct for deviations of linearity between measured intensities and elemental concentrations, especially if the measurement conditions are simplified as much as possible by using a high intensity monochromatic primary excitation and employing an energy dispersive detector for the measurements.

Powerful methodologies are now available for modeling by *ab initio* Monte Carlo (MC) simulation both the beam optics and the beam–sample interaction within the sample and the detector (Vincze et al., 1999a; Vincze et al., 1999b; Vincze et al., 1999c) (see also Subchapter 6.1). The combined use of nonlinear least squares deconvolution of X-ray spectra and the methodology for modeling the beam optics and beam–sample interaction allow the optimum use of all spectral information, including the use of Rayleigh and multiple Compton scattering.

A number of background effects that were never fully evaluated in XRF appear in the SR spectra, the most important being electron Bremsstrahlung in sample and detector. Electron Bremsstrahlung is able to generate fluorescence radiation from low Z elements and, hence, appears to be a quite important factor in quantitative analysis that was never systematically taken into account for quantitative analysis. It appeared, for instance, in the simulation of experimental data that ca. 50% of the fluorescence radiation of bulk Si in a 250 μm thick Si wafer irradiated with 27 keV results indirectly from the 25.16 keV photoelectrons generated by the beam in the sample, rather than directly through impinging X-rays (Vincze et al., unpublished results). Also, specific spectral artefacts complicate the spectra, e.g. resonant Raman scattering which creates peak-like structures in the spectra at excitation energies near the absorption edge, thus complicating spectral deconvolution (Gel'mukhanov and Agren, 1999).

A systematic comparison of experimentally obtained spectral data with modeling results based on MC simulations can pinpoint specific artefacts

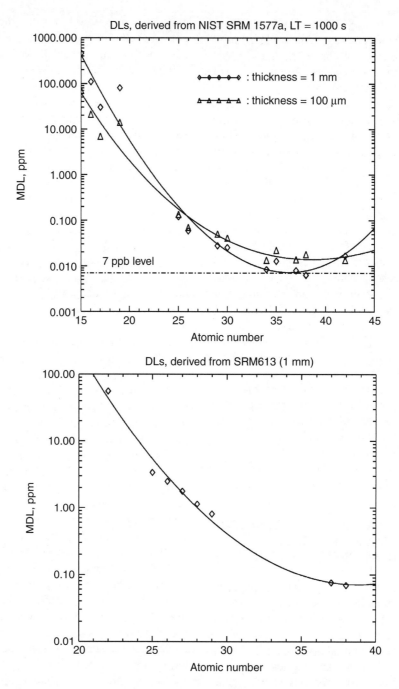

Figure 5.4.2 Relative detection limits with ID 18F using a 100 component compound reflective lens set at 21 keV at 2 μm × 2 μm in biological material (NIST SRM 1577a, bovine liver) and NIST 613 glass SRM 613 (live time measurement of 1000 s). Adapted from Somogyi *et al.* (2001)

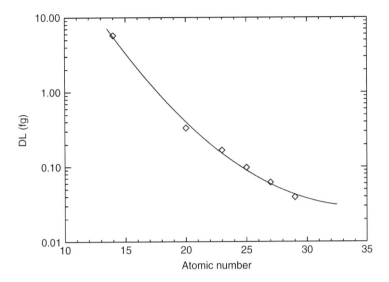

Figure 5.4.3 Absolute detection limits with ESRF (ID 13), 21 keV monochromatic radiation at $2\,\mu m \times 2\,\mu m$ (live time measurement of 100 s). Adapted from Somogyi *et al.* (2001)

and show ways to correct for them in order to bring eventually the measurement accuracy in line with the accuracy of the physical constants characterising the X-ray sample–interaction processes. Ultimately the quality of results that can be reached with this method depends on the accuracy of available physical constants, mainly the cross-sections for X-ray interactions in the most favorable circumstances (5 % for K radiation of elements in the range $Z = 20\text{--}50$, 10–15 % for L radiation between $Z = 50\text{--}80$) (Elam *et al.*, 2002).

A reasonable goal of all such efforts is to reach an average accuracy of quantification in the range of ca. 5 % for elements above atomic number 12 determined through the K and L fluorescence radiation. Table 5.4.1 shows a few results obtained for the standard reference material (SRM) NIST SRM 1832 thin glass (see Figure 6.1.7, Subchapter 6.1 for a comparison of the experimental and MC simulated spectra). Deviations of the results from the certified concentrations are in the 2–4 % range. For more heterogeneous samples such the SRM 1577a (bovine liver) experimental data and certified results correspond only to within 12 % for some elements such as Fe (Table 5.4.2). Systematically repeated measurements on reference materials allow the determination of the homogeneity level of samples and the determination

Table 5.4.1 Analysis of SRM 1832 thin glass standard

Element	Concentration (%)		Relative deviation (%)
	Certified	Measured	
Ca	12.1	12.6	+4.1
V	2.8	2.74	−2.1
Mn	2.8	2.88	+2.9
Co	0.61	0.63	+3.3
Cu	1.50	1.44	−4.0

Table 5.4.2 Certified and calculated concentrations for the analysis by iterative MC calculation of NIST SRM 1577 bovine liver

Element	Concentration (ppm)		Relative deviation (%)
	Certified	Measwed	
Ca	(123)	120	−2
Mn	10.3 ± 1	10.8	+5
Fe	270 ± 20	303	+12
Cu	193 ± 10	183	−5
Zn	130 ± 10	138	+6
Se	1.1 ± 0.1	1.2	+9
Rb	18.3 ± 1	18.4	+0.5
Sr	(0.14)	0.15	+7
Mo	(3.2)	3.4	+6

of the minimum sample size required to reach a given measurement accuracy. The results show that at this confidence limit the minimum sample mass to be analysed is of the order of 10 ng for most elements, which corresponds to a beam size of $3 \times 3 \mu m^2$. The large deviation between the certified and measured concentration for Fe in SRM 1577a (Table 5.4.2) can be attributed by such measurements to inhomogeneities of the elemental distribution in the reference material. Indeed the minimum sample mass for obtaining 5 % reproducibility amounts to 32 µg for this elements, which requires a $400 \times 400 \mu m^2$ beam instead of the small beam used (Kempenaers et al., 2002).

Improvements in quantitative μ-XRF analysis are now taking place in a European Union project of the 5th Framework Programme. The project 'Micro-XRF' investigates the comparability of multi-elemental XRF analysis down to sub-ppm concentration levels with microscopic lateral resolution with the ID18F instrument and with a μ-XRF instrument built at the ANKA SR source (Karlsruhe Research Centre, Germany) (Simon et al., 2003). The two XRF instruments together with selected other 'microbeam' analytical techniques (X-ray photoelectron spectroscopy, microscopic proton induced X-ray emission and Rutherford backscattering, secondary ion mass spectrometry) are used for the determination of the microscopic heterogeneity of available certified reference materials (Kempenaers et al., 2002). Within the project the analytical characteristics of microscopic techniques are compared and feasibility studies are undertaken for the production of reference materials for microanalysis.

5.4.4 X-RAY ABSORPTION METHODS

In the applications of XAS (also called X-ray absorption fine structure spectroscopy, XAFS) the energy dependence of the inner shell photoelectric cross is exploited to increase specificity of measurements or to obtain information on the chemical environment (speciation analysis) (see Subchapter 5.3). Extended X-ray absorption fine structure analysis (EXAFS) provides information on the number, the atomic number and the distance of neighboring atoms. The technique is based on irradiation with a highly monochromatic X-ray beam of tunable energy and scanning over an absorption edge of an element of interest while recording either the absorption of the beam (absorption XAS), the fluorescence radiation produced (fluorescence XAS) or another shell dependent phenomenon. XANES measures the position of the edge and characterisation can be achieved by exploiting specific features of the X-ray absorption spectrum. Recently, the combination of μ-XRF with spatially resolved XAS became an important tool for speciation in environmental and geological materials and for the study of processes in chemical species transformation. Most μ-XAS applications are performed in the XANES fluorescence detection mode as a 'fingerprinting' technique.

Recently there were a number of demonstrations on the use of spatially resolved speciation to distinguish valence states of the elements Cr, Mn (Zaw et al., 2002), Fe (Dyar et al., 2002), Zn (Manceau et al., 2000), U and Pu (Cutler et al., 2001; Salbu et al., 2001), etc. in geological, cosmological and environmental studies (Hsiao et al., 2002; Pinzani et al., 2002) especially for the determination of redox state, solution complex formation, and sorption on mineral phases or natural organic components, finally the bioavailability of metal compounds. Such information is essential for risk assessment, management and reduction of hazards associated with elemental release of contaminants. Bertsch and Hunter (2001) cover the literature on the subject until 2000.

5.4.5 COMPUTERISED MICROTOMOGRAPHY (CMT)

CMT at high spatial resolution with absorption or fluorescence radiation from a spatially confined SR beam is based on the systematic measurement of the beam as it is impinging on the sample (Simionovici et al., 2000; Rau et al., 2001). For obtaining three-dimensional information X-ray tomography exploits one of the weaknesses of μ-XRF, the penetrative nature of the impinging X-rays. It is now possible to carry out CMT with spatial resolutions as low as about $1 \mu m^3$ and with more than 10^9 voxels. This provides possibilities for nondestructive observation of the interior of the sample for the study of shape, density and composition, e.g. in inclusions, interior pore structures, buried phases or other features within the sample.

For CMT measurements the sample, in addition to being rastered, is also rotated over 180° through the incorporation of a sample rotation stage. A two-dimensional elemental distribution map in the horizontal sample plane can then be obtained using a reconstruction procedure based on filtered back projection. A systematic repetition of the process at other planes eventually reconstructs the entire three-dimensional image. Imaging is possible with the white spectrum as well as with monochromatic radiation (Larson et al., 2002).

The technique is now in full development, with major applications being in inorganic matrix composites, transport phenomena in porous media, the study of calcified tissues and fatigue cracks in materials (Stock, 1999; Maire et al., 2001). Techniques are being developed to speed up the measurement process, combining fast detector systems, high speed data networks and parallel computing systems to a few minutes (Wang et al., 2001).

Major problems are the complexity (and length) of calculations when sample self-absorption is taken into account. In principle it is possible to combine XAS and CMT with μ-XRF. The complexity of XAS can be diminished by obtaining the fluorescence or absorption information with two closely spaced excitation energies that are characteristic for the valence states of different ions in the sample (Yamamoto et al., 2000).

5.4.6 MICRO X-RAY DIFFRACTION

The measurement of an XRD pattern over a complex sample provides information on the variation of its crystallographic structure. Micro XRD is becoming common on many X-ray microprobes (Rindby et al., 1997; Riekel, 2000). (See Figure 6.1.6 in Subchapter 6.1 for its incorporation in the ID 18F beamline.) Micro XRD maps showing the repartition at every impact point of several crystallographic states can be obtained together with the maps of elemental information and can assist considerably in the characterisation of complex samples.

5.4.7 IMAGING

Several hard X-rays imaging techniques greatly benefit from the coherence of the beams delivered by the modern SR sources. Phase imaging is directly related to the small angular size of the source as seen from one point of the sample. Phase radiography and tomography are instrumentally very simple. They are often used in the 'edge detection' regime, where the jumps of density are clearly observed. Recently a more quantitative approach has been developed, which provides a three-dimensional density mapping of the sample ('holo-tomography'). The combination of diffraction topography and phase-contrast imaging constitutes a powerful tool that can help in monitoring the areas of a sample to be selected for further chemical or structural analysis (Baruchel et al., 2000).

The coherence of X-ray radiation does not bring anything directly useful for analysis but the production of a coherent and divergent X-ray beam allows the observation of small features in objects both in two and three dimensions. X-Ray tomography can provide three-dimensional density and chemical distributions of such structures with submicrometer resolution (Larson et al., 2002). Such techniques are still in development.

5.4.8 CONCLUSIONS

Within the field of microanalytical techniques SR based μ-XRF emerges as an important new methodology for the characterisation and analysis of diverse materials. The simultaneous application of μ-XRF with XAS, CMT and XRD greatly enhances the capabilities for the study of microscopically heterogeneous materials providing three-dimensional nondestructive information and speciation. The gains in brilliance of SR sources over the past decade and advances in focusing optics during the same period provide spatial resolution at the sub-μm range while maintaining a high sensitivity.

Thanks to the full understanding of the interaction processes of X-rays with matter (Cesareo, 2000), computer techniques, particularly those based on MC simulation, are able to predict the spectral response without limitations in approximations or idealisations of the sample geometry. Hence, they allow methods for calibration and correction for radiation absorption, thus opening up the way to perform reliable quantitatitve analysis.

Scanning techniques are wasteful ways and time consuming to generate two- and three-dimensional maps. The first attempts appeared recently for performing elemental analysis with a full-field XRF microscope on the basis of measurements with charge-coupled device photon counting systems. As demonstrated by Ohigashi *et al.* (2002), it appeared possible to obtain a spatial resolution of 10 μm with a field of view of 200 μm and an energy resolution of 350 eV FWHM.

Exhaustive recent reviews of applications of synchrotron based μ-XRF and related techniques are available in the literature (Schulze and Bertsch, 1995; Potts *et al.*, 2000; Revenko, 2000; Bertsch and Hunter, 2001; Larson *et al.*, 2002).

General recent reviews of micro-XRF and related methods also provide numerous examples of the advantages of the methodology (Chevallier *et al.*, 1998; Chevallier *et al.*, 1999; Ellis *et al.*, 1998; Haberkorn and Beck, 2000; Szaloki *et al.*, 2002). A comparison of different excitation modes is given by Gigante and Gonsior (2000).

REFERENCES

Adams, F. *Encyclopedia of Analytical Chemistry* (Ed. R.A. Meyers), Chichester, Wiley, 2000, Vol. 15, pp. 13636–13644.
Baruchel, J., Cloetens, P., Hartwig, J., Ludwig, W., Mancini, L., Pernot, P. and Schlenker, M. *J. Synchrotron Rad.* **7**, 196–201 (2000).
Bertsch, P.M. and Hunter, D.B. *Chem. Rev.* **101**, 1809–1842 (2001).
Bilderback, D.H. and Hoffman, S.A. *Science* **263**, 201–204 (1994).
Cesareo, R. *Riv. Nuovo cim.*, **23**(7), 1–231 (2000).
Chevallier, P., Firsov, A., Populos, P. and Legrand, F. *J. Phys.* **8**, 407–412 (1998).
Chevallier, P., Populos, P. and Firsov, A. *X-ray Spectrom.* **28**, 348–351 1999.
Cutler, J.N., Jiang, D.T. and Remple, G. *Can. J. Anal. Sci. Spectrosc.* **46**, 130–135 (2001).
Dyar, M.D., Lowe, E.W., Guidotti, C.V. and Delaney, J.S. *Am. Mineral.* **87**, 514–522 (2002).
Elam, W.T., Ravel, B.D. and Sieber, J.R. *Radiat. Phys. Chem.* **63**, 121–128 (2002).
Ellis, A.T., Kregsamer, P., Potts, P.J., Streli, C., West, M. and Wobrauschek, P. *J. Anal. At. Spectrom.* **13**, 209–232 (1998).
Falkenberg, G., Clauss, O., Swiderski, A. and Tschentscher, T. *Nucl. Instrum. Methods* **467**, 737–740 (2001).
Gel'mukhanov, F. and Agren, H. *Phys. Rev. Lett.* **312**, 91–140 (1999).
Gigante, G.E. and Gonsior, B. *Fresenius. J. Anal. Chem.* **368**, 644–648 (2000).
Haberkorn, R. and Beck, H.P. *Mikrochim. Acta* **133**, 51–58 (2000).
Hsiao, M.C., Wang, H.P., Wie, Y.L., Chang, J.E. and Jou, C.J. *J. Hazard. Mater.* **91**, 301–307 (2002).
Janssens, K., Vekemans, B., Adams, F., Van Espen, P. and Mutsaers, P. *Nucl. Instrum. Methods B* **109/110**, 179–185 (1996).
Janssens, K.H., Adams, F.C. and Rindby, A. (Eds) *Microscopic X-ray Fluorescence Analysis*, Wiley, Chichester, 2000.
Kempenaers, L., Janssens, K., Vincze, L., Vekemans, B., Somogyi, A., Drakopoulos, M., Simionovici, A. and Adams, F. *Anal. Chem.* **74**, 5017–5026 (2002).
Larson, B.C., Yang, W., Ice, G.E., Budai, J.D. and Tischler, J.Z. *Nature* **415**, 887–890 (2002).
Lengeler, B. *Naturwissenschaften* **88**, 249–260 (2001).
Maire, E., Buffiere, J.Y., Salvo, L., Blandin, J.J., Ludwig, W. and Letang, J.M. *Adv. Eng. Mater.* **3**, 539–546 (2001).
Manceau, A., Lanson, B., Harge, J.C., Musso, M., Eybert-Berard, L. Hazemann, J.-L., Chateigner, D. and Lanble, G. *Am. J. Sci.* **300**, 289–343 (2000).
Margaritondo, G., Tromba, G., Hwu, Y. and Grioni, M. *Phys. Low-dimen. Struct.* **12**, 39–54 (1998).
Newville, M., Sutton, S., Rivers, M and Eng, P. *J. Synchrotron Radiat.* **6**, 353–355 (1999).
Ohigashi, T., Watanabe, N., Yokosuka, H., Aota, T., Takano, H., Takeuchi, A. and Aoki, S. *J. Synchrotron Radiat.* **9**, 128–131 (2002).
Pfeiffer, F., David, C., Burghammer, M., Riekel, C. and Salditt, T. *Science* **297**, 230–234 (2002).
Pinzani, M.C.C., Somogyi, A., Simionovici, A.S., Ansell, S., Steenari, B.M. and Lindqvist, O. *Environ. Sci. Technol.* **36**, 3165–3169 (2002).
Potts, P.J., Ellis, A.T., Holmes, M., Kregsamer, P., Streli, C., West, M. and Wobrauchek, P. *J. Anal. At. Spectrom.* **15**, 1417–1442 (2000).
Rau, C., Weitkamp, T., Snigirev, A., Schroer, C.G., Tummler, J. and Lengeler, B. *Nucl. Instrum. Methods A* **467**, 929–931 (2001).
Revenko, A.G. *Indust. Lab.* **66**, 637–652 (2000).

Riekel, C. *Rep. Progr. Phys.* **63**, 233–262 (2000).

Rindby, A., Engstrom, P., Janssens, K. and Osan, J. *Nucl. Instrum. Methods B* **124**, 591–604 (1997).

Salbu, B., Krekling, T., Lind, O.C., Oughton, D.H., Drakopoulos, M., Simionovici, A., Snigireva, I., Snigirev, A., Weitkamp, T., Adams, F., Janssens, K. and Kashparov, V.A. *Nucl. Instrum. Methods A* **467**, 1249–1252 (2001).

Schulze, D.G. and Bertsch, P.M. *Adv. Agron.* **55**, 1–66 (1995).

Simionovici, A., Chukalina, M., Schroer, C., Drakopoulos, M., Snigirev, A., Snigireva, I., Lengeler, B., Janssens, K. and Adams, F. *IEEE Trans. Nucl. Sci.* **47**, 2736–2740 (2000).

Simon, R., Buth, G. and Hagelstein, M. *Nucl. Instrum. Methods B* **199**, 554–558 (2003).

Smith, J.V. *Analyst* **23**, 1231–1245 (1995).

Somogyi, A., Drakopoulos, M., Vincze, L., Vekemans, B., Camerani, C., Janssens, K., Snigirev, A. and Adams, F. *ESRF Highlights* 2002, 96–97 (2001a).

Somogyi, A., Drakopoulos, M., Vincze, L., Vekemans, B., Camerani, C., Janssens, K., Snigirev, A. and Adams, F. *X-ray Spectrom.* **30**, 242–252 (2001b).

Somogyi, A., Drakopoulos, M., Vekemans, B., Vincze, L., Simionovici, A. and Adams, F. *Nucl. Instrum. Methods B* **199**, 559–564 (2003).

Stock, S.R. *Int. Mater. Rev.* **44**, 141–164 (1999).

Szaloki, I., Torok, S.B., Injuk, J. and Van Grieken, R.E. *Anal. Chem.* **74**, 2895–2917 (2002).

Van Grieken, R.E. and Markowicz, A.A. (Eds) *2002 Handbook of X-ray Spectrometry*, 2 Edn, Marcel Dekker, New York, 2002.

Vekemans, B., Vincze, L., Somogyi, A., Drakopoulos, M., Kempenaers, L., Simionovici, A. and Adams, F. *Nucl. Instrum. Methods B* **199**, 396–401 (2003).

Vincze, L., Janssens, K., Vekemans, B. and Adams, F. *J. Anal. At. Spectrom.* **14**, 529–533 (1999a).

Vincze, L., Janssens, K., Vekemans, B. and Adams, F. *Spectrochim. Acta, Part B* **54**, 1711–1722 (1999b).

Vincze, L., Janssens, K., Vekemans, B. and Adams, F. *J. Anal. At. Spectrom.* **14**, 529–533 (1999c).

Vincze, L., Wei, F., Proost, K., Vekemans, B., Janssens, K., He, Y., Yan, Y. and Falkenberg, G. *J. Anal. At. Spectrom.* **17**, 177–182 (2002a).

Vincze, L., Somogyi, A., Osán, J., Vekemans, B., Török, S., Janssens, K. and Adams, F. *Anal. Chem.* **74**, 1128–1135 (2002b).

Wang, Y.X., De Carlo, F., Mancini, D.C., McNully, I., Tieman, B., Bresnahan, J., Foster, I., Insley, J., Lane, P., von Laszewski, G., Keselmann, C., Su, M.H. and Thiebaux, M. *Rev. Sci. Instrum.* **72**, 2062–2068 (2001).

Yamamoto, K., Watanabe, N., Takeuchi, A., Takano, H., Aota, T., Fukuda, M. and Aoki, S. *J. Synchrotron Radiat.* **7**, 34–39 (2000).

Zaw, M., Szymczak, R. and Twining, J. *Nucl. instrum. Methods B* **190**, 856–859 (2002).

5.5 High-energy X-ray Fluorescence

I. NAKAI
Tokyo University of Science, Tokyo, Japan

5.5.1 INTRODUCTION

Traditionally, the X-ray fluorescence (XRF) analysis of heavy elements is based on their L series spectral lines. This is because the K lines of these elements are difficult to excite by commercially available XRF spectrometers. Photoelectric absorption can only occur if the energy of the photon is equal or greater than the binding energy of the electron. The energy of the K absorption edge of an element increases with the atomic number of the element. For example, to measure the K fluorescence line of uranium, an X-ray photon with energy of 115.62 keV is necessary to eject an electron from the K shell of uranium. In addition, an X-ray detector capable of measuring high-energy X-rays is required.

Since the vacancy can be in any of the three subcells, L1, L2, or L3, the L line XRF spectrum is more complex than the K line spectrum. A typical energy-dispersive X-ray fluorescence (EDXRF) spectrum of a multicomponent material in energy regions of less than 20 keV is, therefore, usually crowded with an overlap of the K, L, and M emission lines of the component elements. In contrast, the XRF spectrum above 20 keV contains only K lines, and the spectrum becomes simple. Therefore, it is expected that the use of the K lines would be ideal for the analysis of heavy elements of atomic number $Z \geqq 45$ (=Rh Kα_1 = 20.12 keV).

This subchapter first presents a brief review of work done in this field, and then introduces characteristic features and analytical performance of the high-energy XRF technique developed through those studies. The potential capability of this technique will be demonstrated through practical examples in several fields where high-energy XRF is useful and therefore promising.

5.5.2 REVIEW OF STUDIES ON HIGH-ENERGY XRF ANALYSIS

5.5.2.1 LABORATORY X-RAY SOURCES

Harada and Sakurai[1] reported in detail on the advantages of the use of high-energy X-rays in EDXRF analysis based on their laboratory data. They constructed a sealed X-ray tube (W anode) system with a high-voltage power supply (160 kV and a maximum load of 1.6 kW for normal focus and 0.64 kW for fine focus) for high-energy XRF analysis. The detection limit of iodine concentration was found to be 8 ppm and an absolute amount of 3 µg, based on the use of a preliminary beam filter technique (80 kV, 18.5 mA). The corresponding values obtained through the use of white excitation (80 keV, 0.6 mA) were 40 ppm and 15 µg. Monochromatic excitation significantly reduced the background signal compared with white X-ray excitation, and is favorable for trace element analysis.[1]

Recently, PANalytical began marketing a high-energy EDXRF spectrometer, Epsilon 5[2] (Almelo, The Netherlands). It is equipped with a 600 W X-ray tube with a Gd anode operating in a range of

25 kV to 100 kV and a liquid nitrogen-cooled high-resolution solid state Ge detector. The XRF spectrometer features a three-dimensional (Cartesian) polarizing optical geometry and 15 programmable polarizing targets. Backgrounds can be an order of magnitude lower than traditional two-dimensional optics resulting in much lower detection limits, down to sub-ppm levels.

5.5.2.2 FIRST- AND SECOND-GENERATION SYNCHROTRON RADIATION SOURCES

Heavy element analysis using K line spectra was first studied utilizing either first-generation synchrotron radiation (SR) light sources such as VEPP-4 in Novosibirsk, USSR[3-5] and HASYLAB in Hamburg, Germany,[6] or a second-generation light source at NSLS in Brookhaven.[7] White X-ray excitation or relatively weak monoenergetic beams with energy of less than 75 keV have been utilized. Chen et al.[7] reported minimum detection limits (MDLs) of 6 ppm (La) to 26 ppm (Lu) for a counting time of 3600 s when using a wiggler beam. Baryshev et al.[3] reported an MDL of 50 ppm for rare earth elements. Janssens et al.[6] reported the use of lead-glass capillaries for the microfocusing of highly energetic (0–60 keV) synchrotron radiation. With this great improvement in the analytical sensitivity of the focusing optics, they obtained superb detection limits of 1–10 fg/0.8–2 ppm for 100 μm silicate samples for elements from Mn ($Z = 25$) to Gd ($Z = 64$) using their K lines with a counting time of 1000 s.

5.5.2.3 THIRD-GENERATION SR LIGHT SOURCES

Third-generation synchrotron light sources incorporate insertion devices, wigglers, and undulators. The primary characteristics of third-generation light sources are their extremely high brilliance and high-energy X-rays. The former characteristic is already well appreciated in relation to the construction of X-ray microscope or microanalysis systems with a beam size of less than 1 μm[8-10] and that of TXRF systems with fg sensitivity.[11-13] However, the utilization of high-energy X-rays from third-generation light sources in XRF analyses had remained a challenge until our first application in the forensic analysis of arsenic in a murder case in Wakayama city aroused attention in 1998.[14] We used 116 keV X-rays for the first time as an excitation source in the XRF analysis of crime-related materials. A wiggler beam-line at SPring-8 was suitable for producing high-energy X-rays up to 300 keV. This work revealed that high-energy XRF is powerful analytical tool in a variety of scientific fields as well as for forensic analysis.[15]

5.5.3 INSTRUMENTS FOR HIGH-ENERGY XRF

Figure 5.5.1 compares the detection efficiency of the Ge solid state detector (SSD) as a function of X-ray energy with those of the Si(Li) and Si drift detectors. As can be seen in Figure 5.5.1, the efficiency of the Si(Li) SSD and that of the Si drift detector, which are usually used in an energy-dispersive (ED) spectrometer, decrease drastically at above 20 keV with an increase in X-ray energy. In contrast, the Ge detector maintains high efficiency for the energy of the K lines of heavy elements. Therefore, the Ge SDD is suitable for high-energy EDXRF analysis. In wavelength-dispersive (WD) XRF, the energy resolution becomes poorer at a lower 2θ angle of analyzing crystal, namely at higher X-ray energies. In practical terms, the difference in resolution between the ED and WD spectrometers is therefore insignificant for high-energy XRF analysis.

Examples of experimental systems for high-energy EDXRF analysis in laboratories and in SR facilities are schematically illustrated in Figure 5.5.2(a) and 5.5.2(b). At the National Institute for Materials Science, the high-energy XRF system (Figure 5.5.2a) is composed of a sealed W tube (Comet MXR-160) with a high-voltage generator and its controller (Gulmay CP-160 and MP-1).[1] The maximum tube voltage is 160 kV and maximum loading is 1.6 kW for a normal focus (1.5 mm × 1.5 mm) and 0.64 kW for a fine focus (0.4 mm × 0.4 mm). A shielding box is necessary

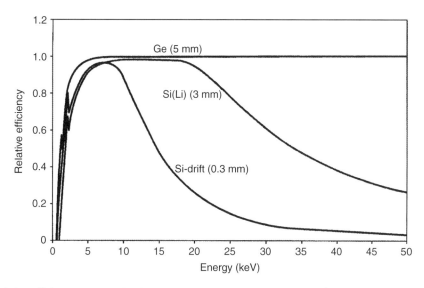

Figure 5.5.1 Relative efficiency vs energy for Ge SSD, Si(Li) SSD and Si drift detectors[1]

for conducting experiments. The box is made of 4-mm thick lead plate sandwiched between steel plates, and 5-mm thick lead plate is also placed at the position of direct beam irradiation. The beam size is limited by the $x-y$ slits (W–Ni alloy, 3 mm thick). The detection system consists of a Ge detector (Canberra, CT, USA, PSR505), a spectroscopy amplifier (Canberra 2021, shaping time 4 s), a multichannel analyzer (NAIG E-553, 562A, 563A), and a computer (NEC, Tokyo, Japan, PC9801RA).

As a laboratory spectrometer, the polarizing optics adopted in the EDXRF spectrometer, Epsilon 5, demonstrated better performance than conventional nonpolarizing optics in high-energy XRF analysis. Figure 5.5.3 is a schematic diagram of the optics. The use of targets of different materials enables the optimization of the excitation source specifically for analyte elements of interest. The primary beam from the Gd anode first irradiates a polarizing target placed along the first axis. After scattering at 90°, the X-rays travel along the second axis to the sample. The XRF signals are measured by the Ge detector placed along the third axis. This Cartesian geometry eliminates the X-ray tube spectrum by polarization, thereby reducing the spectral background.

The first high-energy XRF experiment utilizing monochromatic X-rays with energy higher than 100 keV was conducted at beam line BL-08 W of SPring-8 (Figure 5.5.2b).[14,15] Monochromatic X-rays of 116 keV were obtained from a doubly bent Si(400) monochromator utilizing the high-energy X-rays from an elliptical multipole wiggler as an excitation source. The energy resolution, $\Delta E/E$, was 1.25×10^{-3} at 115 keV, and the photon flux was 10^{12} photons/s.[16,17] The EDXRF analysis system consisted of an XY automatic stage (SIGMA KOOKI, Saitama, Japan), a pure Ge SSD (Canberra GUL0055p), a spectroscopy amplifier (Canberra 2021, shaping time 12 µs), and a multichannel analyzer (Seiko EG&G, Tokyo, Japan, MCA7700). The range of the multichannel analyzer was adjusted to 102.4 keV.

5.5.4 PERFORMANCE OF HIGH-ENERGY XRF

5.5.4.1 CHARACTERISTIC ANALYTICAL FEATURES

Harada and Sakurai[1] compared the sensitivity of the following three-excitation techniques of high-energy XRF analysis utilizing conventional laboratory X-ray sources: i.e. white excitation (80 kV, 0.9 mA), primary beam filter (80 kV, 0.6 mA, filter 40 mm Al plate), and secondary

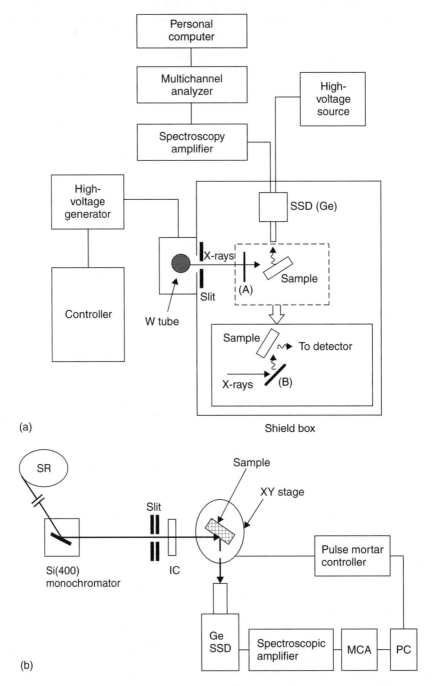

Figure 5.5.2 Schematic diagram of the experimental setup for high-energy XRF analyses. (a) Laboratory spectrometer. In the primary beam filter technique, a filter (A) is interposed between the tube and the sample. In the secondary target technique, the part enclosed in broken lines is replaced by the system below, where (B) represents the secondary target.[1] (b) SR spectrometer at BL-08 W of SPring-8

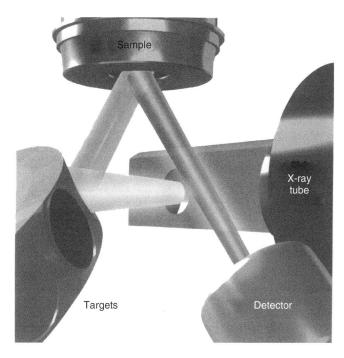

Figure 5.5.3 Three-dimensional polarization optics adopted in Epsilon 5

Table 5.5.1 Comparison of the results from the three excitation techniques. The detection limit is given here as three times the statistical uncertainty of the background[1]

Method	White excitation	Primary-beam filter	Secondary target
Conditions of excitation			
Tube voltage (kV)	80	80	150
Tube current (mA)	0.6	18.5	10.6[a]
Signal (counts/300 s)	5367	5403	5444[b]
Background (counts/300 s)	17050	786	843[b]
S/B	0.31	6.9	6.5
Irradiation area (mm × mm)	6 × 7	6 × 7	15 × 15
Detection limit of iodine			
Concentration (μg g^{-1})	40	8	8
Absolute amount (μg)	15	3	13

[a] Limited by the X-ray generator's maximum loading.
[b] Compensated by the tube current.

target techniques (150 kV, 3 mA, Er target). The sample was the environmental reference material, sargasso (NIES No. 9), which contains a low concentration of iodine (520 μg/g). The results are summarized in Table 5.5.1. When white excitation was applied, a high background was caused by the Compton scattering of incident X-rays. In contrast, with the primary beam filter technique, an absorbing material of 40 mm thick Al plate filtered out the low-energy part of the primary beam. This technique reduced the background at the iodine Kα line, and the signal-to-background (S/B) ratio was improved to 6.9. For measurements using the secondary-target technique, a 3 mm thick Er plate was employed as a secondary target. This technique reduced the background around the iodine Kα line, improving the S/B more than 20 times that of the conventional method.

A combination of secondary target, primary beam filter, and three-dimensional polarizing optical geometry adopted by Epsilon 5 greatly reduced the scattered X-ray tube spectrum and enhanced the sensitivity of the high-energy XRF. With these optics and a CsI secondary target, the lower limit of detection (LLD) of Cd in plastic was reported to be 0.53 ppm for a measurement time of 100 s, Gd anode X-ray tube operated at 100 kV, 4.6 mA. Here, the LLD was calculated according to the equation:

$$\text{LLD} = 3 \times C/I_p \times (I_b/t_b)^{1/2}$$

where I_b is the background count and I_p is the signal (number of counts) produced by the concentration C (the certified value reported in the reference) of the element with measurement time t_b (live time). The sensitivity changes with the voltage of the X-ray tube, as shown in Figure 5.5.4.

There are several superior features of synchrotron radiation in high-energy XRF analysis: the SR X-rays are fully polarized small parallel beams, and brilliant monochromatic high-energy X-rays are obtained from a combination of the wiggler source and monochromator. These features provide the most suitable X-ray source for high-energy XRF. Figure 5.5.5 shows an XRF spectrum of a metamict mineral (a variety of uraninite UO_2) excited by 116 keV SR X-rays, which shows the Kα peak of uranium. Thus, we have confirmed that this system can analyze all heavy elements up to uranium by their K lines. The analytical performance of this method was clarified by analyzing some standard samples. Figure 5.5.6 shows a typical XRF spectrum of a bulk geological standard sample (JG-1: granite, counting time of 500 s). The certified element concentrations range from 54.7 ppm for Zr down to 1.7 ppm for Er and W (Table 5.5.2). As is shown in Figure 5.5.6, tungsten as well as various rare earth elements give the distinct peaks of the K lines. Here, the value of MDL was calculated according to the following equation:

$$\mathrm{MDL} = 3 \times C/I_p \times (I_b)^{1/2}$$

Table 5.5.2 shows the values of MDL calculated from Figure 5.5.6 for Fe, Rb, Sr, Zr, Cs,

Table 5.5.2 Minimum detection limit (MDL) values calculated from the XRF spectrum of JG-1 sample in Figure 5.5.6 for a measurement time of 500 s[15]

	Contents (ppm)	I_p (counts)	I_b (counts)	MDL (ppm)
Fe	2.02[a]	1557	366	0.097[a]
Rb	181	577	281	30.8
Sr	184	719	258	19.2
Zr[b]	108	395	293.5	54.7
Cs	10.2	280	181	4.2
Ba	462	7205	354.5	3.8
La	23	535	355.5	7.2
Ce	46.6	520	86	3.0
Nd	20	862	154.5	1.1
Sm	5.1	136	45	1.1
Gd	3.7	108	42.5	1.1
Dy	4.6	110	41	1.3
Er	1.7	86	515	1.1
Yb	2.7	125	61	1.0
Hf	3.5	268	98.5	0.6
W	1.7	737	199.5	0.1

[a] wt%.
[b] Calculated by using Kβ line.

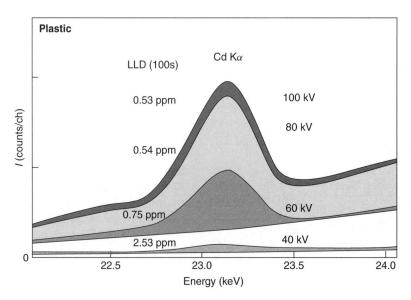

Figure 5.5.4 Cd Kα XRF spectra. Sensitivity (LLD for 100 s) changes with X-ray tube voltage

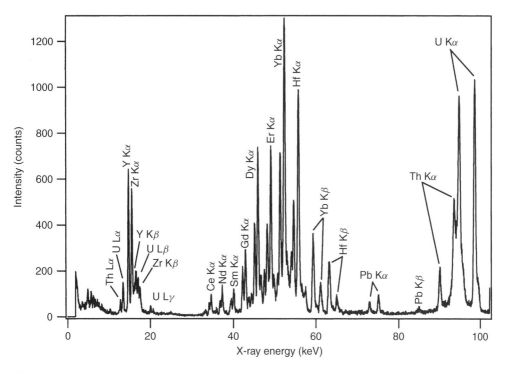

Figure 5.5.5 SRXRF spectrum of a metamict mineral (a variety of uraninite, UO_2). The Oddo–Harkins law is clearly observed

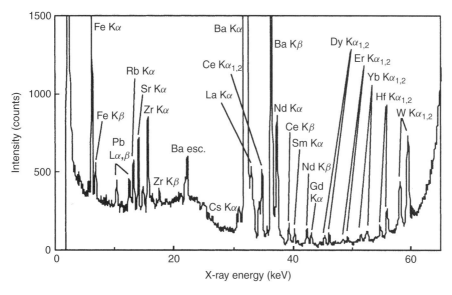

Figure 5.5.6 A typical XRF spectrum of geological samples (JG-1, granite rock standard reference sample) excited by 116 keV X-rays and a measurement time of 500 s[15]

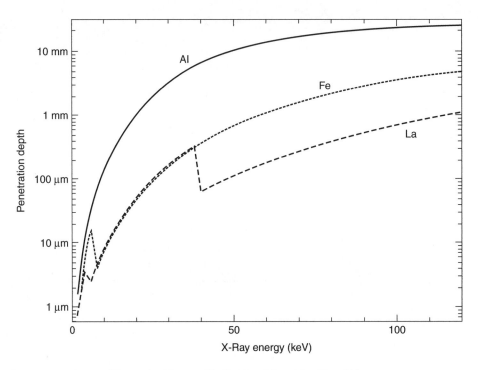

Figure 5.5.7 Penetration depths of X-rays for Al ($Z = 13$), Fe ($Z = 26$) and La ($Z = 57$)[1]

Ba, La, Ce, Nd, Sm, Gd, Dy, Er, Yb, Hf, and W together with the certified concentration of each element. In this sample, the calculated MDL for W was 0.1 ppm, which is the lowest MDL value in Table 5.5.2. For the higher-energy region, Compton scattering caused a high background, which degrades the signals of low-concentration elements. In contrast, the MDL increased with decreases in energy, which was due to the excitation efficiency. Since the energy of the excited X-ray was 116 keV, which is far from the absorption edge of Fe (ca. 7 keV), there was a much poorer detection limit for Fe (0.097 wt%) than for the other heavy elements (Table 5.5.2). The high background region resulting from the Compton scattering can be changed by increasing or decreasing the excitation X-ray energy.

An advantage of the use of high-energy X-rays in XRF analysis is that they have high transmission power. The absorption coefficients of the elements sharply decrease with an increase in the energy of the X-rays. The penetration depths of X-rays for Al, Fe, and La as a function of X-ray energy are given in Figure 5.5.7. This figure shows that penetration depth increases with an increase in X-ray energy. For example, the mass absorption coefficients of the JG1 sample were calculated for X-rays with energies of 0.011 nm (=113 keV), 0.021 nm (=59 keV, which is equal to the energy of the W $K\alpha_1$ line), 0.062 nm (=20 keV), and 0.148 nm (=8.4 keV, which is equal to that of the W $L\alpha_1$ line) to be 0.16, 0.28, 3.05, and 37.87, respectively, based on data in the literature.[18] These values clearly indicate that the absorption coefficient of JG1 sample (3.05) for the 20 keV X-rays, typical energy for conventional SR-XRF analysis, is almost 20 times larger than that (0.16) for the 113 keV X-rays. The high-energy X-rays, therefore, have a greater penetration depth and are favorable for obtaining the bulk chemical composition of a sample. In addition, the absorption coefficient of the JG1 sample (37.87) for the W $L\alpha_1$ line is 135 times larger than that (0.28) for the W $K\alpha_1$ line. This suggests that quantitative analysis using the K lines of the heavy

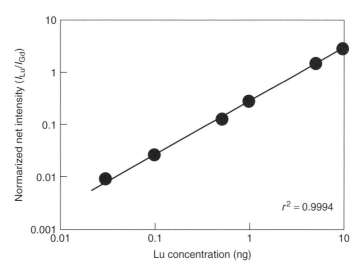

Figure 5.5.8 Calibration curves for the determination of Lu. The normalized X-ray intensity of Lu was plotted against the absolute amount of Lu (ng). The XRF intensity of the internal standard of Gd was used for normalization[15]

elements requires less absorption correction than that using the L lines of those elements.

5.5.4.2 DETECTION LIMIT MEASURED FOR DROPLET SAMPLES

The detection limit of the SR-XRF technique is largely affected by the degree of elastic and Compton scattering relative to the XRF signal from the analyzed elements. The background counting rate is critical for a bulk sample, and the limit of the counting rate of the detector becomes a dominant factor in determining the detection limit. In fact, we must reduce the X-ray intensity by opening the gap of the wiggler to 60 mm in order to measure the bulk sample of JG1. However, the gap should be at its minimum (40 mm) for measuring such items as a tiny glass flake of SRM612 or a liquid droplet on Mylar™ film. The detection limit in the determination of rare earth elements by a calibration curve technique using a liquid droplet sample on a Mylar™ film was evaluated. This method of sample preparation should give almost the lowest MDL when compared with other methods except for the use of total reflection geometry. Lu was used as the target element, and Gd was selected as an internal standard. The net intensities of Lu were normalized by those of Ga and were plotted against the absolute amount of Lu. The calibration curve thus obtained from the spectrum with a counting time of 1000 s is shown in Figure 5.5.8. The data show excellent linearity from a Lu level of 10 ng down to 30 pg. The estimated MDL value at the lowest end is 16 pg. These results suggest that a quantitative trace analysis would be promising for high-energy XRF analysis with proper correction of the matrix effect.

5.5.5 APPLICATION OF HIGH-ENERGY XRF

From the above observations, we found that high-energy XRF analysis enables us to determine all heavy elements with sufficient sensitivity. The minimum detection limit of the current analysis is at a sub-ppm level. Accordingly, high-energy XRF analyses will likely become a powerful tool for analyzing environmental samples, archaeological samples, forensic samples, geochemical samples, and high-tech materials containing heavy elements such as rare earth elements. It is expected that the utilization of high-energy X-rays will open new application fields for X-ray fluorescence analyses.

Several examples of applications of this technique are shown below and the potential advantages of this technique are elucidated.

5.5.5.1 ANALYSES OF RARE EARTH ELEMENTS

In modern industry, rare earth elements are the most important elements used in high-tech devices employing lasers, as well as magnetic, fluorescence, and superconducting materials. Therefore, the chemical analysis of rare earth elements is important in their industrial use. Thus far, however, XRF analysis of rare earth elements is difficult to accomplish using low-energy X-rays. Development of a sensitive, nondestructive method for the total analysis of all rare earth elements had been expected. Figure 5.5.9 compares the XRF spectra of standard SRM612 glass samples excited by (a) 116 keV SR- X-rays and by (b) Pd Kα X-rays with a tube voltage of 40 kV. The XRF spectra (a) and (b) were obtained from measurements for 1000 s and 300 s, respectively. The nominal trace element concentration of SRM612 glass is 50 mg/kg (=50 ppm) for each of the 61 elements that have been added to the glass support matrix with the following composition: 72 % SiO_2, 12 % CaO, 14 % Na_2O, and 2 % Al_2O_3. Figure 5.5.9(a) shows that more than 30 heavy elements are clearly detectable, and the peak of each rare earth element is clearly separated in the spectrum. In contrast, the L line peaks of the rare earth elements cannot be recognized in Figure 5.5.9(b). The problem in the analysis of rare earth elements excited by conventional X-ray sources such as Pd Kα X-rays is that the L lines of the rare earth elements appear from 4.650 keV for La Lα to 9.938 keV for Lu Lβ_2. The K line spectra of the transition metals from Ti (4.508 keV for Kα) to Cu (8.040 keV for Kα) can overlap in the same energy region of the spectrum. Practical samples often contain these transition elements as major components, and disturb the analysis of the trace heavy elements by L lines. In fact, this SRM612 sample contains Ti, V, Cr, Mn, Fe, Co, Ni, and Cu. Therefore, it is practically impossible to analyze the rare earth elements of this sample using the L lines, as can be seen in Figure 5.5.9(b). In addition, without such light elements, possibly 14 rare earth elements (from La to Lu except for Pm) and 42 peaks (Lα, Lβ_1, and Lβ_2 lines for each element if we neglect the Lγ line) are present in the XRF spectrum within the small energy region of 5.288 keV (=9.938−4.650). In contrast, in the energy region above 20 keV, there are no peaks other than the K lines of the heavy elements. Moreover, the energy difference between La Kα_1 (33.4418 keV) and LuKα_1 (52.3889 keV) is ca. 19 keV, which is large enough to distinguish each rare earth element under the resolution of a pure Ge SSD, as is demonstrated in Figure 5.5.9(a).

Certified values for the rare earth elements have not been reported for SRM612 glass, but information values (in ppm) are given as follows: Ag(21), Ba(41), La(36), Ce(39), Nd(36), Sm(39), Eu(36), Gd(39), Dy(35), Er(39), and Yb(42). The intensity of the peak of each element in Figure 5.5.9(a) appears to be qualitatively consistent with these information values. This suggests that the present technique is applicable for the nondestructive total determination of rare earth elements as well as for other heavy elements, such as Hf and Ta, at the ppm level.[19]

5.5.5.2 ENVIRONMENTAL APPLICATIONS

Heavy elements are often environmentally toxic. Recently, many countries have initiated government regulations to restrict severely the maximum permissible concentrations of potentially toxic heavy metals in soils and in sludges used on the land. The heavy elements include Cr, Ni, Cu, Zn, Ag, Cd, Mo, and Pb and their limit values are sub-ppm to a few hundred ppm. Regulation of the concentration of Cd, Hg, As, and Pb in plastic has also become an important issue in the electric industry. Conventional multi-element analysis techniques, such as ICP-AES (inductively coupled plasma-atomic emission spectroscopy), require decomposition of the samples into solutions, which are time-consuming treatments. Therefore, the nondestructive feature of the XRF technique is quite

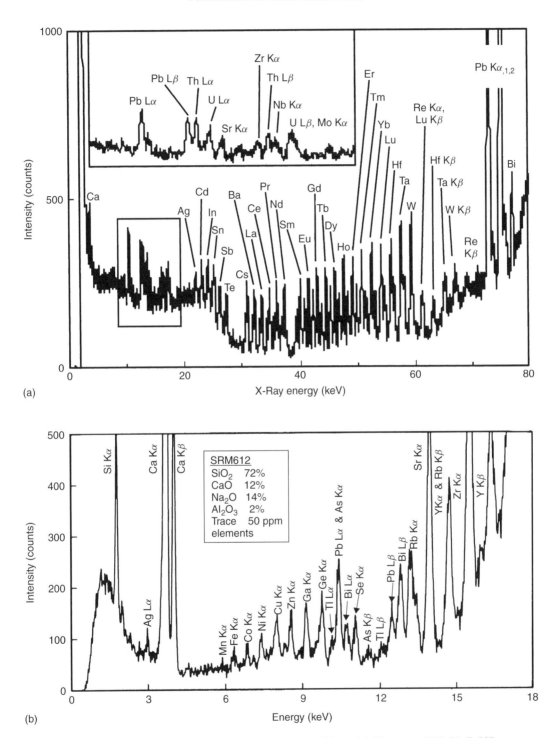

Figure 5.5.9 XRF spectrum of NIST SRM612 glass. Measurement conditions: (a) SR source, 116 keV, GeSSD, measurement time 1000 s[15]; (b) excitation source Pd Kα, 40 kV, 1 mA, Si drift detector, measurement time 300 s

Table 5.5.3 Accuracy and detection limit of the measurement and limit value for heavy metals in soil according to EU Directive 86/278/EEC

Element	Calibration range (ppm)	Calibration rms (ppm)	GSS-1[a]		GSD-7[a]		WT-H (sludge)[a]		LLD 100 s	LLD 30 min	Limit value (mg/kg ppm)
			Certified	Measured	Certified	Measured	Certified	Measured			
As	0.23–412	1.39	33.5	34.7	84	84	146	146	1.5	1.5	50
Cd	0.03–55	0.63	1.05	0.92	4.30	4.39	55	56	1	0.4	1–3
Cr	4.8–1340	17.44	62	58	122	119	1340	1256	4	4	–
Cu	4.1–3140	10.53	21	20.9	38.0	36.0	3140	3106	2.5	2.5	50–140
Mo	0.09–92	0.93	1.40	1.22	1.40	1.02	78	74	0.7	0.7	4
Ni	1.6–1140	5.42	20.4	19.7	53	54	1140	1147	4	4	30–75
Pb	4.4–2290	7.57	98	89	350	358	2290	2278	2.5	2.5	50–300
Zn	16–6360	10.98	680	679	238	246	6360	6359	2	2	150–300

[a] GSS-1 and GSD-7 are geochemical reference materials (Institute of Geophysical and Geochemical Prospecting, People's Republic of China. WT-H is sewage sludge reference materials.

attractive, and high-energy XRF is expected to provide a solution to these problems.

Common heavy metal contaminants in soils and sludges were analyzed with a commercially available laboratory spectrometer, Epsilon 5. A series of soil and rock standards were used for calibration. The soil samples were analyzed in the form of pressed powder pellets. Approximately 12 g of a mixture of sample and wax/styrene additive was pressed into 36-mm diameter pellets. The measurement time was 200 s (live time). The accuracy of the results is presented in Table 5.5.3. Twenty consecutive measurements of a sample demonstrated relative standard deviations better than 4% at the 24 ppm level (i.e. 25 ± 1 ppm). Typical detection limits for heavy metals in soil are also given in Table 5.5.3. For most elements, the LLDs calculated for 100 s are well within the requirement laid down by the EU soil directive as shown in Table 5.5.3 except for Cd, which required a counting time of 30 min.

5.5.5.3 ARCHAEOLOGICAL APPLICATIONS

Trace element components of a material often reflect its origin. Heavy elements in particular are useful as fingerprint elements in provenance analyses of archaeological samples, as heavy elements are trace elements in nature and exhibit unique geochemical behavior because of their large ionic radii and relatively high valency. To date, neutron activation analysis has often been used for analyzing heavy elements. However, destructive sample preparation is necessary for this method because the sample becomes radioactive after the analysis. This precludes the analysis of precious samples such as those in museums. In contrast, no beam-induced damage of samples was observed in high-energy XRF analysis, and this technique appears to be truly nondestructive, making it suitable for analyzing archeological samples and works of art. Here, this technique is applied to reveal the locality of Old Kutani chinaware.[20]

Kutani chinaware was first produced in the late 17th century in Kaga Province, which, today, is Ishikawa Prefecture in Japan. In 1710, however, after half a century of continuous production, the kiln was suddenly closed. Pottery from this early period is known as Old Kutani, and is extremely precious. However, it was thought that Old Kutani china might come from Arita, another world-famous area of porcelain production in Japan since the 17th century. Therefore, identification of Old-Kutani and Arita is an important problem–almost a mystery–in Japanese art history. The high-energy SRXRF analysis of porcelain clay pottery is expected to reveal the origin of the source materials of museum-grade specimens.

The reference samples consist of chinaware excavated from old kilns dated 17th to 19th century in the Kutani, Arita, and Fukuyama districts of Japan. Several original samples of museum-grade dishes have also been nondestructively analyzed (Figure 5.5.10). They were very precious so-called Old-Kutani and Arita pieces, which were borrowed from several collectors and artists. This was the first nondestructive analysis of museum-grade samples of Old Kutani china.

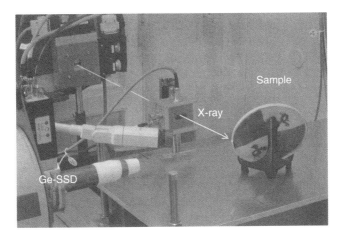

Figure 5.5.10 Photograph showing nondestructive SRXRF analysis of a museum-grade dish (Old Kutani). The measurement was made at BL-08 W of SPring-8

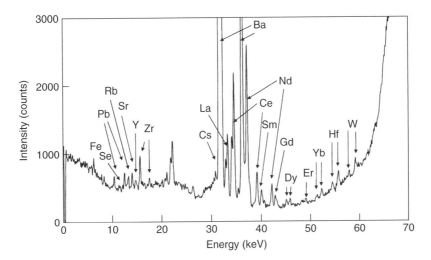

Figure 5.5.11 SRXRF spectrum of a fragment of old Kutani chinaware excavated from an old kiln dated 17th century

The analysis confirmed that 116 keV X-ray irradiation caused practically no damage to the samples. Therefore, this technique is suitable for the nondestructive characterization of precious historical samples, as is demonstrated in Figure 5.5.10. An XRF spectrum of the excavated fragment of Old Kutani ware is shown in Figure 5.5.11. It shows that tungsten as well as various rare earth elements give distinct fluorescence peaks of K lines and, therefore, the spectrum is rich with information. The XRF peak intensities of these heavy elements were used as the parameters of some statistical treatments for the provenance analysis. It was found that the Ba/Ce–Nd/Ce plot shown in Figure 5.5.12 is the most useful to estimate the origin of the chinaware. The result shows that Kutani and Arita chinaware can be clearly distinguished using this plot.[20,21] It was found that the analytical data of some museum-grade samples were located in the region of the Kutani chinaware and that some were from that of the Arita chinaware. The data suggest that the former samples were produced using potter's clay of the Kutani area and that the latter were those of

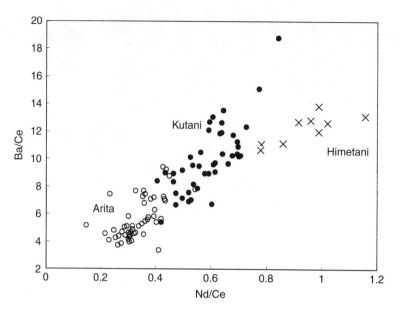

Figure 5.5.12 Ba/Ce vs Nd/Ce plot showing three clusters of analytical data. Samples were fragments of chinaware excavated from old kilns of Arita, Kutani, and Himetani

the Arita area. Thus, high-energy XRF has opened new fields of application for nondestructive analyses of historical samples. Detailed results will be published elsewhere.

5.5.5.4 FORENSIC APPLICATIONS

Originally, we first developed high-energy XRF utilizing 116 keV SR to conduct a scientific investigation to aid in the solving of an arsenic murder that occurred in Wakayama City on 25 July 1998.[14,15] Four people were killed and 63 people suffered arsenic poisoning after eating curry stew served at a summer festival in a small town in Japan. A trace amount of arsenic was attached to the crime-related substances. It was expected that the use of high-energy X-rays from SPring-8 as an excitation source for XRF would be suitable for distinguishing the difference in the origin of the trace amount of the arsenic samples, which were produced commercially. Arsenic compounds often contain impurities of Sb and Bi, whose excitation energies are 30.5 and 90.6 keV, respectively. Therefore, the high-energy X-rays from SPring-8 were required as an excitation source for detecting the ppm levels of Sb and Bi in the trace amounts of the samples. We had analyzed various arsenic oxides of different industrial origins for comparison. The sample was placed on a Mylar^{TM} film on a plastic holder, which was set on the automatic XY stage. The analysis point was irradiated by the laser and monitored by a video camera.

XRF spectra of various arsenic oxides were successfully measured. The examples, which were produced in China and Mexico, are given in Figure 5.5.13(a) and 5.5.13(b), respectively. The former contains Sn, Sb, and Bi while the latter contains Sb and Bi. It was found that the trace heavy element compositions are distinct from each other, reflecting the different places of production. Our analysis revealed that the arsenic oxide in the curry was the same product as the arsenic powder found in the defendant's house. These data were presented to the Wakayama District Public Prosecutors Office. This was the first successful application of SR in forensic analysis. The defendant in the case was tried in Wakayama District Court and found guilty.

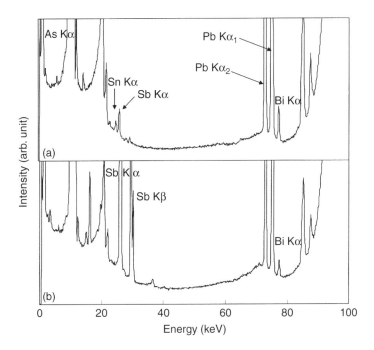

Figure 5.5.13 XRF spectra of arsenic trioxide, As_2O_3, produced in (a) China and (b) Mexico[14]

The judge sentenced the defendant to death in December 2002. The defendant has appealed the conviction to a higher court.

After this experience, we examined several samples from the perspective of forensic analysis. Through our studies, the following samples were found to be suitable for the application of high-energy XRF techniques. Identification of a tiny glass fragment from a hit-and-run car accident and automobile paint,[22] gunshot residues,[23] cement, etc. Only a tiny sample as small as $100\,\mu m\,\phi$ is necessary, and the analysis is truly nondestructive. Through these studies, we have demonstrated that high-energy XRF is a powerful tool in the forensic identification of material evidence based on trace heavy element compositions. This technique has already come into routine use by the Forensic Science Laboratory (headed by Dr T. Ninomiya) of the Hyogo Prefecture Police Headquarters to solve several important criminal cases occurring all over Japan. These cases had been found difficult to solve by conventional analytical techniques. Several practical examples are introduced by T. Ninomiya in this book (Subchapter 7.4).

5.5.5.5 GEOLOGICAL AND GEOCHEMICAL APPLICATIONS

In geochemistry, a chondrite-normalized rare earth element (REE) pattern is an important indication of the origin of a sample. The high-energy XRF technique enables direct measurement of the REE pattern as stated above. In addition, we can obtain two-dimensional information from a small region of a sample by using X-ray microbeams obtained by collimators, capillary optics,[6] or Fresnel zone plate optics.[10,24]

The cosmic abundance of heavy elements ($Z > 30$) is extremely low and they exhibited unique geochemical behavior during the earth's formation because of their large ionic radii and relatively high valence state. Therefore, analysis of trace heavy elements is an important subject for earth and planetary sciences including geochemistry. Here, the elements of the Pt group play an especially important role in the evolution of the earth. The potential ability of the high-energy XRF technique in the analysis of Pt group elements was examined. Measurements were made at BL-08 W,

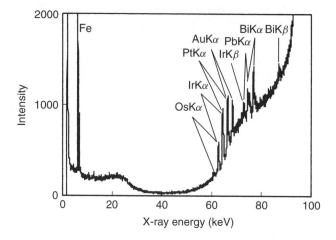

Figure 5.5.14 XRF spectrum of iron meteorite (octahedrite)[25]

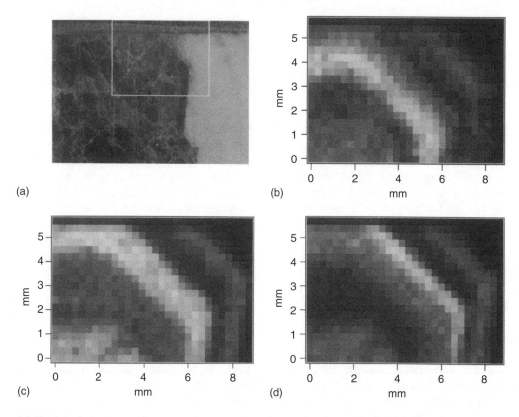

Figure 5.5.15 (a) Optical microscope image of a garnet sample. XRF imaging of (b) Ce, (c) Gd and (d) Yb measured for the area indicated by the white square in (a). White to black corresponds to the highest to lowest XRF intensity[26]

SPring-8 utilizing the same experimental setup as stated above. An example of the XRF spectrum of octahedrite is shown in Figure 5.5.14.[25] It is remarkable that trace amounts of Re, Os, and Ir were clearly detected. The advantage of the high-energy SRXRF technique is the nondestructive, two-dimensional chemical imaging of the trace heavy elements. The distribution of each REE and other heavy elements in garnet was analyzed.[26] The sample showed a zoning of trace REEs, which was formed during the growth of the crystal. The optical microscope image of the sample is given in Figure 5.5.15(a). The distributions of Ce, Gd, and Yb in the garnet are shown in Figure 5.5.15(b), 5.5.15(c) and 5.5.15(d), respectively. These results show that the zoning position of each element shifts outside the garnet crystal with an increase in the atomic number. Assuming that the rare earth elements exist in trivalent ions and are 12 coordinated in the garnet crystals, Figure 5.5.15(b), 5.5.15(c) and 5.5.15(d) suggest that the larger ions (Ce ion, in this case) were positively incorporated into the garnet crystal at an early stage of growth, while the smaller ions (Yb ion, in this case) are not incorporated into the crystal until a late stage of growth.

The advantages of high-energy XRF over conventional analytical techniques used in geology and geochemistry are summarized as follows. The identification of the XRF peaks is more straightforward and measurement is easy compared with secondary ion mass spectroscopy (SIMS), and enables us to analyze trace heavy elements with two-dimensional resolution. We can analyze geological samples as large as 100 kg in weight and 1 m in length. If we used focusing optics such as a Fresnel zone plate, the spatial resolution could reach 1 µm or less. This spatial resolution is close or even better than an electron microprobe, by which trace element analysis is difficult.

ACKNOWLEDGEMENTS

The author greatly appreciates the help of Dr Yasuko Terada of SPring-8 for her kind assistance in the high energy XRF experiments at SPring-8, and in the preparation of some of the figures used in this text. He acknowledges the permission that PANalytical has given him to use the diagrams (Figures 5.5.1, 5.5.3 and 5.5.4). The analytical data (Table 5.5.3) were supplied by the courtesy of Dr Simon Milner of PANalytical B.V.

REFERENCES

1. Harada, M. and Sakurai, K. K-line X-ray fluorescence analysis of high-Z elements. *Spectrochim. Acta B* **54**, 29–39 (1999).
2. *XRF Globe* 1–2, 12–13 (2003).
3. Baryshev, V. B., Gil'bert, A. E., Koz'menko, O. A., Kulipanov, G. N. Zolotarev, K. V. Determination of the concentrations and distributions of rare-earth elements in mineral and rock specimens using the VEPP-4 synchrotron radiation. *Nucl. Instrum. Methods Phys. Res., Sect. A* **261**, 272–278 (1987).
4. Dar'in, A. V. and Bobrov, V. A. Measurement of rare earth element content in rock standards by XFA method with use of synchrotron radiation from the storage ring VEPP-4. *Nucl. Instrum. Methods Phys. Res., Sect. A* **261**, 292–294 (1987).
5. Khvostova, V. P. and Trunova, V. A. Samples for X-ray fluorescence analysis using synchrotron radiation. *Nucl. Instrum. Methods Phys. Res., Sect. A* **261**, 295–300 (1987).
6. Janssens, K., Vincze, L., Vekemans, B., Adams, F., Haller, M. and Knochel, A. Use of lead-glass capillaries for micro-focusing of highly-energetic (0–60 keV) synchrotron radiation. *J. Anal. At. Spectrom.* **13**, 339–350 (1998).
7. Chen, J. R., Chao, E. C. T., Back, J. M., Minkin, J. A., Rivers, M. L., Sutton, S. R., Cygan, G. L., Grossman, J. N. and Reed, M. J. Rare earth element concentrations in geological and synthetic samples using synchrotron X-ray fluorescence analysis. *Nucl. Instrum. Methods Phys. Res., Sect. B* **75** (1–4), 576–581 (1993).
8. Snigirev, A., Snigireva, I., Engstroem, P., Lequien, S., Suvorov, A., Hartman, Ya, Chevallier, P., Idir, M. and Legrand, F. Testing of submicrometer fluorescence microprobe based on Bragg–Fresnel crystal optics at the ESRF. *Rev. Sci. Instrum.* **66**, 1461–1463 (1995).
9. Kagoshima, Y., Takai, K., Ibuki, T., Hashida, T., Yokoyama, Y., Yokoyama, K., Takeda, S., Urakawa, M., Miyamoto, N., Tsusaka, Y. and Matsui, J. Formation of X-ray microbeam using Ta phase zone plate and its application to scanning X-ray microscope-III. *SPring-8 User Exp. Rep.* No. 5, 436 (2000).
10. Suzuki, Y., Takeuchi, A., Takano, H., Ohigashi, T. and Takenaka, H. Diffraction-limited microbeam with Fresnel zone plate optics in hard X-ray regions. *Jpn. J. Appl. Phys.* **40**, 1508–1510 (2001).
11. Wobrauschek, P., Gorgl, R., Kregsamer, P., Streli, C. Pahlke, S. Fabry, L., Haller, M., Knochel, A. and

Radtke, M. Analysis of Ni on Si-wafer surfaces using synchrotron radiation excited total reflection X-ray fluorescence analysis. *Spectrochim. Acta* **B52**, 901–906 (1997).
12. Ortega, L., Comin, F., Formoso, V. and Stierle, A. Trace element analysis on Si wafer surfaces by TXRF at the ID32 ESRF undulator beamline. *J. Synchrotron Radiat.* **5**(3), 1064–1066 (1998).
13. Sakurai, K., Eba, H., Inoue, K. and Yagi, N. Wavelength-dispersive total-reflection X-ray fluorescence with an efficient Johansson spectrometer and an undulator X-ray source: detection of 10^{-16} g-level trace metals. *Anal. Chem.* **74**, 4532–4535 (2002).
14. Nakai, I., Terada, Y., Itou, M. and Sakurai, Y. X-ray fluorescence analysis of heavy elements by using $K\alpha$ X-ray fluorescent lines excited by high energy X-rays. *SPring-8 User Exp. Rep. (JASRI)* No. 3, 88 (1999).
15. Nakai, I., Terada, Y., Ito, M. and Sakurai, Y. Use of highly energetic (116 keV) synchrotron radiation for X-ray fluorescence analysis of trace rare-earth and heavy elements. *J. Synchrotron Rad.* **8**, 360–362 (2001).
16. Hara, M. High energy inelastic scattering (BL08W). *SPring-8 Ann. Rep. 1998* **53** (1999).
17. Sakurai, Y., Hiraoka, N., Ito, M., Ohta, T. and Sakai, N. Performance of a high-resolution Compton scattering spectrometer for heavy elements at BL08W. *SPring-8 User Exp. Rep.* No. 3, 80 (1999).
18. Sasaki, S. X-ray absorption coefficients of the elements (Li to Bi, U). *KEK Rep. 90* **16**, 1–142 (1990).
19. Noma, T., Takada, K., Mukaide, T., Terada, Y., Nakai, I. Application of high-energy X-ray fluorescence analysis for fluorites. *SPring-8 User Exp. Rep. (JASRI)* No. 8, 64 (2002).
20. Nakai, I., Terada, Y., Yamato, S., Yamana, K., Itou, M. and Sakurai, Y. Development and application of high energy X-ray fluorescence technique for provenance analysis of archaeological samples. *SPring-8 User Exp. Rep. (JASRI)* No. 4, 69 (1999).
21. Nakai, I., Terada, Y., Yamato, S., Yamana, K., Miura, Y., Itou, M. and Sakurai, Y. Application of high-energy X-ray fluorescence analysis for archaeometric analysis of old Kutani china wares. *SPring-8 User Exp. Rep. (JASRI)* No. 5, 128 (2000).
22. Ninomiya, T., Nakanishi, T., Muratsu, S., Saitoh, Y., Shimoda, O., Watanabe, S., Nishiwaki, Y., Matsushita, T., Suzuki, S., Suzuki, Y., Ohta, H., Kasamatsu, M., Nakai, I. and Terada, S. Elemental analysis of a trace of paint chip using SR-XRF. *SPring-8 User Exp. Rep. (JASRI)* No. 7, 67 (2001).
23. Nakai, I., Terada, Y. and Ninomiya, T. Forensic application of synchrotron radiation X-ray fluorescence analysis. *Proc. 16th Meet. Int. Assoc. Forensic Sci.* 29–34 (2002).
24. Tamura, S., Yasumoto, M., Kamijo, N., Suzuki, Y., Awaji, M., Takeuchi, A., Takano, H. and Handa K. Development of multilayer Fresnel zone plate for high-energy synchrotron radiation X-rays by DC sputtering deposition. *J. Synchrotron Rad.* **9**, 154–159 (2002).
25. Terada, Y., Miura, Y., Nakai, I., Takahashi, Y., Itou, M. and Sakurai, Y. Development and application of new X-ray fluorescence technique for earth and planetary materials using high energy X-ray. *SPring-8 User Exp. Rep. (JASRI)* No. 7, 72 (2001).
26. Terada, Y., Nakai, I., Miura, Y., Takahashi, Y., and Kato, Y. XRF imaging of the heavy elements in geological samples using high energy X-rays. *SPring-8 User Exp. Rep. (JASRI)* No. 8, 60 (2002).

5.6 Low-energy Electron Probe Microanalysis and Scanning Electron Microscopy

S. KUYPERS
VITO (Flemish Institute for Technological Research), Mol, Belgium

5.6.1 INTRODUCTION

The distinction between an electron probe microanalyser and a scanning electron microscope has always been clear, at least to the microanalysis community:

- An electron probe microanalyser is an analysis tool equipped with at least three, preferably five, wavelength-dispersive spectrometers providing high spectral resolution. A high beam current is required and can be provided by a conventional electron source. Electron image quality is rather poor.
- A state-of-the-art scanning electron microscope is an imaging tool equipped with a field emission gun providing an electron beam with a very high brightness and is therefore capable of working at high magnifications. The preferred attachment for elemental analysis is an energy-dispersive spectrometer, because the relatively low current in the electron beam does not allow efficient use of a wavelength-dispersive spectrometer.

Low-energy electron probe microanalysis (EPMA) and scanning electron microscopy (SEM) are highly complementary techniques, essential in materials R&D. Wavelength-dispersive spectrometry (WDS) in an electron probe microanalyser is a powerful tool for quantitative near-surface analysis and hitherto unrivalled for quantitative analysis of ultra-light-element-based thin films on substrates. A scanning electron microscope with a field emitter source allows high resolution imaging at reduced beam energies, where specimen charging can be reduced or eliminated. The flexibility, the ease-of-use and the availability of tools for local elemental analysis, make low-energy SEM a tool equal to none for studying surface morphology on a nanoscale.

Recent developments in instrumentation, such as high-sensitivity wavelength-dispersive spectrometers, the bolometer and high-current field emitter sources, will allow to combine high spectral resolution and high lateral resolution in one scanning electron beam instrument. Only if this can be done without compromise, will the distinction between EPMA on the one hand and SEM with accessories for elemental analysis on the other hand, disappear completely. Meanwhile, the total cost of a 'hybrid' instrument as compared to two separate instruments can be expected to be such that many laboratories will allow for some compromise in the near future.

In this subchapter, the possibilities and limitations of low-energy EPMA and of low-energy SEM, as performed in two separate instruments, are discussed. The potential of the two techniques is illustrated with recent examples related to the development of ultra-light-element based coatings for sliding wear applications, membranes for ultrafiltration and packaging materials for meat.

5.6.2 SOFT X-RAYS IN PRACTICE

5.6.2.1 GENERAL

For a review on quantitative electron probe microanalysis touching on all aspects, we refer to the textbook by Scott *et al.* (1995). A good account of the practical use of soft X-rays in microanalysis was given by Pouchou (1996). In this paper it was pointed out that the present definition of the soft X-ray range as extending from 1 keV down to 100 eV, is largely due to the limits of spectrometers and absorption correction models in the early days of microanalysis. On the other hand, even with today's instrumentation and software, accurate analysis in the soft X-ray range requires special care and is still a challenge for many applications. Three main reasons can be cited for using soft X-rays:

- Analysis of ultra-light elements (Be, B, C, N, O, F), in which case of course the soft K lines are the only characteristic lines available.
- Analysis with increased surface sensitivity by reducing the electron beam energy, in which case one should select the L rather than the K lines for medium-Z elements and the M rather than the L lines for high-Z elements.
- Minimisation of fluorescence effects, because of low fluorescence contributions in the soft X-ray lines.

There are a number of (potential) experimental problems to be overcome and precautions to be taken when soft X-ray lines in general and ultra-light elements in particular are analysed with wavelength-dispersive spectrometers; the more so when thin coatings and/or low concentrations are involved:

- Carbon contamination at the point of impact of the electron beam, due to the presence of hydrocarbons in the vacuum; its influence on the results can be substantial (Willich and Bethke, 1993); the best way to control it seems to be a gas-jet combined with a cold finger, in particular for lengthy measurements (Willich and Bethke, 1993; Bastin and Heijligers, 1986).
- Insulating specimens can be a serious source of problems; coating with carbon is not always the best choice; good results have e.g. been obtained with sputtered gold (Pouchou, 1996) and aluminium (see below) films; applying identical conductive coatings on reference materials (RMs) and unknowns simultaneously is not obvious and not really essential as long as the different conductive coatings applied are taken into account in the matrix correction.
- The choice of RMs should be ideally such that the electrical conductivity of RMs and unknowns is comparable; in practice one often has to use what is (commercially) available and make the best of it.
- Dedicated monochromators are required; the newer synthetic multilayers have advantages over the conventional lead stearate; the peak count rates are higher and because of their somewhat poorer resolution they are less sensitive to peak shape alterations (Bastin and Heijligers, 1990).
- Peak shape differences between RMs and unknown can be substantial in the case of boron and carbon peaks and should be taken into account by measuring peak areas rather than peak heights; in practice one can use the so-called area-to-peak factor (APF) concept (Bastin and Heijligers, 1990).
- The matrix correction program is of course essential; software for the analysis of multilayer films, based on reliable $\varphi(\rho z)$ models, where φ represents the ionization and ρz the mass depth in the sample, was made commercially available some eight to ten years ago (Bastin and Heijligers, 1990; Pouchou and Pichoir, 1990; Bastin *et al.*, 1993) and has been further refined since then.
- The lack of reliable reference methods for the analysis of ultra-light elements, especially in coatings, emphasises the unique character and the importance of EPMA, but at the same time it prevents the validation of EPMA for this important application.

5.6.2.2 IN PRACTICE

Our interest in soft X-rays is primarily for the analysis of ultra-light-element-based thin coatings. Our microprobe is a Jeol JXA-8621 with three wavelength-dispersive spectrometers (six monochromators) and one energy-dispersive spectrometer with a Be window. Two monochromators are for ultra-light element analysis: a conventional lead stearate crystal (NSTE; $2d = 10.04$ nm) and a synthetic W/Si multilayer crystal (LDE; $2d = 6$ nm). A gas-jet for eliminating carbon contamination is not available on our instrument. Instead we use a combination of a trap at the entrance to the diffusion pump (baffle) and a metal plate just beneath the objective lens, both liquid-nitrogen cooled, when analysing ultra-light elements. With this system we never detected carbon contamination within the time frame of the measurements.

Until recently only ZAF-correction software was available on-line on our system and data had to be fed manually into separate software for further analysis with $\varphi(\rho z)$-based methods. We could not take into account peak areas. Matrix correction was done with Strata 5.0 (SAMx), which uses the models developed by Pouchou and Pichoir (1990).

Most of our RMs were commercially available and were coated by the manufacturer with a protective carbon coating prior to delivery. The thicknesses of the protective carbon coatings on the different ultralight-element RMs were estimated using the Strata program and were found to be in the range 10–25 nm. The measured values were taken into account in the matrix correction program.

Ultra-light-element-based coatings for analysis with EPMA were mostly deposited on silicon wafers. The substrate signals were taken into account for matrix correction. Although the coatings are insulators, charging was only a problem when the coating thickness exceeded the excitation depth. In such cases reasonable results were obtained by depositing an aluminium coating approximately 10 nm thick prior to analysis. Results for a series of B–N–Si:H coatings obtained by varying different deposition parameters are shown in Table 5.6.1. The thicknesses of the aluminium layers were derived from Strata assuming a density of 2.7 g/cm^3. The results look acceptable, perhaps deceivingly so, considering the fact that peak shapes were not taken into account, while the boron content does vary over a wide range. The carbon and oxygen content might (partly) be due to contamination from the aluminium sputtering process.

In Section 5.6.4.1, more experimental details are given for EPMA applied to thin coatings in the boron–nitrogen–carbon (BNC) composition triangle.

5.6.3 LOW-ENERGY SEM AND HIGH-RESOLUTION IMAGING

5.6.3.1 GENERAL

Low-energy SEM is defined as SEM at beam energies below 5 keV. An excellent and practical introduction to the subject can be found in the paper

Table 5.6.1 Compositions of B–N–Si:H coatings, obtained with EPMA at 8 keV and 100 nA. The coating thicknesses were obtained independently from EPMA. An aluminium coating was deposited to achieve conductivity in EPMA. Its thickness was derived from the measured intensity. Reproduced by permission of Springer-Verlag Wien

Sample	B (at%)	N (at%)	Si (at%)	C (at%)	O (at%)	Total wt%	Thickness (nm)	Thickness Al (nm)
GLC917	13.9	49.2	33.8	2.1	1.0	97.3	2800	10
GLC919	7.6	48.0	40.3	2.8	1.3	97.8	3200	15
GLC920	10.8	57.8	27.9	2.3	1.2	97.6	3400	15
GLC921	4.7	58.9	32.7	2.5	1.3	99.7	4000	14
GLC932	30.8	52.1	9.6	1.5	6.0	101.9	1460	16
GLC933	29.0	54.5	10.8	1.4	4.3	102.3	1580	15
GLC934	25.1	56.9	14.3	1.1	2.6	100.1	1620	10
GLC935	19.4	59.2	17.6	1.9	1.9	97.6	2300	14

by Joy and Joy (1996), while a more theoretical background is provided by Reimer (1993).

In principle, low-energy SEM is an excellent approach to achieving high-resolution imaging (e.g. Goldstein *et al.*, 1992), because the electron–specimen interaction volume decreases rapidly with beam energy. The problem is the electron optics: at low beam energies the source brightness will be reduced and one may be left with a degraded probe size or insufficient beam current. This can be overcome by using a high-brightness source, i.e. a field-emitter tip, rather than a more conventional tungsten hairpin or LaB_6 filament. A modern field-emission-gun SEM (FEG-SEM) allows imaging at intermediate (10 000 to 100 000×) and even high (>100 000×) magnifications, with beam energies well below 5 keV.

Advantages of SEM at low beam energies are:

- Charging of nonconductive specimens can be reduced or eliminated; this is achieved by working close to or at the beam energy ('E2') where no net charging occurs, i.e. where the number of electrons emitted by the specimen equals the number of electrons received; for many materials of technological importance (semiconductors, ceramics, polymers) E2 is in the range of a few keV; a method to determine E2 experimentally is described by Joy and Joy (1996).
- The extent of radiation damage is reduced (not necessarily the radiation damage as such).
- Better topographic contrast is obtained.
- Surface sensitivity is increased.
- Image resolution in backscattered electron (BSE) mode is improved.
- X-ray resolution is improved.

Disadvantages are:

- Electron-optical performance is reduced.
- Overvoltage for X-ray microanalysis is (too) low.

5.6.3.2 IN PRACTICE

Conventional SEM is usually done in the range 15–25 keV. This range of electron beam energies is a bad choice in all respects. Charging of nonconductive specimens under these conditions can be severe. In extreme cases of charging the electron beam is reflected off the specimen and scans the microscope chamber. The 'fisheye lens' view of the specimen chamber of our JSM-6340F, shown in Figure 5.6.1, was obtained in this way. It reveals the liquid-nitrogen-cooled anticontamination plate just below the objective lens, the in-lens secondary electron detector (SED), the retracted backscattered electron detector (BSED), the chamber camera and the end cap of the EDS detector. Not visible is the conventional secondary electron detector or 'lower electron detector' (LED) below the objective lens.

The Jeol JSM-6340F is a digital FEG-SEM with a cold field emission electron source and a semi-in-lens type objective lens (OL) (Nakagawa, 1994; Yamamoto *et al.*, 1996; Yamamoto *et al.*, 1999). The semi-in-lens type OL induces a strong magnetic field on the sample. The field rolls up almost all the secondary electrons into the OL, where they are guided towards the in-lens SED by means of a set of accelerating and retarding electrodes. This is schematically represented in Figure 5.6.2. An important consequence is that the relative contribution of backscattered electrons to the LED image is much larger than for an out-of-lens type OL.

Images of metal oxide particles and of the surface of a polymer-based membrane shown in Figure 5.6.3, illustrate the importance of carefully selecting an appropriate beam energy. Here, the effect on image quality and information depth when the beam energy is reduced from a conventional 20 keV to a relatively low 5 keV are dramatic. Figure 5.6.4 demonstrates the high-resolution capabilities of a modern FEG-SEM at low beam energies. The images are from aluminium surfaces anodised under different conditions. They were obtained in the JSM-6340F at 5 keV, working distance (WD) 5–6 mm, original magnification 100 000×; approximately 1.5–2 nm of Pt–Pd coat was applied by sputtering. Note that instead of trying to achieve dynamic charge balance conditions at E2, we are working at slightly higher beam energies and have applied a very thin

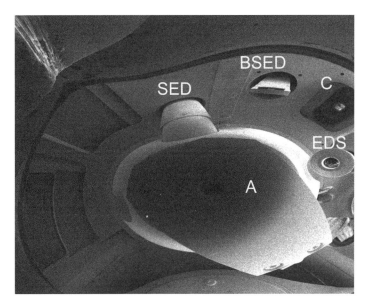

Figure 5.6.1 'Fisheye lens' view of the specimen chamber of the JSM-6340F seen from the specimen and obtained as a result of extreme specimen charging. We see the anti-contamination plate (A) just below the objective lens, the in-lens secondary electron detector (SED), the retracted backscattered electron detector (BSED), the chamber camera (C) and the end-cap of the EDS detector (EDS). Reproduced by permission of Springer-Verlag, Wien

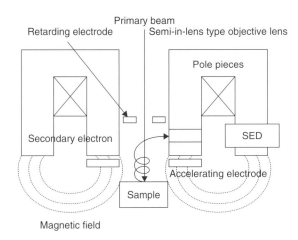

Figure 5.6.2 Schematic representation of the semi-in-lens type objective lens of the JSM-6340F, with accelerating and retarding electrodes and in-lens SED. Courtesy of Jeol Ltd. Reproduced by permission of Springer-Verlag Wien

metallic coating. It is our experience that a sufficiently low beam energy (very often 5 keV), combined with a suitable metallic coating, allows to obtain excellent high-resolution images of nonconductive specimens. We usually apply 1–2 nm of Pt–Pd (80–20) by sputtering (Cressington 208HR). When applied in this thickness range the Pt–Pd coating is 'invisible' to the JSM-6340F and hence does not introduce any observable artefacts. This way of working is very often more efficient than working under dynamic charge balance conditions, especially for heterogeneous surfaces where E2 changes with the position of the beam on the sample.

5.6.4 APPLICATIONS

5.6.4.1 COMPOSITION OF ULTRA-LIGHT-ELEMENT-BASED COATINGS – THE BNC TRIANGLE

Ultra-light-element-based coatings within the BNC composition triangle have been studied intensively in recent years for a wide range of applications (wear protection, optics, electronics). Crystalline phases such as β-C_3N_4, diamond and c-BN seem attractive for wear protection applications because of their extreme hardness. When prepared in the amorphous state these materials have the potential

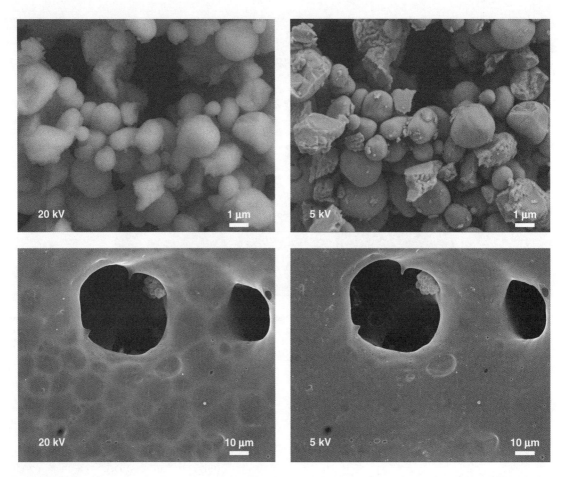

Figure 5.6.3 Effect of reducing the electron beam energy from 20 keV to 5 keV. Top: Metal oxide grains (ZrO_2–MgO) imaged with the lower electron detector (LED). At 20 keV the grain boundaries are vague and all surface detail is lost. Bottom: polymer-based membrane imaged with the in-lens SED. At 20 keV the polymer network below the surface is also imaged. Reproduced by permission of Springer-Verlag, Wien

to retain to some extent the excellent properties of the crystalline phases. Amorphous coatings can be deposited at lower temperatures and will inherently possess smooth surfaces, which makes them particularly useful for sliding wear applications. Moreover, amorphous ternary coatings $B_xN_yC_z$ can be prepared in a wide composition range, allowing material properties to be tailored for specific applications. We have published results for thin films prepared on the C–N and B–N axes, as well as for films prepared across the whole BNC triangle (Dekempeneer *et al.*, 1994, 1995, 1996a, 1996b). In each case the correlation between deposition parameters and structural, mechanical and tribological properties was investigated. All $B_xN_yC_z$:H coatings were deposited with radio frequency plasma assisted chemical vapour deposition (RF PACVD), starting from CH_4–N_2–B_2H_6 (H_2 diluted) gas mixtures.

The B, C, N and impurity-O content of coatings deposited on silicon was measured with EPMA, while the H content was measured with elastic recoil detection analysis (ERDA). Coating thicknesses were in the range 200 nm – 2.5 μm. There were attempts to set up nuclear reaction analysis (NRA) as a reference method for determining the B, C, N and O content of some of the coatings. However, reliable NRA results were never

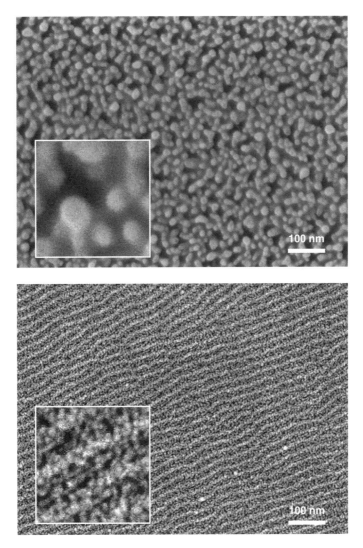

Figure 5.6.4 High resolution images of anodised aluminium surfaces after different anodisation treatments. The insets are $100 \times 100\,\text{nm}^2$. The images were obtained at 5 keV with the SED in the JSM-6340F, working distance 5–6 mm, original magnification $100\,000\times$; approximately 1.5–2 nm of Pt–Pd coat. Reproduced by permission of Springer-Verlag, Wien

obtained. This was very unfortunate because they would have allowed to employ the coatings analysed with NRA as additional RMs for EPMA and, most importantly, to assess and validate EPMA for quantitative analysis of $B_xN_yC_z$:H films.

EPMA was carried out in the Jeol JXA-8621. Samples were typically $20 \times 20\,\text{mm}^2$. They were glued to the sample holder with carbon tape. Per sample 6–9 points were analysed. Reference materials were B, C, h-BN, SiO_2 and Si; all commercially available. The electron beam energy was chosen in the range 5–10 keV; probe current was 100 nA. Analysing crystals were NSTE ($2d = 10.04\,\text{nm}$) for B and C, LDE1 (W/Si; $2d = 6\,\text{nm}$) for N and O, PET ($2d = 0.8742\,\text{nm}$) for Si. Peak heights were measured. Systematic use was made of the liquid-nitrogen-cooled baffle and anti-contamination device. A gas-jet was not available. There were no indications for carbon contamination within the time frame of the

Table 5.6.2 Experimental data for C-rich films in the BNC composition triangle, grown with PACVD. Low total weight percentages for EPMA emphasise the importance of using area-to-peak factors for boron analysis. Reproduced by permission of Springer-Verlag Wien

Sample	Gas flow (sccm)			EPMA results					H content (at%)	Thickness (nm)	Density (g/cm^3)
	CH$_4$	B$_2$H$_6$	N$_2$	C (at%)	B (at%)	N (at%)	O (at%)	Total wt%			
GLC812	9	1	–	97.0	2.8	–	0.2	100.5	32	430	1.49
GLC813	8	2	–	93.7	4.7	–	1.6	99.1	32	470	1.38
GLC814	7	3	–	90.6	7.5	–	1.9	97.1	31	485	1.43
GLC815	6	4	–	89.2	10.2	–	0.6	98.3	32	500	1.31
GLC817	5.7	3.2	1.1	80.3	10.4	7.9	1.4	95.2	36	480	1.43
GLC818	5.3	2.3	2.4	76.3	7.1	13.2	3.4	94.4	30	380	1.27
GLC819	5	1.5	3.5	77.8	4.8	16.4	1.0	93.7	31	380	1.30
GLC821	5	–	5	81.8	–	17.4	0.9	100.3	31	175	1.11
GLC822	4.5	5.5	–	83.0	16.6	–	0.4	94.8	37	790	1.69
GLC823	3	7	–	76.6	22.4	–	1.1	92.5	37	640	1.69

measurements. Matrix correction on the data was done off-line with Strata 5.0 thin film correction software, using the XPP model. The measured Si signal from the substrate and the actual thickness of the protective carbon coatings on the RMs was taken into account. In Table 5.6.2 results are shown for a number of carbon-rich B$_x$N$_y$C$_z$:H coatings with thicknesses in the range 170–800 nm. Electron beam energy was 8 keV for all analyses. The thicknesses were obtained from step height measurements and were fed into Strata to obtain an estimate of the film densities. It is apparent from Table 5.6.2 that the total weight percentage decreases with increasing B content. Most probably this is due to not taking into account differences in B peak shape between RMs and unknowns. The presence of nitrogen seems to add to the effect. The results emphasise the importance of using area-to-peak factors for the analysis of B. It can be assumed that the reported results underestimate the B content of the coatings.

5.6.4.2 SURFACE CHARACTERISTICS OF POLYMER BASED MEMBRANES FOR ULTRAFILTRATION

SEM is generally accepted as an indispensable tool for polymer-membrane research. Its power lies in the direct visualisation of the membrane pore structure. However, membranes for ultrafiltration (UF) have dense boundary layers with surface pores in the nanometer range, which is beyond the resolving power of conventional SEM, even at high electron beam energies. Transmission electron microscopy (TEM) has the required resolution, but the sample preparation (thin foil cross-section or replica) is tedious and likely to introduce artefacts. A solution is offered by FEG-SEM, which combines high-resolution imaging at low beam energies with all the advantages of conventional SEM. FEG-SEM has been successfully applied to different types of commercially available UF membranes (Kim et al., 1990, 1991; Kim and Fane, 1994).

In the early 1990s, VITO developed new UF membranes loaded with inorganic (metal oxide) fillers (Doyen et al., 1990a). The membranes consist of a polymer network (mostly polysulfone) and an inorganic filler (originally zirconium oxide). They possess a dense boundary layer (skin) on top of a porous support (matrix). Originally these membranes were developed for use as separators in electrochemical systems. Applications in the field of ultrafiltration were envisaged when it was discovered that the transport properties vary with the inorganic filler content (Doyen et al., 1990b). The permeability was found to increase significantly with increasing metal oxide content. The observed flux behaviour was explained by assuming that the metal oxide particles are present in the skin of the membrane where they modify skin morphology and/or skin thickness. Direct microscopic observations were required to support this assumption.

Feasibility tests with transmission electron microscopy (TEM), atomic force microscopy (AFM) and FEG-SEM were set up. The path of

TEM was soon abandoned: ultramicrotomy did not produce useful thin sections, mainly because of the presence of large (relative to the skin thickness) inorganic grains that were torn out and damaged the section during cutting. AFM posed no problems of sample preparation, but the apparent roughness of the membrane surfaces complicated image interpretation, it proved extremely difficult to obtain an overview over larger sample areas, and there was the suspicion of imaging artefacts (pores elongated in one direction). The first FEG-SEM results (on coated samples) looked very promising and it was decided to introduce this technique in UF membrane development and optimisation. Different experiments were set up in which membrane casting parameters were varied and the effect on membrane performance studied. In all these experiments low-energy SEM played a key role in linking membrane performance with membrane morphology (Kuypers et al., 1995; Genné et al., 1996; Aerts et al., 2000).

Most of the FEG-SEM work was carried out in instruments with a cold field emitter source (JSM-6400F, JSM-6320F, JSM-6340F) at electron beam energies of 5–10 keV and working distances of 3–5 mm. Our standard conditions on these UF membranes with the JSM-6340F are: 5 keV, WD 3 mm, in-lens detector, 1.5 nm Pt–Pd, magnification 100 000×. To our experience a metallic coating suited for high magnification work greatly increases experimental efficiency, without introducing artefacts within the resolution limits of the instrument. Initially, observations were also made on uncoated surfaces using electron beam energies of 1–3 keV. There was little or no charging, but (minor) beam damage inflicted to the polymer surface and the moderate performance of the instrument at the lowest voltages, resulted in insufficient image detail for accurate pore measurements. After the skin surfaces were coated with 1.5–2 nm of Pt or Pt–Pd, the electron beam did no longer inflict observable damage to the polymer surface. It has been reported that in certain cases Pt is to be preferred over Cr for high resolution imaging of clean UF membranes (Kim and Fane, 1994).

The image in Figure 5.6.5 is representative for the skin surface of UF membranes with metal oxide filler. In general, surfaces with well-resolved pores of 3–50 nm are observed. The distribution of pores over the surface is uniform. Apart from pores the image also reveals bright areas of several hundreds of nanometres in diameter, the positions of which are independent of the pore

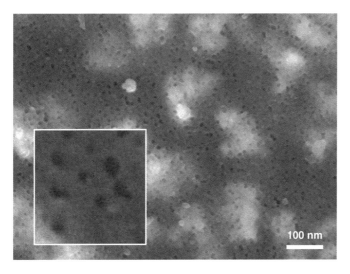

Figure 5.6.5 Ultrafiltration membranes. Surface morphology of a polysulfone based membrane containing 85 wt% of inorganic filler, imaged at 5 keV in the JSM-6340F. The bright regions are due to metal oxide beneath the surface. The inset is 100 × 100 nm². Reproduced by permission of Springer-Verlag, Wien

positions. These areas correspond to conglomerates of metal oxide particles just below the skin surface. For meaningful quantitative analysis of the pore structure, at least six SE images were taken for each membrane, at 100 000× magnification, thus covering an area of about 6 μm². The selected regions were verified to be representative for the membrane under study by scanning larger regions of the surface. The images were analysed for pore size, pore distribution, porosity and pore density (pore density is the number of pores per unit surface area irrespective of their size; porosity is obtained from the ratio of the area of the pores to the corresponding total area). Quantitative data on how pore density and porosity vary with a casting parameter could easily be derived from the corresponding images (Kuypers et al., 1995).

5.6.4.3 FOULING OF POLYMER MEMBRANES FOR REVERSE OSMOSIS

Fouling (blocking, clogging) of a membrane results in a gradual decrease of membrane flux with time and is obviously undesirable. Fouling can be caused by: (1) adsorption/deposition of solute on the surface of the membrane and/or within the membrane; (2) gradual, irreversible changes of the polarised layer. Low-energy SEM has been successfully applied for studying fouling mechanisms (cf. Kim et al. (1992) and references therein). We were asked to contribute to the analysis of fouled membranes for reverse osmosis. The flux through these membranes had dropped to unacceptably low values only a few hours after start up of an industrial installation. To allow for SEM investigation of the fouled membrane, a membrane module was cut open and pieces of typically 30 × 30 mm² were cut from the membrane at several positions along the surface. The pieces were dried in air at 40 °C. Together with the freshly fouled membrane, pieces of an unused membrane and of a membrane used in a pilot installation, both of the same type as the fouled membrane, were prepared for investigation. The pilot membrane had shown a normal flux decline over several months of use, typical for gradual fouling. Pieces of all membranes were cut into samples of typically 10 × 10 mm² for examination of the surface; or broken under liquid nitrogen for examination of the cross-section. In all cases they were coated with approximately 2 nm of Pt–Pd. Images (Figure 5.6.6) were obtained in the JSM-6340F under the following conditions: 5 keV, WD 16 mm, in-lens detector for surface imaging, lower detector for cross-sections. EDS spectra were obtained at 20 keV over a sample area of 24 × 18 μm². The pilot membrane reveals a contamination cake with many cracks. The fouled membrane has irregularly shaped particles on its surface, containing elements strange to the membrane. This contamination is not dense or uniform and cannot explain the drastic drop of the membrane flux. However, when the morphology of particle-free regions of the fouled membrane is compared with the morphology of the unused membrane, it is obvious that a contaminant has been adsorbed on the surface. In EDS the high sulfur signal, typical for the unused membrane, is strongly suppressed on the fouled membrane, while the carbon signal has increased. This strongly suggests a contamination with an organic fluid such as oil, which is known to be lethal for membrane performance. EDS of the pilot membrane reveals the elements expected from the process. The cross-sections give important additional information. The cross-sections of the unused and the pilot membrane are very similar. The contamination on the pilot membrane is superficial. Not so for the fouled membrane, where the morphology has changed micrometres deep below the surface, indicating that a contaminant has not only been adsorbed on the surface, but has penetrated the membrane as well.

FEG-SEM/EDS has revealed the nature of the fouling and has indicated a possible source (organic fluid). The next challenge is to track down and remove the source of fouling.

5.6.4.4 SURFACE CHARACTERISTICS OF CASINGS FOR MEAT

Meat in general, and sausages in particular, are very often wrapped in a casing that sticks firmly

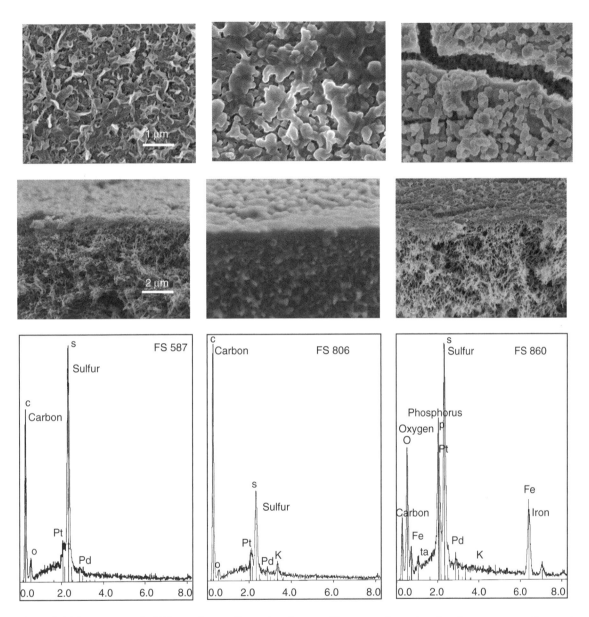

Figure 5.6.6 Reverse osmosis (RO) membranes. Comparison between (from top to bottom) surface images, cross-sections and EDS-spectra of (from left to right) an unused membrane, a fouled membrane and a fouled pilot membrane, all of the same type. Reproduced by permission of Springer-Verlag Wien

to the meat, protects it and keeps it from going taint. The casing can be natural (animal intestines) or synthetic (cellulose-based). The production of a good casing is not trivial and production parameters have to be kept within close ranges to assure its quality. One can easily understand that the condition of the inner surface of the casing, which will be in contact with the meat, is extremely important. The inner surface is treated to assure a proper degree of adhesion to the meat. If the adhesion is poor, the casing is not effective; if the adhesion is too strong, the meat will stick to the

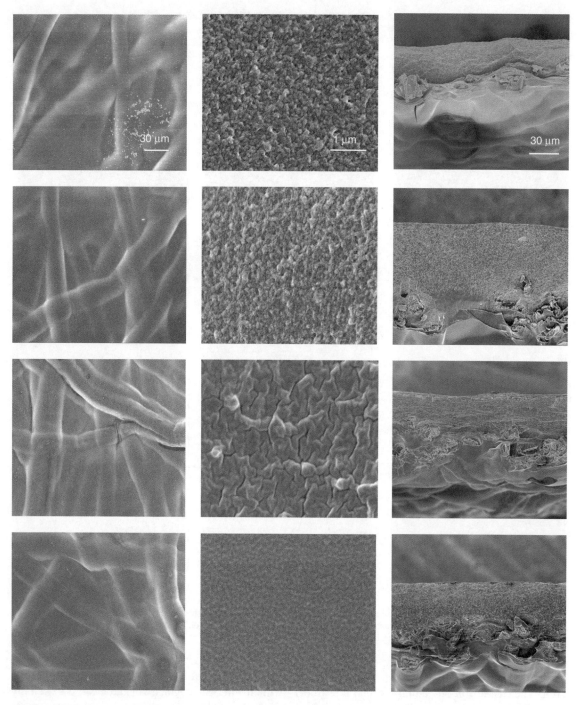

Figure 5.6.7 Casings for meat. Comparison between images of the inner surface at low (left) and high (middle) magnification and of the cross-sections (right), after four different treatments of the inner surface. The surface images were obtained at 1.5 keV. Reproduced by permission of Springer-Verlag, Wien

casing when this is removed prior to consumption and the consumer will not be happy.

On several occasions we were asked to evaluate the impact of different surface treatments on the morphology and/or the composition of the inner surfaces by means of FEG-SEM and X-ray photoelectron spectroscopy (XPS), respectively. An example of a morphological study of the inner surfaces of four casings is shown in Figure 5.6.7. A Pt–Pd coating with a thickness of approximately 1.5–2 nm was applied prior to observation in the SEM. Even then the beam energy had to be reduced to 1.5 keV to avoid charging. The working distance for surface imaging was 3 mm. The in-lens detector of the JSM-6340F was used. The surfaces suffered beam damage resulting in reduced topography when acquiring images in slow scan mode. Representative images were obtained by averaging in fast scan mode. At low magnification the surfaces look identical, but at high magnification the differences are obvious and might explain the different behaviour of the products. It would be extremely difficult to obtain this information in any other way.

ACKNOWLEDGEMENTS

The author is indebted to his colleagues of the Materials Technology and Process Technology Groups of VITO. Special thanks are due to Mrs Hong Chen and Mr Raymond Kemps of VITO's Centre for Materials Advice and Analysis; skilled operators of EPMA and FEG-SEM, respectively.

REFERENCES

Aerts, P., Genné, I., Kuypers, S., Leysen, R., Vankelecom, I.F.J. and Jacobs, P.A. Polysulfone–aerosil composite membranes – Part 2. The influence of the addition of aerosil on the skin characteristics and membrane properties. *J. Membr. Sci.*, **178**, 1–11 (2000).

Bastin, G.F., Dijkstra, J.M., Heijligers, H.J.M. and Klepper, D. In-depth profiling with the electron probe microanalyzer. *Microbeam Anal.*, **2**, 29–43 (1993).

Bastin, G.F. and Heijligers, H.J.M. Quantitative electron probe microanalysis. *X-ray Spectrom.*, **15**, 135–141 (1986).

Bastin, G.F. and Heijligers, H.J.M. Quantitative electron probe microanalysis of ultralight elements (Boron–Oxygen). *Scanning*, **12**, 225–236 (1990).

Dekempeneer, E.H.A., Meneve, J., Kuypers, S. and Smeets, J. Microstructure and mechanical properties of a-$B_{1-x}N_x$:H films prepared by r.f. PACVD. *Surf. Coat. Technol.*, **74/75**, 399–404 (1995).

Dekempeneer, E.H.A., Meneve, J., Kuypers, S. and Smeets, J. Tribological properties of r.f. PACVD amorphous B–N–C coatings. *Thin Solid Films*, **281/282**, 331–333 (1996a).

Dekempeneer, E.H.A., Meneve, J., Smeets, J., Kuypers, S., Eersels, L. and Jacobs, R. Structural, mechanical and tribological properties of plasma-assisted chemically vapour deposited hydrogenated C_xN_{1-x}:H films. *Surf. Coat. Technol.*, **68/69**, 621–625 (1994).

Dekempeneer, E.H.A., Wagner, V., van IJzendoorn, L.J., Meneve, J., Kuypers, S., Smeets, J., Geurts, J. and Caudano, R. Tribological and structural properties of amorphous B–N–C coatings. *Surf. Coat. Technol.*, **86/87**, 581–585 (1996b).

Doyen, W., Leysen, R., Mottar, J. and Waes, G. New composite tubular membranes for ultrafiltration. *Desalination*, **79**, 163–179 (1990b).

Doyen, W., Proost, R. and Leysen, R., Method for preparing a semi-permeable membrane. *European Patent,* 0 241 995 B1 (1990a).

Genné, I., Kuypers, S. and Leysen, R. Effect of the addition of ZrO_2 to polysulfone based UF membranes. *J. Membr. Sci.*, **113**, 343–350 (1996).

Goldstein, J.I., Newbury, D.E., Echlin, P., Joy, D.C., Romig, A.D., Lyman, C.E., Fiori, C. and Lifshin, E. *Scanning Electron Microscopy and X-ray Microanalysis*, Chapter 4, Plenum Press, New York, 1992.

Joy, D.C. and Joy, C.S. Low voltage scanning electron microscopy. *Micron*, **3–4**, 247–263 (1996).

Kim, K.J., Dickson, M.R., Fane, A.G. and Fell, C.J.D. Electron microscopy in synthetic polymer membrane research. *J. Microsc.*, **162**, 403–413 (1991).

Kim, K.J. and Fane, A.G. Low voltage scanning electron microscopy in membrane research. *J. Membr. Sci.*, **88**, 103–114 (1994).

Kim, K.J., Fane, A.G. and Fell, C.J.D. Quantitative microscopic study of surface characteristics of ultrafiltration membranes. *J. Membr. Sci.*, **54**, 89–102 (1990).

Kim, K.J., Fane, A.G., Fell, C.J.D. and Joy, D.C. Fouling mechanisms of membranes during protein ultrafiltration. *J. Membr. Sci.*, **68**, 79–91 (1992).

Kuypers, S., Genné, I. and Leysen, R. Surface characteristics of Zirfon composite ultrafiltration membranes. *J. Microsc.*, **177**, 313–319 (1995).

Nakagawa, S., Development of JSM-6320F scanning microscope. *Jeol News*, **31E/1**, 36–38 (1994).

Pouchou, J.-L. Use of soft X-rays in microanalysis. *Mikrochim. Acta (Suppl.)*, **13**, 39–60 (1996).

Pouchou, J.-L. and Pichoir, F. Surface film X-ray microanalysis. *Scanning*, **12**, 212–224 (1990).

Reimer, L. *Image Formation in Low-Voltage Scanning Electron Microscopy*, TT12, SPIE, Bellingham, 1993.

Scott, V.D., Love, G. and Reed, S.J.B. *Quantitative Electron-Probe Microanalysis*, Ellis Horwood, New York, 1995.

Willich, P. and Bethke, R. Electron probe microanalysis of submicron coatings containing ultralight elements. *Microbeam Anal.*, **2**, 45–52 (1993).

Yamamoto, Y., Yamada, A., Kazumori, H., Negishi, T. and Saito, M. Application of semi-in-lens FE-SEM for chargeless observation. *Jeol News*, **34E/1**, 47–49 (1999).

Yamamoto, Y., Yamada, A., Miyokawa, T. and Tamura, N. Development of a high resolution semi-in-lens digital field emission scanning electron microscope: JSM-6340F. *Jeol News*, **32E/1**, 39–41 (1996).

5.7 Energy Dispersive X-ray Microanalysis in Scanning and Conventional Transmission Electron Microscopy

E. VAN CAPPELLEN

FEI Company, Hillsboro, OR, USA

5.7.1 INTRODUCTION

X-Ray microanalysis did not really become a common technique on transmission electron microscopes (TEMs) until the second half of the 1970s. Designers had to wait until large lithium drifted silicon (Si(Li)) solid-state detectors became available and it remained to be seen that these liquid nitrogen cooled devices would not impair the fundamental TEM specifications because of their weight and the vibrations caused by the constantly boiling nitrogen. Until the 1990s most laboratories would make a clear distinction between a dedicated high-resolution TEM without analytical capabilities and an analytical TEM with a significantly wider pole-piece gap and thus reduced resolution specification. The latter had the capability of accepting an energy dispersive X-ray microanalysis system abbreviated to EDX or EDS (energy dispersive X-ray microanalysis or energy dispersive spectroscopy). Whenever both high-resolution TEM imaging and EDX data were needed, it was up to the scientist to try to find the same area of a sample on two different scopes. The 1990s saw this distinction fade and although column manufacturers still offer different pole-pieces with different gaps and resolution specifications, the reason no longer is the incompatibility between high-resolution TEM and EDX but it is just a trade-off between the available space between the pole-pieces for other applications needing high tilt or an exotic specimen holder and resolution. Ultra-high-resolution microscopes can now safely be fitted with an EDX system.

Besides hardware refinements, the 1970s and 1980s were also rich in software developments, not least in the field of quantitative X-ray microanalysis. It is probably fair to say that by the end of the 1980s EDX had matured into a relatively easy-to-use technique, compatible with most commercially available TEMs. Single point analysis, line scans and X-ray maps were by then relatively straightforward to obtain. If one criticism is to be voiced, it is the fact that commercially available software packages are clearly derived from their counterparts developed for scanning electron microscopy analysing bulk samples instead of thin specimens. Apart from the sometimes-erroneous terminology, many non-expert users have been misled by the quantitative analysis outputs that industry wide are given with an incredible precision of two digits after the decimal point of atomic or weight percentages. Systematic errors and ill-defined correction schemes for absorption and fluorescence have led to erroneous results.

The 1990s have not really addressed this issue but at the same time a clear shift from quantitative analysis back to qualitative analysis was noticeable albeit in a totally different ballgame than in the early days. Field emission gun (FEG) technology capable of producing much smaller electron probes had been around since the 1970s but had never seen widespread use on TEMs because of their complexity and ultra-high vacuum requirements. In the course of the last decade, however, the thermally assisted Schottky FEG with less stringent vacuum needs and a constant beam current became the norm on high-end TEMs. One of the benefits is that these microscopes can produce much smaller probes and as a consequence the bulk of EDX research shifted back from quantitative to qualitative analysis but this time on a much smaller scale: the nanometer or even smaller. When objects or features become so small, a two-dimensional grain boundary for instance, detection limits and/or concentration gradients became the ultimate pursuit. The technique's spatial resolution improved dramatically, so much that the term 'microanalysis', probably too well established by now, should be replaced by 'nanoanalysis'.

The 1970s brought along a new technique on TEMs: EELS or electron energy loss spectroscopy. EELS measures the characteristic energy-loss of the primary electron due to the ionization of a target atom. In that respect, EELS and EDX look at the same ionization events, the difference being that EELS measures the primary event while EDX looks at a secondary product of the ionization, the emitted X-ray. The first generation of spectrometers was the so-called serial EELS (SEELS) because the energy-loss range is acquired sequentially. SEELS only enjoyed a moderate success because it was time consuming and the interpretation of the spectra was all but trivial. In the mid 1980s the PEELS or parallel electron energy loss spectrometer was introduced with a 1024 diode array capable of acquiring a whole energy interval at once. It became fashionable to add a PEELS to a TEM column and soon the decline of EDX on TEMs was predicted because of EELS's significantly better energy resolution and far greater collection efficiency. This prediction never came true and actually EDX on TEMs has never been stronger than right now. This subchapter about EDX on TEMs is not only meant to be a tribute to the technique, but also tries to explain this revival.

5.7.2 BRIEF HISTORICAL OVERVIEW OF HARDWARE DEVELOPMENTS

For a complete review on this matter we refer to the textbooks edited by Williams and Carter (1996a,b) and Joy et al. (1986).

The first attempts to fit a WDX (wavelength dispersive X-ray microanalysis) on a TEM date back to the early 1960s. Only by the end of the decade was a real 100 kV TEM successfully equipped with two twin WDX spectrometers. A specially designed mini-lens allowed EMMA-4 to have almost comparable TEM performance to other contemporary 100 kV TEMs. Despite obvious advantages to be able to combine TEM imaging and diffraction with chemical analysis, the cumbersome WDX systems had only encountered a very moderate success because of their slowness. Also, the bulky set-up of a WDX system was not compatible with the new emerging field of high-resolution TEM (HR-TEM) imaging. Phase contrast lattice imaging rapidly became the most compelling reason to purchase a TEM. This is probably the reason why WDX equipped TEMs did not enjoy the same success as their electron-probe and SEM (scanning electron microscope) counterparts.

When the Si(Li) solid-state detector combined with a multichannel analyser became available X-ray microanalysis really tookoff on TEMs. Because a whole energy range can be acquired simultaneously the new technique was called EDX (energy dispersive X-ray microanalysis) or EDS (energy dispersive spectroscopy). Not withstanding the fact that EDX systems have energy resolutions about ten times worse than WDX spectrometers and as such are not well suited for light element detection, EDX systems did far better because of their smaller dimensions and larger solid-angles. Because of bad light element performance caused by peak overlap and because of inherent fragility, windowless detectors never really broke through. Instead, diamond windows with a detection down to sodium and later the so-called ultra-thin windows with a theoretical detection down to boron became the industry standard. Beam diameters on TEMs in these days were of the order of several hundreds of nanometers and analysis of submicron areas showed a tremendous potential and was rightfully called 'microanalysis'.

The 1970s saw the development of the cold field emission gun (CFEG) with a source brightness of at least a thousand times that of the best thermionic source, lanthanum hexaboride (LaB_6). This high brightness, a consequence of the small size of the emitter, translates into improved beam coherence and much smaller electron probes. This technology enabled the development of the scanning transmission electron microscope (STEM) in which a small intense electron probe is scanned on a thin, TEM-like sample. Instead of detecting the secondary or backscattered electrons as in a scanning electron microscope (SEM), the STEM detectors are beneath the specimen, where different types of 'transmitted' electrons can be used for imaging. Just as a TEM, a STEM can yield bright field (BF) and dark field (DF) images. Although different in nature and thus yielding slightly different information, STEM imaging initially was not the decisive argument in favor of the new instrument, especially because STEM images require longer exposure times and are much noisier than their TEM counterparts. The major reason to develop STEM was indeed its natural compatibility with EDX. The smaller probe sizes allowed EDX analysis of much smaller features and beam control enables line scans and X-ray maps to be generated just like in a SEM but with a phenomenally better resolution because of the absence of the pear-shaped interaction volume present in bulk samples. It is fair to say that it is exactly these EDX and EELS analytical capabilities that drove the development of the so-called dedicated STEM.

Over time two tendencies emerged, resolution-wise STEM imaging caught up with TEM imaging and TEMs were fitted with scanning coils and later with field emission guns (FEGs) which enabled TEMs to operate in STEM mode. Although originally the STEM mode was not intended as a real competitor for the dedicated STEM, but just to boost the EDX capabilities of the TEM, the scope changed when the only manufacturer of dedicated STEMs went out of business. Simultaneously it also became clear that there are some advantages of doing STEM on a TEM column because of the post-specimen lenses. One can look at a highly magnified image of the probe and although the correlation with the probe shape at the sample level is not perfect because of the spherical aberration of the lower pole-piece, it is a good approximation and allows to roughly focus and stigmate the STEM image before even switching to STEM mode. They also allow changing the virtual distance between the sample and the STEM detectors so that with one fixed size annular detector different minimum collection angles can be selected yielding different types of dark field images (see below). Figure 5.7.1 shows a 2002 model of a 200 kV STEM/TEM capable of the highest TEM and STEM resolution and equipped with all the analytical techniques, an EDX detector and an electron energy filter for PEELS and energy filtered TEM (EFTEM).

5.7.3 SIGNALS GENERATED IN A THIN TEM SPECIMEN

This section is not intended to review the electron-beam – thin sample interactions in detail, the intention is to discuss without the use of formulae the different signals that are generated within a thin sample and that are used in a (S)TEM. This will help us understand the importance of EDX in transmission electron microscopy. For a complete description, please refer to the literature (Joy et al., 1986; Egerton, 1996; Williams et al., 1996a,b).

A high-energy incident electron can go through a thin foil without interacting with it or it can have a phonon interaction with a sub-eV energy loss too small to be measured. Such electrons are located on the optical axis and are obviously not very interesting from a chemical standpoint, as they carry no (measurable) information. From an imaging point of view this on-axis signal is called the BF signal since a hole (vacuum) shows up bright. Both TEM and STEM modes yield about the same information.

Electrons can also undergo different types of elastic scattering: Bragg diffraction and Rutherford scattering. The Bragg diffracted electrons are

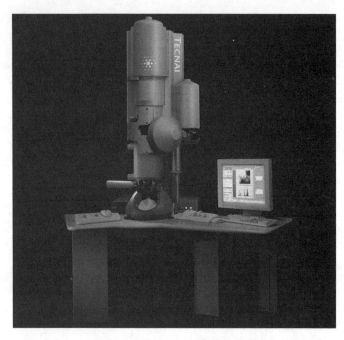

Figure 5.7.1 Modern STEM/TEM capable of combining the highest STEM and TEM resolution. It is equipped with EDX and an energy filter for PEELS and EFTEM. (Courtesy of FEI Company)

those that are diffracted by the regular array of lattice planes in a crystalline structure and therefore carry crystallographic information. A crystal will generate discrete diffracted beams whereas an amorphous material will produce diffuse rings. Imaging with Bragg diffracted electrons is called DF imaging as vacuum will show up dark. In the TEM mode one or several Bragg diffracted beams can be selected with the objective aperture, whereas in STEM mode, an annular solid-state detector averages over all Bragg diffracted electrons. This is called annular dark field (ADF) imaging and is equivalent to conical dark field imaging in TEM. Further away from the optical axis, where the intense Bragg diffracted electron signal fades, the Rutherford scattered electrons, those that have been scattered by the sample's nuclei, can be recorded. Figure 5.7.2 schematizes the location of Bragg diffracted and Rutherford scattered electrons around the optical axis. Rutherford scattered electrons carry atomic mass or Z-contrast, just as the backscattered electrons in the SEM (also Rutherford scattering) and since they also are recorded with an annular detector, but with a

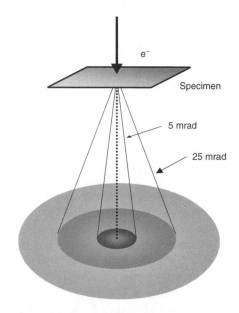

Figure 5.7.2 Transmitted electrons can be found in three zones: the optical axis and just around hosts the electrons that have not interacted with the sample and the inelastically scattered electrons; an annular region between 5 and 25 mrad (at 200 kV) where the Bragg diffracted electrons are located; the outer region only contains Rutherford scattered electrons, which carry Z information

larger minimum acceptance angle, it is referred to as high angle annular dark field (HAADF) imaging. Figure 5.7.3(a) is an example of a HAADF image and shows that this mode is particularly suited for analytical microscopy because the Z-contrast image greatly helps to determine where to acquire point spectra. In the case of elemental maps it is also useful to correlate the Z-contrast image with the EDX or PEELS elemental maps or even with line profiles.

Far more interesting chemically, are the inelastically scattered electrons as most lose a characteristic amount of energy when ionizing the target atoms. Ionized atoms will relax by isotropic emission of either an Auger electron or an X-ray photon. In general Auger emission is more likely except for K shell ionization of elements with an atomic mass exceeding 30. The problem with Auger electrons is that because of their high matter interaction cross-section only those emitted within

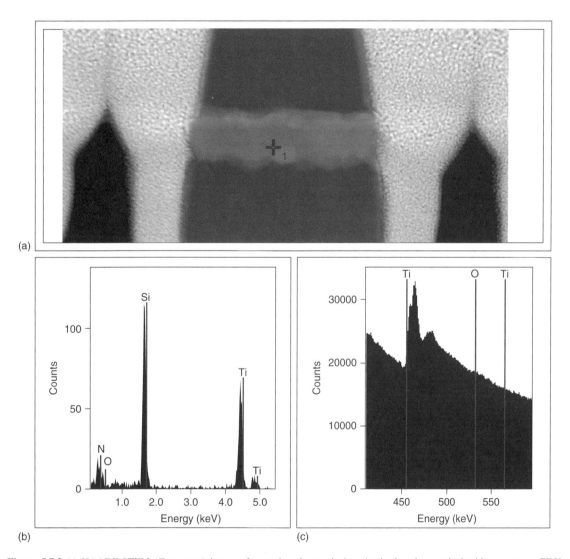

Figure 5.7.3 (a) HAADF STEM (Z-contrast) image of a semiconductor device. At the location marked with a cross, an EDX (b) and an EELS (c) spectrum were acquired. Notice the huge difference in signal intensities and background shapes. (Courtesy of Y.C. Wang, FEI Company)

a few ångstroms from the surface will escape unaltered which makes it a surface analysis method only. Auger spectroscopy also requires ultra-high vacuum to avoid interaction of the Auger electrons with the residual molecules. As a consequence Auger electrons are not used in a TEM. However in a TEM (or STEM) there are two ways of looking at the inelastic scattering processes, one is to measure the actual energy loss of the primary electron (PEELS) and the other is to measure the energy of the emitted X-ray (EDX). Figure 5.7.3(b) and 5.7.3(c) show a comparison between an EDX and a PEELS spectrum from the same sample.

Because of the relatively small energy losses that can be measured (maximum 2000 eV loss above which the signal is too small), and the high initial accelerating voltage (at least 200 kV), conservation of energy and momentum dictates small scattering angles, a few milliradians maximum. So inelastically scattered electrons are located close to the optical axis and are superimposed on the BF signal (Figure 5.7.2).

5.7.4 ELECTRON ENERGY LOSS SPECTROSCOPY VERSUS ENERGY DISPERSIVE X-RAY MICROANALYSIS

This section, while mentioning the similarities, will primarily focus on the dissimilarities of EDX and EELS. It is mostly the differences that explain the modern trend to combine both techniques on one instrument and being able to perform both simultaneously on a sample has proven to be invaluable.

When looking for chemical information, EDX and EELS both analyse the same physical phenomenon: the ionization process of the target atoms. In the case of EELS the energy of primary electron that has lost energy because of the ionization of an atom is measured whereas in the case of EDX a secondary product, an emitted X-ray photon because of the relaxation process, is detected. Because both techniques basically look at the same events, one could argue that one spectroscopic technique should fulfill all the needs. The main difference between them lies in detection efficiency and energy resolution.

The previous paragraph alluded to the fact that inelastic scattering angles are very small because of energy and momentum conservation and as a consequence most of the inelastic signal can be detected in an on-axis spectrometer. It is not uncommon to detect 90 to 95 % of the core-loss events. In this respect the EDX case is entirely different. The EDX spectrum is builtup of characteristic X-ray photons related to the de-excitation of the target atoms and Bremsstrahlung or white radiation that results from primary electrons slowingdown in the Coulomb fields in the sample. Bremsstrahlung obeys scattering laws and is found in a forward peaked thorus around the optical axis. Emission of characteristic X-ray photons however happens isotropically since it is a secondary process that has forgotten its past. There is no preferential direction and the optical axis no longer has a special status. Modern EDX set-ups have 30 mm^2 Si(Li) detectors which at a practical distance translates into an effective detection solid angle of about 0.1 sterrad. A whole sphere, spans $4\pi (= 12.6$ sterrad), so only about 0.8 % of the generated events can possibly be detected. Many attempts, such as bigger or multiple detectors, have been made to maximize the collection angle, but because X-rays are nearly impossible to focus and because of the limited available space between the pole pieces, the difference with EELS will remain huge. Furthermore, characteristic X-ray emission competes with Auger electron emission for the de-excitation process of atoms. In fact, unless these are K-shell ionizations of heavy atoms ($Z > 30$), Auger emission is far more likely. For light elements ($Z < 11$) the likelihood of X-ray emission drops below 10 % for K shells. As a rule of thumb it is not exaggerated to state that EELS detects at least 1000 times more events than EDX. Herein lies the main difference between EDX and EELS.

The other striking difference is the energy resolution. The energy resolution of solid-state detectors, whether Si(Li) or Ge is around 130 eV. Other systems based on microcalorimeters have resolutions of a couple eV, but so far there are no plans to develop these systems for transmission microscopes. The EELS spectrum has a resolution mainly determined by the energy distribution of the source, which depends on the type of source but also on the beam intensity. When electrons are squeezed into a condenser crossover they tend to repel each other, which increases the energy spread. This is the so-called Boersh effect. All in all the following energy resolutions can realistically be achieved: 1.2 eV for a LaB_6 source, 0.7 eV for a thermally assisted field emission gun and 0.5 eV for a CFEG. FEI Company has recently launched a 'monochromator' which further reduces the energy resolution below 0.1 eV. This means that EELS has an energy resolution that is two to three orders of magnitude better than EDX.

It is precisely because of the far superior detection efficiency and energy resolution of EELS that about 20 years ago it was predicted that EELS spectroscopy would replace EDX on transmission microscopes. We will now highlight the reasons why this has not happened.

One of the biggest problems EELS faces is multiple scattering. As soon as multiple scattering events start to occur, the information in the EEL spectrum starts to scramble and the signal to background ratio deteriorates rapidly. This means that in practice the samples need to be significantly thinner than one mean free path of the electrons in the specimen (1λ). Lambda is primarily a function of the sample and the accelerating voltage of the microscope. To a certain extent the spectrometer acceptance angle also has an influence, but for practical purposes we will assume λ to be about 100 nm at 200 kV. This means that the TEM sample has to be less than, say, 50 nm in order to do EELS, which is not always practical.

The signal to background ratio in general is the other major EELS problem and is obvious from the comparison between Figure 5.7.3(b) and Figure 5.7.3(c) showing and EDX and EELS spectrum of the same specimen. Because of the far superior detection geometry, the EEL spectrum clearly has a much better signal to noise ratio but its signal to background ratio cannot compete with the EDX spectrum because of the large exponentially decaying background. Whatever algorithm is used to do quantitative analysis and whether it is EELS or EDX, the first step needs to be the background subtraction to extract the sample specific signal. In the EELS case this is a major cause for uncertainties whereas in the EDX case, even a bad job will still yield a realistic result. The way background subtraction is done in EELS is by extrapolating pre-edge background under the edge and in that respect, the high energy resolution of an EEL spectrum does not help as edge overlap or even edge proximity will make any extrapolation impossible. Just as in EDX, EELS quantification is done by a k-factor ratio technique (see next paragraph). However in the case of EELS the k-factors are not only functions of the chemical species, but also of the specific chemical binding and the acquisition parameters such as convergence angle and acceptance angle (Egerton, 1996). As a consequence experimental k-factors are very difficult to obtain and are only practical in a limited number of cases. The problems of background subtraction and experimental k-factor determination make it practically impossible to obtain accurate quantitative EELS results in a timely manner. This makes EDX the technique of choice for quantitative analysis.

Because of the higher energy resolution and the large dynamic signal range because of the strongly decaying background, EEL spectra usually only cover a limited energy range, which complicates qualitative analysis of truly unknown samples. It is therefore probably fair to say that EDX is also used for determining which elements are present. Fast, qualitative analysis and relatively easy quantitative analysis are the main reasons why EDX has not and will not disappear on transmission microscopes, as the output of an EDX analysis will often be the starting point of a more detailed EELS study.

Because of its superior detection geometry and therefore better signal to noise ratio, EELS is much better for detecting minor constituents or even trace elements. For elements exhibiting a steep edge, a so-called white edge, detection limits can be as low as the 10 to 100 ppm range, whereas EDX is limited to 1000 ppm. The lateral analytical resolution is defined by the primary excitation volume defined by the size of the electron probe and subsequent beam broadening in the sample. As such the best practical resolution is determined by the smallest probe yielding sufficient current to generate a decent EELS or EDX spectrum in a reasonable amount of time. Because of its better detection geometry, EELS can be done with the smallest available probes whereas EDX cannot. Practically speaking this translates into 0.2 nm EELS resolution and 0.4 nm EDX on most modern field emission microscopes.

Another consequence of its better than 1 eV energy resolution, is that the EELS applications go far beyond just identifying and maybe quantifying the chemical species present in the sample. We refer to the book by Egerton (1996) for a detailed review of all EELS applications, but in a nutshell we will just mention some major EELS applications so as to better understand the need to have both techniques EDX and EELS available on the same instrument. The 'energy loss near edge structure' (ELNES) is a fingerprint of the chemical binding and band structure. In other words, just by looking at the edge fine structure enables to distinguish between different compounds. As an example, the Si L edge signature can distinguish pure Si from SiO_2 or Si_3N_4 and the C K edge is different for amorphous carbon, graphite and diamond. The so-called chemical shift allows the valence state of an element to be determined by simply measuring the energy shift of a core-loss edge. Last but not least, the ratio of the total spectrum intensity over the zero-loss peak (electrons that have not interacted with the sample) is related to the sample thickness divided by λ, the mean free path of the electrons. Provided λ can be calculated or experimentally measured on a sample with known thickness, the logarithm of the whole spectrum intensity divided by the zero-loss peak yields a quick value for sample thickness as a function of λ, a bonus for many other applications such as a conventional absorption correction for a quantitative EDX analysis.

The combination of EDX and EELS offers one of the most powerful analytical tools and a recent trend has been the acquisition of so-called 'spectrum images'. A spectrum image is a STEM image with behind each pixel a full-size spectrum. This can be an EDX spectrum or an EEL spectrum or better both can be acquired simultaneously (Figure 5.7.4). A spectrum image is a data cube (or two) with the third axis being energy. The advantage of spectrum imaging is that the whole experiment is canned and can be archived. Further data mining, such as determining the composition of a particle or extracting a line profile across a grain boundary can be done even after the sample has been destroyed. A spectrum image can of

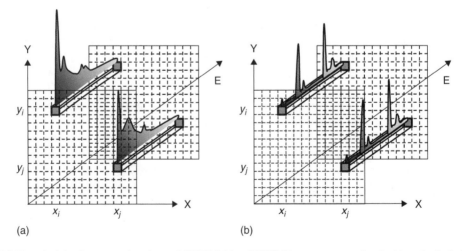

Figure 5.7.4 The principle of spectrum imaging: a full EELS (a) and EDX (b) spectrum associated with each pixel of the STEM image. Data mining is done off-line after the acquisition is completed

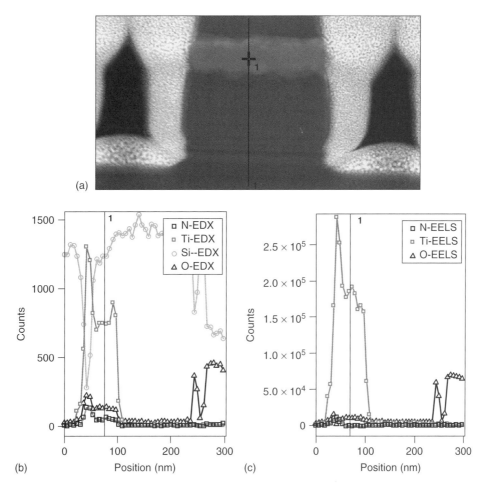

Figure 5.7.5 (a) HAADF STEM (Z-contrast) image of a semiconductor device. Along the vertical line a one-dimensional spectrum image with 100 points and 300 nm long was acquired. A posteriori the EDX (b) and EELS (c) line profiles were extracted. The cursor (1) is located in a titanium silicide layer also shown on the profiles. (Courtesy of Y.C. Wang, FEI Company)

course be one-dimensional. Figure 5.7.5 shows an example of a one-dimensional spectrum image. The line profiles were extracted off-line.

5.7.5 QUANTITATIVE ANALYSIS IN EDX

As explained in the previous paragraph, the single most important reason why EDX has not disappeared from the TEM scene, as predicted 20 years ago, is its capability of doing quantitative analysis. Therefore a review of the different quantitative techniques seems pivotal and special attention will be given to the recently developed correction schemes which all have in common the desire to eliminate the problem of unknown parameters.

Basic quantitative EDX still relies on the Cliff–Lorimer ratio equation relating the intensities of the characteristic X-rays above background (denoted I_A, I_B, etc.) to the compositions (denoted C_A, C_B, etc.):

$$\frac{C_A}{C_B} = k_{AB}\left(\frac{I_A}{I_B}\right) \quad (5.7.1)$$

where k_{AB} is the Cliff–Lorimer k-factor. There are two fundamental limitations associated with

Equation (5.7.1). First the k-factors are not universal and vary from instrument to instrument and second the Cliff–Lorimer equation assumes that there is no significant X-ray absorption in thin TEM-compatible samples. Often this assumption is not realistic especially when both low- and high-energy peaks are used, as the low-energy line will be more absorbed. Methods to overcome both these limitations have been developed over the years, but all are based on the Cliff–Lorimer equation that remains the basis for quantification more than 25 years after its initial publication by Cliff and Lorimer (1975).

5.7.5.1 ABSORPTION AND FLUORESCENCE

Characteristic X-rays are generated isotropically in the thin TEM sample. Only those emitted in the small solid angle seen by the EDX detector are eligible for detection provided they are not absorbed in the sample on their way out. If all characteristic X-rays would be absorbed with the same probability, quantification would not be affected, as the intensities are ratioed. However absorption is a function of the energy and low energy peaks will be more strongly absorbed than high energy ones. A quantitative analysis without any kind of absorption correction will therefore penalize light elements. The magnitude of the error will depend on the specimen thickness and density as well as the take-off angle, which is the angle between the sample surface and the detector axis. When X-ray absorption is no longer negligible, the composition is no longer simply proportional to the intensity and Equation (5.7.1) becomes:

$$\frac{C_A}{C_B} = k_{AB} \left(\frac{I_A}{I_B}\right) \times \left[\frac{(\mu/\rho)_{sp}^A}{(\mu/\rho)_{sp}^B}\right]$$

$$\times \left\{\frac{1 - \exp[-(\mu/\rho)_{sp}^B \rho t \csc\theta]}{1 - \exp[-(\mu/\rho)_{sp}^A \rho t \csc\theta]}\right\}$$

(5.7.2)

where $(\mu/\rho)_{sp}^A$ and $(\mu/\rho)_{sp}^B$ are the mass absorption coefficients of the characteristic X-ray lines A and B, ρ and t are, respectively, the density and thickness of the specimen at the beam position and θ is the X-ray take-off angle. Straightforward correction of absorption requires ρ, t and θ, which is also uncertain unless the sample is perfectly flat and horizontal at zero tilt.

Primary X-radiation, generated by the electron beam, will partially be absorbed in the sample and this in turn generates secondary or fluorescence emission. Although this phenomenon can also alter the peak ratio of two elements, it is negligible in TEM-like samples except for 'bulky' samples and even then only when the energy of a strong characteristic peak is just above the ionization threshold of another element. For most of the periodic table, this happens when two elements have an atomic number difference of two. This translates into an exceptionally large fluorescence yield. It has been demonstrated by simulations (Van Cappellen et al., 1990) that fluorescence strongly depends on the shape of the sample and not so much by the orientation of the specimen in the electron beam. It should also be realized that the secondary excitation volume is orders of magnitude larger than the primary one. It goes without saying that a fluorescence correction of a point analysis is almost impossible as it is virtually impossible to determine the shape of a sample in the millimeter range, which is the size of the secondary excitation volume.

5.7.5.2 THE PARAMETERLESS CORRECTION METHOD

The parameterless correction method, first presented at the ICXOM 10 conference in 1983 (Van Cappellen et al., 1984) and extensively reviewed in 1990 (Van Cappellen, 1990) was also independently called the extrapolation method in 1986 (Horita et al., 1986). The parameterless correction method got its name because it performs absorption and fluorescence corrections without requiring external parameters such as sample thickness and density; absorption coefficients and fluorescence yield coefficients. The price to be paid is that not one but several spectra are needed for one analysis. The spectra are taken at different sites on the specimen with different thickness. The sample has to be homogeneous in the analysed area.

A characteristic line in an EDX spectrum can be considered as being composed of primary and secondary characteristic radiation, both attenuated due to absorption in the specimen. Both components can be expressed as polynomials in T, the mass-thickness of the foil. The intensity of a characteristic line tends monotonically to zero for vanishing specimen thickness. Consequently the polynomial in T will have no constant term. Hence the ratio of two net intensities such as in the Cliff–Lorimer Equation (5.7.1) is a polynomial with a constant term. This term times the proper k-factor yields the mass concentration ratio of the elements under consideration. From the different spectra taken at different thickness, uncorrected concentrations are calculated with Equation (5.7.1). These are plotted versus the foil thickness and a least square fit extrapolates the concentrations at zero thickness.

Zero thickness-extrapolated concentrations are free from absorption effects as absorption is a two-dimensional phenomenon confined to the optical axis – detector axis plane. For any arbitrary specimen geometry the absorption path towards the detector tends to zero when the sample thickness vanishes. The same conclusion does not apply for fluorescence because here the phenomenon is truly three-dimensional. One can imagine that only one electron hits the tip of a wedge-shaped specimen (where thickness is zero) and that only one primary X-ray photon is generated. This photon could travel into the wedge and produce a secondary photon that eventually ends-up being detected. In this hypothetical case no primary radiation is measured whereas one secondary photon is captured. The fact that fluorescence is not completely extrapolated away was confirmed via simulations (Van Cappellen et al., 1990), however it was also shown with a worst case scenario that the absolute systematic error on the extrapolated concentrations was below 0.25 wt%, which is smaller or at least comparable to the overall statistical error.

In order to be 'parameterless', the sample thickness or mass-thickness used as the parameter for the extrapolation, needs to be replaced by an internal thickness measure. This substitute must monotonically tend to zero for vanishing specimen thickness so as to ensure the same extrapolation values as with the true thickness. A valid candidate for this internal measure is characteristic radiation. A net peak or better the sum of net peaks for increased statistics are suitable.

The parameterless correction or extrapolation method corrects for absorption and fluorescence, irrespective of the shape of the specimen. Besides this unquestionable advantage and also in contrast with more conventional approaches, the procedure also provides the final results with an accuracy figure. This statistical error which typically lies between 0.2 and 1.0 at wt%, easily allows to distinguish between good and bad experiments. A bad k-factor will introduce a systematic error, however accurate k-factors can be measured with accuracy estimates with the parameterless correction method.

5.7.5.3 THE ZETA-FACTOR METHOD

A variant of this approach is the recent ζ-factor method which allows k-factor determinations without prior knowledge of the mass-thickness of the sample (Watanabe et al., 1996). For a TEM-like sample, it is safe to assume that the ionization cross-sections do not change with depth. Most incident electrons will not undergo inelastic scattering and the average energy of the incident beam can be considered constant. Consequently the intensity of a characteristic peak is proportional to the local mass-thickness ρt. Inversely the mass-thickness can be expressed as a function of the generated intensity:

$$\rho t = \zeta \left(\frac{I}{C} \right) \quad (5.7.3)$$

where ζ is the proportionality factor connecting the characteristic intensity I and C is the weight fraction. The zeta-factor (ζ) is independent of the sample composition as the characteristic intensity is normalized by the weight fraction. Equation (5.7.3), or the proportionality between mass-thickness and intensity, is valid on condition that absorption in the sample is negligible.

Assuming that a characteristic line of element B is not significantly absorbed, Equation (5.7.3) becomes:

$$\rho t = \zeta_B \left(\frac{(I_B)_m}{C} \right) \quad (5.7.4)$$

where $(I_B)_m$ is the measured intensity of element B. Substituting (5.7.4) into (5.7.2) eliminates ρt

$$\frac{C_A}{C_B} = k_{AB} \left(\frac{I_A}{I_B} \right) \times \left[\frac{(\mu/\rho)_{sp}^A}{(\mu/\rho)_{sp}^B} \right]$$

$$\times \left\{ \frac{1 - \exp[-(\mu/\rho)_{sp}^B (\csc \theta) \zeta_B (I_B)_m / C_B]}{1 - \exp[-(\mu/\rho)_{sp}^A (\csc \theta) \zeta_B (I_B)_m / C_B]} \right\}$$

(5.7.5)

Equation (5.7.5) no longer requires the knowledge of the local mass-thickness. The determination of the zeta-factor is done simultaneously with the k-factor on a standard sample with known composition. For a known sample Equation (5.7.5) only contains two unknowns: k_{AB} and ζ_B. These are extracted using the parameterless correction method. With the k- and zeta-factors determined, quantification through Equation (5.7.5) and the boundary condition that the sum of all concentrations should be one becomes straightforward. When using the zeta-factor technique, it is paramount to use the same beam current as for the k- and zeta-factor determination. It should also be realized that an iterative procedure is required since Equation (5.7.5) contains mass absorption coefficients, which are composition dependent.

The main advantage of the zeta-factor approach is that once the k- and zeta-factors for the different elements are determined through a parameterless or extrapolation method, only one spectrum is needed for a quantitative analysis. Moreover mass-thickness and thus thickness if density is known can also be retrieved through Equation (5.7.3). This approach is extremely well suited to determine quantitative line scans and even X-ray maps.

The same approach had already been used to analyse second-phase precipitates in copper/zinc/aluminum shape memory alloys (Van Cappellen et al., 1987). Because of the highly absorbed Al K line, some kind of absorption correction is needed, but because the precipitates are preferentially etched during sample preparation, it is virtually impossible to apply a conventional correction. An analysis of the surrounding matrix with the parameterless correction method allows determining the exact k-factors and provided the beam current is stable, it is also possible to calibrate the mass-thickness as function of the net intensities in the same way as the zeta-factor approach.

5.7.5.4 QUANTITATIVE CHEMICAL MAPPING

Because of the amount of data and the tediousness of a conventional absorption correction, real quantitative maps only showed up recently. Williams et al. (1998) presented a few years ago real quantitative maps using the zeta-factor approach. The maps show interfacial segregation with a spatial resolution of better than 5 nm. Williams is keen to point out that while line scans are very useful, they are very selective and operator-biased because of the choice that has to made of where to acquire the line-scan. Quantitative X-ray maps have been a standard procedure on bulk sample both in the SEM and the electron probe microanalyser (EPMA) for many years now. Because X-ray generation in bulk samples is significantly more important than in thin specimens and sample thickness is not an issue, quantitative maps can be acquired in a timely manner and absorption and fluorescence corrections can be applied on the fly or off-line depending on the computing power. The main problem with X-ray maps on bulk samples is the poor spatial resolution which depending on the initial beam energy is of the order of 0.5 to 1.0 μm. For most materials science problems, such as interfacial segregation, this is by far insufficient.

X-Ray spatial resolution when using thin TEM-like samples is mainly determined by the electron probe size and some beam broadening occurring in the foil. The primary excitation volume is orders of magnitude smaller than in the bulk case. An immediate consequence is that the X-ray count rate is dramatically reduced which means that probe intensity and X-ray collection angle have to be maximized in order to achieve realistic acquisition times. A FEG is key to high resolution X-ray mapping. A FEG is capable of producing a probe current of 1 nA with a probe size of 1 nm (FWHM). A way to maximize the detection solid angle is to use two EDX detectors, but even with everything optimized, the trade-off will always be spatial resolution against acquisition time. Besides testing the operator's patience, long acquisition times can also be detrimental to the sample because of beam damage. Also drift can become an issue and this ruins the spatial resolution if no measures are taken to compensate for this. A way to compensate for drift is to acquire intermediate images at fixed intervals to measure the drift through a cross-correlation procedure and to use this information to move the beam accordingly. More sophisticated systems assume the drift to be linear in between measurements and have the beam follow the drift between two images.

When generating a quantitative map, the first step is to subtract the background under the characteristic peaks in each pixel. This yields the net intensities, which can be converted into weight concentrations by the k-factor. Depending on the elements and the sample's thickness, the zeta-factor approach must be used to correct for absorption. Williams *et al.* (1998) carried out a full zeta-factor analysis to generate the example shown in Figure 5.7.6. It is an investigation of the boundary segregation of Cu in an Al–4 wt% Cu alloy. The alloy had been aged to cause segregation to the grain boundaries. A high magnification analysis was performed from an edge-on grain boundary with the following parameters: 2 M magnification, 64 × 64 pixels and a dwell time of 200 ms per pixel. The probe size was estimated to be below 1 nm, the current was 0.5 nA and the pixel size was 0.625 nm. The ADF image (Figure 5.7.6a) was acquired with a small camera length to increase the minimum acceptance angle to reduce diffraction contrast and increase Z-contrast (atomic number contrast). The bright diagonal line in Figure 5.7.6(a) reveals a layer of higher atomic number. A quantitative Cu map (Figure 5.7.6b) was generated from the background subtracted Cu K and Al K lines that are nothing else but the net intensity maps, using a k-factor measured from stoichiometric θ-phase (Al_2Cu) particles, which were distributed throughout the material. The quantitative Cu map (Figure 5.7.6b) confirms that Cu has segregated to the boundary, but the extra information is that right on the boundary the Cu concentration is about 10 wt% compared to the average matrix concentration of 3 wt%. Granted, the image is noisy, but statistical meaningful data may be extracted by averaging over many pixels. The matrix composition was determined to be 2.6 ± 0.1 wt% by sampling 1600 pixels. This is close to the equilibrium solubility of 2.75 wt% Cu in Al at 475 °C, the used homogenizing temperature.

Line profiles can be extracted from the image as shown in Figure 5.7.6(c). To obtain a statistical meaningful profile, the profile was integrated

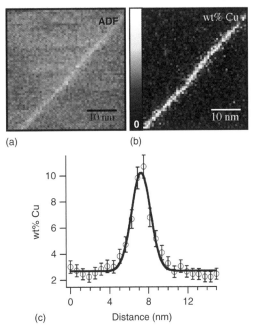

Figure 5.7.6 (a) High magnification HAADF image showing the increase in average atomic number at a grain boundary. (b) Quantitative Cu map from the same boundary. The gray scale goes from 0 to 18 wt%. (c) Compositional profile extracted from (b). (Reprinted from *Mikrochim. Acta [Suppl.]*, **15**, Williams, D., Watanabe, M. and Carpenter, D., Thin film analysis and chemical mapping in the analytical electron microscope, 49–57, Copyright (1998), with permission from Springer-Verlag)

parallel to the boundary over 30 pixels, giving it an effective width of about 20 nm. The line profile reveals the degree of local boundary Cu enrichment, and shows a spatial resolution of composition variation of about 4 nm. This is expected to improve with thinner specimens, although thinner samples would generate less X-rays exacerbating the signal problem. The error bars on the profile are based on counting statistics and represent the 95 % confidence limits.

Williams et al. (1998) also performed a lower magnification analysis to visualize more than one boundary at the time. X-Ray maps of Cu K and Al K were gathered at 240 kV and 256 × 256 pixels with a dwell time of 40 ms, for a total frame time of about 1 h. The BF image (Figure 5.7.7a) shows diffraction contrast, from which it can be seen that most of the boundaries are not oriented in the ideal configuration parallel to the beam. Figure 5.7.7(b) is the corresponding quantitative Cu concentration map, calculated using an experimentally determined k-factor from nearby θ-phase particles.

5.7.5.5 QUANTITATIVE ANALYSIS OF IONIC COMPOUNDS

Another absorption correction based on a 'known' property of the sample was presented in 1994 (Van Cappellen and Doukhan, 1994). Just like the parameterless correction method and the zeta-factor method it does not require specimen thickness, X-ray take-off angle and specimen density. Modern ultra-thin window X-ray detectors are capable to detect light elements such as nitrogen (N K: 392 eV) and oxygen (O K: 532 eV) but even in the thinnest possible samples an absorption correction will be necessary. Absorption does not only occur in the sample itself but also in surface layers such as contamination or on purpose evaporated carbon or metal conductive layers. Also the ultra-thin detector window cannot be neglected in the case of nitrogen or oxygen. The correction procedure described in Van Cappellen and Doukhan (1994) was originally developed for ionic compounds (oxides, ceramics, minerals, etc.) and is based on the principle of electroneutrality or, in other words, the sum of all anions and cations times their respective valence states must cancel out.

The ionic compound method uses the conventional absorption correction software available on all commercially available systems and based on Equation (5.7.2). The exponentials of the absorption correction factor A of Equation (5.7.2) can be expanded in specimen thickness and up to the first order in t equals:

$$A \approx 1 + 1/2[(\mu/\rho)_{sp}^A - (\mu/\rho)_{sp}^B] \operatorname{cosec} \theta \times \rho t \quad (5.7.6)$$

Whenever the difference between the mass-absorption coefficients is small the absorption correction will be close to unity, which is the same as saying that when both characteristic peaks are absorbed the same way, no correction is needed. However with light elements such as O and N $[(\mu/\rho)_{sp}^A - (\mu/\rho)_{sp}^B]$ is always significant and even for a very thin sample the amplitude of the correction will be substantial. A new parameter called the 'mean mass-absorption length' is defined as:

$$\Delta \tau = 1/2 \operatorname{cosec} \theta \times \rho t \quad (5.7.7)$$

When considering a perfect plane parallel sample, $\Delta \tau$ corresponds to the mass-distance through which an X-ray photon generated in the middle

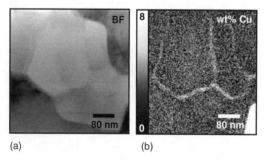

Figure 5.7.7 (a) Bright field image of several grains. (b) Quantitative Cu map at a lower magnification than Figure 5.7.6, showing the Cu distribution in several grain boundaries. The lower intensity from the Cu-rich boundaries is due to the fact that the boundaries are generally not parallel to the electron beam. (Reprinted from Mikrochim. Acta [Suppl.], **15**, Williams, D., Watanabe, M. and Carpenter, D., Thin film analysis and chemical mapping in the analytical electron microscope, 49–57, Copyright (1998), with permission from Springer-Verlag)

of the foil has to propagate in the sample before escaping in the direction of the detector. As mentioned before, the X-ray generation probability is uniform throughout the thin film and thus $\Delta\tau$ also represents the average mass-distance the X-ray photons have to travel in the sample on route to the detector. For all other sample geometries $\Delta\tau$ has no straightforward meaning, but still represents the mass-absorption length that when used in a conventional absorption correction yields the correct concentrations. Equation (5.7.2) reduces to:

$$\frac{C_A}{C_B} = k_{AB}\left(\frac{I_A}{I_B}\right) \times \{1 + [(\mu/\rho)_{sp}^A - (\mu/\rho)_{sp}^B]\Delta\tau\} \quad (5.7.8)$$

In appearance, Equation (5.7.8) is a linear equation in $\Delta\tau$, but since the mass-absorption coefficients $(\mu/\rho)_{sp}^A$ and $(\mu/\rho)_{sp}^B$ are concentration dependent, the relation is quadratic. Mass-absorption coefficients in a particular compound are linear combinations of the values in pure element targets and the coefficients are the weight fractions of the different elements. Therefore in practice, concentrations are obtained through an iterative procedure.

The ionic compound method does not attempt to 'measure' $\Delta\tau$, but instead uses random numbers. Only one value for $\Delta\tau$ will yield a composition that will be electroneutral. Three arbitrarily chosen values for $\Delta\tau$ are used to process the spectrum to be quantified. One spectrum yields three sets of concentrations. This unequivocally defines a parabola per analysed element showing how the concentrations of these elements change as function of the mean mass-absorption length $\Delta\tau$. Each curve is then multiplied by the valence state of the element it represents. The sum of all cation curves yields a parabola showing how positive charge varies with $\Delta\tau$ and a similar curve for the negative charge is obtained by adding the anions. These two 'positive' and 'negative' curves intersect in one point, the only value of $\Delta\tau$ for which the condition of electroneutrality can be met. The real value of $\Delta\tau$ now being available, the correct composition of the sample can be calculated by simply reprocessing the spectrum a fourth time with the exact value for $\Delta\tau$.

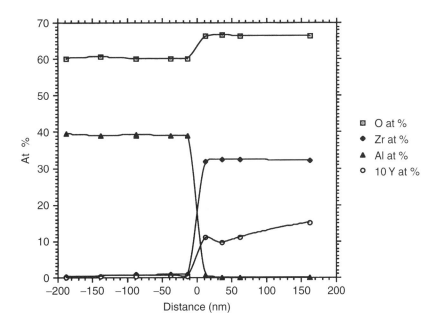

Figure 5.7.8 Diffusion profiles of O, Zr, Al and Y across an Al_2O_3/Y-TPZ interface. The Y profile is multiplied by a factor 10. (Reprinted from *Ultramicroscopy*, **53**, Van Cappellen, E. and Doukhan, J. C., Quantitative transmission X-ray microanalysis of ionic compounds, 343–349, Copyright (1994), with permission from Elsevier Science)

5.7.5.6 REAL WORLD APPLICATIONS

Two problems using the ionic compound correction method to obtain truly quantitative data are reviewed to demonstrate the power of quantitative EDX in transmission microscopy (Van Cappellen and Doukhan 1994). The first example is the accurate quantification of diffusion profiles in fine-grained ceramic composites. The analysed material is an Al_2O_3 reinforced (20 vol%), yttria-stabilized (3 mol% Y_2O_3), tetragonal zirconia polycrystalline composite (Al_2O_3/Y-TZP). The purpose is to investigate mutual diffusion of Al and Zr in neighboring grains after superplastic deformation. Figure 5.7.8 shows that Al does not penetrate in ZrO_2 grains whereas Zr slightly diffuses into the Al_2O_3 crystals. Noteworthy is that on both sides and away from the interface the correct O concentrations are obtained (60 and 66.7 at%). Oxygen, the only common element on both sides, is used, as the ratio element in the Cliff–Lorimer ratio technique, confirming that contrary to some beliefs, strongly absorbed elements can be used as the ratio element. The k-factors are measured experimentally on well-characterized standards using the parameterless correction method. The relative errors on k_{AlO} and k_{ZrO} are approximately 2%, whereas for k_{YO}, the error is believed as high as 5%, which is not a problem as Y is only a minor constituent.

The second example is the study of a complex garnet [$(Mg, Ca, Fe)_3Al_2Si_3O_{12}$]/*ortho*-pyroxene [$(Mg, Fe)_2Si_2O_6$] interface. The concentration profiles obtained with the ionic compound correction scheme (Figure 5.7.9) are consistent with the rules of crystal chemistry and show that the interface contains a certain amount of silica especially on the *ortho*-pyroxene (OPX) side of the interface, a conclusion that would have been difficult to draw otherwise. In this sample, not only the interface is altered, but also the garnet phase is contaminated with monoxide layers. This can be concluded from the following facts: first the measured Si concentration is well under 15 at%, second there is a systematic excess of 2+ valence elements, and third the O/Si ratio far exceeds four. At 5 μm from the interface, the O/Si ratio equals 4.41 whereas on clean garnet samples the measured ratio consistently is within 1% of the theoretical value. It is not possible from the X-ray spectra to determine which monoxide is present, but the concentration at 5 μm from the interface is consistent with: 89 at% of $(Mg, Ca, Fe, Mn)_{3.0}(Al,Cr)_{1.9}Si_{3.0}O_{12.1}$ plus 11 at% of (Mg, Ca, Fe, Mn)O. Noteworthy in this sample is that the OPX phase is pure. Also at 5 μm from the interface, the measured OPX composition is (minor elements are omitted): $(Mg_{1.48} Fe_{0.48} Al_{0.10}) [Si1_{.85} Al_{0.10}] O_{5.96}$. The sum of the so-called M_1 and M_2 sites, the elements between the round brackets equals 2.06 instead of 2.0 and the tetragonal site elements between the square brackets add up to 1.95 instead of 2.0. Even more impressive are the O index of 5.96 and the O/Si ratio of 3.05, which theoretically should be 6 and 3. The k-factors for this study are also measured experimentally with the parameterless correction method. All cations have k_{XSi}s with relative accuracies of better than 3% and k_{OSi}, which was measured on different compounds, has a statistical accuracy of better than 1%. With an estimated take-off angle of 34° and the known densities of garnet and *ortho*-pyroxene (3.95 and 3.4 g/cm^3, respectively) it is also possible to calculate the thickness profile (Figure 5.7.10).

5.7.6 CONCLUSION

More than ever EDX on (S)TEMs is an extremely powerful technique in materials science that surely did not vanish in favor of EELS as predicted 20 years ago. The main reasons are its relative ease-of-use and the possibility of truly quantifying the data. The examples discussed in the previous paragraph are not the most trivial ones, but they clearly show that with some care, quantification can be accurate and precise. As a matter of fact, all modern analytical (scanning) transmission microscopes are equipped with at least an EDX system whereas not all of them have electron energy loss spectrometers.

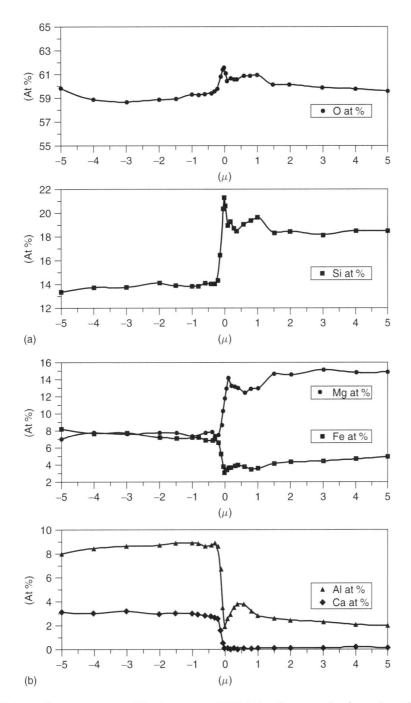

Figure 5.7.9 Diffusion profiles across a garnet (a)/*ortho*-pyroxene (OPX, b) interface, extending from −5 to +5 μm. The oxygen and silicon curves clearly show that the OPX side of the interface is contaminated over a range of about 1 mm, with silica (SiO_2). (Reprinted from *Ultramicroscopy*, **53**, Van Cappellen, E. and Doukhan, J. C., Quantitative transmission X-ray microanalysis of ionic compounds, 343–349, Copyright (1994), with permission from Elsevier Science)

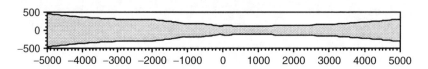

Figure 5.7.10 Calculated thickness profile perpendicular to the garnet/OPX interface. The vertical and horizontal scales are identical and are in nm. (Reprinted from *Ultramicroscopy*, **53**, Van Cappellen, E. and Doukhan, J. C., Quantitative transmission X-ray microanalysis of ionic compounds, 343–349, Copyright (1994), with permission from Elsevier Science)

REFERENCES

Cliff, G. and Lorimer, W. The quantitative analysis of thin specimens. *J. Microsc.*, **103**, 203–207 (1975).

Egerton, R. F. *Electron Energy-loss Spectroscopy in the Electron Microscope*, 2nd Edition, Plenum Press, New York, 1996.

Horita, Z., Sano, T. and Nemoto, M. Determination of the absorption-free kANi factors for quantitative microanalysis of nickel base alloys. *J. Electron Microsc.*, **35**, 324–334 (1986).

Joy, D. C., Romig, Jr, A. D. and Goldstein, J. I. *Principles of Analytical Electron Microscopy*, Plenum Press, New York, 1986.

Van Cappellen, E., Van Dyck, D., Van Landuyt, J. and Adams, F. A parameterless method to correct for X-ray absorption and fluorescence in thin film microanalysis. *J. Phys. (Suppl. C2)*, **45**, 411–414 (1984).

Van Cappellen, E., Van Landuyt, J. and Adams, F. X-ray microanalysis of second-phase precipitates in copper/zinc/aluminium shape memory alloys based on the parameterless correction method. *Anal. Chim. Acta*, **195**, 257–263 (1987).

Van Cappellen, E. The parameterless correction method in X-ray microanalysis. *Microsc. Microanal. Microstruct.*, **1**, 1–22 (1990).

Van Cappellen, E., Deblieck, R. and Van Dyck, D. On the secondary X-ray emission induced by electron irradiation in thin samples. *Microsc. Microanal. Microstruct.*, **1**, 127–140 (1990).

Van Cappellen, E. and Doukhan, J. C. Quantitative transmission X-ray microanalysis of ionic compounds. *Ultramicroscopy*, **53**, 343–349 (1994).

Watanabe, M., Horita, Z. and Nemoto, M. Absorption correction and thickness determination using ζ factor in quantitative X-ray microanalysis. *Ultramicroscopy*, **65**, 187–198 (1996).

Williams, D. B. and Carter, C. B. *Transmission Electron Microscopy, Volume 1: Basics*, Plenum Press, New York, 1996a.

Williams, D. B. and Carter, C. B. *Transmission Electron Microscopy, Volume 4: Spectrometry*, Plenum Press, New York, 1996b.

Williams, D., Watanabe, M. and Carpenter, D. Thin film analysis and chemical mapping in the analytical electron microscope. *Mikrochim. Acta (Suppl.)*, **15**, 49–57 (1998).

5.8 X-Ray Absorption Techniques

J. KAWAI
Kyoto University, Kyoto, Japan

5.8.1 INTRODUCTION

X-Rays are absorbed by matter and the intensity is attenuated. The degree of absorption (absorbance) depends on the wavelength of the X-rays as well as the thickness, density, atomic number, and the local structure of the absorber. Figure 5.8.1[1] shows typical X-ray fluorescence spectra measured by an energy dispersive X-ray fluorescence (EDXRF) spectrometer, a Shimadzu EDX-700 desktop spectrometer,[2] which has Zr, Al, Ti, Ni, and polymer filter to absorb part of the primary X-rays. The X-ray tube is a Rh anode tube. The sample is a 1000 ppm cadmium standard solution for atomic absorption spectrometry, which is in a sample cell with a Mylar film window. Without a Zr filter, the Cd Kα peak (23.2 keV) cannot be observed because the Rh Kβ peak (23.1 eV) overlaps. The Cd Kα peak becomes observable with the use of the Zr filter. The Zr K absorption edge is at 18.0 keV, which corresponds to the sharp edge at 18 keV in the spectrum in Figure 5.8.1. The Zr filter effectively absorbs the Rh Kα and Kβ X-rays, which are not used to excite the Cd X-ray fluorescence, but the high energy continuum X-rays go through the filter to excite the Cd. This kind of automatic spectrometer has now a quantitative analysis computer program including the effect of X-ray absorber filter with an automatic change of the filter. The computer code also includes the effect of 6 μm thickness Mylar window X-ray absorption and air absorption of the X-ray path in the spectrometer. We can successfully obtain the Cd concentration of the sample solution. The filter technique is also widely used in PIXE (particle induced X-ray emission), radioisotope X-ray analysis,[3] and X-ray absorption spectroscopy using the X-ray fluorescence yield method.

The interaction between X-rays and matter is classified into absorption, elastic scattering, and inelastic scattering. The basic physical formula of X-ray absorption are described elsewhere.[4-6] The physical constants, such as X-ray absorption coefficients or equivalently the imaginary part of the atomic scattering factor are listed elsewhere.[4-9] The relation among the imaginary part of the atomic scattering factor, the imaginary part of the refractive index, the linear absorption coefficient, and the mass absorption coefficient, are concisely described in the literature.[4,6,7] The measurement and interpretation of the X-ray absorption coefficients as the change of the X-ray energy are called the X-ray absorption spectrometry or spectroscopy. Since most of the materials analysis techniques using the X-ray absorption phenomena have already been described in the *Encyclopedia of Analytical Chemistry*[4] the present subchapter is devoted to the concise summary and addenda to the Encyclopedia.

X-Ray absorption spectroscopy is referred to as XANES (X-ray absorption near edge structure) and EXAFS (extended X-ray absorption fine structure) (Figure 5.8.2).[10] XANES refers to the fine structure about 50 eV around the edge. EXAFS is used for the oscillating fine structure from 50 to 1000 eV above the absorption edge. The Fourier transform of this EXAFS oscillation yields a

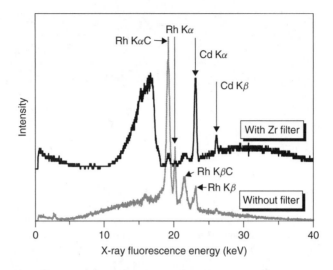

Figure 5.8.1 X-Ray fluorescence spectra of cadmium 1000 ppm solution with and without Zr filter. Taken from Furuya et al.[1] Reproduced by permission of *Adv. X-ray Chem. Anal. Japan*

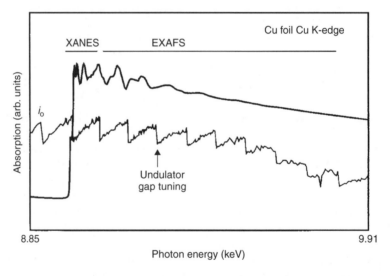

Figure 5.8.2 K-edge EXAFS spectra of Cu foil measured using an undulator beamline. Taken from Oyanagi et al.[10] Incident X-ray intensity (i_0) is also shown, which has saw-teeth like structure due to the undulator gap change. The EXAFS spectrum (heavy solid line) after being normalized with respect to the incident X-ray intensity has a smooth oscillation but no discontinuity. Reproduced by permission of Electrotechnical Laboratory

radial distribution function of the X-ray absorbing atom. The XANES spectra represent the electronic structure of the conduction band. XAFS (X-ray absorption fine structure) is the generic name for XANES and EXAFS. XAS (X-ray absorption spectroscopy) is less used than XAFS, but has a similar meaning.

5.8.2 BASICS FOR XAFS

XAFS has a long history since de Broglie and a summary is given by Lytle.[11] The 10th International Conference on XAFS was held in Chicago in 1998,[12] the 11th at Ako, Japan in 2000,[13] and the 12th at Malmö, Sweden in 2003 (near to

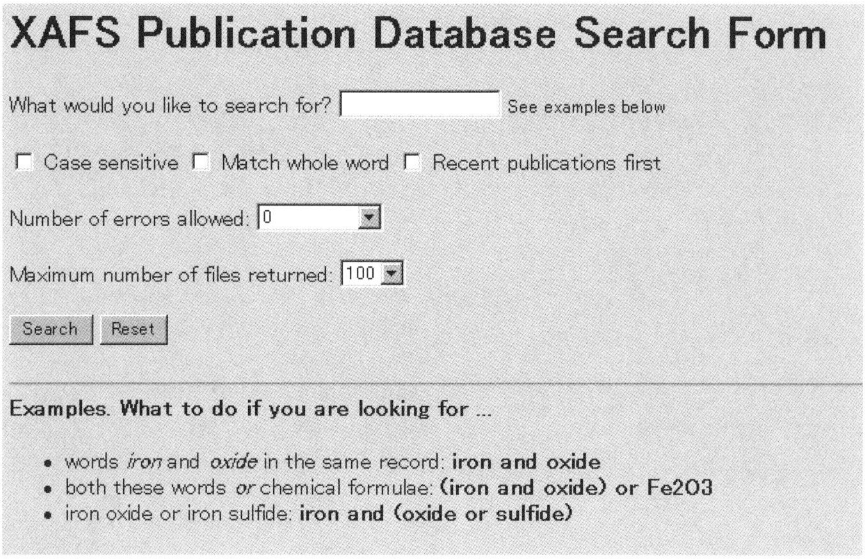

Figure 5.8.3 Web page of XAFS publication database. Reproduced from http://ixs.csrri.iit.edu/

the Swedish synchrotron radiation facility MAX lab).[14] Other earlier international conferences and their proceedings are listed elsewhere.[4] We can obtain information on the activity of the XAFS Society as well as analysis programs and other scientific issues through The International XAFS Society web page.[15] The XAFS Publications Database is freely accessible (Figure 5.8.3). The fundamental review of X-ray spectrometry is published in even years in a journal,[16] and X-ray absorption spectroscopy is included.

XAFS is classified into two different spectroscopies, XANES and EXAFS, as described above. EXAFS is a structural analysis method, and thus it has similarities to X-ray diffraction data analysis, such as the requirement for standardization methods. The multiple scattering method is the major method of EXAFS analysis.[17,18] XANES, which is also called NEXAFS (near edge X-ray absorption fine structure), is analysed by molecular orbital or band structure calculations, and is similar to the soft X-ray emission/fluorescence line shape analysis.[19-24] The basic measurement method is common to these two spectroscopies, EXAFS and XANES. The monochromatized incident X-ray intensity (I_0) is measured, and the X-ray intensity transmitted through a specimen (I) is measured. The value $-\log(I/I_0)$ is plotted versus the incident X-ray photon energy.

5.8.3 EXTREME CONDITIONS

The measurement of XAFS is now a routine experiment for local structure or electronic structure characterizations. Therefore techniques to measure the XAFS spectra under extreme conditions, such as high pressure and/or high temperature, have been developed.

With a combination of an atomic absorption spectrometer and XAFS spectrometer, Nakai et al.[25] measured a 2000 °C flame nebulized from $Cu(NO_3)_2$ solution at various heights of the flame. The spectra showed that the chemical state of Cu was different with the change of the position in the flame (Figure 5.8.4); It went from divalent to atomic like, according to the height in the flame.

High temperature solid phase (3000 K),[26] high temperature liquid phase (3000 K),[27] and both high temperature (1650 °C) and high pressure (2000 bar) supercritical fluids[28] have been measured. The effect of temperature was interpreted through the Debye–Waller factor[29] and anharmonic vibration.[30,31] Isotope effect, though

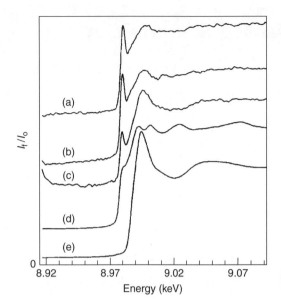

Figure 5.8.4 Cu K XANES spectra of Cu(NO$_3$)$_2$ solution in flame at various heights from an atomic absorption burner head. Taken from Nakai et al.[25] (a) Flame at 6 mm height; (b) 3 mm; (c) 1.8 mm; (d) Cu metal foil; (e) Cu(NO$_3$)$_2$ solution itself. Reproduced by permission of Elsevier

this was not measured under an extreme condition, was observable for the line broadening of the absorption peak as shown in Figure 5.8.5, which was an effect due to the molecular vibration.[32]

5.8.4 COMBINATION WITH SCANNING TUNNELING MICROSCOPES

When the X-rays are absorbed, secondary quanta will be created by the energetic X-ray photons. Probing these secondary quanta using a scanning tunneling microscope (STM) or scanning capacitance microscope,[33-38] high spatial resolution microscopes will be realized in the near future, though the present spatial resolution is only 1 mm or less. Figure 5.8.6 shows a schematic diagram of the experimental set-up for STM-XAFS.[35] Capacitance measurements can be performed both with scanning and without scanning mode as shown in Figure 5.8.7.[38] We can change the bias potential in these tunneling or capacitance probe methods. Though the details of the physical processes have not yet been clarified, the change of bias potential will change the probing site on the surface or position in the valence band, and we will obtain novel

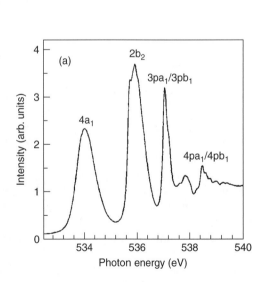
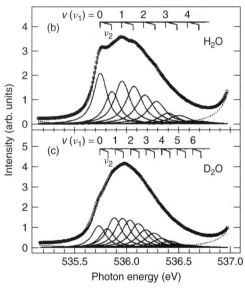

Figure 5.8.5 Oxygen K edge XAFS spectra of water. Reproduced from Hiraya et al.[32] (a) Total ion yield O 1s absorption spectrum of H$_2$O; (b) 2b$_2$ peak of H$_2$O; (c) 2b$_2$ peak of D$_2$O. Vibration fine structures are different due to the isotope effect

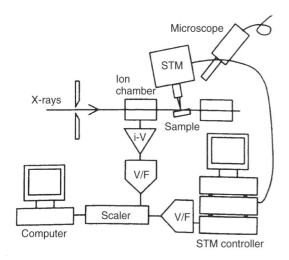

Figure 5.8.6 Experimental set-up of a scanning tunneling microscope under synchrotron radiation. Taken from Tsuji et al.[35] Reproduced by permission of John Wiley & Sons, Ltd

information on the electronic and local structures with an atomic scale using these techniques.

5.8.5 COMBINATION WITH OPTICAL LUMINESCENCE

When X-rays are absorbed, visible light is emitted from the sample. This is called the X-ray excited optical luminescence (XEOL). Its intensity depends on the absorbance, but does not directly depend on it, because of very complicated

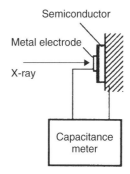

Figure 5.8.7 Experimental set-up for the capacitance XAFS measurement. Taken from Ishii[38]

relaxation processes.[4] Its wavelength depends on the analyte. Figure 5.8.8 shows a decay spectrum of $Y_3Al_5O_{12}$:Ce measured with single bunch synchrotron radiation as an input,[39] which was regarded as a delta function in the time domain. When a narrow time window is selected in the decay line shape, for recording the XAFS spectra, some optical relaxation process corresponding to the time delay can be probed, because some optical processes are delayed depending on its transition probability, or in other words, lifetime of an excited state. When a narrow wavelength window is selected in the optical luminescence spectra, a chemical species can be probed. By selecting these time and wavelength windows at the same time, a novel characterization method will be realized. Porous materials,[40] nanoparticles[41] and optically

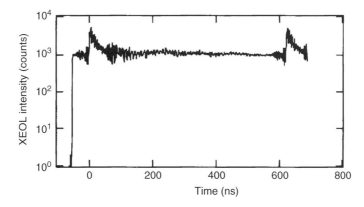

Figure 5.8.8 Decay curve of $Y_3Al_5O_{12}$:Ce optical emission measured with 8 keV X-ray irradiation using single bunch operation mode at the Photon Factory. Taken from Hayakawa et al.[39] Reproduced by permission of John Wiley & Sons, Ltd

active materials[42] are suitable targets for the XEOL method. The line shape measured by the luminescence yield is usually broader than those of the total electron yield XAFS spectra.[43] The XEOL-XAFS can easily be performed using a conventional grating optical spectrometer (either by a step scan type or a position sensitive detector type) at the synchrotron radiation beam line. The detection limit is low compared with the transmission method. Though the thickness or concentration of a film specimen should be adjusted to yield a strong XAFS modulation in the conventional X-ray transmission measurement, the XEOL is measurable even for bulk samples.

5.8.6 COMBINATION WITH X-RAY FLUORESCENCE

When the stronger incident X-rays are absorbed, the stronger X-ray fluorescence is emitted. Therefore the X-ray fluorescence intensity is a measure of absorbance. This type of measurement is called the X-ray fluorescence yield (XFY) method. The XFY method has been used with an X-ray detector such as a photodiode or a proportional counter without wavelength selection. The X-ray detector is placed close to the sample and the total X-ray fluorescence intensity is measured. Jaklevic et al.[44] pointed out in 1977 that the XFY method was quite sensitive down to 10^{19} atoms/cm^2 concentrations. This value has become several orders of magnitude lower after 25 years.

De Groot pointed out that, by the use of a wavelength dispersive X-ray fluorescence spectrometer, spin-state selective XAFS spectra could be obtained.[24] The $K\beta'$ XFY and $K\beta_{1,3}$ XFY spectra of early transition metal compounds yield the minority and the majority spin state XAFS spectra separately.

Muramatsu et al.[45] measured XFY-XAFS spectra by two parts of an X-ray fluorescence spectrum using a grating spectrometer (Figure 5.8.9). Figure 5.8.10 shows the XANES spectra of

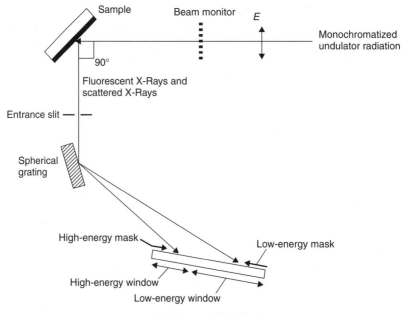

Figure 5.8.9 Experimental set-up of $L\alpha$ and $L\beta$ selective X-ray fluorescence yield method at BL-8.0 at ALS. Taken from Muramatsu et al.[45] To put a blind mask on a part of the position sensitive detector makes it possible to measure the XAFS spectra of a certain X-ray fluorescence line yield method. This is also electrically possible, because the output of a position sensitive counter is usually pulse height signal dependent on the position

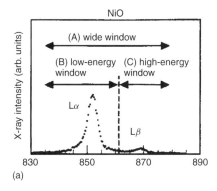

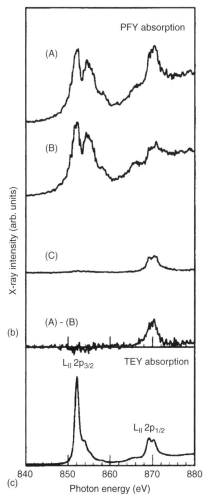

Figure 5.8.10 (a) Nickel L X-ray fluorescence spectrum. (b) L absorption spectra of (A) Lα,β X-ray fluorescence yield, (B) Lα yield and (C) Lβ yield XAFS spectra. (c) Total electron yield spectrum. From Muramatsu.[46]. Reproduced by permission of *Adv. X-ray Chem. Anal. Japan*

total XFY, Ni Lα and Lβ partial X-ray fluorescence yields, the difference of these two, and the total electron yield (TEY).[46] X-Ray fluorescence spectra show chemical shift due to the difference in effective charges. Therefore the XAFS spectra yielded by the chemical-shifted X-ray fluorescence peaks can be obtained. Izumi *et al.*[47] measured site-selective XANES spectra utilizing the chemical shift of Cu Kα_1 between Cu on ZnO and Cu metal in a catalyst. The signal from a position sensitive X-ray detector in an X-ray fluorescence spectrometer has a relation; electric voltage to the position. Therefore we can obtain the partial X-ray fluorescence yield XAFS spectra by the use of discrimination of electric pulses using an electric circuit, without using a mechanical mask on the position sensitive detector, which was used by Muramatsu *et al.*[45]

The resonant Raman scattering peak is observable when the incident X-ray energy is just below the absorption edge. The remarkable feature of the X-ray Raman peak is that the line width is narrower than the X-ray fluorescence, though the transition from the Raman peak to the X-ray fluorescence is continuous. Utilizing this phenomenon, XAFS spectra free from lifetime broadening can be obtained.[48–50]

5.8.7 PHOTOCONDUCTIVE AND ELECTROCHEMICAL MEASUREMENTS

The electrical conductance of a solid changes with the change of the wavelength of the incident X-rays. This is because the valence electrons are excited into the conduction band and thus the density of conduction electrons changes, or, an atom, whose core electron is photoionized, behaves as an impurity in the metal. The response of conductance is thus sometimes positive and sometimes negative.[51] Compared with solid samples, liquids are suitable to measure the conductivity with the change of the incident X-ray wavelength. Sham and Holroyd[52] used a liquid cell with electrodes, and measured the change of the conductivity of $(CH_3)_4Sn$ in trimethylpentane near the Sn K edge. The behavior of conductivity was positive for low concentration (0.07 M) but negative for high concentration (0.10 M). This behavior also depends on the structure of the sample liquid cell. Though the interpretation is very complicated, we can get information on the liquid or solution using XAFS spectra.

Many kinds of electrochemical cell or *in situ* cell for EXAFS measurements were proposed and are summarized by Sharpe et al.[53] Yamaguchi et al.[54] used an *in situ* electrochemical cell designed by Heineman's group[55] to measure unstable chemical species synthesized during a redox reaction and thus very difficult to stabilize them without applying the electric potential. They measured XAFS spectra of a Mn^{IV}/Mn^{III} system at several points in a cyclic voltammogram.

Nakai et al.[56,57] measured the XAFS spectra of a lithium battery electrode during charge–discharge processes using an *in situ* cell as shown in Figure 5.8.11.[56] They observed the chemical shift of the strongest absorption peak (the so-called white line) during the charge–discharge process, emergence of pre-edge structure, which means the change of coordination structure, and the change of atomic distances from the Fourier transform of the EXAFS data.

The stainless steel corrosion process was analysed using an electrochemical cell by the measurement of EXAFS.[58]

5.8.8 TOTAL REFLECTION XAFS

Heald et al.[59] proposed measuring a buried $CuAl_2$ layer in an Al(100 nm)/$CuAl_2$(5 nm)/Cu(100 nm) multilayer using the glancing angle incident XAFS. Adjusting the glancing angle, the signal from a buried layer could be maximized due to the interference effect between the incident and reflected X-rays. Recently, Shirai et al.[60] and Kawai et al.[61] developed the total reflection XAFS method. Most total reflection papers have been cited in Kawai et al.[62]

Figure 5.8.12[63] shows the schematic probing and information depth of total reflection XAFS. The TEY method is more surface sensitive than the X-ray fluorescence yield method. Figure 5.8.13[64] shows the use of polarization of the incident X-rays for the determination of the orientation of adsorbates on a substrate. Sample holders[65,66] and an X-ray detector[67] suitable for the measurement of polarized total reflection XAFS were developed. Polarization total reflection XAFS has been used to study structure transformation of a Pt cluster on

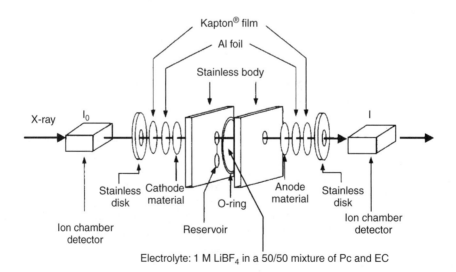

Figure 5.8.11 Schematic illustration of an *in situ* electrochemical cell for transmission XAFS measurements. Taken from Nakai et al.[56] Reproduced by permission of Elsevier

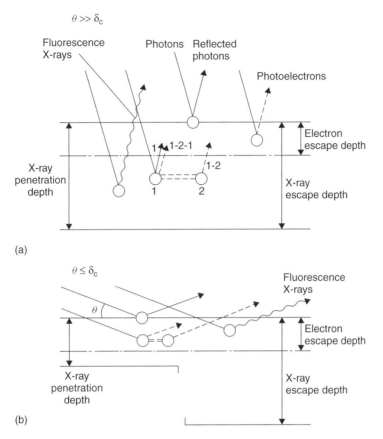

Figure 5.8.12 Schematic representation of normal (a) and surface sensitive (b) X-ray fluorescence excitation. Taken from Oyanagi et al.[63] The critical angle of X-ray total reflection is denoted by δ_c. Total reflection geometry can reduce the probing depth by several orders of magnitude, achieving a surface-sensitive excitation. Reproduced by permission of International Union of Crystallography

an Al_2O_3 substrate[68] and molybdenum oxide on a single crystal.[69]

The X-rays emitted from a bending magnet of a synchrotron light source is polarized in such a way that the electric vector is parallel to the horizontal plane. Thus the electric dipole moment in this plane in a sample can be easily excited and thus the X-rays are strongly absorbed when a column of atoms is present in this direction. Figure 5.8.14[70] shows π and σ orbital components in Cu K XANES spectra of an oxide superconductor single crystal by adjusting a crystal axis to the electric vector of the incident X-rays. This experiment was not a total reflection, but similar to the total reflection method in such a way that both methods used the polarization of X-rays. When L_3 edge spectra are measured with polarized X-rays, the orientation of the empty d band can be determined,[71] because the 2p–3d electric dipole transition is dominant over the 2p–4s transition. The polarization dependent XANES spectra of the $L_{2,3}$ edge of a TiO_2 single crystal are also observable.[72]

Different thermal treatment of a Nb/Al interface was studied by the total reflection EXAFS.[73] Total electron yield XANES under a total reflection condition was measured for Cr thin films on Fe.[74] The small change of the incident glancing angle makes it possible to change the probing depth. This will be a powerful tool to characterize the depth-selective analysis of multilayer samples.

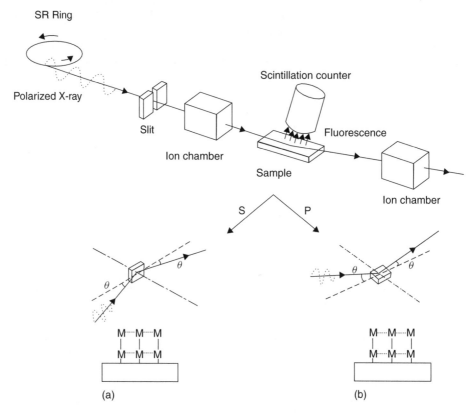

Figure 5.8.13 Schematic diagram of polarized total reflection X-ray fluorescence XAFS, with the molecular adsorbates on the surface. Reproduced by permission of Elsevier

The total reflection X-ray method is suitable for the characterization of liquid surfaces, e.g. a metal stearate Langmuir monolayer on a metal ion aqueous solution.[75] Figure 5.8.15 shows a schematic illustration of the experimental set-up.[76] Usually the conversion electron yield method using He gas, a schematic illustration of which is shown in Figure 5.8.16, is used to measure the spectra.[77] Applications of XANES spectroscopy to solutions have been summarized by Sakane.[78]

X-Ray reflectivity is an alternative method to measure the XAFS spectra. When the absorption is stronger, the reflectivity becomes weaker, and vice versa. Therefore the X-ray reflectivity intensity is a replica of the XAFS spectra.[79,80] An anodic silver oxide film treated *in situ* electrochemically is measured by the reflectivity method.[81]

The total reflection experimental set-up also reduces the self-absorption effect,[82,83] as shown in Figure 5.8.17.[83] The self-absorption effect heavily smears the XAFS spectra when the X-ray fluorescence yield method is used.

5.8.9 ELECTRON AND OTHER SECONDARY YIELD METHODS AND THEIR SURFACE SENSITIVITY

When X-rays are impinging on a surface and absorbed, then photoelectrons, consequently Auger electrons or X-ray fluorescence, and secondary electrons are emitted. These electrons are generated both in deep places where the incident X-rays reach, as well as in the shallow surface region. Therefore the detection of these electrons, or alternatively the measurement of the sample electric current (this method is called the TEY method), was thought to be bulk sensitive. Recently it has

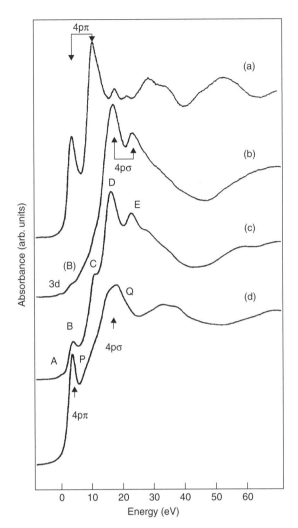

Figure 5.8.14 Polarized Cu K edge XANES spectra of a single crystal of Sr_2CuO_3. Taken from Kosugi et al.[70] (a) The electric vector is parallel with the single crystal axis b; (b) parallel with a; (c) powder; and (d) Cu_2O powder

however been clarified that the electrons emitted from the sample surface, or the sample electric current, are both surface sensitive. This evidence has been independently found by several researchers.[61,84–86] The mechanism and difference of the TEY (sample electric current), secondary electrons, and Auger electrons are described in Kawai[4] and a detailed analysis is reported in Erbil et al.[87] The signal to background ratio of the total reflection X-ray TEY method was estimated by Zheng and Gohshi.[88] The conversion electron yield method, under atmospheric conditions[89,90] was compared with the TEY method, using the total reflection X-ray method in order to change the probing depth.[91] The surface sensitivity of the TEY method, where the majority of the detected electrons are so-called secondary electrons (kinetic energy is less than 50 eV), is approximately 2 nm, which is comparable with the photoelectron inelastic mean free path. The ion yield method is no longer a difficult technique because a channeltron electron multiplier can be used.[92]

Energy dispersive incident X-rays (described later in Figure 5.8.33) on the surface are used as the excitation source. In this experiment, the emitted Auger electrons are detected by a position sensitive electron detector after the energy selection by a hemispherical electron energy analyzer, as shown in Figure 5.8.18.[93] An Auger electron yield XANES spectrum can be obtained with one shot. A surface sensitive XANES spectrum can be thus quickly measured. These electron yield methods are not suitable for measuring insulators. However, as shown in Figure 5.8.19,[94] with the use of an electron flood gun as well as an Ar ion gun, and also with an electron yield detector equipped with an electron energy filter to avoid the effect of low energy electrons from the flood gun, the electron yield XAFS spectra become observable for insulators.[95]

The coincidence spectra are observable by the use of e.g. Auger electron yield and XFY. This kind of coincidence method is useful to clarify the electron transition process in a single atom. We can still obtain chemical information on both the surface and bulk separately for environmental samples such as aerosol or flyash, where the electron yield (surface sensitive) and XFY (bulk sensitive) are measured in one scan.[4]

5.8.10 STRONG FIELD EFFECTS

XAFS are measured for a sample with the application of a strong field, such as laser light or a magnetic field. With a combination of an external field, we can obtain a new chemical information on the sample.

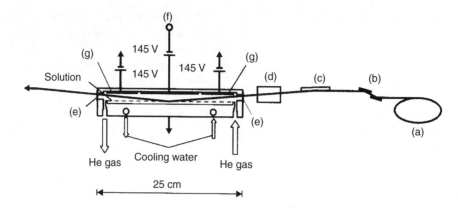

Figure 5.8.15 Schematic diagram of total reflection TEY XAFS experiment. Taken from Tanida.[76] (a) Synchrotron storage ring, (b) monochromator, (c) mirror to remove the higher order harmonic X-rays, (d) ion chamber to monitor the incident X-ray intensity, (e) a polyethylene window, (f) a wire electrode for measuring electrons from sample solution, (f) to amplifier from a collector electrode, and (g) a fringe electrode. Reproduced by permission of H. Tanida

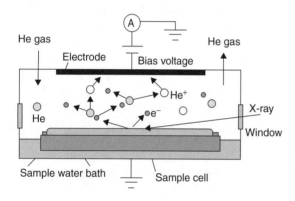

Figure 5.8.16 Mechanism of conversion electron yield method using helium gas. Taken from Harada et al.[77] Reproduced by permission of Japan Society for Analytical Chemistry

Figure 5.8.20 shows a schematic illustration of the transition in the pump-probe XAFS method.[96] Using strong laser irradiation, most of the valence electrons are excited and X-ray absorption spectra of the optically activated state can be measured.

The EXAFS measurement of a glassy sample during laser irradiation has been reported.[97] The EXAFS data of a glassy sample show that the structure is changed temporarily or permanently by the laser irradiation.

We can obtain a magnetic property of a sample by the measurement of XAFS spectra in a strong magnetic field as shown in Figure 5.8.21.[98] The X-rays from an undulator in a synchrotron

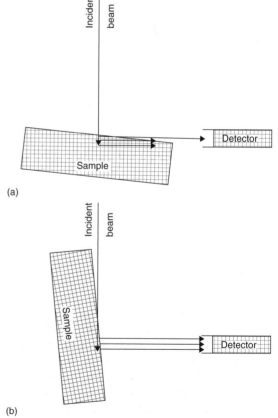

Figure 5.8.17 Sketch of the geometry among incident beam, fluorescent beam, and detector at (a) normal incidence and (b) grazing incidence. Taken from Pfalzar et al.[83]

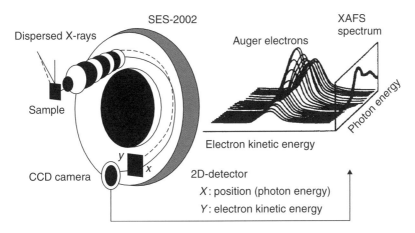

Figure 5.8.18 Schematic diagram for the energy dispersive NEXAFS. Taken from Amemiya et al.[93] The horizontal position at the sample surface corresponds to the photon energy. Reproduced by permission of Japan Society of Applied Physics

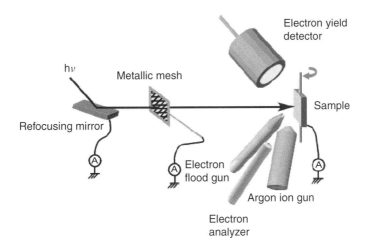

Figure 5.8.19 Schematic drawing of the experimental set-up for an insulator of which charge is neutralized by a flood gun. Taken from Tanaka et al.[94] Reproduced by permission of Japan Society for Analytical Chemistry

radiation facility can be polarized in such a way that the electric vector of the incident X-rays are circulated clockwise or counter-clockwise with respect to the magnetic field applied to the sample. This is also achieved by a phase retarder, such as a λ/4 wavelength slab. The absorption coefficient is slightly different for these two polarizations of the incident X-rays. The transmitted X-ray intensity as well as the X-ray fluorescence intensity show the X-ray magnetic circular dichroism (XMCD). Figure 5.8.21 shows an experimental set-up. Either the helicity or magnetic pole is to be changed, and the resulting difference in X-ray absorbance is measured. This is a Faraday effect in the X-ray region. One of the most important fundamentals of the interpretation of the change in absorption spectra due to the magnetic field is a sum rule proved by Thole et al.;[99] general expressions are given by van der Laan[100] and Ankudinov et al.;[101] how to use the sum rule is given in Tobin et al.[102] Recently a monoatomic wire was characterized using XMCD.[103]

Yokoyama et al.[104] studied carbon monoxide adsorbed on nickel and cobalt thin layers using

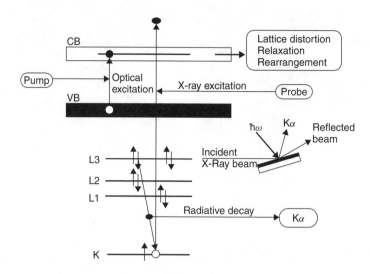

Figure 5.8.20 Principle of pump and probe XAFS. Taken from Oyanagi et al.[96] An electron in a valence band (VB) is excited into a conduction band (CB) by laser light pumping, and the XAFS spectra of such a system are measured using the XFY method. Reproduced by permission of International Union of Crystallography

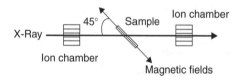

Figure 5.8.21 Schematic diagram of an experimental layout for the measurement of magnetic circular dichroism X-ray absorption fine structure. Taken from Nakamura et al.[98]

XMCD and clarified the interaction between the magnetization film and CO unoccupied orbital π^*, which had orbital magnetic moments.

The XMCD experiment will be useful, when combined with the grazing incident technique, for the characterization of thin multilayer films made by changing the parameters (temperature, pressure, concentration, etc.) in a processing method.

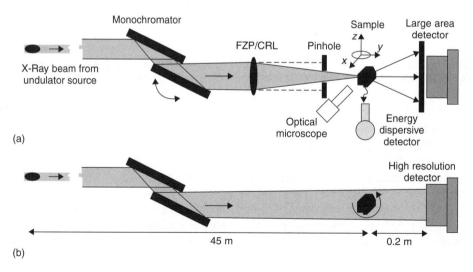

Figure 5.8.22 Experimental set-up for (a) μ-X-ray absorption, μ-X-ray diffraction and μ-X-ray fluorescence and (b) μ-tomography. Taken from Salbu et al.[105] Reproduced by permission of Elsevier

5.8.11 MICROSCOPY

Microscopy and imaging of various samples using the X-ray absorption techniques are summarized in Figure 5.8.22;[105] the techniques used are the microbeam scanning technique (the sample is moved in the real experiment) and tomography technique. Transmission-type micro-imaging without an X-ray optical lens has been widely used. Microscopes using a soft X-ray lens (Figure 5.8.23[106]) have also been extensively used.[107,108] The scanning microbeam method is usually combined with μ-X-ray diffraction[109–112] or X-ray fluorescence[113,114] beamlines in a synchrotron radiation facility.

The principle of X-ray microtomography is shown in Figure 5.8.22(b), which also utilizes the X-ray absorption effect. The X-ray absorption spectra can be obtained by a synchrotron microtomography set-up.[115] Figure 5.8.24[105] shows μ-X-ray absorption spectroscopic tomography of a U fuel particle collected from Chernobyl soils. The image of a cross-section of the fine particle can be numerically reconstructed. Oxidation states of the particle can be determined from the chemical shift of the XANES spectra. The spatial resolution of laboratory microtomography is not better than synchrotron radiation microtomography, but it is now less than 0.1 μm.[116] Figure 5.8.25 shows the inner structure of a lithium electric battery measured by laboratory microtomography developed by Hirakimoto.[117] Desktop microtomography has also been developed.[118]

XANES microscopes have been used to study geological and environmental samples,[119–122] bio and organic materials which show phase separation,[123–125] and adsorbates which are studied using X-ray linear dichroism microscopy.[126] The microbeam analysis method has the potential to be used as a screening method in combinatorial chemistry using an integrated microchemical chip.[127]

5.8.12 EELS, ELNES AND EXELFS

Electron energy loss spectroscopy (EELS) is a technique used in transmission electron microscopy (TEM), where electrons transmit through a thin film and energy absorbed by the thin film during transmission is measured by an electron energy analyser. EELS is classified into energy loss near edge structure (ELNES) and extended energy loss fine structure (EXELFS). This classification is similar to XANES and EXAFS. When the electron energy is measured in the forward direction, the

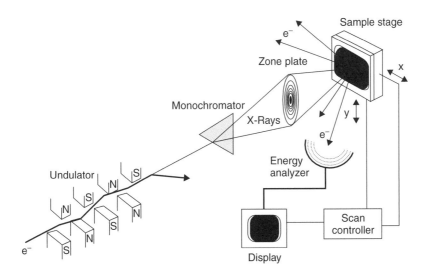

Figure 5.8.23 Schematic representation of the μ-NEXAFS and XPS beamline. Taken from Ade et al.[106] Reproduced by permission of Elsevier

Figure 5.8.24 Tomographic reconstruction and computerised slicing of the three-dimensional image demonstrating an inhomogeneous distribution of U within a Chernobyl particle measured using the beamline shown in Figure 5.8.22. Taken from Salbu et al.[105]. Reproduced by permission of Elsevier

selection rule of an electron transition in an atom is similar to that of an optical (X-ray) transition, i.e. the electric dipole selection rule. Thus the resulting EELS spectra can be interpreted in the same way as the X-ray absorption spectra.[128–130] The presence of the extended fine structure in EELS spectra was found by Leapman et al.[131] and they coined the term 'EXELFS'. EELS measurements are suitable for low energy loss, such as light elements (B, C, N, and O) and soft X-ray lines (L edge of Si and transition metals[132–134]). This is because an overlap of the spectra is avoidable among different elements for low energy losses. EELS has the potential to detect a single atom spectrum of even a biological structure,[135] because the resolution of a transmission electron microscope becomes as high as an atomic level spatial resolution.

5.8.13 EXEFS

Very weak fine structures are always found at the low energy side of strong characteristic X-ray fluorescence or X-ray emission lines. The intensity is usually less than 1 % of the Kα main line. These fine structures are called the 'radiative Auger satellites'. Kawai et al.[136,137] pointed out that the overall line shape of the radiative Auger satellites is similar to that of XAFS spectra and the Fourier transform of the radiative Auger satellites yields a radial distribution function.[136] They coined 'EXEFS' for the use of the radiative Auger satellites to measure XAFS spectra.[137] The EXEFS method is now occasionally used with an EPMA (electron probe microanalyser)[138–140] for chemical state mapping or imaging, and microarea chemical state analysis. Figure 5.8.26 shows the mapping

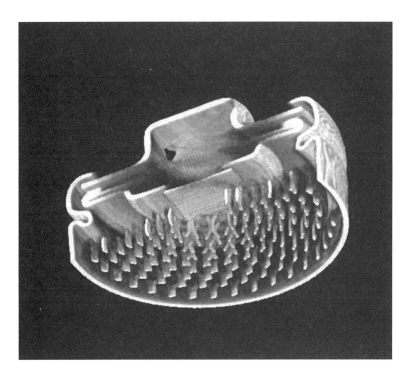

Figure 5.8.25 An inner structure of a lithium electric battery measured by a microtomography developed by Hirakimoto.[117] Reproduced by permission of Shimadzu Corp.

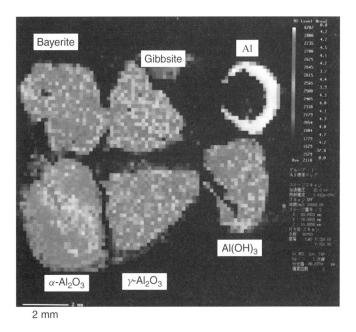

Figure 5.8.26 Chemical state imaging of aluminum standard chemicals using the EXEFS method. Taken from Watanabe et al.[140] Reproduced by permission of International Union of Crystallography

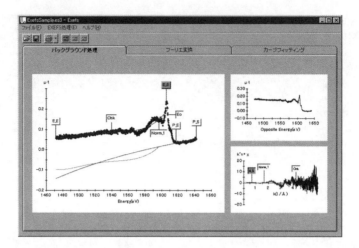

Figure 5.8.27 EXEFS background subtraction window of EXEFA analysis computer software. Taken from Taguchi.[141] The spectrum shown is a real α-Al_2O_3 radiative Auger satellite spectrum. The resulting k-χ spectrum shows an oscillating fine structure. Reproduced by permission of *Adv. X-ray Chem. Anal. Japan*

of aluminum oxides, α- and γ-Al_2O_3, bayerite and gibbsite.[140] These aluminum oxides contain similar elements, thus the discriminations among these different phases are not achieved only by the elemental mapping using the characteristic X-ray lines. However, with the use of EXEFS we can obtain an image due to the difference in phase.

Recently a wavelength dispersive X-ray fluorescence spectrometer was equipped with an EXEFS analysis computer program as shown in Figure 5.8.27.[141] Basic studies on the radiative Auger spectra compared with the Auger electron spectra[142] have become active after the emergence of the EXEFS method. The reason why the EXEFS spectra coincide with the XAFS spectra was at the first stage not clear, but an exact theory has been proposed using a T-matrix expansion;[143] the phase shift is slightly different from the true EXAFS analysis. EXEFS in X-ray fluorescence spectrometry is used to investigate environmental samples because such samples are sometimes very difficult to insert into an ultrahigh vacuum chamber at the synchrotron radiation facility.[144]

5.8.14 DAFS

Diffraction anomalous fine structure (DAFS) is an alternative method to measure the XAFS spectra proposed by Stragier *et al.*[145] The imaginary part of the X-ray scattering factor is closely related to the X-ray absorption coefficient, and thus the measurement of the X-ray diffraction peak by changing the X-ray energy reveals an EXAFS-like oscillation. This method has diffraction peak selectivity, which means that the site or phase selective information is obtainable. The measurement of Ti K XAFS spectra in $BaTiO_3$ is always interfered with by the overlap of the close Ba L_3 edge, but this interference is avoidable by using the DAFS method.[146]

5.8.15 THEORY AND INTERPRETATION OF XAFS SPECTRA

Though the contribution from many electron excitations in XAFS spectra was suggested more than 20 years ago,[147] this effect was not seriously considered until recently. These multiple-excitation phenomena were studied by Ito *et al.*[148] for Br^- in organic solvents and by Magnuson *et al.*[149] for nickel metal. These more-than-one electron excitation signals are a source of spurious noise in the analysis of XAFS spectra[150].

'Atomic XAFS' has been said to exist in XAFS spectra. It is another origin of a step-like structure in EXAFS spectra. Usually EXAFS oscillation is due to the condensed matter effect; the outgoing photoelectron wave interferes with the backscattered waves by its neighboring atoms. Therefore, the EXAFS signal is not observable for an isolated single atom. However, the shape of the potential of a single atom itself has the effect to backscatter the electron wave and some kind of interference pattern with longer period than the EXAFS oscillation is observable. The structures observed in an isolated single atom EXAFS are interpreted both from the view point of double-electron excitation and from atomic XAFS; the existence of 'atomic XAFS' is still controversial.[151,152]

The relativistic effect is also an important factor for the analysis of XAFS spectra. One of the long-standing theoretical problems was that the absence of a sharp absorption line (i.e. white line) in the L_2 XANES of Pt metal, though L_3 has a sharp absorption line. This problem has been solved by the inclusion of the relativistic effect.[153]

All kinds of computer calculation programs of XAFS spectra are accessible through the International XAFS Society (http://ixs.csrri.iit.edu/). One of the most widely used programs is the FEFF series.[154] The newest version is FEFF8. FEFF is a computer program for *ab initio* multiple scattering calculations of EXAFS and XANES spectra for a cluster molecule. The FEFF8 code includes relativistic Dirac–Fock atomic calculation, self-energies, a fully relativistic cross-section, polarization dependence, and many other features.

Another class of computation code is the molecular orbital (MO) method expressed by a linear combination of atomic orbitals (LCAOs). The discrete variational $X\alpha$ (DV-$X\alpha$) method is most convenient in the LCAO-MO method, while the multiple scattering $X\alpha$ (MS-$X\alpha$) method is the basis of the FEFF and other multiple scattering methods. The DV-$X\alpha$ method can calculate the core-hole relaxation state easily. This is the reason why the DV-$X\alpha$ method is extensively used. The LCAO-MO method including the DV-$X\alpha$ method is only applicable to XANES (below 50 eV above the absorption edge) spectra, though the multiple scattering methods are applicable to both XANES and EXAFS spectra. Jiang and Ellis[155] calculated the XANES spectra of various Co compounds using the DV-$X\alpha$ method. The molecule SF_6 has been a reference material for more than 30 years[156,157] in experiments as well as a check for a computer code.[158,159] The measured spectra are satisfactorily reproduced by the DV-$X\alpha$ calculation for the SF_6 molecule. XANES spectra of transition metal complexes,[160,161] silicon oxides[129] and oxyanions[162] are well interpreted by using the DV-$X\alpha$ method.

The core-hole effect is an important factor to interpret mixed valence compounds. According to the calculation of Suzuki *et al.*,[163] the number of 3d electrons in the late transition metal compounds increases by one electron after the creation of a core hole. Contrary to this, that of the early transition metal compounds does not change before and after the creation of the core hole, as is shown in Figure 5.8.28. Therefore the interpretation of the valency or oxidation number of transition metal compounds is very controversial, especially for Co compounds.[164]

'White line' is the name given to the strongest sharp lines in XANES spectra, because these lines develop white when the X-ray spectra are recorded on photographic film. The intensity ratio of the white lines is closely related to the occupancy of valence orbitals, as shown in Figure 5.8.29.[165] The white lines in oxygen K absorption spectra of transition metal oxides (Figure 5.8.30[166]) are also interpreted in a similar way, though the charge-transfer effect[163] and other effects[167] should be included.

The pre-edge structure is a good index of the chemical environment of the X-ray absorbing atom. It is strong for tetragonally coordinated transition metals, but weak for octahedrally coordinated transition metals. The pre-edge structure is sometimes assigned to the 1s–3d electric quadrupole transition, but the relation to $K\beta_5$[168–170] should be studied more extensively.

5.8.16 STANDARDIZATION

The standardization of EXFAS spectra and their numerical processings are important in the practical

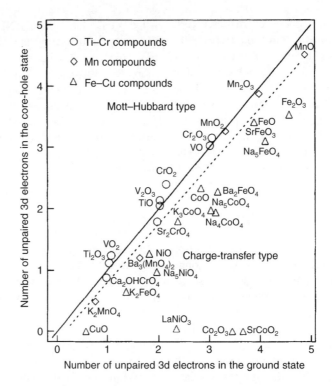

Figure 5.8.28 Plot of the calculated number of unpaired 3d electrons in the ground state and the $1s^{-1}$ hole state. Taken from Suzuki et al.[163] The compounds whose spin states are zero both in the ground state and in the core hole state are omitted. The effect of $1s^{-1}$ and $2p^{-1}$ are approximately the same. Reproduced by permission of Elsevier

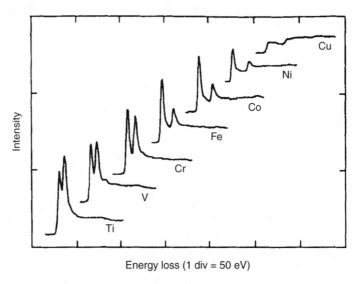

Figure 5.8.29 $L_{2,3}$ edge spectra of 3d transition metals. Taken from Pearson et al.[165]

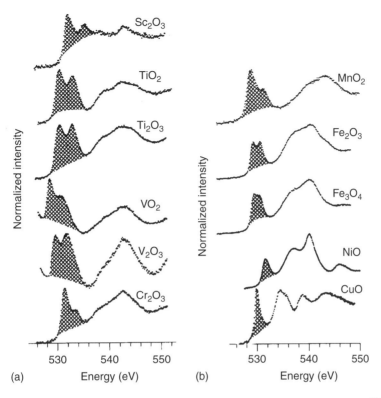

Figure 5.8.30 Oxygen 1s X-ray absorption spectra of transition metal oxides, taken from de Groot et al.[166] The shaded area is assigned to the oxygen p character in the transition metal 3d band

use of EXAFS spectroscopy. This is similar to X-ray diffraction data standardization. The EXAFS spectra measured at many synchrotron beamlines and measured by laboratory spectrometers should be compared with each other. The white line intensity is related to the coherence of the incident X-rays.[171] The problem of the standardization of EXAFS spectra has been discussed with comparison of calculation codes[172] and standard spectra.[173] The International XAFS Society has a committee for the standardization of EXAFS measurements and data analysis. The committee report can be read at http://ixs.csrri.iit.edu/. The report includes: (1) user controlled parameters at beamlines; (2) data collection methods such as transmission, fluorescence, or electron yield methods; (3) detectors used; (4) data analysis programs; (5) error assessment; (6) standardized error reporting procedure; (7) sample preparation; (8) data processing; and (9) modeling.

Many methods have been devised for the analysis of XAFS data. Taguchi and White[174] proposed the use of a crystal structure database to simplify the modeling problem in EXAFS analysis. A rapid calculation method using parallel computation has been proposed.[175] The second derivative of the XANES spectra reveals a doublet structure when the raw spectra themselves look similar.[176]

Cross-correlation analysis of binary mixtures (Co_3O_4 and $CoAl_2O_4$) of EXAFS spectra is used to obtain a quantitative analysis of each component.[177] The EXAFS analysis of Co_3O_4 and $CoAl_2O_4$ has no problem, but XANES analysis should include the charge-transfer effect and spin flip phenomena due to the creation of a core hole.[163,164] Many kinds of empirical parameters are proposed to classify the complicated Cu and Co XANES spectra.[178,179]

Diffraction peaks sometimes contaminate the EXAFS spectra as is shown in Figure 5.8.31.[180]

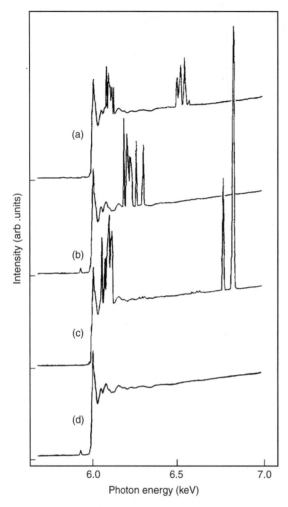

Figure 5.8.31 (a–c) Cr K edge raw X-ray fluorescence yield XAFS spectra of ruby. Taken from Emura and Maeda.[180] Diffraction peaks interfere with the analysis of spectra. (d) A diffraction-free spectrum numerically synthesized from spectra (a) and (b). Reproduced by permission of American Institute of Physics

Using a numerical method, a diffraction-free EXAFS spectrum (Figure 5.8.31d) is available.

5.8.17 INSTRUMENTATION AND SPECTROMETER

A typical synchrotron beamline for XAFS measurement (Figure 5.8.32) can be found in a beamline manual,[181] where a sample is inserted between two ion chambers. To scan the incident X-ray energy, the double-crystal monochromator is step-scanned. A mirror effectively excludes the higher order X-ray reflection. The step-scanning method needs time to scan, and thus energy-dispersive geometry is developed to measure the XAFS spectra by one shot as shown in Figure 5.8.33.[182]

Laboratory EXAFS spectrometers are commercially available[183,184] as shown in Figure 5.8.34, which looks like an X-ray diffractometer. The laboratory EXAFS machines use rotating anode high power X-ray tubes such as reported by Sakurai[185] and recently a parabolic cylinder mirror has been used.[186] The commercially available spectrometers are usually modified versions of spectrometers developed in academic institutions.[187–189] The specimen in these laboratory XAFS spectrometers is moved with a change in the incident X-ray energy but the X-ray tube is fixed. Therefore, complicated experiments using a cryogenic cooler or high temperature furnace are very difficult to perform. To overcome these difficulties, a new spectrometer where the sample is fixed and the X-ray tube moves has been developed.[190]

A soft X-ray synchrotron beamline for XAFS measurements looks like an electron spectrometer, such as an ESCA (electron spectroscopy for chemical analysis) instrument. This is because the soft X-ray beamline (less than 1 keV) requires a windowless system due to the strong absorption coefficients of any type of X-ray windows. Therefore the sample chamber is directly connected to the synchrotron main ring without a window. Consequently the samples allowed into the sample chamber are very limited because of the ultra high vacuum system. However, recently using a thin film such as BN, liquid samples have been measurable[191] at ALS (Advanced Light Source).

5.8.18 SUMMARY

X-ray absorption techniques are used in commercially available film thickness process monitors, which are used in plating, printed circuit and magnetic disk processes. Though these process monitors are not spectrometers, because only thin film

Figure 5.8.32 Ritsumeikan University SR center; 1 m diameter synchrotron radiation facility. Several beamlines are for XAFS spectroscopy. Taken from http://www.ritsumei.ac.jp/se/d11/index-e.html

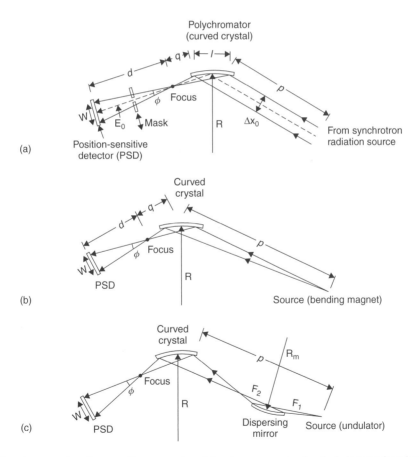

Figure 5.8.33 Focusing geometry of energy-dispersive set-up (a) and arrangements of optical elements for a bending magnet (b) and an undulator (c). Taken from Oyanagi et al.[182] Dispersed X-rays with different energies are focused at a sample position. The undulator band width is optimized to match a typical EXAFS scan range, 1 keV. Reproduced by permission of Electrotechnical Laboratory

Figure 5.8.34 A typical in-laboratory EXAFS spectrometer. Taken from the RIGAKU catalogue[183]

filters are used to select the X-ray wavelengths, they are used in a variety of industries.

On the other hand, X-ray absorption spectrometers are used in both laboratories and synchrotron facilities for basic science, such as surface science,[192,193] cluster chemistry[194] and electron correlation in transition metal oxides.[167] Other applications are in electrodes,[195,196] biological samples,[197] catalysts,[198] especially *in situ* reaction conditions,[199,200] industrial materials, such as flat panel displays,[201] geological samples[202] and environmental analysis.[203–205] An unknown chemical, called 'manganese blue', has been characterized by Mn L XANES and concluded to be a mixture of $Mn^{4+}:Mn^{6+} = 1:1$, which was believed to be Mn^{5+} before the XANES analysis.[206]

A simple method for fabricating uniform and large area X-ray absorption filters from powder materials has been published.[207]

The X-ray absorption techniques described in the present section are summarized using three key words, namely probe, signal and field. The probes used at present are X-rays and electrons; the X-rays are sometimes polarized and sometimes totally reflected. The detected signals are transmitted X-rays, X-ray fluorescence, electrons, electric currents, and many other measurable signals such as ions, reflectivity, and energy selected X-rays or electrons. The number of fields are becoming remarkably greater, such as high temperature, high pressure, low temperature, *in situ* chemical reaction, strong magnetic field, applying an electric potential, short measurement time, and plasma state, using the timescale selection such as coincidence or delay. New opportunities are proposed from time to time.[208] After the absorption of X-rays, the de-excitation processes we consider are usually X-ray fluorescence and Auger, but nuclear excitation by electron transition (NEET) has been actively studied recently.[209–211] The NEET is, in other words, a nuclear Auger process. The nuclear energy level is excited by the inner shell electron transition. Recently, it was found that the γ-radiation rate was accelerated by the absorption of X-rays above the absorption edge. This kind of X-ray absorption can be applied to shorten the lifetime of radioisotopes.

One of the shortcomings of X-ray absorption spectroscopy was that absorption spectra of all the elements were not measurable using one beamline.

This shortcoming has been overcome in many beamlines at many synchrotron facilities and X-ray absorption spectroscopy has become a powerful tool for the analysis of practical samples.

REFERENCES

1. Furuya, Y., Manabe, S. and Kawai, J. *Adv. X-ray Chem. Anal. Japan* **33**, 345–362 (2002).
2. http://www.shimadzu.co.jp/
3. Piorek, S. Radioisotope-excited X-ray analysis, in *Handbook of X-ray Spectrometry*, 2nd Edition, revised and expanded (Eds R. Van Grieken and A. A. Markowicz), Dekker, New York, 2002, pp. 464–465.
4. Kawai, J. Absorption techniques in X-ray spectrometry, in *Encyclopedia of Analytical Chemistry* (Ed. R. A. Meyers), Wiley, Chichester, 2000, pp. 13288–13315.
5. Markowicz, A. A. X-Ray Physics, in *Handbook of X-ray Spectrometry*, 2nd Edition, Revised and Expanded (Eds R. Van Grieken and A. A. Markowicz), Dekker, New York, 2002, pp. 1–94.
6. Attwood, D. *Soft X-rays and Extreme Ultraviolet Radiation, Principles and Applications*, Cambridge University Press, Cambridge, 2000.
7. Henke, B. L., Gullikson, E. M. and Davis, J. C. Atomic Data Nucl. Data Tables **54**, 181–342 (1993).
8. Thompson, A., Lindau, I., Attwood, D., Pianetta, P., Gullikson, E., Robinson, A., Howells, M., Scofield, J., Kim, K.-J., Underwood, J., Kirz, J., Vaughan, D., Kortright, J., Williams, G. and Winick H. *X-Ray Data Booklet* Lawrence Berkeley National Laboratory, LBNL/PUB-490 Rev. 2, 2001; http://xdb.lbl.gov/.
9. http://www.esrf.fr/computing/expg/subgroups/theory/DABAX/dabax.html
10. Oyanagi, H., Ishii, M., Lee, C.-H., Saini, N. L., Kuwahara, Y., Saito, A., Hashimoto, H., Izumi, Y., Kiri, M., Jinno, M. and Ueno, Y. *Bull. Electrotech. Lab.* **62**, 407–418 (1998).
11. Lytle, F. W. *J. Synchrotron Rad.* **6**, 123–134 (1999).
12. Hasnain, S. S., Helliwell, J. R. and Kamitsubo, H. (Eds) *J. Synchrotron Rad.* **6**, 121–807 (1999).
13. Hasnain, S. S., Kamitsubo, H. and Mills, D. M. (Eds) *J. Synchrotron Rad.* **8**, 47–1014 (2001).
14. Swedish synchrotron radiation facility MAX lab (http://www.maxlab.lu.se/).
15. http://ixs.csrri.iit.edu/
16. Szalóki, I., Török, S. B., Ro, C.-U., Injuk, J. and Van Grieken, R. E. *Anal. Chem.* **72**, 211R–233R (2000).
17. Lee, P. A., Citrin, P. H., Eisenberger, P. and Kincaid, B. M. *Rev. Mod. Phys.* **53**, 769–806 (1981).
18. Rehr, J. J. and Albers, R. C. *Rev. Mod. Phys.* **72**, 621–654 (2000).
19. Bonnelle, C., *X-ray Spectroscopy, Annual Report C*, The Royal Society of Chemistry, London, 1987, pp. 201–272.
20. Meisel, A., Leonhardt, G. and Szargan, R. *X-Ray Spectra and Chemical Binding*, Springer-Verlag, Berlin, 1989.
21. Kondratenko, A. V. and Neiman, K. M. *Kvantovaya Khimiya i Spektroskopia Vysokovozbuzhdennykh Sostoyanii, Koordinatsionnye Soedineniya Perekhodnykh Metallov*, Izdatel'stvo NAUKA, Novosibirsk, 1990.
22. Stöhr, J. *NEXAFS Spectroscopy*, Springer-Verlag, Berlin, 1992.
23. Kurmaev, E. Z. *Crit. Rev. Solid State Mater. Sci.* **23**, 65–203 (1998).
24. De Groot, F. *Chem. Rev.* **101**, 1779–1808 (2001).
25. Nakai, I., Terada, Y., Nomura, M. and Uchida, T. *Physica B* **208&209**, 209–211 (1995).
26. Filipponi, A. and Di Cicco, A. *Nucl. Instrum. Methods Phys. Res. B* **93**, 302–310 (1994).
27. Landron, C., Hennet, L. and Thiaudiere, D. *Anal. Sci.* **17** (Suppl.), i165–i168 (2001).
28. Tamura, K., Inui, M. and Hosokawa, S. *Rev. Sci. Instrum.* **70**, 144–152 (1999).
29. Rennert, P. *Jpn. J. Appl. Phys.* **32** (Suppl. 32-2) 79–82 (1993).
30. Yokoyama, T., Kobayashi, K., Ohta, T. and Ugawa, A. *Phys. Rev. B* **53**, 6111–6122 (1996).
31. Miyanaga, T. and Fujikawa, T. *J. Phys. Soc. Jpn.* **67**, 2930–2937 (1998).
32. Hiraya, A., Nobusada, K., Simon, M., Okada, K., Tokushima, T., Senba, Y., Yoshida, H., Kamimori, K., Okumura, H., Shimizu, Y., Thomas, A.-L., Millie, P., Koyano, I. and Ueda, K. *Phys. Rev. A* **63**, 042705 (2001).
33. Tsuji, K. and Hirokawa, K. *Jpn. J. Appl. Phys.* **34**, L1506–L1508 (1995).
34. Tsuji, K., Hasegawa, Y., Wagatsuma, K. and Sakurai, T. *Jpn. J. Appl. Phys.* **37**, L1271–L1273 (1998).
35. Tsuji, K., Wagatsuma, K., Sugiyama, K., Hiraga, K. and Waseda, Y. *Surf. Interface Anal.* **27**, 132–135 (1999).
36. Ishii, M. *Physica B* **308–310**, 1153–1156 (2001).
37. Japanese Patent P2001-50911A (2001).
38. Ishii, M. *Phys. Rev. B* **65**, 085310 (2002).
39. Hayakawa, S., Hirose, T., Yan, L., Morishita, M., Kuwano, H. and Gohshi, Y. *X-Ray Spectrom.* **28**, 515–518 (1999).
40. Coulthard, I. and Sham, T. K. *Solid State Commun.* **110**, 203–208 (1999).
41. Zhang, P., Kim, P. S. and Sham, T. K. *J. Electron Spectrosc. Relat. Phenom.* **119**, 229–233 (2001).
42. Naftel, S. J., Yiu, Y. M., Sham, T. K. and Yates, B. W. *J. Electron Spectrosc. Relat. Phenom.* **119**, 215–220 (2001).
43. Sham, T. K., Jiang, D. T., Coulthard, I., Lorimer, J. W., Feng, X. H., Tan, K. H., Frigo, S. P., Rosenberg, R. A., Houghton, D. C. and Bryskiewicz, B. *Mater. Res. Soc. Symp. Proc.* **281**, 525–530 (1993).
44. Jaklevic, J., Kirby, J. A., Klein, M. P., Robertson, A. S., Brown, G. S. and Eisenberger, P. *Solid State Commun.* **23**, 679–682 (1977).
45. Muramatsu, Y., Grush, M. M. and Perera, R. C. C. *Phys. Rev. B* **61**, R2393 (2000).

46. Muramatsu, Y. *Adv. X-ray Chem. Anal. Japan* **32**, 147–159 (2001).
47. Izumi, Y., Kiyotaki, F., Nagamori, H. and Minato, T. *J. Electron Spectrosc. Relat. Phenom.* **119**, 193–199 (2001).
48. Loeffen, P. W., Pettifer, R. F., Müllender, S., van Veenendaal, M. A., Röhler, J. and Sivia, D. S. *Phys. Rev. B* **54**, 14877–14880 (1996).
49. Etelaniemi, V., Hamalainen, K., Manninen, S. and Suortti, P. *J. Phys.: Condens. Matter* **4**, 879–886 (1992).
50. Kotani, A. and Shin, S. *Rev. Mod. Phys.* **73**, 200–243 (2001).
51. Hu, T. D., Xie, Y. N., Qiao, S., Hai, Y., Jin, Y. L. and Xian, D. C. *Phys. Rev. B* **50**, 2216 (1994).
52. Sham, T. K. and Holroyd, *Phys. Rev. B* **39**, 8257–8262 (1989).
53. Sharpe, L. R., Heineman, W. R. and Elder, R. C. *Chem. Rev.* **90**, 705–722 (1990).
54. Yamaguchi, T., Mitsunaga, T., Yoshida, N., Wakita, H., Fujiwara, M., Matsushita, T., Ikeda, S. and Nomura, M. *Adv. X-ray Chem. Anal. Japan* **25**, 159–166 (1994).
55. Dewald, H. D., Watkins II, J. W., Elder, R. C. and Heineman, W. R. *Anal. Chem.* **58**, 2968–2975 (1986).
56. Nakai, I., Shiraishi, Y. and Nishikawa, F. *Spectrochim Acta B* **B54**, 143–149 (1999).
57. Terada, Y., Yasaka, K., Nishikawa, F., Konishi, T., Yoshio, M. and Nakai, I. *J. Solid State Chem.* **156**, 286–291 (2001).
58. Kimura, M., Kaneko, M. and Suzuki, T. *J. Synchrotron Rad.* **8**, 487–489 (2001).
59. Heald, S. M., Tranquada, J. M. and Chen, H. *J. Phys. (Paris)* **48**, C8-825-830 (1986).
60. Shirai, M., Asakura, K. and Iwasawa, Y. *Chem. Lett.* **15**, 247–254 (1992).
61. Kawai, J., Hayakawa, S., Kitajima, Y., Suzuki, S., Maeda, K., Urai, T., Adachi, H., Takami, M. and Gohshi, Y. *Proc. Japan Acad.* **69** (Ser. B), 179–184 (1993).
62. Kawai, J., Hayakawa, S., Kitajima, Y. and Gohshi, Y. *Spectrochim. Acta B* **54**, 215–222 (1999).
63. Oyanagi, H., Shioda, R., Kuwahara, Y. and Haga, K. *J. Synchrotron Rad.* **2**, 99–105 (1995).
64. Chun, W. J., Shirai, M., Tomishige, K., Saskura, K. and Iwasawa, Y. *J. Mol. Catal. A: Chem.* **107**, 55–65 (1996).
65. Shirai, M., Nomura, M., Asakura, K. and Iwasawa, Y. *Rev. Sci. Instrum.* **66**, 5493–5498 (1995).
66. Oyanagi, H. *J. Synchrotron Rad.* **5**, 48–53 (1998).
67. Chun, W.-J., Asakura, K. and Iwasawa, Y. *J. Synchrotron Rad.* **3**, 160–162 (1996).
68. Asakura, K., Chun, W.-J., Shirai, M., Tomishige, K. and Iwasawa, Y. *J. Phys. Chem.* **101**, 5549–5556 (1997).
69. Asakura, K. and Iijima, K. *J. Electron Spectrosc. Relat. Phenom.* **119**, 185–192 (2001).
70. Kosugi, N., Tokura, Y., Takagi, H. and Uchida, S. *Phys. Rev.* **41**, 131–137 (1990).
71. Bianconi, A., Della Longa, S., Li, C., Pompa, M., Congiu-Castellano, A., Udron, D., Flank, A. M. and Lagarde, P. *Phys. Rev. B* **44**, 10126–10138 (1991).
72. Harada, Y., Kinugasa, T., Eguchi, R., Matsubara, M., Kotani, A., Watanabe, M., Yagishita, A. and Shin, S. *Phys. Rev. B* **61**, 12854–12859 (2000).
73. d'Acapito, F., Mobilio, S., Cikmacs, P., Merlo, V. and Davoli, I. *Surf. Sci.* **468**, 77–84 (2000).
74. Nagoshi, M., Okude, N. and Kobayashi, K. *Surf. Interface Anal.* **30**, 472–474 (2000).
75. Watanabe, I., Tanida, H. and Kawauchi, S. *J. Am. Chem. Soc.* **119**, 12018–12019 (1997).
76. Tanida, H. Studies on halide ion solvation by photoelectron emission spectroscopy and extended X-ray absorption fine structure, Doctoral Thesis, Osaka University (1995).
77. Harada, M., Okada, T. and Watanabe, I. *Anal. Sci.* **17** (Suppl.), i1207–i1209 (2001).
78. Sakane, H. *Anal. Sci.* **17** (Suppl.), i131–i134 (2001).
79. Borthen, P. and Strehblow, H.-H. *Phys. Rev. B* **52**, 3017–3019 (1995).
80. Filatova, E. and Lukyanov, V. *J. Electron Spectrosc. Relat. Phenom.* **79**, 63–66 (1996).
81. Hecht, D., Borthen, P. and Strehblow, H.-H. *J. Electroanal. Chem.* **381**, 113–121 (1995).
82. Terada, S., Yokoyama, T., Sakano, M., Imanishi, A., Kitajima, Y., Kiguchi, M., Okamoto, Y. and Ohta, T. *Surf. Sci.* **414**, 107–117 (1998).
83. Pfalzer, P., Urbach, J.-P., Klemm, M., Horn, S., denBoer, M. L., Frenkel, A. I. and Kirkland, J. P. *Phys. Rev. B* **60**, 9335–9339 (1999).
84. Abbate, M., Goedkoop, J. B., de Groot, F. M F, Grioni, M., Fuggle, J. C., Hofmann, S., Petersen, H. and Sacchi, M. *Surf. Interface Anal.* **18**, 65–69 (1992).
85. Vogel, J. and Sacchi, M. *J. Electron Spectrosc. Relat. Phenom.* **67**, 181–188 (1994).
86. Schroeder, S. L. M., Moggridge, G. D., Ormerod, R. M., Rayment, T. and Lambert, R. M. *Surf. Sci.* **324**, L371–L377 (1995).
87. Erbil, A., Cargill III, G. S., Frahm, R. and Boehme, R. F. *Phys. Rev. B* **37**, 2450–2464 (1988).
88. Zheng, S. and Gohshi, Y. *Anal. Sci.* **13**, 997–1001 (1997).
89. Zheng, S. and Gohshi, Y. *Anal. Sci.* **13**, 1003–1005 (1997).
90. Hayakawa, S., Noda, J. and Gohshi, Y. *Spectrochim. Acta B* **54**, 235–239 (1999).
91. Zheng, S., Hayakawa, S. and Gohshi, Y. *J. Electron Spectrosc. Relat. Phenom.* **87**, 81–89 (1997).
92. Shimada, H., Matsubayashi, N., Imamura, M., Sato, T., Yoshimura, Y., Hayakawa, T., Toyoshima, A., Tanaka, K. and Nishijima, A. *Rev. Sci. Instrum.* **66**, 1780–1782 (1995).
93. Amemiya, K., Kondoh, H., Nambu, A., Iwasaki, M., Nakai, I., Yokoyama, T. and Ohta, T. *Jpn. J. Appl. Phys.* **40**, L718–L720 (2001).
94. Tanaka, T., Bando, K. K., Matsubayashi, N., Imamura, M., Shimada, H., Takahashi, K. and Katagiri, G. *Anal. Sci.* **17** (Suppl.), i1077–i1079 (2001).

95. Tanaka, T., Bando, K. K., Matsubayashi, N., Imamura, M. and Shimada, H. *J. Electron Spectrosc. Relat. Phenom.* **114–116**, 1077–1081 (2001).
96. Oyanagi, H., Kolobov, A. and Tanaka, K. *J. Synchrotron Rad.* **5**, 1001–1003 (1998).
97. Chen, G., Jain, H., Khalid, S., Li, J., Drabold, D. A. and Elliott, S. R. *Solid State Commun.* **120**, 149–153 (2001).
98. Nakamura, T., Shoji, H., Nanao, S., Iwazumi, T., Kishimoto, S. and Isozumi, Y. *Phys. Rev. B* **62**, 5301–5304 (2000).
99. Thole, B. T., Carra, P., Sette, F. and van der Laan, G. *Phys. Rev. Lett.* **68**, 1943–1946 (1992).
100. Van der Laan, G. *J. Phys. Soc. Jpn.* **63**, 2393–2400 (1994).
101. Ankudinov, A. and Rehr, J. J. *Phys. Rev. B* **51**, 1282–1285 (1995).
102. Tobin, J. G., Waddill, G. D., Jankowski, A. F., Sterne, P. A. and Pappas, D. P. *Phys. Rev. B* **52**, 6530–6541 (1995).
103. Gambardella, P., Dallmeyer, A., Maiti, K., Malagoli, M. C., Eberhardt, W., Kern, K. and Carbone, C. *Nature* **416**, 301–304 (2002).
104. Yokoyama, T., Amemiya, K., Yonamoto, Y., Matsumura, D. and Ohta, T. *J. Electron Spectrosc. Relat. Phenom.* **119**, 207–214 (2001).
105. Salbu, B., Krekling, T., Lind, O. C., Oughton, D. H., Drakopoulos, M., Simionovici, A., Snigireva, I., Snigirev, A., Weitkamp, T., Adams, F., Janssens, K. and Kashparov, V. A. *Nucl. Instrum. Methods Phys. Res. A* **A467–468**, 1249–1252 (2001).
106. Ade, H., Smith, A. P., Zhang, H., Zhuang, G. R., Kirz, J., Rightor, E. and Hichcock, A. *J. Electron Spectrosc. Relat. Phenom.* **84**, 53–72 (1997).
107. Urquhart, S. G., Hichcock, A. P., Smith, A. P., Ade, H. W., Lidy, W., Rightor, E. G. and Michell, G. E. *J. Electron Spectrosc. Relat. Phenom.* **100**, 119–135 (1999).
108. Ade, H. and Urquhart, S. G. NEXAFS spectroscopy and microscopy of natural and synthetic polymers, in *Chemical Applications of Synchrotron Radiation Part I*, (Ed. T. K. Sham, World Scientific, Singapore 2002, pp. 285–355.
109. Ice, G. E. *X-Ray Spectrom.* **26**, 315–326 (1997).
110. Ice, G. E. and Sparks, C. J. *Ann. Rev. Mater. Sci.* **29**, 25–52 (1999).
111. Larson, B. C., Yang, W., Ice, G. E., Budai, J. D. and Tischler, J. Z. *Nature* **415**, 887–890 (2002).
112. Cargill III, G. S. *Nature* **415**, 844–845 (2002).
113. Osán, J., Török, S. and Rindby, A. Environmental and biological applications of μ-XRF, in *Microscopic X-ray Fluorescence Analysis* (Eds K. H. A. Janssens, F. C. V. Adams and A. Rindby), Wiley, Chichester, 2000, pp. 315–346.
114. Hayakawa, S., Tohno, S., Takagawa, K., Hamamoto, A., Nishida, Y., Suzuki, M., Sato, Y. and Hirokawa, T. *Anal. Sci.* **17** (Suppl.), i115–i117 (2001).
115. Jones, K. W. Application in the geological sciences, in *Microscopic X-ray Fluorescence Analysis* (Eds K. H. A. Janssens, F. C. V. Adams and A. Rindby), Wiley, Chichester, 2000 pp. 247–289.
116. Hirakimoto, A. *Anal. Sci.* **17** (Suppl.), i123–i125 (2001).
117. Hirakimoto, A. Personal communication (2002).
118. Gurker, N., Nell, R., Seiler, G. and Wallner, J. *Rev. Sci. Instrum.* **70**, 2935–2949 (1999).
119. Bajt, S., Sutton, S. R. and Delaney, J. S. *Geochim. Cosmochim. Acta* **58**, 5209–5214 (1994).
120. Nakai, I., Numako, C., Hayakawa, S. and Tsuchiyama, A. *J. Trace Microprobe Tech.* **16**, 87–98 (1998).
121. Tonner, B. P., Droubay, T., Denlinger, J., Meyer-Ilse, W., Warwick, T., Rothe, J., Kneedler, E., Pecher, K., Nealson, K. and Grundl, T. *Surf. Interface Anal.* **27**, 247–258 (1999).
122. Matsunaga, M., Fukuda, K., Kato, Y. and Nakai, I. *Resource Geol.* **50**, 75–81 (2000).
123. Kawai, J., Takagawa, K., Fujisawa, S., Ektessabi, A. and Hayakawa, S. *J. Trace Microprobe Tech.* **19**, 541–546 (2001).
124. Smith, A. P., Laurer, J. H., Ade, A. W., Smith, S. D., Ashraf, A. and Spontak, R. J. *Macromolecules*, **30**, 663–666 (1997).
125. Smith, A. P., Ade, A. W., Balik, C. M., Koch, C. C., Smith, S. D. and Spontak, R. J. *Macromolecules* **33**, 2595–2604 (2000).
126. Zharnikov, M. and Neuber, M. *Surf. Sci.* **464**, 8–22 (2000).
127. Xiang, X.-D. *Annu. Rev. Mater. Sci.* **29**, 149–171 (1999).
128. Leapman, R. D., Grunes, L. A., Fejes, P. L. and Silcox, J. Extended core edge fine structure in electron energy loss spectra, in *EXAFS Spectroscopy: Techniques and Applications* (Eds B. K. Teo and D. C. Joy), Plenum, New York, 1981, pp. 217–239.
129. Tanaka, I., Kawai, J. and Adachi, H. *Phys. Rev. B* **52**, 11733–11739 (1995).
130. Wu, Z., Seifert, F., Poe, B. and Sharp, T. *J. Phys.: Condens. Matter* **8**, 3323–3336 (1996).
131. Leapman, R. D. and Cosslett, V. E. *J. Phys. D: Appl. Phys.* **9**, L29–L32 (1976).
132. Leapman, R. D., Grunes, L. A. and Fejes, P. L. *Phys. Rev. B* **26**, 614–635 (1982).
133. Guerlin, Th, Sauer, H., Engel, W. and Zeitler, E. *Phys. Status Solidi. A* **150**, 153–161 (1995).
134. Poe, B., Seifert, F. and Wu, Z. *Phys. Chem. Miner.* **24**, 477–487 (1997).
135. Leapman, R. D. and Rizzo, N. W. *Ultramicroscopy* **78**, 251–268 (1999).
136. Kawai, J., Hayashi, K. and Awakura, Y. *J. Phys. Soc. Jpn.* **66**, 3337–3340 (1997).
137. Kawai, J., Hayashi, K. and Maeda, K. *Adv. X-ray Anal.* **42**, 83–90 (2000).
138. Kawai, J. and Takahashi, H. *Surf. Interface Anal.* **31**, 114–117 (2001).
139. Takahashi, H., Harrowfield, I., MacRae, C., Wilson, N. and Tsutsumi, K. *Surf. Interface Anal.* **31**, 118–125 (2001).

140. Watanabe, T., Kawano, A., Ueda, K., Umesaki, N. and Wakita, H. *J. Synchrotron Rad.* **8**, 334–335 (2001).
141. Taguchi, T. *Adv. X-Ray Chem. Anal. Japan* **33**, 299–305 (2002).
142. Abrahams, I., Kövér, L., Tóth, J., Urch, D., Vrebos, B. and West, M. *J. Electron Spectrosc. Relat. Phenom.* **114–116**, 925–931 (2001).
143. Fujikawa, T. and Kawai, J. *J. Phys. Soc. Jpn.* **68**, 4032–4036 (1999).
144. Kawai, J. and Tohno, S. *J. Trace Microprobe Tech.* **19**, 497–507 (2001).
145. Straiger, H., Cross, J. O., Rehr, J. J. and Sorensen, L. B. *Phys. Rev. Lett.* **69**, 3064–3067 (1992).
146. Ravel, B., Bouldin, C. E., Renevier, H., Hodeau, J.-L. and Berar, J.-F. *Phys. Rev. B* **60**, 778–785 (1999).
147. Salem, S. I., Little, D. D., Kumar, A. and Lee, P. L. *Phys. Rev. A* **24**, 1935–1938 (1981).
148. Ito, Y., Mukoyama, T., Emura, S., Takahashi, M., Yoshikado, S. and Omote, K. *Phys. Rev. A* **51**, 303–308 (1995).
149. Magnuson, M., Wassdahl, N., Nilsson, A., Föhlisch, A., Nordgren, J. and Mårtensson, N. *Phys. Rev. B* **58**, 3677–3681 (1998).
150. Holland, B. W., Pettifer, R. F., Pendry, J. B. and Bordas, J. *J. Phys. C: Solid State Phys.* **11**, 633–642 (1978).
151. Ramaker, D. E. and O'Grady, W. E. *J. Synchrotron Rad.* **6**, 800–802 (1999).
152. Filipponi, A. and Di Cicco, A. *Phys. Rev. B* **53**, 9466–9467 (1996).
153. Tamura, E., van Ek, J., Froba, M. and Wong, J. *Phys. Rev. Lett.* **74**, 4899–4902 (1995).
154. Zabinsky, S. I., Rehr, J. J., Ankudinov, A., Albers, R. C, and Eller, M. J. *Phys. Rev. B* **52**, 2995–3009 (1995).
155. Jiang, T. and Ellis, D. E. *J. Mater. Res.* **11**, 2242–2256 (1996).
156. Best, P. E. *J. Chem. Phys.* **47**, 4002–4006 (1967).
157. LaVilla, E. E. *J. Chem. Phys.* **57**, 899–909 (1972).
158. Nakamatsu, H., Mukoyama, T. and Adachi, H. *Chem. Phys.* **143**, 221–226 (1990).
159. Nakamatsu, H., Mukoyama, T. and Adachi, H. *Chem. Phys.* **95**, 3167–3174 (1991).
160. Yamashita, S., Fujiwara, M., Kato, Y., Yamaguchi, T., Wakita, H. and Adachi, H. *Adv. Quantum Chem.* **29**, 357–371 (1997).
161. Matsuo, S., Sakaguchi, N., Obuchi, E., Nakano, K., Perera, R. C. C., Watanabe, T., Matsuo, T. and Wakita, H. *Anal. Chem.* **17**, 149–153 (2001).
162. Nakamatsu, H. *Chem. Phys.* **200**, 49–62 (1995).
163. Suzuki, C., Kawai, J., Adachi, H. and Mukoyama, T. *Chem. Phys.* **247**, 453–470 (1999).
164. Suzuki, C., Kawai, J., Tanizawa, J., Adachi, H., Kawasaki, S., Takano, M., and Mukoyama, T. *Chem. Phys.* **241**, 17–27 (1999).
165. Pearson, D. H., Ahn, C. C. and Fultz, B. *Phys. Rev. B* **47**, 8471–8478 (1993).
166. De Groot, F. M. F., Grioni, M., Fuggle, J. C., Ghijsen, J., Sawatzky, G. A. and Petersen, H. *Phys. Rev. B* **40**, 5715–5723 (1989).
167. De Groot, F. M. F. *J. Electron Spectrosc. Relat. Phenom.* **62**, 111–130 (1993).
168. Sugiura, C. and Yamasaki, H. *Jpn. J. Appl. Phys.* **32**, 1135–1141 (1993).
169. Kawai, J., Nakamura, E., Nihei, Y., Fujisawa, K. and Gohshi, Y. *Spectrochim. Acta.* **45B**, 463–479 (1990).
170. Török, I., Papp, T., Pálinkás, J., Budnar, M., Mühleisen, A., Kawai, J. and Campbell, J. L. *Nucl. Instrum. Methods Phys. Res. B* **114**, 9–14 (1996).
171. Dunn, J. H., Arvanitis, D., Baberschke, K., Hahkin, A., Karis, O., Carr, R. and Mårtensson, N. *J. Electron Spectrosc. Relat. Phenom.* **113**, 67–77 (2000).
172. Vaarkamp, M., Dring, I., Oldman, R. J., Stern, E. A. and Koningsberger, D. C. *Phys. Rev. B* **50**, 7872–7883 (1994).
173. Li, G. G., Bridges, F. and Booth, C. H. *Phys. Rev. B* **52**, 6332–6348 (1995).
174. Taguchi, T. and White, P. *Acta Cryst.* **B58**, 358–363 (2002).
175. Bouldin, C., Sims, J., Hung, H., Rehr, J. J. and Ankudinov, A. L. *X-Ray Spectrom.* **30**, 431–434 (2001).
176. Yamamoto, T., Tanaka, T., Matsuyama, T., Funabiki, T. and Yoshida, S. *Solid State Commun.* **111**, 137–142 (1999).
177. Hoffmann, D. P., Proctor, A., Fay, M. J. and Hercules, D. M. *Anal. Chem.* **61**, 1686–1693 (1989).
178. Mishra, A., Pandey, D., Dubey, R., Kekre, P. A. and Kumar, A. *X-ray Spectrom.* **29**, 161–165 (2000).
179. Katare, R. K., Joshi, S. K., Shrivastava, B. D., Pandeya, K. B. and Mishra, A. *X-ray Spectrom.* **29**, 187–191 (2000).
180. Emura, S. and Maeda, H. *Rev. Sci. Instrum.* **65**, 25–27 (1994).
181. Nomura, M. KEK-PF report.
182. Oyanagi, H., Sasaki, S., Hashimoto, H., Jinno, M. and Ueno, Y. *Bull. Electrotech. Lab.* **61**, 395–401 (1997).
183. RIGAKU catalogue.
184. TECHNOS catalogue.
185. Sakurai, K. *Adv. X-ray Anal.* **39**, 149–153 (1997).
186. Taguchi, T. *Anal. Sci.* **17** (Suppl.), i139–i141 (2001).
187. Tohji, K., Udagawa, Y., Kawasaki, T. and Masuda, K. *Rev. Sci. Instrum.* **54**, 1482–1487 (1983).
188. Tohji, K., Udagawa, Y., Kawasaki, T. and Mieno, K. *Rev. Sci. Instrum.* **59**, 1127–1131 (1988).
189. Kamijo, N., Kageyama, H., Kaeabata, K., Nishihagi, K., Uehara, Y. and Taniguchi, K. In-laboratory XAFS facility for low Z elements: application to the local structure study of amorphous alumina, in *X-Ray Absorption Fine Structure* (Ed. S. S. Hasnain), Ellis Horwood, Chichester, 1991, pp. 613–615.
190. Taguchi, T., Harada, J., Kiku, A., Tohji, K. and Shinoda, K. *J. Synchrotron Rad.* **8**, 363–365 (2001).
191. Matsuo, S. Correlation between local structures and chemical behaviors of some transition metal compounds as studied by X-ray absorption spectroscopy, Doctoral Thesis, Fukuoka University (2001).

192. Citrin, P. H. *Surf. Sci.* **299/300**, 199–218 (1994).
193. Yagi, S., Takenaka, S., Yokoyama, T., Kitajima, Y., Imanishi, A. and Ohta, T. *Surf. Sci.* **325**, 68–74 (1995).
194. Flank, A. M., Karnatak, R. C., Blancard, C., Esteva, J. M., Lagarde, P. and Connerade, J. P. *Z. Phys. D: At., Mol. Clusters* **21**, 357–366 (1991).
195. Grush, M. M., Horne, C. R., Perera, R. C. C., Ederer, D. L., Cramer, S. P., Cairns, E. J. and Callcott, T. A. *Chem. Mater.* **12**, 659–664 (2000).
196. Ohzono, H., Kouno, M., Miyake, H. and Ohyama, H. *Anal. Sci.* **17** (Suppl.), i135–i137 (2001).
197. Wang, H., Peng, G., Miller, L. M., Scheuring, E. M., George, S. J., Chance, M. R. and Cramer, S. P. *J. Am. Chem. Soc.* **119**, 4921–4928 (1997).
198. Yoshida, T., Tanaka, T., Yoshida, H., Funabiki, T., Yoshida, S. and Murata, T. *J. Phys. Chem.* **99**, 10890–10896 (1995).
199. Bando, K. K., Bihan, L. L., Yasuda, H., Sato, K., Tanaka, T., Dumeignil, D., Imamura, M., Matsubayashi, N. and Yoshimura, Y. *Anal. Sci.* **17** (Suppl.), i127–i130 (2001).
200. Ichikuni, N., Maruyama, H., Bando, K. K., Shimazu, S. and Uematsu, T. *Anal. Sci.* **17** (Suppl.), i1193–i1196 (2001).
201. Stohr, J., Samant, M. G. and Luning, J. *Synchrotron Rad. News* **14**(6), 32–38 (2001).
202. Fukuda, K., Matsunaga, M., Kato, Y. and Nakai, I. *J. Trace Microprobe Tech.* **19**, 509–519 (2001).
203. Kawai, J., Hayakawa, S., Esaka, F., Zheng, S., Kitajima, Y., Maeda, K., Adachi, H., Gohshi, Y. and Furuya, K. *Anal. Chem.* **34**, 1526–1529 (1995).
204. Yasoshima, M., Matsuo, M. and Takano, B. *Anal. Sci.* **17** (Suppl.), i1557–i1560 (2001).
205. Hirabayashi, M. and Matsuo, M. *Anal. Sci.* **17** (Suppl.), i1581–i1584 (2001).
206. Kawai, J., Mizutani, Y., Sugimura, T., Sai, M., Higuchi, T., Harada, Y., Ishiwata, Y., Fukushima, A., Fujisawa, M., Watanabe, M., Maeda, K., Shin, S. and Gohshi, Y. *Spectrochim. Acta B* **55**, 1385–1395 (2000).
207. Wong, J. *Nucl. Instrum. Methods Phys. Res.* **224**, 303–307 (1984).
208. Ohta, T. *J. Electron Spectrosc. Relat. Phenom.* **92**, 131–137 (1998).
209. Collins, C. B., Davanloo, F., Rusu, A. C., Iosif, M. C., Zoita, N. C., Camase, D. T., Hicks, J. M., Karamian, S. A., Ur, C. A., Popescu, I. I., Dussart, R., Pouvesle, J. M., Kirischuk, V. I., Strilchuk, N. V., McDaniel, P. and Crist, C. E. *Phys. Rev. C* **61**, 054305 (2000).
210. Ahmad, I., Dunford, R. W., Esbensen, H., Gemmell, D. S., Kanter, E. P., Rütt, U. and Southworth, S. H. *Phys. Rev. C* **61**, 051304 (2000).
211. Collins, C. B., Zoita, N. C., Rusu, A. C., Iosif, M. C., Camase, D. T., Davanloo, F., Emura, S., Uruga, T., Dussart, R., Pouvesle, J. M., Ur, C. A., Popescu, I. I., Kirischuk, V. I., Strilchuk, N. V. and Agee, F. J. *Europhys. Lett.* **57**, 677–682 (2002).

Chapter 6

New Computerisation Methods

6.1 Monte Carlo Simulation for X-ray Fluorescence Spectroscopy

L. VINCZE, K. JANSSENS, B. VEKEMANS and F. ADAMS
University of Antwerp, Antwerp, Belgium

6.1.1 INTRODUCTION

With the tremendous development of computer technology – virtually beyond recognition – in the past few decades, computer simulation has become a standard tool in solving particle transport problems, approaching in importance the traditional experimental and theoretical scientific methods. Computer simulation traces back its origin to the work of Neumann, Ulam and Metropolis in the late 1940s, when they applied a mathematical technique they called 'Monte Carlo analysis' to solve certain nuclear shielding problems which were either too expensive to solve experimentally or too complicated for analytical treatment (Metropolis, 1985). The feasibility of applying analytical techniques, such as the Boltzmann transport equation, for solving photon (or other particle) transport problems becomes very limited when concerning the description of real systems in which the transport phenomena occur. These realistic situations, possibly involving complicated geometries with complex boundaries, complex source distributions or different kinds of particles, most often cannot be represented with a mathematical model that can be treated by analytical techniques. In this respect, Monte Carlo (MC) simulation has become an attractive solution for complicated particle-transport problems.

A summary of the most important MC codes for photon-transport problems with a detailed comparison of the employed photon scattering models has been given by Fernández (Fernández *et al.*, 1993; Fernández, 1997). Important general purpose MC codes capable of treating electromagnetic radiation transport problems are EGS4 (Nelson *et al.*, 1985; Bielajew *et al.*, 1994), MCNP (Briesmeister, 1993), ITS (Halbleib, 1992) and GEANT4 (Apostolakis, 1999). Especially the low-energy extensions of EGS4 to its photoelectric and Compton/Rayleigh scattering and electron-treatment model make it an appropriate choice for simulating X-ray fluorescence (XRF) experiments (Namito *et al.*, 1998; Hirayama *et al.*, 2000; Namito and Hirayama, 1999, 2000). Similarly to dedicated MC models, developed to simulate XRF spectroscopy experiments (Vincze *et al.*, 1995; Vincze, 1995; Ao *et al.*, 1997a; Evans *et al.*, 1998), these extensions include K and L fluorescence from multielement samples, Compton and Rayleigh scattering of linearly polarized photons from bound (atomic) electrons as well as Doppler broadening in Compton scattering.

A MC simulation of the complete response of an energy dispersive (ED) XRF spectrometer is interesting from various points of view. With respect to quantification, even simple MC models, which only take the major fluorescent lines of each element into account and dispense with simulation of scattering phenomena altogether, are useful. For example, quantification can be achieved by establishing the X-ray intensities corresponding to a number of standard samples via MC simulation. The composition of the simulated standard can be chosen to be close to that of the unknown samples so that simple calibration relations can be employed. A more complete MC simulation of an EDXRF spectrometer also covers scattering of the primary radiation and includes second and higher order effects such as the enhancement of fluorescent lines by higher energy fluorescent or scattered radiation. As phenomena that contribute to the background of EDXRF spectra can be accounted for (e.g. low energy multiple scattering tails of the scatter peaks in the case of monochromatic excitation and the scattering of the primary spectrum in the case of polychromatic excitation), it is possible to 'predict' the complete spectral response of an EDXRF spectrometer (Vincze et al., 1999a, 1999b).

A significant advantage of the MC simulation based quantification scheme compared to other methods, such as fundamental parameter (FP) algorithms, is that the simulated spectrum can be compared directly to the experimental data in its entirety, taking into account not only the fluorescence line intensities, but also the scattered background of the XRF spectra. This is coupled with the fact that MC simulations are not limited to first or second order approximations and to ideal geometries.

In the past, MC models have been applied in an iterative manner to quantification of XRF data corresponding to homogeneous or simple heterogeneous samples in much the same way as other quantification algorithms based on the fundamental parameter approach. Relative deviations in the range of 2–15% have been achieved by the MC quantification scheme, depending on the analysed element and sample type (Vincze, 1995). Errors are mostly due to the uncertainties in the physical constants (cross-sections, fluorescence yields, transition probabilities, etc.) applied in the simulations.

Important applications of MC simulations include the optimization/characterization of *in vivo* XRF analysis systems in the field of medical science. Al-Ghorabie reported on the use of EGS4 to aid the design of a 90° geometry polarized XRF system for the measurement of cadmium concentration in deep body organs such as the kidney (Al-Ghorabie, 1999) and compared the performance of EGS4 and MCNP for modeling *in vivo* XRF systems (Al-Ghorabie et al., 2001).

Hugtenburg et al. have described the use of EGS4 to model the measurement of *cisplatin* uptake with *in vivo* X-ray fluorescence (Hugtenburg et al., 1998) while Lewis et al. used the same code to aid the design of a polarized source for *in vivo* XRF analysis (Lewis et al., 1998). The use of a dedicated XRF code CEARXRF for *in vivo* XRF was demonstrated by Lee et al. who studied the possibility of combining K and L XRF methods for *in vivo* bone lead measurements (Lee et al., 2001). The same code was used for the optimization of *in vivo* XRF analysis methods for bone lead for a ^{109}Cd source based K and for an X-ray tube based L XRF spectroscopy system (Ao et al., 1997b). O'Meara et al. used MC simulation models to improve the *in vivo* XRF measurement of renal mercury and uranium in bone (O'Meara et al., 1998, 2000).

While a number of models exist for solving photon-transport problems for both unpolarized and (linearly) polarized incident beams, their direct usability is often limited when dealing with the case of conventional (X-ray tube based) or synchrotron radiation XRF on multi-element samples due to the lack of a sufficiently detailed combination of photoelectric/fluorescence and scattering models for low X-ray energies. The computer model discussed in this subchapter is optimized specifically for calculating the results of synchrotron radiation induced X-ray fluorescence (SRXRF) experiments not only for homogeneous but also for heterogeneous samples. Figure 6.1.1 shows the generic experimental arrangement that can be simulated.

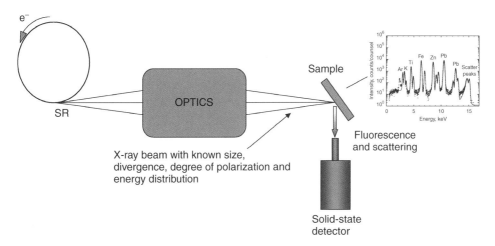

Figure 6.1.1 Generic experimental set-up for synchrotron radiation induced X-ray fluorescence studies for which the MC code by Vincze et al. (Vincze et al., 1995) is optimized

Detailed description of this code and validation for X-ray tube excited XRF setups is given in Vincze (1995), the case of monochromatic synchrotron excitation with primary photon energies below 20 keV (linearly polarized beam) is described in Vincze et al. (1995) and the case of high energy, linearly polarized primary beams (60–100 keV energy range) is discussed in Vincze (1999a,b).

In what follows, we will briefly describe the general framework of MC models which are capable of predicting the spectral response of EDXRF spectrometers using either conventional or synchrotron radiation sources. Next, applications of MC simulation are presented in the field of quantitative XRF analysis, calculation of multiple scattering contribution to XRF spectra as well as the simulation of XRF tomography experiments.

6.1.2 GENERAL FRAMEWORK OF THE MC SIMULATION FOR XRF SPECTROSCOPY

A MC code which models photon–matter interactions, simulates the fate of individual photons, from the point where they enter the volume of interest with a certain direction and energy to the stage where they are either absorbed by a sample atom or emerge from the sample and are optionally detected. For a typical XRF experiment, the simulation code must consider the three most important interaction types in the X-ray energy range of 1–100 keV, i.e. (i) photoelectric absorption followed by XRF or Auger electron emission, (ii) Rayleigh and (iii) Compton scattering. The simulation of these interactions allows the complete spectral response of materials subjected to X-ray excitation in the above energy range to be built up.

A simulated photon can be represented by five parameters: its energy (E), propagation vector **k**, degree of linear polarization p, the plane of polarization with respect to a given reference plane and its initial weight factor W. The latter corresponds to e.g. the intensity of the photon beam at the particular energy E in case of a polychromatic incident beam. As the photon undergoes a given interaction, some or all of these photon parameters change and these changes are simulated on the basis of the appropriate probability density functions (pdfs) derived from basic atomic data (cross-sections, fluorescence yields, emission rates) characterizing the photon–matter interaction processes. During the simulation, the photon is ray-traced within the volume of interest, that is, within the solid angle accepted by the detector.

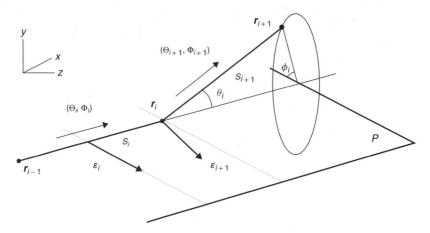

Figure 6.1.2 The elemental step of the simulation of the photon trajectory. S_i is the distance between two subsequent interactions occurring at r_{i-1} and r_i respectively. At r_i an interaction with polar angle θ_i and azimuth angle ϕ_i occurs, after which the angles Θ_{i+1} and Φ_{i+1} characterize the direction of propagation in the laboratory system and ε_{i+1} denotes the electric field vector of the photon after the ith interaction

The trajectory of each photon is modeled as consisting of a number of straight steps. The basic calculation step is illustrated in Figure 6.1.2.

6.1.2.1 SELECTION OF STEP LENGTH

A basic element of a simulation is the selection of distance between the subsequent interactions by generating the so-called step-length values in a way that the produced (pseudo) random sequence satisfies the well-known exponential attenuation law in a statistical sense. Suppose, we have a volume of interest including a heterogeneous sample and its surrounding medium, in which the different phases are separated by distances s_j in the direction of propagation of the photon (Figure 6.1.3). These s_j distances are calculated by the geometry module of the simulation after each interaction for the given sample topology (Vincze *et al.*, 1999c). The phases of the simulated volume of interest are characterized by their linear absorption coefficients $\{\mu_j\}_{j=1}^n$ at the particular photon energy E.

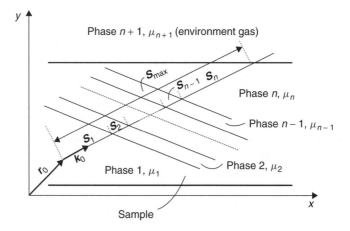

Figure 6.1.3 Illustration of the pathlength calculations in two dimensions. The steplength S between the subsequent interactions at a given photon energy is determined by the pathlength values s_i within the different sample phases

Once the path length values $\{s_j\}_{j=1}^n$ within the various phases are known, the (pseudo) random step length S_i, defining the location of the next interaction, is calculated as follows. First, the largest index m is found, for which the inequality holds:

$$\sum_{j=1}^{m} \mu_j s_j < -\ln(1 - RP_{\text{abs}}) \text{ with}$$

$$P_{\text{abs}} = 1 - \exp\left(-\sum_{j=1}^{n} \mu_j s_j\right) \quad (6.1.1a)$$

As a second step, S_i is calculated according to:

$$S_i = \sum_{j=1}^{m}\left(1 - \frac{\mu_j}{\mu_{m+1}}\right) s_j - \frac{1}{\mu_{m+1}} \ln(1 - RP_{\text{abs}})$$
(6.1.1b)

where R is a uniform random number. It is clear from the equation above, that for an infinitely thick homogeneous material ($P_{\text{abs}} = 1$) in which $\mu_j = \mu = $ constant, Equation (6.1.1b) reduces to the well-known choice of step-length:

$$S_i = -\frac{1}{\mu}\ln(1-R) \equiv -\lambda \ln R \quad (6.1.2)$$

where the λ term is the absorption length of the photon at energy E_i at the beginning of this step. Note that Equation (6.1.1b) does not allow the photon to escape from the volume of interest even in case of low-absorption material, which is an important improvement in terms of the efficiency of the model. After applying this basic variance reduction technique, preventing the loss of photons from the volume of interest, the weight fraction of the photon must be updated with the probability that the photon remains (i.e. absorbed) within the volume of interest:

$$W = W \times P_{\text{abs}} \quad (6.1.3)$$

The initial step ($i = 0$) starts from where the photon enters the volume of interest (i.e. the solid angle seen by the detector), along the original direction of propagation of the photon.

In the laboratory coordinate system XYZ, the direction of propagation is described by two angles (Θ_i, Φ_i). At the end of each step, coordinates $\vec{r}_i = (x_i, y_i, z_i)$, an interaction with a particular type of sample atom (with atomic number Z) occurs.

6.1.2.2 SELECTION OF ATOM TYPE

Once the location of the simulated interaction is determined from Equation (6.1.1a) and (6.1.1b), the next step is to choose the atom type for the subsequent interaction. In case N different atomic species constitute the sample material in this location, each present with a weight fraction of w_i and mass absorption coefficient μ_i^* an interaction with an atom of species $1 \leq k \leq N$ (atomic number Z_k) is chosen by means of a random number R so that:

$$\sum_{i=1}^{k-1} w_i m_i \leq R < \sum_{i=1}^{k} w_i m_i \text{ with } m_i = \frac{\mu_i^*}{\sum_{j=1}^{N} w_j \mu_j^*}$$
(6.1.4)

6.1.2.3 SELECTION OF INTERACTION TYPE

For photons with energy in the range 1 to 100 keV, three interaction types are important: the photoelectric effect, Rayleigh (elastic) and atomic Compton (inelastic) scattering. These interactions can take place respectively with probability $\tau_Z(E_i)/\mu_Z(E_i)$, $\sigma_{R,Z}(E_i)/\mu_Z(E_i)$ and $\sigma_{C,Z}(E_i)/\mu_Z(E_i)$ where $\mu_Z = \tau_Z + \sigma_{R,Z} + \sigma_{C,Z}$.

Accordingly, the interaction type for atomic number Z is chosen as follows:

$$\text{If} \begin{cases} 0 \leq R < \dfrac{\tau_Z}{\mu_Z} & \text{Photoelectric effect} \\[6pt] \dfrac{\tau_Z}{\mu_Z} \leq R < \dfrac{\tau_Z + \sigma_{R,Z}}{\mu_Z} & \text{Rayleigh scattering} \\[6pt] \dfrac{\tau_Z + \sigma_{R,Z}}{\mu_Z} \leq R < 1 & \text{Compton scattering} \end{cases}$$
(6.1.5)

Values of the photoelectric τ_Z, the Rayleigh $\sigma_{R,Z}$ and Compton $\sigma_{C,Z}$ scattering cross-sections can be calculated by using compiled data libraries, such as the Evaluated Photon Data Library, '97 Version (EPDL97) by Cullen, Hubbell and Kissel (Cullen et al., 1997).

Depending on the type of interaction, the energy and/or the direction of propagation of the photon is changed. The change in direction in the local coordinate system attached to the photon is described by a polar angle θ_i and azimuth angle ϕ_i, as shown in Figure 6.1.2.

The orientation in the laboratory system (Θ_{i+1}, Φ_{i+1}) of the next segment (if any) of the photon trajectory (from $\mathbf{r}_i$ to $\mathbf{r}_{i+1}$) can then be calculated and the coordinates of the next interaction point established:

$$x_{i+1} = x_i + S_i \sin \Theta_{i+1} \cos \Phi_{i+1}$$
$$y_{i+1} = y_i + S_i \sin \Theta_{i+1} \sin \Phi_{i+1} \quad (6.1.6)$$
$$z_{i+1} = z_i + S_i \cos \Theta_{i+1}$$

When the photon is not absorbed at $\mathbf{r}_{i+1}$ (Figure 6.1.2) and this location is still inside the volume of interest, a specific interaction is simulated (see below the simulation of various interaction types) after which the next segment of the trajectory is calculated. In case the photon has escaped from the sample and has a direction within the detector solid angle Ω_{Det}, the content of the appropriate channel (corresponding to the final photon energy) is incremented in the equivalent of a multichannel analyser (MCA) memory. In this way, the energy distribution of the photons just before they enter the detector crystal is obtained.

6.1.2.4 SIMULATION OF PHOTOELECTRIC EFFECT

This type of interaction causes the original photon to be absorbed. Depending on the relative magnitude of the shell-specific contributions to $\mu_Z(E_i)$, in a particular shell s (K, L_I, L_{II}, L_{III}, ...) a vacancy is created with the ejection of a photoelectron. In case of an L shell excitation, the finite probability of non-radiative transitions of electrons between L sub-shells (L_i, L_j) is accounted for by the so-called Coster–Kronig transition probabilities $f_{i,j}$.

A choice between fluorescence or Auger electron emission is made on the basis of the fluorescence yield $\omega_{Z,s}$ of the shell in question. Values for $\omega_{Z,s}$ and $f_{i,j}$ are taken from Krause (Krause, 1979). In the case of Auger electron production, the trajectory is terminated; for simplicity, it is assumed that the X-ray production due to Auger and photoelectrons is negligible compared to that by direct fluorescence. Alternatively, in the trajectory calculation, the photon path can be simply continued along a random direction (θ_i, ϕ_i).

The probability of a fluorescent line l originating from a shell s of atom Z can be written as:

$$P_{Z-sl} = \omega_{Z,s} \times \frac{\tau_{Z,s}}{\mu_Z} \times p_{Z-sl} \quad (6.1.7)$$

Accordingly, by means of a uniform random number R, a particular line l of a specific shell s of atom Z can be chosen which satisfies the condition:

$$\sum_{s=0}^{S} \sum_{l=0}^{\ell} P_{Z-sl} < R \leq \sum_{s=0}^{S} \sum_{l=0}^{\ell+1} P_{Z-sl} \quad (6.1.8)$$

If no (S, ℓ) combination can be found which satisfies Equation (6.1.8), the photon is considered to be absorbed (non-radiative relaxation of the atom) and the trajectory is terminated. In the case of radiative relaxation, since the characteristic radiation is emitted isotropically (i.e. $P(\theta, \theta + d\theta) = 1/2 \sin \theta d\theta$ and $P(\phi, \phi + d\phi) = d\phi/2\pi$), the angles θ and ϕ (Figure 6.1.4) can be selected using:

$$\cos \theta = 2R - 1$$
$$\phi = 2\pi R \quad (6.1.9)$$

The energy of the fluorescent photon is chosen on the basis of the relative emission rates P_{Z-sl} of the various characteristic lines l corresponding to shell s. Table 6.1.1 lists the various lines the program implemented by Vincze et al. takes into account (Vincze, 1995). A maximum of 40 fluorescent lines can be considered for each element.

As the fluorescent photons in subsequent steps are treated in exactly the same manner as the primary photons, higher order phenomena such as the scattering of fluorescent photons or enhancement effects are covered by this scheme.

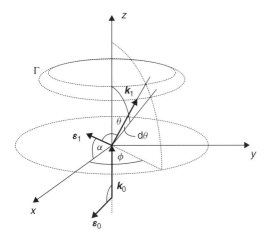

Figure 6.1.4 The coordinate system attached to the photon used to describe the scattering phenomena. The photon is traveling along the Z-axis before the interaction (propagation vector $\mathbf{k}_0$) and its net electric field vector $\boldsymbol{\varepsilon}_0$ is parallel with the X-axis. After the scattering, characterized by the scattering angle θ and azimuth angle ϕ (indices dropped for clarity), the propagation and net electric field vectors of the photon change to $\mathbf{k}_0$ and $\boldsymbol{\varepsilon}_0$, respectively. From Vincze et al. (1999b), reproduced by permission of Elsevier Science Ltd

Table 6.1.1 Simulated fluorescent lines in the MC model implemented by Vincze (1995)

Shell	Transition	Siegbahn notation
K	K-L$_3$, K-L$_2$	K$_{\alpha 1}$, K$_{\alpha 2}$
	K-M$_3$, K-N$_{2,3}$, K-M$_2$	K$_{\beta 1}$, K$_{\beta 2}$, K$_{\beta 3}$
L$_1$	L$_1$-M$_2$, ..., L$_1$-M$_5$	L$_{\beta 4}$, ..., L$_{\beta 9}$
	L$_1$-N$_2$, ..., L$_1$-N$_5$	L$_{\gamma 2}$, ..., L$_{\gamma 11}$
L$_2$	L$_2$-M$_1$, L$_2$-M$_4$	L$_\eta$, L$_{\beta 1}$
	L$_2$-N$_1$, L$_2$-N$_4$	L$_{\gamma 5}$, L$_{\gamma 1}$
	L$_2$-O$_1$, L$_2$-O$_4$	L$_{\gamma 8}$, L$_{\gamma 6}$
L$_3$	L$_3$-M$_1$, L$_3$-M$_4$, L$_3$-M$_5$	L$_{\alpha 3}$, L$_{\alpha 2}$, L$_{\alpha 1}$
	L$_3$-N$_1$, L$_3$-N$_4$, L$_3$-N$_5$	L$_{\beta 6}$, L$_{\beta 1,5}$, L$_{\beta 2}$
	L$_3$-O$_1$, L$_3$-O$_{4,5}$	L$_{\beta 7}$, L$_{\beta 5}$

6.1.2.5 SIMULATION OF SCATTERING INTERACTIONS

In the case where a scattering interaction (Compton or Rayleigh scattering) is selected by Equation (6.1.5), the change in direction of the photon is sampled on the basis of the appropriate differential scattering cross-sections $d\sigma/d\Omega$, which characterize the angular distribution of the scattered photons. A detailed description of the treatment of the scattering of linearly polarized X-rays is given by several authors (Namito, 1993; Vincze, 1995; Matt et al., 1996; Vincze et al., 1999b). The treatment of general beam polarization state, including elliptically polarized X-ray beams, is given by Fernández (Fernández, 1995a,b, 1996, 1997, 1998a, 1998b, 1999, 2000; Fernández et al., 1998).

In the following paragraphs, the approach adopted by Vincze et al. (Vincze et al., 1999b) for a linearly polarized photon beam is described, which can be easily implemented in MC calculations. The local coordinate system is chosen in such a way that the photon beam (having an initial propagation vector $\mathbf{k}_0$) travels along the Z-axis prior to the interaction and its (net) electric field vector $\boldsymbol{\varepsilon}_0$ is parallel with the X-axis (Figure 6.1.4). After the scattering event, the new direction of photon propagation in the attached coordinate system is characterized by the (unit) propagation vector $\mathbf{k}_1$ and the (net) electric field vector $\boldsymbol{\varepsilon}_1$.

In case of linearly polarized radiation having a degree of polarization p with respect to the reference plane XZ (Figure 6.1.4), the expressions for the Rayleigh ($d\sigma_R/d\Omega$) and Compton ($d\sigma_C/d\Omega$) differential scattering cross-sections are, respectively, given by:

$$\frac{d\sigma_R}{d\Omega}(\theta, \phi, E) = \frac{d\sigma_T}{d\Omega}(\theta, \phi) F^2(x, Z)$$
$$= \frac{r_e^2}{2}[2 - \sin^2\theta(1 - p \quad (6.1.10)$$
$$+ 2p\cos^2\phi)]F^2(x, Z)$$

$$\frac{d\sigma_C}{d\Omega}(\theta, \phi, E) = \frac{d\sigma_{KN}}{d\Omega}(\theta, \phi, E) S(x, Z)$$
$$= \frac{r_e^2}{2}\left(\frac{E}{E_0}\right)^2\left[\frac{E}{E_0} + \frac{E_0}{E} - \sin^2\theta \right.$$
$$\left. \times (1 - p + 2p\cos^2\phi)\right]S(x, Z)$$
$$(6.1.11)$$

where $d\sigma_T/d\Omega$ denotes the Thomson and $d\sigma_{KN}/d\Omega$ the Klein–Nishina differential scattering cross-sections (Hanson, 1986). $F(x, Z)$ and $S(x, Z)$ are the atomic form factor and the incoherent scattering function, respectively (Hubbell et al., 1975), for an element with atomic number Z, $x(\text{Å}^{-1}) = \sin(\theta/2)E$ (keV)/12.39 is the momentum transfer of the photon and r_e is the classical electron radius.

The probability that a photon is scattered within a finite solid angle characterized by the angles $(\theta, \theta + d\theta; \phi, \phi + d\phi)$ is calculated by the following equation:

$$f(\theta, \phi, E)\,d\theta\,d\phi = \frac{1}{\sigma(E)} \frac{d\sigma}{d\Omega}(\theta, \phi, E) \sin\theta\,d\theta\,d\phi \quad (6.1.12)$$

where $f(\theta, \phi, E)$ is the probability density function of this process at energy E. The integration of Equation (6.1.12) over ϕ in the interval $[0, 2\pi]$ gives the probability that the scattered photon is found in a spherical section Γ characterized by $(\theta, \theta + d\theta)$ (Figure 6.1.4):

$$\begin{aligned} f(\theta, E)\,d\theta &= \frac{1}{\sigma(E)} \left(\int_0^{2\pi} \frac{d\sigma}{d\Omega}(\theta, \phi, E)\,d\phi \right) \sin\theta\,d\theta \\ &= \frac{2\pi}{\sigma(E)} \left(\frac{d\sigma}{d\Omega} \right)_U (\theta, E) \sin\theta\,d\theta. \end{aligned} \quad (6.1.13)$$

As a result of the integration over ϕ, the polarization dependence drops out (Equation 6.1.5), which means that the probability of scattering into the region characterized by $(\theta, \theta + d\theta)$ is independent from the degree of linear polarization p. As a consequence of this, the scattering angle θ can be sampled on the basis of the differential scattering cross-sections for unpolarized radiation.

By defining the cumulative distribution function $F(\theta)$ for the scattering angle at a fixed photon energy E as:

$$F(\theta, E) = R = \int_0^\theta f(t, E)\,dt \quad (6.1.14)$$

the scattering angle can be numerically sampled by $\theta = F^{-1}(R, E)$, where R is a uniformly distributed random number in the interval $[0,1]$. (See Vincze 1995) for the details of the selection of θ for both Rayleigh and Compton scattering for the unpolarized case. In practice, the $F^{-1}(R, E)$ inverse cumulative distribution function is pre-calculated for a well-defined energy-random number grid for all elements in the atomic number range of $1 \leq Z \leq 92$. An example of such pre-calculated 'scattering surface' is shown for Rayleigh scattering on Fe in Figure 6.1.5. The

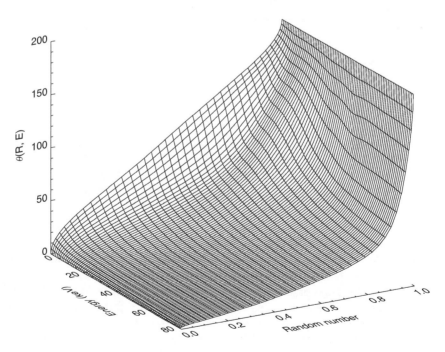

Figure 6.1.5 Pre-calculated scattering (polar) angle $\theta(R, E)$ surface in the energy range of 0 to 80 keV for coherent scattering by an Fe atom

scattering angle θ can be chosen by means of a simple bilinear interpolation scheme for a particular (R, E) combination.

Once the scattering angle θ is generated from the appropriate cumulative distribution functions for unpolarized radiation, the azimuth angle ϕ, which has a non-uniform distribution in the linearly polarized case, can be calculated as follows. At a given scattering angle θ, the probability P' that the azimuth angle falls within the interval $(\phi, \phi + d\phi)$ is given by:

$$P'(\phi, \phi + d\phi) = \frac{1}{2\pi} \frac{A - \sin^2\theta(1 - p + 2p\cos^2\theta)}{A - \sin^2\theta} d\phi \quad (6.1.15)$$

where $A = 2$ for Rayleigh and $A = E/E_0 + E_0/E$ for Compton scattering. It can easily be seen that the special case of $p = 0$ (unpolarized beam) yields a uniform distribution which is independent of ϕ:

$$P'(\phi, \phi + d\phi) = \frac{1}{2\pi} d\phi. \quad (6.1.16)$$

The term A for Compton scattering can be derived from the well-known expression for the Compton energy. The energy E' after a Compton scattering event of a photon with energy E_0 (without considering Doppler broadening) is determined by the scattering angle θ as:

$$E' = E_0 \left[1 + \frac{E_0}{m_e c^2}(1 - \cos\theta) \right]^{-1} \quad (6.1.17)$$

where m_e is the rest mass of the electron and c is the speed of light.

Using Equation (6.1.15), the cumulative distribution function for the azimuth angle at a fixed scattering angle θ can be written as:

$$F(\phi) = R = \frac{1}{2\pi} \left[\phi - \frac{p \sin^2\theta}{A - \sin^2\theta} \sin(2\phi) \right] \quad (6.1.18)$$

The azimuth angle for a particular random number R can then be selected by the numerical solution of $\phi = F^{-1}(R)$.

The modeling of a scattering event at a given photon energy E_0 can be summarized as follows. First, the scattering angle θ is sampled on the basis of the angular distribution $f(\theta, E)$ described by Equation (6.1.13). The corresponding inverse cumulative distribution function (shown in Figure 6.1.5 for Fe as an example) can be pre-calculated for both Compton and Rayleigh scattering events for each element present in the sample in order to accelerate the simulation by reducing on-line calculations.

After the scattering angle is selected, the corresponding energy of the scattered photon E_1 is determined as follows. For Rayleigh (elastic) scattering during which the energy of the scattered photon is preserved $E_1 = E_0$ is chosen; for Compton scattering Equation (6.1.17) is used to determine the energy of the inelastically scattered photon.

In the case of Compton scattering, so far no Doppler has been included in the calculations. This effect is caused by the non-zero momentum of the scattering atomic electrons. In order to take this effect into account, the energy E_1 after Compton scattering is chosen according to:

$$E_1 = E_0 \left(\frac{E_0}{E'} - \frac{2p_z}{m_e c} \sin\frac{\theta}{2} \right)^{-1} \quad (6.1.19)$$

where E_0 is the photon's incident energy, E' is the energy calculated by Equation (6.1.17) without Doppler broadening and E_1 is the final energy of the scattered photon after considering the Doppler effect.

The second term in the above equation refers to the influence of the momentum p_z of the scattering electron on the energy-transfer during the Compton scattering, which can be calculated according to:

$$p_z = q \frac{m_e e^2}{2\varepsilon_0 h} \quad (6.1.20)$$

where q is the reduced momentum of the scattering electron, ε_0 the vacuum permittivity, e the electron charge and h Planck's constant. Biggs et al. (Biggs et al., 1975) provide numerical values for the probability density function $f_Z(q)$ for every atomic number Z. Whenever necessary, a suitable q value is chosen based on the distribution $f_Z(q)$ by means of a uniform random number. (See Vincze (1995) for a detailed description of the method.)

Once the polar (scattering) angle θ and azimuth angle ϕ which define the new direction of propagation of the photon, are determined, the degree of polarization corresponding to this direction can be obtained as:

$$p(\theta,\phi) = \frac{\sqrt{[p_0\cos 2\phi(\cos^2\theta+1)-\sin^2\theta]^2 + 4p_0^2\sin^2 2\phi\cos^2\theta}}{A-\sin^2\theta(1+p_0\cos 2\phi)} \quad (6.1.21)$$

where p_0 is the degree of linear polarization before scattering.

The net electric field vector corresponding to the new direction, defined by (θ, ϕ) in the local coordinate system is given by:

$$\varepsilon_1 = \frac{1}{\sqrt{1-\sin^2\theta\cos^2\phi}} \begin{pmatrix} 1-\sin^2\theta\cos^2\phi \\ -\sin^2\theta\sin\phi\cos\phi \\ -\sin\theta\cos\theta\cos\phi \end{pmatrix} \quad (6.1.22)$$

Using Equations (6.1.10)–(6.1.22) we obtain all the necessary parameters which describe the photon after a particular scattering event: θ and ϕ, defining the direction of propagation after the event, the energy E_1 as well as the degree of linear polarization p and polarization plane of the photon.

6.1.2.6 VARIANCE REDUCTION

The general concept of variance reduction may be found in various standard works. A detailed discussion on the theoretical background of variance reduction techniques in particle transport problems has been given recently by Milgram (2001). In order to escape a formidable or even impractical amount of simulation work, it is profitable to reformulate the original problem in such a way that the statistical uncertainty in the answers is reduced. Fishman (1996) describes variance reduction in the following terms: 'von Neumann, Ulam and others recognized that one could modify the standard Monte Carlo method in a way that produced a solution to the original problem with specified error bound at considerably less cost, in terms of computing time, than directly generating the random tour that corresponded to the original problem', and then goes on to equate the concept with the term 'efficiency-improving technique', or variance reduction.

In case of our present XRF MC model, the optimized code essentially follows the basic random walk of photons within the simulated sample, including the simulation of higher-order phenomena. However, at each interaction point $\mathbf{r}_i$, the probability of every possible pathway for that photon to travel from $\mathbf{r}_i$ to a point on the detector surface (assuming no other interactions along the way) is calculated. This point is selected randomly on the detector surface. Each pathway is defined to consist of (i) an interaction process during which the direction of propagation (and energy) of the photon is altered so that it travels towards the detector and (ii) the path the photon must travel to finally reach the detector. Scatter-type interactions simply involve a scattering interaction with a sample atom; fluorescent-type of conversion consists of a photo-ionization followed by the emission of a fluorescent photon in the appropriate direction.

In general, the probability $P_{c,Z}$ of a pathway resulting in detection which involves an interaction with a sample atom of type Z and a conversion of type c can be written as:

$$P_{c,Z} = P_{c,Z}^{(\text{conv})} \times P_{c,Z}^{(\text{dir})} \times P_{c,Z}^{(\text{esc})} \quad (6.1.23)$$

where $P_{c,Z}^{(\text{conv})}$ denotes the probability for the particular conversion (interaction) process to occur, $P_{c,Z}^{(\text{dir})}$ the probability of the photon to change its direction over the appropriate angles (θ_i, ϕ_i) within the detector solid angle and $P_{c,Z}^{(\text{esc})}$ the probability to escape from the sample when it is traveling in this direction.

During the simulation of a photon history, prior to simulating the next interaction on the trajectory, the probabilities of all possible pathways (and the corresponding energies) leading to detection are calculated, as explained above. These pathways with probability $P_{c,Z}$ can be thought of as 'fractional photons' with a final weight fraction equal to $W \times P_{c,Z}$ which travel towards the detector.

The arrival of each fractional photon is recorded by adding the weight $P_{c,z}$ to the content of the appropriate channel of the MCA memory. Only then, the next step of the current photon's trajectory is calculated.

6.1.2.7 DETECTOR RESPONSE FUNCTION

In order to compare the simulation results with experimental data, the simulated distributions need to be convoluted with a so-called detector response function. The model proposed by He *et al.* (1990) was used for this purpose, with values of the empirical parameters appropriate for the detector that was employed. Next to Gaussian peak broadening, the detector response function considers spectral artifacts such as escape peaks, short- and long-term exponential tails of the Gaussian peaks and flat continuum from zero to the full peak energy. An additional artifact being considered is the low-energy Compton escape contribution, which is clearly visible in case of detected X-rays in the 60–100 keV energy range.

6.1.3 SIMULATION VERSUS EXPERIMENT

A very important step in the development of any MC simulation is its validation, which can be done either by comparison of results from various simulation codes (e.g. Al-Ghorabie *et al.*, 2001) or comparing simulated data directly to experimental results (Fajardo *et al.*, 1998; Vincze *et al.*, 1999b). The latter validation method is discussed in this section.

A recently established experimental set-up, called ID18F, at the European Synchrotron Radiation Facility (ESRF) for microscopic XRF is shown in Figure 6.1.6 (Somogyi *et al.*, 2001). At this instrument the above-discussed MC simulation coupled with a sophisticated spectral deconvolution software provides a no-compromise, general solution for the XRF quantification problem (Vekemans *et al.*, 1995, 2002).

Two examples of simulated and experimental XRF spectra measured at this beamline are shown in Figure 6.1.7 for a NIST SRM1832 thin glass calibration standard and for a NIST SRM1577b Bovine Liver biological reference material. The spectral distributions correspond to monochromatic excitation by a 21 keV microbeam of $3 \times 14 \mu m^2$. Relative deviations between simulated and experimental fluorescent line intensities of 2–4 % for the glass standard and 1–12 % for the more heterogeneous biological standard have been achieved (Vekemans *et al.*, 2002), which illustrates the high potential of the method as an XRF quantification tool.

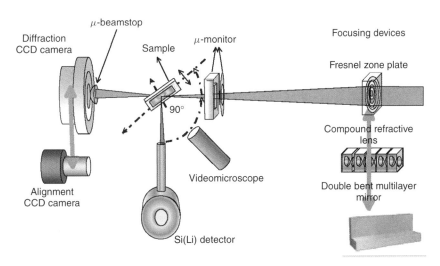

Figure 6.1.6 Schematic illustration of the micro XRF end-station ID18F at the ESRF. Reproduced by permission of John Wiley & Sons from Somogyi *et al.*, 2001

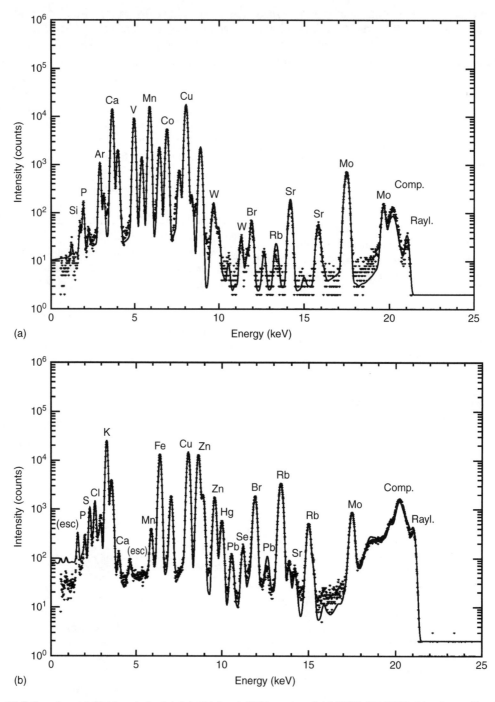

Figure 6.1.7 Experimental (dots) and simulated (solid lines) XRF spectra of (a) NIST SRM1832 thin glass calibration and (b) NIST SRM 1577b Bovine Liver standard, corresponding to 21 keV excitation at the ESRF ID18F end-station

For higher incident energies, typical simulated spectral distributions for Al (4 mm thickness) and Cu (10 mm thickness) samples, irradiated with a high-energy monochromatic beam of linearly polarized X-rays, are shown in Figure 6.1.8. These spectral distributions were collected at the high-energy scattering beamline BW5 of HASYLAB (Hamburg, Germany) (Schneider, 1995). Here the energy distribution of the scattered synchrotron beam is compared with the simulated equivalent in the case of the two sample matrices. The energy of the incident beam is 80 keV and the degree of linear polarization is estimated to be 90–91 % (Vincze, 1999b). The figure also shows the individually calculated contributions of the various scattering orders, i.e. the spectral components which are generated by single, double and higher order multiple scattering. The calculation was done up to the 6th scattering order for Al and the 4th for Cu, above which the contribution of multiple scattering is negligible to the total scattered distribution.

By summing up these scattering orders one obtains the full spectral response of the irradiated sample which can be compared directly with experimental data. The residual values (Figure 6.1.9) derived from the simulated and experimental data show deviations which are in general below the statistical uncertainties of the experimental spectrum in the multiple Compton scatter region of 45–65 keV. In Figure 6.1.9, the statistical uncertainties are indicated as dashed curves, corresponding to 3σ confidence limits. In these figures, σ is the relative standard deviation per channel content.

In the region of the primary/multiple Compton scattering the agreement between simulation and experiment is considered to be satisfactory in view of the uncertainties of the individual physical constants involved in the calculations, the uncertainties of the degree of polarization of the primary beam and that of the angles involved in the excitation/detection geometry.

In Figure 6.1.8(b), next to the contribution of Rayleigh and Compton scattering, the Cu K lines can also be observed at 8.05 and 8.90 keV. The ratio of these fluorescent lines to the scatter peaks is correctly calculated by the simulation code indicating that the assumed degree of linear polarization ($p = 0.91$) is a good estimate in the case of the primary beam available at beamline BW5 of HASYLAB. It is interesting to note that the elevated spectral background in the energy region of 5–20 keV around the Cu Kα and Cu Kβ fluorescent lines is also correctly modeled by the simulation code. This background contribution is mostly attributed to the effect of high energy photons escaping from the detector after Compton scattering within the Ge crystal, depositing part of their energy during the inelastic scattering process. As estimated from simulation studies by the current MC code, the magnitude of such Compton escape for a high-purity Ge (HPGe) detector is about 2–3 % at incident X-ray energies of 80 keV.

6.1.3.1 VARIATION OF THE DETECTION GEOMETRY

Figure 6.1.10 illustrates the change in the spectral distribution of the scattered radiation as the angle of detection (and therefore also the scattering angle) is varied. The experimental and simulated scattered distributions correspond to scattering angles of 112°, 90° and 62°, obtained from a 2 mm thick polypropylene disk irradiated by an 80 keV incident beam. In the case of the simulated distributions, the higher order scatter contributions are also shown up to the 6th interaction order. Note, that in the case of the simulation an arbitrary number of subsequent interactions can be modeled, however, beyond the 5–6th order the final result does not change in a significant way. Next to the evident change in energy of the primary Compton peak (65.8, 69.2 and 73.9 keV corresponding to scattering angles of 112°, 90° and 62°, respectively), the structure of the multiply scattered Compton continuum also clearly changes with the scattering angle. These changes are correctly taken into account by the simulation. Especially the first- and second-order scatter distributions are influenced by the observation angle while this dependence gradually diminishes for the higher orders. This is understandable, as the multiply scattered radiation field becomes more and more isotropic.

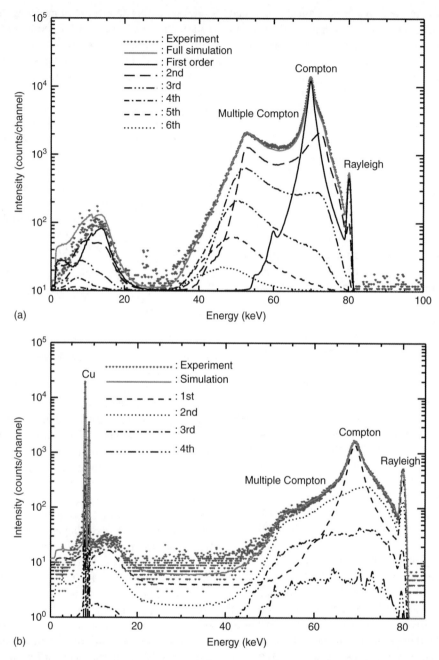

Figure 6.1.8 Comparison of experimental (dots) and simulated (solid curves) scattered distributions from (a) an Al disk with a thickness of 4 mm and a diameter of 25 mm and (b) a Cu disk with a thickness of 10 mm. The experiment was done using an 80 keV incident beam with a degree of linear polarization of about 0.91. Next to the full simulations, the various scattering orders are also shown up to the 6th order. Reproduced by permission of Elsevier Science Ltd from Vincze et al. (1999b)

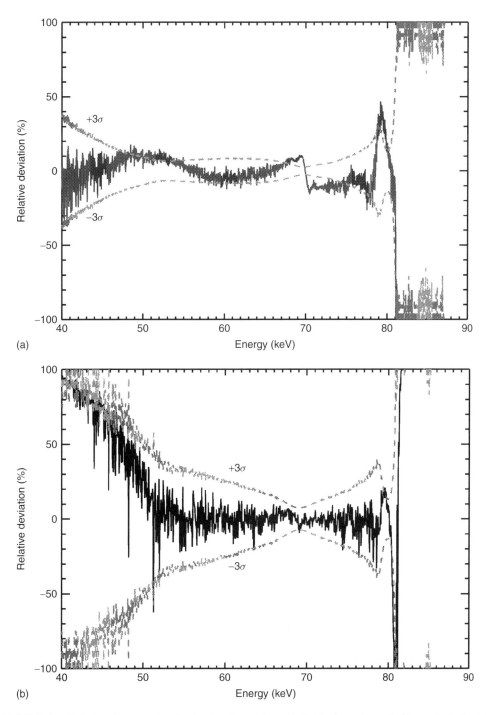

Figure 6.1.9 Relative deviations between the simulated and experimental distributions shown in Figure 6.1.8 for (a) Al and (b) Cu. Here, $\sigma = 1/\sqrt{N}$ denotes the relative deviation corresponding to the statistical uncertainty of the measured channel content N. Reproduced by permission of Elsevier Science Ltd from Vincze et al. (1999b)

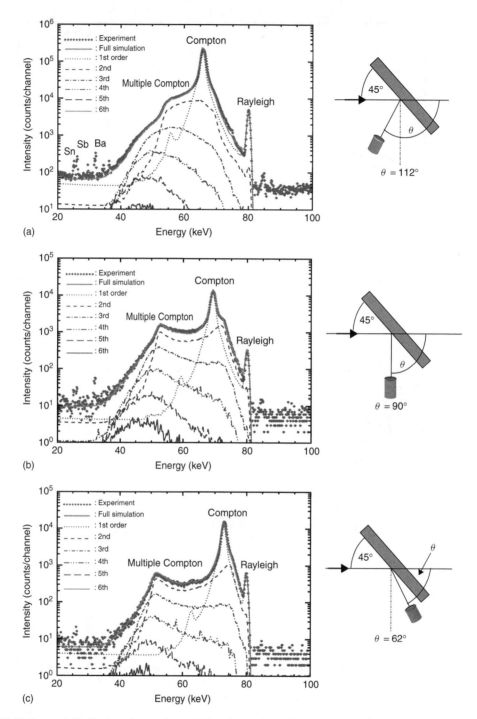

Figure 6.1.10 Scattered distributions from a 2 mm thick polypropylene disk measured using various detector positions. The position of the detector determines the scattering angle θ to be (a) 112°, (b) 90° and (c) 62°. The experimental data are indicated by the dots and the simulations are shown by the solid line. Next to the full simulations, the various multiple scattering orders are shown. Reproduced by permission of Elsevier Science Ltd from Vincze *et al.* (1999b)

6.1.3.2 ESTIMATION OF DEGREE OF LINEAR POLARIZATION

By means of the simulation model, it is possible to estimate the degree of linear polarization of the primary beam from the ratio of known fluorescent line intensities to the primary and multiple Compton intensity. The fluorescence to Compton scatter ratio for given and fixed experimental conditions (fixed incident beam divergence, detector solid angle) and sample is determined by the degree of linear polarization of the incoming beam, and this ratio is very sensitive to even slight changes of p under 90° scattering in the plane of polarization.

Figure 6.1.11 shows simulated XRF spectra, calculated for different linear polarization values for a 420 μm thick polypropylene standard having known trace-element composition (Ca, Fe, Zn). In the simulations, instrumental conditions of the ID18F beam line of the ESRF were assumed. The experimentally obtained spectral distribution at this instrument is also indicated on this graph (dots), corresponding to an excitation energy of 21.1 keV.

From the comparison of spectra normalized to the Zn Kα line, it is clear that the low intensity of the Compton scatter peak obtained experimentally requires a degree of linear polarization which is higher than 99%. The effective degree of linear polarization of the primary beam available at ID18F of the ESRF is estimated to be about 99.7% by this simple method, which is expected for an instrument installed at a high-brilliance third generation SR source.

A major source of uncertainty of this simple method, next to uncertainties in the physical constants used in the simulation, is the possible error of the concentration of the employed reference line (Zn Kα in this case) in the test sample. Next to fluorescence/Compton ratio, the multiple-to-primary Compton ratio can also be used for the estimation of the degree of linear polarization for energetic monochromatic synchrotron beams (Vincze *et al.*, 1999b). Recently, They *et al.* have described a polarimeter for synchrotron photon beams based on the measurement of the angular distribution of Compton intensities (They *et al.*, 2001).

6.1.3.3 PREDICTION OF XRF DETECTION LIMITS FOR RARE EARTH ELEMENTS

As an application of the simulation code, the improvement of detection limits (DLs) is calculated for K-line based XRF analysis of rare earth

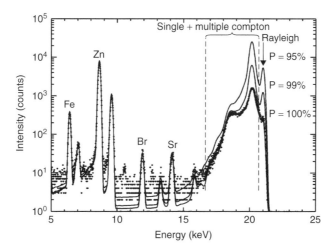

Figure 6.1.11 Calculated spectral distributions corresponding to the spectral response of a polypropylene standard of 420 μm thickness for various degree of linear polarization of the incident beam. The experimentally recorded XRF spectrum, measured at the ID18F end-station of the ESRF is shown by the dotted curve. The latter corresponds to a degree of linear polarization of $p > 99.5\%$

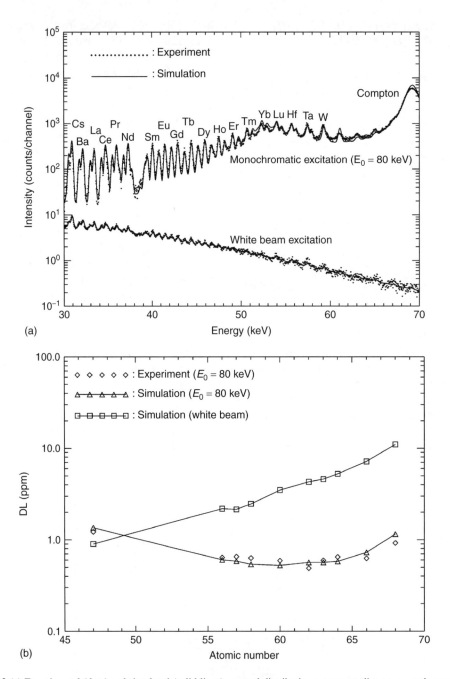

Figure 6.1.12 (a) Experimental (dots) and simulated (solid lines) spectral distributions corresponding to monochromatic (80 keV) and white beam excitation measured from a 3 mm thick NIST SRM 612 glass standard indicating the energy region of REE K fluorescence lines. The simulated and experimental spectra corresponding to the white excitation are scaled down by a factor of 100 for visual clarity (b) Detection limits corresponding to mono (80 keV) and polychromatic (bending magnet, white beam) excitation modes as calculated by the simulation code. The experimentally determined DLs corresponding to the monochromatic primary beam are also shown. Reproduced by permission of Elsevier Science Ltd from Vincze et al. (1999b)

elements (REEs) when an 80 keV monochromatic excitation is employed instead of white incident radiation, such as available at the SRXRF spectrometer installed at Beamline L, HASYLAB (Lechtenberg, 1996). The calculation of DL values requires the reliable estimation of both fluorescence line intensities of interest and the background level which, at the high X-ray energies in question, are determined mainly by single and multiple Compton scattering. In the case of monochromatic excitation tuned above the absorption edges of the examined elements, the fluorescence lines of interest are separated in energy from the primary Compton and Rayleigh scatter peaks. Even though at the incident energy of 80 keV the multiple Compton scattering region (35–65 keV) overlaps with the characteristic lines of REE, there is still a significant improvement in terms of peak-to-background ratios when compared to the white beam excitation. In the case of the latter, the fluorescence lines are superimposed on the scattered primary and multiple Compton/Rayleigh continuum without any energy separation.

In the following calculations, the estimation of DL values for the elements in the atomic number range of 47–68 is based on simulated spectra from the glass calibration standard NIST SRM 612 (3 mm thickness) corresponding to the above excitation modes. In Figure 6.1.12(a) simulated and experimental SRXRF spectra of this standard are compared for an 80 keV incident beam and for white beam excitation.

For the simulation of the monochromatic case the parameters of beamline BW5 and for the polychromatic situation the parameters of the bending magnet source of Beamline L of HASYLAB were used. For clarity, the spectrum corresponding to this source was scaled down by a factor of 100. In Figure 6.1.12(b), experimentally determined and calculated DL values for both modes of excitation are compared. All simulated spectra used in these calculations were normalized to an integrated intensity of 1.10×10^6 which was recorded experimentally at Beamline BW5 using a live time of 1000 s, corresponding to a count rate of 1100 s^{-1}. In both simulated cases (i.e. mono- and polychromatic excitation) a degree of polarization of 91 % was assumed.

Overall, a very good agreement is found between experiment and theory. As shown in Figure 6.1.12(b), the DLs in case of the monochromatic excitation are in the range of 0.5 to 1 ppm for the elements Nd ($Z = 60$) to Er ($Z = 68$). These DLs are nearly an order of magnitude lower than those obtained for the polychromatic incident beam assuming the same counting conditions in which case the DLs are situated in the 2–10 ppm range.

It is clear from Figure 6.1.12 that the simulation model discussed in this work can reliably be used for estimating the analytical characteristics of SRXRF spectrometers operating in the incident energy range of 60–100 keV.

6.1.4 MODELING OF XRF ON HETEROGENEOUS SAMPLES: SIMULATION OF XRF TOMOGRAPHY

Many applications of microscopic XRF aim to study the heterogeneity of the analysed materials, i.e. two- or three-dimensional distribution of chemical elements, such as heavy metals in different environmental, geological and biological samples. Owing to its high sensitivity and non-destructive nature, synchrotron radiation based micro XRF computed tomography (XFCT) is one of the emerging methods of providing potentially three-dimensional, quantitative information on the elemental distribution in the probed sample volume with trace-level DLs (Bohic *et al.*, 2001; Simionovici *et al.*, 2001; Vincze *et al.*, 1999c, 2002a). The technique can be easily performed on a regular scanning micro XRF set-up with the addition of a sample rotation stage to the usual *xyz* linear stages (Janssens, 1998).

By performing repeated line scans at a fixed sample-height under a large number of observation angles (i.e. taking different views of the sample) around a given sample axis and recording the emerging fluorescent and scattered signals, the conventional X-ray fluorescent mapping is replaced by a XFCT experiment. In this way one obtains a special representation of the elemental intensity distributions across the investigated cross-section of the sample, called elemental sinograms. The usual two-dimensional elemental distribution in the investigated horizontal sample plane (i.e. the tomographic plane perpendicular to the vertical rotation axis) is obtained by a mathematical reconstruction procedure called filtered backprojection algorithm (Russ, 1995).

In Figure 6.1.13 an example of the type of heterogeneous sample which can be simulated is shown. The horizontal cross-section of the sample is divided into several phases having boundaries of arbitrary polygonal and circular shape. The sample composition does not change in the vertical direction. This invariance along the vertical axis implies that the simulated object is only quasi three-dimensional and so the simulated trajectories are dependent only on the horizontal and not on the vertical coordinates.

Figure 6.1.14 shows XRF microtomography results taken from a phantom sample of three attached glass capillaries filled with 1 % solutions of Mn, Ni and As (Figure 6.1.14a), which were used to verify and validate the simulation code for heterogeneous sample types. The experimental data were collected at HASYLAB, Beam line L. The recorded sinograms (Figure 6.1.14b) correspond to 100 translation steps of 6 μm and 180 rotation steps of 2°, with a live time of 10 s per point.

The reconstruction of the displayed tomographic data sets was done by a filtered backprojection algorithm, without using self-absorption corrections. This results in a clearly poorer

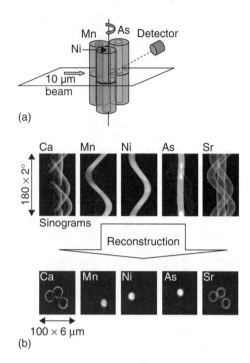

Figure 6.1.14 (a) Phantom sample of three attached glass capillaries filled with 1 % solutions of Mn, Ni and As, which were used to verify and validate the simulation code for XRF microtomography on heterogeneous sample types. The experimental data were collected at HASYLAB, Beam line L. The recorded sinograms and reconstructed images (b) correspond to 100 translation steps of 6 μm and 180 rotation steps of 2°, with a live time of 10 s per point. Reproduced by permission of SPIE from Vincze et al. (2002a)

reconstructed image for Ca Kα ($E_{K\alpha} = 3.69$ keV) when compared to the reconstruction of e.g. Sr Kα ($E_{K\alpha} = 14.16$ keV) even though Ca and Sr have the same spatial distributions (glass walls). Low-energy lines suffer from sample self-absorption, which can be accounted for by appropriate reconstruction algorithms. Several methods exist to correct for the attenuation of both primary and fluorescent X-rays within the sample matrix (Schroer, 2001; Simionovici et al., 2001), however, the use of these algorithms is beyond the scope of the present subchapter.

Figure 6.1.15 shows a comparison of experimental and simulated sinograms and reconstructed images for Ca and Sr (both of which are present in the glass capillary walls), indicating that the

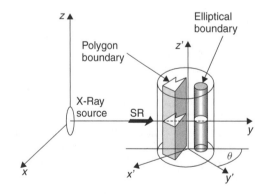

Figure 6.1.13 An example of the modeled heterogeneous sample, containing heterogeneities of polygonal and/or circular boundaries

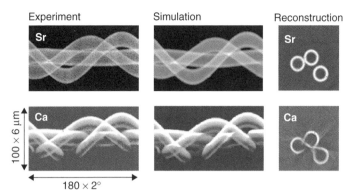

Figure 6.1.15 Comparison of experimental and simulated sinograms as well as reconstructed images for Ca and Sr

simulation code can describe adequately the sample topology represented by the particular phantom sample shown in Figure 6.1.14(a). The fact that both sinograms and reconstructed images can be reproduced from the simulated data set indicates that the numerical model for the heterogeneous system approximates well the topology of the sample in question.

6.1.5 QUANTITATIVE TRACE-ELEMENT ANALYSIS OF INDIVIDUAL FLY ASH PARTICLES

In this section we focus on micro XRF spectroscopy coupled with MC simulation as a useful analytical tool for determining the sample composition down to trace concentration levels. The method is appreciated for its high sensitivity, non-destructiveness and because of its simple relation to the fundamental physics of atom–radiation interaction.

Several papers have appeared on the applicability of microscopic XRF for the analysis of particulate matter without aiming to determine trace element concentrations quantitatively (Török et al., 1994; Rindby et al., 1997). Lankosz and Pella applied recently the fundamental parameter (FP) method for quantitative XRF analysis of individual particles having irregular shapes, but only for major and minor elements (Lankosz and Pella, 1997). The theoretically derived calculations were verified by analysis of standard glass particles with known stoichiometry and an agreement within 3–10 % was found between calculated and nominal concentrations. The measurements were performed using an X-ray microbeam with 177 μm diameter.

Generalized iterative procedures based on FP for quantitative XRF analysis were published (Szalóki et al., 1999). In general, the estimation of the matrix attenuation for the iterative calculation is based on the average atomic number of the sample matrix. The procedure offers a wide ranging extension for quantitative standardless XRF analysis of complex samples, especially for environmental and biological materials. In contrast to these methods, our approach uses the MC technique.

The quantification method presented below is based on MC simulations for dealing with the energy and spatial distribution of the X-ray microbeam, the interaction between the exciting photons and the sample elements taking into account the geometry effects caused by the size and topology of the particles. The applicability of the method is demonstrated on fly ash particles because they are common atmospheric particles having relatively elevated concentrations of some toxic elements (Vincze et al., 2002b).

6.1.5.1 THE QUANTIFICATION ALGORITHM

The concentration values are calculated by an iterative MC simulation method taking into account not

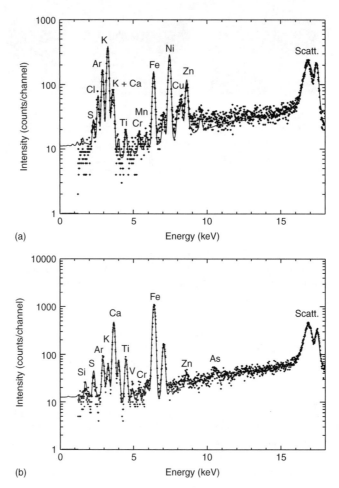

Figure 6.1.16 Measured (dots) and simulated (solid lines) XRF spectra obtained by the X-ray tube based set-up for typical wood (a) and coal (b) fly-ash particles. Reprinted with permission from Vincze et al. (2002b). Copyright 2002 American Chemical Society

only self-absorption within a spherical particle, but also more subtle higher-order effects, such as inter-element enhancement, or fluorescence induced by in-sample scattering. In order to get the initial concentration values, necessary to start the iteration, self absorption was neglected as a first step. The initial concentration of element i, $C_{i,0}$ with detectable X-ray line was calculated as:

$$C_{i,0} = \frac{I_{i,\text{meas}}/S_i}{\sum_{j=m+1}^{n} I_{j,\text{meas}}/S_j} \times \left(1 - \sum_{j=1}^{m} C_j\right) \quad (6.1.24)$$

where n is the total number of elements in the sample, and the first m elements constitute the dark matrix (which can be derived from stoichiometry or known from other, bulk measurements), $I_{i,\text{meas}}$ the experimental intensity and S_i the experimentally determined sensitivity factor for element i.

As a second step, the above detailed MC simulation code is used to iteratively refine the initial concentration values. The iterative procedure to refine the concentration of each element is given by:

$$C_i^{(k+1)} = C_i^{(k)} \cdot \frac{I_{i,\text{meas}}}{I_{i,\text{calc}} \cdot \sum_{j=1}^{n} C_j^{(k)} \cdot \frac{I_{j,\text{meas}}}{I_{j,\text{calc}}}} \quad (6.1.25)$$

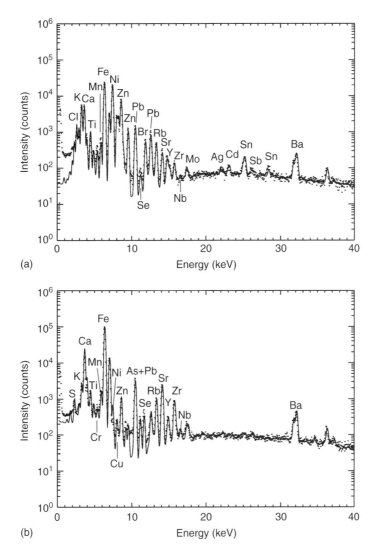

Figure 6.1.17 Experimental (dots) and simulated (solid lines) SRXRF spectra of the individual wood (a) and coal (b) fly-ash particles. Reprinted with permission from Vincze *et al.* (2002b). Copyright 2002 American Chemical Society

where $C_i^{(k)}$ is the calculated concentration of element i after k iteration steps.

The procedure is terminated, if either of the following conditions are met:

$$|C_i^{(k+1)} - C_i^{(k)}| < \varepsilon \qquad (6.1.26)$$

$$\sum_{i=1}^{n} \frac{(I_{i,\text{meas}} - I_{i,\text{calc}})^2}{\sigma_{i,\text{meas}}^2} < \delta \qquad (6.1.27)$$

where $\sigma_{i,\text{meas}} = \sqrt{I_{i,\text{meas}}}$ is the statistical uncertainty of the measured intensity corresponding to element i.

6.1.5.2 QUANTITATIVE RESULTS FOR COAL AND WOOD FLY ASH PARTICLES

In this section an example is shown for the use of MC simulation in the field of quantitative particulate analysis. The trace element composition of 100 individual coal and wood fly ash particles, originating from Hungarian power plants, is determined by the above-described iterative MC quantification method.

In Figure 6.1.16 measured and simulated XRF spectra (corresponding to the final iteration step of the MC calculation) of a typical coal and wood fly ash particle are shown, measured by a mono-capillary based laboratory micro-XRF set-up. This instrument is installed at the University of Antwerp (UIA) making use of a high-power rotating anode Mo X-ray tube (Janssens *et al.*, 1996). The agreement between the measured and simulated spectra is satisfactory for both types of fly-ash particles, both with respect to the fluorescence line and scatter peak intensities.

The trace element content of the fly-ash particles was determined from the micro-SRXRF measurements at HASYLAB Beam line L.

In Figure 6.1.17 two examples of simulated (result of the final iteration) and measured spectra corresponding to the two types of particle are shown. Using the elemental composition of the (low-Z) fly ash matrix, determined from the tube-excited measurements, the trace element content of each individual fly-ash particle could be determined and the results are summarized in Table 6.1.2.

In Table 6.1.3, the average minor and major elemental concentrations and the variation in the particles of the wood and coal fly ash particles determined on the basis of 40 wood and 60 coal fly ash particles are shown. The same major elements could be detected in both types of fly ash and these elements are visible in most of the particles. The major element concentration ratios are different in the two cases: in the case of wood fly ash a higher concentration of S and Cl can be detected (originating from the organic component of the wood) while in case of the coal fly ash the amount of elements characteristic to the mineral such as Si, Ca, Fe is larger. In the case of coal fly ash a higher number of potentially hazardous trace elements, e.g. Cu, As, Pb, are detected. The variation of the elemental concentrations is large for both types of particles showing the highly inhomogeneous nature of the ash material.

Table 6.1.2 Major and minor element concentrations of a typical wood and coal fly ash particle with corresponding experimental and simulated spectra shown in Figure 6.1.5(a). The calculated concentrations were obtained by assuming $\rho = 0.5\,\text{g/cm}^3$ density and 50% carbon residual matrix for the wood fly ash and by assuming $\rho = 2.6\,\text{g/cm}^3$ and $\rho = 1.0\,\text{g/cm}^3$ density values for the coal particle[a]

Element	Coal fly ash calc. conc. (%)	Wood fly ash calc. conc. (%)
O[a]	47	10
Si	26	–
S	8.3	5
Cl	–	7
K	1.7	20
Ca	12.0	2.1
Ti	0.9	0.3
Cr	0.05	0.1
Mn	0.07	0.05
Fe	4.8	1.3
Ni	–	1.8
Cu	–	0.2
Zn	0.05	0.5

[a] Concentration value of oxygen was calculated from stoichiometric considerations.

Table 6.1.3 Trace element concentrations in an individual wood and coal fly ash particle, whose spectra are shown in Figure 6.1.6(a) and 6.1.6(b). In the case of the wood fly ash, a residual matrix consisting of 50% C and a density of $0.5\,\text{g/cm}^3$ was assumed while in the case of the coal fly ash a density of $1.0\,\text{g/cm}^3$ was used

Element	Wood fly ash conc. (ppm)	Coal fly ash conc. (ppm)
Se	21	53
Br	238	<5
Rb	148	170
Sr	100	415
Y	10	14
Zr	22	85
Nb	5	9
Mo	21	21
Ag	28	<5
Cd	117	<5
Sn	161	10
Sb	40	<5
Ba	588	440
Pb	1044	210

6.1.6 CONCLUSIONS AND FUTURE PERSPECTIVES

The use of MC simulation models in the field of EDXRF spectrometry is becoming more and more viable in view of the rapid increase of inexpensive computing power and because of the availability of accurate atomic data for photon–matter interactions. By considering the three most important interaction types in the 1–100 keV energy range (photoelectric effect followed by fluorescence emission, Compton and Rayleigh scattering) such models can be used in a general fashion to predict the achievable analytical characteristics of future (SR)XRF spectrometers and to aid the optimization/calibration of existing instruments.

With an efficient combination of sampling and variance reduction techniques in a given MC model, the complete spectral response of heterogeneous multi-element samples irradiated with a polychromatic (optionally linearly polarized) X-ray beam can be simulated using a CPU time in the order of minutes on a typical personal computer. For monochromatic excitation, the calculation time is in the order of a few seconds.

The code illustrated in this subchapter has been experimentally verified by comparisons of simulated and experimental spectral distributions of samples of various nature, recorded at different SRXRF spectrometers. As the code could reliably predict both the fluorescent and scattered intensities in the measured fluorescent spectra, it has been applied to determine unknown sample characteristics, such as the sample composition. The latter was estimated by the (reverse) iterative use of the code, which provides an alternative for the quantitative analysis of unknown samples next to the traditional FP methods.

The code could also be successfully used to estimate unknown primary beam characteristics, such as the degree of linear polarization, which is an interesting application for the characterization of existing SRXRF instruments.

With respect to the simulation of heterogeneous samples, an example was given for the modeling of XRF tomography experiments. The simulation of such lengthy XRF imaging experiments is extremely important for performing feasibility studies and optimization before the actual measurement is performed.

Potential future development for MC codes specific for EDXRF spectroscopy includes the extension of the simulated energy range below 1 keV and the refinement of the employed physical models serving as the engine for the simulation model. On the one hand, there is a general necessity to select the most up-to-date atomic constants (X-ray cross-sections, fluorescence yields, emission rates, etc.) in order to increase the accuracy of the simulated results. And, on the other hand, to improve the validity of background modeling in the case of highly polarized monochromatic X-ray sources.

One of the important effects, determining the spectral background in the case of XRF experiments performed at third generation synchrotron sources employing highly polarized monochromatic excitation is the bremsstrahlung emission by energetic photoelectrons (and to a lesser extent Auger electrons) produced during photoelectric interactions. The simulation of these processes requires the modeling of electron transport phenomena, which have been implemented only in the most general MC codes for particle transport problems (e.g. EGS4, MCNP). As these general-purpose codes often cannot be used directly for modeling EDXRF spectra efficiently, there is a general need to further develop specific MC codes for XRF spectroscopy.

Next to electron bremsstrahlung, another important effect contributing to the spectral response of samples illuminated with highly monochromatic synchrotron radiation is resonant Raman X-ray scattering, which is currently under implementation in the simulation model described in this subchapter.

REFERENCES

Al-Ghorabie, F. H. H. *Radiat. Phys. Chem.* **55**, 377–384 (1999).
Al-Ghorabie, F. H. H., Natto, S. S. A. and Al-Lyhiani, S. H. A. *Comput. Biol. Med.* **31**, 73–83 (2001).
Ao, Q., Lee, S. H. and Gardner, R. P. *Appl. Radiat. Isot.* **48**, 1403–1412, (1997a).
Ao, Q., Lee, S. H. and Gardner, R. P. *Appl. Radiat. Isot.* **48**, 1413–1423 (1997b).
Apostolakis, J., Giani, S., Maire, M., Nieminen, P., Pia, M. G. and Urban, L. GEANT4 Low Energy Electromagnetic Models for electrons and photons. CERN-OPEN-99-034, preprint, 1999.

Bielajew, A. F., Hirayama, H., Nelson, W. R. and Rogers, D. W. O. National Research Council of Canada Report PIRS-0436 (1994).

Biggs, F., Mendelsohn, L. B. and Mann, J. B. *At. Data Nucl. Data Tables* **16**, 201 (1975).

Bohic, S., Simionovici, A., Snigirev, A., Ortega, R., Deves, G., Heymann, D. and Schroer, C. G. *Appl. Phys. Lett.* **78** (22), 3544–3546 (2001).

Briesmeister, J. F. (Ed.) Los Alamos National Laboratory Report LA-12625-M, (1993).

Cullen, D. E., Hubbell, J. H. and Kissel, L., Lawrence Livermore National Laboratory Report UCRL-50400, Vol. 6, Rev. 5 (1997).

Evans, C. J., Shamsaie, M., Ghara'Ati, H. and Ali, P. A. *Appl. Radiat. Isot.*, **49** 559–560 (1998).

Fajardo, P., Honkimaki, V., Buslaps, T. and Suortti, P. *Nucl. Instr. Methods B* **134**, 337–345 (1998).

Fernández, J. E. *X-ray Spectrom.* **24**, 283–292 (1995a).

Fernández, J. E. *Appl. Radiat. Isot.* **46**, 383–400 (1995b).

Fernández, J. E. *J. Trace Microprobe Tech.* **14**, 489–516 (1996).

Fernández, J. E. *Appl. Radiat. Isot.* **48**, 1635–1646 (1997).

Fernández, J. E. *Radiat. Phys. Chem.* **51**, 383–385 (1998a).

Fernández, J. E. *Appl. Radiat. Isot.* **49**, 83–87(1998b).

Fernández, J. E. *Radiat. Phys. Chem.* **56**, 27–59 1999.

Fernández, J. E. *Interaction of X-rays with matter, in Microscopic X-ray Fluorescence Analysis* (Eds K. H. A. Janssens, F. C. V. Adams and A. Rindby), John Wiley & Sons, Ltd, Chichester, (2000).

Fernández, J. E., Bastiano, M. and Tartari, A. *X-ray Spectrom.* **27**, 325–331 (1998).

Fernández, J. E., Hubbell, J. H., Hanson, A. L. and Spencer, L. V. *Radiat. Phys. Chem.* **41**, 579 (1993).

Fishman, G. S. *Monte Carlo: Concepts, Algorithms, and Applications*, Springer-Verlag, New York (1996).

Halbleib, J. A., Kensek, R. P., Mehlhorn, T. A., Valdez, G. D., Seltzer, S. M. and Berger, M. J. Sandia Report SAND91-1634 (1992).

Hanson, A. L. *Nucl. Instrum. Methods A* **243**, 583–598 (1986).

He, T., Gardner, R. P. and Verghese, K. *Nucl. Instrum. Methods A* **299**, 354–366 (1990).

Hirayama, H., Namito, Y. and Ban, S. KEK Internal Report 2000-3 (2000).

Hubbell, J. H., Veigele, W. J., Briggs, A., Brown, R. T., Cromer, D. T. and Howerton, R. J. *J. Phys. Chem. Ref. Data.* **4**, 471 (1975)

Hugtenburg, R. P., Turner, J. R., Mannering, D. M. and Robinson, B. A. *Appl. Radiat. Isot.* **49**, 673–676 (1998).

Janssens, K., Vekemans, B., Vincze, L., Adams, F. and Rindby, A. *Spectrochim. Acta B* **51** 1661–1678 (1996).

Janssens, K., Vincze, L., Vekemans, B., Adams, F., Haller, M. and Knöchel, A. *J. Anal. At. Spectrom.* **13**, 339–350 (1998).

Krause, M. O. *J. Phys. Chem. Ref. Data* **8**, 307–327 (1979).

Lankosz, M. and Pella, P. A. *X-Ray Spectrom.* **26**, 347–349 (1997).

Lechtenberg, F. *J. Trace Microprobe Tech.* **14** (3), 561–587 (1996).

Lee, S. H., Gardner, R. P. and Todd, A. C. *Appl. Radiat. Isot.* **54**, 893–904 (2001).

Lewis, D. G., Kilic, A. and Ogg, C. A. *Appl. Radiat. Isot.* **49**, 707–709 (1998).

Matt, G., Feroci, M., Rapisarda, M. and Costa, E. *Radiat. Phys. Chem.* **48**, 403–411 (1996).

Metropolis, N. *Lecture Notes Phys.* **240**, 62 (1985).

Milgram, M. S *Ann. Nucl. Energy* **28**, 297–332 (2001).

Namito, Y. and Hirayama, H. *Nucl. Instrum. Methods A* **423**, 238–246 (1999).

Namito, Y. and Hirayama, H. KEK Internal Report 2000-4 (2000).

Namito, Y., Hirayama, H. and Ban, S. *Radiat. Phys. Chem.* **53**, 283–294 (1998).

Nelson, W. R., Hirayama, H and Rogers, D. W. O. The EGS4 Code System Report SLAC-265, Stanford Linear Accelerator Center, Stanford, California (1985).

O'Meara, J. M., Borjesson, J., and Chettle, D. R. *Appl. Radiat. Isot.* **53**, 639–646 (2000).

O'Meara, J. M., Chettle, D. R., McNeill, F. E. and Webber, C. E. *Appl. Radiat. Isot.* **49**, 713–715 (1998).

Osán, J., Szalóki, I., Ro, C.-U. and Van Grieken, R. *Mikrochim. Acta* **132**, 349–355 (2000).

Rindby, A., Engström, P., Janssens, K. and Osán, J. *Nucl. Instrum. Methods Phys. Res. B* **124**, 591–604 (1997).

Russ, J. C. *The Image Processing Handbook*, 2nd edition CRC Press, Boca Raton, FL (1995).

Schroer, C. G. *Appl. Phys. Lett.* **79**, 1912–1914 (2001).

Schneider, J. R. *Synchr. Rad. News* **8**, 26 (1995).

Simionovici, A., Chukalina, M., Gunzler, F., Schroer, C., Snigirev, A., Snigireva, I., Tummler, J., and Weitkamp, T. *Nucl. Instrum. Methods A* **467**, 889–892 (2001).

Somogyi, A., Drakopoulos, M., Vincze, L., Vekemans, B., Camerani, C., Janssens, K., Snigirev, A. and Adams, F. *X-ray Spectrom.* **30**, 242–252 (2001).

Szalóki, I., Lewis, D. G., Benett, C. A. and Kilic, A. *Phys. Med. Biol.* **44**, 1245–1255 (1999).

They, J., Buschhorn, G., Kotthaus, R. and Pugachev, D. *Nucl. Instrum. Methods Phys. Res. A* **467–468**, 1167–1170 (2001).

Török, Sz., Faigel, Gy., Osán, J., Török, B., Jones, K. W., Rivers, M. L., Sutton, S. R. and Bajt, S. *Adv. X-Ray Anal.* **37**, 711–716 (1994).

Vekemans, B., Janssens, K., Vincze, L., Adams, F. and Van Espen, P. *Spectrochim. Acta* **50B**(2), 149–169 (1995).

Vekemans, B., Vincze, L., Somogyi, A., Drakopoulos, M., Simionovici, A. and Adams, F. *Nucl. Instrum. Methods Phys. Res. B* **199**, 396–401 (2003).

Vincze, L., PhD Thesis, University of Antwerp (1995).

Vincze, L., Janssens, K., Adams, F., Rivers, M. L. and Jones, K. W. *Spectrochim. Acta, Part B* **50**, 127–148, (1995b).

Vincze, L., Janssens, K., Vekemans, B. and Adams, F. *J. Anal. At. Spectrom.* **14**, 529–533 (1999a).

Vincze, L., Janssens, K., Vekemans, B. and Adams, F. *Spectrochim. Acta, Part B* **B54**, 1711–1722, (1999b).

Vincze, L., Janssens, K., Vekemans, B. and Adams, F. *Proc. SPIE.* **3772**, 328–337 (1999c).

Vincze, L., Vekemans, B., Szalóki, I., Janssens, K., Van Grieken, R., Feng, H., Jones, K. W. and Adams, F. *Proc. SPIE.* **4503**, 240–248 (2002a).

Vincze, L., Somogyi, A., Osán, J., Vekemans, B., Török, S., Janssens, K. and Adams, F. *Anal. Chem.* **74**, 1128 (2002b).

6.2 Spectrum Evaluation

P. LEMBERGE

University of Antwerp, Antwerp, Belgium

6.2.1 INTRODUCTION

Spectrum evaluation essentially comprises the mathematical procedures to extract relevant information from acquired X-ray spectra. Apart from smoothing and peak search methods, the most important aspect is, beyond doubt, the extraction of the analytically important net peak areas of the element characteristic fluorescence lines.

Spectrum evaluation remains a crucial step in X-ray spectrometry and can be considered as important as sample preparation and quantification. Due to the relatively low resolution of solid-state detectors, spectrum evaluation is certainly more critical to energy-dispersive X-ray spectrometry (ED-XRF) than to wavelength-dispersive X-ray spectrometry (WD-XRF). However, the often-quoted inferior accuracy of ED-XRF can, to a large part, be attributed to errors associated with the evaluation of these spectra. Although ED-XRF lacks the precision of WD-XRF, a correct spectrum evaluation improves the accuracy of ED-XRF to the same or even higher level than WD-XRF. This is also important from an economical perspective, i.e. in many industrial applications the more expensive WD-XRF systems might be replaced or backed-up by cheaper and more versatile ED-XRF instruments. To achieve this, the precision of ED-XRF is to be improved while the analysis time should remain comparable to WD-XRF.

As discussed in the chapter on new detector technologies, the ever-expanding computer and chip industry has also triggered the development of Si-PIN photodiodes and semiconductor drift detectors (SDD) or drift chambers (SDC). The versatility of the Integrated Circuits (IC) production apparatus made it possible to manufacture these detectors of high quality at low cost. Such detectors can handle a much higher count rate than Si(Li) or HPGe/Ge(Li) detectors (Gatti and Rehak, 1984; Murty *et al.*, 1998; Castoldi *et al.*, 2000). At the same time, the introduction of digital signal processing (DSP) further improved the count rate and made it possible to operate solid-state detectors at high dead-times while preserving the peak shape of the characteristic lines in the spectrum (Jordanov *et al.*, 1994). In the early 1970s when the first commercial ED-XRF systems based on Si(Li) detectors became available the maximal count rate was about 10 kcps. Over the past few years, count rates of 50 kcps to 100 kcps are no longer exceptional.

With an increased count rate and hence a better precision, more details (e.g. the Lorentz character of characteristic lines) as well as specific features of the spectrum (e.g. peak tailing and incomplete charge collection) become apparent. The simple fitting model used in the past to evaluate spectra should therefore be improved and extended to describe all the aspects of the spectrum. Until now, such a complex model requiring many parameters demanded too much processing power of the computer attached to the instrument. Today, the availability of inexpensive and powerful personal computers enables the implementation of mature spectrum evaluation packages bringing sophisticated spectrum evaluation within reach.

X-Ray Spectrometry: Recent Technological Advances. Edited by K. Tsuji, J. Injuk and R. Van Grieken
© 2004 John Wiley & Sons, Ltd   ISBN: 0-471-48640-X

The previous subchapter already indicated that an ED spectrum is in fact the original spectrum, as it reaches the detector, convoluted with the response function of this detector. In this respect, spectrum evaluation is also known as spectrum deconvolution. The Monte Carlo (MC) simulation code also relies on an accurate detector response function to deliver, through convolution, a close agreement between simulated and experimental spectrum. The ability to simulate X-ray spectra that agree very well with measured spectra opens the possibility to use them as standards for new quantification methods.

The various methods for spectrum evaluation discussed in this subchapter emphasize ED X-ray spectra. Most of the methods are relevant for X-ray fluorescence (XRF), particle-induced X-ray emission (PIXE) and electron beam X-ray analysis (electron probe X-ray microanalysis (EPXMA), scanning electron microscopy - energy dispersive X-ray analysis (SEM-EDX) and analytical electron microscopy (AEM)). A complete and detailed text on every aspect of spectrum evaluation can be found elsewhere (Van Espen, 2002).

6.2.2 SIMPLE NET PEAK AREA DETERMINATION

For both WD-XRF and ED-XRF, the concentration of an element is proportional to the number of counts under the characteristic X-ray peak of the element, corrected for the continuum. Provided that the resolution is constant, this proportionality extends to the entire net peak height. Due to the inherent nature of the ED-XRF to detect X-rays simultaneously, preference is given to the peak area. In WD-XRF the detection of X-rays is a sequential process. Acquisition of the entire peak profile is very time-consuming and pointless except when detecting low-energy X-rays where the resolution is inadequate due to the limited ability of the crystal to disperse the X-rays. In such a case, peak integration is advisable. For higher X-ray energies, the count rate is measured only at the peak maximum. To utilize the count rate at the 2Θ angle of the peak maximum as the analytical signal, it should be corrected for the continuum. The continuum is estimated at a 2Θ position on the left- and right-hand side of the peak. This straightforward method is sufficiently adequate to obtain both accurate and precise results. For now, improvement of this procedure is not considered necessary.

In ED-XRF, a closely related method to obtain the net area of an isolated peak in a spectrum consists of interpolating the continuum under the peak and summing the channel contents (corrected for the continuum) in a window over the peak. This method can only be used for peaks that are free from interferences while the continuum should be linear over the extent of the region taken into consideration. Because of these restrictions, a simple peak integration method cannot be used as a general tool for the evaluation of ED-XRF spectra and good results are obtained in a limited number of applications only. At present, application of this method has become rare.

6.2.3 LEAST-SQUARES FITTING USING REFERENCE SPECTRA

A much more robust technique is the least-squares fitting method using reference spectra. The reference spectra are measured or calculated spectra of pure compounds. A spectrum of an unknown sample described as a linear combination of pure element spectra can be formulated mathematically as

$$y_i^{\text{mod}} = \sum_{j=1}^{m} a_j x_{ij} \qquad (6.2.1)$$

in which y_i^{mod} stands for the content of channel i in the model spectrum and x_{ij} the content of channel i in the jth reference spectrum. The coefficients a_j represent the contribution of the reference spectra to the unknown spectrum and can be used for quantitative analysis. The values of the a_j coefficients are found via multiple least-square fitting. During the fit the sum of weighted squared differences between the measured spectrum and the model are minimized. This object function, χ^2, is written as

$$\chi^2 = \frac{1}{i_2 - i_1 + 1 - m} \sum_{i=i_1}^{i_2} \frac{1}{\sigma_i^2} (y_i - y_i^{\text{mod}})^2$$

$$= \frac{1}{i_2 - i_1 + 1 - m} \sum_{i=i_1}^{i_2} \frac{1}{\sigma_i^2} \left(y_i - \sum_{i=i_1}^{i_2} a_j x_{ij} \right)^2$$

(6.2.2)

with y_i the content of channel i and σ_i^2 the variance at channel i, normally taken as $\sigma_i^2 = y_i$ (Poisson statistics). Channels i_1 and i_2 are the beginning and the end of the fitting region.

The approach assumes that a measured spectrum can be described as a linear combination of pure element spectra. This assumption holds to some extent for the characteristic lines in the spectrum but is not valid for the continuum. Therefore, prior to the least-squares fitting the background must be removed. A frequently used approach is the application of a digital filter to both unknown and reference spectra. This variant is known as the filter-fit method (Schamber, 1977; Statham, 1978; McCarthy and Schamber, 1981). The continuum can also be estimated by peak stripping or using polynomial functions as discussed later.

The advantage of using reference spectra is the ability to deal with a complex and difficult to model continuum. On the other hand, when net intensities of small peaks in the vicinity of very large peaks are to be derived, the method is less optimal. The filter-fit and related methods are still available in some software packages. Recently, no further research has been done in this area of spectrum evaluation.

6.2.4 LEAST-SQUARES FITTING USING ANALYTICAL FUNCTIONS

A widely used and certainly the most flexible procedure for evaluating complex X-ray spectra, is based on least-squares fitting of spectral data with an analytical function. The method is conceptually simple, but not trivial to implement and use.

6.2.4.1 CONCEPT OF SPECTRUM FITTING

In least-squares fitting of spectral data an algebraic function or fitting model, including analytically important parameters, such as the net areas of the fluorescence lines, is used to describe the measured spectrum. Chi-square, χ^2, is defined as the weighted sum of squares over a certain region of the spectrum, i_1 to i_2, of the differences between the model and the measured spectrum y_i:

$$\chi^2 = \frac{1}{i_2 - i_1 + 1 - m} \sum_{i=i_1}^{i_2} \frac{1}{\sigma_i^2}$$
$$\times [y_i - y(i, a_1, \ldots, a_m)]^2 \quad (6.2.3)$$

again with σ_i^2 the uncertainty of the measured spectrum, and a_j are the parameters of the model. The optimum values of the parameters are those for which χ^2 reaches a minimum. They can be found by setting the partial derivatives of χ^2 to the parameters to zero:

$$\frac{\partial \chi^2}{\partial a_j} = 0 \quad j = 1, \ldots, m \quad (6.2.4)$$

If the model is linear in all the parameters a_j, these equations result in a set of m linear equations in m unknowns a_j, which can be solved algebraically. This is known as linear least-squares fitting. If the model has one or more nonlinear parameters no direct solution is possible and the optimum value of the parameters must be found by iteration. The latter is known as nonlinear least-squares fitting. The selection of a suitable minimization algorithm is important because it determines to a large extent the performance of the method. In most of the software packages the Marquardt–Levenberg algorithm is used (Levenberg, 1944; Marquardt, 1963).

The difficulty with this method is to find an algebraic function that accurately describes the observed spectrum or at least the spectral region of interest. This requires a model describing the continuum, the element characteristic lines and all other features present in the spectrum. Though the response function of the solid-state detector is, to a very good approximation, Gaussian, the deviation from the Gaussian shape needs to be taken into account. An inaccurate model will result in systematic errors, which may lead to gross errors in the estimated peak areas and ultimately the element concentrations. On the other hand, the fitting function may not become too complex and preferably use few parameters. Especially for nonlinear fitting, the location of the χ^2 minimum is demanding on the computing power when a large number of parameters is involved.

During the last decade much of the research in this field has been focused on the development of a model including a Gaussian and some extra functions to describe the deviation from the ideal Gaussian and has also increased the number of parameters. This evolution was not possible until the availability of cheap and powerful PCs. Before, the deviation from the Gaussian peak shape was taken into account numerically. This method is only satisfying when applied for similar detectors. With the availability of detectors different from the original Si(Li) and Ge detectors, spectrum evaluation required more flexible models.

In general, the fitting model consists of two parts; the first part describes the continuum while the second part deals with the element characteristic lines and other peak-like features:

$$y(i) = y_{\text{Cont}}(i) + \sum_{P=1}^{\text{Peaks}} y_P(i) \qquad (6.2.5)$$

where $y(i)$ is the calculated content of channel i.

6.2.4.2 THE CONTINUUM

Several functions can be used to describe the continuum, depending on the excitation conditions and on the width of the fitting region. Since it is almost impossible to construct an acceptable physical model describing the continuum, in most cases some kind of polynomial expression is used. An exception, worthwhile to be mentioned, is the case of electron microscopy where a Brehmsstrahlung continuum is observed. This can be modelled with an exponentially decreasing function according to Kramer's formula, corrected for the absorption by the detector windows and by the sample. To be physically correct, the absorption term must be convoluted with the detector response function, because the sharp edges due to absorption by detector windows or elements present in the sample are smeared out by the finite resolution of the detector.

6.2.5 LINEAR AND EXPONENTIAL POLYNOMIALS

A linear polynomial is defined as

$$y_{\text{Cont}}(i) = a_0 + a_1(E_i - E_0) + a_2(E_i - E_0)^2 \\ + \cdots + a_k(E_i - E_0)^k \qquad (6.2.6)$$

with E_i the energy (in keV) of channel i and E_0 a suitable reference energy, often the middle of the fitting region. The user is allowed to choose the degree of the polynomial, k. $k = 0$ produces a constant while $k = 1$ produces a straight line and $k = 2$ a parabolic continuum. Values larger than $k = 4$ are rarely useful because such high-degree polynomials tend to have physical nonrealistic oscillations. Linear polynomials are used to describe the continuum over a region of 2 to 3 keV wide. Wider regions usually exhibit too much curvature and can be modelled by an exponential polynomial instead:

$$y_{\text{Cont}}(i) = a_0 \exp[a_1(E_i - E_0) + a_2(E_i - E_0)^2 \\ + \cdots + a_k(E_i - E_0)^k] \qquad (6.2.7)$$

where k is the degree of the exponential polynomial. A value $k = 6$ or higher might be necessary to accurately describe a continuum from 2 to 16 keV.

6.2.6 ORTHOGONAL POLYNOMIALS

Steenstrup (1981) proposed the use of orthogonal polynomials to model the continuum as demonstrated by evaluating energy-dispersive X-ray diffraction spectra. The spectrum is fitted using orthogonal polynomials, and the weights of the least-squares fit are iteratively adjusted so that only channels belonging to the continuum are included in the fit. The method is generally applicable and can be implemented as an algorithm that needs little or no control parameters. Vekemans *et al.* (1994) used the method for the unsupervised evaluation of spectra collected with a micro X-ray fluorescence (μ-XRF) setup.

6.2.6.1 THE FLUORESCENCE LINES

The second part of $y(i)$ is the more important one since it represents the actual detector response function. As already mentioned, given that the response function of solid-state detectors is predominantly Gaussian, all mathematical expressions used to describe the fluorescence lines involve this function. As will be discussed later, the intrinsic energy distribution of a characteristic X-ray is of Lorentzian nature. Convolution with the Gaussian broadening function of the detector actually results in Voigtian distribution. However, in most cases the width of the Lorentzian is negligible in comparison with the resolution of the detector so that a Gaussian peak shape model is adequate (Wilkinson, 1971; Gunnink, 1977).

6.2.7 A SINGLE GAUSSIAN

A Gaussian peak is characterized by three parameters: position, width and area. The first approximation to the profile of a single peak is given by

$$\frac{A}{\sigma\sqrt{2\pi}} \exp\left[-\frac{(x_i - \mu)^2}{2\sigma^2}\right] \quad (6.2.8)$$

where A is the peak area (counts), σ is the width of the Gaussian expressed in channels, and μ the location of the peak maximum. The full width at half-maximum (FWHM) is related to σ by the factor $2\sqrt{2\ln 2}$ or FWHM = $2\sqrt{2\ln 2}\sigma$.

To describe part of a measured spectrum, the fitting function must contain as many of such Gaussian functions as there are peaks. For 10 elements and 2 peaks (Kα and Kβ) per element, 60 parameters need to be optimized. It is highly unlikely that a nonlinear least-square will successfully reach a global minimum for the χ^2. This problem is overcome by writing the fitting function in a different way.

6.2.8 ENERGY AND RESOLUTION CALIBRATION

A first step is to drop the idea of optimizing the position and width of each peak independently. In X-ray spectrometry, the energies of the fluorescence lines are known with an accuracy of 1 eV or better. The pattern of peaks observed in a spectrum is directly related to elements present in the sample. Based on those elements, we can predict all X-ray lines that constitute the spectrum and their energies. Obviously, a peak function is therefore best written in terms of energy rather than channel number. Defining 'zero' as the energy at channel 0 and expressing the spectrum 'gain' in eV per channel, the energy of channel i is given by:

$$E_i = \text{zero} + \text{gain } i$$

The resolution or peak width S_{jk} of an X-ray line at energy E_{jk} is given by

$$S_{jk} = \left[\left(\frac{\text{noise}}{2\sqrt{2\ln 2}}\right)^2 + \varepsilon \text{ Fano } E_{jk}\right]^{\frac{1}{2}} \quad (6.2.9)$$

in which 'noise' is the detector system's electronic contribution to the peak width (typically 80 to 100 eV FWHM) with the factor $2\sqrt{2\ln 2}$ to convert to σ units. Fano is the Fano factor ($\sim$0.114 for Si(Li) detectors) and ε the energy required to produce an electron–hole pair in the solid-state detector (3.85 eV for Si(Li) detectors).

Taking into account the energy and resolution calibration, the Gaussian can be written as

Gaussian (E_i, E_{jk})

$$= \frac{\text{gain}}{S_{jk}\sqrt{2\pi}} \exp\left[-\frac{(E_i - E_{jk})^2}{2S_{jk}^2}\right] \quad (6.2.10)$$

in which gain/($S_{jk}\sqrt{2\pi}$) is required to normalize the Gaussian so that the sum over all channels is unity. Instead of optimizing the position and width of each peak the parameters of the energy and resolution parameters are optimized, reducing the dimensionality of the problem. In the case of 10 elements with two peaks each, the number of parameters to be optimized is now 24. More importantly, all the information available in the spectrum is now used to estimate zero, gain, noise and Fano, and thus the positions and widths of all peaks. In this way, small overlapping peaks exhibiting low counting statistics can be estimated much more accurately provided that some well-defined peaks are available in the fitting region (Nullens *et al.*, 1979).

6.2.9 RESPONSE FUNCTION FOR AN ELEMENT

To further reduce the number of fitting parameters, entire elements are modelled rather than single peaks. This way, a number of lines belonging together in a logical way, such as the Kα_1 and Kα_2 of a doublet or all the K lines of an element, are fitted as one group, with one area parameter A representing the total number of counts in all the lines within the group. This is represented by:

$$y_P(i) = \sum_{j=1}^{ng} A_j \left[\sum_{k=1}^{np(j)} R_{jk} \text{ Gaussian } (E_i, E_{jk}) \right] \quad (6.2.11)$$

In the above equation, R_{jk} is the relative intensity of an individual line k within a peak group j, E_i is the energy of channel i, E_{jk} is the energy of the kth line of peak group j. The inner summation runs over all lines within the group $np(j)$ with $\sum_k R_{jk} = 1$. R_{jk} values are available in the literature. The outer summation runs over all the groups specified in the spectral region of interest.

6.2.10 DEVIATION FROM THE IDEAL GAUSSIAN PEAK SHAPE

As new energy-dispersive detection systems are able to handle much higher count rates, very large peaks can be obtained in a few seconds while this used to take several minutes or even hours. When such large peaks are fitted with a Gaussian, the deviation from the pure Gaussian shape becomes significant.

Figure 6.2.1 shows part of the spectrum of a V thin film standard. The almost flat continuum is higher at the low-energy side of the Kα and Kβ peaks than at the high-energy side. Also, at the low-energy side the peak is broader, this is the so-called tail which results in an asymmetric peak shape.

As indicated previously, the observer peak shape is partially caused by the non-ideal behaviour of the detector. For low-energy lines ($<10\,\text{keV}$), incomplete charge collection and other detector artifacts result in the observed tailing. For lines of higher energy, the tailing is a result of Compton scattering in the detector. Part of the deviation from the Gaussian shape is attributed to phenomena occurring in the sample; X-ray satellite transitions such as KLM radiative Auger transition on the low-energy side of Kα peak as well as KMM transitions on the low-energy side of the Kβ (Van Espen *et al.*, 1980; Campbell and Wang, 1991; Campbell, 1996; Campbell *et al.*, 1997; Campbell *et al.*, 1998).

Failure to account for the deviation from the Gaussian peak shape causes a number of problems when fitting X-ray spectra. A small peak located on the tail of a larger one cannot be fitted accurately, resulting in large systematic errors for the small peak. Because the least-squares method seeks to minimize the difference between observed and fitted spectrum, the tail might become filled up with peaks of elements that are not really present. Also, the continuum over the entire range of the spectrum becomes difficult to describe. To a certain extent, the least-square method might use the polynomial, used to describe the continuum, to correct for the tailing of the characteristic peaks.

6.2.11 NUMERICAL PEAK SHAPE CORRECTION

A simple but efficient method to account for the peak tailing is a numerical one (Van Espen *et al.*, 1977). The deviation from the Gaussian shape is stored as a table of numerical values, representing the difference of the observed shape and the pure

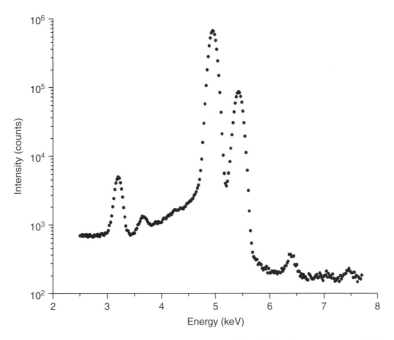

Figure 6.2.1 Part of the spectrum of a V thin film standard detected with a Si(Li) detector. For excitation a Rh X-ray tube is used in combination with a 0.05 mm thick Rh filter. The analysis time was 15 000 s. Main peaks in the spectrum are the V Kα and Kβ peaks, the peaks located at $\sim$3.1 keV and $\sim$3.6 keV are the escape peaks. The shelf and tail features are clearly visible

Gaussian. The table extends from zero energy up to the high-energy side of the Kβ peak and is normalized to the area of the Kα peak. The deviation is obtained from pure element spectra having high counting statistics (area K$\alpha \sim 10^7$ counts). Preferably, thin films are used to keep the continuum as low as possible and to avoid absorption effects. The area, position, and width of all peaks in the spectrum are determined by fitting Gaussians on a constant continuum over the full width at tenth maximum (FWTM) of the peaks. The Gaussian contributions are then stripped from the spectrum. The resulting non-Gaussian part is further smoothed and subsequently used as a numerical peak-shape correction. The fitting function for the characteristic lines is now given by

$$y_P(i) = \sum_{j=1}^{ng} A_j \left\{ R_{jK\alpha}[\text{Gaussian}(E_i, E_{jK\alpha}) + C_i] \right.$$
$$\left. + \sum_{k=2}^{np(j)} R_{jk} \text{Gaussian}(E_i, E_{jk}) \right\} \quad (6.2.12)$$

where C_i is the numerical peak shape correction at channel i. Values in the table are interpolated to account for the difference between the energy scale of the correction and the actual energy calibration of the spectrum.

A major advantage of this method is its computational simplicity and the fact that no extra parameters are required in the model. However, it is quite difficult and laborious to obtain good experimental peak-shape corrections and they are, to some extent, detector dependent. Also, since the peak shape correction is only related to the area of Kα, the peak shape correction for Kβ becomes underestimated when strong differential absorption takes place. In addition, it is nearly impossible to use this method to describe the L lines. Certainly, numerical peak shape correction meant a breakthrough in the early years of ED spectrum evaluation, enabling an efficient way to correct for the deviation from the ideal Gaussian peak shape. Nowadays, it is replaced by more flexible methods carrying out the correction through analytical functions.

6.2.12 MODIFIED GAUSSIANS

A number of analytical functions have been proposed to account for the true line shape. Nearly all of them include a flat shelf and an exponential tail, both convoluted with the Gaussian response function. The original function was first introduced by Philips and Marlow (1976) to describe the peak shape observed in γ-ray spectra. Later on, other authors adopted, extended and improved this function (Jorch and Campbell, 1977; Gardner and Doster, 1982; Yacout et al., 1986; Vekemans et al., 1994).

To account for the deviation from the Gaussian peak shape, the Gauss function, Gaussian (E_i, E_{jk}) in Equation (6.2.12) is replaced by

$$F(E_i, E_{jk}, f_{jk}^S, f_{jk}^T, \gamma_{jk})$$
$$= \text{Gaussian } (E_i, E_{jk}) + f_{jk}^S \text{ shelf } (E_i, E_{jk})$$
$$+ f_{jk}^T \text{ tail } (E_i, E_{jk}, \gamma_{jk}) \quad (6.2.13)$$

in which Gaussian (E_i, E_{jk}) is the Gaussian part given earlier while the shelf and tail functions denoted by shelf (E_i, E_{jk}) and tail $(E_i, E_{jk}, \gamma_{jk})$ are given by.

$$\text{Shelf } (E_i, E_{jk}) = \frac{\text{gain}}{2E_{jk}}$$
$$\times \text{erfc} \left(\frac{E_i - E_{jk}}{S_{jk}\sqrt{2}} \right) \quad (6.2.14)$$

$$\text{Tail } (E_i, E_{jk}) = \frac{\text{gain}}{2\gamma_{jk} S_{jk} \exp\left(-\frac{1}{2\gamma_{jk}^2}\right)}$$
$$\times \exp\left(\frac{E_i - E_{jk}}{\gamma_{jk} S_{jk}} \right)$$
$$\times \text{erfc} \left(\frac{E_i - E_{jk}}{S_{jk}\sqrt{2}} + \frac{1}{\gamma_{jk}\sqrt{2}} \right) \quad (6.2.15)$$

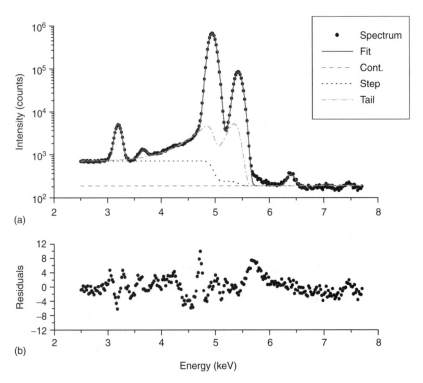

Figure 6.2.2 The same spectrum of V as shown in Figure 6.2.1 but now the spectrum is fitted with a modified Gaussian. Also the escape peaks are fitted. This is done with a simple Gaussian, the fraction of counts ending up in the escape peaks can be calculated from theory. The continuum is fitted with a first-degree linear polynomial, i.e. a constant value

In these equations, S_{jk} represents the spectrometer resolution and γ_{jk} is the broadening of the exponential tail. The parameters f_{jk}^S and f_{jk}^T describe the fraction of the photons that end up in the shelf and the tail respectively. In some studies, the exponential function is slightly different from the one presented here. Those authors use β_{jk} instead of the term $\gamma_{jk}S_{jk}$. Here, β_{jk} is defined as a multiple, γ_{jk}, of S_{jk}. In this way, the parameter γ_{jk} becomes independent of the resolution of the detector. However, both β_{jk} and $\gamma_{jk}S_{jk}$ may be used to describe the slope of the exponential tail.

Figure 6.2.2 represents the fitted spectrum of V (already shown in Figure 6.2.1) and demonstrates that a fitting model based on a modified Gaussian describes the data well. A χ^2 value of 7.1 was obtained which is very good taking into account that the intensity at the maximum of V Kα is nearly 10^6 counts. A discrepancy clearly seen at the high-energy side of the V Kβ peak is due to the Lorentzian character of the peaks.

Figures 6.2.3(a) and 6.2.3(b) show the effect of using simple Gaussians instead of modified Gaussians. Here, the spectrum of a brass sample (NIST SRM 1106) is fitted. When simple Gaussians are used, the Ni Kα peak is clearly overestimated since it is used to fill up the tail of the Cu lines. When modified Gaussians are used, a realistic fit is obtained.

In the above equations, erfc is the complement of the error function erf and serves to round off the exponential and shelf terms to avoid non-physical sharp edges. (Actually, the erfc component results from convoluting an exponential tail or shelf with the Gaussian detector response function.) Erf and erfc are given by the following expressions

$$\mathrm{erfc}(x) = 1 - \mathrm{erf}(x) = 1 - \frac{2}{\sqrt{\pi}} \int_0^x e^{-t^2} dt$$

$$= \frac{2}{\sqrt{\pi}} \int_x^{+\infty} e^{-t^2} dt \qquad (6.2.16)$$

and can be calculated via series expansion.

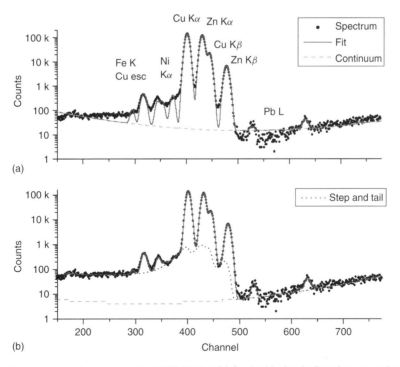

Figure 6.2.3 Part of the spectrum of a brass sample NIST SRM 1106 fitted with simple Gaussians (a) and modified Gaussians (b). The continuum is fitted with a third degree polynomial in both cases. When only simple Gaussians are used, the Ni peaks are overestimated and the continuum is not correctly fitted. Both the Ni peaks and the continuum are used to compensate for the tailing of the Cu and Zn K lines. Reproduced by permission of international centre for diffraction data

The parameters f_{jk}^S, f_{jk}^T and γ_{jk} vary smoothly with the X-ray energy of the characteristic peak. Campbell *et al.* (1985, 1987) conducted many experiments employing monoenergetic X-ray beams that were obtained from a curved crystal monochromator. Deriving the values of the different parameters and relating them to the X-ray energy showed, evidently, no difference between $K\alpha$ and $K\beta$ lines. As already shown by Wielopolski and Gardner (1979), this is not the case when real K X-rays are examined. It is found that the parameters are not unique but take on different values for the $K\alpha$ and $K\beta$ series. This is due to the appearance of radiative phenomena (e.g. radiative Auger effects) in the vicinity of the $K\alpha$ and $K\beta$ signals.

Fitting real X-ray spectra with modified Gaussians dramatically increases the number of parameters that need to be optimized for each peak. Besides the area, A, also the tail parameters, f_{jk}^S, f_{jk}^T and γ_{jk} are now involved. In our example of 10 elements with two peaks each, the number of parameters has again increased to 84! In practice less parameters are needed since not all the peaks in the spectrum have such intensities that peak shape correction through modified Gaussians are required. To decrease the number of parameters, the energy dependence of the shelf and tail parameters can be used. In this way, functions describing the energy dependence can again reduce the total number of parameters, similar to the energy and resolution calibration.

6.2.13 REPLACING THE GAUSSIAN BY A VOIGTIAN

The fact that a characteristic X-ray has an intrinsic energy distribution of Lorentzian nature might influence the fitting results. Two aspects are important: the Lorentzian width itself and the tailing of the Lorentzian function. For the energy region usually investigated (1 to 15 keV) the contribution of the Lorentzian width (1 to 5 eV) is negligible in comparison with the FWHM of the detector (100 to 300 eV). As indicated previously, the convolution of the Lorentzian function with the Gaussian broadening function of the detector, resulting in a Voigt function is usually not taken into account. Instead, only a Gaussian function is used. If the energy of the characteristic photon is higher than 15 keV, the contribution of the Lorentzian width increases rapidly, e.g. for Sn ($K\alpha = 25.27$ keV) the Lorentzian width is in the order of 11 eV, for Ba ($K\alpha = 32.19$ keV) 16 eV and for W ($K\alpha = 59.31$ keV) 45 eV. In the case of L lines the effect is even more important since the Lorentzian width of these lines is larger at lower energies, e.g. 6 eV for W $L\alpha_1$ ($E = 8.398$ keV) and 12 eV for W $L\beta_3$ ($E = 9.819$ keV) (Campbell and Papp, 1995). If such lines are fitted, the Lorentzian broadening cannot be neglected. Another more important feature of the Lorentzian distribution is its tailing. Unlike a Gaussian, the Lorentzian function falls to zero much less rapidly. While the peak broadening becomes apparent only at characteristic photon energies higher than 15 keV, the tailing effect is also observable for photon energies below 15 keV. Therefore, replacing the Gauss by a Voigt profile results in a more accurate detector response function (Campbell and Wang, 1992).

The Voigtian is the convolution of a Lorentzian with a Gaussian distribution and can be written as

$$V(E) = \int_{-\infty}^{+\infty} L(E')G(E - E')dE' \quad (6.2.17)$$

in which E is the energy along the convoluted spectrum. $L(E')$ and $G(E - E')$ are the Lorentzian distribution and the Gaussian distribution, given by the following expressions:

$$L(E') = \frac{\frac{\Gamma}{2\pi}}{(E' - E_{jk})^2 + \left(\frac{\Gamma}{2}\right)^2} \quad (6.2.18)$$

$$G(E_i - E') = \frac{\text{Gain}}{S_{E'}\sqrt{2\pi}} \exp\left[-\frac{(E_i - E')^2}{2S_{E'}^2}\right] \quad (6.2.19)$$

in which Γ is the Lorentzian width of characteristic line of energy E_{jk} and $S_{E'}$ is the Gaussian width at energy E'. Substitution in the expression representing the convolution gives:

$$V(E_i) = \int_{-\infty}^{+\infty} \frac{\Gamma}{S_{E'} 2\pi\sqrt{2\pi}} \frac{\text{gain}}{(E' - E_{jk})^2 + \left(\frac{\Gamma}{2}\right)^2}$$

$$\times \exp\left[-\frac{(E_i - E')^2}{2S_{E'}^2}\right] dE' \quad (6.2.20)$$

Rearranging and adding $E_i - E_i$ gives

$$V(E_i) = \int_{-\infty}^{+\infty} \frac{1}{2\sqrt{2}} \frac{\Gamma}{S_{E'}} \frac{1}{\pi\sqrt{\pi}}$$

$$\times \frac{\text{gain}}{[(E_i - E_{jk}) - (E_i - E')]^2 + \left(\frac{\Gamma}{2}\right)^2}$$

$$\times \exp\left[-\frac{(E_i - E')^2}{2S_{E'}^2}\right] dE' \quad (6.2.21)$$

Introducing in this expression the following substitutions

$$x = \frac{E_i - E_{jk}}{S_{E'}\sqrt{2}} \quad (6.2.22)$$

$$y = \frac{E_i - E'}{S_{E'}\sqrt{2}} \quad (6.2.23)$$

$$a = \frac{1}{2\sqrt{2}} \frac{\Gamma}{S_{E'}} \quad (6.2.24)$$

gives for the Voigt profile, symbolized as $V(x, a)$

$$V(x, a) = \int_{-\infty}^{+\infty} \frac{a\sqrt{2}S_{E'}}{\pi\sqrt{\pi}}$$

$$\times \frac{\text{gain}}{[\sqrt{2}S_{E'}(x - y)]^2 + (\sqrt{2}S_{E'}a)^2}$$

$$\times \exp(-y^2) dy \quad (6.2.25)$$

or

$$V(x, a) = \frac{\text{gain}}{\sqrt{2\pi}S_{E'}} K(x, a) \quad (6.2.26)$$

where $K(x, a)$ is in general known as the Voigt function

$$K(x, a) = \frac{a}{\pi} \int_{-\infty}^{+\infty} \frac{\exp(-y^2)}{(x - y)^2 + a^2} dy \quad (6.2.27)$$

The Voigt function can be written as the real part of the complex error function:

$$W(z) = e^{-z^2}\left(1 + \frac{2i}{\sqrt{\pi}} \int_0^z e^{t^2} dt\right) = e^{-z^2}$$

$$\text{erfc}(iz) = K(x, a) + iL(x, a) \quad (6.2.28)$$

where $z = x + ia$. $L(x, a)$ is expressed as follows:

$$L(x, a) = \frac{1}{\pi} \int_{-\infty}^{+\infty} \frac{(x - y)\exp(-y^2)}{(x - y)^2 + a^2} dy \quad (6.2.29)$$

Various computational procedures have been published for the evaluation of the Voigt function, $K(x, a)$. They are based upon numerical expansions in different regions of the x, a space. Procedures are known that calculate only the real part of the complex probability integral while others evaluate the complete probability function $W(z)$ (Armstrong, 1967; Humlicek, 1982). The calculation of the complete probability function has the advantage that the imaginary part can be used for the evaluation of the partial derivatives of the Voigt function. Schreier (1992) gives an overview of the available computational procedures. Until now, the best algorithm is the one developed by Poppe and Wijers (1990a, 1990b), which is an improved version of Gautschi's algorithm (Gautschi, 1969). The accuracy of this algorithm is 14 significant digits throughout most of the complex plane while the speed is comparable with other, less accurate algorithms.

Figure 6.2.4 shows the same spectrum as Figures 6.2.1 and 6.2.2 but now a Voigtian is used in combination with a shelf and tail. The χ^2 value drops to 3.7 and the misfit at 5.8 keV is now absent.

To conclude this section, Figures 6.2.5 and 6.2.6 show the fitted spectra of NIST SRM 1155 and NIST SRM 2710, a stainless steel and contaminated soil sample, respectively. χ^2 values of 1.46 and 4.0 are obtained. Notice that almost every characteristic peak in the spectrum of the soil sample is fitted in the same run.

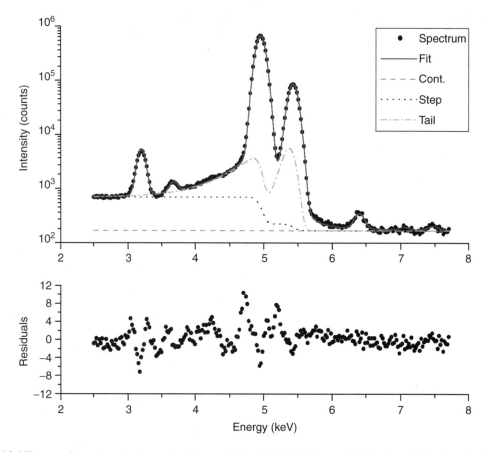

Figure 6.2.4 The same V spectrum is shown as in Figures 6.2.1 and 6.2.2. This time, a Voigtian based fitting function including shelf and tail is used. The skirts of the Lorentzian distribution of the V K lines, clearly seen at the basis of the V Kβ peak, are now fitted. This was not the case in Figure 6.2.3. The standardized residuals do not show a systematic deviance anymore around 5.7 keV

6.2.14 CONVOLUTION OF MC SIMULATED SPECTRA

The previous subchapter gives a detailed overview of the MC technique as applied to the simulation of XRF spectra. Simulated X-ray spectra can be used for quantitative analysis as well as to study the behaviour and performance of spectrum processing methods. The know-how gained via MC simulations can also be used for the design and optimization of new instrumental set-ups. In either case, the simulated spectrum is one as seen by an ideal detector with infinite resolution or, put in another way, it is the spectrum as it impinges on the detector surface. This spectrum still needs to be convoluted with the detector response function to obtain the familiar pulse-height spectrum. Therefore, apart from a sophisticated MC simulation code, the method also requires an accurate detector response function.

We already identified peak tailing as being a detector related phenomenon. Obviously, the feature is not taken into account during the MC simulation step. During the convolution all the counts belonging to a characteristic peak are redistributed over a Gaussian (or Voigtian) as well as the tail and the shelf. Numerical peak shape correction cannot be used here because it would imply adding extra counts to the system and altering the peak ratios. The sole plausible

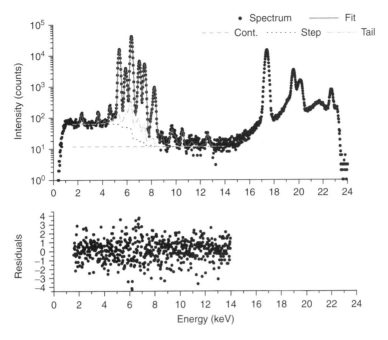

Figure 6.2.5 The entire spectrum of a stainless steel sample NIST SRM 1155. The region between 1 keV and 14 keV is fitted. The major peaks are fitted with modified Voigtians. To fit the continuum a first degree linear polynomial is used. Notice that for this energy region the continuum is nearly absent

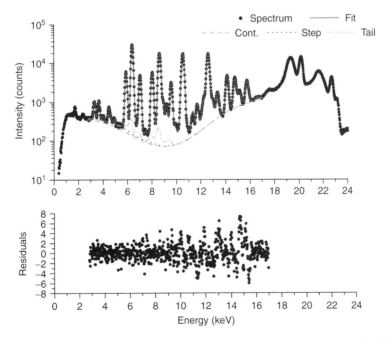

Figure 6.2.6 Spectrum of a contaminated soil sample NIST SRM 2710. All characteristic peaks are fitted simultaneously. The continuum is modelled with a sixth degree exponential polynomial. The L lines of Pb are fitted with Voigtians without shelf and tail. The high intensity K lines are fitted with Voigtians including shelf and tail functions

method is to use a detector response function based on modified Gaussians. To obtain a perfect agreement between simulated and experimental spectra it is necessary to derive the values of the parameters f_{jk}^S, f_{jk}^T and γ_{jk} for the detector system with which the experimental spectra are obtained. The characteristic element lines, continuum and, if present, the elastically and inelastically scattered characteristic lines of the source are all convoluted with the same detector response function. In addition, also escape peaks and sum peaks are taken into consideration.

An example is given in Figure 6.2.7 where the spectrum of the NIST SRM 1106 brass sample is simulated and convoluted with the appropriate detector response function. The experimentally measured spectrum is also shown. The standardized residuals show that the agreement is nearly perfect.

6.2.15 PARTIAL LEAST-SQUARES REGRESSION

As discussed throughout this subchapter, the aim of spectrum evaluation is to derive the net intensities of the characteristic element lines. In a next step, these net peak intensities are used to determine the constituent concentrations using one of the empirical, semi-empirical or fundamental quantification methods. The procedure of spectrum evaluation followed by an independent quantification has proven to work very well but it remains a rather complicated and time-consuming process requiring a high degree of experience and knowledge from the operator. This makes it very hard to automate the process. Moreover, the spectrum evaluation method will fail when the encountered peak shapes differ too much from the employed peak model. Also, any structures in the spectrum not taken

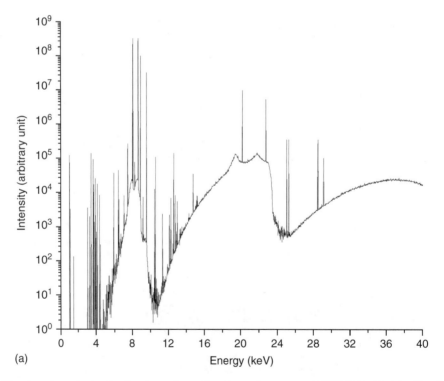

(a)

Figure 6.2.7 The plots show the simulated and experimentally measured spectrum of NIST SRM 1106. (a) Simulated spectrum before convolution with the detector response function. This is the spectrum as seen by an ideal detector with infinite resolution and detector efficiency equal to 1. (b) Simulated spectrum after convolution with a detector response function based on modified Gaussians and corrected for the detector efficiency. Together with the simulation the real spectrum is also shown. The standardized residuals show that the agreement is almost perfect when the uncertainty due to counting statistics is taken into account

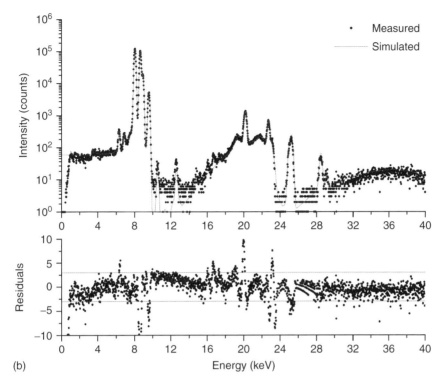

Figure 6.2.7 (*continued*)

into account by the fitting model (e.g. Compton peaks) might pose problems. During the past few years new approaches based on multivariate calibration have appeared in the field of quantitative X-ray spectrometry. In what follows we will concentrate on partial least-squares regression (PLS), a technique comprehensively written about in the chemeometrics literature. PLS showed to be very useful with quantitative infrared (IR) and ultraviolet (UV) spectrometry where the spectral information is less selective compared to ED-XRF. This selectivity is probably the only reason why PLS had not yet been applied to ED-XRF or WD-XRF. An important advantage of PLS is its ability to take care of the spectrum evaluation and quantification in one single step.

6.2.15.1 THEORY

PLSR is a multivariate calibration method able to relate element concentrations directly to the measured XRF spectra (or portions of them). The spectral variables are collected as a matrix $\mathbf{X}$ with a row number equal to the number of samples and a column number equal to the number of channels in the spectra. The $\mathbf{Y}$ matrix consists of the concentrations with a number of rows equal to the number of samples and the number of columns equal to the constituents of interest. The relationship can then be written in the form:

$$\mathbf{Y} = \mathbf{XB} + \mathbf{F} \qquad (6.2.30)$$

in which $\mathbf{F}$ is the matrix containing the residuals (variation not described by the model, e.g. misfit or noise). The regression coefficients $\mathbf{B}$ can be calculated in several ways. One widespread method is multiple linear regression (MLR) in which the least squares solution is given by:

$$\mathbf{B} = (\mathbf{X'X})^{-1}\mathbf{X'Y} \qquad (6.2.31)$$

However, when the number of x-variables exceeds the number of samples and/or when there is a high degree of correlation (also known as collinearity) among the variables, the resulting solution is not stable. Mathematically, this means the inverse of covariance matrix $\mathbf{X'X}$ does not exist. This is the case for ED-XRF where the number of channels largely exceeds the number of samples and where the intensities of neighbouring channels in each peak and peaks of the same element are very strongly correlated.

The PLSR method handles this collinearity problem by compressing the $\mathbf{X}$ data matrix $\mathbf{X} = [\mathbf{x}_1, \mathbf{x}_2, \ldots, \mathbf{x}_p]$ containing p spectrum channels for n samples, into a number of A orthogonal latent variables or scores $\mathbf{T} = [\mathbf{t}_1, \mathbf{t}_2, \ldots, \mathbf{t}_A]$. In theory, there are as many latent variables as there are original samples (or variables, depending on which of the two is smaller). In practice, only the most significant, carrying most of the variance in the data, are used. In the end, a number of latent variables usually much smaller than the original number of variables p are retained. The scores $\mathbf{T}$ are employed to fit a set of n observations to m dependent concentration variables $\mathbf{Y} = [\mathbf{y}_1, \mathbf{y}_2, \ldots, \mathbf{y}_p]$. Since the latent variables are orthogonal (linearly independent from each other), the inverse can easily be obtained, solving the problem of collinearity. When one PLS model is built per constituent (PLS1), the vectors $\mathbf{t}$ are easily found via singular value decomposition of the matrix $\mathbf{X'yy'X}$. If more than one constituent is modelled with the same PLS model (PLS2), the derivation of the $\mathbf{t}$ vectors is conceptually less straightforward.

The PLSR model can be considered as consisting of two outer relations and an inner relation. There is an outer relation for the $\mathbf{X}$ matrix:

$$\mathbf{X} = \mathbf{TP'} + \mathbf{E} = \sum_{a}^{A} t_a p'_a + \mathbf{E} \quad (6.2.32)$$

As well as for the $\mathbf{Y}$ matrix:

$$\mathbf{Y} = \mathbf{UQ'} + \mathbf{F} = \sum_{a}^{A} u_a q'_a + \mathbf{F} \quad (6.2.33)$$

In which $\mathbf{P}$ and $\mathbf{Q}$ are the loadings of the $\mathbf{X}$ and $\mathbf{Y}$ variables block, respectively, $\mathbf{E}$ and $\mathbf{F}$ are matrices containing residuals. A represents the number of latent variables retained in the PLSR model. The loadings describe how the original $\mathbf{X}$ and $\mathbf{Y}$ variables are related to the scores $\mathbf{T}$ and $\mathbf{U}$. The inner relation is written as:

$$\mathbf{u}_a = \mathbf{b}_a \mathbf{t}_a \quad (6.2.34)$$

In essence, the inner relation is a least squares fit between the $\mathbf{X}$ block scores and $\mathbf{Y}$ block scores. When all the necessary scores and loadings are calculated the final PLSR model can be written as:

$$\mathbf{Y} = \mathbf{TBQ'} + \mathbf{F} \quad (6.2.35)$$

Figure 6.2.8 illustrates these relationships graphically.

As explained, the PLS method compresses the original variables into a number of latent variables. Validating the PLS model essentially concerns the selection of the optimal number of PLS dimensions/components. In addition, the validation method provides a value for the prediction error enabling the assessment of the predictive capacity of the model.

The determination of the optimum number of PLS components is mainly done by calculation of the root mean squared error (RMSE)

$$\text{RMSE} = \sqrt{\frac{\sum_{i=1}^{n} (\hat{y}_i - y_i)^2}{n}} \quad (6.2.36)$$

in which n denotes the number of observations, y is the given (or 'true') value of the constituent of interest and $\hat{y}$ the concentration predicted by the PLS model.

Given a certain data set, the RMSE values are calculated for different numbers of components included in the PLS model. Normally the RMSE reduces with increasing number of PLS components until a minimum or constant value is reached and the corresponding number of components is regarded as optimal. The prediction error is composed of two contributions, the remaining interference error and the estimation error. The former is the systematic error due to unmodelled interference in the spectral data and the latter is caused by random measurement noise of various kinds or by

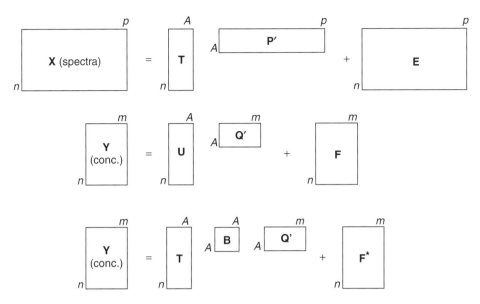

Figure 6.2.8 Graphical representation of the outer and inner models of PLS with n the number of calibration samples, p the number of channels in the spectrum, A the number of latent variables retained and m the number of constituents modelled. Reproduced by permission of John Wiley & Sons, Ltd

systematic error not relevant for the modeled analyte y. The two contributions to the prediction error have opposite trends with increasing complexity of the PLS model. The interference error decreases with an increased modelling of the systematic variance by including more PLS components. However, at the same time the statistical uncertainty error increases. Including too few PLS components results in underfitting and one risks that important phenomena are not modelled. Including too many components results in overfitting and this is equivalent to the modelling of noise.

Calculation of the RMSE value is mostly done in either of two ways; leave one out cross validation (LOO-CV) or by means of a separate data set. The LOO-CV technique is generally used when only a limited number of samples are available to set up and validate a calibration. Each sample is excluded once from the data set and a PLS model is built based on the remaining samples. Once built, the excluded sample is predicted and the deviation stored. This is done until all samples have been excluded once. The RMSE value is the mean of the stored deviations. In the case of a separate data set or prediction set, the PLS model is built from the so-called training set and used to predict the samples of the prediction set.

Geladi and Kowalski (1986) published a tutorial on PLSR and its relation to other regression methods. A standard work on PLS and multivariate calibration in general is the book by Martens and Næs (1989). For a thorough discussion of the theoretical and statistical aspects of PLS, we refer to articles by Manne (1987), Lorber *et al.* (1987), Höskuldsson (1988), de Jong (1993), Phatak *et al.* (1992), Burnham *et al.* (1996) and Ter Braak and de Jong (1998).

6.2.15.2 APPLICATION

The PLS method is illustrated with the analysis of cement containing CaO, SiO_2, Al_2O_3, SO_3 and Fe_2O_3. Cement analysis is one of the major applications of XRF and most cement production plants maintain a quality control programme to assure that the final product is of constant quality and meets its preset specifications. Without exception, completely automated WD-XRF set-ups are used to determine the constituents of the product. In this example a cheap ED-XRF instrument equipped with a low power Cu anode and a sealed neon gas proportional counter with an energy resolution at Mn Kα of 700 eV is used. Due to the very low resolution of the instrument compared to Si based solid-state detectors, a high degree of overlap exists between the characteristic peaks in the spectrum. In such a case, spectrum evaluation will fail. The example will show that for these kind of applications PLS is well suited. More on the analysis of cement with PLS and low-resolution ED-XRF can be found in an article published by Lemberge *et al.* (2000).

The purpose of this PLS calibration is to determine the concentration of the most abundant component, CaO. In total, 14 certified cement standards (601A Cement, Japan Cement Association, Research and Development Laboratory, Tokyo, Japan) are measured. Table 6.2.1 gives the composition of the samples used in the PLS calibration. Sample preparation consisted of pressing 5 g of cement powder into an aluminium sample cup without use of a binder or diluent. Spectra were recorded with a tube voltage of 12 kV and a measurement time of 150 s. The recorded spectra consist of 2048 channels and span an energy range going from 0 to 8 keV. Figure 6.2.9 shows the spectrum for sample 1 with composition given in Table 6.2.1. Since there is only one constituent of interest, a PLS1 model for CaO is built. Due to the limited number of cement standards available it is not possible to split the data in a training and prediction set. Therefore we shall use LOO-CV to calculate the RMSE value and to determine the optimal number of PLS components.

In Figure 6.2.10 the RMSE is plotted as a function of the number of latent variables. In this case, the minimum is located at the first latent variable and the error slightly increases when more latent variables are included in the model. The minimum RMSE is equal to 0.29 % (m/m) and a PLS model with one latent variable is retained. Figure 6.2.11 compares the true CaO concentration with the predicted concentrations for the data set. The PLS model predicts the CaO concentrations very accurately in the range of 49 to 66 % (m/m). The regression coefficients of the PLS model show what parts of the spectrum (spectral variables or channels) are used by the PLS model to predict the CaO concentrations. The regression coefficients are plotted in Figure 6.2.12 and show that, as expected, the channels corresponding to the Ca signal contribute most to the PLS model. In addition, there is also a contribution at the position of the Si peak but unlike the positive contribution for Ca the one for Si is negative. This can be explained by observing Table 6.2.1 showing that the CaO concentration is to a large extent complementary to the SiO_2 concentration. When there is less CaO in the cement the SiO content is increased and vice versa. This information is also used by the PLS model to achieve better results. At the same time it makes the model only valid for this kind of sample and hence, the model becomes less robust.

To build an accurate PLS model, a large number of standards spanning the concentration range of each element or constituent of interest is necessary. This allows the PLS model to take into account all interactions occurring. Compared to the FPM or semi-empirical methods relying on theoretical principles, the PLS method is empirical and relies entirely on the data given

Table 6.2.1 Most abundant constituents of the analysed cement samples (601A Cement, Japan Cement Association, Research and Development Laboratory, Tokyo, Japan). In addition to the tabulated constituents, the samples also contain MgO, Na_2O, K_2O, TiO_2. P_2O_5 and MnO are at concentration levels below 1 % (m/m)

Sample no.	Concentrations [% (m/m)]				
	CaO	SiO_2	SO_3	Al_2O_3	Fe_2O_3
1	63.13	22.09	2.22	5.23	3.01
2	63.66	21.02	1.83	5.29	2.86
3	65.12	20.49	3.13	4.54	2.37
4	65.41	20.69	2.58	4.69	2.80
5	64.93	20.32	2.97	5.04	2.99
6	65.14	20.51	2.55	4.88	2.71
7	63.11	22.42	2.33	4.18	4.01
8	63.36	23.09	1.85	3.73	4.00
9	61.07	23.03	2.03	6.25	2.38
10	58.33	24.36	1.92	7.35	2.24
11	54.31	26.18	1.93	8.91	1.81
12	54.46	25.99	1.19	9.16	2.01
13	54.76	25.77	2.09	8.56	2.02
14	49.25	29.56	1.32	10.70	1.33

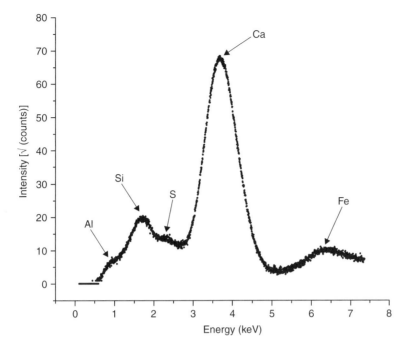

Figure 6.2.9 Low-resolution ED-XRF spectrum of cement sample 1 of Table 6.2.1. Due to the low resolution (FWHM at Mn Kα of 700 eV) a high degree of overlap exists between the element characteristic peaks. Reproduced by permission of John Wiley Sons, Ltd

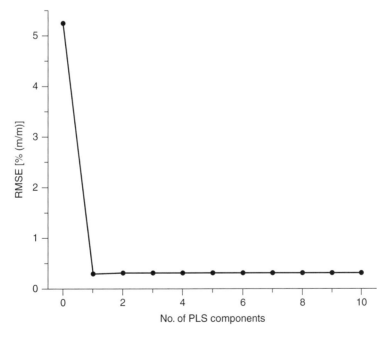

Figure 6.2.10 RMSE curve for CaO obtained via LOO-CV. The minimal value is already obtained with one PLS component (also known as latent variable)

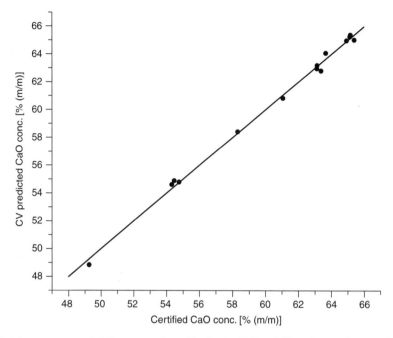

Figure 6.2.11 Predicted versus measured CaO concentrations. Clearly, the PLS model based on one latent variable only accurately predicts the CaO concentration. Reproduced by permission of John Wiley & Sons, Ltd

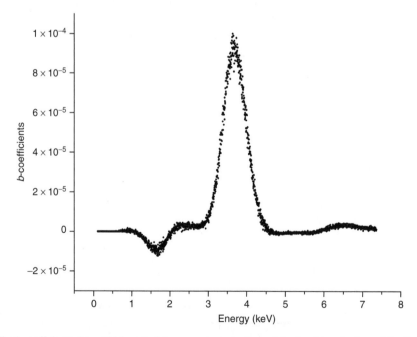

Figure 6.2.12 The b-coefficients show which part of the cement spectra is used for the determination of the CaO concentration. Large positive values are located at the position of the Ca peak. Evidently the Ca peak contains most of the information to predict the CaO concentration. Next to the Ca peak also the region of the Si peak is used. The negative values around 1.5 keV indicate an inverse relationship between the Si concentration and the Ca concentration. This is explained by the specificity of the cement composition. Reproduced by permission of John Wiley & Sons, Ltd

to it during the calibration. The large number of calibration samples required and the empirical aspect related to it, is certainly the major drawback of PLS for its application in XRF. However, with the availability of excellent MC simulation codes this problem can be overcome by using simulated spectra. Virtually any composition in accordance with an experimental design can then be chosen. Another drawback remains, due to matrix effects encountered with XRF spectrometry, the relationship between intensities of element characteristic lines and the element concentrations are in part nonlinear. PLS, being in essence a linear method able to handle only a certain degree of nonlinearity is probably less suited for XRF than for IR or UV spectrometry.

Wang et al. (1990) applied the PLS method for the analysis of nickel alloy samples employing WD-XRF. The authors used the instrument in scan mode to acquire a spectrum of each sample. Each spectrum consisted of 261 channels and the total measuring time was 7830 s (30 s per channel). Adams and Allen (1998) reported on the application of PLS to WD-XRF in combination with variable selection for the analysis of iron-based alloys, copper-based alloys and geological samples. Swerts et al. (1993) were the first to apply PLS in combination with ED-XRF for the analysis of sulfur–graphite mixtures. The PLS method was used both for the analysis of sulfur and the explanation of various artifacts observed in the spectra. Later, Urbanski and Kowalska (1995) applied the PLS method to various low-resolution ED-XRF analyses. Lemberge and Van Espen (1999) used the PLS method for the determination of Ni, Cu and As in liquid samples. They showed that taking the square root of the data improves the PLS model and that the PLS method extracts information from the scattered excitation radiation to describe the matrix effects. PLS has also been applied for EPXMA and μ-XRF analysis of glass samples (Lemberge et al., 2000). The PLS method seemed to perform better for EPXMA than for XRF because of smaller matrix effects and hence a reduced nonlinearity effect in the intensity–concentration relationship for EPXMA.

6.2.16 FUTURE PERSPECTIVES

Though the development of modified Gaussian and Voigtian based detector response functions have been going on for over 10 years, it was not possible to implement them as practical and user-friendly spectrum evaluation software. The reason for this is two-fold. The shelf and tail functions add extra parameters to the fitting algorithm and though only peaks with high counting statistics require these functions, it becomes more difficult to find a global minimum in the χ^2 space. Preferably, the software should decide automatically which peaks require an extended peak model. Next to this, ordinary PCs were not powerful enough to fit a spectrum containing many peaks using only modified Gaussians or Voigtians. However, new PCs deliver massive amounts of processing power. Also, the relationship of the shelf and tail parameters as a function of energy is intensively studied. Most probably this research will result in physically meaningful functions describing the relationships, further decreasing the number of fitting parameters. With this knowledge, robust and accurate spectrum evaluation software is possible. The availability of such software will renew interest in ED-XRF. In the near future, new high performance ED-XRF instruments will become available. Together with recent advances in hardware the precision of such instruments will improve considerably. Detection limits tube-excited ED-XRF instruments will improve from the lower p.p.m. range at the moment to sub p.p.m. during the next few years. Hence, ED-XRF will become an important competitor of inductively coupled plasma (ICP) instruments, especially for solid samples.

PLS and other multivariate methods might prove important for instruments operating without supervision, e.g. portable instruments used in the field or on-line systems. In such dedicated applications the type of samples are usually of the same type, e.g. soil analysis, alloy sorting and the PLS method will be more than adequate to obtain accurate results. PLS can also be applied to the analysis of images obtained with micro X-ray spectroscopy. The image consists of pixels and in each pixel a complete X-ray spectroscopy spectrum is recorded. For a small image of 50 × 50 pixels this results in 2500 spectra and all of them need to be deconvoluted individually. With PLS, the data processing will be much faster provided that appropriate calibration standards are available.

REFERENCES

Adams, M.J. and Allen, J.R. Variable selection and multivariate calibration models for X-ray fluorescence spectrometry. *J. Anal. At. Spectrom.* **13**, 119–124 (1998).

Armstrong, B.H. Spectrum line profiles: the Voigt function. *J. Quant. Spectrosc. Radiat. Transfer* **7**, 61–88 (1967).

Burnham, A.J., Viverbos, R. and MacGregor, J.F. Frameworks for latent variable multivariate regression. *J. Chemom.* **10**, 31–45 (1996).

Campbell, J.L. Si(Li) detector response and PIXE spectrum fitting. *Nucl. Instrum. Methods B* **109/110**, 71–78 (1996).

Campbell, J.L. and Papp, T. Atomic level widths for X-ray spectrometry. *X-Ray Spectrom.* **24**, 307–319 (1995).

Campbell, J.L. and Wang, J.X. Improved model for the intensity of low-energy tailing in Si(Li) X-ray spectra. *X-Ray Spectrom.* **20**, 191–197 (1991).

Campbell, J.L. and Wang, J.X. Lorentzian contributions to X-ray lineshapes in Si(Li) spectroscopy. *X-Ray Spectrom.* **21**, 223–227 (1992).

Campbell, J.L., Cauchon, G., Lépy, M.-C., McDonald, L., Plagnard, J., Stemmler, P., Teesdale, W.J. and White, G. A quantitative explanation of low-energy tailing features of Si(Li) and Ge X-ray detectors, using synchrotron radiation. *Nucl. Instrum. Methods A* **418**, 394–404 (1998).

Campbell, J.L., Maxwell, J.A., Papp, T. and White, G. Si(Li) detector lineshapes: contributions from atomic physics and detector properties. *X-Ray Spectrom.* **26**, 223–231 (1997).

Campbell, J.L., Millman, B.M., Maxwell, J.A., Perujo, A. and Teesdale, W.J. Analytic fitting of monoenergetic peaks from Si(Li) X-ray spectrometers. *Nucl. Instrum. Methods B* **9**, 71–79 (1985).

Campbell, J.L., Perujo, A. and Millman, B.M. Analytic description of Si(Li) spectral lineshapes due to monoenergetic photons. *X-Ray Spectrom.* **16**, 195–201 (1987).

Castoldi, A., Gatti, E., Guazzoni, C., Longoni, A., Rehak, P. and Strüder, L. The controlled-drift detector. *Nucl. Instrum. Methods A* **439**, 519–528 (2000).

de Jong, S. SIMPLS: An alternative approach to partial least squares regression. *Chemom. Intell. Lab. Syst.* **18**, 251–263 (1993).

Gardner, R.P. and Doster, J.M. Treatment of the Si(Li) detector response as a probability density function. *Nucl. Instrum. Methods* **198**, 381–390 (1982).

Gatti, E. and Rehak, P. Semiconductor drift chamber – an application of a novel charge transport scheme. *Nucl. Instrum. Methods A* **225**, 608–614 (1984).

Gautschi, W. Algorithm 363: Complex error function. *Commun. ACM* **12**, 635 (1969).

Geladi, P. and Kowalski, B.R. Partial least-squares regression: a tutorial. *Anal. Chim. Acta* **185**, 1–17 (1986).

Gunnink, R. An algorithm for fitting Lorentzian-broadened, K-series X-ray peaks of the heavy elements. *Nucl. Instrum. Methods* **143**, 145–149 (1977).

Höskuldsson, A. PLS regression methods. *J. Chemom.* **2**, 211–228 (1988).

Humlicek, J. Optimized computation of the Voigt and complex probability function. *J. Quant. Spectrosc. Radiat. Transfer* **27**, 437 (1982).

Jorch, H.H. and Campbell, J.L. On the analytic fitting of full energy peaks from Ge(Li) and Si(Li) photon detectors. *Nucl. Instrum. Methods* **143**, 551–559 (1977).

Jordanov, V.T., Knoll, G.F., Huber, A.C. and Pantazis, J.A. Digital techniques for real-time pulse shaping in radiation measurements. *Nucl. Instrum. Methods A* **353**, 261–264 (1994).

Lemberge, P. and Van Espen, P.J. Quantitative energy-dispersive X-ray fluorescence analysis of liquids using partial least-squares regression. *X-Ray Spectrom.* **28**, 77–85 (1999).

Lemberge, P., De Raedt, I., Janssens, K., Wei, F. and Van Espen, P. Quantitative analysis of 16–17th century archaeological glass vessels using PLS regression of EPXMA and μ-XRF data. *J. Chemom.* **14**, 751–763 (2000).

Lemberge, P., Van Espen, P. and Vrebos, B. Analysis of cement using low-resolution energy-dispersive X-ray fluorescence and partial least-squares regression. *X-Ray Spectrom.* **29**, 297–304 (2000).

Levenberg, K. A method for the solution of certain non-linear problems in least squares. *Quart. Appl. Maths.* **2**, 164–168 (1944).

Lorber, A., Wangen, L.E. and Kowalski, B.R. A theoretical foundation for the PLS algorithm. *J. Chemom.* **1**, 19–31 (1987).

Manne, R. Analysis of two partial-least-squares algorithms for multivariate calibration. *Chemom. Intell. Lab. Syst.* **2**, 187–197 (1987).

Marquardt, D.W. An algorithm for least-squares estimation of non-linear parameters. *J. Soc. Ind. Appl. Math.* **11**, 431–441 (1963).

Martens, H. and Naes, T. *Multivariate Calibration*, Wiley, Chichester, 1989.

McCarthy, J.J. and Schamber, F.H. NBS Special Publication 604, 1981.

Murty, V.R.K., Winkoun, D.P. and Devan, K.R.S. On the comparison of performance of freoelectric cooled Si(Li) and Si-PIN Peltier cooled detectors. *Radiat. Phys. Chem.* **51**, 459–460 (1998).

Nullens, H., Van Espen, P. and Adams, F. Linear and non-linear peak fitting in energy-dispersive X-ray fluorescence. *X-ray Spectrom.* **8**, 104–109 (1979).

Phatak, A., Reilly, P.M. and Pemlidis, A. The geometry of 2-block partial least squares regression. *Commun. Statist. Theory Meth.* **21**, 1517–1553 (1992).

Philips, G.W. and Marlow, K.W. Automatic analysis of Gamma-ray spectra from Germanium detectors. *Nucl. Instrum. Methods* **137**, 525–536 (1976).

Poppe, G.P.M. and Wijers, C.M.J. More efficient computation of the complex error function. *ACM Trans. Math. Software* **16**, 38–46 (1990a).

Poppe, G.P.M. and Wijers, C.M.J. Algorithm 680: evaluation of the complex error function. *ACM Trans. Math. Software* **16**, 47 (1990b).

Schamber, F.H. in *X-ray Fluorescence Analysis of Environmental Samples* (Ed. T. Dzubay), Ann Arbor, Michigan, 1977 pp. 241–257.

Schreier, F. The voigt and complex error function: a comparison of computational methods. *J. Quant. Spectrosc. Radiat. Transfer* **48**, 743–762 (1992).

Statham, P.J. Pitfalls in linear and non-linear profile-fitting procedures for resolving severely overlapped peaks. *X-Ray Spectrom.* **7**, 132–137 (1978).

Steenstrup, S.J. A simple procedure for fitting a background to a certain class of measured spectra. *Appl. Crystallogr.* **14**, 226–229 (1981).

Swerts, J., Van Espen, P. and Geladi, P. Partial least squares determination in the energy-dispersive X-ray fluorescence determination of sulfur–graphite mixtures. *Anal. Chem.* **65**, 1181–1185 (1993).

Ter Braak, C.J.F. and de Jong, S. The objective function of partial least squares regression. *J. Chemom.* **12**, 41–54 (1998).

Urbanski, P. and Kowalska, E. Application of partial least-squares calibration methods in low-resolution EDXRS. *X-Ray Spectrom.* **24**, 70–75 (1995).

Van Espen, P. in *Handbook of X-ray spectrometry: Second Edition, Revised and Expanded* (Eds R.E. Van Grieken and A.A. Markowicz), Marcel Dekker, New York, 2002 pp. 239–339.

Van Espen, P., Nullens, H. and Adams, F. A computer analysis of X-ray spectra. *Nucl. Instrum. Methods* **142**, 243–250 (1977).

Van Espen, P., Nullens, H. and Adams, F. An in-depth study of energy-dispersive X-ray spectra. *X-Ray Spectrom.* **9**, 126–133 (1980).

Vekemans, B., Janssens, K., Vincze, L. Adams, F. and Van Espen, P. Analysis of X-ray spectra by iterative least squares (AXIL): new developments. *X-ray Spectrom.* **23**, 278–285 (1994).

Wang, Y., Zhao, X. and Kowalski, B.R. X-ray fluorescence calibration with partial least-squares. *Appl. Spectrosc.* **44**, 998–1002 (1990).

Wielopolski, L. and Gardner, R.P. Development of the detector response function approach in the least-squares analysis of X-ray fluorescence spectra. *Nucl. Instrum. Methods* **165**, 297–306 (1979).

Wilkinson, D.H. Breit–Wigner viewed through Gaussians. *Nucl. Instrum. Methods* **95**, 259–264 (1971).

Yacout, A.M., Gardner, R.P. and Verghese, K. A semi-empirical model for the X-ray Si(Li) detector response function. *Nucl. Instrum. Methods A* **243**, 121–130 (1986).

Chapter 7

New Applications

7.1 X-Ray Fluorescence Analysis in Medical Sciences

J. BÖRJESSON[1,2] and S. MATTSSON[1]

[1] Malmö University Hospital, Malmö, Sweden and [2] County Hospital, Halmstad, Sweden

7.1.1 INTRODUCTION

7.1.1.1 ANALYSIS OF ELEMENT CONCENTRATION

Many elements are essential for the function of the human body. Others are toxic. There is thus a need to control their levels in human organs and tissues. This is especially important for occupationally exposed subjects. Moreover, it is essential to increase our knowledge of relations between observable toxic effects and element concentrations in man and his environment. Monitoring and basic occupational/environmental research rely on measurements directly in humans as well as of samples from humans and the environment. The same technique can also be used to follow, either a pure element or an element part of a molecule, after they have been administered to patients for diagnostic and therapeutic purposes. This subchapter focuses on recent advances in *in vivo* X-ray fluorescence (XRF) methods and their applications since 1995, including examples of *in vitro* use of the technique in the medicine field.

In Vitro **Analysis**

In vitro XRF is primarily used for laboratory analysis of, e.g. metals, minerals, samples of the environment, food, body fluids and tissue specimen (Szalóki *et al.*, 2000). Element concentrations in blood and urine are readily analysed by, e.g. chemical techniques (AAS, atomic absorption spectrophotometry; ICP-MS, inductively coupled plasma mass spectrometry, etc.) and atomic/nuclear techniques (XRF; PIXE, particle induced X-ray emission, etc.). A disadvantage of the PIXE method, compared with XRF, is that PIXE may destroy the sample due to very high local energy absorption (Amokrane *et al.*, 1999). However, with proper conditions fulfilled, even μ-PIXE using a scanning proton microprobe (SPM), can be used to detect most elements above $Z \approx 10$ with detection limits of $1–10\,\mu g/g$. A recent application is the identification of atoms in proteins (Garman, 1999).

Under certain conditions, element concentrations in blood and urine can be used to predict the concentration of the same element in an organ or tissue. In general, however, these relationships are not simple or well-known. The relations are influenced by, e.g. endogenous and ongoing external exposure, individual kidney function and variability in kinetics.

Faeces and hair as well as biopsy and autopsy specimens have also been used to estimate the exposure and the amount of the element retained in the body. A tissue biopsy provides information on the element concentration. However, it is invasive, involves a risk and may not be possible to repeat. In addition, the concentration in the biopsy may not necessarily represent the mean concentration in the organ. Thus, there is an interest to develop *in vivo* methods.

In Vivo Analysis

The two main non-invasive *in vivo* methods are XRF and neutron activation analysis (NAA). Reviews of *in vivo* XRF can be found in the literature (Börjesson *et al.*, 1998; Bradley and Farquharson, 1999; Chettle, 1999; McNeill and O'Meara, 1999) as can reviews of *in vivo* NAA (Sutcliffe, 1996).

Neutron activation analysis is possible for elements, which have a high cross-section for neutron capture. The technique has proven useful for *in vivo* studies of nitrogen, calcium, cadmium and aluminium. Many *in vivo* NAA studies have been concerned with prompt neutron activation of nitrogen in order to determine the whole-body content of protein or of cadmium in kidneys and liver (McNeill and Chettle, 1998; Kadar *et al.*, 2000; O'Meara *et al.*, 2001a).

Although useful for non-invasive quantification, both *in vivo* XRF and *in vivo* NAA have limitations. The attenuation of photons used to excite the element and of emitted characteristic X-rays from the atom imposes a limit to which element that practically can be measured with XRF. In *in vivo* KXRF, elements must have a Z higher than ≈ 25. For elements with $Z = 25-45$, studies are limited to the most superficial tissues of the body.

The first *in vivo* XRF application was the non-invasive measurements of natural *iodine* in the thyroid (Hoffer *et al.*, 1968). Later, a technique to measure iodine in human tissue *in vivo* from X-ray contrast agents was reported (Grönberg *et al.*, 1983). Measurements of *lead in vivo* began in 1971 (Ahlgren *et al.*, 1976, Ahlgren and Mattsson, 1979) and since then a number of alternative XRF techniques have been developed (Somervaille *et al.*, 1985; Wielopolski *et al.*, 1989; Gordon *et al.*, 1993). Descriptions of systems

Table 7.1.1 Application, principal measurement site(s), K absorption energy and characteristic X-ray energies of elements subject to *in vivo* XRF (Börjesson, 1996 and references therein). A question mark indicates that the application or measurement site is more or less speculative and based on scarce information

Element (Z)	Application	*In vivo* measurement site(s)	K absorption energy (keV)	K_α X-ray energies (keV)
Iron (26)	Medical (splinters, disease)	Eye, skin	7.11	6.39, 6.40
Copper (29)	Medical (splinters, disease)	Eye, skin	8.98	8.03, 8.05
Zinc (30)	Medical (splinters, disease)	Eye, skin	9.66	8.62, 8.64
Strontium (38)	Natural abundance	Bone	16.11	14.10, 14.17
Cadmium (48)	Occupational, environmental	Kidneys, liver	26.71	22.98, 23.17
Iodine (53)	Natural abundance, medical (X-ray contrast)	Thyroid, blood	33.17	28.32, 28.62
Xenon (54)	Medical (cerebral blood flow)	Brain	34.56	29.46, 29.78
Barium (56)	Medical (X-ray contrast)	Lungs	37.44	31.82, 32.19
Platinum (78)	Medical (cytotoxic agent)	Kidneys, liver, tumours	78.40	65.12, 66.83
Gold (79)	Medical (anti-rheumatic agent)	Kidneys, liver, bone joints	80.73	66.99, 68.81
Mercury (80)	Occupational, environmental	Kidneys, liver?, thyroid?, bone?	83.10	68.89, 70.82
Lead (82)	Occupational, environmental	Bone	88.00	72.80, 74.97
Bismuth (83)	Medical (treatment of ulcus duodeni, cytotoxic agent?)	Stomach?, intestines?, brain tumour?	90.53	74.81, 77.11
Thorium (90)	Medical (previously used X-ray contrast)	Liver, spleen	109.65	89.96, 93.35
Uranium (92)	Nuclear weapons industry, 'war', 'crime' (uranium covered ammunition)	Bone, lung	115.61	94.66, 98.44

for measurements of *cadmium, mercury, gold, platinum, uranium* and a number of other elements have been given through the years (Ahlgren and Mattsson, 1981; Bloch and Shapiro, 1981; Christoffersson and Mattsson, 1983; Jonson *et al.*, 1985; 1988; Börjesson *et al.*, 1993; Shakeshaft and Lillicrap, 1993; Börjesson *et al.*, 1995; O'Meara *et al.*, 1997). Table 7.1.1 summarizes elements, which have been subjects to *in vivo* XRF.

The *In Vivo* and *In Vitro* Situations – Differences

At first glance one may think that there is no great difference between the *in vivo* and *in vitro* XRF situations. However, some factors are more prominent for *in vivo* XRF. First, volumes are considerably larger and situated deeply in the body. Attenuation of incoming and characteristic X-rays is very pronounced, although the effect is also seen at *in vitro* measurements (Streli, 1995). For example, characteristic cadmium X-rays are attenuated by 98 % when passing through 50 mm of water, i.e. a typical distance between the skin and the kidney surface. Moreover, matrix effects can not be decreased by, e.g. drying.

Second, the measurement time is limited for two reasons. One is the time that the subject can lay or sit still. The other is the absorbed dose to the subject. The irradiated volumes are however quite small, thus, the energy imparted and mean absorbed dose to the whole body are low compared with other diagnostic procedures in medicine. For example, for an absorbed dose to the skin of the finger of 3 mGy, the mean absorbed dose to the whole body, at a finger-bone lead measurement, is two to four orders of magnitude lower than that of an ordinary chest X-ray examination (Börjesson, 1996).

7.1.2 *IN VITRO* APPLICATIONS OF XRF IN MEDICINE

This section describes some recent applications of *in vitro* XRF for measurement of heavy metals and other elements. It is not intended to completely cover all improvements and applications, rather to give an overview of some areas of research.

7.1.2.1 MEASUREMENT OF HEAVY METALS AND OTHER ELEMENTS

Mercury

Hair mercury correlates with blood mercury at the time when the hair is formed. Toribara (2001) used a scanning XRF method to investigate the mercury level in 1 mm portions of a single hair and could in an exact way determine when a fatal mercury accident had happened (Figure 7.1.1).

During dental intervention, amalgam particles may become absorbed in the oral mucosa; an 'amalgam tattoo'. Small amounts of mercury are believed to find their way to the blood. Forsell *et al.* (1998) found correlations between inflammatory effects and the mercury content in tissue biopsies in the surroundings of tattoos.

Autoclave sterilisation of extracted teeth, which are used for educational purposes, possesses a potential health hazard. A LXRF study of mercury on the inside of the autoclave bag suggested that mercury is transported to a vapour phase during the procedure and that vapour escapes and is finally released into the room (Parsell *et al.*, 1996).

Chromium and Lead

Small airway epithelial cells may be targets for chromium-induced lung cancer (Singh *et al.*, 1999). Phagocytosed lead chromate particles and intracellular lead inclusion bodies were observed by electron microscopy and confirmed by XRF (beam size $0.5\,\mu m \times 0.5\,\mu m$). Interaction of chromium and lead with DNA may be involved in lead–chromate carcinogenesis.

It is well-known that heavy metal compounds, e.g. lead chromate, may leach from crystal glass and ceramic glazes and imply a health risk. XRF techniques may be valuable to separate ceramic

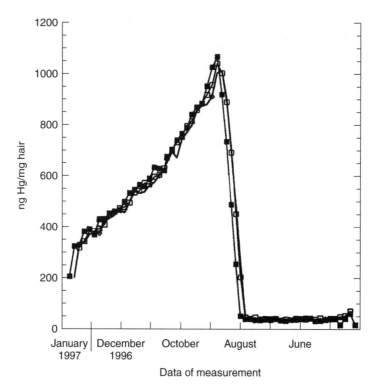

Figure 7.1.1 XRF analysis of hair collected 31 January 1997 (Toribara, 2001). In August 1996, a fatal mercury accident occurred. Reproduced by permission of Arnold Publishers

household utensils with high concentrations of toxic elements from those with low levels (Anderson et al., 1996).

In situ measured lead in remaining, scattered bones, recovered 1993 from a purported campsite of the 1845 Franklin expedition[1] was seen to range up to 1800 µg/g bone mineral (Keenleyside et al., 1996; Figure 7.1.2). Correlations between lead at different bone sites were used to identify missing bones, i.e. bones could be associated to the same individual. Improperly soldered tin containers were believed to have been a major source of lead exposure for the expedition members.

[1] The purpose of the expedition was to map out the North-West Passage from Europe to Asia, a sea route linking the Atlantic and Pacific Oceans. It lies above the Arctic Circle between Canada and Greenland and the Arctic itself. The sea in this region is frozen over for most of the year. Temperatures in winter fall below −50 °C. During the winter of 1846–1847, the ships became trapped in thick ice. After leaving the ships and trying to find help the crew broke up in small groups but finally all men died.

Iodine

In animals and man iodine is concentrated in the thyroid gland as a constituent of thyroid hormones. Iodine deficiency may lead to neurological syndromes, e.g. cretinism. Liu et al. (2001) showed that iodine supplementation to iodine-deficient rats improved thyroid hormone metabolism, whereas the trace elements bromine in brain (increased) and zinc (decreased), manganese (decreased) and copper (decreased) in erythrocytes, were affected by the iodine supplementation. The data stimulate further studies of the role of trace elements in the central nervous system (CNS) and their possible links to neurological defects. Moreover, Majewska et al. (2001) showed that the *zinc* concentration in TXRF (total reflection XRF) measured samples of the thyroid was lower in patients with thyroid cancer (23 µg/g) compared with patients with Grave's disease (42 µg/g). In blood samples, the reverse relation was obtained. Regarding whether elevated

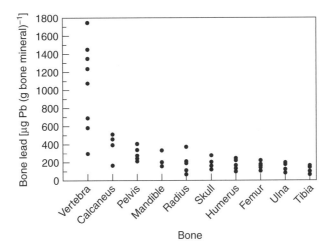

Figure 7.1.2 Bone lead content for all Franklin expedition bones measured (Keenleyside *et al.*, 1996). Reproduced by permission of Academic Press

blood zinc concentration is an indicator of thyroid cancer is a question for further studies.

Most X-ray contrast agents contain iodine. Such substances can also be used as markers for glomerular filtration rate (GFR). The contrast agent Iohexol has been compared with the most commonly used GFR marker for clearance measurements, ^{51}Cr-EDTA (ethylenediaminetetraacetic acid). Both substances were injected in patients and subsequent blood samples were analysed for iodine with XRF technique (Brändström *et al.*, 1998) and for ^{51}Cr with γ-spectrometry. The observed strong correlation between Iohexol and ^{51}Cr-EDTA clearances implied compatibility between GFR markers. Measurement of urographic iodine contrast media can also be used to estimate the residual renal function and the efficiency of haemodialysis (Sterner *et al.*, 2000).

Thorium

Thorotrast (25 % thorium oxide), another contrast medium, was used from the late 1920s until about 1950. After i.v. injection, high concentrations of thorium were seen in spleen (10–50 mg/g), liver (1–10 mg/g), bone marrow and lymph node. With respective biological and physical half-lives of 400 and 1.4×10^{10} years, internal organs are irradiated by α-particles for the rest of the subject's life. Increased occurrences of liver cancer, liver cirrhosis and leukaemia have been reported. Using a μ-beam XRF technique (beam Ø160 µm, Si(Li) detector for thorium L_α line) the microdistribution of thorium was seen to form conglomerates in the liver, whereas it more looked like a belt in the spleen (Muramatsu *et al.*, 1999). XRF can give complementary information to results gained by γ-spectrometry and α-autoradiography.

Platinum

Platinum, in the chemical form of cisplatin, may be used to cure cancer. The drug has been in use for about 30 years and it has shown very good results in the treatment of, e.g. testicular carcinoma. Cisplatin's adverse effects, however, limit the amount which can be administered to patients. In animals, treated with cisplatin for tumours, increased levels of trace elements, e.g. iron, copper, zinc, ruthenium and bromine, in addition to platinum, was observed in the kidneys and liver and suggested that platinum toxicity is a result of the overall accumulation of trace elements in these organs and not only high platinum levels (Shenberg *et al.*, 1994). Administered amounts of selenite reduced the trace element levels, hence, selenium might have a future as chemoprotector. It has also been shown that rubidium, bromine, selenium, zinc, copper and iron concentrations in blood, liver, kidneys, colon, and skin were significantly different in tissue samples obtained from mice inoculated with tumours and normal mice (Feldstein et al., 1998). The rubidium level in a tumour was one order of magnitude higher than in normal tissue.

Calcium

Coronary heart disease (CHD) is common, especially in the Western countries. Inverse relations have been found between health district standardised mortality ratio (SMR) and the mean hair calcium concentration (MacPherson and Bacsó, 2000; Figure 7.1.3). Part of the explained variance was environmental factors believed to influence calcium metabolism, e.g. water hardness and number of sunshine hours. For example, the south-east part of England had the highest hair calcium, the hardest water and the most sunshine hours and the lowest mortality from CHD. The converse was true of Scotland. Additionally, the authors point out that confounding socio-economic conditions may have been more beneficial to the south-east part than to Scotland, thus reducing the SMR.

Synchrotron radiation XRF may be a suitable method to study the interface between bone and biomaterials, i.e. substitutes for autogenous bone grafts, with regard to mineral content. Circular defects (Ø 4 mm), made in tibias of rabbits, were augmented with a composite of hydroxyapatite (HA) granules. The calcium distribution, studied with µ-XRF line scans, suggested that calcium phosphate compositions, e.g. HA, have interesting properties as bone substitutes (Liljensten et al., 2000).

Nickel

In order to detect allergen metals, XRF is used in dentistry and dermatology. Suzuki (1995) reported

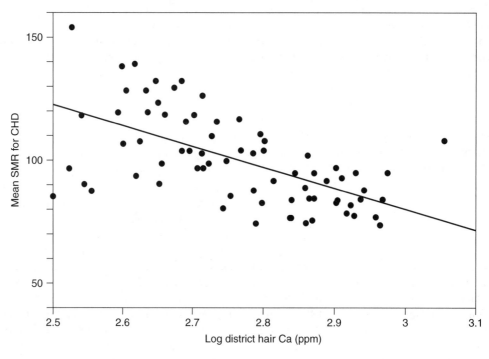

Figure 7.1.3 The relationship between the mean district SMR for CHD and log district hair calcium concentration (MacPherson and Bacsó, 2000). Reproduced by permission of Elsevier Science

successful allergen elimination in patients with metal allergy from dental restoration work. Detected nickel and gold in skin specimens, taken from lesions in pierced earlobes, suggested that small metal fragments remain in lesions for a long time, even after the studs have been removed (Suzuki, 1998). This causes irritation and cutaneous reactions.

Bromine

The determination of extracellular water, using ^{82}Br dilution technique, has drawbacks, e.g. radiation dose and the short physical half-life of ^{82}Br. An approach with administered stable bromine detected with ^{109}Cd sources and a Si(Li) detector *in vitro* indicated that measurements of blood bromine may be substituted with bromine in urine and saliva, which imply less discomfort to the patient (Zaichick, 1998a,b).

Palladium

The increasing pollution of palladium in the environment, as well as in the body of living species is discussed controversially. Sures *et al.* (2001) showed, with TXRF technique, that automobile catalyst emitted palladium is bioavailable for aquatic animals. Also, the production and recycling of palladium-containing materials, as well as the use of palladium in dental restorative alloys, may be a source of toxic/allergic reactions.

Palladium in body fluids can be analysed with ICP-MS, however, for environmental samples, e.g. road dust, the technique is less suitable (Messerschmidt *et al.*, 2000). Therefore, a TXRF method was developed for analysis of palladium and gold in environmental samples (detection limits 2.5 ng/l and 2.0 ng/l, respectively). The method is sensitive but time consuming, hence it was suggested to serve as 'gold standard'.

Gold

Mechanisms of drowning were studied using an immersion fluid, containing small gold tracers (12–48 µm), which were introduced into the airways of rats (Bajanowski *et al.*, 1998). Microanalyses after the animals had died showed that small diameter tracers had penetrated the intercellular gaps of the alveolar epithelium, while larger tracers were incorporated into the epithelial and endothelial cells. An active post-mortem transport proceeded in alveolar pneumonocytes and macrophages and functioned for a time period even after death and cessation of circulation.

Multielement Studies

Improved element-specific (iron, cobalt, copper and zinc), detection in capillary electrophoresis, with monochromatic 10 keV synchrotron radiation XRF was seen to be superior compared with ICP-MS and PIXE methods (Mann *et al.*, 2000). The present detection limit was ≈0.5 ng and further development seems possible.

Synchrotron radiation XRF has also been used to study the myocardial blood flow by means of heavy element microspheres (Mori *et al.*, 1995). Elements studied were bromine, yttrium, zirconium, niobium, iodine, and barium and 20 and 50 keV SR energy was used.

Multielement studies of human placenta have indicated that most element concentrations are the same, regardless of the mother's state ('normal', undernourished, smoker or hypertension) (Meitín *et al.*, 1999). However, strontium levels for the latter three states were markedly lower than for the 'normal' state.

7.1.3 SYSTEM DEVELOPMENT OF *IN VIVO* XRF IN MEDICINE

7.1.3.1 *PHOTON SOURCES GEOMETRIES*

Photons from an X-ray tube or from a γ-emitting radionuclide source may be used to excite elements *in vivo*. In some cases it is possible to find radionuclide sources with suitable energy, negligible or filterable emission of other γ- and β-radiation, high yield and high specific activity at a reasonable cost. Sources used or proposed for *in vivo* studies are among others ^{57}Co, ^{99}Tc^m, ^{109}Cd, ^{133}Xe and ^{241}Am. An advantage is the stable output, compared with the X-ray tube output, which may vary to some extent over time and thus need to be continuously monitored by a separate detector. A radionuclide source is also compact and transportable. On the other hand, it may have a short half-life, be expensive and give a low photon fluence rate.

A limitation of the XRF technique is the amount of scattered photons that are detected. Since the human body mainly consists of low Z elements, incoherent scattering is the dominating photon interaction process. Scattering in low Z matrices leads to a broad and large background in measured pulse height distributions. The concentration of the element of interest is generally low, on the order of μg/g, which implies a low signal-to-background ratio as the small characteristic X-ray peaks are superimposed on a high background distribution of scattered radiation. Thus, it is essential to improve this ratio by increasing the net signal and/or decreasing the background. One should, however, be aware that incoherently scattered photons can be used constructively. For most biological materials, the amount of *incoherently* scattered photons is proportional to the mass of the analysed volume, thus, a normalisation can be made. Moreover, *coherent* scattering can be used for bone mineral measurements and for normalisation to bone mineral content when lead in bone is determined.

Polarised Photons from an X-ray Tube

Background reduction may be achieved by using partly plane-polarised photons, e.g. produced by incoherent scattering of primary photons in 90°. Other methods to produce a beam of plane-polarised photons are synchrotron radiation (SR), thin-anode transmission X-ray tubes and Bragg diffraction. However, incoherent scattering has been described as one of the best ways to improve the ratio between fluorescent and scattered radiation for low Z matrices (Heckel, 1995).

Briefly, the detector is put in the plane defined by the electric vector of the polarised photon. A reduced fluence rate of scattered photons will result if the detector axis is parallel to the electric vector. The differential collision cross-section ($d_e\sigma/d\Omega$) ratio for the photon fluence rate in the directions of minimum and maximum scatter is 1.6% at a photon energy of 100 keV (Evans, 1955). However, collimation and multiple scattering largely reduce the benefit of polarisation, e.g. an *in vivo* method for platinum had a ratio of 40% (Jonson et al., 1988).

The tube potential significantly impacts the detection limit since the number of photons, capable of exciting the element increases, and the overall number of photons also increases when the voltage of the tube is increased (Figure 7.1.4). Theoretical analyses of system parameters for a renal mercury measurement set-up indicate that a decreased detection limit is within reach (O'Meara et al., 2000). An improvement might be possible through a higher accelerating potential (250 kVp) than was used by Börjesson et al. (1995). However, this has to be tested experimentally.

Development of tube design leads to considerable changes in spectral shape and possibly to lower MDCs. The feasibility of a nearly monochromatic X-ray tube, with a gold anode and a secondary target of tantalum for measuring cadmium *in vivo*, was studied by Börjesson et al. (2000). However, characteristic tantalum X-rays (56–67 keV) and a low fluence rate of bremsstrahlung did not offer an improved detection limit. In the future, interchangeable targets may facilitate a better match between the studied element's absorption edge and the primary photon energy. A recent Monte Carlo simulation of a monochromatic X-ray tube seems promising (O'Meara et al., 2002) for decreasing the detection limit.

Radionuclide Sources

Besides the aforementioned *in vivo* methods for iodine and lead, developed during the 1960s and 1970s (Hoffer et al., 1968; Ahlgren et al., 1976; Ahlgren and Mattsson, 1979), another KXRF

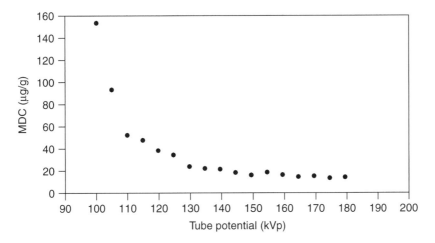

Figure 7.1.4 The minimum detectable concentration (MDC) for mercury in the kidney for different X-ray tube voltages (kVp) (Börjesson, 1996)

technique based on a ^{109}Cd source was developed for bone lead measurements. In the beginning, an annular source was used (Somervaille et al., 1985) but was later replaced by a point-source, placed in front of a large area detector. This resulted in an improved precision (Gordon et al., 1993).

A third technique, uses low photon energy LXRF from bone lead excited from a ^{109}Cd source or an X-ray tube (Ao et al., 1997b; Rosen, 1997; Figure 7.1.5). Methodological differences in KXRF and the 'short-sighted' LXRF presumably imply that the KXRF and LXRF methods provide complementary information. Recently, Todd (2002a,b) reviewed the LXRF method and showed that predicted bone lead and measurement uncertainty were influenced by the choice of linear attenuation coefficient, with which to correct for overlying tissue. Inter-individual variability in body composition, methodological uncertainty in the ultrasound measurement of overlying tissue thickness as well as discrepancy between the site of LXRF and the site of ultrasound measurement were also important factors. Interference from lead in non-bone tissues may not be negligible. A combination of radionuclide KXRF and tube LXRF has been proposed, however, Todd et al. (2002a) found that the variability of the method was large enough to give concerns regarding the application of this method at all for *in vivo*

use. Feasibility studies of various sources (^{99}Tc^m, ^{133}Xe, etc.) for measurements of platinum and gold have also been presented, but will not be reviewed here.

7.1.3.2 DETECTOR AND ELECTRONICS

It is essential to optimise the detector for various elements studied as illustrated by the reduced (50%) detection limit when switching from a germanium detector to a silicon detector for *in vivo* cadmium measurements (Nilsson et al., 1990). The combination of a large-area detector and a fast analogue-to-digital converter (ADC) has also proven advantageous. The new 'clover-leaf' detector is also of special interest (Nie et al., 2002). It consists of several detectors with separate sets of electronics. Consistency between simulations and experiments and a greatly reduced detection limit (70% lower than with standard detector) were reported for bone lead KXRF.

Furthermore, digital spectroscopy systems may offer important advantages, due to their increased throughput of pulses from photon interactions in the detector (Fleming, 1998, 1999). Instead of conventional pulse shaping circuits, the new spectrometer filters and shapes the pulses digitally.

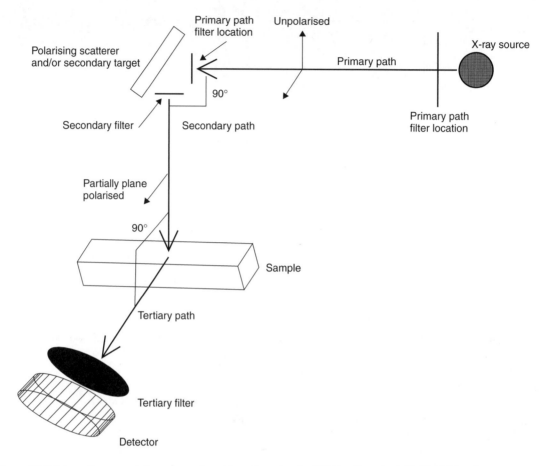

Figure 7.1.5 Schematic representation of experimental configuration for LXRF of bone lead (Todd, 2002b). The sketch shows the X-ray source, scatterer with filters, sample and detector with filter. Notice the state of polarisation at various positions along the beam line. Reproduced by permission of IOP Publishers

Improved energy resolution, detection limit and precision, the latter up to 27%, have been recorded (Bateman *et al.*, 2000). The detection limit for tibia bone lead was 2.2 μg/g bone mineral, corresponding to about 1.3 μg/g wet bone (Fleming and Forbes, 2001). The *in vivo* performance is to be explored.

Portable XRF systems, with generators producing 35 kV and 0.1 mA, and thermoelectrically cooled cadmium zinc telluride detectors are today available on the market. Primarily, they seem applicable to *in vitro* studies, e.g. for measurement of mercury and lead contamination of surfaces, at a detection limit of 0.1 mg/cm^2, but may have a future potential for *in vivo* XRF.

7.1.3.3 MONTE CARLO SIMULATION

The detection limit *in vivo* is often higher or comparable with the concentration of the element in the subject. Thus, improving the sensitivity by modifications of the set-up parameters are desirable. The generally large number of parameters (e.g., tube voltage, collimation, angles, filters, detector) that can be changed makes this time consuming and costly. However, modern computers make it possible to simulate the measurement situation, using a Monte Carlo code. Several such codes are now available and need only small modifications. We will briefly review some recent publications in the field. Some codes account for Compton momentum broadening and photon polarisation (Tartari et al., 1991; Fernandez et al., 1993; Tartari et al., 1999). It has to be stressed that the ultimate test of a simulated result is to apply the optimised system to an in vivo measurement to verify that the optimisation works in reality.

Wielopolski et al. (1983, 1989) used an X-ray tube for bone lead LXRF analysis. Later, Ao et al. (1997a,b) and Gardner et al. (1999a,b) developed the CEARXRF code to improve three set-ups with KXRF (^{109}Cd source) and LXRF (^{109}Cd and X-ray tube source). The lowest detection limit was found for a polarised X-ray tube source. For the LXRF technique, L_α and L_β X-rays provide a method for minimising the skin thickness effect. The code may be used for tasks such as the proposed combination of KXRF and LXRF (see section on 'Radionuclide sources'). It was also shown that the normalisation of bone lead to coherent scatter improves the accuracy for the KXRF system, in accordance with Somervaille et al. (1985) and Bradley et al. (1999).

The code developed by O'Meara et al. (1997, 1998a) for simulation of source-excited in vivo XRF of heavy metals showed agreement in most parts of the spectrum, whereas, in the low energy tail of Compton scattered photons, agreement was less. Simulations of a ^{57}Co source in a backscatter geometry (Figure 7.1.6), to measure bone uranium showed that the uranium concentration in tibia was insensitive to variations in source–sample geometry, thickness of overlying tissue and tibia size.

O'Meara et al. (1999, 2001b) also presented simulation results of the original set-up with the ^{57}Co technique (Ahlgren and Mattsson, 1979) to measure finger-bone lead. As for the ^{109}Cd set-up, the ^{57}Co may also benefit from using the coherent scatter peak normalisation, as an alternative to how it presently is made by radiographs of the subject's finger, which increase the absorbed dose.

Al-Ghorabie (2000) used the EGS4 Monte Carlo code to simulate a measurement situation with a ^{133}Xe source in a backscatter geometry, and found detection limits of 15–60 µg/g for cadmium KXRF in kidneys situated at a depth of 30–60 mm. Further optimisation and in vivo results are expected.

A version of the EGS4 code (Kilic, 1995) has also been used to optimise a set-up with a polarised source (X-ray tube) for in vivo XRF of platinum (Lewis et al., 1998). Good agreement between simulation and experiment was observed (Figure 7.1.7). Another version of EGS4, also intended to model platinum in vivo measurements, was presented by Hugtenburg et al. (1998). In contrast, analytical methods, e.g. the fundamental parameter method, for describing the in vivo KXRF of platinum measurement situation have been suggested (Szaloki et al., 1999).

7.1.3.4 SPECTRUM ANALYSIS

At most in vivo XRF measurements, a small net signal is analysed in the presence of a large background of scattered photons. More or less complex fitting algorithms may be applied for extraction of the net peak area. The commonly used straight line approximation (least-squares method) of the background under the peak is easy to apply and works well in situations with a 'straight' background. However, this is often not the case and the background may be better described with a polynomial.

There are also nonlinear regression methods that model both background and signal, e.g. the Marquardt algorithm and sequential quadratic programming. These methods find the optimal fit by iteratively changing the parameters in small steps, starting with initial data supplied by the user. The algorithm may, for example, be the sum of a polynomial background and Gaussian shaped peaks. Iterative methods are sensitive to the supply of 'correct' seed values at the start of fitting, otherwise the procedure may capsize.

For those interested in spectrum analysis and fitting, especially regarding 'outliers', the publication by Reich (1992) is recommended. The author discusses, among other things, the use of absolute values of the deviations instead of the squared deviations. The proposed technique might produce more stable fits in spectra with outliers. These outliers have a destructive effect on the whole fit as

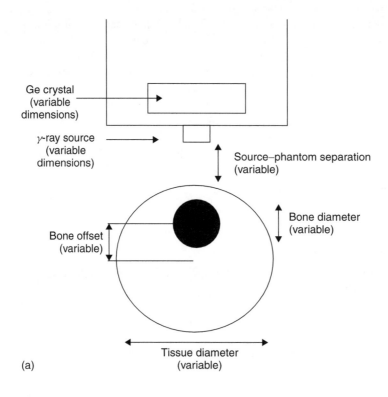
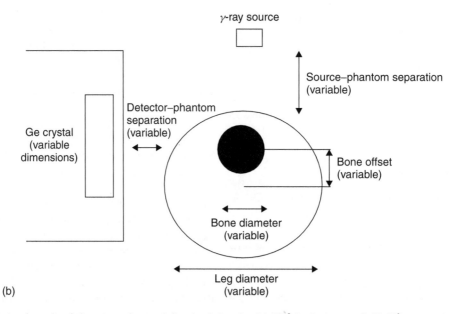

Figure 7.1.6 A schematic of the geometries used for simulating the (a) 180° backscatter and (b) 90° scatter systems. Note the alignment of the detector with the centre of the bone cylinder (O'Meara *et al.*, 1998a). Reproduced by permission of IOP Publishers

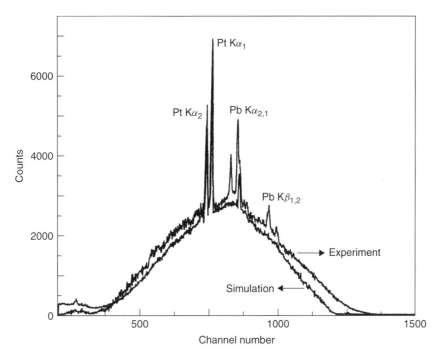

Figure 7.1.7 Simulated and measured XRF spectra using a 1000 μg/g platinum solution (Lewis *et al.*, 1998)

their impact become very strong due to the squaring when minimising the deviations. In the end, it is up to the individual scientist to determine the method to use (Kondrashov *et al.*, 2000). Moreover, it is wise to check the accuracy and precision of the fitting procedure by, e.g. fitting a large number of background distributions (no phantom) and determine the mean net number and the coefficient of variation.

New algorithms for fitting data have evolved and one of them uses a least moduli approach. Compared with the least-squares method the 'new' approach is more effective for 'strong' peaks on a low background, i.e. calibrations peaks, whereas the least-squares method is more effective for a weak peak situated on a large background, i.e. the *in vivo* situation (Kondrashov *et al.*, 2000). Moreover, a doublet deconvolution technique for correcting the peak of coherently scattered photons for the $K_{\beta 2}$ signal in the bone lead set-up was suggested by Kondrashov and Rothenberg (2001a). Kondrashov and Rothenberg (2001b) presented mathematical proof that the estimate of bone lead uncertainty in measurements with ^{109}Cd needs revision. Authors also suggested a reduced detector–target distance to improve the measurement. A shorter distance may, however, lead to pile-up of pulses and this effect has been studied by simulations (Gardner *et al.*, 1999a).

7.1.3.5 CALIBRATION

Calibration of an *in vivo* measurement technique for quantitative analysis of element tissue concentration is not an easy task. Often there is no 'gold standard' and phantoms have to be used. One may also make an *in vivo* XRF measurement and chemically determine the element concentration in a biopsy of the measured volume. Another variant is to measure the concentration *in situ*, using XRF, in autopsy material and then analyse the material by, e.g. AAS. Yet another way is to use animal organs, e.g. animal kidneys immersed in water for simulation of *in vivo* XRF of kidney cadmium. One must, however, always be aware of the strengths and weaknesses the actual calibration method has. In the last example, clearly visible horse kidneys, at an accurately known depth in water, is not the same as a kidney at an unknown depth in a non-transparent human being.

In the following section we will limit the discussion to calibration of bone lead systems. Regarding the phantom, it has traditionally been made of silica paraffin wax and plaster-of-Paris, a powdery, slightly hydrated, calcium sulfate made by calcining gypsum. However, phantoms of polyurethanes and calcium carbonates are claimed to be much more uniform in density and composition and to better describe the *in vivo* situation regarding scatter, attenuation, positional dependency and dead-time loss (Spitz *et al.*, 2000). Another phantom, a synthetic apatite matrix was proposed by Todd (2000a). Todd concluded that the effect on the coherent conversion factor, which converts between calibration standard and human bone, from impurities in plaster, coherent scatter from non-bone tissues and the subject's measurement geometry were all of minor effect. Furthermore, validation using non-human bones may introduce 'new' uncertainties due to changes in spectral shape (Todd *et al.*, 2001d).

Contamination of subjects, measurement systems and phantoms has occurred, and may happen again, unless measures are taken to prevent it. For example, measurement of bone lead on the grounds of a lead plant demands that the measurement room is clean from environmental lead. It means that workers are to enter the room without bringing lead with them and that lead on skin at the measurement position is carefully removed. The effects of contamination may be considerable (Todd, 2000b; Todd *et al.*, 2000b).

7.1.3.6 DETECTION LIMIT

Various definitions of the MDC have been presented (Currie *et al.*, 1994; Todd, 2002b). For example, the MDC may be defined as

$$\text{MDC} = 3 * \frac{C\sqrt{N_{bg}}}{N_{net}}$$

where N_{bg} is the background counts under the characteristic X-ray peaks, N_{net} the net counts in the peaks and C the element concentration. N_{bg} and N_{net} linearly depend on the measurement time, thus the MDC is inversely proportional to the square root of the measurement time. The N_{net} varies, according to, e.g. attenuation of the emitted characteristic X-rays. Hence, the MDC varies strongly with depth (Figure 7.1.8). Photon attenuation in lateral direction results in MDC variability for a given dorsal kidney depth. It is

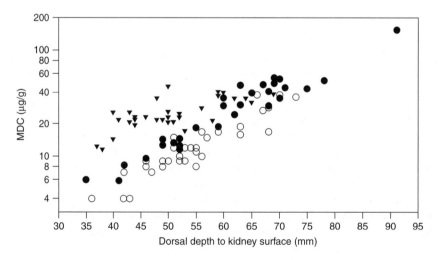

Figure 7.1.8 The MDC for cadmium (● and ○) and mercury (▼) in the kidney cortex plotted as a function of the dorsal kidney depth (Börjesson, 1996)

important that all data are used and not discarded in the analysis. For example, excluding all data points with concentrations under the detection limit will lead to errors in the calculation of mean values and standard deviations (Kim *et al.*, 1995).

7.1.3.7 PRECISION AND ACCURACY

It is essential that the measurement presents a result close to the 'true' value. If the *in vivo* XRF measurement may be repeated with the same set-up, a low precision may be acceptable as long as the mean value of the series of measurements is accurate. However, if only a single measurement is possible, which is usually the case in *in vivo* XRF, precision must be high. Accuracy and precision are discussed from a KXRF bone lead point of view.

The high precision for repeated measurements at the same bone site and the bone lead heterogeneity, i.e. potential differences in bone lead at various sites of the studied bone, need to be controlled. In 1995, Hoppin *et al.*, after measuring with a ^{109}Cd KXRF technique, concluded that it '... may not yet be a useful diagnostic tool for individual subjects, but it may be of great use to environmental scientists trying to characterise long-term lead exposure and dose in the general population or specific subpopulations.' Hoppin *et al.* (2000) later demonstrated a considerable variability in bone lead for measurement locations only 10 mm from the centre of the tibia. This may, according to the authors, limit the interpretation of bone lead.

On a group level, the ^{109}Cd KXRF technique is capable of revealing bone lead differences of 5 μg/g bone mineral (i.e. 2.9 μg/g wet bone; conversion factor from bone mineral to wet bone 1/1.72) or less (McNeill, 1999; McNeill *et al.*, 1999). However, the individual measurement uncertainty is significantly affected by age, sex, and subject obesity (Figure 7.1.9). Women have poorer precision than men because of smaller bone mass. Obese subjects have tissue overlay thicknesses that attenuate the signal, thus, precision can be poor (>10 μg/g bone mineral; >5.8 μg/g wet bone). A precision better than 2 μg/g bone mineral (1.2 μg/g wet bone) may be necessary for meaningful evaluations of individual measurements (Bradley and Farquharson, 1999). The uncertainty at measurements in young individuals was shown to be marginally worse than in adults (Todd *et al.*, 2001a). This is important since measurement in children and adolescents is of special interest.

Ideally, the uncertainty assigned to an individual bone lead measurement would be the standard deviation of multiple measurements made in the subject. Regarding ^{109}Cd KXRF measurements, Todd *et al.* (2000a) studied factors that might explain why the observed uncertainty underestimates (up to 18 % deviation) the standard deviation

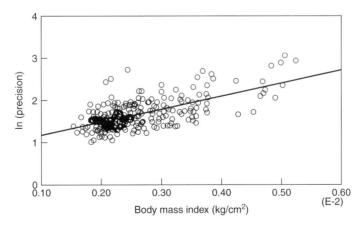

Figure 7.1.9 The natural logarithm of the precision against body mass index for female subjects (McNeill *et al.*, 1999a). Reproduced by permission of IOP Publishers

of replicate tibia lead measurements. Changes over time, e.g. source decay, degradation in detector resolution and efficiency, may partly be responsible for the observed discrepancy. Repositioning of the sample, as suggested by Hoppin *et al.* (2000), seemed however not to add to the uncertainty.

Macroscopically, lead is fairly homogeneously distributed within the skeleton. Microscopically, lead-to-bone mineral concentration ratio is homogenous in the endosteum but peaked in the periosteum region (Jones *et al.*, 1990; Schidlovsky *et al.*, 1990). Bone lead in cadaver legs show higher concentrations (5–8 µg/g bone mineral; 2.9–4.7 µg/g wet bone) in surface compared with core tibia bone (Todd *et al.*, 2001b,e). These findings will have an impact on the evaluation of KXRF and LXRF since the former technique samples more than just the superficial 1–2 mm of the tibia. Thus, it is not expected that the results of the two techniques should agree. Moreover, comparison with AAS showed that the KXRF method is likely to overestimate the core lead level by about 5–8 µg/g bone mineral (2.9–4.7 µg/g wet bone) (Todd *et al.*, 2002b).

Besides tibia and calcaneus, the two most widely used sites for ^{109}Cd KXRF measurements, also the patella has been used. The question whether orientation of the patella affects the measurement was looked into by Todd *et al.* (2001c), who reported no obvious effect on lead concentration, whereas uncertainty was significantly changed. Contributions from non-patellar lead in distal femur, proximal tibia and synovium could not be excluded.

Aro *et al.* (2000) validated the ^{109}Cd KXRF by initial measurements in patella and tibia in cadaveric legs with surrounding tissues intact. KXRF was repeated after all soft tissue had been stripped. The bone was then isolated and measured by ICP-MS. Results showed strong correlations between KXRF and ICP-MS bone lead. The study indicated that KXRF measurement is not influenced by interference from surrounding soft tissue.

The relationships between indicators of lead exposure have mainly been studied by linear regression methods. However, there are basic underlying assumptions associated with this technique, e.g. that the uncertainty in the X values is zero, which as a rule is not true. For example, the relationship between bone lead (Y) and blood lead (X) assumes that blood lead is determined with a 100 % accuracy, which is not realistic. Thus, the regression line would be different if we would change the X and Y variables. To take into account the uncertainty in both variables one may use 'structural analysis' (Brito *et al.*, 1999).

7.1.4 SUMMARY OF *IN VIVO* APPLICATIONS OF XRF IN MEDICINE

7.1.4.1 LEAD

Lead has found a range of modern applications in which occupational exposure may occur (Tables 7.1.2 and 7.1.3). For the general

Table 7.1.2 Sources of exposure and uptake of cadmium, mercury and lead (Börjesson, 1996, references therein and papers within this review)

Element	Exposure		Uptake (%)	
	Occupationally exposed	General public	Gastrointestinal	Inhalation
Cadmium	Production and recycling of Ni-Cd batteries. Use of alloys, solders, pigments, fertilisers	Food, smoking	≈5	0.1–50
Mercury	Mercury and gold mining. Chloralkali and thermometer factories. Dental amalgam handling	Food, dental amalgam	<1 (metallic mercury)	≈80 (metallic mercury vapour)
Lead	Production and recycling of lead batteries, ammunition, paint, plumbing, cable cover, radiation shield	Food, water, lead in gasoline	15–20 (adults, higher in children)	10–60

Table 7.1.3 The estimated biological half-life and tissue levels for cadmium, mercury and lead. Note that multiple half-life components may be found in, e.g. blood, and this mirrors the excretion of an element from various organs and tissues

Element	Biological half-life (d = days, y = years)			Typical levels in tissues (µg/g)	
	Blood	Urine	Tissue	Occupationally exposed	General public
Cadmium	75–130 d and 7–16 y	No studies found	26 y (whole body) 1 d and 30–60 d (lungs), 10–30 y (kidney), 5 y (liver)	Up to 600 (kidney), Up to 120 (liver)	10–50 (kidney), ≈1 (liver) (Swedish figures)
Mercury	3–4 d and 45 d	40–90 d	42–58 d (whole body), 64 – 180 d (kidney), 21 d (brain) (Long-term components not known)	Up to 70 (kidney), <2 (brain), <1 (liver), up to 100 (thyroid gland), up to 70 (pituitary gland), ?? (bone)	<1 (kidney), <0.03 (brain), <2 (pituitary gland)
Lead	1 d, 30–50 d, 1–5 y and 13 y	150 d and probably a longer component	5–30 y (cortical/trabecular bone) 30 d (soft tissue)	>200 (bone)	3–10 (bone) (Swedish figures)

population, exposure mainly comes from e.g. ingested food, drinking water and inhalation of car exhausts (leaded petrol). Moreover, lead poisoning from lead-based paint in homes is claimed to be the number one environmental disease among children in the USA (Holmes, 1995). Presence of the paint is implicated either directly or indirectly in elevated blood lead levels. KXRF and LXRF techniques, for rapid on-site testing of lead paint, quickly present a result expressed in mg of lead per cm^2, whereas a chemical test kit only tells whether there is lead in the paint or not (Schmehl et al., 1999).

Lead accumulates in bone. The skeleton contains >90 % of the body burden of lead, of which 70 % is in the cortical bone. This bone type constitutes 80 % of the skeleton, the rest being trabecular bone. Cortical bone lead linearly increases with age, whereas trabecular bone lead increases until the age of about 50 years, after which it levels off or even decreases. Observed differences in lead accumulation between subjects from Canada, Scandinavia and England have been observed (Roy et al., 1997) but the exact cause of differences was unclear.

Significantly higher tibia and blood lead concentrations in adults who, as children, were exposed to lead, were presented by McNeill et al. (2000) (Figure 7.1.10). Children, living in the vicinity of lead plants, may have dramatically

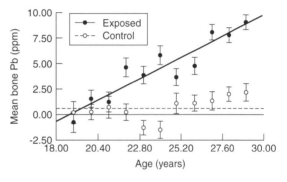

Figure 7.1.10 Bone lead concentrations versus age for exposed and control subjects (McNeill et al., 2000). Reproduced by permission of BMJ Publishing Group

increased bone lead concentrations (Todd et al., 2001a). Controls had tibia lead levels of around 1 µg/g bone mineral, whereas in exposed children the average level was 39 µg/g bone mineral.

Lead also accumulates in teeth. An in vivo technique (^{109}Cd) to measure lead in intact teeth was developed by Zaichick and Ovchjarenko (1998) and Zaichick et al. (1999), (Figure 7.1.11). Lead concentrations in permanent teeth of teenagers and adults did not exceed 3 µg/g; in accordance with literature data (Skerfving et al., 1999). Another XRF technique (^{57}Co) revealed elevated lead concentrations, caused by environmental pollution, in shed teeth of Russian and Chinese children (Bloch et al., 1998).

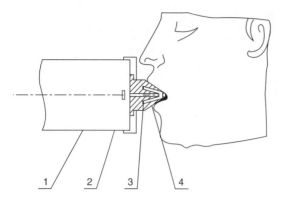

Figure 7.1.11 Schematic diagram of the *in vivo* XRF of lead in teeth (Zaichick and Ovchjarenko, 1996). 1, Detector casing; 2, Si(Li) detector; 3, annular ^{109}Cd radionuclide source; 4, shielding and collimation. Reproduced by permission of Marcel Dekker, Inc

Bone lead XRF has mostly been applied to occupationally exposed subjects. Retired smelters may have much higher bone lead levels than active smelters. This is ascribed to the long exposure duration and high exposure levels during earlier employment periods, in combination with the slow bone lead excretion.

A positive correlation between blood and bone lead concentrations is often found for retired smelters, confirming the skeleton as an endogenous source of lead. The fraction of blood lead that comes from the skeleton may be up to 70%. In active workers, the influence from ongoing exposure masks the relation between blood and bone lead.

Re-measurements of bone lead in smelter workers have indicated a shorter bone lead half-life, about 5 years, in younger persons (<40 years) compared with subjects older than 40 years ($t_{1/2}$ = 14 years) (Brito *et al.*, 2001). The latter figure is consistent with the bone lead half-life estimated from repeated bone lead KXRF in retired workers during two decades (Nilsson *et al.*, 1991). Thus, current metabolic models may have to be adjusted for the subject's age. In another study, Brito *et al.* (2000) noted decreased tibia bone lead concentrations over a 5 year period, presumably due to improved industrial hygiene measures. The estimated tibia lead half-life was 9–15 years. Surprisingly, a corresponding decrease in calcaneus bone lead was not observed, although it is believed to have a shorter half-life. The observation needs attention.

The accumulation of lead in bone is probably affected by factors that have an impact on the calcium metabolism. Thus, a high body burden of lead may result in an increased endogenous exposure, and thus a risk, for persons with fractures, bone diseases, thyreotoxicosis, pregnancy or general osteoporosis, as observed in women after menopause.

Recent studies indicate that lead is mobilised from the maternal skeleton during pregnancy and the postpartum period. Skeletal turnover may be >10% and in some cases possibly >30% (Gulson *et al.*, 1999). However, the changes in bone lead would be unlikely to be detected by current bone XRF methods. The most influencing factors on maternal plasma lead seem to be maternal bone lead stores, air lead exposure and recent cooking with lead-glazed ceramics (Chuang *et al.*, 2001). The breastfeeding practice also seems to be an important predictor of blood lead concentration (Tellez-Rojo *et al.*, 2002). Women who exclusively breastfed their infants had blood lead levels that were increased by 1.4 µg/dl and women who practised mixed feeding had levels increased by 1.0 µg/dl, in relation to those who had stopped lactation. These results support the hypothesis that lactation is related to the amount of lead released from bone.

Apart from the acute threat of being killed by a gunshot, retained lead bullets tend to increase blood lead and the situation is probably aggravated if a bone fracture is caused by the gunshot (McQuirter *et al.*, 2001).

KXRF indicated a correlation between uric acid level and bone lead, whereas no obvious association between gouty arthritis and lead, at environmental exposure levels was seen (Shadick *et al.*, 2000).

For the reader interested in kinetics/toxicology we refer to papers published within the last decode (Cake *et al.*, 1996; Börjesson *et al.*, 1997b; Chettle *et al.*, 1997; Fleming *et al.*, 1997; Bergdahl *et al.*, 1998; Hernandez-Avila *et al.*, 1998; Elreedy *et al.*, 1999; Fleming *et al.*, 1999; Schwartz *et al.*,

1999; Skerfving et al., 1999; Suarez et al., 1999; Ambrose et al., 2000; Olsson et al., 2000; Erfurth et al., 2001; Markowitz and Shen, 2001; McQuirter et al., 2001; Schütz et al., submitted).

Low-level lead exposure in early development and childhood may be linked with disturbances in physical and mental development (Needleman et al., 1996). Central and peripheral neurological effects were found in young adults some 20 years after childhood environmental lead exposure (Stokes et al., 1998). However, an association between neurological outcomes and bone lead was absent, due to a presumably low precision of the bone lead measurement. Furthermore, data suggested that cognitive function (verbal memory and learning, visual memory, etc.) may progressively decline due to past occupational exposure (Schwartz et al., 2000). An increased tibia lead of 16 μg/g was equivalent to 5 more years of age.

7.1.4.2 CADMIUM

Occupational cadmium exposure takes place in various sites of industry (Tables 7.1.2 and 7.1.3). The total uptake of cadmium in the normal population may typically range 0.5–3 μg/day. The natural cadmium content in soil varies and agricultural use of phosphate fertilisers and sewage sludge lead to soils with increased cadmium content. Acid precipitation facilitates mobilisation of cadmium in soil. Hence, elevated cadmium levels in vegetables and animal food may be anticipated. This eventually results in an increased human exposure (Chan et al., 2000). Increasing levels of cadmium in humans have been observed during the 20th century (Drasch, 1983). In kidney specimens from 1897–1939 (year of death), subjects from the normal population of that period had 50 times lower cadmium concentrations compared with specimens from the 1980s. However, kidney cadmium, as measured with KXRF in groups of non-smoking farmers with high and low pH in their drinking water did not differ between groups (Nilsson et al., 2000). Neither did two groups with high and low blood cadmium levels differ in their kidney cadmium. On the other hand, for the pooled group of farmers a correlation was seen between urinary cadmium and decreasing drinking water pH.

Cadmium is virtually absent in new-borns, but the element's long biological half-life may lead to a considerable increase in the body burden during life. Kidney cadmium increases progressively with age up till 50–60 years and then tends to decline, whereas a decline is not observed for liver cadmium (Torra et al., 1995). The renal cortex cadmium concentration in non-occupationally exposed subjects generally ranges 10–100 μg/g (Friis et al., 1998; Barregård et al., 1999a; Benedetti et al., 1999).

Cadmium in kidneys and liver is readily measured using KXRF, but many studies have also been made using in vivo NAA. Both techniques have advantages and drawbacks regarding absorbed dose, sensitivity, radiation shielding, etc. The polarised KXRF system is devoted to measurements not only in occupationally exposed subjects but also in the general population. NAA and XRF studies have shown that the relation between kidney and liver cadmium is broadly consistent, regardless of method used (Börjesson et al., 1997a).

The toxic effects of cadmium exposure include acute and chronic poisoning (Järup et al., 1998; Skerfving et al., 1999). The kidney is considered the critical organ for chronic poisoning and an established damage seems irreversible and may progress even though the exposure has ceased. Low-level exposure may be linked with tubular proteinuria, i.e. an increased excretion of proteins, at a considerably lower U–Cd, 1.0 nmol/mmol creatinine, than earlier expected (Järup et al., 2000). Low-level exposure may also be associated with osteoporosis (Staessen et al., 1999; Alfvén et al., 2000) and end-stage renal disease (Hellström et al., 2001).

7.1.4.3 MERCURY

Mercury is naturally introduced into the biosphere by degassing from the earth's crust and the oceans. It is also released by combustion of fossil fuels, cremation of people with amalgam fillings and by a number of other human activities (Tables 7.1.2

and 7.1.3). In chloralkali plants, mercury cathodes may be used for the production of caustic soda and chlorine gas from brine. Another application is the primitive extraction of gold in mining, using mercury amalgamation. In general, the occupational exposure levels may be about 10–20 times higher for chloralkali and thermometer workers than for dental personnel. Food and dental amalgam dominate the mercury exposure for the general population (Barregård et al., 1995).

Absorbed mercury is quickly distributed to most regions of the body with the kidneys as the major depot. The kidneys and the central nervous system may be regarded as risk organs. For a review of mercury kinetics/toxicology see, e.g. Skerfving et al. (1999).

Our group seems to be the first to have published in vivo KXRF measured kidney mercury concentrations in occupationally exposed subjects (Börjesson et al., 1995). The average concentration was 24 μg/g (maximum 54 μg/g; Figure 7.1.12). The results were in accordance with literature data on chemically measured kidney mercury concentrations (Barregård et al., 1999b). Mercury measured in liver, thyroid, tibia and bone were below the respective detection limits.

Repeated measurements in subjects over a few months non-occupational exposure indicated a decrease in kidney mercury, although measurements contained substantial uncertainty. The estimated mean biological half-life was 115 days (range 55–173 days; Figure 7.1.13), thus longer than that reported from whole-body counter measurements after short-term exposure (Rahola et al., 1973; Hursh et al., 1976). However, long-term occupational exposure may lead to a considerable accumulation of mercury in the kidneys, thus, components with long half-lives may possibly be observed.

7.1.4.4 IRON

Those who suffer from thalassaemia, an inherited chronic blood disorder, are unable to produce sufficient amounts of haemoglobin, hence regular blood transfusions are required (Bradley and Farquharson, 1999). A build-up of residual iron in liver, heart and other organs may in turn lead to coronary problems, etc. A KXRF system (X-ray tube using 20 kV and 20 mA and a germanium detector; Figure 7.1.14) was developed for indirect determination of organ iron stores via measurements on the skin (Farquharson and Bradley, 1999; Farquharson et al., 2000). A quasi-monoenergetic beam of 8.4 keV, which is just above the absorption edge of Fe (7.11 keV), produced a detection limit of

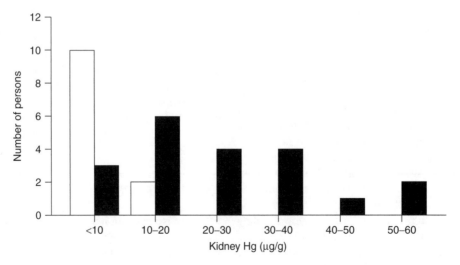

Figure 7.1.12 Estimated mercury concentration in the kidneys (Kidney Hg) of occupationally exposed workers (closed bars) and referents (open bars) (Börjesson, 1996)

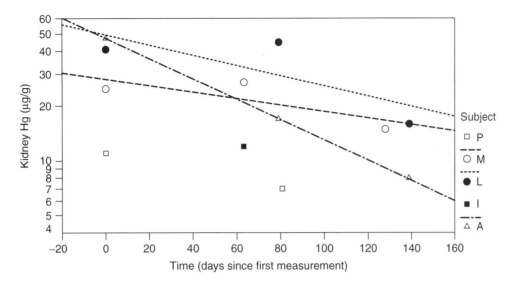

Figure 7.1.13 Retention of mercury in kidneys (Kidney Hg; log scale) of occupationally exposed subjects, during a period of non-occupational exposure. Straight lines fitted for subjects having initial kidney Hg above the detection limit (Börjesson, 1996)

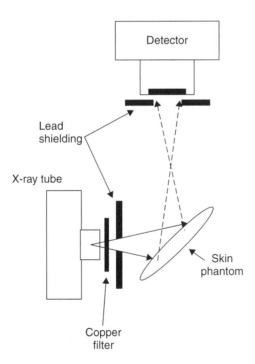

Figure 7.1.14 Schematic diagram of the experimental set-up for *in vivo* iron measurement (Bradley and Farquharson, 1999). Reproduced by permission of John Wiley & Sons, Ltd

≈15 μg/g. This implies a step forward in being able to measure iron *in vivo*. A strong correlation between skin and liver iron concentrations was demonstrated in rats.

7.1.4.5 IODINE

In an *in vivo* K-XRF study, a mean iodine concentration of 325 μg/g was noted in healthy persons, whereas decreased concentrations were found in euthyroid goitre patients and hyperthyroid patients with focal functional autonomy or Graves' disease (Reiners *et al.*, 1996). The method was reported to be well suited for individual follow-up studies because of its sensitivity, high reproducibility and low radiation exposure.

7.1.4.6 PLATINUM

A method for non-invasive determination of the platinum concentration in tumours is desirable in order to collect more information on cisplatinum uptake, retention and metabolism in tumours and risk organs. A plane polarised XRF system for platinum measurement in head and neck tumours was developed and optimised by Monte Carlo methods (Ali *et al.*, 1998a,b; Kadhim *et al.*, 2000). A radiotherapy treatment unit operated at 220 kV, a filter and a bi-layer polariser produced a detection limit of 5.6 μg/g for a tumour depth of 20 mm and a skin dose of 3 mGy. This system is similar to that developed by Jonson *et al.* (1985) and used for platinum measurements in tumours and kidneys after cisplatinum therapy.

7.1.4.7 GOLD

Gold salt is used in the treatment of rheumatoid arthritis. *In vivo* XRF of gold has been made by at least two groups. The Bath group (Shakeshaft and Lillicrap, 1993) put their efforts into a system based on a ^{153}Gd source and a 32 mm diameter germanium detector placed in 90° geometry. The kidney gold levels in the 12 patients measured were broadly consistent with those reported by Börjesson *et al.* (1993). There was no strong correlation between measured concentrations and administered amounts of gold.

7.1.4.8 URANIUM

Bone uranium has been measured *in vivo* by O'Meara *et al.* (1997, 1998b) using a ^{57}Co source in a backscatter geometry. The MDC of 20 μg uranium per g bone mineral in tibia and results from 10 measured subjects indicated that the current technique is not sensitive enough for measurements in occupationally exposed subjects, in which bone uranium seems to be at least an order of magnitude less.

7.1.5 FUTURE ASPECTS AND POTENTIALS

This subchapter reviews some of the methods used for quantification of various elements using *in vitro* and *in vivo* methods. Nondestructive *in vitro* XRF methods are well established and the detection limits imply that low concentrations of many elements can be measured in samples of minerals, food, body fluids, tissue specimen, etc. Issues of medical interest include studies of heavy elements in blood, urine, sweat, skin, hair, teeth, biopsies of tissues and organs, breathing air and drinking water. Carcinogenic and anti-tumour substances as well as elements causing allergic reactions have been studied.

Due to a decreased use of lead in petrol and decontamination of lead-based paint, although taking time and with delayed effects, lead concentrations may decrease over the coming decades. On the other hand, elements like palladium and cadmium may become more important due to the use of automobile catalysors and long-term use of fertilisers in modern agriculture, respectively. Epidemiological studies of elements like calcium may be of interest in order to establish presumed relations with coronary failure. Ancient bones, studied in connection with prehistoric finds of e.g. pottery, give us ideas of lead intake a long time ago.

Progress in the *in vivo* field has produced original techniques for measuring the toxic elements cadmium, mercury and lead. For all three elements the *in vivo* studies made so far have supplied us with important data on element concentrations and kinetics. This information is of great value to researchers in the environmental and occupational area. Lead is the one that has been most studied at a number of research centres around the world. Studies of bone lead in relation to occupational exposure, hormonal alterations and fractures are important.

XRF is the only known *in vivo* method to quantify bone lead. The XRF technique can be utilised for determining the bone lead concentration and its relation to more traditional indices of lead exposure (lead in blood and urine). In cases of suspected lead intoxication and in more basic studies, determinations of lead in bone are important. At lead intoxication the technique gives hints on how much of the lead that has been retained in bone, showing the risk of future endogenous exposure from bone lead.

XRF set-ups for cadmium and mercury are still limited but their number will presumably increase. The cadmium method has the lowest MDC and may already be used for measurements in the general population. For mercury the situation is more troublesome, since only subjects with very high exposure may have detectable levels. However, the MDC may be improved by changes in the set-up; a process guided by Monte Carlo simulation. Moreover, novel techniques to study iron, iodine, platinum, gold and uranium have been presented and applied to volunteers or patients. Some of the methods' significance needs further evaluation.

The use of fast computers and simulation programs may greatly help to study the influence of parameters changed in efforts to optimise methods. The new generation of synchrotron radiation sources needs to be investigated further, also in connection with *in vivo* applications, although the practical aspects of measurements need to be considered. Improvements in radiation sources, e.g. semi-monoenergetic X-ray tubes and filtered beams, are worth looking into in more detail. The use of new electronics and detector configurations have already proven valuable.

For both *in vivo* and *in vitro* methods, calibration procedures need to be rigorous and phantoms need to be constructed with careful considerations. Detection limits must be stated clearly and the effects of a matrix or overlying tissue need to be controlled for. Figures for accuracy and precision should be mentioned, preferably for the *in vivo* situation rather than just for the phantom measurement. Intercomparison programs, with the elements in various types of structures, should be stimulated.

REFERENCES

Ahlgren, L., Lidén, K., Mattsson, S. and Tejning, S. X-ray fluorescence analysis of lead in human skeleton *in vivo*. *Scand. J. Work Environ. Health* 1976;**2**:82–6.

Ahlgren, L. and Mattsson, S. An X-ray fluorescence technique for *in vivo* determination of lead concentration in a bone matrix. *Phys. Med. Biol.* 1979;**24**:136–45.

Ahlgren, L. and Mattsson, S. Cadmium in man measured *in vivo* by X-ray fluorescence. *Phys. Med. Biol.* 1981;**26**:19–26.

Alfvén, T., Elinder, C.G., Carlsson, M.D., Grubb, A., Hellström, L., Persson, B., Pettersson, C., Spång, G., Schütz, A. and Järup, L. Low-level cadmium exposure and osteoporosis. *J. Bone Miner. Res.* 2000;**15**:1579–86.

Al-Ghorabie, F.H. Evaluation of ^{133}Xe for X-ray fluorescence analysis of cadmium *in vivo*: a Monte Carlo study. *Radiat. Environ. Biophys.* 2000;**39**:141–5.

Ali, P.A., Al-Hussany, A.F., Bennett, C.A., Hancock, D.A. and El-Sharkawi, A.M. Plane polarized X-ray fluorescence system for the *in vivo* measurement of platinum in head and neck tumours. *Phys. Med. Biol.* 1998a;**43**:2337–45.

Ali, P.A., Bennet, C., El-Sharkawi, A.M. and Hancock, D.A. Optimisation of a polarised X-ray source for the *in vivo* measurement of platinum in head and neck tumours. *Appl. Radiat. Isot.* 1998b;**49**:647–50.

Ambrose, T.M., Al-Lozi, M. and Scott, M.G. Bone lead concentrations assessed by *in vivo* X-ray fluorescence. *Clin. Chem.* 2000;**46**:1171–8.

Amokrane, A., Gamaz, F., Bouchagra, T. and Heitz, C. Possibilities for the determination of trace elements in blood by PIXE and NAA. In *Proceedings of the European Conference on Energy Dispersive X-ray Spectrometry EDXRS-98* (Fernandez, J.E and Tartari, A., eds), Editrice Compositori, Bologna, 1999, pp. 191–6.

Anderson, D.L. and Cunningham, W.C. Nondestructive determination of lead, cadmium, tin, antimony, and barium in ceramic glazes by radioisotope X-ray fluorescence spectrometry. *J. AOAC Int.* 1996;**79**:1141–57.

Ao, Q., Lee, S.H. and Gardner, R.P. Development of the specific purpose Monte Carlo code CEARXRF for the design and use of *in vivo* X-ray fluorescence analysis systems for lead in bone. *Appl. Radiat. Isot.* 1997a;**48**:1403–12.

Ao, Q., Lee, S.H. and Gardner, R.P. Optimization of *in vivo* X-ray fluorescence analysis methods for bone lead by simulation with the Monte Carlo code CEARXRF. *Appl. Radiat. Isot.* 1997b;**48**:1413–23.

Aro, A., Amarasiriwardena, C., Lee, M.L., Kim, R. and Hu, H. Validation of K X-ray fluorescence bone lead measurements by inductively coupled plasma mass spectrometry in cadaver legs. *Med. Phys.* 2000;**27**:119–23.

Bajanowski, T., Brinkmann, B., Stefanec, A.M., Barckhaus, R.H. and Fechner, G. Detection and analysis of tracers in experimental drowning. *Int. J. Legal Med.* 1998;**111**:57–61.

Barregård, L., Sällsten, G. and Järvholm, B. People with high mercury uptake from their own dental amalgam fillings. *Occup. Environ. Med.* 1995;**52**:124–8.

Barregård, L., Svalander, C., Schütz, A., Westberg, G., Sällsten, G., Blohmé, I., Molne, J., Attman, P.O. and Haglind, P. Cadmium, mercury, and lead in kidney cortex of the general Swedish population: a study of biopsies from living kidney donors. *Environ. Health Perspect.* 1999a;**107**:867–71.

Barregård, L., Sällsten, G. and Conradi, N. Tissue levels of mercury determined in a deceased worker after occupational exposure. *Int. Arch. Occup. Environ. Health* 1999b;**72**:169–73.

Bateman, S.N., Pejovic-Milic, A., Stronach, I.M., McNeill, F.E. and Chettle, D.R. Performance appraisals of digital spectroscopy systems for the measurement of bone lead. *Appl. Radiat. Isot.* 2000;**53**:647–50.

Benedetti, J.L., Samuel, O., Dewailly, E., Gingras, S. and Lefebvre, M.A. Levels of cadmium in kidney and liver tissues among a Canadian population (province of Quebec). *J. Toxicol. Environ. Health A* 1999;**56**(3):145–63.

Bergdahl, I.A., Strömberg, U., Gerhardsson, L., Schütz, A., Chettle, D.R. and Skerfving, S. Lead concentrations in tibial and calcaneal bone in relation to the history of occupational lead exposure. *Scand. J. Work Environ. Health* 1998;**24**:38–45.

Bloch, P. and Shapiro, I.M. An X-ray fluorescence technique to measure the mercury burden of dentists *in vivo*. *Med. Phys.* 1981;**8**:308–11.

Bloch, P., Shapiro, I.M., Soule, L., Close, A. and Revich, B. Assessment of lead exposure of children from K-XRF measurements of shed teeth. *Appl. Radiat. Isot.* 1998;**49**:703–5.

Börjesson, J. Studies of cadmium, mercury and lead in man. The value of X-ray fluorescence. Thesis, Lund University, Malmö (1996).

Börjesson, J., Alpsten, M., Huang, S., Jonson, R., Mattsson, S. and Thornberg, C. *in vivo* X-ray fluorescence analysis with applications to platinum, gold and mercury in man – experiments, improvements and patient measurements. In *Human Body Composition. In vivo Methods, Models, and Assessment* (Ellis, K.J. and Eastman, J., eds), Plenum Publishing, New York, 1993, pp. 275–80.

Börjesson, J., Barregård, L., Sällsten, G., Schütz, A., Jonson, R., Alpsten, M. and Mattsson, S. *In vivo* XRF analysis of mercury: the relation between concentrations in the kidney and the urine. *Phys. Med. Biol.* 1995;**40**:413–26.

Börjesson, J., Bellander, T., Järup, L., Elinder, C.G. and Mattsson, S. *In vivo* analysis of cadmium in battery workers versus measurements of blood, urine, and workplace air. *Occup. Environ. Med.* 1997a;**54**:424–31.

Börjesson, J., Mattsson, S., Strömberg, U., Gerhardsson, L., Schütz, A. and Skerfving, S. Lead in fingerbone: a tool for retrospective exposure assessment. *Arch. Environ. Health.* 1997b;**52**:104–12.

Börjesson, J., Mattsson, S. and Alpsten, M. Trace element concentrations studied *in vivo* using X-ray fluorescence analysis. *Appl. Radiat. Isot.* 1998;**49**: 437–45.

Börjesson, J., Olsson, M. and Mattsson, S. Feasibility of a fluorescent X-ray source for *in vivo* X-ray fluorescence measurements of kidney- and liver cadmium. *Ann. N. Y. Acad. Sci.* 2000;**904**:255–8.

Bradley, D.A., Kissel, L. and Pratt, R.H. Elastic photon scattering and normalisation of *in vivo* XRF analyses of Pb in bone. In *Proceedings of the European Conference on Energy Dispersive X-ray Spectrometry EDXRS-98* (Fernandez, J.E. and Tartari, A., eds), Editrice Compositori, Bologna, 1999, pp. 231–4.

Bradley, D.A. and Farquharson, M.J. XRF and the *in vivo* evaluation of toxicological metals. *X-Ray Spectrum.* 1999;**28**:270–4.

Brändström, E., Grzegorczyk, A., Jacobsson, L., Friberg, P., Lindahl, A. and Aurell, M. GFR measurement with iohexol and 51Cr-EDTA. A comparison of the two favoured GFR markers in Europe. *Nephrol. Dial. Transplant* 1998;**13**:1176–82.

Brito, J.A.A., Fleming, D.E.B. and Chettle, D.R. Structural analysis of relationships amongst bone lead, cumulative blood lead and current blood lead. In *Proceedings of the European Conference on Energy Dispersive X-ray spectrometry EDXRS-98* (Fernandez, J.E. and Tartari, A., eds), Editrice Compositori, Bologna, 1999, pp. 235–40.

Brito, J.A., McNeill, F.E., Chettle, D.R., Webber, C.E. and Vaillancourt, C. Study of the relationships between bone lead levels and its variation with time and the cumulative blood lead index, in a repeated bone lead survey. *J. Environ. Monit.* 2000;**2**:271–6.

Brito, J.A., McNeill, F.E., Stronach, I., Webber, C.E., Wells, S., Richard, N. and Chettle, D.R. Longitudinal changes in bone lead concentration: implications for modelling of human bone lead metabolism. *J. Environ. Monit.* 2001;**3**:343–51.

Cake, K.M., Bowins, R.J., Vaillancourt, C., Gordon, C.L., McNutt, R.H., Laporte, R., Webber, C.E. and Chettle, D.R. Partition of circulating lead between serum and red cells is different for internal and external sources of lead. *Am. J. Ind. Med.* 1996;**29**:440–5.

Chan, D.Y., Black, W. and Hale, B. Bioaccumulation of cadmium from durum wheat diets in the livers and kidneys of mice. *Bull. Environ. Contam. Toxicol.* 2000;**64**: 526–33.

Chettle, D.R. X-ray fluorescence and neutron activation for measuring trace elements *in vivo*. Abstract at 4th topical meeting on industrial radiation and radioisotope measurement applications (IRRMA), Raleigh, NC, 1999.

Chettle, D.R., Fleming, D.E.B., McNeill, F.E. and Webber, C.E. Serum (plasma) lead, blood lead, and bone lead. *Am. J. Ind. Med.* 1997;**32**:319–20.

Christoffersson, J.-O. and Mattsson, S. Polarised X-rays in XRF-analysis for improved *in vivo* detectability of cadmium in man. *Phys. Med. Biol.* 1983;**28**:1135–44.

Chuang, H.Y., Schwartz, J., Gonzales-Cossio, T., Lugo, M.C., Palazuelos, E., Aro, A., Hu, H. and Hernandez-Avila, M. Interrelations of lead levels in bone, venous blood, and umbilical cord blood with exogenous lead exposure through maternal plasma lead in peripartum women. *Environ. Health Perspect.* 2001;**109**:527–32.

Currie, L.A. and Svehla, G. Nomenclature for the presentation of results of chemical analysis (IUPAC Recommendations 1994). *Pure Appl. Chem.* 1994;**66**:595–608.

Drasch, G. *Die anthropogene Blei- und Cadmiumbelastung des Menschen – Untersuchungen an Skelett- und Organmaterial*, Habil.-Schrift, München, 1983.

Elreedy, S., Krieger, N., Ryan, P.B., Sparrow, D., Weiss, S.T. and Hu, H. Relations between individual and neighbor-hood-based measures of socioeconomic position and bone lead concentrations among community-exposed men: the Normative Aging Study. *Am. J. Epidemiol.* 1999;**150**:129–41.

Erfurth, E.M., Gerhardsson, L., Nilsson, A., Rylander, L., Schütz, A., Skerfving, S. and Börjesson, J. Effects of lead on the endocrine system in lead smelter workers. *Arch. Environ. Health* 2001;**56**: 449–55.

Evans, R.O. *The Atomic Nucleus* McGraw-Hill, New York, 1955, pp. 677–84.

Farquharson, M.J. and Bradley, D.A. The feasibility of a sensitive low-dose method for the *in vivo* evaluation of Fe in skin using K-shell x-ray fluorescence (XRF). *Phys. Med. Biol.* 1999;**44**:955–65.

Farquharson, M.J., Bagshaw, A.P., Porter, J.B. and Abeysinghe, R.D. The use of skin Fe levels as a surrogate marker for organ Fe levels, to monitor treatment in cases of iron overload. *Phys. Med. Biol.* 2000;**45**: 1387–96.

Feldstein, H., Cohen, Y., Shenberg, C., Klein, A., Kojller, M., Maenhaut, W., Cafmeyer, J. and Cornelis, R. Comparison between levels of trace elements in normal and cancer inoculated mice by XRF and PIXE. *Biol. Trace Elem. Res.* 1998;**61**:169–80.

Fernandez, J.E., Hubbell, J.H., Hanson, A.L. and Spencer, L.V. Polarization effects on multiple scattering gamma transport. *Radiat. Phys. Chem.* 1993;**41**:579–630.

Fleming, D. Human lead metabolism: chronic exposure, bone lead and physiological models. PhD Thesis McMaster University, Hamilton, Ontario, Canada (1998).

Fleming, D.E.B. Construction of a K X-ray fluorescence bone lead system. Abstract at 4th topical meeting on industrial radiation and radioisotope measurement applications (IRRMA), Raleigh, NC, USA, 1999.

Fleming, D.E. and Forbes, T.A. Calibration and characterization of a digital X-ray fluorescence bone lead system. *Appl. Radiat. Isot.* 2001;**55**:527–32.

Fleming, D.E., Boulay, D., Richard, N.S., Robin, J.P., Gordon, C.L., Webber, C.E. and Chettle, D.R. Accumulated body burden and endogenous release of lead in employees of a lead smelter. *Environ. Health Perspect.* 1997;**105**:224–33.

Fleming, D.E., Chettle, D.R., Webber, C.E. and O'Flaherty, E.J. The O'Flaherty model of lead kinetics: an evaluation using data from a lead smelter population. *Toxicol. Appl. Pharmacol.* 1999;**161**: 100–9.

Forsell, M., Larsson, B., Ljungqvist, A., Carlmark, B. and Johansson, O. Mercury content in amalgam tattoos of human oral mucosa and its relation to local tissue reactions. *Eur. J. Oral. Sci.* 1998;**106**:582–7.

Friis, L., Petersson, L. and Edling, C. Reduced cadmium levels in human kidney cortex in Sweden. *Environ. Health Perspect.* 1998;**106**:175–8.

Gardner, R.P., Lee, S.H. and Todd, A.C. Error analysis and design of the XRF measurement of *in vivo* lead in the tibia with the Monte Carlo code CEARXRF. In *Proceedings of the European Conference on Energy Dispersive X-ray Spectrometry EDXRS-98* (Fernandez, J.E. and Tartari, A., eds), Editrice Compositori, Bologna, 1999a, pp. 203–16.

Gardner, R.P., Todd, A.C. and Lee, S.H. A combined K and L XRF method for *in vivo* bone lead measurement. Abstract at 4th topical meeting on industrial radiation and radioisotope measurement applications (IRRMA), Raleigh, NC, USA, 1999b.

Garman, E. Leaving no element of doubt: analysis of proteins using microPIXE. *Structure Fold Des.* 1999;**7**:R291–9.

Gordon, C.L., Chettle, D.R. and Webber, C.E. An upgraded ^{109}Cd K X-ray fluorescence bone lead measurement. In *Human Body Composition. in vivo Methods, Models, and Assessment* (Ellis, K.J. and Eastman, J., eds), Plenum Publishing, New York, 1993, pp. 285–8.

Grönberg, T., Sjöberg, S., Almén, T., Golman, K. and Mattsson, S. Noninvasive estimation of kidney function by X-ray fluorescence analysis: elimination rate and clearance of contrast media injected for urography in man. *Invest. Radiol.* 1983;**18**:445–52.

Gulson, B.L., Pounds, J.G., Mushak, P., Thomas, B.J., Gray, B. and Korsch, M.J. Estimation of cumulative lead releases (lead flux) from the maternal skeleton during pregnancy and lactation. *J. Lab. Clin. Med.* 1999;**134**:631–40.

Heckel, J. Using Barkla polarised X-ray radiation in energy dispersive X-ray fluorescence analysis (EDXRF). *J. Trace Microprobe Tech.* 1995;**13**: 97–108.

Hellström, L., Elinder, C.G., Dahlberg, B., Lundberg, M., Järup, L., Persson, B. and Axelson, O. Cadmium exposure and end-stage renal disease. *Am. J. Kidney Dis.* 2001;**38**: 1001–8.

Hernandez-Avila, M., Smith, D., Meneses, F., Sanin, L.H. and Hu, H. The influence of bone and blood lead on plasma lead levels in environmentally exposed adults. *Environ. Health Perspect.* 1998;**106**:473–7.

Hoffer, P.B., Jones, W.B., Crawford, R.B., Beck, R. and Gottschalk, A. Fluorescent thyroid scanning: a new method of imaging the thyroid. *Radiology* 1968;**90**: 342–4.

Holmes, M.C. Identifying lead sources by X-ray fluorescence spectrum analysis. *J. Tenn. Med. Assoc.* 1995;**88**:191–2.

Hoppin, J.A., Aro, A.C., Williams, P.L., Hu, H. and Ryan, P.B. Validation of K-XRF bone lead measurement in young adults. *Environ. Health Perspect.* 1995;**103**: 78–83.

Hoppin, J.A., Aro, A., Hu, H and Ryan, P.B. Measurement variability associated with KXRF bone lead measurement in young adults. *Environ. Health Perspect.* 2000;**108**: 239–42.

Hugtenburg, R.P., Turner, J.R., Mannering, D.M. and Robinson, B.A. Monte Carlo methods for the *in vivo* analysis of cisplatin using X-ray fluorescence. *Appl. Radiat. Isot.* 1998;**49**:673–6.

Hursh, J.B., Clarkson, T.W., Cherian, M.G., Vostal, J.J. and Vander Mallie, R. Clearance of mercury (Hg-197, Hg-203) vapor inhaled by human subjects. *Arch. Environ. Health* 1976;**31**: 302–9.

Järup, L., Berglund, M., Elinder, C.G., Nordberg, G. and Vahter, M. Health effects of cadmium exposure – a review of the literature and a risk estimate. *Scand. J. Work Environ. Health.* 1998;**24**(Suppl):1–51.

Järup, L., Hellström, L., Alfvén, T., Carlsson, M.D., Grubb, A., Persson, B., Pettersson, C., Spång, G., Schütz, A. and Elinder, C.G. Low level exposure to cadmium and early kidney damage: the OSCAR study. *Occup. Environ. Med.* 2000;**57**:668–72.

Jones, K.W., Schidlovsky, G., Burger, D.E., Milder, F.L. and Hu, H. Distribution of lead in human bone: III. Synchrotron X-ray microscope measurements. In *Advances in In Vivo Body Composition Studies* (Yasumura, S., Harrison, J.E., McNeill, K.G., Woodhead, A.D. and Dilmanian, F.A., eds), Plenum Press, New York, 1990, pp. 281–6.

Jonson, R., Mattsson, S. and Unsgaard, B. *In vivo* determination of platinum concentration after cisplatin therapy of testicular carcinoma. In *Recent Advances in Chemotherapy* (Ishigami, J., ed.), The University of Tokyo Press, Tokyo, 1985, pp. 1222–4.

Jonson, R., Mattsson, S. and Unsgaard, B. A method for *in vivo* analysis of platinum after chemotherapy with cisplatin. *Phys. Med. Biol.* 1988;**33**: 847–57.

Kadar, L., Albertsson, M., Areberg, J., Landberg, T. and Mattsson, S. The prognostic value of body protein in patients with lung cancer. *Ann. N. Y. Acad. Sci.* 2000;**904**:584–91.

Kadhim, R., al-Hussany, A., Ali, P.A., Hancock, D.A. and el-Sharkawi, A.M. *In vivo* measurement of platinum in the kidneys using X-ray fluorescence. *Ann. N. Y. Acad. Sci.* 2000;**904**:263–6.

Keenleyside, A., Song, X., Chettle, D.R. and Webber, C.E. The lead content of human bones from the 1845 Franklin expedition. *J. Archeolog. Sci.* 1996;**23**: 461–5.

Kilic, A. A theoretical and experimental investigation of polarised X-rays for the *in vivo* measurement of heavy metals. PhD Thesis, University of Wales (1995).

Kim, R., Aro, A., Rotnitzky, A., Amarasiriwardena, C. and Hu, H. K x-ray fluorescence measurements of bone lead concentration: the analysis of low-level data. *Phys. Med. Biol.* 1995;**40**:1475–85.

Kondrashov, V.S. and Rothenberg, S.J. How to calculate lead concentration and concentration uncertainty in XRF *in vivo* bone lead analysis. *Appl. Radiat. Isot.* 2001a;**55**:799–803.

Kondrashov, V.S. and Rothenberg, S.J. One approach for doublet deconvolution to improve reliability in spectra analysis for *in vivo* lead measurement. *Appl. Radiat. Isot.* 2001b;**54**:691–4.

Kondrashov, V.S., Rothenberg, S.J., Sajo-Bohus, L., Greaves, E.D. and Liendo, J.A. Increasing reliability in gamma and X-ray spectral data analysis: least moduli approach. *Nucl. Instrum Methods* 2000:**A446**: 560–8.

Lewis, D.G., Kilic, A. and Ogg, C.A. Computer aided design of a polarised source for *in vivo* X-ray fluorescence analysis. *Appl. Radiat. Isot.* 1998;**49**:707–9.

Liljensten, E.L., Attaelmanan, A.G., Larsson, C., Ljusberg-Wahren, H., Danielsen, N., Hirsch, J.M. and Thomsen, P. Hydroxyapatite granule/carrier composites promote new bone formation in cortical defects. *Clin. Implant. Dent. Relat. Res.* 2000;**2**(1):50–9.

Liu, N.Q., Xu, Q., Hou, X.L., Liu, P.S., Chai, Z.F., Zhu, L., Zhao, Z.Y., Wang, Z.H. and Li, Y.F. The distribution patterns of trace elements in the brain and erythrocytes in a rat experimental model of iodine deficiency. *Brain Res. Bull.* 2001;**55**:309–12.

MacPherson, A. and Bacsó, J. Relationship of hair calcium concentration to incidence of coronary heart disease. *Sci. Total Environ.* 2000;**255**:11–9.

Majewska, U., Braziewicz, J., Banas, D., Kubala-Kukus, A., Kucharzewski, M., Waler, J., Gozdz, S. and Wudarczyk, J. Zn concentration in thyroid tissue and whole blood of women with different diseases of thyroid. *Biol. Trace Elem. Res.* 2001;**80**:193–9.

Mann, S.E., Ringo, M.C., Shea-McCarthy, G., Penner-Hahn, J. and Evans, C.E. Element-specific detection in capillary electrophoresis using X-ray fluorescence spectroscopy. *Anal. Chem.* 2000;**72**:1754–8.

Markowitz, M.E. and Shen, X.M. Assessment of bone lead during pregnancy: a pilot study. *Environ. Res.* 2001;**85**:83–9.

McNeill, F. *in vivo* precision in ^{109}Cd K X-ray fluorescence bone lead measurements: Implications for populations studies. Abstract at 4th Topical Meeting on Industrial Radiation and Radioisotope Measurement Applications (IRRMA), Raleigh, NC, 1999.

McNeill, F.E. and Chettle, D.R. Improvements to the *in vivo* measurement of cadmium in the kidney by neutron activation analysis. *Appl. Radiat. Isot.* 1998;**49**:699–700.

McNeill, F.E. and O'Meara, J.M. The *in vivo* measurement of trace heavy metals by K X-ray fluorescence. *Adv. X-ray Anal.* 1999;**41**:910–21.

McNeill, F.E., Stokes, L., Chettle, D.R. and Kaye, W.E. Factors affecting *in vivo* measurement precision and accuracy of ^{109}Cd K x-ray fluorescence measurements. *Phys. Med. Biol.* 1999;**44**:2263–73.

McNeill, F.E., Stokes, L., Brito, J.A., Chettle, D.R. and Kaye, W.E. ^{109}Cd K x ray fluorescence measurements of tibial lead content in young adults exposed to lead in early childhood. *Occup. Environ. Med.* 2000;**57**:465–71.

McQuirter, J.L., Rothenberg, S.J., Dinkins, G.A., Manalo, M., Kondrashov, V. and Todd, A.C. The effects of retained lead bullets on body lead burden. *J. Trauma.* 2001;**50**:892–9.

Meitín, J.J., de la Fuente, F. and Mendoza, A. Some achievements on EDXRF analysis in Cuba. Applications in industry, medicine and environment. In *Proceedings of the European Conference on Energy Dispersive X-ray Spectrometry EDXRS-98* (Fernandez, J.E. and Tartari, A. eds), Editrice Compositori, Bologna, 1999, pp. 351–8.

Messerschmidt, J., von Bohlen, A., Alt, F. and Klockenkämper, R. Separation and enrichment of palladium and gold in biological and environmental samples, adapted to the determination by total reflection X-ray fluorescence. *Analyst* 2000;**125**:397–9.

Mori, H., Chujo, M., Haruyama, S., Sakamoto, H., Shinozaki, Y., Uddin-Mohammed, M., Iida, A. and Nakazawa, H. Local continuity of myocardial blood flow studied by monochromatic synchrotron radiation-excited X-ray fluorescence spectrometry. *Circ Res.* 1995;**76**:1088–100.

Muramatsu, Y., Ishikawa, Y., Yoshida, S. and Mori, T. Determination of thorium in organs from Thorotrast patients by inductively coupled plasma mass spectroscopy and X-ray fluorescence. *Radiat Res.* 1999;**152**:S97–S101.

Needleman, H.L., Riess, J.A., Tobin, M.J., Biesecker, G.E. and Greenhouse, J.B. Bone lead levels and delinquent behaviour. *JAMA* 1996;**275**: 363–9.

Nie, H., Chettle, D., Stronach, I., Arnold, M., Huang, S., McNeill, F. and O'Meara, J. A study of MDL improvement for the in-vivo measurement of lead in bone. Abstract at 5th International Topical Meeting on Industrial Radiation and Radioisotope Measurement Applications, Bologna, 2002.

Nilsson, U., Ahlgren, L., Christoffersson, J.-O. and Mattsson, S. Further improvements of XRF analysis of cadmium *in vivo*. In *Advances in In Vivo Body Composition Studies* (Yasumura, S., Harrison, J.E., McNeill, K.G., Woodhead, A.D. and Dilmanian, F.A., eds), Plenum Press, New York, 1990, pp. 297–301.

Nilsson, U., Attewell, R., Christoffersson, J.-O., Schütz, A., Ahlgren, L., Skerfving, S. and Mattsson, S. Kinetics of lead in bone and blood after end of occupational exposure. *Pharmacol. Toxicol.* 1991;**68**:477–84.

Nilsson, U., Schütz, A., Bensryd, I., Nilsson, A., Skerfving, S. and Mattsson, S. Cadmium levels in kidney cortex in Swedish farmers. *Environ. Res.* 2000;**82**:53–9.

Olsson, M., Gerhardsson, L., Jensen, A., Börjesson, J., Schütz, A., Mattsson, S. and Skerfving, S. Lead accumulation in highly exposed smelter workers. *Ann. N. Y. Acad. Sci.* 2000;**904**:280–3.

O'Meara, J.M., Chettle, D.R., McNeill, F.E. and Webber, C.E. The feasibility of measuring bone uranium concentrations *in vivo* using source excited K X-ray fluorescence. *Phys. Med. Biol.* 1997;**42**:1109–20.

O'Meara, J.M., Chettle, D.R., McNeill, F.E., Prestwich, W.V. and Svensson, C.E. Monte Carlo simulation of source-excited *in vivo* X-ray fluorescence measurements of heavy metals. *Phys. Med. Biol.* 1998a;**43**:1413–28.

O'Meara, J.M., Chettle, D.R., McNeill, F.E. and Webber, C.E. *In vivo* X-ray fluorescence (XRF) measurement of uranium in bone. *Appl. Radiat. Isot.* 1998b;**49**: 713–5.

O'Meara, J.M., Chettle, D.R. and McNeill, F.E. The validity of coherent scatter peak normalisation of X-ray intensities detected during *in vivo* XRF measurements of metals in bone. *Adv. X-ray Anal.* 1999;**41**:932–40.

O'Meara, J.M., Börjesson, J. and Chettle, D.R. Improving the *in vivo* X-ray fluorescence (XRF) measurement of renal mercury. *Appl. Radiat. Isot.* 2000;**53**:639–46.

O'Meara, J.M., Blackburn, B.W., Chichester, D.L., Gierga, D.P. and Yanch, J.C. The feasibility of accelerator-based *in vivo* neutron activation analysis of nitrogen. *Appl. Radiat. Isot.* 2001a;**55**:767–74.

O'Meara, J.M., Börjesson, J., Chettle, D.R. and Mattsson, S. Normalisation with coherent scatter signal: improvements in the calibration procedure of the ^{57}Co-based *in vivo* XRF bone-Pb measurement. *Appl. Radiat. Isot.* 2001b;**54**:319–25.

O'Meara, J.M., Börjesson, J., Chettle, D.R. and McNeill, F.E. Further developments of an *in vivo* polarised X-ray fluorescence mercury measurement system. Abstract at 5th International Topical Meeting on Industrial Radiation and Radioisotope Measurement Applications, Bologna, 2002.

Parsell, D.E., Karns, L., Buchanan, W.T. and Johnson, R.B. Mercury release during autoclave sterilization of amalgam. *J. Dent. Educ.* 1996;**60**:453–8.

Rahola, T., Hattula, T., Korolainen, A. and Miettinen, J.K. Elimination of free and protein-bound ionic mercury (^{203}Hg^{2+}) in man. *Ann. Clin. Res.* 1973;**5**:214–9.

Reich, J.-G. *Curve Fitting and Modelling for Scientists and Engineers*, McGraw-Hill, New York, 1992.

Reiners, C., Sonnenschein, W., Caspari, G., Yavuz, A., Ugur, T., Leder-bogen, S. and Olbricht, T. Non-invasive measurement of thyroidal iodine content (TIC) by X-ray fluorescence analysis (XFA). *Acta Med. Austriaca* 1996;**23**(1–2):61–4.

Rosen, J.F. Clinical applications of L-line X-ray fluorescence to estimate bone lead values in lead-poisoned young children and in children, teenagers, and adults from lead-exposed and non-lead-exposed suburban communities in the United States. *Toxicol. Ind. Health.* 1997;**13**:211–8.

Roy, M.M., Gordon, C.L., Beaumont, L.F., Chettle, D.R. and Webber, C.E. Further experience with bone lead content measurements in residents of southern Ontario. *Appl. Radiat. Isot.* 1997;**48**:391–6.

Schidlovsky, G., Jones, K.W., Burger, D.E., Milder, F.L. and Hu, H. Distribution of lead in human bone: II Proton microprobe measurements. In *Advances in In Vivo Body Composition Studies* (Yasumura, S., Harrison, J.E., McNeill, K.G., Woodhead, A.D. and Dilmanian, F.A., eds), Plenum Press, New York, 1990, pp. 275–80.

Schmehl, R.L., Cox, D.C., Dewalt, F.G., Haugen, M.M., Koyak, R.A., Schwemberger Jr, J.G. and Scalera, J.V. Lead-based paint testing technologies: summary of an EPA/HUD field study. *Am. Ind. Hyg. Assoc. J.* 1999;**60**:444–51.

Schwartz, B.S., Stewart, W.F., Todd, A.C. and Links, J.M. Predictors of dimercaptosuccinic acid chelatable lead and tibial lead in former organolead manufacturing workers. *Occup. Environ. Med.* 1999;**56**:22–9.

Schwartz, B.S., Stewart, W.F., Bolla, K.I., Simon, P.D., Bandeen-Roche, K., Gordon, P.B., Links, J.M. and Todd, A.C. Past adult lead exposure is associated with longitudinal decline in cognitive function. *Neurology* 2000;**55**: 1144–50.

Schütz, A., Olsson, M., Jensen, A., Gerhardsson, L., Börjesson, J., Mattsson, S. and Skerfving, S. Lead in finger-bone, whole-blood, plasma and urine in lead smelter workers – extended exposure range. *Int. Arch. Occup. Environ. Health* (submitted).

Shadick, N.A., Kim, R., Weiss, S., Liang, M.H., Sparrow, D. and Hu, H. Effect of low level lead exposure on hyperuricemia and gout among middle aged and elderly men: the normative ageing study. *J. Rheumatol.* 2000;**27**:1708–12.

Shakeshaft, J. and Lillicrap, S. An X-ray fluorescence system for the determination of gold *in vivo* following chrysotherapy. *Br. J. Rad.* 1993;**66**:714–7.

Shenberg, C., Boazi, M., Cohen, J., Klein, A., Kojler, M. and Nyska, A. An XRF study of trace elements accumulation in kidneys of tumor-bearing mice after treatment with cis-DDP with and without selenite and selenocistamine. *Biol. Trace Elem. Res.* 1994;**40**:137–49.

Singh, J., Pritchard, D.E., Carlisle, D.L., Mclean, J.A., Montaser, A., Orenstein, J.M. and Patierno, S.R. Internalization of carcinogenic lead chromate particles by cultured normal human lung epithelial cells: formation of intracellular lead-inclusion bodies and induction of apoptosis. *Toxicol. Appl. Pharmacol.* 1999;**161**:240–8.

Skerfving, S., Bencko, V., Vahter, M., Schütz, A. and Gerhardsson, L. Environmental health in the Baltic region – toxic metals. *Scand. J. Work Environ. Health.* 1999;**25**(Suppl. 3): 40–64.

Somervaille, L.J., Chettle, D.R. and Scott, M.C. In vivo measurement of lead in bone using X-ray fluorescence. *Phys. Med. Biol.* 1985;**30**: 929–43.

Spitz, H., Jenkins, M., Lodwick, J. and Bornschein, R. A new anthropometric phantom for calibrating *in vivo* measurements of stable lead in the human leg using X-ray fluorescence. *Health Phys.* 2000;**78**:159–69.

Staessen, J.A., Roels, H.A., Amelianov, D., Kuznetsova, T., Thijs, L., Vangronsveld, J. and Fagard, R. Environmental exposure to cadmium, forearm bone density, and risk of fractures: prospective populations study. Public health and environmental exposure to cadmium (PheeCad) study group. *Lancet* 1999;**353**:1140–4.

Sterner, G., Frennby, B., Månsson, S., Ohlsson, A., Prutz, K.G. and Almén, T. Assessing residual renal function and efficiency of hemodialysis – an application for urographic contrast media. *Nephron* 2000;**85**:324–33.

Stokes, L., Letz, R., Gerr, F., Kolczak, M., McNeill, F.E., Chettle, D.R. and Kaye, W.E. Neurotoxicity in young adults 20 years after childhood exposure to lead: the Bunker Hill experience. *Occup. Environ. Med.* 1998;**55**:507–16.

Streli, C. Light element trace analysis by means of TXRF using synchrotron radiation. *J. Trace Microprobe Tech.* 1995;**13**:109–18.

Suarez, A.M., Sajo-Bohus, L., Greaves, E., Scott, M.C., Somervaille, L.J., Green, S., Foglietta, L.M., Avila, T. and Loaiza, F. Study in-vivo of the lead concentration in the tibia, using X-ray fluorescence. Abstract at 4th Topical Meeting on Industrial Radiation and Radioisotope Measurement Applications (IRRMA), Raleigh, NC, 1999.

Sures, B., Zimmermann, S., Messerschmidt, J., von Bohlen, A. and Alt, F. First report on the uptake of automobile catalyst emitted palladium by European eels (*Anguilla anguilla*) following experimental exposure to road dust. *Environ Pollut.* 2001;**113**:341–5.

Sutcliffe, J. A review of *in vivo* experimental methods to determine the composition of the human body. *Phys. Med. Biol.* 1996;**41**:791–834.

Suzuki, H. Nickel and gold in skin lesions of pierced earlobes with contact dermatitis. A study using scanning electron microscopy and X-ray fluorescence. *Arch. Dermatol. Res.* 1998;**290**:523–7.

Suzuki, N. Metal allergy in dentistry: detection of allergen metals with X-ray fluorescence spectroscope and its application toward allergen elimination. *Int. J. Prosthodont.* 1995;**8**:351–9.

Szaloki, I., Lewis, D.G., Bennett, C.A. and Kilic, A. Application of the fundamental parameter method to the *in vivo* X-ray fluorescence analysis of Pt. *Phys. Med. Biol.* 1999;**44**(5):1245–55.

Szaloki, I.I., Torok, S.B., Ro, C.U., Injuk, J. and Van Grieken, R.E. X-ray spectrometry. *Anal Chem.* 2000;**72**:211R–233R.

Tartari, A., Baraldi, C., Felsteiner, J. and Casnati, E. Compton scattering profile for *in vivo* XRF techniques. *Phys. Med. Biol.* 1991;**36**:567–78.

Tartari, A., Casnati, E., Baraldi, C., Fernandez, J.E. and Felsteiner, J. Comments on the paper 'Monte Carlo simulation of source-excited *in vivo* X-ray fluorescence measurements of heavy metals'. *Phys. Med. Biol.* 1999;**44**:L3–6.

Tellez-Rojo, M.M., Hernandez-Avila, M., Gonzalez-Cossio, T., Romieu, I., Aro, A., Palazuelos, E., Schwartz, J. and Hu, H. Impact of breastfeeding on the mobilization of lead from bone. *Am. J. Epidemiol.* 2002;**155**:420–8.

Todd, A.C. Coherent scattering and matrix correction in bone-lead measurements. *Phys. Med. Biol.* 2000a;**45**:1953–63.

Todd, A.C. Contamination of *in vivo* bone-lead measurements. *Phys. Med. Biol.* 2000b;**45**:229–40.

Todd, A.C. L-shell X-ray fluorescence measurements of lead in bone: theoretical considerations. *Phys. Med. Biol.* 2002a;**47**:491–505.

Todd, A.C. L-shell X-ray fluorescence measurements of lead in bone: system development. *Phys. Med. Biol.* 2002b;**47**:507–22.

Todd, A.C., Carroll, S., Godbold, J.H., Moshier, E.L. and Khan, F.A. Variability in XRF-measured tibia lead levels. *Phys. Med. Biol.* 2000a;**45**:3737–48.

Todd, A.C., Ehrlich, R.I., Selby, P. and Jordaan, E. Repeatability of tibia lead measurement by X-ray fluorescence in a battery-making work-force. *Environ. Res.* 2000b;**84**: 282–9.

Todd, A.C., Godbold, J.H., Moshier, E.L. and Khan, F.A. Patella lead X-ray fluorescence measurements are independent of sample orientation. *Med. Phys.* 2001c;**28**: 1806–10.

Todd, A.C., Moshier, E.L., Carroll, S. and Casteel, S.W. Validation of X-Ray fluorescence-measured swine femur lead against atomic absorption spectrometry. *Environ. Health. Perspect.* 2001d;**109**:1115–1119.

Todd, A.C., Buchanan, R., Carroll, S., Moshier, E.L., Popovac, D., Slavkovich, V. and Graziano, J.H. Tibia lead levels and methodological uncertainty in 12-year-old children. *Environ. Res.* 2001a;**86**:60–5.

Todd, A.C., Carroll, S., Godbold, J.H., Moshier, E.L. and Khan, F.A. The effect of measurement location on tibia lead XRF measurement results and uncertainty. *Phys. Med. Biol.* 2001b;**46**:29–40.

Todd, A.C., Carroll, S., Geraghty, C., Khan, F.A., Moshier, E.L., Tang, S. and Parsons, P.J. L-shell X-ray fluorescence measurements of lead in bone: accuracy and precision. *Phys. Med. Biol.* 2002a;**47**:1399–1419.

Todd, A.C., Parsons, P.J., Tang, S. and Moshier, E.L. Individual variability in human tibia lead concentration. *Environ. Health. Perspect.* 2001e;**109**:1139–1143.

Todd, A.C., Parsons, P.J., Carroll, S., Geraghty, C., Khan, F.A., Tang, S. and Moshier, E.L. Measurements of lead in human tibiae. A comparison between K-shell X-ray fluorescence and electrothermal atomic absorption spectrometry. *Phys. Med. Biol.* 2002b;**47**:673–687.

Toribara, T.Y. Analysis of single hair by XRF discloses mercury intake. *Hum. Exp. Toxicol.* 2001;**20**:185–8.

Torra, M., To-Figueras, J., Rodamilans, M., Brunet, M. and Corbella, J. Cadmium and zinc relationships in the liver and kidney of humans exposed to environmental cadmium. *Sci. Total. Environ.* 1995;**170**:53–7.

Wielopolski, L., Vartsky, D. and Cohn, S.H. *In vivo* elemental analysis utilizing XRF techniques. *Neurotoxicology* 1983;**4**:173–6.

Wielopolski, L., Rosen, J.F., Slatkin, D.N., Zhang, R., Kalef-Ezra, J.A., Rothman, J.C., Maryanski, M. and Jenks, S.T. *In vivo* measurement of cortical bone lead using polarised X rays. *Med. Phys.* 1989;**16**:521–8.

Zaichick, V. Estimation of extracellular water by means of stable bromine and X-ray fluorescence analysis. *Appl. Radiat. Isot.* 1998a;**49**:635.

Zaichick, V. X-ray fluorescence analysis of bromine for the estimation of extracellular water. *Appl. Radiat. Isot.* 1998b;**49**:1665–9.

Zaichick, V.Y. and Ovchjarenko, N.N. *In vivo* X-ray fluorescent analysis of Ca, Zn, Sr and Pb in frontal tooth enamel. *J. Trace Microprobe Tech.* 1996;**14**:143–52.

Zaichick, V. and Ovchjarenko, N. *In vivo* X-ray fluorescence for estimation of essential and toxic trace elements in teeth. *Appl. Radiat. Isot.* 1998;**49**:721.

Zaichick, V., Ovchjarenko, N. and Zaichick, S. *In vivo* energy dispersive X-ray fluorescence for measuring the content of essential and toxic trace elements in teeth. *Appl. Radiat. Isot.* 1999;**50**:283–93.

7.2 Total Reflection X-ray Fluorescence for Semiconductors and Thin Films

Y. MORI
Wacker-NSCE Corp., Yamaguchi, Japan

7.2.1 INTRODUCTION

For many years, only a few materials have been used in the manufacturing of silicon semiconductor devices: SiO_2 for gate oxide, Si_3N_4 for capacitors, polycrystalline Si for electrodes, and Al for wiring. In the past decade, the simple shrinkage of unit transistors in large-scale integrated (LSI) circuits increased manufacturing difficulty because of the physical restrictions imposed by the smaller size. Consequently, a large number of alternative elements have been tested and actually adopted to keep pace with the reduced size. Some examples include ZrO_2 and HfO_2 for gate oxides, Ta_2O_5 and $(Ba,Sr)TiO_3$ (BST) for capacitors, $SrRuO_3$ (SRO) for electrodes, and Cu for wiring.[1] At the same time, in order to reduce the cost of LSI devices, large-diameter silicon wafers were adopted by leading-edge semiconductor manufacturers. Shipment of 200-mm ϕ wafers is mainstream at present, and the shipment of 300-mm ϕ wafers is increasing.[2] In the semiconductor manufacturing process using such large-diameter wafers, controlling the film composition as well as reducing undesirable contaminants over the surface is increasingly significant for stabilizing yield.

Because of the technological transitions mentioned above, two characteristics have become the keys to current semiconductor analysis. These key characteristics are (1) the expansion of the number of analysable elements to evaluate the new materials and (2) the capability of distribution analysis for large-diameter wafers. High throughput and ease of operation are also indispensable in the highly competitive semiconductor industry. X-Ray analyses such as the methods of total-reflection X-ray fluorescence (TXRF) spectrometry and X-ray reflectivity (XRR) meet these requirements, and significant improvements to these techniques have been seen in recent years.

In this subchapter, progress in the industrial TXRF technique will first be shown. The use of TXRF for semiconductor analysis started at the end of the 1980s, and came into popular use in the 1990s. Today, more than 300 TXRF spectrometers are installed in this industry worldwide, meaning that almost all leading-edge semiconductor factories have introduced TXRF. Since the main purpose of TXRF is trace contamination analysis, improvements in detection ability as well as reliability will be discussed. In addition, XRF and XRR/XRF analysers for the characterization of thin films made from new materials will be introduced.

7.2.2 IMPROVEMENTS IN TXRF INSTRUMENTATION

7.2.2.1 EXPANSION OF ANALYSABLE ELEMENTS

The conventional TXRF uses W Lβ (9.67 keV) as an excitation source to analyse the elements such as Cr and Fe that are critical to the properties of LSIs. As a number of new metals have been introduced or are being tested as alternative materials in recent LSI manufacturing, the W Lβ source ceased to be satisfactory due to its narrow excitation window ($_{16}$S – $_{30}$Zn by K lines). Although some TXRF manufacturers used Mo Kα (17.45 keV) as an alternative source, they could not analyse materials such as Zr, Mo, and Ru that have high K-absorption edge energy. To analyse these new elements, new types of excitation sources have been introduced. One type is Ag Kα (22.11 keV) excitation, which can excite the K shells of up to Ru. TXRF with a sealed Ag X-ray tube is actually used to evaluate the cleaning efficiency of Ru.[3] Figure 7.2.1 is a sample spectrum of 10^{12} atoms cm^{-2} Ru on a silicon wafer.

Another interesting approach is to utilize continuous X-rays that had at one time been considered to be useless in semiconductor TXRF. Figure 7.2.2 is a sample spectrum of 5×10^{11} atoms cm^{-2} Mo on a silicon wafer excited by ca. 22 keV X-rays that are monochromatized from continuous X-rays. Because of the relatively high intensity of continuous X-rays from the rotating anode and the improvements in the multilayer monochromator, the system shows a detection limit of ca. 1.5×10^{10} atoms cm^{-2} for Mo. This system is also applicable to Ru analysis.

Besides the heavy elements mentioned above, the analysis of light elements (Na, Mg, Al) by TXRF has been an issue of interest from the beginning of semiconductor application. Because conventional TXRF systems with medium-energy excitation sources such as W Lβ or Mo Kα do not have enough excitation efficiency for light elements such as Na and Al, the detection limit was poor. The low detection ability of light elements is one of the major drawbacks of TXRF when compared with atomic absorption spectrophotometry (AAS) and inductively coupled plasma mass spectrometry (ICPMS). Several researchers attempted to improve the capability of light element analysis. One such system is a straight-TXRF with W Mα (1.78 keV) excitation, which is now commercially available.[4] The monochromatic W Mα, generated by a W rotating anode, can effectively excite light elements up to Al. In addition, the energy of W Mα is lower than the Si K absorption edge (1.84 keV),

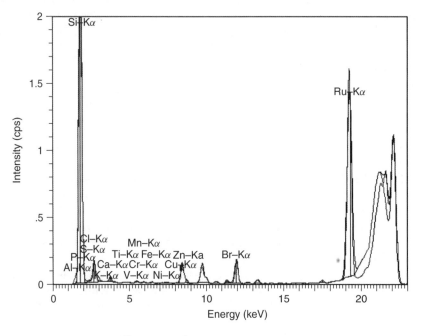

Figure 7.2.1 Sample TXRF spectrum of 10^{12} atoms cm^{-2} Ru on a silicon wafer excited by Ag Kα (22.11 keV). Courtesy of Technos Corp., Japan

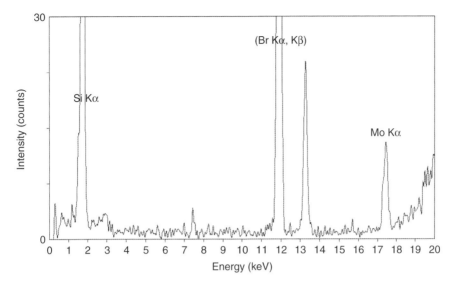

Figure 7.2.2 Sample TXRF spectrum of 5×10^{11} atoms cm^{-2} Mo on a silicon wafer. The excitation source is a monochromatic ca. 22 keV X-ray from continuous X-rays generated by an Au rotating anode (35 kV – 255 mA)

so the spectral background caused by Si Kα from the substrate is reduced. The detection limit of this system, however, is not very good (i.e. 10^{11}–10^{12} atoms cm^{-2} level) because the W Mα emission is not strong and the sensitivity of the solid state detector (SSD) in the low energy region is not high enough to allow determination at the 10^9 atoms cm^{-2} level. To achieve a higher level of sensitivity on this system, a combination of chemical preconcentration was examined.[5–7] The results will be discussed in another subsection.

7.2.2.2 LOW BACKGROUND OPTICS

The lower limit of detection (LLD) of TXRF is generally expressed by the equation

$$\text{LLD} = 3 I_{\text{BG}}^{1/2} C_{\text{STD}} / I_{\text{STD}} \qquad (7.2.1)$$

where I_{BG} is the background count of a blank sample, C_{STD} is the nominal concentration of a standard sample, and I_{STD} is the fluorescent X-ray intensity of the standard sample. At the initial stage of TXRF development for semiconductor application, increasing I_{STD} by intensifying the primary X-ray was a common effort to improve LLD. At first, a rotating anode was introduced instead of a sealed tube, and then an artificial multilayer monochromator was installed. In recent TXRF machines, however, the intensification of the primary X-ray is becoming impractical, because of the increasing dead time of the detection system and the impurity peaks caused by imperfections of the artificial multilayers.[4] Therefore, the next strategy to improve LLD should be the reduction of I_{BG}, and some significant improvements have been achieved.

The first improvement was the introduction of a dual-crystal monochromator. Because an artificial multilayer is not a perfect monochromator, parts of white X-rays sometimes passed through to cause impurity peaks in the background spectrum of single-crystal systems. To reduce the intensity of these impurity peaks, a dual-multilayer monochromator was proposed.[4] Figure 7.2.3 compares the blank spectra for single- and dual-crystal optics of Au Lβ excitation. Impurity peaks in the former system, which appeared at around 6 keV and 8.4 keV, disappeared in the dual-crystal system. Accordingly, the detection ability has improved substantially.

The second improvement is the implementation of an x–y stage instead of the traditional

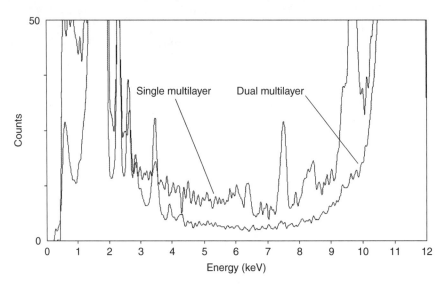

Figure 7.2.3 Comparison of blank spectra for single- and dual-crystal optics of Au Lβ excitation TXRF

$r-\theta$ stage. Because a silicon wafer is a single crystal, the diffraction of irradiated primary X-rays occurs at a certain azimuthal angle against the wafer, raising the overall background of the spectrum. During mapping measurements on an $r-\theta$ controlled stage, the diffraction cannot be avoided because the azimuthal angle is not selectable.[8] Hence, the necessity of an $x-y-\theta$ stage had been suggested,[9] and such a stage was actually implemented.[4] Because of the additional third axis, any arbitrary azimuthal angle can be set at any measurement spot on the sample to avoid the diffraction in mapping analysis.

The third improvement is in the SSD. Spurious peaks originating in impurities in the detection system had been known to raise the background level to degrade the LLD. The principal cause of such spurious peaks were found to be the impurities in the Be window,[9,10] but this is not the only source according to our research. Now, a manufacturer has reported the almost complete removal of impurity peaks. Figure 7.2.4 compares the impurity peaks of Fe for conventional and improved detectors.

The fourth improvement is concerned with the alternative excitation source. Recently, cross-contamination by Cu has become a critical issue in the Cu wiring process in manufacturing high-speed LSI devices, and trace Cu detection is becoming increasingly important. The traditional W Lβ excitation TXRF, however, includes a background problem in trace Cu detection. An escape peak inevitably appears at 7.93 keV in the W Lβ excitation-Si(Li) SSD system, which is unfortunately very close to Cu Kα (8.04 keV). The escape peak not only raises the background of Cu but also makes the peak separation of small Cu Kα difficult, resulting in a degradation of the ability to detect Cu. Au Lβ (11.44 keV) might be one solution to this problem while retaining the ability to detect other elements such as Fe and Ni.[11,12] Figure 7.2.5 compares the blank spectra of W Lβ and Au Lβ excitation. Since the escape peak in Au Lβ excitation appears at 9.70 keV, no interference with Cu Kα is observed. The ability to detect trace Cu by W Lβ and Au Lβ excitation was experimentally compared (Figure 7.2.6). In this experiment, identical 5-point mapping analyses were performed with the two systems for 6-level Cu-contaminated wafers. In W Lβ excitation, although the linearity reaches the level of 10^9 atoms cm^{-2}, the dispersion of 5-point mapping is large, because of interference by the escape peak. In comparison, the data dispersion in Au Lβ excitation is very small even at the lower level of 10^9 atoms cm^{-2}.

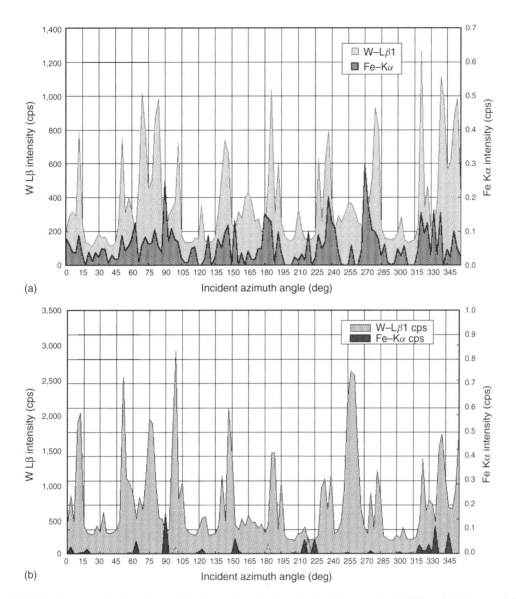

Figure 7.2.4 Incident azimuth dependence of Fe impurity peak intensity for (a) conventional SSD and (b) improved SSD. Courtesy of Technos Corp., Japan

7.2.3 TXRF WITH CHEMICAL PRECONCENTRATION

7.2.3.1 AUTOMATION OF PRECONCENTRATION

The potential of TXRF with chemical preconcentration (generally called vapor phase decomposition (VPD)-TXRF) was pointed out at an early stage of semiconductor-oriented TXRF.[13] The enrichment factor, [wafer area]/[detector view area] ratio, equals about two orders of magnitude, resulting in an ultra-low detection limit comparable to VPD-ICPMS. Until the middle of the 1990s, however, the chemical preconcentration was performed manually by persons with special technical skills, so routine VPD-TXRF analyses

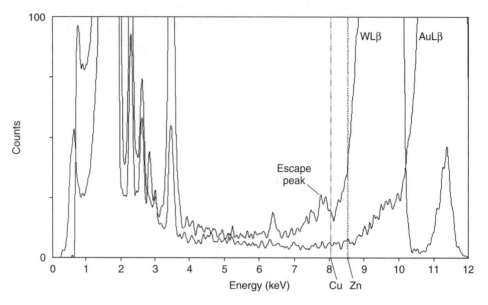

Figure 7.2.5 Comparison of blank wafer spectra for W Lβ and Au Lβ excitation. The applied power is 9 kW for each. Courtesy of Rigaku Corp., Japan

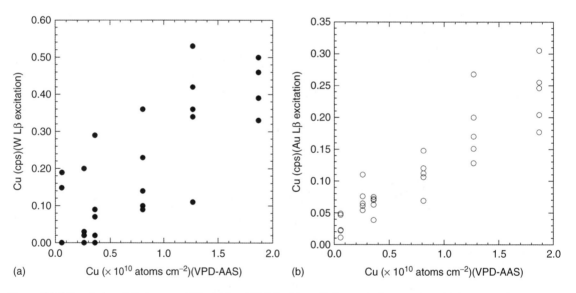

Figure 7.2.6 Correlation of Cu between VPD-AAS and TXRF with (a) W Lβ excitation and (b) Au Lβ excitation

for semiconductor process characterization were actually difficult to implement. To solve this problem so as to meet the coming age of large-diameter wafers, the automation of VPD had been steadily developing. Now, such instruments are commercially available.[14–16] Figure 7.2.7 is an example of one of those instruments. The basic function of such instruments is as follows: loader/unloader for wafer cassette, VPD reaction chamber(s) to make the wafers hydrophobic, a scanning unit to collect the surface impurities into a small droplet, and a stage for drying the droplet to deposit a

Figure 7.2.7 Photograph of an automatic VPD system (WSPS, Wafer Surface Preparation System). Courtesy of GeMeTec Corp., Germany

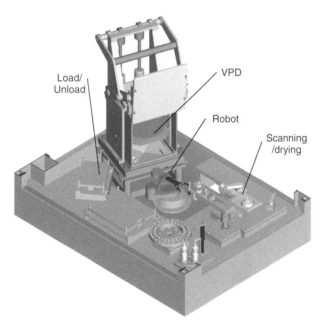

Figure 7.2.8 Overview of an automatic VPD system integration. Courtesy of SES Corp., Japan

residue on a wafer. These combined processes take place in a clean draft chamber, and each sample wafer is sequentially transferred from unit to unit by a robotic system (Figure 7.2.8). The automatic VPD instrument significantly reduced the labor and time required for chemical preconcentration. In addition, isolation from the laboratory environment achieved a drastic reduction in unintentional contamination, and the automatic robotic motions improved the reproducibility of the overall preconcentration process,[17] resulting in a higher degree of reliability of the analytical data for trace metal contamination.

7.2.3.2 LIGHT ELEMENT ANALYSIS

As mentioned above, the capability of light element analysis by straight-TXRF with W Mα excitation is not satisfactory for critical applications. To improve the detection limit, the combination of chemical preconcentration was examined. Figure 7.2.9 is a set of sample spectra for straight- and VPD-TXRF.[7] Because of the concentrating effect, the latter shows clear Na Kα and Al Kα peaks. The detection limits of Na and Al reach 3×10^{10} atoms cm^{-2} and 2×10^9 atoms cm^{-2} for 150 mm ϕ wafers, respectively. These detection limits correspond to ca. 8×10^9 atoms cm^{-2} and 5×10^8 atoms cm^{-2}, respectively, when converted to those of leading-edge 300 mm ϕ wafers. This level of detection is satisfactory for conducting semi-quantitative analysis of current semiconductor surfaces. For Na, however, ca. one-order improvement of the detection limit may still be desirable to meet critical use.

7.2.4 STANDARDIZATION OF TXRF AND OTHER METHODS

7.2.4.1 CROSS-CHECK ACTIVITIES

Along with the improvement of LLD, the standardization of TXRF analysis for semiconductors has been attracting more attention in recent years. On the background of previous two cross-check works,[18,19] ISO/TC201/WG2 was organized to establish international standards of TXRF measurement in 1993, primarily for the semiconductor industry. After an international round-robin test (RRT) and its data analysis, the first international standard, ISO14706:2000 ('Surface chemical analysis–Determination of surface elemental contamination on silicon wafers by total-reflection X-ray fluorescence (TXRF) spectroscopy') was published. The ISO/TC201/WG2 then decided to standardize chemical preconcentration methods for VPD-TXRF as its second major task, and the work started in 1998. This work consisted of two international RRTs. In the first RRT, the experimental conditions were left to each participant's discretion, which brought about poor interlaboratory reproducibility. The lesson learned from the first RRT contributed to the conduct of the second RRT. In the second RRT, both the VPD-TXRF and VPD-wet (e.g. ICPMS) methods were tested, and basic experimental conditions and calculation procedures were carefully provided by the organizer. For example, internal addition of V or Sc was applied to normalize the fluorescence intensities between different dried residues, and 2% HF + 2% H$_2$O$_2$ was specified as the scanning solution. Figure 7.2.10 summarizes the results of the first and second RRT for Ni.[20] The careful control of experimental conditions and calculation procedures in the second RRT accomplished significant

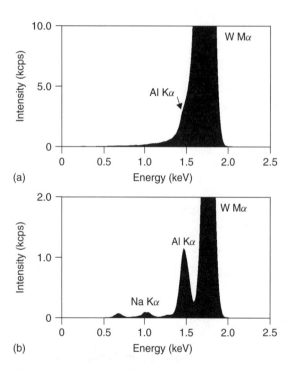

Figure 7.2.9 (a) Straight-TXRF spectrum of a wafer on which 1×10^{11} atoms cm^{-2} Na and Al are added and (b) spectrum after VPD preconcentration for the same wafer.[6] The excitation source is W Mα. Reproduced by permission of John Wiley & Sons, Ltd

STANDARDIZATION OF TXRF AND OTHER METHODS

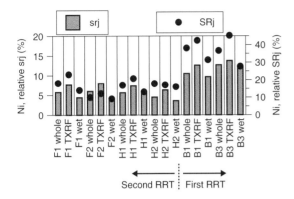

Figure 7.2.10 Summary of the international RRT results conducted by ISO/TC201/WG2. 'SRj' (right axis) and 'srj' (left axis) indicate interlaboratory reproducibility and intralaboratory repeatability, respectively. B1 to H2 are the sample names, and the B series are of the first RRT, while the F and H series are of the second RRT. 'wet' means the result of VPD-AAS or ICPMS, and 'whole' means the overall dispersion of the VPD-TXRF and VPD-wet method.[20] Reproduced by permission of The Japan Society of Applied Physics

improvements in repeatability and reproducibility. Based on the results of the second RRT, an ISO draft was prepared by ISO/TC201/WG2 in 2001, and the final document is scheduled to be published in 2004.

7.2.4.2 STANDARD SAMPLE ISSUE

There are many error factors in TXRF quantification.[21] One of the critical factors is the depth profile of the analyte element; the fluorescent X-ray intensity in TXRF is highly sensitive to the depth profile of the analyte. Figure 7.2.11 schematically demonstrates this fact. Two types of depth profiles for the same amount of analyte are assumed: (a) a near-surface analyte; and (b) an implanted analyte. The primary X-rays attenuate as they penetrate into the substrate, as illustrated in

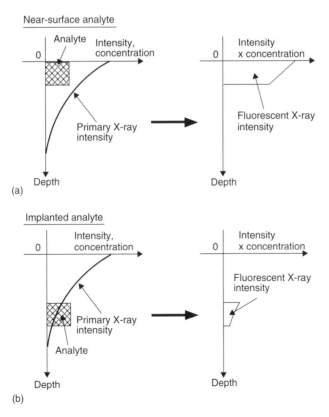

Figure 7.2.11 Schematic illustrations explaining the differences in the intensity of X-ray fluorescence for two types of analytes with different depth profiles

Figure 7.2.11 on the left. The intensity of the fluorescent X-ray is proportional to the integration of the product of concentration and the excitation X-ray intensity along the depth, which is illustrated in the figures on the right. The differences of the areas indicate that different depth profiles give different fluorescent X-ray intensities even though the amount of analyte is the same. Such an effect was actually found in standard samples for TXRF. Figure 7.2.12 shows the angle scans of Ni for two spincoat[22] standard samples prepared by following the same process. Although their targeted concentrations were the same, their angle scans were apparently different. At 0.10°, which is the typical measurement angle in actual use, the difference in fluorescent X-ray intensity is more than double. Similar differences were observed for standard microdrop[23] samples. These findings imply that controlling the depth profile at nanometer-level resolutions in physisorption is difficult, and a method that employs chemisorption was proposed. The method is named 'Immersion in Alkaline Hydrogen Peroxide Solution' (IAP).[24,25] This method utilizes a mixture of ammonia, hydrogen peroxide, and water, which is a very common cleaning solution for removing particles from silicon wafers. If metal ions are contained in this solution, they adsorb onto the surface of silicon wafers,[26] and the IAP method makes use of this chemisorption. In this method, cleaned silicon wafers are immersed in the solution that is intentionally doped with a certain amount of metal ions such as Fe, Ni, and Zn. A schematic illustration of the reactions in the solution is shown in Figure 7.2.13. During immersion, the SiO_2 formation by hydrogen peroxide and the etching of SiO_2 by ammonia balance each other out, continuously leaving ca. 1 nm SiO_2 layer, and metal ions adsorb to the SiO_2 layer based on chemical equilibrium. The 'adsorption isotherms,' the

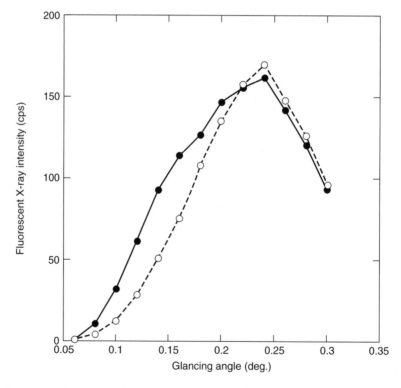

Figure 7.2.12 Angle scan profiles of two spincoat samples (Ni, 5×10^{13} atoms cm^{-2}). Reproduced by permission of The Discussion Group of X-Ray Analysis Japan

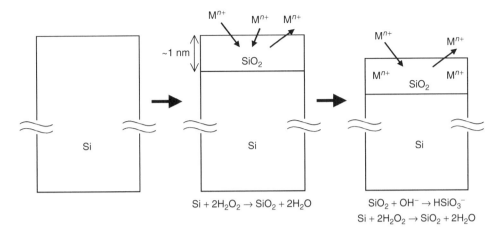

Figure 7.2.13 Schematic models of surface reactions on a silicon wafer in alkaline hydrogen peroxide solution

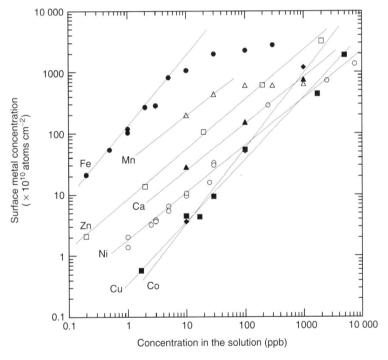

Figure 7.2.14 Adsorption isotherms of several metal ions in alkaline hydrogen peroxide solution (2.2 M NH_3 and 1.4 M H_2O_2, 80 °C, 10 min). Reproduced by permission of The Discussion Group of X-Ray Analysis Japan

amount adsorbed versus the concentration of dissolved metal at a fixed temperature, are shown in Figure 7.2.14 for several metals.[25] The IAP method can be applied to these important elements in the range of at least 10^9 to 10^{13} atoms cm^{-2}. In addition, this method can also be applied to Al and Mg, although they are omitted here because they cannot be analysed with ordinary TXRF. It should be mentioned, however, that alkaline metals (Na and K) and some heavy metals (Cr, W,

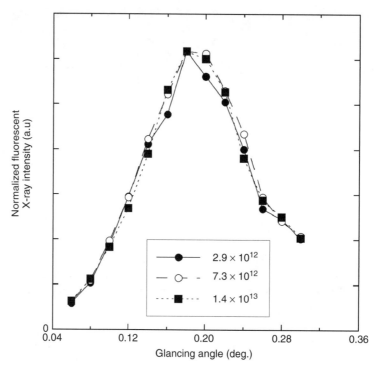

Figure 7.2.15 Comparison of angle scan profiles for three IAP wafers (Ni, different concentrations). Reproduced by permission of The Japan Society for Analytical Chemistry

and Ta) were not adsorbed. The reproducibility of the depth profiles was examined by measuring the angle scans of the analyte elements on IAP wafers. Figure 7.2.15 compares the angle scans for IAP wafers that have different concentrations of Ni.[24] The angle scan profiles agreed well, indicating that the depth profile is independent of the adsorbed concentration. Figure 7.2.16 compares the angle scans between three elements – Fe, Ni, and Zn.[24] The results agreed well with each other, indicating that the depth profile is independent of the element. From a standpoint of use as a standard sample, uniformity of adsorption is also a critical factor. Table 7.2.1 shows the spatial uniformity of metal adsorption evaluated by conducting 9-point TXRF mapping. The uniformity was typically 10 % or less by relative standard deviation (RSD), which is comparable to that of traditional spincoat wafers.[22]

Table 7.2.2 lists the in-batch uniformity of adsorbed concentration. In this experiment, nine wafers were immersed in a single solution at one time, and the wafer-to-wafer uniformity was evaluated by analysis with TXRF or AAS. The dispersion was very small – less than ca. 6 % by RSD. Good in-batch uniformity, as well as spatial uniformity, is advantageous in standard or cross-check samples for the contamination analysis of semiconductor surfaces. The maximum number of IAP wafers made from a single solution was

Table 7.2.1 Summary of spatial uniformity test for IAP wafers

Element	Concentration (atoms cm^{-2})	Relative standard deviation (%)
Fe	9.0×10^{11}	7.7
	4.5×10^{12}	3.5
	1.7×10^{13}	4.0
Ni	3.5×10^{11}	12.3
	3.3×10^{12}	19.6
	1.0×10^{13}	4.3
Zn	7.3×10^{11}	9.3
	3.0×10^{12}	3.4
	6.0×10^{12}	4.8

Reproduced by permission of The Japan Society for Analytical Chemistry.

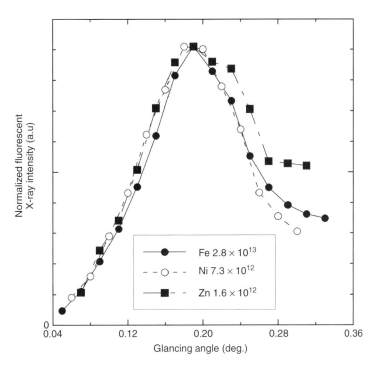

Figure 7.2.16 Comparison of angle scan profiles for IAP wafers on which Fe, Ni, or Zn was adsorbed. Reproduced by permission of The Japan Society for Analytical Chemistry

Table 7.2.2 Wafer-to-wafer uniformity of adsorbed metal concentration for IAP wafer, prepared in each single batch

Element	Concentration (atoms cm^{-2})	Relative standard deviation (%)
Fe	4.5×10^{12}	0.96
Ni	1.6×10^{12}	5.9
Zn	3.0×10^{12}	5.4

Reproduced by permission of The Japan Society for Analytical Chemistry.

typically limited to 25 (i.e. one cassette), but the sequential reuse of a single solution for more than one cassette of wafers increased the maximum number to more than 50.[27] Along with spincoat samples, the IAP wafers were used as crosscheck standard samples in the ISO/TC201/WG2 international RRT mentioned above.[20]

7.2.5 FILM ANALYSIS

As discussed in the Introduction, many new elements are being introduced or tested as alternative materials for recently developed semiconductor devices. Usually, the elements are deposited on the wafer surface to form a thin film layer. Controlling the chemical composition, thickness, and density of each layer is very important for stabilizing the device properties and enhancing the yield. X-Ray analysis is a suitable means of control because of its advantages of nondestructive and mapping capabilities, among others. As for chemical composition, conventional XRF is applicable. In XRF, an $r-\theta$ stage is thought not to be suitable for the mapping analysis of certain kinds of films such as BST, or $(Ba,Sr)TiO_3$, because of diffraction, and an $x-y-\theta$ stage was introduced to avoid diffraction.[28] The conventional XRF, however, cannot be used for stacked films which include a common element. BST deposited on SRO, or $SrRuO_3$, is one such example. To analyse such films, two-angle grazing incidence XRF (GIXRF) was proposed.[29] At first, only the top layer is analysed by conducting a low-angle measurement, and then all the layers are analysed by a high-angle measurement.

The composition of the lower layer can be calculated by subtraction. As for film thickness, both XRF and XRR are applicable. In the XRF method, the fluorescent intensity is converted to film thickness by applying calibration curves, whereas in the XRR method, the thickness is calculated from the oscillation data of reflectivity, as illustrated in Figure 7.2.17. Of the two methods, XRR is more convenient because the thickness can be determined without the use of reference standard samples. In addition, advanced theoretical fitting to the XRR oscillation data enables simultaneous determination of stacked layers.[29] Besides the film thickness, this method can directly determine the density of the first layer from the critical angle at which the reflectivity shows the first inflection. A combined method of GIXRF and XRR for the semiconductor industry was developed for this purpose.[29] A schematic illustration of the instrument is shown in Figure 7.2.18. Many applications using this system are proposed including insulators (TaO_2, etc.), wirings (Cu, Al + Cu, etc.), electrodes

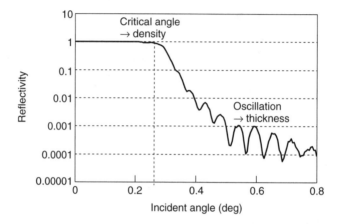

Figure 7.2.17 Sample curve of incidence angle dependence of X-ray reflectivity. Courtesy of Technos Corp., Japan

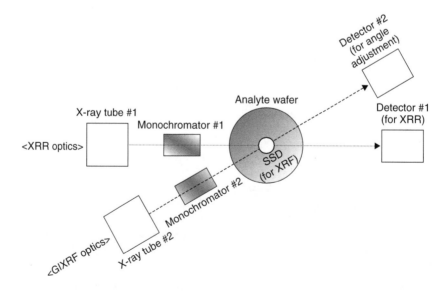

Figure 7.2.18 Schematic illustration of the X-ray optics of an XRR/XRF film analyser (SMAT210, Technos Corp)

(W-Si, Mo-Si, etc.), barrier films (Ti/Ti-N, etc.), and dielectrics (BST on SRO).

ACKNOWLEDGEMENTS

The author wishes to acknowledge and thank GeMeTec Japan Corp., Rigaku Corp., SES Corp. and Technos Corp. (in alphabetical order) for kindly providing us with their data and original illustrations for this contribution.

REFERENCES

1. Front End Processes, in *International Technology Roadmap for Semiconductors*, 2001 Edition, p. 26; http://public.itrs.net/Files/2001ITRS/FEP.pdf.
2. Front End Processes, in *International Technology Roadmap for Semiconductors*, 2001 Edition, p. 6; http://public.itrs.net/Files/2001ITRS/FEP.pdf.
3. Futase, T., Itoh, M. and Katsuyama, K. Wafer back side and edge cleaning to remove metal contamination (4). Removal of Ru at wafer edge. *Extended Abstracts (The 48th Spring Meeting 2001)*, The Japan Society of Applied Physics and Related Societies, Meiji University, 2001, p. 831 (29a-D-6).
4. Funabashi, M., Utaka, T. and Arai, T. Improvement of Total Reflection X-ray Fluorescence (TXRF) spectrochemical analysis for silicon wafers. *Spectrochim. Acta B*, **52**, 887–899 (1997).
5. Funabashi, M., Matsuo, M., Kawada, N., Yamagami, M. and Wilson, R. Enhanced analysis of particles and vapor phase decomposition droplets by total-reflection X-ray Fluorescence. *Spectrochim. Acta B*, **54**, 1409–1426 (1999).
6. Yamagami, M., Nonoguchi, M., Yamada, T., Shoji, T., Utaka, T., Mori, Y., Nomura, S., Taniguchi, K., Wakita, H. and Ikeda, S. Analysis of light elements on Si wafer by vapor-phase decomposition/total reflection X-ray fluorescence. *Bunseki Kagaku*, **48**, 1005 (1999).
7. Yamagami, M., Nonoguchi, M., Yamada, T., Shoji, T., Utaka, T., Nomura, S., Taniguchi, K., Wakita, H. and Ikeda, S. VPD/TXRF analysis of trace elements on a silicon wafer. *X-ray Spectrom.*, **28**, 451–455 (1999).
8. Yakushiji, K., Ohkawa, S., Yoshinaga, A. and Harada, J. Main peak profiles of total reflection X-ray fluorescence analysis of Si (100) wafers excited by monochromatic X-ray beam W-Lβ (I). *Jpn. J. Appl. Phys.*, **31**, 2872–2876 (1992).
9. Yakushiji, K., Ohkawa, S., Yoshinaga, A. and Harada, J. Main peak profiles of total reflection X-ray fluorescence analysis of Si (100) wafers excited by monochromatic X-ray beam W-Lβ (II). *Jpn. J. Appl. Phys.*, **32**, 1191–1196 (1993).
10. Kozono, S., Itoh, T., Yoshinaga, A., Ohkawa, S. and Yakushiji, K. Trace analysis of a beryllium window for a solid state detector system by inductively coupled plasma mass spectrometry. *Anal. Sci.*, **10**, 477–480 (1994).
11. Yamada, T., Shoji, T., Funabashi, M., Utaka, T., Arai, T. and Wilson, R. Tungsten analysis with a total reflection X-ray fluorescence spectrometer using a three crystal changer. *Adv. X-Ray Chem. Anal. Jpn.*, **26s**, 53–56 (1995).
12. Yamada, T., Matsuo M., Kohno, H. and Mori, Y. Sensitive detection of trace copper contamination on a silicon wafer by total reflection X-ray fluorescence using W-Lβ or Au-Lβ excitation source. *Spectrochim. Acta B*, **56**, 2307–2312 (2001).
13. Huber, A., Rath, H. J., Eichinger, P., Bauer, T., Kotz, L. and Staudigl, R. Sub-ppm monitoring of transition metal contamination on silicon wafer surfaces by VPD-TXRF. *ECS Proceedings*, PV88-20, The Electrochemical Society, Pennington, NJ, 1988, pp. 109–112.
14. Pahlke, S., Kotz, L., Ehmann, T., Eichinger, P. and Huber, A. WSPS: wafer surface preparation system. A novel modular automated method capable of the ultratrace analytical inspection of 300 mm silicon wafer surfaces. *ECS Proceedings*, PV98-1, The Electrochemical Society, Pennington, NJ, 1998, pp. 1524–1525.
15. http://www.gemetec.com/
16. http://www.ses-corp.co.jp/en/
17. Fabry, L., Pahlke, S., Kotz, L., Wobrauschek, P. and Streli, C. Novel methods of TXRF analysis for silicon wafer surface inspection. *Fresenius J. Anal. Chem.*, **363**, 98–102 (1999).
18. Hockett, R. S., Ikeda, S. and Taniguchi, T. TXRF round robin results. *ECS Proceedings*, PV92-12, The Electrochemical Society, Pennington, NJ, 1992, pp. 324–337.
19. UC Standardization Committee. UC Standard: Test method for measuring surface contamination on silicon wafers by total reflection X-ray fluorescence spectroscopy, *Ultra Clean Technology*, **8**, 44–82 (1996).
20. Horie, S. *et al.* Progress of ISO standardization activity on TXRF method −1: Report on RRT2 (Collecting method of surface metal). *Extended Abstracts (The 48th Spring Meeting 2001)*, The Japan Society of Applied Physics and Related Societies, Meiji University, 2001, p. 839 (31a-D-1).
21. Mori, Y. and Uemura, K. Error factors in quantitative total reflection X-ray fluorescence analysis. *X-Ray Spectrom.*, **28**, 421–426 (1999).
22. Hourai, M., Naridomi, T., Oka, Y., Murakami, K., Sumita, S., Fujino, N. and Shiraiwa, T. A method of quantitative contamination with metallic impurities of the surface of a silicon wafer. *Jpn. J. Appl. Phys.*, **27**, L2361–L2363 (1988).
23. Kondo, H., Ryuta, J., Morita, E., Yoshimi, T. and Shimanuki, Y. Quantitative analysis of surface contaminations

of Si wafers by total-reflection X-ray fluorescence. *Jpn. J. Appl. Phys.*, **31**, L11–L13 (1992).
24. Mori, Y., Shimanoe, K. and Sakon, T. A standard sample preparation method for the determination of metal impurities on a silicon wafer by total reflection X-ray fluorescence spectrometry. *Anal. Sci.*, **11**, 499–504 (1995).
25. Mori, Y. and Shimanoe, K. Standard sample preparation for the analysis of several metals on silicon wafer. *Anal. Sci.*, **12**, 141–143 (1996).
26. Mori, Y., Uemura, K., Shimanoe, K. and Sakon, T. Adsorption species of transition metal ions on silicon wafer in SC-1 solution. *J. Electrochem. Soc.*, **142**, 3104–3109 (1995).
27. Mori, Y. and Uemura, K. Multi-batch preparation of standard samples from a single doped solution for cross-checking in surface metal analyses of silicon wafers. *Anal. Sci.*, **16**, 987–989 (2000).
28. Funahashi, M., Kuraoka, M., Fujimura, S., Kobayashi, H., Kohno, H. and Wilson, R. BST thin film evaluation using X-ray fluorescence and reflectivity methods. *Adv. X-Ray Anal.*, **42**, 109–118 (1999).
29. Terada, S., Furukawa, H., Murakami, H. and Nishihagi, K. A grazing incidence X-ray fluorescence analysis of the composition of $(Ba,Sr)TiO_3$ (BST), and $SrRuO_3$ (SRO) stacked films. *Adv. X-Ray Anal.*, **43**, 504–509 (2000).

7.3 X-Ray Spectrometry in Archaeometry

A. ZUCCHIATTI

Istituto Nazionale di Fisica Nucleare, Genova, Italy

7.3.1 INTRODUCTION

Any human artefact bears the history of its making, use, and conservation. Fossils as well bear the history of their birth, of their life, of the aggression of nature after their death. Reading this history is the ultimate goal of all the scholars and sciences that operate in the broad environment of the cultural heritage. Part of the history is visible. The expert eyes of an art historian, trained by thousands of examinations and supported by a sound knowledge of an historic period, can detect in the form of bodies, in the shades of colours, in the length and thickness of brush-strokes, the unmistakable signature of a master. A large part of the history is however hidden, lying below the surface, trapped in the texture of the object or deeper in the constituent atoms and molecules. The preparatory drawing of a painting, the artist's second thoughts and remakes, the technology of preparing a colour, of firing pottery, of soldering metals, the signs of climate in a fossil, all require instruments to be read. These are the foundations of archaeometry.

In principle any materials science analytical technique can be transferred to archaeometry if it can meet a few decisive requirements: it should be nondestructive, flexible to accommodate a variety of artefacts and samples and to analyse different object structures, fast, accurate and hopefully quantitative. Techniques using an X-ray beam as the analytical probe or detecting X-rays as the analysis product meet almost generally these requirements.

The use of X-rays in the broad domain of cultural heritage is almost as old[1] as the discovery of this kind of radiation and has progressed in coincidence with the technical and instrumental developments that have marked X-ray production and detection. X-Rays allow the characterization of an ancient material, in terms of the elemental composition (X-ray fluorescence, XRF; proton-induced X-ray emission, PIXE; scanning electron microscopy with energy-dispersive spectrometry, SEM-EDS; synchrotron radiation – X-ray fluorescence, SR-XRF), of the constituent minerals (X-ray diffraction, XRD; SR-XRD; small-angle X-ray scattering, SAXS), of the oxidation state of the atomic species (X-ray absorption near-edge spectroscopy, XANES). Looking at the recent literature we observe the constant progress (40–50 % in 2001, source *Archaeometry*) of techniques like XRF, XRD, SEM-EDS that can be performed with bench or even portable (XRF) equipment. Their principles are well established, and widely covered in textbooks[2,3] or review articles.[4] The extended use and technical development of *traditional* techniques deserve recording on their own. However, looking at recent developments, it is the progress of portable instrumentation (including X-ray based) that has brought the best benefits increasing the accessibility to artefacts and improving considerably the quality of *in situ* analyses.

Accelerator based techniques represent a rapidly evolving alternative to bench instrumentation. In terms of elemental sensitivities (ppb level in SR-XRF) they can compete with neutron activation analysis (NAA) and inductively coupled plasma

(ICP) and offer space-resolved elemental and chemical analysis and an extended accessibility.[5] Ion beam analyses in archaeometry[6] have been between 5 and 10 % of the total in recent renowned topical conferences, an appreciable figure in a field dominated, quite obviously, by the analysis of modern industrial materials. Applications of SR, have been until recently limited to the occasional use of a general-purpose beam line. However, the outstanding performance of second and third generation synchrotrons now begins to be used in a more systematic way even in dedicated facilities, like the newly established archaeometry laboratory at Daresbury.[7] Ancient valuable samples are far from being ideal for the accelerator analyst. They often impose stringent exposure conditions to avoid any form of beam damage and an experimental geometry which conflicts with the need of optimising the measurement sensitivity. Important recent evolutions have concerned the methodology with the aim of assuring quantitative well-controlled characterization of ancient artefacts, and of describing their texture.

7.3.2 ACCESSIBILITY OF TECHNIQUES: PORTABLE SYSTEMS

Artefacts of a historical and artistic nature necessarily attract the attention of art historians and curators so that the opportunity for instrumental investigations is normally inhibited. If the curator has a personal view of what is considered acceptable damage levels, this can typically lead to a prohibition on sampling and consequently the impossibility of meaningful laboratory analyses.

Portable XRF equipment[8] has been a traditional resource in archaeometry. Although limited in sensitivity and in many cases only qualitative, it can nevertheless give a rather complete view of the object conservation state, assist the restoration process and guide the sampling to the areas where a precise quantitative characterization can give the most useful information to restorers and art historians.[9] In some cases (XRF of metal alloys) appropriate algorithms make possible the quantitative analysis[10] to characterize an object in detail. Even if a certain number of radioactive sources can be used to produce the primary X-ray,[10] X-ray tubes are normally preferred for these have higher intensity and are not bound to transport regulations. Several portable XRF systems have managed to combine sensitivity and portability. A light (5 kg) system based on a 30 kV, 0.1 mA, air-cooled tungsten tube and on a Peltier-cooled SDD (silicon drift diode) assures[11] minimum detection limits (MDLs) of 0.15 % for Fe Kα and 0.2 % for Sn Lα. A high power (60 kV, 4 mA), water-cooled tungsten tube coupled to a HPGe detector, liquid nitrogen (LN2) cooled, weighs four times more but has MDLs of 0.1 % for Fe and Pb and 0.01 % for Ag, Sn, Sb.[9] Good MDLs have been reported for an air-cooled rhodium tube operated at 50 kV and 0.35 mA coupled to an LN2 cooled Si(Li) detector and capable of detecting down to 0.05 % of Fe, 0.01 % of Ni, 0.5 % of As (0.01 % in samples with low Pb content), 0.5 % of Zn, 0.001–0.006 % of Ag, Sn or Sb.[12]

A recent device[13,14] makes use of a radioactive alpha source which, combined with a portable X-ray detector and electronics, constitutes a compact and versatile field instrument for quantitative PIXE analysis at the 1 % level. Since the source activity must remain below law enforced limits, the price paid is evidently the increase of MDLs, and the limited penetration depth of an alpha particle compared to a proton. Nevertheless even limited information can be conveniently exploited as proved in several campaigns.[15–18] The source holder is made of Mylar and has a conical shape (Figure 7.3.1) so as to produce a concentration of alpha particles on the target. The irradiated surface is of the order of 1 cm^2. Alpha particles, emitted at about 5 MeV, before reaching the target will cross a sealing epoxy resin layer, one of the kapton windows and a negligible air layer, which is replaced by a helium flow in the most recent configuration. The whole system is very compact and can be used in close contact to the sample, for minimum energy loss and maximum detection efficiency. ^{210}Po has a half-life of 138.4 days and the source initial activity is 16.6×10^7 Bq/g. Only 3.7×10^7 Bq of ^{210}Po is

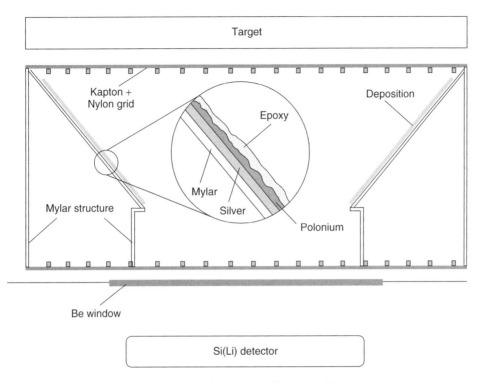

Figure 7.3.1 A schematic cross-section view of the polonium source with target and detector

sufficient to produce in 30 min peaks, which gives rise to a statistical error below 5 %.[19]

7.3.2.1 X-RAY ENERGY DISPERSIVE DETECTORS

The remarkable performance of portable systems is due also to the development of energy dispersive spectrometers. (Table 7.3.1). Today they offer resolution and efficiency comparable to that of previous laboratory instruments with, in addition, compactness and, in the case of SDD detectors,[11] also high speed. In analysis one wants the lowest possible MDL (Equation 7.3.1) and therefore tries to obtain the largest possible integrated source intensity Q, detector solid angle Ω, efficiency ε and the smallest possible FWHM resolution.

$$w_Z^{MDL} \propto \frac{\sqrt{FWHM_R}}{\sqrt{\Omega Q \varepsilon}} \quad (7.3.1)$$

In the case of art objects the source intensity might be limited by the risk of damage and the solid angle subtended by the solid surface, which configuration could be quite complex around the detector. In a very general way we can assume that the space available to the detector is a cone of full aperture $\Delta\theta$ as in Figure 7.3.2. In such a case, the maximum solid angle will be obtained when the angular sector is entirely covered by the cryostat and this corresponds to a value:

$$\Delta\Omega_{MAX} = \frac{S}{(D+t)^2} = \frac{f\pi\left(\frac{d}{2}\right)^2}{\left[\left(\frac{d}{2}\right)\text{ctg}\frac{\Delta\vartheta}{2} + g\frac{d}{2}\right]^2}$$

$$= \frac{f\pi\left(\frac{d}{2}\right)^2}{\left(\frac{d}{2}\right)^2\left(\text{ctg}\frac{\Delta\vartheta}{2} + g\right)^2}$$

$$= \pi\frac{f}{\left(\text{ctg}\frac{\Delta\vartheta}{2} + g\right)^2} \quad (7.3.2)$$

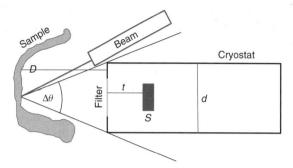

Figure 7.3.2 A schematic view of a typical beam–sample–detector set-up for the analysis of an irregularly shaped artefact

where d is the detector diameter, S the active area, t the semiconductor–filter distance, $g = 2t/d$, $f = 4S/\pi d^2$. Given $\Delta\theta$ the solid angle depends only on f and g, which are critical parameters given in Table 7.3.1.

7.3.3 SYNCHROTRON RADIATION IN ARCHAEOMETRY

The advantage that SR has in comparison to even the most powerful X-ray tube is an intensity 10^{10} times higher (Figure 7.3.3a), a continuous spectrum (Figure 7.3.3b) rich in hard components (40–100 keV), low background[20] and good emittance. These allow a broad range of preparations (ESRF has 48 beamlines, Hasylab 43, Daresbury 44) that can produce on the sample high brilliance monochromatic microbeams. Since the energy is tunable, the penetration depth and the excitation cross-sections are widely controllable. SR offers superior spatial and angular resolution, high XRF sensitivity from low to high Z elements and a broad range of complementary techniques (μ-XRF, μ-XRD and μ-XANES). Let us consider two of the several SR arrangements, which have been used recently in archaeometry.[21–25]

The SR-XRF microprobe set-up at beamline L in Hasylab is used for simultaneous multi-elemental analysis of micro-samples. The experimental set-up is schematically shown in Figure 7.3.4. The beam[26,27] is collimated by a system of 3 mm thick tungsten cross slits to a size of $30 \times 30\,\mu\text{m}^2$ and passes an optional absorber to reduce the low energy part of the white spectrum, if a better sensitivity is wanted for higher Z elements. To further reduce the beam spot straight capillaries or ellipsoidal capillaries can be used giving a minimum beam diameter of $2\,\mu\text{m}$. The sample is mounted on an $XYZ\theta$ table with reproducible positioning of about $0.5\,\mu\text{m}$ and $0.1°$. Fluorescence X-rays are collected by a $30\,\text{mm}^2$, 5 mm thick HPGe detector covering a solid angle of 10^{-3} Sr. The sample is aligned to the beam with a resolution of $3\,\mu\text{m}$ by means of a long distance zoom microscope, coupled to a charge-coupled device (CCD) camera. The synchrotron beam is

Table 7.3.1 Parameters of recent solid state detectors

Detector	Cryostat diameter (mm)	Active area (mm²)	f [%]	Filter distance t (mm)	g (%)	Be (μm)	Resol. FWHM (eV)	Active Depth of S (μm)
1 Si(Li)	25.4	12.6	2.5	8	63	8	164[a]	3490
2 Si(Li)	25.4	78	15.5	6	47.2	25.4	184[a]	5470
3 Si-PIN Peltier cooled	14	25	16.2	1.52	21.7	25.4	280[b]	500
4 SDD Peltier cooled	19.2	10	3.5	1	10.5	8	175[c]	300
5 HPGe	22	95	25	5	45.5	25.4	155[d]	5000
6 UltraLeGe	25.4	100	19.7	7	55.1	25.4	145[d]	5000
7 Si(Li)	25.4	80	15.8	7	55.1	25.4	180	5000
8 Si(Li) without internal collimator	19	80	28.2	3	31.6	25	150[e]	5000
9 Si(Li)	25.4	80	15.8	5.5	43.3	25	150[e]	5000

[a] Quoted FWHM resolution with 6 μs pulses at 3–4 kHz.
[b] Quoted FWHM resolution with 6 μs pulses and rate below 3 kHz.
[c] Quoted FWHM resolution with 0.5 μs pulses at 50 kHz.
[d] Quoted FWHM resolution with 4 μs pulses and rate below 10 kHz.
[e] Quoted FWHM resolution with 40 μs pulses at 1 kHz.

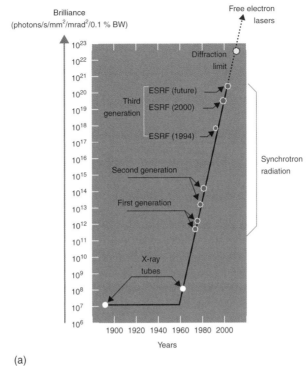

(a)

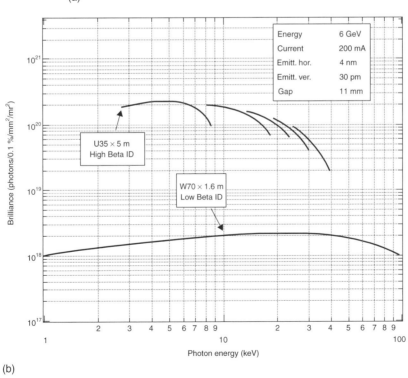

(b)

Figure 7.3.3 Comparison of X-ray sources brilliance (a) and some possible selections of the SR spectrum at the European Synchrotron Radiation Facility (ESRF) (b). Courtesy of the ESRF information office

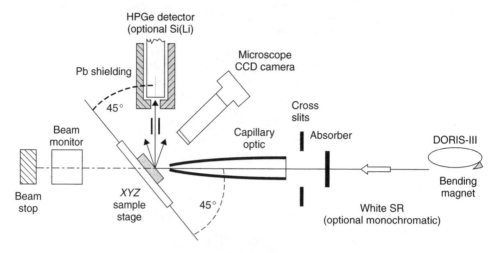

Figure 7.3.4 A schematic layout of the micro-XRF beam of line L at Hasylab. Courtesy of Dr Gerald Falkenberg

monitored after the cross slits and after the sample by ionisation chambers and is finally stopped in a lead block. The fluorescent excitation of K lines is used for element identification and quantification from $Z = 19$ (K) to $Z = 82$ (Pb). MDLs are down to 0.1–1 ppm for $19 < Z < 50$ and 1–10 ppm for $Z > 50$.[28]

A beamline used for SR-XRD on ancient powders is BM16 of ESRF. The optical arrangement receives from a bending magnet 4 mrad of white SR in the horizontal plane, which is focused in the vertical direction by a curved mirror, set at grazing angle. The residual FWHM vertical divergence is around 12 µrad. After the mirror the beam energy is selected by a double (Si-111) crystal monochromator. Routine operation is without further focusing since focusing produces an increase of the vertical divergence at the sample location and therefore worsens the angular resolution. At the sample location the beam is approximately 5 mm wide. The energy range is from 5 keV to 40 keV. The diffractometer[29] consists of nine NaI detectors, spaced by 2°, and preceded by a Ge-111 crystal analyser, that makes diffraction peaks free from aberrations. The crystals are scanned at the same time collecting nine high resolution diffraction patterns, between 1° and 40° in 2θ with a step of 0.001–0.004°. The system is mechanically very stable, accurate to ±1 arcsec and sensitive to displacements of 5×10^{-5}°. The instrumental contribution to the peaks FWHM is around 0.003° in 2θ. Peak positions are accurate and reproducible to $1–5 \times 10^{-4}$°. Flat specimens or capillaries can be accommodated in the set-up. The use of a capillary spinning around the diffractometer axis eliminates possible preferred-orientation effects in the sample. The beam energy and the capillary diameter can be selected to minimize absorption even when high Z elements are contained in the sample, hence the diffracted intensities are also very accurate. The complete analysis of an archaeological powder would take about 4 h.

7.3.4 PROBING THE TEXTURE OF ANCIENT MATERIALS

The surface and the in-depth texture are a highly distinctive feature of all ancient artefacts. Measuring the composition distribution down to the µm scale, as possible with the most recent particle and X-ray beams, can be the key to the understanding of the object making, of its use and of corrosion processes. The invasive or microinvasive sampling of cross-sections of materials (Figure 7.3.5) gives access to a broad range of nondestructive complementary analyses like SEM, TEM (transmission electron microscopy), µ-XRF, µ-XANES, µ-PIXE to describe the details (elements, oxidation states, inclusions, mineral phases)

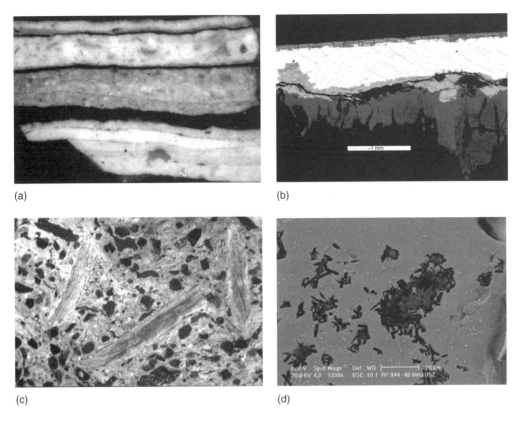

Figure 7.3.5 Some examples of the texture of ancient artefacts as seen in polished cross-sections. (a) A multilayered painting on a German fourteenth century polychrome wooden sculpture. © C2RMF photo S. Colinart. (b) The cross-section of a fourth century bronze vase showing superficial corrosion layers of even very large depth. © C2RMF photo B. Mille. (c) A portion of a fourth century pottery with well evident mineral grains and crushed sea shells. © C2RMF photo A. Leclair. (d) A nineteenth century blue glaze shows the presence of unfused *thenard* blue grains. © C2RMF photo A. Bouquillon

of the object composition at the sub-micron level. Whenever the object integrity is strictly imposed, radiography (X-ray, neutron) is capable of detecting structural details[30] with spatial resolution of the order of 1 mm and can be used for the investigation of whole objects (Figure 7.3.6). With fully nondestructive procedures elemental surface maps in the 10 μm range and depth profiles in the 100 μm range are now possible.

7.3.4.1 RECENT ADVANCES IN PIXE DEPTH PROFILING

Rutherford backscattering spectroscopy (RBS) is the most appropriate, nondestructive, Ion beam analysis (IBA) technique for element depth profiling. Whenever RBS is not applicable[31] the analysis of fluorescence X-rays can be used, despite a poorer depth resolution, to detect the ordering of layers and have a clue of their thickness, especially when each layer may be characterized by a different most abundant (key) element and when the elements to profile have medium to high Z.

A methodology using 68 MeV protons[32,33] has been demonstrated on a set of test paint layers, prepared by the Kunsthistorisches Museum in Vienna, in resemblance of sixteenth–seventeenth century Italian and Flemish paintings and characterized by pigments containing mostly medium and high Z elements (Cu, Hg, Pb). The penetration depth of 68 MeV protons in matter is of the order of a few

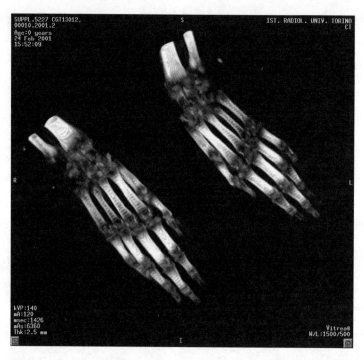

Figure 7.3.6 A three-dimensional CAT scan of Egyptian mummy hands, from a study of the Radiology Department of the University and 'Le Molinette' hospital of Turin (Italy). Courtesy of Soprintendenza al Museo Delle Antichità Egizie – Ministero per i Beni e le Attività Culturali. Imaging by Dr Federico Cesarani

mm and the X-ray production cross-section is 100 times larger than at 3 MeV, so that still 10 % of the Kα X-rays of lead will reach the detector from 3 mm below the surface. For several elements the different absorption of two X-ray lines in their passage through the matrix gives a yield ratio (e.g. Pb Lα/Pb Kα) related, in layered samples, to the average depth at which the element is present. The depth can be quoted as an equivalent CaCO$_3$ thickness since this material should match the average absorption coefficients of the real matrix. The unperturbed peak intensity ratios are extracted from thin pure foils of Cu, Ag, Au and Pb. Two samples were prepared on a Cu backing with a slightly different layers sequence: Au foil, cinnabar (HgS), yellow ochre, lead-tin yellow (Pb$_2$SnO$_4$), azurite (2CuCO$_3$.Cu(OH)$_2$) and, in the second sample, an extra lead-white (PbCO$_3$.2Pb(OH)$_2$) layer after the gold foil. As seen in Figure 7.3.7 the sequence is well reproduced, which is more significant information than the layer's actual thickness (doubled by the method!) since the latter changes considerably with the brush strokes. The layer marking elements can be used to attempt an identification of the pigment; however, in the case of two different layers, characterized by the same major element, this method produces only one average depth and the two levels cannot be disentangled in the sequence.

Another way of probing a material at different depths consists of using on the same spot a set of different beam energies.[34] In gold alloys several conditions are met that make possible even the quantitative determination of a surface layer thickness. An example[5,31] concerns the so-called *tumbaga*, a Cu–Ag–Au alloy, rich in copper (up to 40 %), produced by pre-Columbian goldsmiths with processes that reduced the copper content at the surface thus giving the alloy a colour very close to that of pure gold and a durable protection from corrosion. At a given proton energy, $E_p = 1.6$ MeV, the Cu Kα/Au Lβ yield ratio of 7.8 will be compatible (Figure 7.3.8) with a homogeneous alloy made of Cu 40 %, Au 54 %, Ag 6 %, but

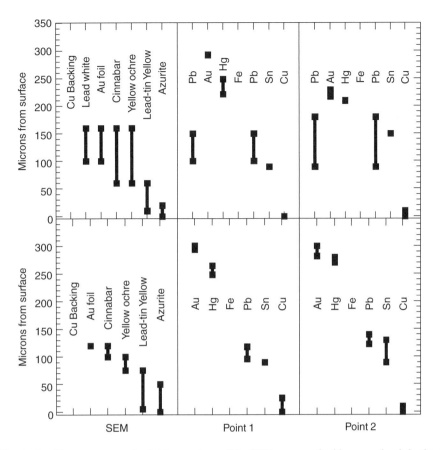

Figure 7.3.7 The depth of two sequences of paint layers observed by SEM, compared with conventional depths deduced from 68 MeV PIXE analyses. Two points per sample were analysed. Constructed with the data of Denker and Maier[32]

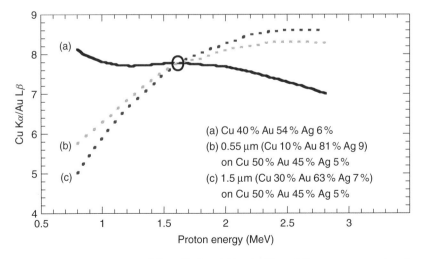

Figure 7.3.8 The behaviour with proton energy of CuKα/AuLβ yield ratio. Three different curves are given for possible layered structures in the examined sample. Reworked from the data of Demortier and Ruvacalba-Sil[32]

also with a 0.55 μm surface layer of Cu 10 %, Au 81 %, Ag 9 % on top of a homogeneous bulk material containing Cu 50 %, Au 45 %, Ag 5 % or even with a 1.5 μm surface layer of Cu 30 %, Au 63 %, Ag 7 % on top of Cu 50 %, Au 45 %, Ag 5 %. The ambiguity can be resolved by measurements at other energies around 1.6 MeV since the yield ratio will follow a different curve according to the layer's structure. Since in metal alloys, there is no beam damage the count rate can be kept high enough to obtain in a short time statistical errors lower (1 %) than the separation of the curves. The outermost layer composition is determined by PIXE analysis with 1 MeV alpha particles and is used as the starting point for the calculation of depth profiles. The Cu, Au, Ag concentrations measured on a *tumbaga* pendant of 44 mm diameter, show (Figure 7.3.9) very clearly the progressive decrease of gold from the surface to a bulk material that contains 28 % of copper.

7.3.4.2 EXTERNAL MICROBEAMS FOR PIXE

Charged particle microbeams extracted in air have a spatial resolution of the order of 10 μm, larger than the 1 μm obtainable in vacuum, but their flexibility gives access to microstructures (inclusions, composition gradients, ...) in a large variety of samples including valuable artefacts, polished sections, fragments. A very effective one is that developed at the ALGAE accelerator of the Louvre museum,[35] based on the nuclear microprobe system of Oxford Microbeams coupled to a special design of the beam-line exit nozzle and to an appropriate choice of the exit window. The window material is silicon nitrate (Si_3N_4), which can be produced in 0.1 μm films, and assures good resistance to pressure, to moderate mechanical shocks and to radiation damage. The Si_3N_4 window is set at 45° to the beam. A collimated exit provided with a 8 μm kapton foil gives access to a Peltier-cooled Si(Li) detector to count the Si 1.740 keV X-rays as a beam monitor (Figure 7.3.10a). The window is highly stable under the particle beam, and this monitor is quite reliable. A specific brass collimator suppresses the contribution of beam halo to the beam monitoring. Figure 7.3.10(b) represents in detail the 110 mm long nozzle. A 3 MeV proton beam emerges from the window with an energy loss of 8.5 keV and an energy straggling of 4.5 keV, while a 3 MeV alpha beam emerges with an energy loss of 94 keV and an energy straggling of 31 keV. These figures make possible RBS and NRA and obviously PIXE

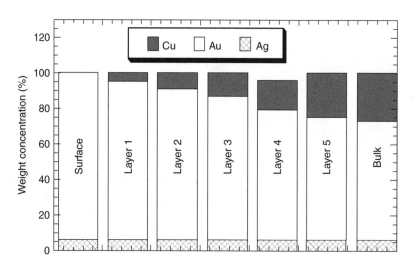

Figure 7.3.9 The concentration of Cu, Au, Ag in different layers of a *tumbaga* pendant. Constructed from the data of Demortier and Ruvacalba-Sil[54]

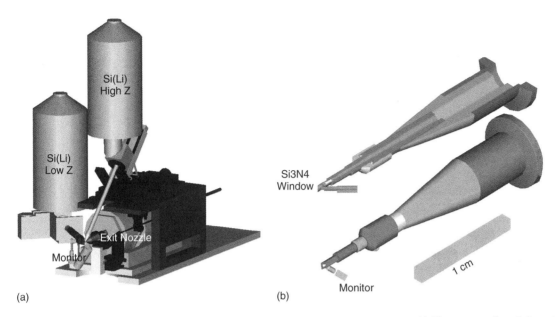

Figure 7.3.10 (a) The layout of the ALGAE microbeam showing the two detectors system. (b) The cross-section of the exit nozzle. Shown are the Si_3N_4 window, the anti-halo collimator and the exit to the Peltier-cooled Si(Li) detector acting as a beam monitor. Courtesy of B. Moignard C2RMF

that is not as sensitive to possible alterations of the beam quality. The spatial resolution has been quantified by scanning, as a reference, with protons, deuterons and alphas a calibrated copper grid. It is 10, 20 and 50 μm FWHM for proton, deuteron and alpha beams, respectively.

7.3.4.3 PIXE MICRO-MAPPING OF FLINT TOOLS

Of great interest is the information on the wear mechanism of archaeological flint tools. Its archaeological importance is in the presence of remnants belonging to the worked material that allows the anthropologists to classify the tools and gather information on the human activities in the area of finding. Flint is a mixture[36] of quartz spherules, averaging 12.5 μm in diameter, embedded in a microfibrous chalcedony cement. The edge chipping that was used to manufacture cutting tools is essentially a fracture of the chalcedony cement and goes around the spherules giving rise to a microscopically rough surface. The tool's surface is modified by continuous use and a polish appears at the cutting edges. The origin of this polish has been debated at length.[37] The microanalysis of test tools has added essential information to the debate. The chipped edges have been implanted with Cu marker ions at a depth of only 0.1 μm with a dose of 5×10^{16} ions/cm² for a resulting Cu Kα count rate of 15 Hz. After intense use on bone, linear microscans of the edge region (both parallel and perpendicular to the edge profile) were performed. The results prove that no loss of the Cu marker is experienced, while the edge area is enriched in Ca. Due to the low Cu implantation depth compared to the spherules diameter it seems therefore excluded that the use of the tool results in ejection of the quartz spherules from the surface. The formation of polish should therefore be due to the deposition of material from the worked body in the interstices of the flint tool.

The micro-mapping of a set of archaeological Mesolithic flints[38] reveals again a deposition of external materials confined in the flint edge and characteristic of the worked object: bone, wood, skin, meat as in the diagrams of Figure 7.3.11.[39] The accumulation of Ca, S, and to a minor extent P and K on the tool edge (Figure 7.3.11) indicates

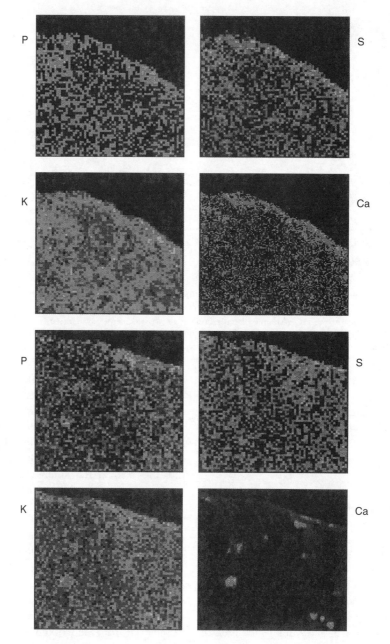

Figure 7.3.11 PIXE micro-mapping of two 1.2 × 1.2 mm² areas of two archaeological flint tools. The tool that produced the top maps was probably used for cutting bone while the other one was used for hide scraping. Reproduced from Smit et al.[39] With permission from the European Commission

that the tool was used for cutting hard material like bone. The absence of Ca but the increase of K and P (Figure 7.3.11) might indicate that the tool was used for hide scraping.

7.3.5 RADIATION DAMAGE

It can never be stressed enough that the nondestructive nature of the X-ray based techniques has been the key of their expansion in the domain of cultural heritage. It is extremely important to guarantee safe procedures throughout the analysis process to control in a quantitative way the beam–sample interaction.

7.3.5.1 THE RADIATION DAMAGE ON GLAZES

The irradiation of pottery glazes with ion beams is known to produce chromatic alterations visible by the naked eye even after a short exposure. The beam can induce atomic dislocations leading to the formation of peroxide groups (the coupling of two oxygen atoms) and centres (sites where an oxygen bound is lost). The absorption spectrum of the medium and consequently the colour of the irradiated sample could change. Although self-annealing takes place within hours or days the effect is unwanted and must be controlled. Chromatic alterations have been quantified for the first time[40,41] in white and blue glazes of the Italian renaissance. Digital macro-images of the irradiated area, are analysed with commercial software and each pixel is assigned the so-called *Lab* colorimetric coordinates (L, a, b), which aim at reproducing the human perception of colours. L is the lightness and may vary from 0 (totally black surface) to 100 (perfectly white surface). a and b account, respectively, for the red–green and yellow–blue balances. The *Lab* coordinates define a vector **E** corresponding to a given colour in the colorimetric space and the chromatic distance between two points is represented by:

$$|\Delta \mathbf{E}| = \sqrt{\Delta a^2 + \Delta b^2 + \Delta L^2}. \qquad (7.3.3)$$

The colorimetric analysis was performed for a series of rings of increasing radius even beyond the visible beam spot (2×10^{-3} cm^2) well into the non-irradiated unaltered surface. The colour change ΔE, with respect to the outermost ring, was measured as a function of the radius for a broad range of beam currents and accumulated charges. A safety value $\Delta E = 1$ was assumed since the human eye can only appreciate $\Delta E \geq 1$. Original renaissance and modern white or blue glazes have been studied with similar results. In white glazes both the amount of the damage and its surface extension increase with the accumulated charge, as shown in Figure 7.3.12 for a beam current of 4 pA. The maximum variation, $\Delta E \approx 12$, is reached close to the beam axis (smallest radius) at an accumulated charge of about 1 nC. Afterwards the colour change saturates at the spot centre while a further increase of the accumulated charge produces a growth of the ΔE values at large radii, in other words an increase of the FWHM of the radial distribution (Figure 7.3.13).

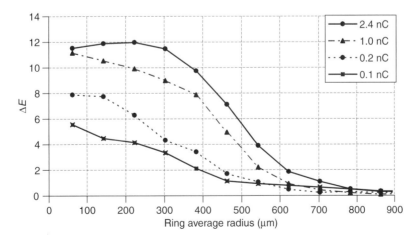

Figure 7.3.12 The radial distribution of the colour change in white glazes as a function of the collected charge for a beam current of 4 pA. Constructed from the data of Migliori[40]

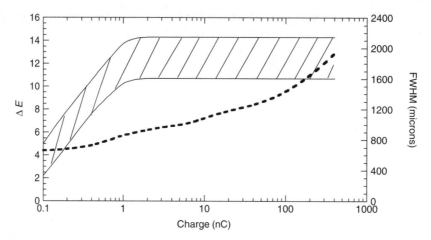

Figure 7.3.13 The maximum ΔE colour change of white glazes at the spot centre as a function of the total collected charge for different choices of the beam current (dashed band). The FWHM of the colour distribution (dotted curve). Reworked from the data of Migliori[40]

Table 7.3.2 Safe limits for IBA of some ancient materials

Material	Beam	Energy (MeV)	Current	Spot	Current density	Safe limit	Ref.
Glaze	Proton	3	4 pA	0.2 mm²	20 pA/mm²	0.1 nC	40
Glaze	Proton	4	0.2 nA	1 mm²	0.2 nA/mm²	10 nC	52
Paper	Proton	4	1 nA	4 mm²	0.25 nA/mm²	5 nC	34
Paper	Proton	3.5	–	–	10–150 pA/mm²	12 nC/mm²	53
Paper	Proton	2.5	–	–	10–150 pA/mm²	8 nC/mm²	53
Paint layers	Proton	4	0.5 nA	1 mm²	0.5 nA/mm²	30 nC	34
Paint layers	Proton	68	0.1–1 pA	0.5 mm²	0.2–2 pA/mm²	200 nC	54
Glass	Proton	2	3.5 nA	12 µm²	300 pA/µm²	2.4 µC	55

In blue glazes the behaviour of the colour change is similar. However ΔE values higher than 30 have been reached at a charge of 2.4 nC, still without signs of saturation. The perception of damage on an irradiated glaze limits therefore the total integrated charge to values below 0.1 nC. This must be delivered to the sample in controlled conditions[40] operating with currents of 1–2 pA for 50–100 s.

Limits for safe IBA of various materials have been reported in the literature and are summarized in Table 7.3.2.

7.3.5.2 DAMAGE WITH X-RAY PRIMARY BEAMS

X-Rays deposit energy in a sample over a volume much larger than in the case of particles at equal penetration depth. The effects on the object appearance and integrity should be negligible. Organic materials could be darkening and become fragile earlier than other materials. However, no damage has been observed,[41] whatever the examined material, in years of practice in XRF even at tube settings of 2.5 kW and 60 kV with most measurements taken at 0.5 mW and 50 kV and 30 s irradiation.

The high photon flux in SR facilities deserves some attention. In the SR-XRF analysis of parchments care must be taken[25] to assess the significance of ionisation and the creation of free radicals which have effects on the sample integrity, especially those with high moisture content. In ink studies with SR-XRF with a 0.5 × 1 mm² polychromatic SR beam the energy deposited in 300 s runs on paper is of the order of 15 µW/cm² almost a factor 100 below solar maximum irradiation[42] and therefore the creation of radicals should be negligible.

7.3.6 RECENT PIXE AND μ-PIXE APPLICATIONS

7.3.6.1 STUDY OF RENAISSANCE GLAZED TERRACOTTA SCULPTURES

Glazed, coloured terracottas, are very distinctive of the Italian artistic Renaissance and enrich the collections of the world major museums. Art historians are confronted by a rich and composite ensemble of hundreds of objects. PIXE can identify the elemental content as a fingerprint of each glaze and help to identify similarities amongst objects and to strengthen hypotheses on artist attribution. Appropriate analytic procedures[42] have been implemented to deal with the non-homogeneous nature of Renaissance glazes, to treat consistently the three kinds of samples available (polished cross-sections, fragments and whole sculptures), to correlate data from two facilities (ALGAE[43] and INFN Florence[44]), to estimate non statistical errors which are between 3 and 8% for many elements and exceed 10% only for low Z elements (Na,Al) or for elements well below 1% in weight.

The nature of samples requires some considerations. Plaster, dust or paint, could create a surface layer on whole sculptures and even more on fragments, which are sampled for evident reasons from peripheral 'dirty' areas. The extra layer could exhaust a non-negligible fraction of the total proton range and therefore increase the concentration of contaminant elements, such as Al, K, and Ca in comparison to a polished SEM section. Other discrepancies could be due to the glaze texture. For example the high amounts of CaO found by PIXE on a micro-sample were explained by SEM images. The interface, characterized by the development of newly formed crystals of calcium silicate was exceptionally thick,[42] more than 100 μm, due to local over-firing inside the Renaissance kiln. A large number of measurements on the same artefact are therefore recommended to avoid possible misinterpretations. Even anomalous analyses can be taken to further statistical treatment since they give complementary information on the artefact manufacture and conservation.

Almost 700 PIXE measurements have been performed on 54 artefacts. Detailed comparisons can be made on groups of objects similar in shape, style and use. The sculpture department of the Louvre museum hosts two pairs of kneeling angels used as candleholders, stylistically quite different (catalogue numbers C52-C53 and RF1533-RF1534). The first is typical of the Buglioni's or some minor Florentine workshop, the second seems convincingly a product of the Della Robbia 'bottega' executed by an apprentice rather than by the masters (the ageing Luca or Andrea).

The glazes have been analysed on the sculptures as well as on cross-sections. All are lead-siliceous glazes, slightly enriched in K_2O and CaO, opacified by tin oxide and coloured with different metallic oxides: cobalt for blue, manganese for purple, lead antimoniate for yellow, copper for green. The microstructure of the glazes (SEM images) is quite typical: a homogeneous glassy matrix (~150 μm thick for both pairs) with unmelted quartz grains or feldspar, heterogeneous re-partition of tin oxides crystals and bubbles. The glazes are very similar for the two angels of the same pair and their compositions are compatible with the Renaissance recipes published in old ceramics treatises. If the blue glazes are considered, the two pairs differ sharply (Figure 7.3.14). Cobalt is usually associated with other metallic elements that reflect the origin of the ores as well as the different methods of purification of the metal. Cobalt is associated with Ni, Cu, Fe for the pair RF1533-RF1534, whereas appreciable amounts of As_2O_3 are measured in the blue glazes of the other pair. Studies aiming at tracing the ancient commercial routes of cobalt in Europe[37,45,46] suggest a compositional change of raw material at the boundary of fifteenth–sixteenth century, with increased amounts of As, probably due to a different refinement process of the mineral ores. The two pairs of angels seem to fit in this scheme: the Della Robbia bottega was founded around 1440 while the

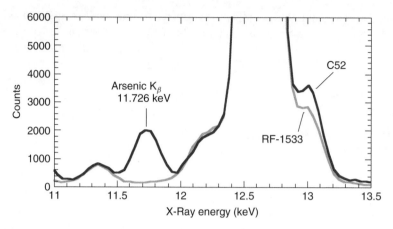

Figure 7.3.14 The X-ray spectra of samples C52 and RF1533 show clearly the discriminating presence of the As Kβ line in one of them

Buglioni's were active from the beginning of the sixteenth century.

7.3.7 RECENT APPLICATIONS OF SYNCHROTRON RADIATION

7.3.7.1 ANALYSIS OF EGYPTIAN COSMETICS WITH SR-XRD

Cosmetic recipes and details of make-up manufacturing in ancient Egypt have been revealed thanks to powder SR-XRD.[47,48] In a series of experiments performed at line BM16 of ESRF, a large number of cosmetics, used in Ancient Egypt and conserved in the Louvre museum, have been analysed showing how great was the variety of compositions using lead compounds and how advanced was at that time (2100–1100 BC) the know-how in chemical synthesis.[49,50]

The Rietveld refinement of powder diffraction patterns (Figure 7.3.15) has identified four lead-based main phases: the black galena (PbS)

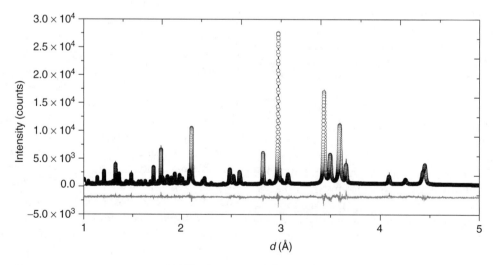

Figure 7.3.15 The XRD intensity pattern of sample E11047 of Egyptian cosmetic powder (black curve), and Rietveld refinement. Open circles represent the calculated diagram and the experimental-calculated difference is in grey. Courtesy of Ph. Walter C2RMF

and three white products, cerussite ($PbCO_3$), phosgenite ($Pb_2Cl_2CO_3$) and laurionite PbOHCl. Galena (greyish black) and cerussite are well known lead ores mined in antiquity along the Red sea coast. They were mixed to give a variety of grey shades. Laurionite and phosgenite (both white) are synthetic compounds produced by wet chemistry, according to recipes that were documented by Greco-Roman authors, and were added quite surely because of their remedial effects. The study has been extended into the analysis of the Bragg line profiles, which are influenced, by the microstructure, size and distortions of the mineral grains. The profile of archaeological galena (PbS) was compared with that of a geological galena hand-ground with a pestle and mortar and then selected through a 63–125 μm sieve. The lattice distortion is found higher in the archaeological powder than in the geological powder. The archaeological powder has been finely ground: the SEM observations reveal a heterogeneous assembly of small cubes ranging from 20 μm to 150 μm long, with a significant fraction of smaller grains. Other archaeological galena powders show that they were finely ground by the Egyptians and sorted as a function of size, to obtain either a black mat powder, or grey powders with metallic overtones. Already 4000 years ago, people wanted their cosmetics to do more than simply highlighting the eyes!

7.3.7.2 XRF AND XANES MICRO-MAPPING OF CORRODED GLASSES

The problem of glass corrosion in ancient Roman glass fragments[22,51] has been studied with SR using both μ-XRF and μ-XANES on the same sample. The severe glass alteration manifests itself, on a microscopic scale, as a series of corrosion layers at progressive depths down to 400 μm, accompanied by a precipitation crust at the surface and separated by interstitial cavities. Such an alteration seemed due to a more complex process than a continuous leaching of alkaline ions, like Na and Ca and their replacement by other ions brought in by the humid environment. Rather, several cycles of leaching and layer separation seemed to have taken place. This hypothesis was confirmed by μ-XRF scans along the depth of the corrosion layers[51] and by μ-PIXE measurements.[22] The migratory patterns of major and minor elements, which are related to the size of the ions and their chemical bonds in the glass network, were reconstructed. While the Ca leaching is evident with a concentration that decreases by a factor 10, some of the elements (K, Ti, Fe, Br) are enriched and uniformly distributed over the corrosion layer. Most trace elements are also enriched in the corrosion layers, except those (Sn, Zr) which are part of the glass network and remain almost constant. Manganese has a more differentiated profile, with maxima and minima within the corroded area. During the mechanical separation of the leached layer from the glass bulk, one expects in the interstices some precipitation of Mn, possibly accompanied by the concentration of trace elements brought in by ground water. This has been clarified by the μ-XANES technique. The K absorption edge of Mn shifts upwards by several eV if the oxidation state of Mn changes from 2+ to 4+. Selective mapping of Mn can be done by looking at the SR-XRF maps performed at two beam energies: 6.550 keV, which is only just above the absorption edge of Mn^{2+} and 6.564 keV where also Mn^{4+} is above the absorption edge. By difference, the chemical Mn^{4+} (MnO_2) map is obtained (Figure 7.3.16). MnO_2 is concentrated only in cavities that have separated two subsequent corrosion layers formed in the repetitive leaching sequence.

7.3.8 CONCLUSIONS

There is growing attention to the problems of study and conservation of our cultural heritage. Many scientific initiatives, on an international scale, have promoted new technologies. The three actions of the European Cooperation in the field of Scientific and Technical Research (COST) concerning 'Application of ion beam analysis to art or archaeological objects', 'Advanced artwork restoration

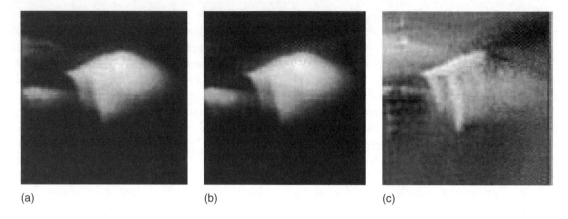

Figure 7.3.16 Fluorescence micro-maps on leached Qumram glass, performed at two primary beam settings at the Mn^{4+} crest (a) and above the Mn^{2+} edge (b). By difference, the map of MnO_2 (c) is obtained. Reworked from Jannsen et al.[51] by permission of John Wiley & Sons, Ltd

and conservation methods using laser technology', 'Non-destructive analysis and testing of museum objects' and the network of 'Laboratories on Science and Technology for the Conservation of European Cultural Heritage' (LabS TECH) are good examples of research promotion. X-Ray based techniques continue to be one of the fundamental tools of archaeometry research. A much larger availability of facilities and several technological advances have produced an impressive number of scientific examinations of artefacts with X-rays in dedicated laboratories, in multi-disciplinary centres and *in situ*, making archaeometry a very dynamic field today and an even greater research opportunity for tomorrow.

ACKNOWLEDGEMENTS

I would like to thank the many colleagues who have helped me with comments, reprints, documents, data and photos. I am pleased to acknowledge the warm and friendly hospitality of the colleagues at the Centre de Recherche et Restauration des Musées de France (C2RMF) during my many visits to the Louvre. Finally, I would like to dedicate this work to the late Prof. Friedel Sellschop who first addressed my attention to IBA and applied sciences.

REFERENCES

1. *Wilhelm Roentger Museum. Conrad Roentgen, una scoperta che ha cambiato il mondo*. Catalogue of the special exhibition of the Roentgen Museum, Remscheid, Germany, 1995.
2. R.E. Van Grieken and A.A. Markowicz (Eds), *Handbook of X-ray Spectrometry*, Dekker, New York, 1993.
3. R. Jenkins, *X-ray Fluorescence Spectrometry*, 2nd edition, John Wiley & Sons, Ltd, New York, 1999.
4. D.E. Cox, in *Handbook of Synchrotron Radiation* (Eds G.S. Brown and D.E. Moncton), North Holland, Amsterdam, pp. 155–200, 1991.
5. G. Demortier and A. Adriaens (Eds), *Ion Beam Study of Art and Archaeological Objects*, Eur. Comm. EUR19218, 2000.
6. M.A. Respaldiza and J. Gomez-Camacho (Eds), *Application of Ion Beam Analysis to Arts and Archeometry*, Secretariado de Pubblicaciones, Universidad de Sevilla, 1996.
7. M. Pantos, Synchrotron Radiation Special Interest Group, http://srs.dl.ac.uk/arch and links therein.
8. E.T. Hall, F. Scwezer and P.A. Toller, *Archaeometry* **15**, 53–78, 1973.
9. M. Ferretti, M. Miazzo and P. Mo'ioli, *Stud. Conserv.* **42**, 241–246, 1997.
10. M. Milazzo and C. Cicardi, *Archaeometry* **40-2**, 351–360, 1998.
11. A. Longoni, C. Fiorini, P. Leutenegger, S. Sciuti, G. Fonterotta and L. Strudel and P. Lechner, *Nucl. Instrum. Methods* **A409**, 407–409, 1998.
12. J. Lutz and E. Pernicka, *Archaeometry* **38-2**, 313–323, 1996.
13. G. Pappalardo, J. de Sanoit, A. Musumarra, G. Calvi and C. Marchetta, *Nucl. Instrum. Methods* **B109-110**, 214–217, 1996.

14. G. Pappalardo, *Nucl. Instrum. Methods* **B150**, 576–580, 1999.
15. G. Pappalardo, M. Bicchieri, C. Giuliani, L. Pappalardo and F.P. Romano, *Boll. Acc. Gioiea Sci. Nat.* **33**, 415–421, 2000.
16. A. Carastro, L. Pappalardo and F.P. Romano, *Boll. Acc. Gioiea Sci. Nat.* **33**, 1655–173, 2000.
17. L. Pappalardo, *Boll. Acc. Gioiea Sci. Nat.* **33**, 422–430, 2000.
18. C. Aurisicchio, S.N. Cesaro, G. Pappalardo, L. Pappalardo and F.P. Romano, *Boll. Acc. Gioiea Sci. Nat.* **33**, 5–14, 2000.
19. B. McConnel and L. Pappalardo, *Boll. Acc. Gioiea Sci. Nat.* **33**, 399–414, 2000.
20. L. Vincze, K. Janssens, F. Adams and K.W. Jones, *Spectrochim. Acta* **B50**, 1481–1500, 1995.
21. W. Kockelmann, E. Pantos and A. Kirfel, in *Radiation in Art and Archeometry* (Eds D.C. Creagh and D.A. Bradley), Elsevier Amsterdam, pp. 361–377, 2000.
22. K. Janssens, A. Aerts, L. Vincze, F. Adams, C. Yang, R. Utui, K. Malmqvist, K.W. Jones, M. Radtke, S. Garbe, F. Lechtenberg, A. Knöchel and H. Wouters, *Nucl. Instrum. Methods* **B109–110**, 690–695, 1996.
23. E.S. Friedman, Y. Sato, A. Alatas, C.E. Johnson, T.J. Wilkinson, K.A. Yener, B. Lai, G. Jennings, S.M. Mini and E.E. Alp, *Adv. X-Ray Anal.* **42**, 151–160, 2000.
24. R.R. Martin, T.K. Sham, G. Wong Wou, K.W. Jones and H. Feng, *X-Ray Spectrum.* **30**, 338–341, 2001.
25. T.J. Wess, M. Drakopoulos, A. Snigirev, J. Wouters, O. Paris, P. Fratzel, M. Collins, J. Hiller and K. Nielsen, *Archaeometry* **43**, 117–129, 2001.
26. M. Haller and A. Knöchel, *J. Trace Microprobe Tech.*, **14–3**, 461–466, 1996.
27. F. Lechtenberg, S. Garbe, J. Bauch, D.B. Dingwell, F. Freitag, M. Haller, T.H. Hansteen, P. Ippach, A. Knöchel, M. Radtke, C. Romano, P.M. Sachs, H.U. Schmincke and H.J. Ullrich, *J. Trace Microprobe Tech.* **14–3**, 561–587, 1996.
28. K. Janssens, L. Vincze, B. Vekemans, C.T. Williams, M. Radtke, M. Haller and A. Knöchel, *J.Anal. Chem.* **363**, 413–420, 1998.
29. A.N. Fitch, *Mater. Sci. Forum* **228–231**, 219–222, 1996.
30. F. Cesarani, M.C. Martina, A. Ferraris, M.C. Cassinis, R. Grilletto, R. Boano, A.M. Donadoni Roveri and G. Gandini, in *COST Int. Works Non-Destruct. Anal. Conserv./Restorat. Museum Objects*, Genova, Unpublished, 2001.
31. G. Demortier and J.L. Ruvacalba-Sil, *Nucl. Instrum. Methods* **B118**, 352–358, 1996.
32. A. Denker and K.H. Maier, *Nucl. Instrum. Methods* **B161–163**, 704–708, 2000.
33. A. Denker and K.H. Maier, in *Ion Beam Study of Art and Archaeological Objects* (Eds G. Demortier and A. Adriaens), Eur. Comm. EUR19218, pp. 81–83, 2000.
34. C. Neelmeijer, A. Wagner and H.P. Schram, *Nucl. Instrum. Methods* **B118**, 338–345, 1996.
35. T. Calligaro, J.C. Dran, E. Ioannidou, B. Moignard, L. Pichon and J. Salomon, *Nucl. Instrum Methods* **B161–163**, 328–333, 2000.
36. M. Christensen, T. Calligaro, S. Consigny, J.C. Dran, J. Salomon and Ph. Walter, *Nucl. Instrum. Methods* **B136–138**, 869–874, 1998.
37. B. Gratuze, I. Soulier, M. Blet and L. Vallauri, *Rev. Archeom.* **20**, 77–94, 1996.
38. Z. Smit, G.W. Grime, S. Petru and I. Rajta, *Nucl. Instrum. Methods* **B150**, 565–570, 1999.
39. Z. Smit, G.W. Grime and S. Petru, in *Ion Beam Study of Art and Archaeological Objects* (Eds G. Demortier and A. Adriaens), Eur. Comm. EUR19218, pp. 67–71, 2000.
40. A. Migliori, *Misura di Basse correnti di fascio per lo studio del danneggiamento in misure PIXE di campioni ceramici rinascimentali*, Thesis, University of Florence, 2001.
41. M. Chiari, A. Migliori and P.A. Mandò, *Nucl. Instrum. Methods* **B188**, 151–155, 2002.
42. M. Mantler and M. Schreiner, *X-ray Spectrum.* **29**, 3–17, 2000.
43. A. Bouquillon, J. Castaing, J. Salomon, A. Zucchiatti, F. Lucarelli, P.A. Mando', P. Prati, G. Lanterna and M.G. Vaccari, in *Fundamental & Applied Aspects Of Modern Physics* (Eds S.H. Connel and R. Tegen), Word Scientific, Singapore, pp. 441–448, 2000.
44. H. Mommsen, Th. Beier, H. Dittmann, D. Heimermann, A. Hein, M. Rosenberg, M. Boghardt, E.M. Manebutt-Benz andH. Halbey, *Archaeometry* **38–2**, 347–357, 1996.
45. T. Calligaro, J.C. Dran, J.P. Poirot, G. Querrè, J. Salomon and J.C. Zwan, *Nucl. Instrum. Methods* **B161–163**, 769–774, 2000.
46. P.A. Mandò, *Nucl. Instrum. Methods*, **B85**, 51–57, 1992.
47. B. Gratuze, I. Uzonyi, Z. Elekes, A.Z. Kiss and E. Mester, in *Ion Beam Study of Art and Archaeological Objects* (Eds G. Demortier and A. Adriaens), Eur. Comm. EUR19218, pp. 50–53, 2000.
48. J. Hartwig, *Verre* **7**, 40–47 2001.
49. Ph. Walter, P. Martinetto, G. Tsoucaris, R. Breniaux, M.A. Lefebvre, G. Richard, J. Talabot and E. Dooryhee, *Nature* **397**, 483–484, 1999.
50. Ph. Walter, C. Ziegler, P. Martinetto and J. Talabot, *Techne* **9–10**, 9–18, 1999.
51. P. Martinetto, E. Dooryhee, M. Anne, J. Talabot, G. Tsoucaris and Ph.Walter, *ESRF Newslett.* **32**, 15–18, 1999.
52. P. Martinetto, M. Anne, E. Dooryhee, G. Tsoucaris and Ph. Walter, in *Radiation in Art and Archeometry* (Eds

D.C. Creagh and D.A. Bradley), Elsevier Amsterdam, pp. 297–316, 2000.
53. K. Janssens, G. Vittiglio, I. Deraedt, A. Aerts, B. Vekemans, L. Vincze, F. Wei, I. Deryck, O. Schalm, F. Adams, A. Rindby, A. Knöchel, A. Simionovici and A. Snigirev, *X-ray Spectrom.* **29**, 73–91, 2000.
54. M. Mader, D. Gramole, F. Hermann, C. Neelmmeijer, M. Schreiner and G. Woisetschlager, *Nucl. Instrum. Methods* **B136–138**, 863–868, 1998.
55. H. Cheng, D. Yanfang, H. Wunquau and Y. Fujia, *Nucl. Instrum. Methods* **B136–138**, 897–901, 1998.
56. G. Demortier and J.L. Ruvacalba-Sil, in *Ion Beam Study of Art and Archaeological Objects* (Eds G. Demortier and A. Adriaens), Eur. Comm. EUR19218, pp. 105–109, 2000.
57. M. Mosbah and J.P. Durand, *Nucl. Instrum. Methods* **B130**, 182–187, 1997.

7.4 X-Ray Spectrometry in Forensic Research

T. NINOMIYA

Hyogo Prefectural Police Headquarters, Kobe, Japan

7.4.1 INTRODUCTION

In the field of forensic sciences, various analytical techniques are used in order to solve crimes or offenses. Total reflection X-ray fluorescence (TXRF) analysis, which is a variant of X-ray fluorescence (XRF) analysis, was first presented by Yoneda and Horiuchi as a trace elemental analytical technique.[1] TXRF is a very useful technique for trace forensic samples because it is nondestructive, and has high sensitivity for trace elements. Klockenkämper reviewed this technique.[2] In the course of the investigation on forensic elemental analysis, Prange et al. reported the characterization of single fibers as a forensic application case of TXRF[3] and Duwel et al. reported quantitative elemental microanalysis of thermoplastic remains using TXRF.[4] The author has applied TXRF to various forensic samples, for example, cloth,[5] mineral water,[6] pigment,[7] plastic,[8] arsenic material,[9] blood,[10] toner,[11] drugs,[10,12–15,18] lipstick,[10] a copied letter,[11] counterfeit bills,[14,15] vinyl tape,[14,15] semen,[14,15] fingerprints,[14,15] soil,[16] liquor,[16] ivory and mammoth tusk,[17] fiber[19] and gunshot residue.[19] The author has also developed a specific instrument with both TXRF mode and fine beam X-ray analysis mode for forensic samples.[14,15] In addition, a successful application of TXRF to a single fiber concerning a murder case has been reported[20] and synchrotron radiation-elemental mapping analysis of a fingerprint has been reported.[21]

In this paper, recent laboratory TXRF applications and synchrotron radiation-XRF applications at SPring-8 are presented.

7.4.2 FORENSIC APPLICATIONS OF LABORATORY TXRF

7.4.2.1 POISONED FOOD

In 1998, a murder case, which involved poisoning curried food, happened at Wakayama Prefecture and after this triggering case, many food poisoning cases occurred in Japan. TXRF was used to examine suspicious foods and beverages quickly and nondestructively.[22]

A droplet of 1 µl of a canned coffee sample was spotted on a Si wafer and after drying, a residue sample was measured under conditions of 40 kV and 30 mA excitation and a monochromated Mo Kα X-ray was incident onto a sample surface at 0.06° incident angle for 1000 s. Figure 7.4.1(a) shows a TXRF spectrum of a suspicious canned coffee sample A and Figure 7.4.1(b) shows that of a normal canned coffee sample B. Phosphorus and chlorine are clearly detected in Figure 7.4.1(a), while in Figure 7.4.1(b), no phosphorus is detected and a weak peak of chlorine is observed. The peak of silicon is due to the Si wafer. The difference between spectra in (a) and (b) suggests that some poisoning agents containing both phosphorus and chlorine were added to sample A. By detailed chemical analysis using gas chromatography-mass spectrometry (GC-MS), dichlorvos (DDVP) was detected in sample A. DDVP is a type of insecticide with molecular formula $(CH_3O)_2P(O)CH=CCl_2$. The peaks of phosphorus and chlorine in the spectrum in Figure 7.4.1(a) reflect the formula of DDVP.

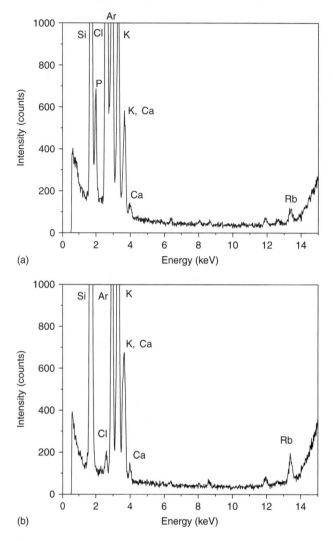

Figure 7.4.1 TXRF spectra of two canned coffee samples: (a) sample A (suspicious); (b) sample B (normal)

This TXRF technique can also be applied to examine the chemical terrorism weapons, sarin, soman, tabun, VX gas and their hydrolysed residues in water, because these chemicals have phosphorus elements in their molecular structures.

7.4.2.2 LIQUOR

Residues of liquors are often discovered as evidential materials at crime scenes. TXRF was used to examine elemental ingredients contained in liquors.

Figure 7.4.2 shows TXRF spectra of four kinds of liquors: sample A (sake, Japanese liquor, alcohol content: 14%); sample B (beer, alcohol content: 5%); sample C (Chinese liquor, alcohol content: 17%); and sample D (red wine, alcohol content: 14%). Each droplet of 10 μl of liquor was spotted on a Si wafer and after drying was measured. X-Ray excitation conditions were 40 kV and 40 mA. Other analytical conditions were the same as for the previous example.

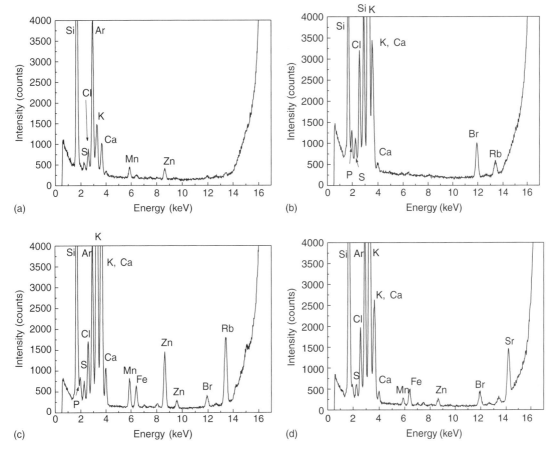

Figure 7.4.2 TXRF spectra of liquors: (a) sample A (Japanese liquor, sake); (b) sample B (beer); (c) sample C (Chinese liquor); (d) sample D (red wine)

On comparing the spectra shown in Figure 7.4.2, the spectral patterns of the four liquors were all different. That is, S, Cl, K, Ca, Mn and Zn were detected as characteristic elements in sample A, P, S, Cl, K, Ca, Br and Rb were detected in sample B, P, S, Cl, K, Ca, Mn, Fe, Zn, Br and Rb were detected in sample C and S, Cl, K, Ca, Mn, Fe, Zn, Br and Sr were detected in sample D. The peak of silicon was due to the Si wafer. As a result, the four liquors can be distinguished from each other on the basis of TXRF data.

7.4.2.3 BRANDY

It was reported elsewhere[14,15] that a counterfeit VSOP cognac had been proved to contain sulfur abundantly by TXRF and to be circular dichroism (CD) inactive by CD analysis, while genuine VSOP cognacs had no sulfur and were CD active.

Elemental analyses of four kinds of genuine brandies were examined by TXRF.

Each droplet of 10 μl of brandy was spotted on a Si wafer and the TXRF measurement conditions were the same as for the case of liquors. TXRF spectra of four genuine brandy samples (A, B, C and D) are shown in Figure 7.4.3.

As a characteristic element, Cu was detected in common in Figure 7.4.3(a)–(d) and no sulfur could be detected in Figure 7.4.3(a)–(d). On the other hand, the spectrum of a red wine sample shown in Figure 7.4.2(d) gave no Cu peaks although many other elements (S, Cl, K, Ca, Mn,

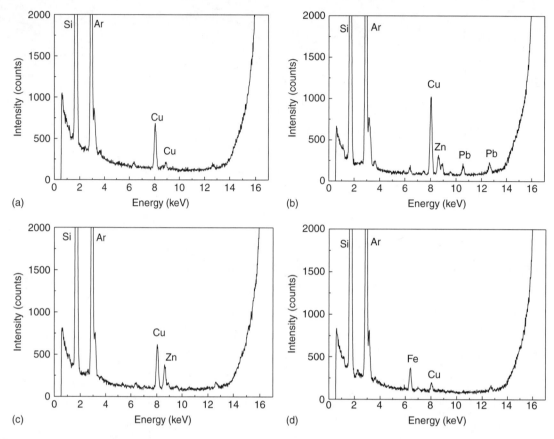

Figure 7.4.3 TXRF spectra of brandies: (a) sample A; (b) sample B; (c) sample C; (d) sample D

Fe, Zn, Br and Sr) were observed in the spectrum. It is well known that brandy is made by gentle distillation of a large amount of wine. Therefore, the Cu in the four brandies might originate from the distillation procedure. Also, compared with the spectra shown in Figure 7.4.3(a)–(d), every spectral pattern is different from each other. That is, Cu is observed as a main element in sample A and Cu, Zn and Pb are detected in sample B and both Cu and Zn are detected in sample C, and both Fe and Cu are detected in sample D. The lead in Figure 7.4.3(b) is supposed to be derived from the glass bottle.

7.4.2.4 WASTE WATER

Figure 7.4.4 shows the TXRF spectrum of 1 μl of waste water from a factory. Sulfur and nickel were detected clearly as shown in Figure 7.4.4 and the nickel concentration was estimated to be 170 mg/l by quantitative analysis of TXRF. In this factory, recovery of nickel was usually in operation using a NiS precipitation process. However, the precipitation process could not function perfectly and could not recover efficiently the valuable nickel resources from the waste water. As a result, the waste water still contained valuable nickel after the precipitation process.

7.4.2.5 SEAL INK

In Japan, traditionally seal impressions are often used as certification marks instead of signatures. Nowadays various kinds of seal inks are used in Japan and they appear vermilion in appearance,

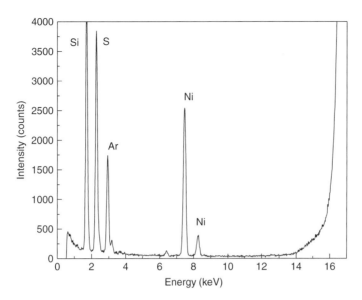

Figure 7.4.4 TXRF spectrum of waste water

while the elements contained in them are often different. In disputed documents seal impressions are often examined. Even if counterfeit seal impressions have been used, genuine original seal inks can be differentiated from illegal-used seal inks using TXRF.

Figure 7.4.5 shows TXRF spectra of five kinds of seal inks (samples A–E) which are vermilion in appearance. Both samples A and B are low priced, while sample C is moderately priced and both samples D and E are considerably more expensive. The spectral patterns differ from each other. That is, in sample A, Cl, Ti and Zn were detected and in sample B, Pb, S, Ba, Cr and Sr were detected and in sample C, Ca, Fe, Hg and Sr were detected and in sample D, Hg, S, Sb and Pb were detected and in sample E, Hg, S and Ba were detected. Hence, each seal ink could be differentiated from one another by the TXRF data.

7.4.2.6 METHAMPHETAMINE

Abuse of drugs is one of the most serious social issues. Drugs of abuse, such as methamphetamine, amphetamine, heroin, cocaine and other drugs, usually contain trace intrinsic ingredients due to the synthetic and purification processes or methods of smuggling. Those ingredients can be used as tagging factors to indicate clandestine synthetic laboratories or illegal import routes. Impurity profiling analysis has been investigated for the characterization and classification of illicit drug samples. Trace organic ingredient analyses of methamphetamine salts or amphetamine salts have been reviewed by Verweij[23] and Inoue[24] and have been reported by other authors.[25–31] Inorganic ingredient analyses of methamphetamine salts or amphetamine salts by neutron activation analysis (NAA),[32,33] inductively coupled plasma-mass spectrometry (ICP-MS) analysis[34–36] and atomic absorption spectrometry[37] have also been reported. In those methods, they have used sample amounts of 2–50, 10–100 mg, and so on, and detected ppm levels of each trace element. Further, the above methods need complicated pretreatment of samples and are often destructive.

TXRF was applied to examine trace elements in methamphetamine salts and nanogram levels of detection of each trace element in 1 mg of methamphetamine salts has been reported.[18]

Figure 7.4.6 shows TXRF spectra of two seized methamphetamine HCl salts (samples A and B). Both samples A and B were previously proved to

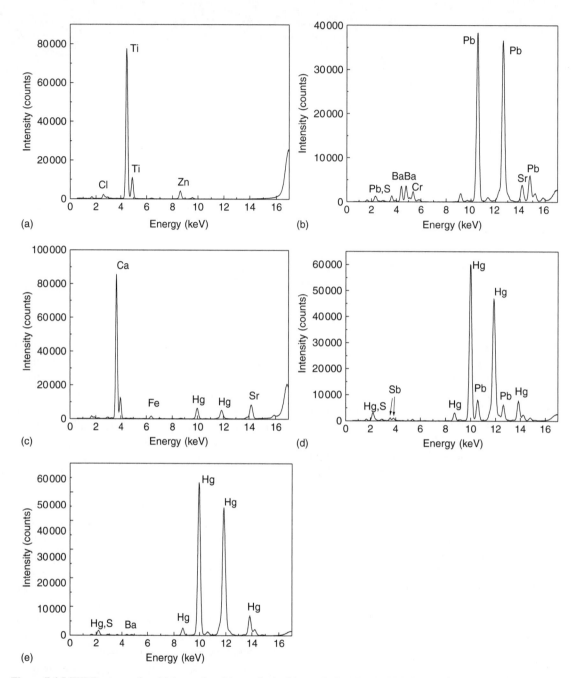

Figure 7.4.5 TXRF spectra of seal ink samples: (a) sample A; (b) sample B; (c) sample C; (d) sample D; (e) sample E

be chemically pure by melting point analyses. In Figure 7.4.6(a), Ca, Fe, Ni, Cu, Zn and Br were detected. Bromine is supposed to be derived from the impurities of the methamphetamine HCl salt and Fe, Ni, Cu and Zn are supposed to be from the metallic vessels or tools used during the synthesis processes or handling. In Figure 7.4.6(b), iodine and mercury are detected as specific elements,

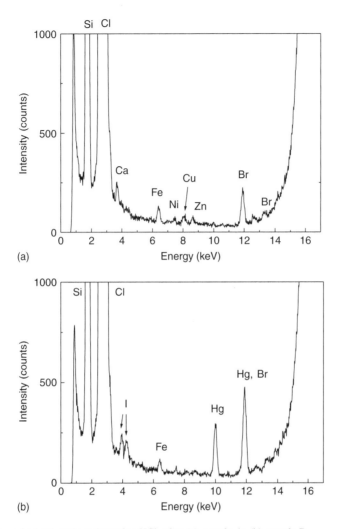

Figure 7.4.6 TXRF spectra of seized methamphetamine HCl salts: (a) sample A; (b) sample B

and these elements are supposed to originate from the chemical reagents in the methamphetamine synthesis processes.

7.4.2.7 METEORITE

On 26 September 1999, a rock-like material penetrated a house in Kobe city. The rock-like material broke into 20 pieces, and these pieces were sent to the forensic laboratory at Hyogo Prefectural Police Headquarters for analysis to confirm whether they were from a meteorite or not. The pieces weighed 135.197 g and were examined by many nondestructive analytical techniques. Figure 7.4.7 shows the grazing incidence XRF spectra of a small fragment (D) from the sample pieces; in fragment D both Fe and Ni are detected clearly. It is known that meteorites contain both Fe and Ni, so that the spectra suggest the sample might be from a meteorite.

On the basis of many kinds of analytical results, including microscopic observation, density, elemental analysis, γ-ray analysis of radioactive nuclides originating from cosmic rays, the pieces were concluded to be meteorite pieces.

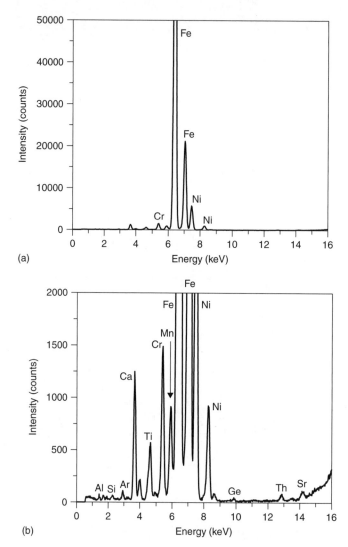

Figure 7.4.7 Grazing incidence XRF spectra of the Kobe meteorite (fragment D). Spectrum (b) is the magnified version of spectrum (a)

The meteorite was clarified to be a carbon-type meteorite, which was the first example in Japan. These fragments have magnetic properties and elemental mapping analysis of a fragment using SR-XRF at SPring-8 has been applied to examine the magnetic structure in the meteorite.[38] In addition, radioisotopes with very short life-time (Ni57, Mg28, Sc47 and K43) were detected in this meteorite.

This meteorite was named 'Kobe' and details have been presented elsewhere.[38]

7.4.3 FORENSIC APPLICATIONS OF SYNCHROTRON RADIATION XRF

7.4.3.1 FLUORESCENT TRACING AGENT

A plastic ball, called a 'color ball' in Japan, contains a fluorescent powder suspension, which may act as a tracing agent for criminals or cars involved in crimes. If a robbery happened in a bank, this ball would be thrown at the criminal or a car leaving the crime scene to trace the criminal or the car. The criminals or cars may be traced if some of the fluorescent powder remained on the road and this might help to investigate the criminal or the car. Synchrotron radiation XRF (SR-XRF) can be applied to forensic trace samples.

As an X-ray source, 10 keV X-rays from undulator radiation at beam line 39XU of SPring-8 was used and the X-rays were shaped to a small beam (100 μm × 100 μm) to analyse the ultra trace of fluorescent powder. The XRF spectrum was measured by a Si(Li) semiconductor detector for 200 s. Each trace powder sample was kept in a small polypropylene bag (6 μm thick film) and was measured. Figure 7.4.8 shows SR-XRF spectra of three kinds of reddish color fluorescent powders. Each spectral pattern is different. Peaks of S and Zn are characteristic in Figure 7.4.8(a), there is a large Ca peak in Figure 7.4.8(b), and peaks of S, Fe and Zn are the main peaks in Figure 7.4.8(c).

SR-XRF is a very useful technique to identify ultra traces of fluorescent powder. Recently this technique was applied successfully to solve a bank robbery case. The SR-XRF spectrum of the fluorescent powder used in the case is shown in Figure 7.4.9.

7.4.3.2 DRUGS OF ABUSE (COCAINE, HEROIN, MARIJUANA AND OPIUM)

Using our laboratory TXRF system, it was very difficult to detect pg amounts of trace elements in drugs of abuse. Ultra trace analysis of seized drugs of abuse (cocaine, heroin, marijuana and opium) using synchrotron radiation total reflection X-ray (SR-TXRF) analysis has been presented elsewhere.[39]

Sample amounts of 1 μl solutions containing 10 μg of drugs (cocaine and heroin) were spotted on Si wafers. A leaflet of marijuana was set directly on a Si wafer and opium in the form of a soft lump was smeared on another Si wafer. These samples were then measured by SR-TXRF. The X-ray source was 10 keV X-rays from undulator radiation at Hyogo Prefecture beam line 24XU (hutch B) of SPring-8 and X-rays were incident onto the sample surface at 0.005° under atmospheric conditions. The TXRF spectrum was measured by a Si (Li) semiconductor detector for 500 s. In these experiments, pg levels of contaminant elements in 1 μl sample could be detected.

Figure 7.4.10(a) and 7.4.10(b) show SR-TXRF spectra of cocaine and heroin, respectively. In Figure 7.4.10(a), a large peak of Cl at 2.6 keV is derived from the cocaine hydrochloride salt and Ca and Zn are supposed to be contaminant elements from the purification procedures of cocaine. In Figure 7.4.10(b), characteristic peaks were observed at 3.9 keV ($L\alpha$), 4.2 keV ($L\beta_1$), 4.5 keV ($L\beta_2$) and 4.8 keV ($L\gamma_1$), attributed to iodine. Infante et al. referred to metal contamination in illicit heroin samples using atomic absorption spectrometry and also reported that calcium was encountered in most of the samples.[37] However, they did not mention the existence of iodine in illicit heroin samples. The origin of iodine in the sample shown in Figure 7.4.10(b) is under investigation.

Marijuana and opium are derived from botanical tissues and SR-TXRF spectra of them were compared. A SR-TXRF spectrum of a chip of marijuana leaflet is shown in Figure 7.4.10(c) and a SR-TXRF spectrum of a smear of black opium is shown in Figure 7.4.10(d). In Figure 7.4.10(c), peaks of Ca at 3.7 keV ($K\alpha$) and 4.0 keV ($K\beta$), a peak of Fe $K\alpha$ at 6.4 keV, a peak of Zn $K\alpha$ at 8.6 keV, a peak of Ti $K\alpha$ at 4.5 keV, a peak of Cl $K\alpha$ at 2.6 keV and a small peak of S $K\alpha$ at 2.3 keV were observed. In Figure 7.4.10(d), a peak of K $K\alpha$ at 3.3 keV and a moderately large peak of S $K\alpha$ at 2.3 keV were observed. No peaks of Ti, Fe or Zn were observed.

From the above results, ultra trace drugs of abuse samples could be differentiated from one another by SR-TXRF, while, in general, black resin-like compounds could hardly be discriminated by microscopic observation. This method can also be applied to other botanical samples.

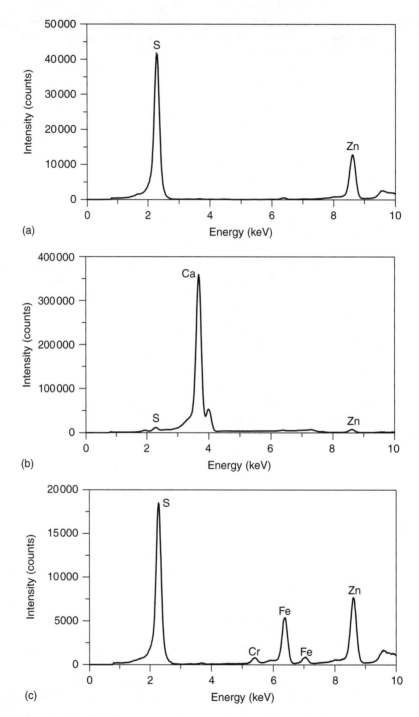

Figure 7.4.8 SR-XRF spectra of reddish fluorescent powders: (a) powder A; (b) powder B; (c) powder C

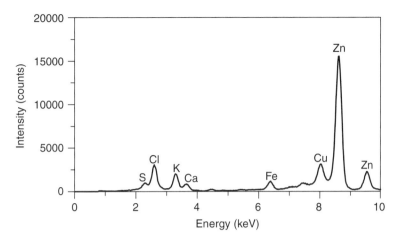

Figure 7.4.9 SR-XRF spectrum of the reddish fluorescent powder D

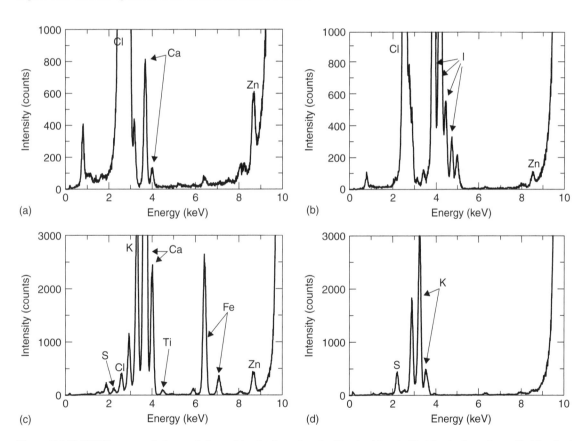

Figure 7.4.10 TXRF spectra of drugs of abuse: (a) seized cocaine A; (b) seized heroin B; (c) seized marijuana C; (d) seized opium D

7.4.4 FORENSIC APPLICATIONS OF HIGH ENERGY SYNCHROTRON RADIATION XRF

7.4.4.1 GUNSHOT RESIDUE

It is known that trace gunshot residue (GSR) remains in an empty cartridge case or on the surface of the hand which held the handgun. Usually, electron probe microanalysis (EPMA) has been utilized to examine GSR particles. However, electron beam excitation is far less sensitive than X-ray excitation on X-ray elemental analysis. Also, EPMA needs high vacuum analytical conditions,

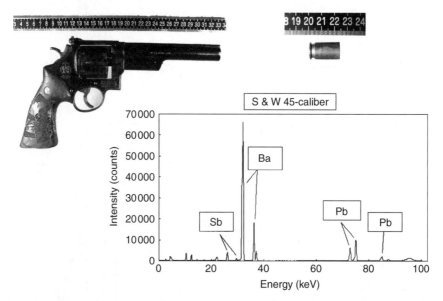

Figure 7.4.11 High energy SR-XRF spectrum of GSR sample A

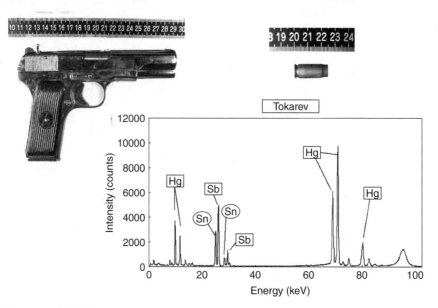

Figure 7.4.12 High energy SR-XRF spectrum of GSR sample B

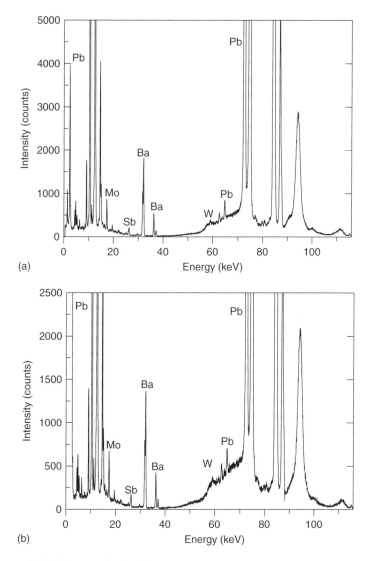

Figure 7.4.13 High energy SR-XRF spectra of paint chips: (a) a trace of paint chip from the scene of the crime; (b) a paint chip of the car concerned

which interrupts detection of a volatile element, for example, mercury from GSR.

High energy SR-XRF at SPring-8 was applied to detect trace elements in GSR under atmospheric conditions. The X-ray source was high energy X-rays (116.4 keV) through a Si (400) monochromator from an elliptic multipole wiggler at beam line 08 W of SPring-8. Traces of two kinds of GSR collected from the empty cartridge cases of guns A and B were mounted on surfaces of Scotch mending tape (No. 810). The SR-XRF spectrum was measured by using a Ge semiconductor detector for 600 s. The results are shown in Figures 7.4.11 and 7.4.12. Each spectral pattern gave a distinct profile: Ba, Sb and Pb were detected in the GSR sample A, while Hg, Sb and Sn were detected in that of GSR sample B. Also, GSR analysis on a hand having shot a gun can be applied using this technique.[40]

7.4.4.2 PAINT CHIP

In hit-and-run cases, it is very important to examine traces of evidential samples obtained at crime scenes. High energy SR-XRF analysis at SPring-8 was applied to analyse traces of paint chips from a hit-and-run case.[41] The X-ray measurement system was the same as for the above GSR case.

Figure 7.4.13(a) shows a high energy SR-XRF spectrum of a trace of a paint chip, which was discovered at a hit-and-run scene. The high energy SR-XRF spectrum of the paint chip from the car concerned is shown in Figure 7.4.13(b). Both spectra resemble each other closely. Microscopic color observation and Fourier transform infrared spectra of the two paint chips confirmed that the trace of paint chip at the scene was of the same quality of the car concerned.

As shown in the above cases, K-series XRF signals of Sn, Sb, Ba, W, Hg and Pb, which are impossible to detect using conventional laboratory XRF instruments, can be detected by high energy SR-XRF at SPring-8.

ACKNOWLEDGEMENTS

The author expresses thanks to all of his coworkers for their contributions and also expresses thanks to Dr S. Hayakawa for helping with small beam SR-XRF experiments and to Dr I. Nakai for helping with high energy SR-XRF experiments. The synchrotron radiation experiments were performed at the SPring-8 with the approval of the Japan Synchrotron Radiation Research Institute (JASRI).

REFERENCES

1. Yoneda, Y. and Horiuchi, T. Optical flats for use in X-ray spectrochemical microanalysis. *Rev. Sci. Instrum.*, **42**, 1069–1070 (1971).
2. Klockenkämper, R. *Total-Reflection X-ray Fluorescence Analysis*. John Wiley & Sons, New York, 1997.
3. Prange, A., Reus, U., Boddeker, H., Fischer, R. and Adolf, F. P. Microanalysis in forensic science–characterization of single textile fibers by total reflection X-ray fluorescence. *Anal. Sci.*, **11**, 483–487 (1995).
4. Duwel, F., Fischer, R., Schonberger, T., Simmross, U. and Weis, D. Quantitative elemental microanalysis of thermoplastic remains using total reflection X-ray fluorescence (TXRF) spectrometry. *Proc. Meet. Int. Assoc. Forensic Sci.* **14**, 1–4 (1996).
5. Ninomiya, T., Nomura, S. and Taniguchi, K. Application of total reflection X-ray fluorescence analysis to forensic samples: chromium detection in fibers. *Memoirs Osaka Electro-Commun. Univ.* **22**, 51–57 (1986).
6. Nomura, S., Ninomiya, T. and Taniguchi, K. Elemental analysis of micro water samples using total reflection X-ray fluorescence spectrometry. *Memoirs Osaka Electro-Commun. Univ.* **24**, 127–138 (1989).
7. Nomura, S., Ninomiya, T. and Taniguchi, K. Trace elemental analysis of titanium oxide pigments using total reflection X-ray fluorescence analysis. *Adv. X-ray Chem. Anal. Jpn.* **19**, 217–226 (1988).
8. Ninomiya, T., Nomura, S. and Taniguchi, K. Elemental analysis of trace plastic residuals using total reflection X-ray fluorescence analysis. *Adv. X-ray Chem. Anal. Jpn.* **19**, 227–235 (1988).
9. Ninomiya, T., Nomura, S., Taniguchi, K. and Ikeda, S. Quantitative analysis of arsenic element in a trace of water using total reflection X-ray fluorescence spectrometry. *Adv. X-ray Anal.* **32**, 197–204 (1989).
10. Ninomiya, T., Nomura, S. and Taniguchi, K. Probing into residual forensic evidences–application of total reflection X-ray fluorescence spectrometry. *J. Surface Sci. Soc. Jpn.* **11**, 189–194 (1990).
11. Ninomiya, T., Nomura, S. and Taniguchi, K. Application of total-reflection X-ray fluorescence spectrometry to elemental toner analysis. *J. Jpn. Soc. Color Mater.* **65**, 176–181 (1992).
12. Nomura, S., Ninomiya, T., Taniguchi, K. and Ikeda, S. Application of total reflection X-ray fluorescence spectrometry to drug analysis. *Adv. X-ray Anal.* **35**, 969–974 (1992).
13. Nomura, S., Ninomiya, T. and Taniguchi, K. Indirect drug analysis using total reflection X-ray fluorescence spectrometry. *Adv. X-ray Chem. Anal. Jpn.* **24**, 113–119 (1993).
14. Ninomiya, T., Nomura, S., Taniguchi, K. and Ikeda, S. Application of GIXF to forensic samples. Proceedings of the 5th workshop on total reflection X-ray fluorescence spectroscopy and related spectroscopical methods held in Tukuba, Japan. *Adv. X-ray Chem. Anal. Jpn.* **26s**, 9–18 (1994).
15. Ninomiya, T., Nomura, S., Taniguchi, K. and Ikeda, S. Application of grazing incidence X-ray fluorescence analysis to forensic samples. *Anal. Sci.*, **11**, 489–494 (1995).
16. Nomura, S., Taniguchi, K. and Ninomiya, T. Application of total reflection X-ray fluorescence analysis to soils and liquors. *Memoirs Osaka Electro-Commun. Univ.* **31**, 45–49 (1996).
17. Shimoyama, M., Nakanishi, T., Hamanaga, Y., Ninomiya, T. and Ozaki, Y. Non-destructive discrimination between elephant ivory products and mammoth tusk

products by glancing incidence X-ray fluorescence spectroscopy. *J. Trace Microprobe Tech.* **16**, 175–181 (1998).
18. Muratsu, S., Fukui, S., Maeda, T., Matsushita, T., Kaizaki, S. and Ninomiya, T. Trace elemental analysis of illicit methamphetamines using total reflection X-ray fluorescence spectroscopy. *J. Health Sci.* **45**, 166–171 (1999).
19. Ninomiya, T. Japanese Patent Application No. 194381 (1993).
20. Nomura, S., Ninomiya, T. and Taniguchi, K. Applications of total reflection X-ray fluorescence spectrometry to trace evidential materials. Abstract of 42nd Annual Denver Conference on Applications of X-ray Analysis, Denver, Colorado, p. 28 (1993).
21. Ninomiya, T., Nomura, S., Nakai, I., Hayakawa, S. and Taniguchi, K. XRF imaging of a fingerprint using synchrotron radiation. *Photon Factory Activ. Rep.* **11**, 43 (1993).
22. Ninomiya, T. Analytical chemistry in human life-criminal investigation and chemical analysis. *Bunseki* **325**(10), 578–585 (2000).
23. Verweij, A. M. A. Impurities in illicit drug preparations: amphetamine and methamphetamine. *Forensic Sci. Rev.* **1**, 1–11 (1989).
24. Inoue, T. Discrimination of abused drug samples by impurity profiling analysis (chemical fingerprint). *Jpn. J. Forensic Toxicol.* **10**, 204–217 (1992).
25. Kobayashi, K., Iwata, Y., Kanamori, T., Inoue, H. and Kishi, T. Analysis of impurities in methamphetamine and impurity profiling. *Rep. Nat. Res. Inst. Police Sci. Res. Forensic Sci.* **53**, 1–9 (2000).
26. Inoue, T., Tanaka, K., Ohmori, T., Togawa, Y. and Seta, S. Impurity profiling analysis of methamphetamine seized in Japan. *Forensic Sci. Int.* **69**, 97–102 (1994).
27. Tanaka, K., Ohmori, T., Inoue, T. and Seta, S. Impurity profiling analysis of illicit methamphetamine by capillary gas chromatography. *J. Forensic Sci.* **39**, 500–511 (1994).
28. Perkal, M., Ng, Y. L. and Pearson, J. R. Impurity profiling of methylamphetamine in Australia and the development of a national drugs database. *Forensic Sci. Int.* **69**, 77–87 (1994).
29. King, L. A., Clarke, K. and Orpet, A. J. Amphetamine profiling in the U.K. *Forensic Sci. Int.* **69**, 65–75 (1994).
30. Jonson, C. S. L. Amphetamine profiling–improvements of data processing. *Forensic Sci. Int.* **69**, 45–54 (1994).
31. Kishi, T., Inoue, T., Suzuki, S., Yasuda, T., Oikawa, T. and Niwaguchi, T. Analysis of impurities in methamphetamine. *Eisei Kagaku* **29**, 400–406 (1983).
32. Kishi, T. Application of neutron activation analysis to forensic chemistry. *Eisei Kagaku* **32**, 335–343 (1986).
33. Kishi, T. Forensic neutron activation analysis–the Japanese scene. *J. Radioanal. Nucl. Chem.* **114**, 275–280 (1987).
34. Kishi, T. Determination of sodium, bromine, palladium, iodine and barium in authentic methamphetamine hydrochloride by inductively coupled plasma-mass spectrometry. *Rep. Nat. Res. Inst. Police Sci. Res. Forensic Sci.* **41**, 256–259 (1988).
35. Suzuki, S., Tsuchihashi, H., Nakajima, K., Matsushita, A. and Nagao, T. Analyses of impurities in methamphetamine by inductively coupled plasma mass spectrometry and ion chromatography. *J. Chromatogr.* **437**, 322–327 (1988).
36. Marumo, Y., Inoue, T. and Seta, S. Analysis of inorganic impurities in seized methamphetamine samples. *Forensic Sci. Int.* **69**, 89–95 (1994).
37. Infante, F., Domiguez, E., Trujillo, D. and Luna, A. Metal contamination in illicit samples of heroin. *J. Forensic Sci.* **44**, 110–113 (1999).
38. Maeda, T., Shimoda, O., Muratsu, S., Shimoyama, M., Misaki, K., Nakanishi, T., Ninomiya, T., Kagoshima, Y., Takai, K., Ibuki, T., Yokoyama, K., Takeda, S., Tsusaka, Y. and Matsui, J. An identification of a Meteorite. *Rep. Nat. Res. Inst. Police Sci. Res. Forensic Sci.* **54**, 11–18 (2001).
39. Muratsu, S., Ninomiya, T., Kagoshima, Y. and Matsui, J. Trace elemental analysis of drugs of abuse using synchrotron radiation total reflection X-ray fluorescence analysis (SR-TXRF). *J. Forensic Sci.* **47**, 944–949 (2002).
40. Ninomiya, T., Muratsu, S., Maeda, T., Shimoda, O., Nakanishi, T., Hashimoto, T., Saitoh, Y., Nakai, I., Terada, Y., Nishiwaki, Y., Marumo, Y., Suzuki, S., Suzuki, Y. and Kasamatsu, M. Trace element characterization of gunshot residues using SR-XRF technique. *SPring-8 User Experiment Report (JASRI)*, No. 5 (2000A), 129 (2000).
41. Ninomiya, T., Nakanishi, T., Muratsu, S., Saitoh, Y., Shimoda, O., Watanabe, S., Nishiwaki, Y., Matsushita, T., Suzuki, S., Suzuki, Y., Ohta, H., Kasamatsu, M., Nakai, I. and Terada, S. Elemental analysis of a trace of paint chip using SR-XRF. *SPring-8 User Experiment Report (JASRI)*, No. 7 (2001A), 67 (2001).

7.5 Speciation and Surface Analysis of Single Particles Using Electron-excited X-ray Emission Spectrometry

I. SZALÓKI[1], C.-U. RO[2], J. OSÁN[3], J. DE HOOG[4] and R. VAN GRIEKEN[4]

[1] University of Debrecen, Debrecen, Hungary, [2] Hallym University, Chun Cheon, Korea, [3] KFKI Atomic Energy Research Institute, Budapest, Hungary and [4] University of Antwerp, Antwerp, Belgium

7.5.1 INTRODUCTION

In the last decade, both the technological background and the data evaluation methods, i.e. X-ray spectra analysis and quantitative determination of sample composition, have been significantly improved for electron probe microanalysis (EPMA). One of the most essential breakthroughs in this field was the appearance of newly designed high-resolution energy-dispersive detectors such as microcalorimeters, as well as commercial silicon-based spectrometers equipped with thin polymer window having weak attenuation for low-energy X-rays. Another important improvement can be found in the recently developed evaluation models for quantitative analysis, which are capable of handling a wide variety of target sample types. The ultimate goal of this scientific effort has been to increase the X-ray detection efficiency, to broaden the energy range of the X-rays to be analysed, to extremely decrease the irradiated and excited volume (Watanabe and Williams, 1999) in the specimen and to obtain maximum information about sample composition and structure by application of adequate quantitative model for fast calculation. The recent development trend in the field of calculation models shows two main directions in quantitative EPMA i.e. conventional $\phi(\rho z)$-based methods and Monte Carlo (MC) simulation of basic elementary interactions between electrons–atoms and photons–atoms.

EPMA equipped with an energy-dispersive detector qualifies to detect simultaneously all the morphology and the constitution elements (within a microscopic size volume) of a sample. This advantageous analytical capability of the EPMA has been successfully utilized in atmospheric aerosol research, where two new techniques have appeared resulting in a significant improvement of the applicability of EPMA in single-particle analysis: (i) grazing-exit EPMA (Tsuji *et al.*, 1999a); and (ii) low-Z EPMA described in this subchapter (Scott and Love, 1999).

One of the principal aspects that limit the efficient application of energy-dispersive X-ray (EDX) detectors lies in the fact that the detection of low-Z elements is hindered by the absorption of characteristic X-rays by the beryllium window of the detectors. This technical difficulty can be avoided by using thin polymer windows instead; their thickness is approximately 200 nm, they are commercially available and have been introduced in routine analysis for several years. The characteristic X-ray lines of the main components, which are mostly low-Z elements ($Z < 9$), undergo extremely strong attenuation while propagating through the particle volume, therefore the estimation of the matrix effect is important. Quantitative determination of low-Z elements is a necessary development for further research of individual particles, firstly because these elements (C, N, O) are abundantly present e.g. in atmospheric particles, and secondly because quantitative information is necessary for speciation of individual microscopic particles; indeed, many environmental particles contain low-Z elements in the form of nitrates, sulfates, oxides or mixtures including a carbon matrix.

7.5.2 EXPERIMENTAL CONDITIONS FOR PARTICLE ANALYSIS

7.5.2.1 TECHNOLOGICAL DEVELOPMENTS IN EPMA FOR PARTICLE ANALYSIS

Electron microscopy has always had to cope with some limitations, not only for particle analysis. Since the development of the first electron microscopes (either called electron probe microanalysers or electron microscopes), researchers have been improving the technology within their instruments. Most of the useful developments were commercialized and put into use for different applications. During the last decades, particle analysis has benefited from the technological improvements in the different components of the electron microscope. However, for practical and financial reasons, new technology has not always been accepted or applied very fast in the whole of the particle analysis community (like in many other fields). In environmental particle analysis, for instance, one could say that most of the research goes into the study of the environmental effects of the particles themselves, and that less time and money is spent on applying the new available techniques. Environmental institutes are, therefore, expected to spend money on new technology when the costs are acceptable, or when the need for better, rapidly available data is high. A contrary example is the semiconductor industry, which has always supported the developers of electron microscopes by offering its semiconductor technology, getting better analytical instruments in return and resulting in mutual, stimulating benefits.

One of the basic parts of an electron microscope is the electron gun. A short discussion about the evolution in the electron gun technology is appropriate in the context of X-ray spectrometry, since electron beam optics not only play an important role in image formation, but also in X-ray analysis. For example, the size of the electron beam determines the resolution of the electron images, as well as the size of the interaction volume in which the X-rays for microanalysis are generated. A detailed discussion on the principles of electron guns and electron beam optics is beyond the scope of this subchapter, but can be found in literature on the basics of electron microscopy. Numerous electron microscopes 'in the field' are still equipped with conventional electron guns based on thermionic emission, requiring a tungsten filament or a lanthanum hexaboride emitter, but much recent research has been done on field-emission sources. Although field-emission guns (FEGs) were developed several decades ago, they are still in the process of gaining more and more acceptance in the particle analysis community. This type of electron gun uses a very small source (e.g. a tungsten hairpin) and requires only simple optics to obtain much narrower and brighter beams (nanometer sizes) than with the thermionic guns (micrometer sizes). This advantage is very useful for high-resolution imaging at low voltages, since thermionic guns show insufficient beam currents and degraded probe sizes under low-voltage conditions, whereas FEGs provide nanometer probes with nanoampere beam currents. The main practical disadvantage, however, lies in the vulnerability of the hairpin emitter to residual gas traces in the vacuum of the specimen chamber. The vacuum should be at least better than 10^{-8} Pa in order to obtain reasonably stable emission, since monolayers or thicker coatings of foreign gas molecules on the tip's surface reduce the electron emission. So, the quality of the vacuum determines the stability of the electron gun and the electron beam. Some methods (e.g. 'flashing') can be used to restore stable emission, but repeated use of these methods tends to blunt the sharp tip of the hairpin, which should, therefore, be replaced after many months of operation. Short-term beam instabilities causing streaks in scanned images may be reduced using feedback circuits for compensation. The quality of the images made with field-emission electron guns is very high, and, therefore, more and more authors mention the application of this type of source for making images of single particles in their publications (Ortner et al., 1998; Ortner, 1999; Höflich et al., 2000; Jones et al., 2000; Laskin and Cowin, 2001). The very narrow probe size could also be very advantageous for the X-ray analysis of small particles, since the electron interaction volume is also considerably smaller. Many authors, however, use field-emission gun scanning electron microscopy (FEG-SEM) mostly for imaging or for qualitative analysis only. One reason might be that the long-term or short-term instabilities of field emission are expected to hamper quantitative X-ray analysis. Other authors, however, have already used FEGs for quantitative particle analysis (Laskin and Cowin, 2001) or to compare the quality of different X-ray detectors for particle analysis (Newbury et al., 1999; Wollman et al., 2000a). FEG-SEM is sometimes referred to as low-voltage scanning electron microscopy (LV-SEM), since it can be used for X-ray analysis at low voltages (still offering small enough probe sizes and high enough beam currents), however, this technique is not used very often. One reason is that the analysts should be able to select beam-accelerating voltages that are sufficiently adequate to excite the characteristic peaks of the elements of interest, which is not evident. First of all, the efficiency of characteristic X-ray generation, and evidently also the analytical range, are strongly dependent on the overvoltage U, which is inevitably low for low-voltage analysis. Secondly, the very unpredictable effect of charging becomes also very important at lower voltages, since its influence on the overvoltage is very critical for X-ray generation (Newbury, 2000). A second reason is that low-voltage X-ray analysis also requires the capability to perform light-element detection, which was and is not always straightforward in all applications (as will be discussed further below). The fact that earlier simulation models for quantitative analysis also did not take into account the different physical behavior of electrons at low voltages, could have been a third reason for many researchers not to use FEG-SEM or LV-SEM for quantitative analysis. However, many changes in the available software are being made, and, for example, the CASINO Monte Carlo program (Hovington et al., 1997), which was adapted for quantitative particle analysis in the presented research, was originally developed for this reason, implementing improved functions for better simulating the electron interactions at low voltages. The interest in low-voltage analysis is growing, so further studies using FEGs for quantitative analysis are to be expected in the future.

The most spectacular technological evolutions can probably be found in the development of X-ray detectors (explained more in detail in other subchapters of this book). Although wavelength-dispersive spectrometers (WDS) are able to record spectra with very high resolution, their applications for particle analysis are rather limited. Due to its low quantum and geometry efficiency, a higher beam current has to be set in order to cause enough electron interactions which result in a suitable amount of detected X-ray signals. Since WDS also requires long measuring times, the exposure of small or volatile particles could cause damage or even total evaporation (Szalóki et al., 2001a). The long measuring times also limit the applicability of WDS, because particle studies mostly involve the analysis of huge amounts of particles, which would require too much time. These disadvantages limit WDS to the analysis of more stable particles, e.g. metal oxides or silicates (Ortner et al., 1998; Ortner, 1999; Höflich et al., 2000).

The improved energy-dispersive (semiconductor) spectrometers (EDS), which can now also detect the X-rays from light elements (with atomic number $Z < 11$), have proven their usefulness for particle analysis. The replacement of the beryllium detector window in the earlier models by a much thinner window has almost become the new standard. The first real attempts to use a different detector window set-up for particle analysis were done either by just removing the beryllium window (windowless mode) or by replacing it with the first versions of thin windows, consisting of a thin foil. The latest thin windows mostly consist of a silicon grid supporting a thin polymer foil, coated with a metal film for opacity. This improved kind of energy-dispersive Si(Li) detectors have been fully commercialized and some recent examples of thin-window electron probe microanalysis (TW-EPMA) can be found in the literature (Diebold *et al.*, 1998; Roth and Okada, 1998; Paoletti *et al.*, 1999; Ro *et al.*, 1999; Szalóki *et al.*, 1999; Newbury, 2000; Osán *et al.*, 2000; Laskin and Cowin, 2001). The advantages of semiconductor EDS over WDS are its better geometry and quantum efficiency, but it always had to cope with many spectral problems (which will be discussed in detail below) due its poorer energy resolution. The fact that many of these problems occur in the low-energy part of the spectra, might seem a drawback to use this technique for quantitative analysis in some applications. However, despite these remaining, but improved, spectral limitations, TW-EPMA currently appears to be the best, commercially available option to perform particle analysis over a broad range of elements (from low-Z to high-Z) in a straightforward, fast way.

Another very recent kind of energy-dispersive detector is the microcalorimeter (sometimes called bolometer), which could hardly have been unnoticed by particle analysts (Wollman *et al.*, 1997; Diebold *et al.*, 1999). These detectors are discussed in detail in subchapter 4.4 of this volume. Microcalorimeters combine the advantages of both WDS and conventional EDS, since their excellent energy resolution (comparable with WDS) allows straightforward identification of closely spaced X-ray peaks in complicated spectra at fast operation times (comparable with EDS). Wollman *et al.* (1998) used a microcalorimeter detector for the analysis of sub-micrometer particles down to 0.1 µm on silicon wafers. Besides proving that microcalorimeter EDS is suitable for this kind of analysis (even in non-optimal cases where the electron beam diameter is larger than the analyzed particle), they also performed chemical shift analysis on particles. Changes in electron binding energies with the chemical environment of atoms result in 'chemical' shifts of peaks in the acquired spectra, providing information on oxidation states. With the X-ray information on chemical bonding, the analysts were able to differentiate between chemical species present in investigated particles. Chemical shift analysis is also possible with WDS, but this was never routinely performed because of the long scanning times involved. Therefore, when the much faster, high resolution microcalorimeters are further improved and commercialized, they would undoubtedly provide significant benefits for particle analysis, certainly if they were combined with FEG-SEM in order to analyze small particles. In order to perform X-ray analysis with FEG-SEM, Newbury *et al.* (1999) equipped a microcalorimeter detector with a polycapillary X-ray optic to increase the solid angle of the instrument by a factor of 300 (dependent on photon energy), since the standard solid angle of the detector was too small to obtain statistically useful spectra with a nanoampere beam within 100–1000 s. The developers often publish the status of the performance of the microcalorimeter (Wollman *et al.*, 2000b), and in the future, efforts to improve the detector resolution, counting rates, low-photon cut-off and the solid angle are to be expected.

A technique, which has been reported to use a different X-ray detector geometry, rather than new detector technology, is grazing exit EPMA (GE-EPMA). In this technique, the X-ray detector (WDS or EDS) is positioned at low take-off or exit angles in order to minimize the effect of the substrate on the acquired spectra, whether by tilting the sample or by moving the detector with a step-motor. Tsuji *et al.* (1999a–c, 2001) have discussed this technique and its advantages in this book, but, in short, it offers a better detection of X-rays coming from very small particles or thin layers. An additional aspect regarding particle analysis is the specific detection of signals coming from the top layers of a particle only. Coated particles or particles with a core-shell structure could be analysed using GE-EPMA at negative exit-angles, if the particles are positioned at the edge of a tilted specimen substrate. By increasing the exit angle in small steps, the particle structure is then revealed layer by layer. Another option for the analysis of structured particles is dual-voltage EPMA (DV-EPMA) in which particles are analysed at two or more different voltages (e.g. 5 and 15 kV), using the change in information depth to study particle layers (Ro *et al.*, 2001a). This technique will be further discussed below in Section 7.5.4.

Regarding sample treatment during analysis, developments have been reported on the use of low-vacuum, environmental or variable pressure SEM (VP-SEM). The difference with SEM can be found in the specimen chamber, which is not under vacuum conditions, but contains a gas mixture in order to perform electron microscopy at higher pressure (Danilatos, 1994). The technology behind VP-SEM depends on a combination of differential pumping and an electron beam transfer system, allowing an electron beam to be formed in vacuum and being transferred to the specimen in the gas medium. The advantage of this technique lies in its possibility to study wet, volatile and even non-conducting samples, since the gas medium preserves the samples from evaporation, and it acts as a charge dissipating means which frees the sample from conductive coatings or chemical treatments. Since these advantages would be able to broaden the range of applications for X-ray analysis in electron microscopy, this technique appears to be very promising. However, most of the applications reported concern visual analysis in morphology studies (using special imaging detectors adapted to the gas medium), since X-ray analysis is still rather difficult. One reason is electron 'skirting', a process in which electrons transfer energy to the gas medium. The energy loss, which is difficult to predict or to quantify, affects all the electron interactions with the specimen. The lack of suitable standards is also reported to be a drawback for X-ray analysis in VP-SEM (Griffin *et al.*, 2000), and, therefore, one can only find examples of particle imaging in the literature. However, VP-SEMs now account for 50% of the market for non-field emission SEMs, so more is to be expected from this technique (Mohan *et al.*, 1998). A recent development, for example, is the combination with FEG-SEM by Kim and Lee (2001).

Another example of enhanced sample treatment can be found in cryogenic SEM (Cryo-SEM) in which the samples are cooled down to low temperatures using special sample stages (Gregory *et al.*, 1998). A stream of nitrogen gas flows through the stage, after it has been cooled down with liquid nitrogen ($-193\,°C$). This technique offers the possibility to study more volatile species, like ammonium sulfate and nitrate. In environmental analysis, these compounds are very important since their abundance in atmospheric particles is very high (Osán *et al.*, 2000; Szalóki *et al.*, 2001a). The reduction of the beam damage effects is reported to be very spectacular, since very small particles (down to $0.3\,\mu m$) can still be analysed. Quantification is still a problem, because it is almost impossible to prevent all particles from evaporating, and since this process is very difficult to predict. The results for several compounds have proven to be quite spectacular, and they will be discussed more in detail below.

Considering the developments mentioned above, one could conclude that different techniques are looking very promising. However, since they are mostly still under development, TW-EPMA with a thermionic electron source and the possibility for cryogenic cooling, offers the most stable, easily available technology for particle analysis over a broad range of sample types. For this reason, the work discussed in this subchapter, has been done using this combination.

7.5.3 QUANTIFICATION MODELS IN EPMA

7.5.3.1 ANALYTICAL MODELS

The ultimate goal in the development of models for quantitative EPMA is to provide an adequate macroscopic description of the basic interactions between the electron beam and the atoms in solid matter in order to solve micro-analytical tasks on various type of target samples: solid bulk materials having rough or flat polished surface, layered thin films, and micro-sized particles. The basic event taking place in the specimen can easily be described: the focused electron beam bombards the sample and the electrons lose a portion of their energy by means of an inelastic scattering process, which yields X-ray emission by de-excited atoms. Through elastic scattering the electrons change their original traveling directions and finally they lose their energy in inelastic events or they leave the sample volume. The characteristic and Bremsstrahlung radiations are partly absorbed by the sample mass, but a certain portion can propagate into the X-ray detector (ED or WD) after they leave the sample volume. The detected intensity of the characteristic X-rays depends on the excitation conditions i.e. the beam current and the electron energy, the sample size and the quantitative composition of the irradiated material. The geometrical shape and sizes of the sample also influence the X-ray flux effectively through absorption and secondary fluorescence excitation effects generated by primary characteristic and Bremsstrahlung radiation. Corresponding to this depiction, the physical kernel of the mathematical models consists of the calculation of: (i) the excitation conditions i.e. the ionization cross sections, the linear energy loss and the backscattering factor of the electrons; (ii) the fluorescence yield or the transition probability of the X-ray lines; and (iii) the absorption and fluorescence correction factors and the spectrometer efficiency function, which allow approximation of the theoretical value of the detected intensities. Based on the calculation mode of the X-ray intensities, the models can be classified into three main groups (Newbury, 1999): (i) classical ZAF correction procedures, where additional X-ray spectra originating from standard samples are required; (ii) standardless calculation procedures that need knowledge of the depth-distribution of the generated X-ray photons [$\phi(\rho z)$ function] for the analysed bulk sample; and (iii) MC simulations. The principal difficulty for empirical or semi-empirical quantitative EPMA is to correctly model the depth-dependence of this X-ray generation function (Brown, 1999). They are constructed on the basis of: (i) theoretical considerations (Pfeiffer et al., 1996); (ii) empirical models (Bastin et al., 1998; Staub, 1998; Bastin et al., 2001) using macroscopic physical parameters; or (iii) direct measurement of the $\phi(\rho z)$ function. Since EPMA is applied to a great variety of material types, sometimes the so-called second order effects (e.g. excitation by Bremsstrahlung) can become important in the numerical calculations of the sample composition (Cazaux, 1996; Pfeiffer et al., 1996). The computer codes written for iterative EPMA quantification sometimes allow the use of different matrix-correction and model parameters suitable for special experimental and sample conditions (Trincavelli et al., 1998; Trincavelli and Castellano, 1999).

7.5.3.2 MC SIMULATION MODELS

The main disadvantage of the conventional analytical models is that they are not flexible enough for various experimental and sample conditions, such as the analysis of microparticles having irregular shapes and heterogeneous compositions. MC simulation-based models are capable of handling any arbitrary geometry conditions (Hu and Pan, 2001) e.g. sample porosity (Sorbier et al., 2000) or rough surfaces (Gauvin and Lifshin, 2000) that influence the X-ray characteristic flux. Therefore, MC simulation can be applied for the analysis of both particles and flat bulk samples or layered films. Other secondary effects such as photon–atom interactions, i.e. photoelectric effect and continuum background generation, can also be involved in the model (Jbara et al., 1997). Using simulation models, unknown parameters such as the $\phi(\rho z)$ function at different X-ray energies, the k-ratio between films and bulk samples, the backscattering factor, etc. can be estimated for different compositions and elementary distributions. In single scattering models all elementary electron–atom collisions are calculated, while in multiple scattering models a great number of events are condensed into one occurrence requiring much shorter computer calculation times (Chan and Brown, 1997a, b). Considering all the possible interaction types, a more reliable assessment for complete detected X-ray spectra can be obtained if angle-dependent Bremsstrahlung simulations are included (Acosta et al., 1999). Such a complex simulation procedure allows the structure of different samples and the reliability of the theoretical model of elementary events used in the MC code to be studied. The fact that classical ZAF or $\phi(\rho z)$ procedures are not suitable for quantitative EPMA of inclusions embedded in a bulk matrix, motivated Hovington et al. (1997) to develop a complex simulation code, called CASINO, specially designed to describe the electron–solid interactions and to generate all the possible recorded electron and X-ray signals in the low energy range of electrons $0.1 < E < 15$ keV. For quantitative analysis of individual particles Ro et al. (1999) modified the original CASINO software in order to enable it to model different particle shapes i.e. spherical, hexahedral and hemispherical, and generate the detected X-ray intensities for a wide range of atomic numbers including low-Z elements ($5 < Z$), which act mostly as the major constitutions in environmental microparticles. They combined the analytical benefits of the simulation method with the extended detection capability of a thin-window Si(Li) detector which is capable of recording low-energy X-rays (0.2 keV $< E$) with reasonable efficiency. Linking this new hardware and software has yielded a flexible micro-analytical technique for quantitative study of aerosol and sediment single particles.

7.5.3.3 ITERATIVE APPROACH TO SOLVE REVERSE MC PROBLEM

For applying the mathematical-physical models (classical or simulated) in quantification problems, they have to be encapsulated into an appropriate iterative process to obtain the concentrations of constituent elements and other sample parameters. A typical example can be found in the literature (Wagner et al., 2001) in which the calculated X-ray flux emitted by embedded particles was calculated using an MC code and in which the size and depth of included particles was determined by means of chi-square minimization between experimental and theoretical values of X-ray intensities. Hu and Pan (2001) developed a reverse MC procedure for thin film analysis based on a limited iteration approach calculating quantitative composition of thin films layered on bulk substrate. The k values and the mass thickness of the films were determined experimentally by independent measurements, i.e. a priori information on the sample parameters was required. The extended CASINO and thin-window detection constitute the basis of a new reverse MC algorithm, which is represented by the EPPROC code designed by Szalóki et al. (2000) using a set of nonlinear equations of the mathematical relationship between the concentrations and the characteristic intensities of the particle elements:

$$I_{i,\text{meas}} = pI_{i,\text{sim}}(C_1, C_2, \ldots, C_n, d, \rho, \ldots)$$

$$i = 1, \ldots, n \text{ and } \sum_{i=1}^{n} C_i = 1 \quad (7.5.1)$$

where n is the number of elements in the sample, $I_{i,\text{meas}}$ and $I_{i,\text{sim}}(C_1, C_2, \ldots, C_n, d, \rho, \ldots)$ are the measured and the simulated intensities of the characteristic lines of the ith element, respectively. Terms $C_1, C_2, \ldots, C_n$ are the concentrations of the particle elements, factor p includes all proportional instrumental parameters (such as the solid angle of the detector or the flux of the excitation electron beam) and $d, \rho, \ldots$ represent the diameter, the density and all other properties of the particle. Although the form of this mathematical expression cannot be given explicitly, it is applicable to model the X-ray intensities from the particle. Practically, the aim of the reverse MC approximation is to find the exact solution of the set of equations (7.5.1). For this reason the following conditions are presumed: (i) the particle material is homogeneous; (ii) all the elements in the sample are detected apart from H, Li, Be and B; (iii) the particle is located on a flat surface having a known composition; and (iv) the shape of the analysed particle is spherical, hemispherical or hexahedral. Szalóki et al. (2000) proved the existence and uniqueness of the mathematical solution of Equation (7.5.1) for a general case, which can be obtained by successive approximation:

$$C_i^{(k+1)} = \frac{C_i^{(k)} I_{i,\text{meas}}}{I_{i,\text{sim}}^{(k)} \sum_{j=1}^{n} C_j^{(k)} I_{j,\text{meas}}/I_{j,\text{sim}}^{(k)}} \quad (7.5.2)$$

where $C_i^{(k+1)}$ is the $(k+1)$th approximation of the ith concentration value. This form contains the normalization of the concentration values based on the assumption that (practically) all the constituent elements are observed. The convergence speed of the numerical approximation depends strongly on the number of simulated electrons and the number of particle elements and the computer performance. Typical convergence of the calculation algorithm is demonstrated by a simple example of a CaSO$_4$ particle for the Ca concentration in Figure 7.5.1. The shape and size of the particle can be determined by means of SEM images, but these estimated parameters sometimes have large uncertainty (Osán *et al.*, 2000). For the estimation of the unknown particle density, the only available possibility is to approximate it by the intensities, to predetermine the type of the particle. In order to avoid the large uncertainties caused by the improper information of the density, the shape and sizes, the influence of the variation of these parameters on the final results of the successive approximations was investigated. Testing the present simulation on the analysis of standard particles showed that the concentration of the particle elements varies only in a relatively narrow range due to the average diameter of the particle as illustrated by Figure 7.5.1, which shows 5–10 % variation of the elements in CaSO$_4$ over the diameter range 400–2000 nm with 10 keV excitation electron energy. The detected X-ray flux is emitted mostly from the volume located directly under the particle surface because of the strong attenuation of the low-energy X-ray lines. An opposite relationship is expected between the particle diameter and the intensity of the substrate lines, because of the

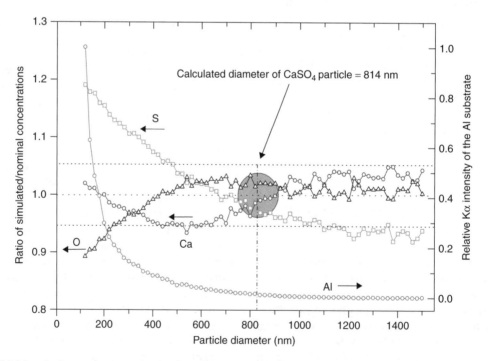

Figure 7.5.1 Result of successive approximation for a CaSO$_4$ standard particle. The left axis is the ratio of simulated and nominal concentration values as a function of the particle diameter; the particle density was chosen to be 2.71 g/cm^3. The right axis is the relative intensity of the Kα radiation emitted by the Al substrate

attenuation effect in the particle volume. Increasing the particle diameter results in a lower number of electrons in the substrate volume; therefore the emitted number of X-ray quanta of the substrate lines is decreased. Because a similar dependence of the intensities on the sample density was also recognized, the measured and simulated substrate intensities were introduced into the iterative algorithm in order to compensate for the improper determination and the uncertainty of the particle diameter and density. Because of the general insensitive behavior of these parameters, the diameter or the density can be varied during the iteration process in order to fit the calculated substrate intensity to the measured one yielding a more realistic result. In practice, after every iteration step of the concentration calculation, an optimization procedure for the diameter is performed based on the minimization of difference between calculated and measured substrate intensities.

7.5.3.4 DETECTION, SAMPLE HOLDER, BEAM DAMAGE EFFECT

Standardless quantitative X-ray spectrometry needs accurate knowledge of the detector efficiency function in the analysed range of X-ray energy. For the conventional semiconductor detectors one can give a simple mathematical expression:

$$\varepsilon(E) = \exp\left(-\frac{\mu_W d_W + \mu_{Si} d_{inSi} + \mu_{con} d_{con} + \mu_{ice} d_{ice}}{\sin\phi}\right)$$
$$\times \left[1 - \exp\left(-\frac{\mu_{Si} d_{acSi}}{\sin\phi}\right)\right] \quad (7.5.3)$$

where μ_W, μ_{Si}, μ_{con}, μ_{ice} and d_W, d_{inSi}, d_{con}, d_{ice}, d_{acSi} are the attenuation function and the thickness of different absorbing layers: window material (W), dead (inSi) and active layer (acSi) of the Si crystal and the conductive layer (con), respectively, and ϕ is the angle between the plane of the detector window and impinging X-ray beam (Szalóki et al., 2001b). For low-energy X-ray detection ($E < 1.5$ keV) the relatively thick (≈ 800–2500 nm) Be window of conventional Si(Li) detectors has to be replaced by materials which consists of other light elements such as H, C, O, N having higher transmission, e.g. very thin windows produced from organic polymers (≈ 200 nm). The exact composition, structure and thickness of the polymer windows are not widely published, however, they have a great importance in the efficiency calculation due to the very short attenuation length in the energy range 200 eV $< E < 1$ keV. To avoid problems due to the unknown parameters, Procop (1999) proposed an indirect method by means of experimental determination of absorption edge parameters for C and O elements involved in the detector window. The main constituent elements of the polymer window material should be H, C, N, O and a support grid or thin layer of metal (Be, Al, Ni, etc.) for opacity. Based on the discrete transmission data of the thin-window (published by the producer of the window), the weight fraction of the elements and the thickness of the absorbing layer can be estimated by an optimization procedure. Taking into consideration the possible kind of materials with different combinations of the possible compositions (such as Formvar, parylene or polyethylene), the optimal composition was found to be 8 % H, 71 % C, 21 % O and the thickness of the window was estimated to be approximately 0.202 μm (Osán et al., 2001a). The optimized window parameters yielded the transmission function for low-Z elements; the calculated curve is plotted in Figure 7.5.2 with values given by the window producer. The detection of low-Z elements (the main constituents of the environmental microparticles) raises new practical questions (Szalóki et al., 2001a) associated with the measurement conditions: the choice of substrate material, the optimization of instrumental conditions, the influence of beam damage effects to the analysis, the spectral and data analysis under automated investigation of huge number of particles, and the classification of particles with appropriate statistical procedure. The polymer foils (e.g. Nuclepore filters) that have been used as ideal substrate materials for elemental analysis in the atomic range of $Z > 12$ with computer-controlled EPMA cannot be used for light element analysis, because the main elements of these foils are carbon

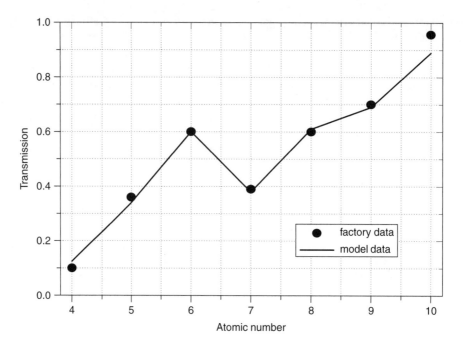

Figure 7.5.2 Fitted transmission model for super atmospheric thin-window of Si(Li) detector. Reproduced from Osán et al. (2001a) by permission of John Wiley & Sons Ltd

and oxygen. They also require conductive coating in order to avoid charging effects induced by electron beam, which is not desirable for low-Z element detection because of the intensive attenuation for low-energy X-rays in the coated layer located on the particle surface and because of the X-ray line interference with lines of the light elements in the particle. In order to minimize the interference from the substrate and the charging effect, metallic substrates are used for aerosol collection and EPMA measurement. Several metallic substrates were investigated, such as Be, Al, Si, and Ag (Szalóki et al., 2001a). If the particles contain the same chemical element as the substrate, the quantification of the chemical element becomes difficult. Sometimes the X-ray peaks from the substrate overlap with those of the chemical elements present in the particles. In that case, the quantitative analysis of the elements shows more uncertainty.

Based on the results reported by Szalóki et al. (2001a), the best substrate for TW-EPMA of individual particles was found to be Be, using low excitation electron accelerate voltages (5–10 keV) and beam currents of 0.5–1 nA. However, other substrate materials (Ag, Al, Si) provide more advantageous analytical conditions for automated analysis in some other aspects. Because the carbon coating on the particle surface disturbs the low-Z detection, the electrical conductivity of the substrate should be high. Better quality of contrast than for the inverse Be image was obtained with Ag substrate for low-Z elements, which resulted in a more reliable estimation of the particle shape and size. The intensive fluorescence peaks of the substrate often disturb the particle spectra, especially in case of the Ag L and M lines. However, in case of the Al K peaks additional information is provided for the iterative correction of the particle size.

The beam damage effect is a considerable difficulty in the electron bombardment of light element compounds, since many of them are very sensitive for the energy impact transferred by electron–atom collisions. In Figure 7.5.3, the beam sensitivity effect is visually illustrated by two secondary

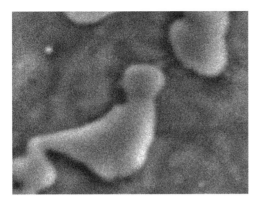 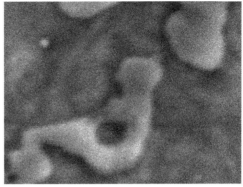

Figure 7.5.3 Visualization of beam damage effect in ammonium nitrate-type particle. Reproduced from Szalóki *et al.* (2001a) by permission of John Wiley & Sons Ltd. The pictures were recorded as SE images after EPMA at room temperature (≈295 K). The measuring time was 10 s, accelerating voltage 10 keV and beam current 0.5 nA

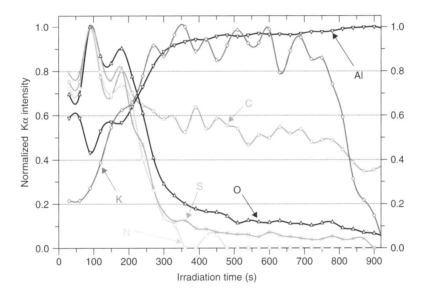

Figure 7.5.4 Experimental evidence is shown for selective beam sensitivity for different species in thin-window EPMA. Reproduced from Szalóki *et al.* (2001a) by permission of John Wiley & Sons Ltd. The O, N and S elements disappear after 250–300 s while K Kα intensity decreases only at 800 s. The higher carbon background intensity between 0.4 and 0.6 may be caused by external contamination originating from the vacuum system of the EPMA device. The scale of the vertical axis is the relative intensity of Kα lines emitted by elements of a volatile aerosol particle located on Al substrate versus irradiation time. The particle diameter is 1.5 μm. The series of individual measurements were carried out sequentially with 30 s recording time for each measurement at ca. ≈81 K sample temperature with 10 kV accelerating voltage and 1 nA beam current. Each set of intensities was normalized to their maximum values

electron (SE) images of an environmental type particle, successively recorded after 10 s irradiation duration at room temperature (∼295 K). This volatilization effect must be considered in quantitative TW-EPMA analysis, because the great variation in detected fluorescence intensity strongly influences the result of the iterative solution of Equation (7.5.1), as demonstrated in Figure 7.5.4. The required reduction of these effects can be achieved by optimizing the measuring conditions (Szalóki *et al.* 2001a), such as choosing the best substrate material, using liquid nitrogen temperature to cool down the sample, minimizing the energy impact on the particle by using low beam currents (0.5–1.0 nA) and as short as possible measuring times 10–50 s.

7.5.3.5 SPECTRUM AND DATA ANALYSIS, AUTOMATED EVALUATION

The main goal of particle analysis is to obtain as much information as possible on the abundance of different types of particles in the analysed sample. For this reason, huge numbers of individual particles must be measured and thus the spectral acquisition time devoted to each individual particle should be minimized. The most important characteristics of spectra collected from individual microparticles using TW-EPMA are the low counting statistics and the strong spectral overlaps under 1 keV. The so-called 'top-hat' filter method has been proven to be very useful in conventional computer-controlled EPMA for the on-line evaluation of spectra, usually having very poor counting statistics (Small, 1998). Because of the strong overlap of the characteristic lines, this method cannot be applied for TW-EPMA spectra under 1 keV. Also the curvature of the background at low energies may cause significant errors. A possible solution for the accurate determination of the characteristic X-ray intensities is the nonlinear least-squares fitting of the spectra. As the elements present in the particle must be included in the fitting model, an automatic evaluation can only be done if supervised by the operator. As peak distortion due to incomplete charge collection is most pronounced for low-energy X-ray lines, the usual Gaussian peak-shape model does not work well, and a tailing and step function must be included (Van Espen and Lemberge, 2000). The nonlinear least-squares fitting procedure (AXIL) was optimized for accurate modeling of the characteristic X-ray peaks and the Bremsstrahlung background below 1 keV (Osán et al., 2001a). The appropriate mass absorption coefficients were taken into account, and the accurate parameters of the detector window were used for modeling the continuum background and the relative abundance of the characteristic lines in the same line group. The spectral distortions caused by incomplete charge collection in the detector crystal were also modeled in the peak functions as tail and step functions. The incomplete charge collection is most pronounced at low energies, and can also cause the energy calibration to become nonlinear in the low-energy range (Joy et al., 1996). Therefore, small deviations from the assumed linear energy calibration were allowed during the fitting procedure. Figure 7.5.5 shows the fitting of a typical organic particle spectrum (algae) collected by TW-EPMA. The contribution of the tailing effect was found to be around 10 % of the total area of the characteristic peaks. The agreement between the fitted functions and the experimental data is good, yielding a quite low χ^2 value. Due to the relatively poor energy resolution of semiconductor energy-dispersive detectors, characteristic L lines of heavier elements can completely overlap with K lines of C, N and O. A possible method for handling strong spectral overlaps is the correction of K X-ray intensities of low-Z element using MC simulated L/K ratios for the overlapping L lines. The correction procedure is described in detail elsewhere (Osán et al., 2001a). The future application of the recently developed tunnel-junction or microcalorimeter detectors could simplify the handling of the ED X-ray spectra in the low-energy range, since their energy resolution is comparable to that of WD systems.

Another possibility for processing X-ray spectra with low energy resolution and poor counting statistics is the application of the partial least-squares (PLS) regression method that provides elemental concentration values without the need of determining the characteristic intensities (Lemberge et al., 2000). The PLS method basically consists of a multivariate linear relation between the concentrations of the constituents and the measured spectral data. An explicit evaluation of the spectrum is not required. The disadvantage of the method is that it requires a large number of standards covering all possible compositions expected in the unknowns. By using MC simulated spectra as standards, major

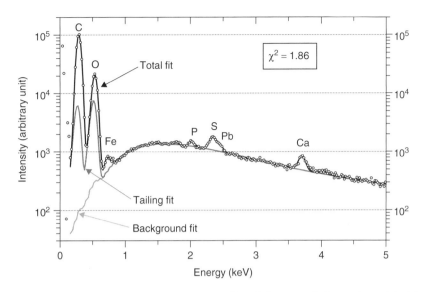

Figure 7.5.5 Least-squares fit of a low-energy X-ray spectrum collected from a typical organic particle (algae, IAEA-413). Reproduced from Osán *et al.* (2001a) by permission of John Wiley & Sons Ltd

elements in brass could be determined with a relative error below 10 % (Van Espen and Lemberge, 2000). After further optimization, the method can be advantageous for thin-window EPMA of individual particles, but the set-up of the simulated standard set requires a priori knowledge about the composition of the particles in the sample to be measured.

The automatic TW-EPMA measurement of a large number of particles in a sample produces a huge amount of spectral and morphological data. For appropriate interpretation of the measured data, quantitative chemical information should be derived from each particle individually. For this reason, the overall procedure for converting the spectral information to concentration values should work automatically if possible. Based on earlier results (Osán *et al.*, 2001a,b; Szalóki *et al.*, 2001a) the most workable procedure for automatic evaluation consists of the determination of the characteristic X-ray intensities by supervised least-squares spectral processing and unsupervised concentration calculation using the reverse MC method. The outline of the automatic evaluation is shown in Figure 7.5.6. Based on the calculated concentration data, chemometric methods should be applied for deriving the most important particle types in the sample. A detailed discussion of the applicability of cluster and principal component analysis for

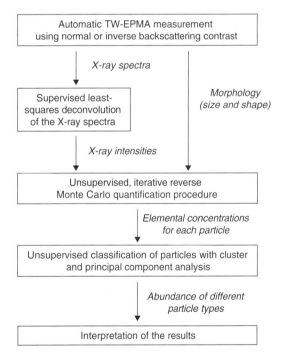

Figure 7.5.6 Outline of the automatic evaluation of the single-particle TW-EPMA data

low-Z EPMA data has been published (Osán et al., 2001b). The combination of TW-EPMA with a reverse MC quantification procedure yields environmentally more relevant particle classes than could be obtained using conventional computer-controlled EPMA. For example, ammonium sulfate and sulfur-containing organic particles can be distinguished, earlier they were classified together as 'sulfur-rich' particles (Osán et al., 2000). Also, the quantitative knowledge of the oxygen concentration makes it possible to distinguish sulfides and sulfates in sediment particles (Osán et al., 2002).

7.5.3.6 ANALYSIS OF STANDARD PARTICLES

Suitable validity for the complete quantitative analytical procedure developed for TW-EPMA is offered by analysis of standard particles of several chemical compounds prepared from pro analysis grade solid chemical compounds. In this test analysis the particles were suspended in n-hexane and dropped by micropipette onto the surface of the metallic foils and dried in air. The quantification method based on MC simulation combined with successive approximation was evaluated by comparing the weight concentrations obtained from EPMA measurements with their nominal weight concentrations. At least 10 independent analyses were performed for each type of standard particles with their diameters varying between 0.5 and 5 μm. The results are summarized as a histogram of relative errors, $\Delta = (C_c - C_n)/C_n$, based on the comparison of the nominal (C_n) and calculated (C_c) concentrations for $CaCO_3$, KNO_3, SiO_2, $NaCl$, $CaSO_4 \cdot 2H_2O$, $BaSO_4$, Fe_2O_3, $(NH_4)_2SO_4$ and NH_4NO_3. Figure 7.5.7 shows the comparison of the error distributions obtained for the reverse MC of particles including light elements and a first-principles standardless method of bulk samples (NIST DTSA X-ray microanalysis software engine; Newbury, 1999). In first-principles standardless analysis, the physical equations for electron-excited X-ray generation in the target, and X-ray absorption in the target and in other components of the spectrometer are used to predict the intensity from a pure element standard for each constituent under the analytical conditions in use (beam energy, beam incidence and X-ray take-off angles and detector solid angles). The sources of the relatively large relative error (95 % of the analyses are in the ±50 % interval) are the uncertainties in the physical parameters (ionization cross-section, mass absorption coefficients, fluorescence yield, detector parameters) and the fact that the sum of the concentrations must be forced to 100 %. When light elements are not analysed directly, i.e. they are not 'visible' and they are only taken into account based on stoichiometry, this normalization can introduce large errors. The distribution of errors for the reverse MC quantification for particle analysis (see Figure 7.5.7) shows that semi-quantitative analysis is possible for a wide range of chemicals. The analytical accuracy of the method is within 5 % relative, which is due to systematic errors for some of the compounds included (crystal water content can be different from the nominal value for hygroscopic particles, uncertainties of the physical parameters, etc.). More details of this study especially for Ag substrate were published (Ro et al., 2001b). The precision, or the statistical uncertainty, of the analytical results (95 % of the analyses fall in the interval of ±30 % relative error) is mainly due to the statistical error of the X-ray intensity determination originated from spectrum fitting procedure and the fluctuations in the shape and size parameters from the idealized models included in the simulation. As all the elements except hydrogen can be detected, the normalization of the concentrations to 100 % is supported. As can be seen in Figure 7.5.7, the performance of the reverse MC method is superior to that of other first-principles standardless methods, even though particles are analysed using a much worse counting statistics (10–20 s counting time vs 100–500 s).

7.5.3.7 CHEMICAL SPECIATION OF SINGLE PARTICLES, HETEROGENEITY

Since the technique can provide quantitative elemental concentrations, except hydrogen, of individual particles, the determination of chemical

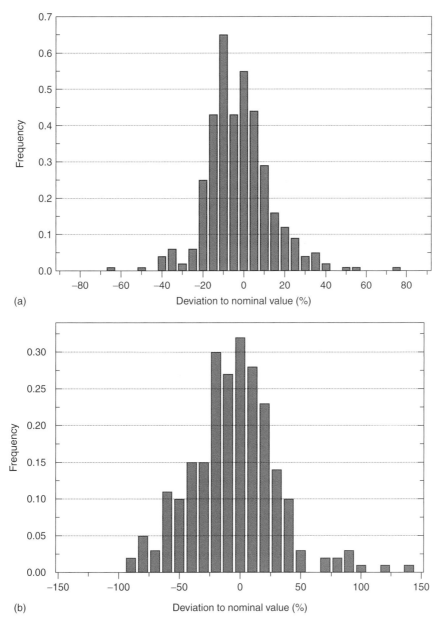

Figure 7.5.7 Distribution of errors with standardless analysis of a broad database of EDS spectra. (a) Reverse MC of particles including light elements (EPPROC/CASINO). (b) First-principles method of bulk samples. (NIST DTSA X-ray microanalysis software engine. Reproduced from Newbury (1999) by permission of Springer-Verlag

species in single particles is possible. This includes 'pure' particles containing only one major chemical species, and internally mixed particles containing two or more chemical species. Furthermore, in optimal cases even 'molar' concentrations of the different chemical species in internally mixed particles can be determined. Because many atmospheric particles contain only one or two chemical species, low-Z EPMA could provide more details about airborne particles of environmental interest.

Some examples and detailed description for this speciation method are presented elsewhere (Ro et al., 2000).

In addition, it is of primary importance to have an analytical tool to distinguish chemical species in the surface region from that of the core region in individual microparticles, because the analysis would allow the direct and more conclusive investigation of the nature of atmospheric reactions, which some airborne particles may experience. For example, sea salt can react with NO_x to produce sodium nitrate particles in the air. Also, the atmospheric reaction between soil particles and SO_x receives considerable attention in the atmospheric environment society. And thus, if gaseous or aqueous NO_x or SO_x species react with solid sea-salt or dust particles in air and if the atmospheric reactions are not completely finished, then it is expected that the product of the reactions would exist in the surface layer and the original solid species in the core region. Therefore, the existence of different atmospheric reactions would be directly proven if we could characterize both regions in individual particles. However, since the analysis volume of individual microparticles is quite small (pg range in mass), quantitative analysis of surface and core regions in individual particles has been a real challenge.

Recently, a methodology based on EDX-EPMA was developed that can analyse chemical species both in surface and core regions of individual particles (Ro et al., 2001a). The idea was to investigate heterogeneous individual particles with different primary electron beam energies, i.e. X-ray photons obtained with different primary electron beam energies carry information on the chemical compositions for different regions in the particles, mainly because of the different excitation volumes according to the energies of the primary electron beam. The excitation volume of the elements is decreased with the decrease of the primary electron beam energy. Figure 7.5.8 shows an example of electron trajectories for different primary electron beam energies. The model particle is a heterogeneous $CaCO_3$–$CaSO_4$ particle with 2 μm gross diameter, while the thickness of the $CaSO_4$ surface layer is 0.25 μm. It is clear that primary

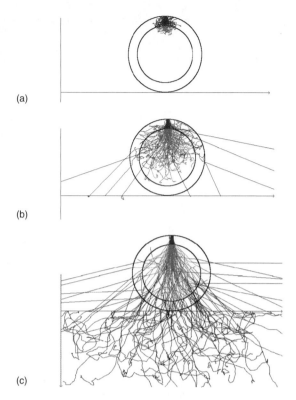

(a)

(b)

(c)

Figure 7.5.8 Simulation of electron trajectories in a heterogeneous $CaCO_3$–$CaSO_4$ spherical particle at (a) 5 kV, (b) 10 kV and (c) 20 kV accelerating voltages. Reproduced from Ro et al. (2001a) by permission of American Chemical Society

electron beams with different energies provide different electron probing regions. At 5 kV electron accelerating energy, the $CaSO_4$ surface region is mostly probed, whereas at 20 kV, many electrons get through the particle and the signal from the surface region is relatively suppressed.

Artificial heterogeneous $CaCO_3$–$CaSO_4$ particles were synthesized, i.e. particles with $CaSO_4$ in the surface region and $CaCO_3$ in the core. X-ray spectra, obtained at 5, 10, 15, and 20 kV electron-accelerating voltages for a heterogeneous $CaCO_3$–$CaSO_4$ spherical particle of 1.5 μm diameter, are shown in Figure 7.5.9. The measured characteristic X-ray intensities for the elements in the particle vary differently with the variation of primary electron beam energies. From the observation of different trends, for different elements, of

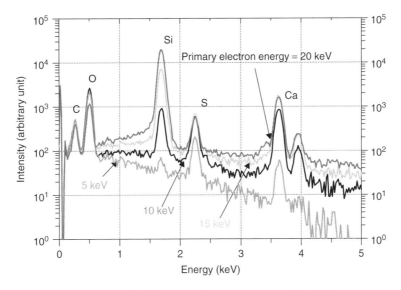

Figure 7.5.9 X-ray spectra obtained from an artificially generated heterogeneous $CaSO_4$–$CaSO_3$ particle at 5, 10, 15, and 20 kV accelerating voltages. Reproduced from Ro *et al.* (2001a) by permission of American Chemical Society

the characteristic X-ray intensity variation according to the variation of electron beam energies, these X-ray spectra certainly contain information on chemical species and heterogeneity of the particle. In Figure 7.5.10, simulated spectra calculated by our modified MC program are shown. The similarity between the simulated and experimental spectra is remarkably obvious. The MC calculation almost perfectly simulates the X-ray intensity variations for the elements according to the variation of the primary electron beam energies.

By the application of the MC calculation, even the thickness of the $CaSO_4$ surface region of the artificially generated $CaCO_3$–$CaSO_4$ particles can be determined. In Figure 7.5.11, ratios of simulated-to-measured intensities with the variation of the $CaSO_4$ surface thickness for a spherical $CaCO_3$–$CaSO_4$ particle of 1.5 μm diameter are shown for a 15 kV primary electron beam energy. For oxygen, the ratios of simulated-to-measured intensities are relatively constant with the variation of the $CaSO_4$ thickness, mainly because of their small compositional differences between the two chemical species. However, the ratios for carbon and sulfur between the simulated and measured intensities are strongly dependent on the thickness of the surface $CaSO_4$ region. Furthermore, the ratios for sulfur decrease as the thickness of $CaSO_4$ region decreases, whereas the ratios for carbon increases as the thickness of the $CaSO_4$ region decreases. For the heterogeneous $CaCO_3$–$CaSO_4$ particles, the sulfur species is in the surface region and carbon is in the core region. Therefore, if the assumed $CaSO_4$ surface thickness for the MC calculation is thicker than the real one, then the calculated intensities are larger than the measured ones for sulfur, whereas they are smaller for carbon. From the result in Figure 7.5.11, the good match between the simulated and measured data is in the range of 160–200 nm thickness of the surface region.

The validity of this technique is investigated more systemically using a more controlled system, i.e. soda-lime glass particles (SPI #2716) coated with carbon by evaporation. The particles were coated with carbon layers consecutively four times, using the standard process for scanning electron microscopes. The estimated carbon thickness for different numbers of carbon layers obtained by the proposed method is tabulated in Table 7.5.1. The obtained results show good agreement between the values obtained at different accelerating voltages, supporting the applicability of the proposed method. Further research is needed

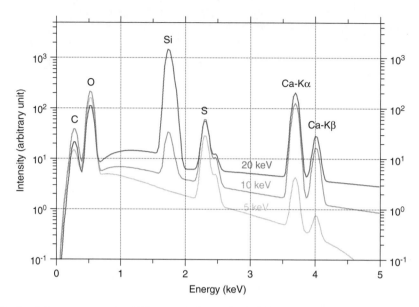

Figure 7.5.10 Simulated X-ray spectra using MC approach for a spherical, heterogeneous $CaCO_3$–$CaSO_4$ particle at different accelerating voltages. Overall size of the particle is ≈1.5 μm in diameter and surface thickness is 0.18 μm. Reproduced from Ro et al. (2001a) by permission of American Chemical Society

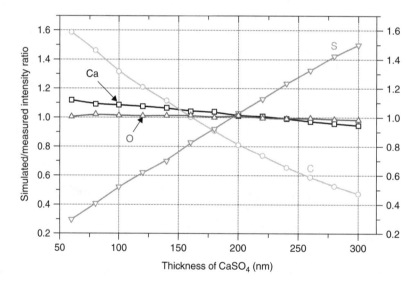

Figure 7.5.11 Dependence of simulated/measured X-ray intensity ratios on $CaSO_4$ surface thickness, at 15 kV acceleration voltages, for a heterogeneous $CaCO_3$–$CaSO_4$ particle. Reproduced from Ro et al. (2001a) by permission of American Chemical Society

to evaluate the accuracy and reproducibility of this approach using well-defined heterogeneous particles and also to optimize the iteration procedure for automatic evaluation of the surface layer thickness of heterogeneous particles. Furthermore, for real atmospheric particles, another complexity is involved in the analysis; the elemental concentrations of chemical compositions as well as the structure of the heterogeneity are not known a priori. It is necessary to find a way to extract information

Table 7.5.1 Estimated carbon thickness for carbon-coated glass particles

Accelerating voltage (kV)	Number of carbon layer depositions			
	1	2	3	4
	Estimated carbon thickness (nm)			
5	35	65	105	135
10	35	70	95	145
15	30	65	95	150
20	25	55	85	140

both on chemical compositions and the heterogeneity of the atmospheric particles from their X-ray spectral data.

7.5.4 APPLICATIONS

7.5.4.1 ENVIRONMENTAL PARTICLES, EXAMPLES

Characterization of Water-insoluble Components of Asian Dust

Low-Z EPMA has been applied to characterize the water-insoluble part of 'Asian Dust' deposited by washout in the form of rainwater during an Asian Dust storm event and collected in Seoul, Korea (Ro *et al.*, 2001b). In addition to the Asian Dust sample, China Loess particles collected in the loess layer in Gansu Province of China and a local soil particle sample collected in the backyard of the Korea Meteorological Administration in Seoul were analysed. In Figure 7.5.12, the observed frequencies of four major water-insoluble chemical species, e.g. aluminosilicates, carbonaceous, $CaCO_3$, and SiO_2 species, in the three samples, are shown. The frequencies are calculated for particles, which contain those species either as single species or mixtures. Particles containing aluminosilicate species are the most abundant, and carbonaceous species the next. The abundance of SiO_2 containing particles is similar between the samples. However, it is quite different for $CaCO_3$ species; the local soil does not have the $CaCO_3$ species, but the Asian Dust and China Loess do. Even though the abundance of particles containing aluminosilicate species is similar between the samples, there are some differences in chemical composition between particle samples originating from China (the Asian Dust and China Loess samples) and the local soil sample. Figure 7.5.13 shows the frequencies of the elements most frequently encountered in aluminosilicate-containing particles. In Asian Dust and China Loess samples,

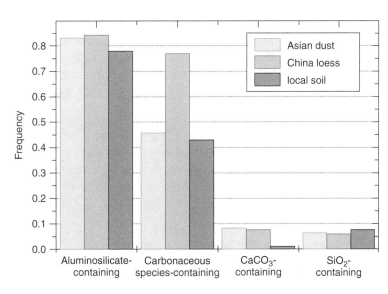

Figure 7.5.12 Frequencies of four major chemical species observed in Asian Dust, China Loess and local soil particle samples

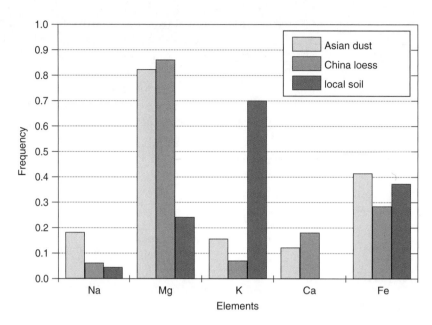

Figure 7.5.13 Frequencies of major elements observed in aluminosilicate-containing particles of Asian Dust, China Loess and local soil particle samples

those aluminosilicate-containing particles also contain almost always Mg. In contrast, K is the most frequently encountered element in aluminosilicate-containing particles of the local soil sample. It implies that the Asian Dust and China Loess samples contain mostly Mg-enriched aluminosilicate minerals, whereas the local soil sample contains K-enriched minerals. Another difference in chemical composition of those aluminosilicate-containing particles is the absence of Ca in the local soil sample. This example demonstrates that the single particle analysis using the low-Z EPMA can provide detailed information on various types of chemical species in the samples and clearly distinguish the particle samples with different sources based on their chemical compositions.

Direct Proof of Nitrate Formation from Sea-salts

The low-Z EPMA was applied for the analysis of aerosol particles collected on 19 June, 1999 at Cheju Island, Korea (Ro *et al.*, 2001c). Cheju Island is an ideal place to study continental and marine influences on aerosols, because it is surrounded by the Korean peninsula, mainland China, Japan and the Yellow Sea and it is also one of the cleanest areas in Korea. Overall, 2888 particles were analysed and one of the most abundant particle types is marine-originated. Using low-Z EPMA, several different types of marine-originated particles were identified, i.e. 'genuine' sea salt particles such as NaCl-containing ones and 'reacted sea salt' particles such as $NaNO_3$- and Na_2SO_4-containing ones. Since low-Z EPMA provides quantitative elemental concentration data for single particles, we could distinguish two different 'genuine' sea-salts (i.e. NaCl particles and particles with NaCl and $MgCl_2$) and also two different nitrate particles (i.e. $NaNO_3$ particles and particles with $NaNO_3$ and $Mg(NO_3)_2$). There is a good possibility that those nitrate and chloride particles which contain only Na, also contain Mg at trace level, because detection limits of ED-EPMA are in the range of 0.1–1 % by weight. However, the average atomic concentrations of Na and Mg elements in all 338 particles containing both $NaNO_3$ and $Mg(NO_3)_2$ species are 14.8 % and 1.9 %, respectively. In contrast, those of Na and

Mg elements in all the NaNO$_3$ particles are 19.8 % and 0.3 %, respectively. No particle identified as NaNO$_3$ and NaCl species contains a Mg content larger than 1 % in atomic fraction. In other words, two types of sea salts were indeed observed; with and without Mg. We do not know exactly why two different types of nitrate and chloride particles were observed, and yet this result shows the powerful applicability of the low-Z EPMA technique for the elucidation of chemical species of aerosol particles.

There are some strong implications that sea salts had reacted with other chemical species before the sampling. For example, the chance to observe 'genuine' sea salt particles is relatively very small (3.1 %), compared to 'reacted sea salt' particles, e.g. NaNO$_3$- and Na$_2$SO$_4$-containing ones (30.2 %). Also, internally mixed particles with NaNO$_3$ and Mg(NO$_3$)$_2$ are abundant (11.9 %). In addition, particles containing Na$_2$SO$_4$ and/or MgSO$_4$ species are also observed (4.6 %). We also observed a significant number of Na- and Mg-containing particles with Cl as well as other chemical species, e.g. (Na, Mg)(NO$_3$, Cl), (Na, Mg)(SO$_4$, Cl) and (Na, Mg)(NO$_3$, SO$_4$, Cl) types, implying that the reactions between sea salts and the other species were not complete so that the particles still have some remnant Cl in them. Most importantly, we calculated the ratio of atomic concentrations between Na and Mg in all the mixture particles containing NaNO$_3$ and Mg(NO$_3$)$_2$, i.e. 338 particles. The ratio is 0.128 with a standard deviation of 0.047, which is remarkably similar to that of seawater (0.122). This result provides strong evidence for the original source of those particles; they were from the sea and reacted with HNO$_3$ in air. The result based on quantitative analysis on Na and Mg concentration in nitrate particles might be the most direct proof on nitrate formation from sea-salt up to now.

Identification of Mine Pollution Particles in River Sediment

Natural water ecosystems are very vulnerable to sedimentary heavy metal pollution. Especially the Tisza and Szamos rivers located in eastern Hungary are frequently polluted mostly due to mining activities in the catchment area. The speciation of heavy metals is necessary for estimating the environmental mobility and bioavailability of these elements. Below, the capability of TW-EPMA for the heavy metal speciation in sediment particles is demonstrated. In March 2000, just a month after the world-widely discussed catastrophic cyanide pollution event, a thousand tons of heavy-metal containing sludge reached the surviving part of the Tisza River due to a tailings dam failure at Baia Borsa, Romania. Sediment samples were collected from the main riverbed of the Hungarian section of Tisza at six sites, a week after the pollution event. All of the collected samples were subjected to bulk XRF prior to the TW-EPMA measurements, and this technique showed that some of the samples collected at the river section between 200–220 km from the pollution site, contained Cu, Zn and Pb at elevated concentrations. The observed concentrations were an order of magnitude higher than the background level for river and lake sediments in Hungary (Szalóki et al., 1999; Weisz et al., 2000). The average population of particles in the sediment samples was investigated using single-particle TW-EPMA, on samples prepared on silver foil using the inverse BSE contrast. As the determination of carbon and oxygen is feasible with this method, the most abundant particle type connected with the mine pollution could be clearly identified as pyrite. Internally mixed particles consisting of aluminosilicates (natural constituents of sediment) and pyrite, zinc sulfide and lead oxide were also detected. The abundance of the pollution particles was below 10 % by number for the most polluted samples. In order to exclusively investigate the heavy-metal containing particles, one sediment sample showing the highest pollution was prepared on silicon wafer for further analysis. Using the normal BSE contrast, only particles having significantly higher average atomic number than that of silicon were selected for automatic measurement. Although a 10 keV electron excitation energy was used, heavy metals could be identified and determined using L and M X-ray lines. Using this measurement setup, pyrite, zinc sulfide, iron–copper sulfide, lead oxide and lead sulfide as well as rare earth (La, Ce, Sn) oxide particle types could be distinguished. Heavy metals were also found to be connected to the aluminosilicate phase in some of the particles. The small size (<2 µm) and the composition of the heavy-metal containing particles in the polluted Tisza sediment indicate their mine tailings origin. Other details on the combination of TW-EPMA and bulk XRF measurements for the identification of mine pollution particles can be found elsewhere (Osán et al., 2002).

REFERENCES

Acosta, E., Llovet, X., Coleoni, E., Salvat, F. and Riveros, J. A. Simulation of X-ray spectra generated by electron impact on solids. *X-Ray Spectrom.*, **28**, 121–127 (1999).

Bastin, G. F., Dijkstra, J. M., Heijligers, H. J. M. PROZA96: an improved matrix correction program for electron probe microanalysis, based on a double gaussian $\varphi(\rho z)$ approach. *X-ray Spectrom.*, **27**, 3–10 (1998).

Bastin, G. F., Dijkstra, J. M., Heijligers, H. J. M. An experimental approach to the determination of the surface ionisation $\varphi(0)$ in electron probe microanalysis. *X-ray Spectrom.*, **30**, 216–229 (2001).

Brown, D. B. The application of depth distribution, $\varphi(\rho z)$ curves to quantitative electron probe microanalysis. *J. Anal. At. Spectrom.*, **14**, 475–478 (1999).

Cazaux, J. Electron probe microanalysis of insulating materials: quantification problems and some possible solutions. *X-ray Spectrom.*, **25**, 265–280 (1996).

Chan, A. and Brown, J. D. Modification to a multiple scattering Monte Carlo model to predict X-ray generation curves. *X-Ray Spectrom.*, **26**, 275–278 (1997a).

Chan, A. and Brown, J. D. Simulation and modelling of thin film $\phi(\rho z)$ curves for electron probe microanalysis. *X-Ray Spectrom.*, **26**, 279–290 (1997b).

Danilatos, G. D. Environmental scanning electron microscopy and microanalysis. *Mikrochim. Acta*, **114–115**, 143–155 (1994).

Diebold, A. C., Lindley, P., Viterralli, J., Kingsley, J., Liu, B. Y. H. and Woo, K.-S. Comparison of the submicron particles analysis capabilities of Auger electron microscopy, time-of-flight secondary ion mass spectrometry, and scanning electron microscopy with energy dispersive X-ray spectroscopy for particles deposited on silicon wafers with 1 µm thick oxide layers. *J. Vac. Sci. Technol.*, **A16**, 1825–1831 (1998).

Diebold, A. C., Wollman, D. A., Hilton, G. C., Irwin, K. D., Martinis, J. M. and Liu, B. Y. H. Applications of microcalorimeter EDS X-ray detectors to particle analysis. *Solid State Phenomena*, **65–66**, 199–202 (1999).

Gauvin, R. and Lifshin, E. Simulation of X-ray emission from rough surfaces. *Mikrochim. Acta*, **132**, 201–204 (2000).

Gregory, C. L., Nullens, H. A., Gijbels, R. H., Van Espen, P. J., Geuens, I. and De Keyzer, R. Automated particle analysis of populations of silver halide microcrystals by electron probe microanalysis under cryogenic conditions. *Anal. Chem.*, **70**, 2551–2559 (1998).

Griffin, B. J., Nockolds, C. E., Philips, M. R. and Remond, G. New needs for imaging and X-ray microanalysis standards: ESEM, CHIME and low voltage microanalysis. *Inst. Phys. Conf.*, **165**, 395 (2000).

Höflich, B. L. W., Wentzel, M., Ortner, H. M., Weinbruch, S., Skogstad, A., Hetland, S., Thomasssen, Y. and Chaschin, V. P. Chemical composition of individual aerosol particles from working areas in a nickel refinery. *J. Environ. Monit.*, **2**, 213–217 (2000).

Hovington, P., Drouin, D. and Gauvin, R. CASINO: A new Monte Carlo code in C language for electron beam interaction – Part I: Description of the program. *Scanning*, **19**, 1–14 (1997).

Hu, Y. and Pan, Y. Method for the calculation of the chemical composition of a thin film by Monte Carlo simulation and electron probe microanalysis. *X-Ray Spectrom.*, **30**, 110–115 (2001).

Jbara, O., Portron, B., Mouze, D., Cazaux, J. Electron probe microanalysis of insulating oxides: Monte Carlo simulations. *X-Ray Spectrom.*, **26**, 291–302 (1997).

Jones, T. P., Williamson, B. J., Bérubé, K. A. and Richards, R. J. Microscopy and chemistry of particles collected on TEOM filters: Swansea, south Wales, 1998–1999. *Atmos. Environ.*, **35**, 3573–3583 (2000).

Joy, D. C., Joy, C. S. and Bunn, R. D. Measuring the performance of scanning electron microscope detectors. *Scanning*, **18**, 533–538 (1996).

Kim, G. M. and Lee, D. H. FEG-SEM investigation of micromechanical deformation processes in ultrafine monospherical SiO_2 particle-filled polymer composites. *J. Appl. Polym. Sc.*, **82**, 785–789 (2001).

Laskin, A. and Cowin, J. P. Automated single-particle SEM/EDX analysis of submicrometer particles down to 0.1 µm. *Anal. Chem.*, **73**, 1023–1029 (2001).

Lemberge, P., Van Espen, P. J. and Vrebos, B. A. R. Analysis of cement using low-resolution energy-dispersive X-ray fluorescence and partial least-squares regression. *X-Ray Spectrom.*, **29**, 297–304 (2000).

Mohan, A., Khanna, N., Hwu, J. and Joy, D. C. Secondary electron imaging in the variable pressure scanning electron microscope. *Scanning*, **20**, 436–441 (1998).

Newbury, D. E. Standardless quantitative electron-excited X-ray microanalysis by energy dispersive spectrometry: what is its proper role? *Microsc. Microanal.*, **4**, 586–597 (1999).

Newbury, D. E. Measures for spectral quality in low-voltage X-ray microanalysis. *Scanning*, **22**, 345–351 (2000).

Newbury, D., Wollman, D., Irwin, K., Hilton, G. and Martinis, J. Lowering the limit of detection in high spatial resolution electron beam microanalysis with the microcalorimeter energy dispersive X-ray spectrometer. *Ultramicroscopy*, **28**, 73–88 (1999).

Ortner, H. M. Sampling and characterization of individual particles in occupational health studies. *J. Environ. Monit.*, **1**, 273–283 (1999).

Ortner, H. M., Hoffmann, P., Stadermann, F. J., Weinbruch, S. and Wentzel, M. Chemical characterization of environmental and industrial particulate samples. *Analyst*, **123**, 833–842 (1998).

Osán, J., Szalóki, I., Ro, C.-U. and Van Grieken, R. Light element analysis of individual microparticles using thin-window EPMA. *Mikrochim. Acta*, **132**, 349–355 (2000).

Osán, J., de Hoog, J., Van Espen, P., Szalóki, I., Ro,C.-U. and Van Grieken, R. Evaluation of energy-dispersive X-ray spectra of low-Z elements from electron-probe microanalysis of individual particles. *X-Ray Spectrom.*, **30**, 419–426 (2001a).

Osán, J., de Hoog, J., Worobiec, A., Ro, C.-U., Oh, K.-Y., Szalóki, I. and Van Grieken, R. Application of chemometric methods for classification of atmospheric particles based on thin-window electron probe microanalysis data. *Anal. Chim. Acta*, **446**, 211–222 (2001b).

Osán, J., Kuruncri, S., Török, S. and Van Grieken, R. X-ray analysis of river sediment of the Tisza (Hungary): identification of particles from a mine pollution event. *Spectrochim. Acta*, **B57**, 413–422 (2002).

Paoletti, L., Diociaiuti, M., De Berardis, B., Santucci, S., Lozzi, L. and Picozzi, P. Characterization of aerosol individual particles in a controlled underground area. *Atmos. Environ.*, **33**, 3603–3611 (1999).

Pfeiffer, A., Schiebl, C. and Wernisch, J. Continuous fluorescence correction in electron probe microanalysis applying an electron scattering model. *X-ray Spectrom.*, **25**, 131–137 (1996).

Procop, M. Estimation of absorbing layer thickness for an Si(Li) detector. *X-Ray Spectrom.*, **28**, 33–40 (1999).

Ro, C.-U., Osán, J. and Van Grieken, R. Determination of low-Z elements in individual environmental particles using windowless EPMA. *Anal. Chem.*, **71**, 1521–1528 (1999).

Ro, C.-U., Osán, J., Szalóki, I., Oh, K.-Y., Kim, H. and Van Grieken, R. E. Determination of chemical species in individual aerosol particles using ultra-thin window EPMA. *Environ. Sci. Technol.*, **34**, 3023–3030 (2000).

Ro, C.-U., Oh, K.-Y., Osán, J., de Hoog, J., Worobiec, A. and Van Grieken, R. Heterogeneity assessment in individual $CaCO_3$–$CaSO_4$ particles using ultra thin window electron probe X-ray microanalysis. *Anal. Chem.*, **73**, 4574–4583 (2001a).

Ro, C.-U., Oh, K.-U., Kim, H., Chun, Y., Osán, J., de Hoog, J. and Van Grieken, R. E. Chemical speciation of individual atmospheric particles using low-Z electron probe X-ray microanalysis: characterizing 'Asian Dust' deposited with rainwater in Seoul, Korea. *Atmos. Environ.*, **35**, 4995–5005 (2001b).

Ro, C.-U., Oh, K.-Y., Kim, H., Kim, Y. P., Lee, C. B., Kim, K.-H., Osan, J., de Hoog, J., Worobiec, A. and Van Grieken, R. Single particle analysis of aerosols at Cheju Island, Korea, using low-Z electron probe X-ray microanalysis: a direct proof of nitrate formation from sea-salts. *Environ. Sci. Technol.*, **35**, 4487–4494 (2001c).

Roth, B. and Okada, K. On the modification of sea-salt particles in the coastal atmosphere. *Atmos. Environ.*, **32**, 1555–1569 (1998)

Scott, V. D. and Love, G. EPMA using low-energy (0.1–1.0 keV) X-rays – an historical perspective. *J. Anal. At. Spectrom.*, **14**, 367–376 (1999).

Small, J. A. The detection of low-intensity peaks in energy-dispersive X-ray spectra from particles. *Scanning*, **20**, 92–98 (1998).

Sorbier, L., Rosenberg, E., Merlet, C. and Llovet, X. EPMA of porous media: a Monte Carlo approach. *Mikrochim. Acta*, **132**, 189–199 (2000).

Staub, P. F. IntriX: a numerical model for electron probe analysis at high depth resolution. Part I. Theoretical description. *X-ray Spectrom.*, **27**, 43–57 (1998).

Szalóki, I., Somogyi, A., Braun, M. and Tóth, A. Investigation of geochemical composition of lake sediments using ED-XRF and ICP-AES techniques. *X-Ray Spectrom.*, **28**, 399–405 (1999).

Szalóki, I., Osán, J., Ro, C.-U., Van Grieken, R. Quantitative characterisation of individual aerosol particles by thin-window electron probe microanalysis combined with iterative simulation. *Spectrochim. Acta B*, **55**, 1015–1028 (2000).

Szalóki, I., Osán, J., Worobiec, A., de Hoog, J. and Van Grieken, R. Optimization of experimental conditions of thin-window EPMA for light element analysis of individual environmental particles. *X-Ray Spectrom.*, **30**, 143–155 (2001a).

Szalóki, I., Szegedi, S., Varga, K., Braun, M., Osán, J. and Van Grieken, R. Efficiency calibration of energy dispersive detectors for application in quantitative X- and γ-ray spectrometry. *X-Ray Spectrom.*, **30**, 49–55 (2001b).

Trincavelli, J. and Castellano, G. MULTI: an interactive program for quantification in EPMA. *X-ray Spectrom.*, **28**, 194–197 (1999).

Trincavelli, J., Castellano, G. and Riveros, J. A. Model for the Bremsstrahlung spectrum in EPMA application to standardless quantification. *X-Ray Spectrom.*, **27**, 27–86 (1998).

Tsuji, K., Spolnik, Z., Wagatsuma, K., Zhang, J. and Van Grieken, R. E. Enhancement of electron-induced X-ray intensity for single particles under grazing-exit conditions. *Spectrochim. Acta*, **B54**, 1243–1251 (1999a).

Tsuji, K., Wagatsuma, K., Nullens, R. and Van Grieken, R. Grazing exit electron probe microanalysis for surface and particle analysis. *Anal. Chem.*, **71**, 2497–2501 (1999b).

Tsuji, K., Nullens, R., Wagatsuma, K. and Van Grieken, R. Elemental X-ray images obtained by grazing-exit electron probe microanalysis (GE-EPMA). *J. Anal. At. Spectrom.*, **14**, 1711–1713 (1999c).

Tsuji, K., Murakami, Y., Wagatsuma, K. and Love, G. Surface studies by grazin-exit electron probe microanalysis (GE-EPMA). *X-Ray Spectrom.*, **30**, 123–126 (2001).

Van Espen, P. and Lemberge, P. ED-XRF spectrum evaluation and quantitative analysis using multivariate and nonlinear techniques. *Adv. X-Ray Anal.*, **43**, 560–569 (2000).

Wagner, H. W., Werner, W. S. M., Störi, H. and Richardson, L. M. Electron probe microanalysis inverse modeling. *Nucl. Instrum. Methods*, **B184**, 450–457 (2001).

Watanabe, M. and Williams, D. B. Atomic level detection by X-ray microanalysis in the analytical electron microscope. *Ultramicroscopy*, **78**, 89–101 (1999).

Weisz, M., Polyák, K. and Hlavay, J. Fractionation of elements in sediment samples collected in rivers and harbors at Lake Balaton and its catchment area. *Microchem. J.*, **67**, 207–217 (2000).

Wollman, D. A., Irwin, K. D., Hilton, G. C., Dulce, L. L., Newbury, D. E. and Martinis, J. M. High-resolution, energy-dispersive microcalorimeter spectrometer for X-ray analysis. *J. Microscopy*, **188**, 196–223 (1997).

Wollman, D. A., Hilton, G. C., Irwin, K. D., Dulcie, L. L., Bergren, N. F., Newbury, D. E., Woo, K.-S., Liu, B. Y. H., Diebold, A. C. and Martinis, J. M. High-resolution microcalorimeter energy-dispersive spectrometer for X-ray microanalysis and particle analysis, in *Characterization and Metrology for ULSI Technology* (Eds D. G. Seiter, A. C. Diebold, W. M. Bullis, T. J. Shaffner, R. McDonald and E. J. Walters), pp. 799–804 (1998).

Wollman, D. A., Nam, S. W., Hilton, G. C., Irwin, K. D., Bergren, N. F., Rudman, D. A., Martinis, J. M. and Newbury, D. E. Microcalorimeter energy-dispersive spectrometry using a low voltage scanning electron microscope. *J. Microscopy*, **19**, 37–44 (2000a).

Wollman, D. A., Nam, S. W., Newbury, D. E., Hilton, G. C., Irwin, K. D., Bergren, N. F., Deiker, S., Rudman, D. A. and Martinis, J. M. Superconducting transition-edge microcalorimeter X-ray spectrometer with 2 eV energy resolution at 1.5 keV. *Nucl. Instrum. Methods*, **A444**, 145–150 (2000b).

Index

Active pixel sensors (APS), 7, 181–190
   and DEPFET detector system, 183
   sideward depletion, 149
   XEUS mission, 181–182
   see also pixel detectors
ADF imaging, 390
Algebraic reconstruction technique (ART), 124–125
Annular dark field (ADF) imaging, 390
APS see Active pixel sensors
APXS technique, 338–339
Archaeometry, 11, 157–158, 533–552
   analytical techniques overview, 533
   EDXRF analysis, 329–334
   PIXE and micro-PIXE applications, 547–548
      terracotta sculptures, 547–548
   portable equipment, 329–334, 534–536
      detectors, 535–536
      energy dispersive spectrometers, 535
   radiation damage, 544–545
      glazes, 545–546
      safe limits, 546
      with X-ray primary beams, 546–547
   synchrotron radiation, 536–538, 548–549
      Egyptian cosmetic analysis, 548–549
      glass corrosion, 549
      micro-mapping, 549
      SR-XRD, 538, 548
      SR-XRF microprobe, 536–538
   texture analysis, 538–539
      beam energies, 540–542
      external microbeams, 542–543
      gold alloys (*tumbaga*), 540–542
      micro-mapping, 543–544
      paint layers, 539–540, 541
      PIXE depth profiling, 539–540
      proton penetration, 539–540
ART see Algebraic reconstruction technique
Atomic scattering factor, 64
Auger electron, 391–392, 414, 415

Bolometer see Cryogenic microcalorimeters

Cadmium
   in fly ashes, 126–129
      concentration levels, 127, 128
      'hot spots', 127, 128
      micro-XAS spectra, 128
      micro-XRF maps, 128
   medical detection, 502, 503, 505, 508, 509
CCDs see Charge-coupled detectors
CDDs see Controlled drift detectors
CFEG see Cold field emission gun
Charge-coupled detectors (CCDs), 133, 137, 140–142
   CAST experiment, 180
   charge transfer efficiency (CTE), 172
   electron emission channelling spectroscopy, 179
   noise, 170–171
   performance, 169–175
   plasma diagnostics, 178–179
   quantum efficiency, 172–173
   quantum optics, 179
   sideward depletion, 141–142, 149
   three-phase MOS, 140–141
   transition radiation, 180
   X-ray microscopy, 178
   XMM satellite, 145–146, 172, 173, 174–175
   see also pn-CCD
Chemical analysis, 2–6
Chemical mapping, 398
CMT, 350–351
Cold field emission gun (CFEG), 389
Collimating optics, 97–98
   divergent angle, 97–98, 99
   intensity gain, 97
Compound refractive lenses see Refractive X-ray lenses
Compton scattering, 10, 254, 441, 443
Computerisation methods, 435–485
   Monte Carlo simulation, 435–461
   spectrum evaluation, 463–485
Computerised microtomography (CMT), 350–351
Controlled drift detectors (CDDs), 7, 137, 163–166
   applications, 164–166
   trigger signal, 163–164
   see also Drift detectors
Cryogenic microcalorimeters, 8, 229–245, 572
   absorber, 230–231
      metals, 230–231
      semiconductors, 231
      superconductors, 231

Cryogenic microcalorimeters (*continued*)
   collecting area, 235–236
      arrays, 235–236, 239
   count rate, 233–235
      time constant, 234, 235
   detector microfabrication, 237–238
   electrothermal feedback, 234–235, 242
   energy resolution, 232–233
      determining factors, 233
   future trends, 241–242
   imaging detectors, 239–240
   lithography, 238
   multiplexers, 238–239
   noise, 232–233, 237, 241–242
   non-ideal effects, 236–237, 240
      absorber decoupling, 236
      compensating mechanisms, 240
      hot-electron effect, 236, 242
      non-ohmic behavior, 236–237
   operating principles, 230
   position sensitive detectors, 239–240
   quantum efficiency, 235
   sensor, 231–232, 240–241
      magnetic, 232, 240, 241
      non-resistive, 232, 240
      thermistor, 231–232
      transistor edge sensors (TES), 231, 232, 236–237
'Cumulative Centre' line, 18, 19

DAFS, 422
Dentistry, 489, 492–493
DEPFET *see* Depleted p-channel field effect transistor
Depleted p-channel field effect transistor (DEPFET), 137, 182–189
   advantages, 183
   clear procedure, 185, 186
   detector system and APS, 183
   detector-amplifier, 142–143, 184–185
   energy resolution, 187, 189
   noise, 187, 189
   performance figures, 185–187, 188
   position resolution, 187–188
   quantum efficiency, 188
   radiation background, 188
   repetitive nondestructive readout (RNDR), 189
   sideward depletion, 184
   system concept, 182–183
   and XEUS mission, 183
Detectors *see* X-ray detectors
Diffraction anomalous fine structure (DAFS), 422
Diodes, 135, 136
Divergent angle, 97–98, 99
Double-multilayer monochromator, 76–77
Drift detectors, 137, 140, 141, 148–161
   controlled drift detector (CDD), 7, 137, 163–166
   energy measurement, 150–152
   energy resolution, 151, 152
   for high-energy XRF, 356
   leakage current, 151–152
   noise reduction, 150, 151, 152
   pn-CCDs, 166–180
   position resolution, 152–153

   principles, 148, 149–150
   sideward depletion, 140, 141
   for X-ray detection, 153–163
   X-ray holography, 161–162
   see also Controlled drift detectors; Silicon drift detectors
DuMond arrangement, 23

EDS, 387, 388
   see also Energy-dispersive X-ray microanalysis (EDX)
EDX *see* Energy-dispersive X-ray microanalysis
EDXRF systems, 2–3, 8–9, 307–341
   applications, 307–308, 321, 328–339
      alloy analysis, 335, 336
      archaeometry, 329–334
         chlorine, 329, 331
         copper, 333, 334
         gold, 329, 331, 333, 334
         lead, 333, 334
         pigments, 329, 330
         sulfur, 329, 330, 331, 333
         titanium, 331
      environmental analysis, 334–335
         lead in paints, 334–335
      Martian soil analysis, 338–339
      soil and rock analysis, 335–337
         air gap correction, 335–336
         grain size, 336
         lead, 336–337
         trace element analysis, 337
   defined, 307
   element identification, 317
   instrumentation, 308–328
   Monte Carlo simulation, 458–459
   portable equipment, 319–328
      9000 XRF Field Analyser, 324–325
      AMPTEK, 321–323
      EDAX, 323–324, 325
      EIS, 319, 321, 322
      Horizon 600 Alloy Sorter, 326, 327
      ICS-4000, 327, 328
      Metallurgist Pro, 326
      Metorex, 327–328, 335
      NITON, 319–321, 323
      Oxford, 326, 327
      Roentec ArtAX, 323, 324
      summarized, 320
      Thermo Measure Tech, 324–326
      Warrington (Lead Star and μ-Lead), 328, 329
      X-MET 880, 970, or 2000, 327–328, 335
      XRF Corporation, 327
   software, 315–319
      'fundamental parameter determination' (FP), 317
      peak area analysis, 317
   spectral lines, 355
   and spectrum evaluation, 463, 483
   X-ray detectors, 313–315
      compared, 317
      cooling, 313, 316, 318
      efficiency, 316
      energy resolution, 313, 314, 315
      features, 313
      types, 313–314

X-ray sources
    radioactive sources, 308–309, 310
    X-ray optics, 310–313
    X-ray tubes, 309–310, 312, 313
EELS see Electron energy loss spectroscopy
Electron energy loss spectroscopy (EELS), 388, 419–420
    applications, 394
    background subtraction, 393
    detection rate and geometry, 392, 393–394
    and EDX compared, 392–395
    'energy loss near edge structure' (ELNES), 394, 419
    energy resolution, 392–393
    extended energy loss fine structure (EXELFS), 419
    multiple scattering, 393
    spectrum image, 394–395
Electron guns, 570–571
    field-emission gun, 376–377, 382, 385, 387–388, 570–571
    thermionic guns, 570
Electron probe microanalysis (EPMA), 9, 103–104, 373–386, 569
    BNC coatings, 377–380
    chemical mapping, 398
    detectors, 572–573
        dual voltage, 573
    grazing exit, 297–302, 572–573
    gunshot residue, 564–565
    particle analysis, 570–590
    polycapillary optics, 103–104
    setup, 379
Electron-excited X-ray emission spectrometry, 569–592
Element imaging, 158–160
Ellipsoidal mirrors, 19–22
    magnification, 21
    performance, 21
    production, 20
    in protein crystallography, 21–22
ENC, 266–267
Energy recovery linac (ERL), 45–46
    brilliance, 46
Energy-dispersive spectroscopy (EDS), 387, 388
    see also Energy-dispersive X-ray microanalysis (EDX)
Energy-dispersive X-ray fluorescence systems see EDXRF systems
Energy-dispersive X-ray microanalysis (EDX), 387–404
    detection rate and geometry, 392, 393–394
    detectors, 569–570, 571–572
    and EELS compared, 392–395
    quantitative analysis, 395–402
        absorption correction, 396, 397, 398, 400
        boundary segregation, 399
        chemical mapping, 398–400
        Cliff-Lorimer Equation, 395–396, 397
        diffusion profiles, 401, 402, 403
        extrapolation method, 396–397
        fluorescence correction, 396, 397, 398
        ionic compounds, 400–401
        line profile, 399–400
        mass-thickness calibration, 397–398
        'mean mass-absorption length', 400–401
        parameterless correction method, 396–397
        thickness profile (garnet/OPX interface), 403, 404
        zeta-factor method, 397–398
    and STEM, 389
    see also Energy-dispersive X-ray spectroscopy; Transmission electron microscopy
Energy-dispersive X-ray spectroscopy (EDS), 103–104, 387–404
    historical overview, 387–388
Energy-dispersive X-ray spectroscopy (EDS)
    see also Energy-dispersive X-ray microanalysis; Transmission electron microscopy
Environmental applications, 5
    EDXRF, 334–335
    high-energy XRF, 364–366
EPMA see Electron probe microanalysis
Epsilon 5, 359
Equivalent Noise Charge (ENC), 266–267
ERL see Energy recovery linac
EXAFS (Extended X-ray absorption fine structure), 405–406, 422–426
    data analysis, 425
    spectra, 422–423
    standardization, 423–426
    and X-ray fluorescence, 422
EXEFS, 420–422
Extended X-ray absorption fine structure see EXAFS

FEG, 376–377, 382, 385, 387–388, 399
FEG-SEM, 571
FEL see Free electron laser
Field-emission gun (FEG), 376–377, 382, 385, 387–388, 399
Field-emission gun scanning electron microscopy (FEG-SEM), 571
Film analysis see Thin-film analysis
Forensic research, 11, 368–369, 553–567
    high-energy synchrotron radiation XRF, 564–566
        gunshot residue, 564–565
        paint chip, 565, 566
    synchrotron radiation XRF, 560–563
        drugs of abuse, 561, 563
        fluorescence tracing, 560–561, 562, 563
    TXRF studies, 553–560
        brandy, 555–556
        liquors, 554–555
        meteorite pieces, 559–560
        methamphetamine, 557–559
        poisoned food, 553, 554
        seal ink, 556–557, 558
        terrorism weapons, 554
        waste water, 556, 557
FP, 317
Free electron laser (FEL), 30, 41–44
    and light power, 42
    and radiation intensity, 41
Front-end electronics, 266–269
    count rate performance, 267
    Equivalent Noise Charge (ENC), 266–267
    feedback capacitance discharge, 267–268
    noise, 266, 267, 268–269
'Fundamental parameter determination' (FP), 317

Gas proportional ionization counter (PC), 195, 196, 197
Gas proportional scintillation counter (GPSC) see Scintillation counter

'Gaussian equivalent FWHM', 18
GIIXD see Grazing-incidence in-plane X-ray diffraction
GPSC see Scintillation counter
Grazing-exit XRS, 8, 293–305
   apparatus, 295
   detector, 304
   electron probe microanalysis (GE-EPMA), 297–302
      particle analysis, 299–301
      setup, 297–298
      surface analysis, 298, 299
      thin-film analysis, 301–302
   exit angle control, 295, 297–298
   fluorescence (GE-XRF), 295–296
   future developments, 303–305
   grazing emission XEF (GE-XEF), 296
   grazing-incidence XRS compared, 293
   light element analysis, 296, 297
   localized surface analysis, 303–304
   particle induced X-ray emission (GE-PIXE), 302–303, 304
      particle analysis, 303
      setup, 302–303
      surface and thin-film analysis, 303
   principles, 293–295
      critical angle, 294
      emission intensities of X-rays, 293–294
      information depth, 294, 295
      refraction of X-rays, 293, 294
   TXRF compared, 293
Grazing-incidence in-plane X-ray diffraction (GIIXD), 25–26
   Goorsky–Tanner technique, 25–26
   in-plane mosaic determination, 26
Grazing-incidence XRS, 8, 277–291
   critical angle, 277
   future developments, 288–290
   impurity analysis, 285
   intensity distribution, 284
   interface analysis, 283–287
   internal X-ray electric field, 284
   microscopic imaging, 287–288, 289
   penetration depth (of X-rays), 277
   standing wave technique, 283, 285, 286, 287
   surface analysis, 283–287
   total reflection XRF, 278–283
   X-ray fluorescence, 278–290
   X-ray reflectometry, 285–287

HAADF, 391
'Halo effect', 108
Hard X-ray multilayers, 72–77
   double-multilayer monochromator, 76–77
   microbeams and microscopy, 72–73
   multilayer-coated grating, 75–76
   reflectivity curves, 72, 73
   structure and composition, 72
   telescopes, 73–75
Hard X-rays, 52–55
HHG see High harmonic generation
High angle annular dark field (HAADF) imaging, 391
High harmonic generation (HHG), 49, 55–59
   short wavelength generation, 55–56
   X-ray source parameters
      conversion efficiency, 57, 58
      laser pulse intensity, 57, 58–59
      polarization, 57, 58
      pulse duration, 56, 57–58
      spectral range, 56–57, 58
High-energy XRF, 9, 355–372
   applications, 363–371
      archaeology, 366–368
      environment, 364–366
      forensic, 368–369
      geology and geochemistry, 369–371
         garnet analysis, 370, 371
         heavy element analysis, 369–371
      Old Kutani chinaware, 366–368
      rare earth elements analysis, 364, 365
      soil analysis, 366
   experimental setup, 356–357, 358
   historical review, 355–356
      laboratory X-ray sources, 355–356
      synchrotron radiation light sources, 356
   instruments, 356–357
      detectors, 356, 357
      monochromatic X-rays, 357
      optics, 357, 359
   performance, 357–363
      absorption coefficients, 362–363
      analytical performance, 360–362
      detection limit, 363
         lower limit (LLD), 359–360
         minimum (MDL), 360–362
      excitation techniques, 357–359
      penetration depths, 362–363
      transmission power, 362–363
   spectral lines, 355
High-resolution X-ray diffraction (HRXRD), 22–25
   configuration, 23
   rocking curve, 23, 24
   semiconductor characterization, 22–23
   spot size, 23, 24
HRXRD see High-resolution X-ray diffraction

Intensity gain, 97

Kirkpatrick–Baez objective, 73, 74

Laser-driven X-ray sources, 6, 49–62
   advances, 49–50
   applications, 50, 52–53, 54–55
   brilliance, 51
   compact coherent, 49–50
   generation mechanism, 53
   hard X-rays, 52–55
   high harmonic generation (HHG), 55–59
   laser technology, 50–52
   parameters, 52
   pulse speed and duration, 50–51, 52, 53, 54
   repetition rates, 51–52
   time-resolved X-ray diffraction, 54–55
   X-ray lasers, 55
Least moduli method, spectrum evaluation, 499
Lenses see Refractive X-ray lenses

Light element analysis
  by TXRF, 281, 282, 283, 518–519
  vapor phase decomposition (VPD-TXRF), 524
Light emission, in FEL, 42
Literature survey, 2–6
  country of origin, 5–6
  language of publication, 5
Lithium drifted silicon (Si(Li)) detectors, 258, 387, 388
Low-energy electron probe microanalysis see Electron probe microanalysis (EPMA)
Low-energy SEM see Scanning electron microscopy (SEM)
Low-voltage scanning electron microscopy (LV-SEM), 571
LV-SEM, 571

Magnetic circular dichroism (MCD), 69–71
MC simulation see Monte Carlo simulation
MCD, 69–71
Medical applications of XRF, 487–515
  element concentration, 487–489
    bromine, 493
    cadmium, 502, 503, 505, 508, 509
    calcium, 492, 508
    chromium, 489
    gold, 493, 508
    iodine, 490–491, 508
    iron, 506–508
    lead, 489–490, 501–502, 503, 509
      accumulation, 503–504
      childhood development, 505
      exposure, 502–503
      in pregnancy and lactation, 504
    mercury, 489, 490, 502, 503, 505–506, 507, 509
    multielement studies, 493
    nickel, 492–493
    palladium, 493, 508
    platinum, 491–492, 508
    thorium, 491
    trace elements, 490
    uranium, 508
  glomerular filtration rate (GFR), 491
  in vitro analysis, 487–488, 489
    PIXE method, 487
  in vivo analysis, 10, 488–489, 493–508
    accuracy, 501–502
    calibration, 499–500
      phantoms, 499, 500
    detectors and detection limit, 495–496, 500–501
    Monte Carlo simulation, 496–497
    neutron activation analysis (NAA), 488
    photons
      from radionuclides, 493, 494–495
      from X-ray tube, 494, 495
      LXRF and KXRF, 495
      polarised, 494
      scattering, 494
    precision, 501–502
    uncertainty, 501–502
  spectrum analysis, 499
  thyroid activity, 490–491
Micro-SRXAS, 126–129
Micro-X-ray analysis, 296
Micro-X-ray diffraction, 351

Micro-X-ray fluorescence computed microtomography (XFCMT), 121–126
  experimental setup, 121
  future developments, 125–126
  ion transport in plants, 121–123
  micrometeorite studies, 123–126
    ART, 124–125
    biological, 124
Micro-X-ray fluorescence computed tomography (XFCT), 453–455
Micro-X-ray sources, 6, 13–27
  application
    GIIXD, 25–26
    HRXRD, 22–25
  focusing optics
    ellipsoidal mirrors, 19–22
    polycapillary optics, 22
  and heat conduction, 13–14
  magnetic focusing, 15–17
  measurement, 17–18
    'pinhole camera', 17
    shadow technique, 17–18
  microfocus generator, 15–17
  and optics, 14–15
  phase contrast imaging, 18–19, 20
  photon collection, 14–15
  spot control, 15–17, 23
  target material, 14
  Wavefront distortion, 19
Micro-XRF, 100–103, 117
  applications, 5
  beam confinement, 343
  and cell analysis, 118–121
  imaging, 287–288
  laboratory, 100, 101
  portable, 100–101, 102
  synchrotron radiation, 102–103
Microbeams see X-ray microbeams
Microcalorimeters see Cryogenic microcalorimeters
Microscopic X-ray fluorescence analysis see Synchrotron radiation micro-XRF
Microtomography
  X-ray absorption techniques, 418, 419, 420, 421
  X-ray fluorescence, 121–126
Monte Carlo (MC) simulation, 10, 435–461
  applications, 436
  Compton scattering, 10, 441, 443
  electrons in scintillation counters, 200, 204, 206
  experimental validation, 445–453
    detection geometry, 447, 450
    detection limits for rare earths, 451–453
    linear polarization, 451
    scatter distributions, 447, 448, 449, 450
    setup, 445
  future trends, 458–459
  general principles, 437–445
    atom type selection, 439
    detector response function, 445
    interaction type selection, 439–440
    photoelectric effect simulation, 440
    photon absorption, 440
    photon trajectory, 437–438
    scattering interactions simulations, 441–444

Monte Carlo (MC) simulation (*continued*)
    azimuth angle, 443
    inverse cumulative distribution function, 442
    scattering angle, 442–443
    scattering event modeling, 443–444
    simulation code, 437
    step length selection, 438–439
    variance reduction, 444–445
  historical summary, 435
  in vivo medical readings, 10, 496–497
  Rayleigh scattering, 10, 441
  spectrum evaluation, 474–476
  trace-element analysis, 455–458
    fly ash particles, 456, 457–458
    quantification algorithm, 455–457
  XRF tomography, 453–455
Multilayer coatings, 7, 63–78
  applications, 63–64
  boundary of total reflection, 65–66
  design, 64–67
  fabrication, 63
  gratings, 75–76
  hard X-ray multilayers, 72–77
  'layer-by-layer' method, 66–67
  material selection, 67
  reflectivity, 64–65
  soft X-ray multilayers, 67–72

Near edge X-ray absorption fine structure *see* XANES
NEET (nuclear excitation by electron transition), 428
NEXAFS (near edge X-ray absorption fine structure) *see* XANES
Noise
  limits CCD performance, 169–171
  in semiconductor detectors, 135–137
NRA, 378–379
Nuclear excitation by electron transition (NEET), 428
Nuclear reaction analysis (NRA), 378–379

Pad detectors, 247
Parabolic compound refractive X-ray lenses *see* Refractive X-ray lenses
Particle analysis, 569–592
  applications, 587–590
    heavy metals, 589–590
    nitrate formation, 588–589
    soil and dust analysis, 587–588
  by EPMA, 299–301, 573–590
  experimental conditions, 570–573
    detectors, 571–573
    electron gun, 570–571
    priorities, 570
    sample treatment, 573
      cryogenic SEM, 573
      variable pressure (VP) SEM, 573
  inclusions in stainless steel, 300–301
  quantification models, 573–587
    analytical models, 573–574
    automatic evaluation, 581
    beam damage effect, 578–580
    beam sensitivity, 579
    CASINO software, 575
    chemical species analysis, 582–584
      surface and core region, 584
    data analysis, 580–582
    detector efficiency, 577
    error distribution, 582, 583
    heterogeneity, 584–587
    iterative approach, 575–577
    MC simulation models, 574–575
    particle diameter calculation, 576–577
    reverse MC procedure, 11, 575–577
    spectrum analysis, 580–582
    substrate material, 577–578
    window parameters, 577
  on silicon wafers, 299–300
Particle/proton-induced X-ray emission (PIXE) analysis, 1, 3–4
  archaeometry, 539–544, 547–548
  grazing-exit, 302–303, 304
  in vitro analysis, 487
  literature survey, 3–4
PC, 195, 196, 197
Phase contrast imaging, 18–19, 20
Photons
  absorption, 143–144
  collection, 14–15
  detection, 146–147, 148
  increase temperature, 229
PIXE *see* Particle-induced X-ray emission analysis
Pixel detectors, 181–190, 247–248
pn-CCD, 141–142, 166–180
  detector quality, 167
  development, 166
  frame store format, 175–176, 178, 180
  low energy response, 167
  p+ strips (shift registers), 167–168, 169, 170
  pixel size, 175–177
  position precision, 176
  readout speed, 175–177
  for ROSITA mission, 175, 178, 179, 180
  XEUS mission, 175, 177, 180
  see also charge-coupled detectors
Polarization, 69–71, 278
Polycapillary optics, 7, 22, 89–110
  applications, 98–108
    electron probe X-ray microanalysis, 103–104
    micro-XRF, 100–103
    wavelength dispersion of X-rays, 104
  collimating optics, 97–98, 106, 108, 311, 314
  divergent angle, 97–98, 99
  intensity gain, 97
  in EDXRF systems, 310–313
  elemental analysis, 98–104
  fabrication, 92
  focusing optics, 90, 93–97
    beam size measurement, 95
    flux density gain, 95–97
    focal distance, 93–95
    focal spot size, 93–95, 108–109
  future trends, 108–109
  micro-XANES, 105–106
  for micro-XRF, 345
  monolithic polycapillary optics, 90, 91

photon trajectory, 92–93
principles, 89–92
simulation tools, 92–93
structural analysis, 104–108
transmission efficiency, 91–92
wavelength dispersion, 104
X-ray diffraction, 106–108
XAFS, 106
Portable EDXRF systems *see* EDXRF systems
Portable equipment for X-ray fluorescence analysis *see* EDXRF systems
Position sensitive semiconductor strip detectors *see* Strip detectors
α-proton X-ray spectrometer 338–339
Proton-induced X-ray emission (PIXE) analysis *see* Particle/proton induced X-ray emission (PIXE) analysis

Quantitative analysis, using EDX, 395–402

Radiative Auger satellites, 420
Radioisotope XRF, 3
γ-ray camera, 162–163
Rayleigh scattering, 10, 254, 441
Readout electronics, 265–273
  analogue scheme, 270, 271, 272–273
  architectures, 269–273
  binary architecture, 269–271
  front-end electronics, 266–269
  multiplexing, 269, 270
  threshold discriminator, 271
  time-over-threshold principle, 270, 271–272
Reciprocity theorem, 294
Refractive X-ray lenses, 7, 111–131
  applications, 117–129
    cell analysis, 118–121
    experimental setup, 117–118
    μ-SRXAS, 126–129
    μ-SRXRF, 126–129
    XFCMT, 121–126
  attenuation, 112, 114
  design, 111, 112–113
    spherical aberration, 113
  imaging, 114–115
  material, 112–113
  microbeams
    background, 116
    knife-edge technique, 116–117
    and nanofocusing lenses, 117
    production, 115–117
    properties, 116–117
  physics, 112–117
    absorption, 112
    refraction, 112
  PINK beam mode, 118, 119, 120
  properties, 113–114
ROSITA mission, 175, 178, 179, 180
Rotating anode generator, 13, 14
  and microfocus sources compared, 21–22
Rotation analyzer, 71
Rotational trapezoidal readout (ROTOR), 163
ROTOR, 163

SASE *see* Self-amplified spontaneous emission system
Scanning electron microscopes, 158–160
Scanning electron microscopy (SEM) (low energy), 373–386
  advantages and disadvantages, 376
  applications
    anodised aluminium, 379
    BNC coatings, 377–380
    composition, 377–380
    fouled membranes, 382, 383
    meat casings, 382–385
    polymer membranes, 380–382
      experimental setup, 381
      feasibility tests, 380–381
      surface morphology, 381–382
  beam energy selection, 376
  high-resolution capabilities, 376, 379
  high-resolution imaging, 375–377
  Jeol JSM-6340F, 376, 377
  metal oxide images, 376–377, 378
  objective lens, 376, 377
  specimen chamber, 376, 377
Scanning transmission electron microscope/microscopy (STEM) *see* Transmission electron microscopy
Schwarzchild optics, 68–69
Scintillation counter (GPSC), 7, 195–214
  electron drift, 202–205, 206
    energy change, 202–205, 206
    velocity, 207, 208
  energy resolution, 209, 213
  Fano factor (F), 201
  material analysis applications, 211–213
  nonlinearity, 198–200, 213
  photodiodes, 211
  photosensors, 210–211
    compact implementation, 210–211
    photomultipliers, 210
  primary electron production, 198–200, 201, 202
    distribution function, 200–201, 203
  secondary scintillation, 206–207
    detection, 207–208
  solid angle compensation, 209
    curved grid technique, 209, 210
    masked photosensor technique, 209, 210
  statistical fluctuations, 200–201
  structure, 195–197, 207–209
    cylindrical geometry, 208
    spherical anode, 208
    uniform electric field, 195–197, 208–209
  tailing effects, 201–202, 203
  transport of electrons, 202–206
  Xe as filling medium, 197–198
    electron cascade, 197–198, 199
Scintillation detectors, 162–163
SDDs *see* Silicon drift detectors
Self-amplified spontaneous emission (SASE) system, 30, 41–45
  brilliance, 44
  electron beam, 42–43
  power production, 42–44
  wavelength region, 44
SEM *see* Scanning electron microscopy

Semiconductor detectors, 133–193
  charge-coupled detectors (CCDs), 137, 140–142
  DEPFET detector-amplifier, 142–143
  diodes, 135, 136
  drift detectors, 137, 140, 141, 148–161
  future developments, 189–190
  gas detectors compared, 134
  germanium and silicon, 134–135
  materials *see* Semiconductor materials
  properties, 134–135
  sideward depletion, 140, 141–142, 149
  signal charge measurement, 135–137
    noise, 135–137
  X-ray detection, 148
  see also Strip detectors
Semiconductor materials, 247, 248, 250–251, 517
  absorption length, 250–251
  total reflection XRF, 517–532
Semiconductor-radiation interaction, 143–144
  photon absorption, 143–144
  radiation entrance window, 144–145
Sideward depletion, 140, 141–142, 149, 184
Silicon drift detectors (SDDs), 7
  art investigation, 157–158
  drift detector drop (SD^3), 155–157
  in EDXRF systems, 314
  element imaging, 158–160
  energy resolution, 156
  entrance window, 153, 154
  γ-ray camera, 162–163
  and grazing-exit XRS, 304
  integrated JFET, 150–151
  layout, 160, 161
  low energy background, 153–155
  portable instrumentation, 158
  potential energy distribution, 154, 155
  properties, 155
  and X-ray fluorescence (XRF) spectrometer, 161
  see also Drift detectors
Silicon microsystems, 148
Silicon strip detectors *see* strip detectors
Single capillary X-ray optics, 7, 79–87
  analytical microscope, 81–86
    aluminum crystal growth, 84–85, 86
    pearl structure, 83
  microbeams, 79–80
  multidimensional analysis, 86–87
  production, 80–81, 82
Soft X-ray multilayers, 67–72
  applications, 68
  focusing, 68–69
  materials selection, 67–68
  microscopy, 68–69
  polarimetry, 69–71
Soft X-rays, 374–375
  applications, 55–56
  coatings analysis, 375
  lasers, 55
  problems and precautions, 374
  reasons for use, 374
Speckle patterns, 39

Spectrum evaluation, 463–485
  detectors, 467, 468
    response function, 474–476
  energy calibration, 467–468
  future applications, 483–484
  Gaussian peak, 467
    modified, 470–472
  Gaussian shape deviation, 468
  group fitting, 468
  least moduli method, 499
  least-squares fitting, 464–465, 468, 497, 499, 580
    filter-fit method, 465
    using analytical functions, 465
      continuum, 466
      linear and non-linear, 465–466
    using reference spectra, 464–465
  Lorentzian distribution, 472–473, 474
  Monte Carlo simulation, 474–476
  partial least-squares regression (PLS), 476–483, 580
    application
      cement analysis, 479–483
      other analyses, 483
    theory, 477–479
      collinearity (correlation) problem, 477–478
      model, 479
      root mean squared error (RMSE), 478–479
      validation method, 478–479
  peak area determination, 464
  peak shape
    deviates from ideal, 468
    modified Gaussian, 470–472
    numerical correction, 468–469
  polynomials, 466–467, 468
    fluorescence lines, 467
  resolution calibration, 467–468
  Voigt profile, 472–474, 475
SPIX, 303
Spot size, 23, 24, 93–95, 108–109
SR *see* Synchrotron radiation
SRXRF *see* Synchrotron radiation induced X-ray fluorescence
SRXRS, 1, 4
STEM *see* Transmission electron microscopy
STJ *see* Superconducting tunnel junctions
Storage rings, 29, 30, 32, 33, 45
  in SR-based micro-SRF, 345
Strip detectors, 8, 137–140, 247–275
  absorption length, 250–251, 253
  basic concept, 248
  biasing methods, 139–140
  diffusion, 250, 251–254
    charge distribution, 252–253
    full width at half maximum (FWHM), 252–253
  double-sided, 138–139
  front-end electronics, 266–269
  lithium drifted silicon Si(Li), 258
  manufacturing technology, 247, 248
  materials, 247, 248, 250–251
    absorption length, 250–251, 253
  non-silicon, 248, 255, 258, 264–265
    coplanar strip structure, 265
    germanium, 264–265
  photon scattering, 254–255, 263–264

readout electronics, 138, 139–140, 265–273
    architectures, 269–273
  silicon, 255, 258–264
    breakdown voltage, 258–259
    coupling capacitors, 260
    dead layer, 263
    double-sided, 261–262
    edge-on, 262–264
    guard ring structure, 263
    layout optimization, 259
    lithium contact layer, 261
    noise, 260
    readout strips, 259, 260, 261
    scattering effects, 263–264
    single-sided, 259–261
    strip isolation techniques, 260, 261
    strip structure, 259–260
  single-sided and double-sided, 248–249
  spatial resolution, 249–258
    diffusion, 250, 251–254
    energy deposition, 255–256, 257
    intrinsic, 250–258
    parallax effects, 256–258
    scattering, 254–256
  strip pitch, 249, 250, 253–254
  structures, 258–265
Superconducting tunnel junctions (STJ), 8, 217–227
  applications, 224–226
  basic functioning, 217–218
  cascade excitation process, 218–220
  cooling systems, 226
  energy resolution, 218–221, 222, 223
  noise performance, 220–221
  one-dimensional-imaging detectors, 218, 223
  quasiparticle excitation, 218–220
  series-junction detectors, 218, 223–224
    absorption efficiencies, 224
  single-junction detectors, 221–223
  statistical fluctuation, 221
  structure, 217–218
Surface analysis
  GE-EPMA, 298, 299
  GE-PIXE, 303
  grazing incidence XRS, 283–287
Surface sensitive particle-induced X-ray analysis (SPIX), 303
Synchrotron radiation
  applications
    in archaeometry, 536–538
    in medicine, 492
  characteristics, 34–40
    beamline performance, 40, 41, 279
    brilliance, 34, 36–37, 38, 44, 45, 46
    coherence, 38–39
    electron beam, 34
    flux and flux spectra, 34–35, 37, 38
    intensity, 34
    light rejection, 39
    photon energy, 35
    polarization, 39, 40
    source degeneracy, 39
    time structure (short pulse production), 39–40
  history, 1, 4–5
  sources, 6, 29–47

energy recovery linac (ERL), 45–46
    first generation, 29
    fourth generation, 30–32, 40–46
      energy recovery linacs, 45–46
      SASE-FEL, 41–44
      storage rings, 45
    listed, 31–32
    for micro-XRF, 346
    second generation, 29, 35
    storage rings, 29, 30, 33, 45
    third generation, 29–30, 32–33
      characteristics, 34–40
      classification, 33
      energy range, 37–38
      undulators, 32–33
    in TXRF spectrometry, 278–279
Synchrotron radiation induced micro-X-ray absorption spectroscopy (micro-SRXAS), 126–129
Synchrotron radiation induced X-ray fluorescence (SRXRF), 126–129
  Monte Carlo simulation, 436–445
Synchrotron radiation micro-XRF, 9, 102–103, 343–353
  accuracy, 347–350
    background effects, 347
    electron Bremsstrahlung, 347
    modeling methodologies, 347
    standards compared, 349–350
  advantages, 344
  applications, 351–352
  beam confinement, 343–344
  computerised microtomography (CMT), 350–351
  imaging, 351
  instrumentation, 344–347
    detection limits, 347, 348, 349
    detectors, 345–346
    optics, 345
    polychromatic v. monochromatic excitation, 344–345
    SR sources, 346
    storage rings, 345
  'Micro-XRF' project, 350
  X-ray absorption methods, 350
Synchrotron radiation X-ray spectrometry (SRXRS), 1, 4

TEM see Transmission electron microscopy
TEY method, 414, 415
Thin-film analysis
  by GE-EPMA, 301–302
  by GE-PIXE, 303
  total reflection X-ray fluorescence (TXRF), 10–11, 529–531
    two-angle grazing incidence XRF (GIXRF), 529–531
    X-ray reflectivity, 530
Thin-window electron probe microanalysis (TW-EPMA), 572
  see also Electron probe microanalysis
Total reflection X-ray fluorescence (TXRF), 1, 4–5, 10–11, 278–283, 517–532
  beam confinement, 343
  with chemical preconcentration, 521–524
  cleanliness, 279, 280
  detection, 518, 519–520, 521, 522
  forensic applications, 553–560

Total reflection X-ray fluorescence (TXRF) (*continued*)
  instrumentation, 517–520
    background optics, 519–520
    continuous X-rays, 518, 519
    detection limit, 518
    elements analysed, 517–519
  light element analysis, 281, 282, 283, 518–519, 524
  monochromator, 278
  parasitic X-rays, 279, 280
  polarized radiation, 278
  semiconductor analysis, 517–532
  standard samples, 525–529
  standardization, 524–529
    angle scan profiles, 526, 528
    cross-check activities, 524–525
    depth profile of analyte, 525–528
    'Immersion in Alkaline Hydrogen Peroxide Solution' (IAP) method, 526–528
    ISO standards, 524, 525
  synchrotron radiation (SR), 278–279
  thin-film analysis, 529–531
  vapor phase decomposition (VPD-TXRF), 521–524
    automation, 521–524
  wavelength-dispersive spectrometer, 281, 282
Transmission electron microscopy (TEM), 9, 387–404
  beam diameters, 388–389
  and electron energy loss spectroscopy (EELS), 388
  generated signals, 389–392
    Bragg scattering, 389–390
    Rutherford scattering, 390–391
  historical developments, 387–389
  inelastic scattering, 391–392
  Si(Li) detectors, 387, 388
  and WDX, 388
  see also Energy dispersive spectroscopy (EDS); Energy dispersive X-ray microanalysis (EDX)
TW-EPMA *see* Thin-window electron probe microanalysis
TXRF *see* Total reflection X-ray fluorescence

Undulators, 32–33, 35–37

Vapor phase decomposition (VPD-TXRF), light element analysis, 524

Wavelength-dispersion of X-rays, 104, 105
Wavelength-dispersive spectrometry (WDS), 281, 373
Wavelength-dispersive X-ray fluorescence (WDXRF), 2–3, 104
  and spectrum evaluation, 463
Wavelength-dispersive X-ray microanalysis (WDX), 388
WDS, 373
WDX, 388
WDXRF *see* Wavelength-dispersive X-ray fluorescence
WFI, 182
Wide field imager (WFI), 182
Windows (detectors), 144–145, 153, 154, 571–572, 577

X-ray absorption fine structure spectrometry *see* X-ray absorption techniques

X-ray absorption near edge structures *see* XANES
X-ray absorption techniques, 9–10, 405–433
  applications, 350, 351, 428
  'atomic XAFS', 422–423
  beamline, 426
  and DAFS, 422
  data analysis, 425
  and EELS, 419–420
  and EXEFS, 420–422
  extreme conditions, 407–408
  history, 406–407
  instrumentation, 426, 427
  liquid surface characterization, 414
  magnetic property determination, 416–418
  measurements, 106
  microscopy and microtomography, 418, 419, 420, 421
  with optical luminescence, 409–410
  photoconductive and electrochemical measurements, 411–412
  polarized X-rays, 413, 415
  with scanning tunneling microscopes, 408–409
  secondary yield methods, 414–415, 417
  self-absorption effect, 414
  spectra, 128, 129
    standardization, 423–426
    theory and interpretation, 422–423, 424, 425
  spectrometer, 426, 428
  strong field effects, 415–418
  TEY method, 414, 415
  total reflection, 412–414, 416
  with X-ray fluorescence, 410–411
  X-ray reflectivity intensity, 414
X-ray analytical microscope, 81–86
  applications, 82–86
  structure, 81–82
X-ray detectors, 7, 133–275, 467, 468, 571–573
  classified, 229
  compared, 313
  count rates, 463
  for EDXRF systems, 313–315, 317
  energy-dispersive spectrometers (EDS), 571–572
  EPMA, 572–573
  gas proportional scintillation counters, 195–214
  microcalorimeter, 229–245, 572
  semiconductor detectors, 133–193, 247–275
  silicon microsystems, 148
  solid state detectors (parameters), 536
  spectrum evaluation, 463
  strip detectors, 247–275
  superconducting tunnel junctions, 217–227
  wavelength-dispersive spectrometers (WDS), 571
X-ray diffraction, 106–108
  GIIXD, 25–26
  high resolution (HRXRD), 22–25
  lyzozyme diffraction patterns, 107–108
  micro X-ray diffraction, 351
X-ray excited optical luminescence (XEOL), 409
X-ray fluorescence (XRF), 1, 2
  applications, 5
  and EXEFS, 420–422
  and gas proportional scintillation counters (GPSC), 211–213
  high energy, 355–372

and interface roughness, 286–287, 288
in medicine *see* Medical applications of XRF
Monte Carlo simulation, 435–461
   detection limits for rare earth elements, 451–453
   experimental validation, 445–453
   tomography, 453–455
   trace-element analysis, 455–458
and silicon drift detectors, 161, 162
trace element *in vivo* analysis, 337
and X-ray absorption techniques, 410–411
and X-ray reflectometry, 285–287
X-ray fluorescence (XRF)
   see also EDXRF systems; Total reflection X-ray fluorescence (TXRF)
X-ray holography, 161–162
X-ray lenses *see* Refractive X-ray lenses
X-ray microanalysis, 9, 387–404
X-ray microbeams
   formation, 79–80, 115–117
   multidimensional analysis, 86–87
   and refractive lenses, 115–117
   single capillary, 80–81
X-ray optics, 7, 63–131
   micro-XRF, 345
   multilayer coatings, 63–78
   polycapillary, 89–110, 310–313
   refractive X-ray lenses, 111–131
   single capillaries, 79–87
X-ray sources, 13–62
   laser-driven, 49–62

micro X-ray sources, 13–27
synchrotron radiation, 29–47
X-ray spectrometry (XRS)
   chemical analysis, 26
   historical aspects, 1–2
   literature survey, 2, 3
   progress in sub-fields, 2
X-ray telescope, 73–75
XAFS (X-ray absorption fine structure spectroscopy) *see* X-ray absorption techniques
XANES (X-ray absorption near edge structures), 405–406, 407
   experiments, 105–106
   microscopes, 419
   spectral analysis, 425
   SR-based micro-SRF, 350
   see also X-ray absorption techniques
XAS *see* X-ray absorption techniques
Xenon, X-ray absorption, 197–198
XEOL, 409
XEUS mission, 175, 177, 180, 188
   and active pixel sensors, 181–182
   and DEPFET, 183
   wide field imager, 182
XFCMT *see* Micro-X-ray fluorescence computed microtomography
XFCT, 453–455
XMM satellite, 145–146, 172, 173, 174–175
XRF *see* X-ray fluorescence
XRS *see* X-ray spectrometry

With kind thanks to A. Griffiths for creation of this index.